# Small AC Generator

## SERVICE MANUAL ■ 3RD EDITION

This manual covers electrical powerplants under 8 kw that generate single-phase alternating current and are driven by air-cooled engines. Models listed herein may be portable or stationary and weigh approximately 300 pounds or less.

# CONTENTS

**Published by**
**INTERTEC PUBLISHING**
P.O. BOX 12901     OVERLAND PARK, KS 66282-2901
Phone: 1-800-262-1954          Fax: 1-800-633-6219
www.intertecbooks.com

# DUAL DIMENSIONS

This service manual provides specifications in both the U.S. Customary and Metric (SI) systems of measurement. The first specification is given in the measuring system perceived by us to be the preferred system when servicing a particular component, while the second specification (given in parenthesis) is the converted measurement. For instance, a specification of "0.011 inch (0.28 mm)" would indicate that we feel the preferred measurement, in this instance, is the U.S. system of measurement and the metric equivalent of 0.011 inch is 0.28 mm.

# CABLE SIZE

Equipment damage can result from low voltage. Therefore, to prevent excessive voltage drop between the generator and the equipment, the cable should be of adequate gauge for the length used. The cable selection chart shown in Fig. A-1 gives the maximum cable length for various gauges of wire that can adequately carry the loads shown. Wattage consumption for typical equipment is shown in Fig. A-2. Refer to these two charts to be sure that the load you are connecting to is within the capacity of the generator and wiring cable. Some electrical devices such as electric motors, incandescent lights and resistance coil-type heaters draw much greater current during start-up than after they are operating. Be sure you use adequate size connecting cables that can carry the required start-up load. Refer to Fig. A-3 for chart showing approximate starting watts required for different types of electric motors.

| CURRENT IN AMPERES | LOAD IN WATTS AT 120 VOLTS | AT 240 VOLTS | MAXIMUM ALLOWABLE CABLE LENGTH #8 WIRE | #10 WIRE | #12 WIRE | #14 WIRE | #16 WIRE |
|---|---|---|---|---|---|---|---|
| 2.5 | 300 | 600 | | 1000 ft. | 600 ft. | 375 ft. | 250 ft. |
| 5 | 600 | 1200 | | 500 | 300 | 200 | 125 |
| 7.5 | 900 | 1800 | | 350 | 200 | 125 | 100 |
| 10 | 1200 | 2400 | | 250 | 150 | 100 | 50 |
| 15 | 1800 | 3600 | | 150 | 100 | 65 | |
| 20 | 2400 | 4800 | 175 ft. | 125 | 75 | 50 | |
| 25 | 3000 | 6000 | 150 | 100 | 60 | | |
| 30 | 3600 | 7200 | 125 | 65 | | | |
| 40 | 4800 | 9600 | 90 | | | | |

Fig. A-1—The cable selection chart shown above gives maximum cable length for various gauges of wire that can adequately carry the loads shown.

| Appliance | Watts | Appliance | Watts |
|---|---|---|---|
| Light bulb | See Bulb | Coffeemakers | 400-700 |
| Fan or heating pad | 40-100 | Electric Drill | 500-1000 |
| Radio | 50-200 | Iron (Hand) | 500-1500 |
| Brooder | 100 and up | Portable Heater | 600-4800 |
| Automatic Washer | 150-1500 | Toasters | 900-1650 |
| Refrigerator | 190-2000 | 6-1/2 inch Hand Saw | 1000-1500 |
| Television | 200-500 | Water Heater | 1000-5000 |
| Vacuum Cleaner | 200-300 | Skillet | 1200 |
| Electric Drill | 225-1000 | 8-1/2 inch Hand Saw | 1500-2000 |
| Hot Plate | 330-1100 | 10 inch Hand Saw | 2000-2500 |

Fig. A-2—The chart above shows typical wattage consumption for various types of electrical devices.

| Motor HP Rating | Approx. Running Watts | APPROXIMATE STARTING WATTS REQUIRED Universal Motors | Repulsion Induction Motors | Capacitor Motors | Split Phase Motors |
|---|---|---|---|---|---|
| 1/8 | 275 | 400 | 600 | 850 | 1200 |
| 1/4 | 400 | 500 | 850 | 1050 | 1700 |
| 1/3 | 450 | 600 | 975 | 1350 | 1950 |
| 1/2 | 600 | 750 | 1300 | 1800 | 2600 |
| 3/4 | 850 | 1000 | 1900 | 2600 | |
| 1 | 1000 | 1250 | 2300 | 3000 | |

Fig. A-3—The chart above shows approximate starting watts required to start-up various types and sizes of electric motors.

# ACME

**ACME NORTH AMERICA CORPORATION**
**5203 West 73rd Street**
**Minneapolis, MN 55435**

| Model | Output-kva* | Voltage | Engine | | Governed Rpm |
| | | | Make | Model | |
|---|---|---|---|---|---|
| AG4200 | .40 | 120/220 | Acme | ALN330 | 3600 |
| AG5500 | .52 | 120/220 | Acme | AT330 | 3600 |
| AG7500 | .75 | 120/220 | Acme | VT94 | 3600 |

*1 kva equals 1000 watts.

# GENERATOR

## MAINTENANCE

Before starting unit, always clean and inspect receptacles. After 50 hours of operation, remove inspection cover and check for dirty windings. Tighten all loose bolts and screws. Inspect wiring for frayed or damaged insulation, tighten all screws on receptacles. Inspect connections to capacitor and make certain they are tight. Clean all ground wires, terminals and connections.

## TROUBLE-SHOOTING

### All Models

If there is no AC voltage output with engine running at specified speed, check condition of circuit breaker. If circuit breaker is set correctly, a capacitor check must be performed.

**NOTE: Even with generator stopped the capacitor will have a retained charge. Do not touch capacitor terminals as electric shock may result. Discharge capacitor by shorting across terminals with a screwdriver or similar tool with an insulated handle.**

Test capacitor using an ohmmeter set to the RX1000 scale. With ohmmeter leads connected to the capacitor terminals, a meter deflection should be seen followed by a slow return to infinity. Reverse the ohmmeter leads and repeat the procedure for the same reading. No meter deflection or continuing continuity indicates an open or shorted capacitor.

If capacitor checks normal, start and run generator at specified speed and check for 2-5 volt output at the 120 volt receptacle. No voltage reading indicates loss of rotor residual magnetism. To restore residual magnetism, operate generator at specified engine speed and, using a 12-volt battery as a power source, "flash" across capacitor terminals by connecting jumper wires from each of the 12-volt battery posts to the capacitor terminals for not more than one second. Polarity is of no concern. Output levels should then build to normal levels.

The generator is equipped with one diode per field winding. Diodes are located on a round heat-sink which is pressed on the ball bearing end of the rotor shaft (Fig. AC5). Diodes can be checked using an ohmmeter set at the lowest scale or a battery-powered test light. Connect one test lead to diode terminal and the other test lead to heat sink and observe ohmmeter reading, then reverse the tester connections and again observe ohmmeter reading. If diode is good, there should be one high reading (infinity) and one low reading (0 or no more than 15 ohms). If using a test light, the light should come on in one direction only. If both readings are the same, diode is defective and must be renewed.

To test stator main windings, remove stator from generator assembly. Use an ohmmeter set at the lowest scale. Ground one ohmmeter lead to metal frame bracket holding stator coils. Connect remaining ohmmeter lead to each individual disconnected winding lead. Ohmmeter reading should be infinite. Reading other than specified indicates field windings are grounded and should be renewed. Connect ohmmeter test leads to coil leads. If meter indicates infinite resistance, field windings are open and should be renewed.

To test rotor windings, remove the rotor from the generator. Use an ohmmeter set at the lowest scale. Connect one ohmmeter lead to coil wire and remaining lead to the metal bracket holding coils. Any ohmmeter reading indicates a short in rotor coil.

## OVERHAUL

Disassembly and reassembly of generator is evident after inspection of unit and reference to Fig. AC5.

# ENGINE

Refer to the appropriate Acme engine section for engine service.

*Fig. AC5—Exploded view of typical Acme generator.*

# COLEMAN POWERMATE, INC.

**125 Airport Road**
**Kearney, NE 68848**

| Model | Rated Output-watts | Voltage | Engine Make | Engine Model | Governed Rpm |
|-------|--------------------|---------|-------------|--------------|--------------|
| PM40-1000 | 800 | 120 | B&S | 80200 | 3600 |
| PM40-1750 | 1400 | 120 | B&S | 80200 | 3600 |
| PM40-1755 | 1400 | 120 | Kawasaki | FA130 | 3600 |

# GENERATOR

### MAINTENANCE

All bearings are sealed and require no lubrication, but should be inspected for smooth operation when generator is disassembled. Inspect unit periodically for condition of receptacles, electrical connections and operation of all switches.

### TROUBLE-SHOOTING

If little or no AC output is generated, check the following: The engine must be in good condition and maintain a governed speed under load of 3600 rpm. Check condition of circuit breakers and make sure all wiring connections are clean and tight. A faulty capacitor, open or shorted diodes, open stator windings, loss of residual magnetism of rotor or open or shorted rotor windings are also possible causes of no generator output.

**NOTE: Even with generator stopped the capacitor should have a retained charge if working properly. Do not touch capacitor terminals as electric shock may result. Discharge capacitor by shorting across terminals with a screwdriver or similar tool with an insulated handle.**

To test capacitor (6—Fig. C1), use an ohmmeter set to the RX10000 scale. Touch leads of ohmmeter to capacitor terminals. If capacitor is good, a meter deflection should be seen followed by a slow return to infinity. No meter deflection or continuing continuity indicates an open or shorted capacitor.

If field coil magnets have lost their residual magnetism, use the following procedure to re-excite the magnetic field. With the generator off, plug in the battery charger cable and reverse the leads to a 12-volt battery. Connect black lead to positive terminal of battery and momentarily (2-3 seconds) touch red lead to negative terminal. Disconnect battery charger cable, start the engine and check generator output, which should then build to normal level.

The stator windings can be checked for shorted or open windings by disconnecting the stator to endbell wiring connector plug. Use an ohmmeter set on RX1 scale and measure resistance between the terminals as follows: To check main winding, measure resistance between black and white leads; specified resistance is 1.05-1.25 ohms. To check excitation winding, measure resistance between two yellow leads; specified resistance is 9.3-9.7 ohms.

To check resistance of rotor coil windings, it will be necessary to first unsolder one side of the rotor diodes from coil winding leads before testing. Measure resistance of each coil using an ohmmeter set on RX1 scale. Specified resistance is 13.1-14.1 ohms. To check diodes, connect ohmmeter test leads to diode terminals and note meter reading. Reverse the tester leads and again note meter reading. There should be resistance in one direction only if diode is good.

If little or no DC output is generated, possible causes could be faulty receptacle, tripped circuit breaker, loose or broken wire, shorted or open stator winding, or shorted or open bridge diode.

To check battery charge winding, disconnect stator to endbell connector plug and measure resistance between two orange leads using ohmmeter set on RX1

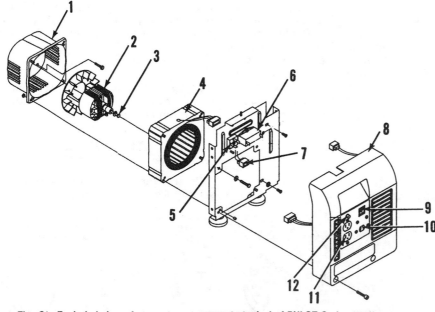

*Fig. C1—Exploded view of generator components typical of PULSE Series "40" generators.*

1. Engine adapter housing
2. Rotor assy.
3. Retaining bolt
4. Stator assy.
5. Bridge diodes
6. Capacitor
7. Connector
8. Endbell
9. Charge indicator light
10. DC receptacle
11. AC receptacle
12. Circuit breaker

scale. Specified resistance is 0.215-0.235 ohms.

To check bridge diodes (5—Fig. C1), disconnect wiring leads and using an ohmmeter, connect one test lead to diode positive terminal and touch other test lead to first one of the AC terminals and then the other. Note ohmmeter reading, then reverse the tester connections and again note reading. Ohmmeter should show either continuity or infinity when each terminal is touched. When leads are reversed, the result should be the opposite. Repeat test sequence with one ohmmeter lead connected to diode negative terminal. If test results are not as outlined above, renew diode assembly.

## OVERHAUL

To disassemble generator, remove endbell retaining bolts, disconnect wiring connector plugs and remove endbell (8—Fig. C1). Unbolt and remove stator bracket and stator assembly (4). Loosen rotor retaining bolt (3), then use a suitable puller to remove rotor from engine crankshaft. Be sure that center bolt of puller is positioned on head of the rotor retaining bolt and that puller legs are positioned on bottom of rotor core laminations. Do not strike the rotor laminations, windings or fan as this will damage the rotor. If necessary, unbolt and remove engine adapter/fan housing (1).

To reassemble generator, reverse the disassembly procedure while noting the following special instructions. Be sure that engine adapter housing (1) is properly aligned on the engine and tighten retaining bolts to 15 ft.-lbs. (20 N·m). Be sure that crankshaft and rotor mating surfaces are clean. Tighten rotor retaining bolt to 10-15 ft.-lbs. (14-20 N·m). Tighten endbell retaining screws to 6-8 ft.-lbs. (8-11 N·m).

# ENGINE

Refer to the appropriate Briggs & Stratton or Kawasaki engine section for engine service.

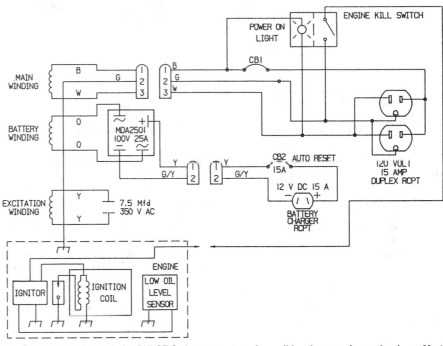

Fig. C3—Wiring schematic for PULSE Series generators. Low oil level sensor is used only on Model PM40-1755.

B. Black
G. Green
O. Orange
W. White
Y. Yellow

# COLEMAN

| Model | Rated Output-watts | Voltage | Engine | | Governed Rpm |
|---|---|---|---|---|---|
| | | | Make | Model | |
| PM45-4002 | 4000 | 120/240 | B&S | 190400 | 3600 |
| PM45-4022 | 4000 | 120/240 | B&S | 190400 | 3600 |
| PM45-5002 | 5000 | 120/240 | B&S | 252400 | 3600 |
| PM45-5022 | 5000 | 120/240 | B&S | 252400 | 3600 |
| PM45-7022 | 7000 | 120/240 | B&S | 422400 | 3600 |

# GENERATOR

### MAINTENANCE

All bearings are sealed and require no lubrication, but should be inspected for smooth operation when generator is disassembled. Inspect unit periodically for condition of receptacles, electrical connections and operation of all switches.

### TROUBLE-SHOOTING

If little or no output is generated, check the following: The engine must be in good condition and maintain a governed speed under load of 3600 rpm.

Check condition of circuit breakers and make sure all wiring connections are clean and tight. Rotor slip rings and brushes must be in good condition. A faulty capacitor, open or shorted diodes, open stator windings, loss of residual magnetism of rotor or open or shorted rotor windings are also possible causes of no generator output.

If field coils have lost their residual magnetism, there will be no excitation voltage produced. To check voltage, remove brush cover (8—Fig. C10) from end bell and start generator. Using a voltmeter set on 250 volts DC scale, touch positive test lead to top or right brush and negative test lead to bottom or left brush. Meter should read minimum of

95 volts for 4000 series generators, 144 volts for 5000 series generators or 150 volts for 7000 series generators. No voltage indicates loss of residual magnetism or an open in excitation circuit. Low voltage indicates short in stator or rotor windings, faulty capacitor or faulty diodes.

If field coil magnets have lost their residual magnetism, use the following procedure to re-excite the magnetic field. Using a 6-volt or 12-volt DC battery, connect battery positive terminal to top or right brush and momentarily (2-3 seconds) touch battery negative lead to bottom or left brush. Recheck generator output voltage and repeat if necessary. If loss of field magnetism is chronic, capacitor (9—Fig. C10) may be defective and should be renewed.

To test capacitor (9—Fig. C10), disconnect wiring from capacitor and short across capacitor terminals with a screwdriver to make sure there is no residual charge in capacitor. Using an ohmmeter set to the RX10000 scale, connect test leads to capacitor terminals. If capacitor is good, a meter deflection should be seen followed by a slow return to infinity. No meter deflection or continuing continuity indicates an open or shorted capacitor.

The diodes (11—Fig. C10) can be tested without removing them after first disconnecting connector plug from the stator. Using an ohmmeter set on RX1 scale, connect one test lead to No. 4 terminal in endbell connector plug (Fig. C11) and other test lead to bare metal of endbell (ground). Note meter reading, then reverse the test lead and again observe meter reading. A good diode will indicate continuity in one direction only. If there is a reading in both directions or no reading in either direction, renew diode assembly. Repeat test procedure for the other diode by connecting test leads to No. 2 terminal (Fig. C11) and ground.

The stator windings can be checked for shorted or open windings by disconnecting the stator to endbell wiring con-

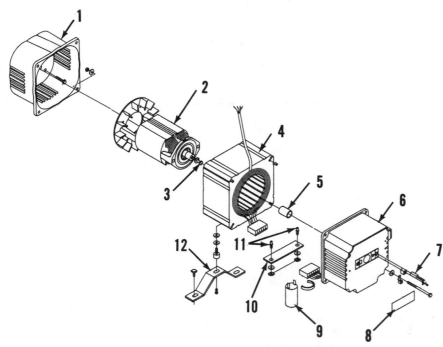

**Fig. C10—Exploded view of generator assembly typical of COMMERCIAL "45" series.**

1. Adapter housing
2. Rotor assy.
3. Retaining bolt
4. Stator assy.
5. Needle bearing
6. Endbell
7. Brush
8. Brush cover
9. Capacitor
10. Heat sink
11. Rectifier diodes
12. Stator mounting bracket

Illustrations courtesy of Coleman Powermate, Inc.

nector plug. Use an ohmmeter set on RX1 scale and measure resistance between the terminals in stator plug (Fig. C11) as follows:

To check power windings on 4000 and 5000 series generators (Fig. C12), measure resistance between black (No. 2 terminal) and white (No. 8 terminal) stator coil leads for power winding (1). Measure resistance between black (No.4 terminal) and white (No.6 terminal) leads for power winding (2). Specified resistance is 0.415-0.435 ohms for both full power windings. Measure resistance between the white/black and black/white leads for power winding (3). Specified resistance is 0.265-0.285 ohms. To check excitation windings, measure resistance between the yellow leads and blue lead; specified resistance is 1.05-1.15 ohms for either winding. Resistance between the two yellow leads should be 2.1-2.3 ohms.

To check power windings on 7000 series generators, measure resistance between the black and green leads and between the white and red leads; specified resistance for either coil is 0.170-0.190 ohms. To check excitation windings, measure resistance between the yellow leads and the blue lead; specified resistance is 1.0-1.1 ohms for either winding. Resistance measured between the two yellow leads should be 2.0-2.2 ohms.

To check rotor coil windings, measure resistance between rotor slip rings using an ohmmeter set on RX1 scale. Specified resistance is 35.9-36.7 ohms for 4000 and 5000 series generators and 46.1-48.5 for 7000 series generators. There should not be continuity between either slip ring and the rotor shaft.

## OVERHAUL

To disassemble generator, first disconnect generator output connector plug from control panel. Remove brush cover (8—Fig. C10) from endbell (6). Remove brush retaining screws and remove brushes and springs. Unbolt and remove endbell (6) and stator assembly (4). Loosen rotor retaining bolt (3), then use a suitable puller to remove rotor (2) from engine crankshaft. Be sure that center bolt of puller is positioned on head of the rotor retaining bolt and that

puller legs are positioned on bottom of rotor core laminations. Do not strike the rotor laminations, windings or fan as this will damage the rotor. If necessary, unbolt and remove engine adapter/fan housing (1).

To reassemble generator, reverse the disassembly procedure while noting the following special instructions. Be sure that engine adapter housing (1) is properly aligned on the engine and tighten retaining bolts to 15 ft.-lbs. (20 N·m). Be sure that crankshaft and rotor mating surfaces are clean. Tighten rotor retaining bolt to 10-15 ft.-lbs. (14-20 N·m). Tighten endbell retaining screws to 6-8 ft.-lbs. (8-11 N·m).

# ENGINE

Refer to the appropriate Briggs & Stratton engine section for engine service.

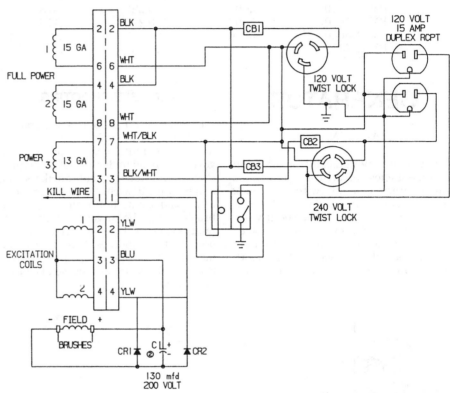

*Fig. C12—Wiring schematic for COMMERCIAL "45" series 4000 and 5000 watt generators.*

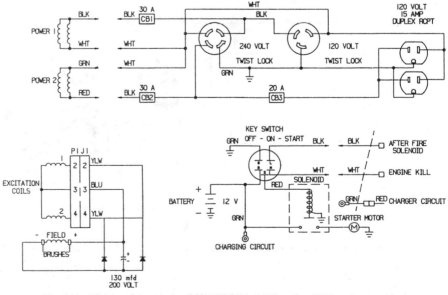

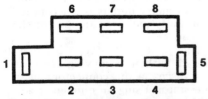

*Fig. C11—View of generator wiring connector plug as viewed from endbell end showing numbered position of wiring leads.*

*Fig. C13—Wiring schematic for COMMERCIAL "45" series 7000 watt generators.*

# COLEMAN

| Model | Rated Output-watts | Voltage | Engine Make | Engine Model | Governed Rpm |
|---|---|---|---|---|---|
| PM52-2000 | 2250 | 120 | Tecumseh | H50 | 3600 |
| PM52-4000 | 4000 | 120/240 | Tecumseh | HM80 | 3600 |
| PM52-4020 | 4000 | 120/240 | Tecumseh | HM80 | 3600 |
| PM52-5000 | 5000 | 120/240 | Tecumseh | HM100 | 3600 |
| PM52-5020 | 5000 | 120/240 | Tecumseh | HM100 | 3600 |
| PM54-2000 | 2250 | 120 | B&S | 130200 | 3600 |
| PM54-4000 | 4000 | 120/220 | B&S | 190400 | 3600 |
| PM54-4020 | 4000 | 120/220 | B&S | 190400 | 3600 |
| PM54-5000 | 5000 | 120/220 | B&S | 252400 | 3600 |
| PM54-5020 | 5000 | 120/220 | B&S | 252400 | 3600 |
| PM54-7020 | 6500 | 120/220 | B&S | 326430 | 3600 |

# GENERATOR

## MAINTENANCE

Inspect unit periodically for condition of receptacles, electrical connections and operation of all switches. Check condition of rotor slip rings and brushes. Brushes should be renewed if worn to 1/4 inch (6.35 mm) or less. All bearings are sealed and require no lubrication, but should be inspected for smooth operation when generator is disassembled.

## TROUBLE-SHOOTING

If little or no output is generated, check the following: The engine must be in good condition and maintain a governed speed under load of 3600 rpm. Check condition of circuit breakers and make sure all wiring connections are clean and tight. Rotor slip rings and brushes must be in good condition. A faulty capacitor, open or shorted diodes, open stator windings, loss of residual magnetism of rotor or open or shorted rotor windings are also possible causes of no generator output.

If field coils have lost their residual magnetism, there will be no excitation voltage produced. To check voltage, remove brush cover (7—Fig. C20) from end bell and start generator. Using a voltmeter set on 250 volts DC scale, touch positive test lead to top or right brush and negative test lead to bottom or left brush. Meter should read minimum of 78 volts for 2000 series generators, 95 volts for 4000 series generators, 144 volts for 5000 series generators or 150 volts for 7000 series generators. No voltage indicates loss of residual magnetism or an open in excitation circuit. Low voltage indicates short in stator or rotor windings, faulty capacitor or faulty diodes.

If field coil magnets have lost their residual magnetism, use the following procedure to re-excite the magnetic field. Using a 6-volt or 12-volt DC battery, connect battery positive terminal to top or right brush and momentarily (2-3 seconds) touch battery negative lead to bottom or left brush. Recheck generator output voltage and repeat if necessary. If loss of field magnetism is chronic, capacitor (8—Fig. C20) may be defective and should be renewed.

To test capacitor (8—Fig. C20), disconnect wiring from capacitor and short across capacitor terminals with a screwdriver to make sure there is no residual charge in capacitor. Using an ohmmeter set to the RX10000 scale, connect test leads to capacitor terminals. If capacitor is good, a meter deflection should be seen followed by a slow return to infinity. No meter deflection or continuing continuity indicates an open or shorted capacitor.

The diodes (10—Fig. C20) can be tested without removing them after first disconnecting connector plug from the stator. Using an ohmmeter set on RX1 scale, connect one test lead to No. 8 terminal in endbell connector plug (Fig. C21) and other test lead to bare

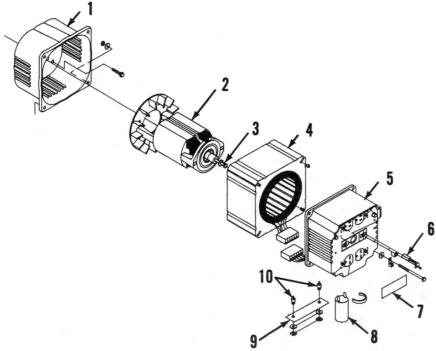

C20—Exploded view of generator typical of MAXA Series "52" and POWERBASE Series "54" generators.

1. Adapter housing
2. Rotor assy.
3. Retaining bolt
4. Stator assy.
5. Endbell
6. Brush
7. Brush cover
8. Capacitor
9. Heat sink
10. Rectifier diodes

Illustrations courtesy of Coleman Powermate, Inc.

metal of endbell (ground). Note meter reading, then reverse the test lead and again observe meter reading. A good diode will indicate continuity in one direction only. If there is a reading in either direction or no reading in either direction, renew diode assembly. Repeat test procedure for the other diode by connecting test leads to No. 6 terminal (Fig. C21) and ground.

The stator windings can be checked for shorted or open windings by disconnecting the stator to endbell wiring connector plug. Use an ohmmeter set on RX1 scale and measure resistance between the terminals in stator plug (Fig. C21) as follows:

To check power windings on 2000 series generators, measure resistance between black and white stator coil leads (Fig. 22); specified resistance is 0.250-0.370 ohms. To measure resistance of excitation windings, measure resistance between the yellow leads and blue lead; specified resistance is 1.250-1.350 ohms for either coil. Resistance between the two yellow leads should be 2.5-2.7 ohms.

To check power windings on 4000, 5000 and 7000 series generators, measure resistance between black (No. 1 terminal) and white (No. 7 terminal) stator coil leads for power coil (1) and between black (No. 3 terminal) and white (No. 5 terminal) leads for power coil (2). Specified resistance for either coil is 0.505-0.525 ohms for 4000 series generators, 0.265-0.285 ohms for 5000 series generators and 0.160-0.180 for 7000 series generators. To check excitation windings, measure resistance between the yellow leads and blue lead. For 4000 and 5000 series generators, specified resistance for either coil is 1.05-1.15 ohms. Resistance between the two yellow leads should be 2.1-2.3 ohms. For 7000 series generators, specified resistance for either coil is 0.850-0.950 ohms. Resistance between the two yellow leads should be 1.700-1.900 ohms.

To check rotor coil windings, measure resistance between rotor slip rings using an ohmmeter set on RX1 scale. Specified resistance is 24.0-24.9 ohms for 2000 series generators, 30.0-31.0 ohms for 4000 series generators, 35.9-36.7 ohms for 5000 series generators and 46.1-48.5 ohms for 7000 series genera-

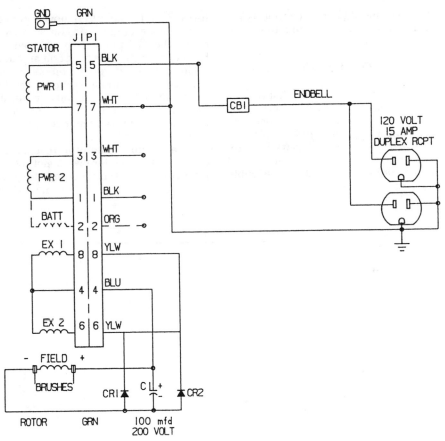

**Fig. C22—Wiring schematic for MAXA Model PM52-2000 and POWERBASE Model PM54-2000 generators.**

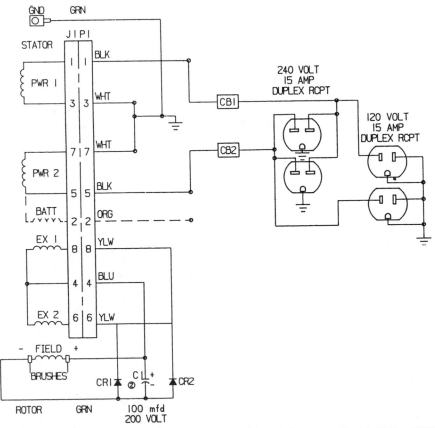

**Fig. C23—Wiring schematic for MAXA Series ''52'' and POWERBASE Series ''54'' 4000, 5000 and 7000 watt generators.**

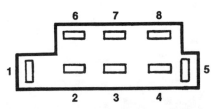

**Fig. C21—View of endbell wiring connector plug showing number location of wiring leads.**

tors. There should not be continuity between either slip ring and the rotor shaft.

### OVERHAUL

To disassemble generator, first disconnect generator output connector plug from control panel. Remove brush cover (7—Fig. C20) from endbell (5). Remove brush retaining screws and remove brushes and springs. Unbolt and remove endbell (5) and stator assembly (4). Loosen rotor retaining bolt (3), then use a suitable puller to remove rotor (2) from engine crankshaft. Be sure that center bolt of puller is positioned on head of the rotor retaining bolt and that puller legs are positioned on bottom of rotor core laminations. Do not strike the rotor laminations, windings or fan as this will damage the rotor. If necessary, unbolt and remove engine adapter/fan housing (1).

To reassemble generator, reverse the disassembly procedure while noting the following special instructions. Be sure that engine adapter housing (1) is properly aligned on the engine and tighten retaining bolts to 15 ft.-lbs. (20 N·m). Be sure that crankshaft and rotor mating surfaces are clean. Tighten rotor retaining bolt to 10-15 ft.-lbs. (14-20 N·m). Tighten endbell retaining screws to 6-8 ft.-lbs. (8-11 N·m).

# ENGINE

Refer to the appropriate Briggs & Stratton or Tecumseh engine section for engine service.

# DAYTON

**DAYTON ELECTRIC MFG. CO.**
**5959 W. Howard St.**
**Chicago, Illinois 60648**

| Model | Output-kw | Voltage | Engine Make | Engine Model | Governed Rpm |
|-------|-----------|---------|------|-------|--------------|
| 1W761A | 1 | 115 | Tecumseh | * | 3600 |
| 1W762A | 1.5 | 115 | B&S | * | 3600 |
| 1W831A | 2.5 | 115/230 | B&S | * | 3600 |
| 1W832A | 3.5 | 115/230 | B&S | * | 3600 |

*Engine model number was not available.

# GENERATOR

## MAINTENANCE

Brushes are accessible after removing end frame (3–Fig. D1-1). Inspect brushes periodically and renew brushes if brush length is less than 5/16 inch (7.9 mm). Inspect slip rings and recondition if required. Bearing (10) is sealed and does not require lubrication but should be inspected whenever end frame is removed.

## TROUBLESHOOTING

If little or no output is generated, check the following: The engine must be in good condition and maintain desired governed speed of 3600 rpm. Brushes and slip rings must be in good condition. All wiring connections must be clean and tight.

If field magnets are too weak to generate voltage, flash the fields as follows: Remove receptacle cover (1–Fig. D1-1) so brushes are exposed. Note that positive brush is on right and negative brush is on left. Run unit; then using a 12 volt battery, momentarily connect positive battery terminal to negative generator brush. Disconnect battery as soon as generator voltage rises. Chronic loss of field magnetism may be due to weak or damaged field magnets and rotor should be renewed.

## OVERHAUL

Disassembly and reassembly of generator is evident after inspection of unit and referral to exploded view in Fig. D1-1 and wiring schematic in Fig. D1-2 or D1-3. Wiring should be marked to aid reconnection. Rotor shaft and engine crankshaft end are tapered and a suitable puller is necessary to dislodge rotor from crankshaft.

# ENGINE

Engine make is listed at beginning of section. Refer to Briggs and Stratton or Tecumseh section for engine service.

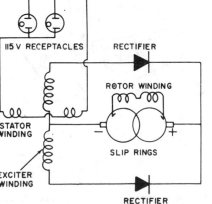

**Fig. D1-1—Exploded view of generator. Note that negative brush holder (14) and rectifier diodes (15) are mounted on backside of plate (5).**

1. Receptacle cover
2. Receptacle
3. End frame
4. Positive brush holder
5. Plate
6. Stator assy.
7. Bolt
8. Lockwasher
9. Washer
10. Bearing
11. Rotor
12. Fan
13. End bell
14. Negative brush holder
15. Diodes

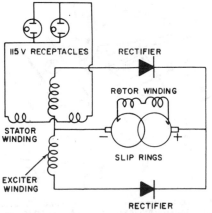

**Fig. D1-2—Wiring schematic for Models 1W761A and 1W762A.**

**Fig. D1-3—Wiring schematic for Models 1W831A and 1W832A.**

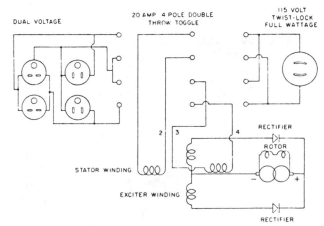

# DAYTON

| Model | Output-kw | Voltage | Make | Model | Governed Rpm |
|-------|-----------|---------|------|-------|--------------|
| | | | | **—Engine—** | |
| 2W183 | 1.2 | 125 | B&S | 80232 | 3600 |
| | | | Tecumseh | H35 | 3600 |
| 2W183A | 1.2 | 125 | B&S | 80232 | 3600 |
| 3W091 | 1.7 | 125 | B&S | 100232 | 3600 |
| | | | Tecumseh | HS40 | 3600 |
| 3W091A | 1.7 | 125 | B&S | 100232 | 3600 |
| 3W092 | 3.0 | 125/250 | B&S | 190432 | 3600 |
| 3W092 | 3.0 | 125/250 | Tecumseh | HM80 | 3600 |
| 3W092A | 3.0 | 125/250 | B&S | 190432 | 3600 |
| 3W093A | 4.0 | 125/250 | B&S | 243432 | 3600 |
| 3W093A | 4.0 | 125/250 | B&S | 251412 | 3600 |
| 3W093A | 4.0 | 125/250 | Tecumseh | HH100 | 3600 |
| 3W093B | 4.0 | 125/250 | B&S | 251412 | 3600 |

# GENERATOR

## MAINTENANCE

Manufacturer recommends inspecting and servicing, if necessary, generator brushes and slip rings once every year. Brushes are accessible after removing end cover plate while the end frame must be removed to service the slip rings. Renew brushes if worn to length of 5/16 inch (7.9 mm) or less.

## TROUBLESHOOTING

If little or no output is generated, check the following: The engine must be in good condition and maintain a governed speed under load of 3600 rpm. On models equipped with a receptacle box attached to the end frame, check fuse or circuit breaker. All wiring connections must be clean and tight. Brushes and slip rings must be in good condition.

If field coil magnets have lost their residual magnetism, use the following procedure: Remove end cover plate and run unit. Using a 6 or 12 volt battery, momentarily connect the negative battery terminal to the generator frame and the positive battery terminal to the positive generator brush. Disconnect battery as soon as generator voltage rises. If loss of field magnetism is chronic, capacitor (9–Fig. D2-1) may be defective and should be renewed.

Models 2W183, 3W091, 3W092 and 3W093A are equipped with diodes (11) which may be checked after removing end frame and unsoldering and removing diodes. Using an ohmmeter, connect ohmmeter leads to diode and note reading; reverse ohmmeter leads and again note reading. Ohmmeter should read very high for one connection and very low for the other connection. If both readings are high or low, diode is faulty and must be renewed.

Models 2W183A, 3W091A, 3W092A and 3W093B are equipped with a rectifier mounted on the end frame as shown in Fig. D2-2. To test rectifier (R), disconnect (AC) leads from rectifier and black lead from positive (+) rectifier terminal. Using an ohmmeter, connect ohmmeter leads to an (AC) terminal and positive (+) terminal, note reading, reverse ohmmeter leads and again note reading. One reading should be high and the other low. If both readings are high or both are low, the rectifier is defective and must be renewed. Repeat test by connecting ohmmeter leads between remaining (AC) terminal and positive (+) terminal. If test readings using remaining (AC) terminal are both high or both

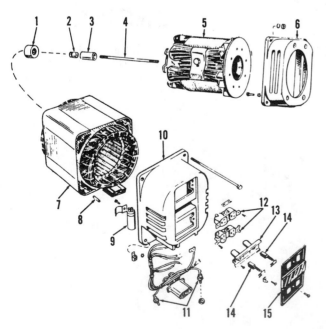

Fig. D2-1—Exploded view of generator used on Models 2W183, 3W091, 3W092 and 3W093A. Other models are similar except a rectifier module is used in place of diodes (11).

1. Roller bearing
2. Special nut
3. Sleeve
4. Stud
5. Rotor
6. Adapter
7. Stator assy.
8. Pin
9. Capacitor
10. End frame
11. Diodes
12. Outlets
13. Brush holder
14. Brushes
15. Cover plate

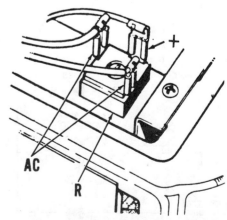

Fig. D2-2—View of rectifier module (R) used on Models 2W183A, 3W091A, 3W092A and 3W093B. Note location of (AC) terminals and (+) positive terminal.

Illustrations courtesy of Dayton Electric Mfg. Co.

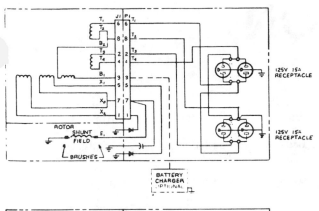

*Fig. D2-3—Wiring schematic for Models 2W183 and 3W091.*

low, rectifier is faulty and must be renewed.

If above tests do not locate malfunction, generator must be disassembled and rotor and stator tested for open and short circuits.

Refer to Figs. D2-3, D2-4 and D2-5 for wiring schematics.

## OVERHAUL

Disassembly and reassembly of alternator is evident after inspection of unit and referral to Fig. D2-1. Refer also to wiring schematic in Fig. D2-3, D2-4 or D2-5.

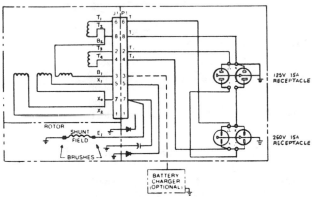

*Fig. D2-4—Wiring schematic for Models 3W092 and 3W093A.*

# ENGINE

Engine make and model are listed at beginning of section. Refer to Briggs & Stratton or Tecumseh engine section for engine service.

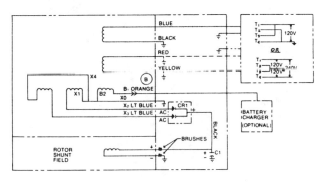

*Fig. D2-5—Wiring schematic for Models 2W183A, 3W091A, 3W092A and 3W093B.*

# DAYTON

| Model | Output-kw | Voltage | Engine Make | Engine Model | Governed Rpm |
|---|---|---|---|---|---|
| 3W013 | 2.0 | 120 | B&S | * | 3600 |
| 3W013A | 2.25 | 120 | B&S | 131232 | 3600 |
| 3W013C | 2.5 | 120 | * | * | 3600 |
| 3W013D | * | * | * | * | 3600 |
| 3W014 | 3.0 | 120/240 | B&S | 170432 | 3600 |
| 3W014I | 3.5 | 120/240 | B&S | 195432 | 3600 |
| 3W014J | * | * | * | * | 3600 |
| 3W014K | 3.75 | 120/240 | * | * | 3600 |
| 3W014L | * | * | * | * | 3600 |
| 3W015 | 4.0 | 120/240 | B&S | * | 3600 |
| 3W015G | 4.5 | 120/240 | B&S | 243432 | 3600 |
| 3W015H | * | 120/240 | * | * | 3600 |
| 3W015I | 4.75 | 120/240 | * | * | 3600 |
| 3W015J | * | 120/240 | * | * | 3600 |
| 3W016 | 6.0 | 120/240 | Wisc. | S12D | 3600 |
| 3W016A | 6.5 | 120/240 | B&S | 326432 | 3600 |
| 3W016B | * | 120/240 | * | * | 3600 |
| 3W017C | 8.0 | 120/240 | * | * | 3600 |
| 3W017D | * | 120/240 | * | * | 3600 |
| 3W056B | 5.0 | 120/240 | B&S | * | 3600 |
| 3W056C | 5.0 | 120/240 | Tec. | OH140 | 3600 |
| 3W057B | 8.0 | 120/240 | B&S | 422431 | 3600 |
| 3W057C | 8.0 | 120/240 | Tec. | OH180 | 3600 |
| 3W181 | 3.0 | 120/240 | B&S | 243432 | 1800 |
| 3W181A | 3.0 | 120/240 | * | * | 1800 |
| 3W181B | 3.0 | 120 | B&S | * | 1800 |
| 3W182 | 4.0 | 120/240 | B&S | 326432 | 1800 |
| 3W182A | 4.0 | 120/240 | B&S | * | 1800 |
| 3W182B | 4.0 | 120/240 | * | * | 1800 |
| 3W182C | 4.0 | 120/240 | * | * | 1800 |

*Specifications not available.

# GENERATOR

## MAINTENANCE

Manufacturer recommends inspecting and servicing, if necessary, brushes, commutator and slip rings after every 100 hours of operation. Brushes and armature are accessible after removing end cover (21–Fig. D3-1). Renew brushes if worn to length of 3/8 inch (9.5 mm) or less. Commutator and slip ring surfaces must be clean with commutator mica separators undercut 1/32 inch (0.8 mm) below copper segments.

## TROUBLESHOOTING

If little or no output is generated, check the following: The engine must be in good condition and maintain desired governed speed under load. Brushes, commutator and slip rings must be in good condition. All wiring connections must be clean and tight.

If field magnets are too weak to generate voltage, flash the fields as follows: Run unit and using a 6 or 12 volt battery, momentarily connect the negative battery terminal to the generator frame and the positive battery terminal to the positive generator brush. Disconnect battery as soon as generator voltage rises. Chronic loss of field magnetism may be due to a faulty capacitor in field circuit.

Condensers (10 and 12–Fig. D3-1) are used to suppress radio interference. A shorted condenser will ground generator output. To check condensers, alternately disconnect condenser leads then note generator output. If generator output improves, then condenser is faulty and must be renewed. If generator output does not improve, then condenser is not faulty and should be reconnected.

To test field coils, disconnect battery, if so equipped, remove end cover (21) then disconnect field coil leads from brush holders. Using an ohmmeter, check resistance between field coil leads as shown in Fig. D3-2. If resistance is infinite, then field coil is open and must be renewed. Connect an ohmmeter lead to one of the field coil leads, then ground other ohmmeter lead to frame as shown in Fig. D3-3. If ohmmeter indicates continuity, then coil is grounded and should be renewed.

To test armature, disconnect battery, if so equipped, remove end cover (21–Fig. D3-1) and all brushes. Using an ohmmeter, check for grounded windings by connecting one ohmmeter lead to armature shaft then touching other ohmmeter lead to each slip ring and each commutator segment. Renew armature if continuity exists between armature shaft and any slip ring or commutator segment. Using an ohmmeter, check

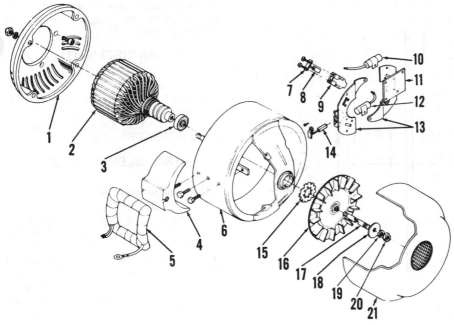

Fig. D3-1 — Exploded view of generator.

1. Adapter
2. Armature
3. Bearing
4. Field magnet
5. Field coil
6. Frame
7. Brush retainer
8. AC brush
9. Brush holder
10. AC condenser
11. Plate
12. DC condenser
13. Brush holder plates
14. DC brush
15. Lockwasher
16. Fan
17. Stud
18. Washer
19. Lockwasher
20. Nut

resistance between front and middle slip rings then between middle and rear slip rings (Models 3W013, 3W013A, 3W013C and 3W013D only have two slip rings). Ohmmeter should indicate little or no resistance between slip rings; otherwise armature windings are open and armature must be renewed. Additional electrical tests may be performed with armature removed from generator.

## OVERHAUL

Disassembly and reassembly of generator is evident after inspection of unit and referral to Fig. D3-1. Refer also to wiring schematics in Figs. D3-5 through D3-10C.

# AUTOMATIC IDLER

### (Demandatrol/Conserver)

## OPERATION

All models except 3W181, 3W181A, 3W181B, 3W182, 3W182A, 3W182B, 3W182C and 1800 rpm models are equipped with an automatic idler system which closes engine throttle when no

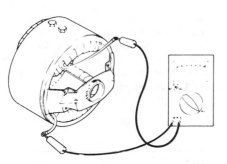

Fig. D3-2 — Connect an ohmmeter to field coil leads as shown and measure resistance. Infinite resistance indicates an open coil.

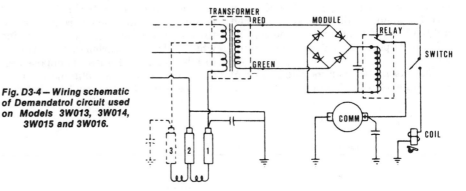

Fig. D3-4 — Wiring schematic of Demandatrol circuit used on Models 3W013, 3W014, 3W015 and 3W016.

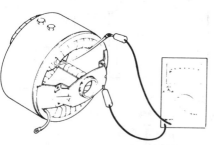

Fig. D3-3 — Connect an ohmmeter to field coil lead and frame. If continuity exists then coil is grounded.

Fig. D3-4A — Wiring schematic of Demandatrol circuit used on Models 3W013A, 3W013C, 3W014I, 3W014K, 3W015G, 3W015I, 3W016A and 3W017C.

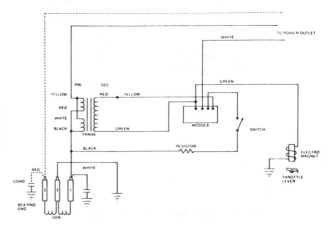

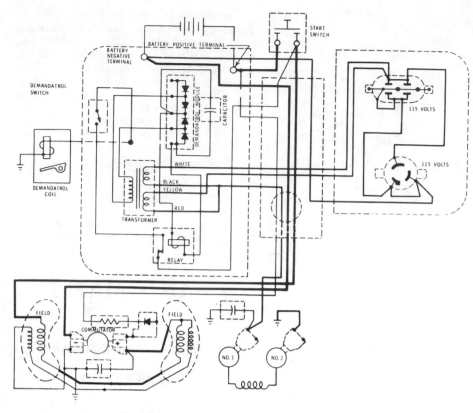

Fig. D3-5 — Wiring schematic for Model 3W013.

tromagnet thereby releasing the throttle and allowing the engine to run at governed speed.

## ADJUSTMENT

The position of the throttle electromagnet is adjustable by turning locknuts so desired engine idle speed may be obtained. Desired engine idle speed is 1800-1900 rpm.

## TROUBLESHOOTING

If engine does not idle under no-load, check for binding and sticking governor and carburetor linkage. Test automatic idler switch to be sure it closes in "AUTO" mode. Open control box and connect an AC voltmeter to small red (yellow on some models) and green secondary wires of transformer. With unit running and a 100 watt load connected to AC outlet, there should be 12-16 volts AC at transformer wires. Renew transformer if desired voltage reading is not obtained. Using an ohmmeter, check resistance of electromagnet coil mounted on engine. Coil resistance should be approximately 70 ohms or coil is defective and electromagnet must be renewed.

On Models 3W013A, 3W014I, 3W015G and 3W016A, check resistor in series with automatic idler switch. On Models 3W013, 3W014, 3W015 and 3W016, check operation of relay in series with automatic idler switch. Relay should close to activate electromagnet.

If the above tests do not locate a malfunction in the automatic idler system, then the automatic idler module should be renewed.

load is imposed on generator. A two-position switch activates system when in "AUTO" mode or disables system when in "MANUAL" mode and engine runs at governed speed regardless of generator load.

The automatic idler module shown in Fig. D3-4 or D3-4A is connected to a transformer in the generator output circuit and senses generator load. if no load is present, the module will route current to the electromagnet on the engine which will pull the carburetor throttle into idle position. If a load is imposed on the generator, the automatic idler module will cut off current to the elec-

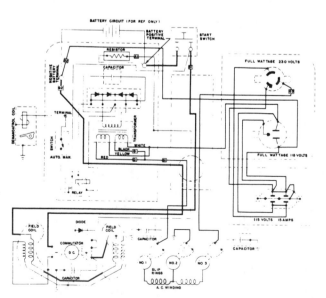

Fig. D3-6 — Wiring schematic for Model 3W014.

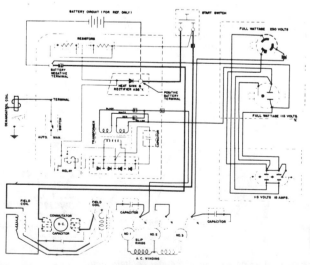

Fig. D3-7 — Wiring schematic for Model 3W015. Model 3W016 is similar.

Illustrations courtesy of Dayton Electric Mfg. Co.

# AUTOMATIC ENGINE START/STOP

### OPERATION

Models 3W056B, 3W056C, 3W057B and 3W057C are equipped with an automatic engine start/stop circuit which senses power outage and automatically starts generator. Power outage is sensed by a start/stop relay which in turn grounds and activates a solenoid in the gas line and activates an overcranking relay. Overcranking relay will stop engine cranking if engine fails to start after a predetermined period of time. When engine starts, generator voltage rises and opens a set of normally closed contacts in a stop cranking relay to disable starter. After engine starts, a transfer delay relay allows engine to reach full speed for approximately 20-45 seconds before load is transferred to generator. When commercial power is restored, start/stop relay is reenergized, generator is disconnected from load, fuel solenoid ground is interrupted and ignition is grounded, thus stopping engine.

# ENGINE

Engine make and model are listed at beginning of section. Refer to Briggs & Stratton, Wisconsin or Tecumseh section for engine overhaul.

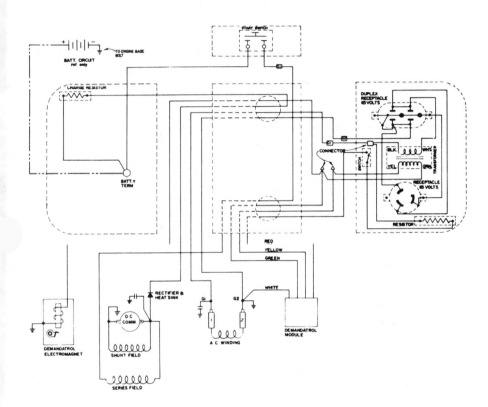

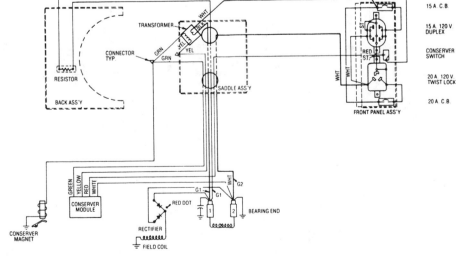

*Fig. D3-8 — Wiring schematic for Model 3W013A.*

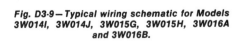

*Fig. D3-9 — Typical wiring schematic for Models 3W014I, 3W014J, 3W015G, 3W015H, 3W016A and 3W016B.*

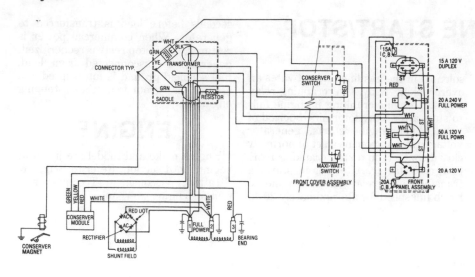

Fig. D3-10 — Wiring schematic for Models 3W181 & 3W181A. Models 3W182 and 3W182A are similar.

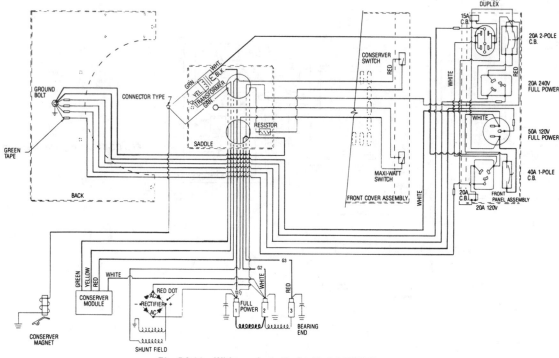

Fig. D3-11 — Wiring schematic for Model 3W014L.

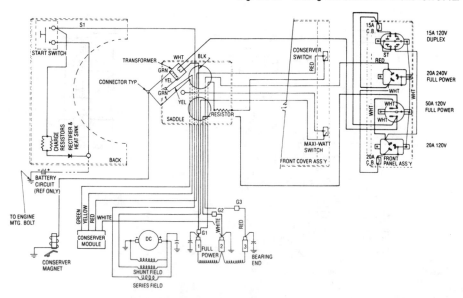

Fig. D3-12 — Wiring schematic for Model 3W015I.

Illustrations courtesy of Dayton Electric Mfg. Co.

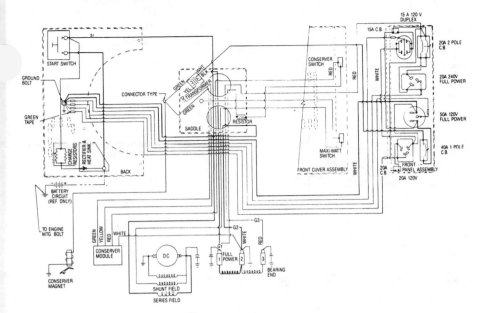

*Fig. D3-13 — Wiring schematic for Model 3W015J.*

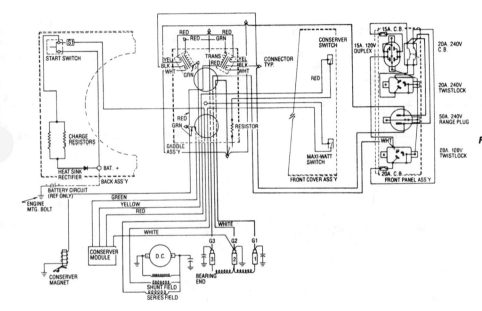

*Fig. D3-14 — Wiring schematic for Model 3W017C.*

*Fig. D3-15 — Wiring schematic for Model 3W017D.*

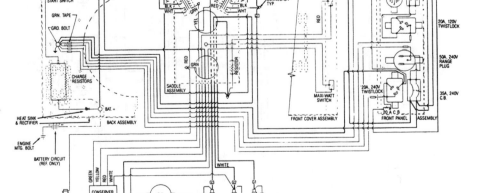

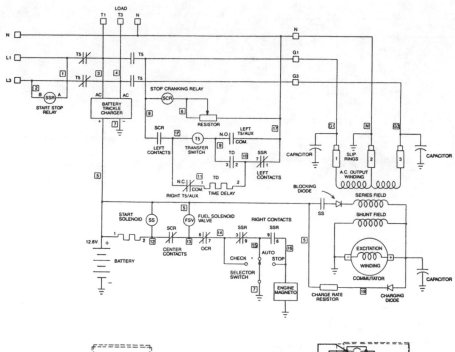

*Fig. D3-16 — Wiring schematic for Models 3W056B, 3W056C, 3W057B and 3W057C.*

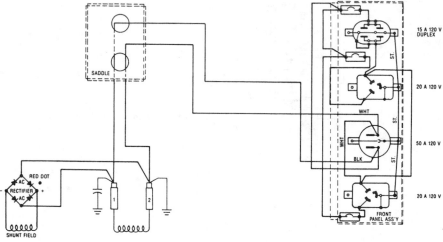

*Fig. D3-17 — Wiring schematic for Model 3W181B.*

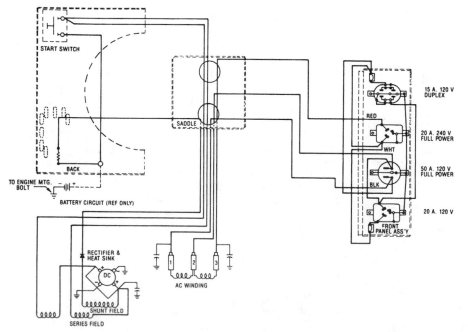

*Fig. D3-18 — Wiring schematic for Model 3W182B.*

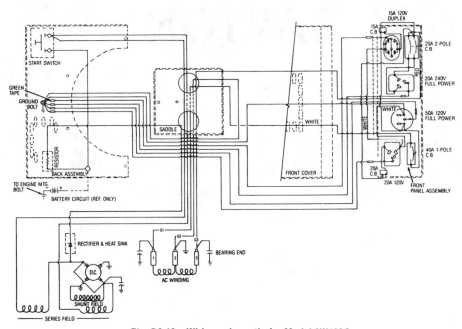

*Fig. D3-19 — Wiring schematic for Model 3W182C.*

# DAYTON

| Model | Output-kw | Voltage | Engine Make | Engine Model | Governed Rpm |
|-------|-----------|---------|------|-------|--------------|
| 3W177 | 2.0 | 120 | B&S | 130232 | 3600 |
| 3W177A | 2.0 | 120 | B&S | 130232 | 3600 |
| 3W177C | 2.0 | 120 | Tec. | H50 | 3600 |
| 3W178 | 3.0 | 120/240 | B&S | 190412 | 3600 |
| 3W178A | 3.0 | 120/240 | B&S | 190412 | 3600 |
| 3W178C | 3.0 | 120/240 | Tec. | HM80 | 3600 |
| 3W179 | 4.0 | 120/240 | B&S | 252412 | 3600 |
| 3W179A | 4.0 | 120/240 | B&S | 252412 | 3600 |
| 3W179B | 4.0 | 120/240 | Tec. | HM80 | 3600 |
| 3W180 | 5.0 | 120/240 | B&S | 252412 | 3600 |
| 3W180A | 5.0 | 120/240 | B&S | 252412 | 3600 |
| 3W180B | 5.0 | 120/240 | Tec. | HM100 | 3600 |
| 3W232 | 7.0 | 120/240 | B&S | 402437 | 3600 |
| 3W232B | 7.0 | 120/240 | B&S | 326437 | 3600 |
| 3W232C | 7.0 | 120/240 | Tec. | OH160 | 3600 |

# GENERATOR

### MAINTENANCE

Brushes (4–Fig. D4-1) and rotor slip rings should be inspected periodically. End frame (3) must be removed for access to brushes and rotor slip rings.

Rotor end bearing (9) is sealed and does not normally require service.

### TROUBLESHOOTING

Determine if circuit breakers are operating properly. Brushes and slip rings must be in good condition. All wiring connections must be clean and tight. The engine must run satisfactorily under load at governed speed of 3600 rpm. Using wiring schematics in Figs. D4-3, D4-4, D4-5, D4-6 and D4-7 along with suitable testing equipment, the faulty component may be located.

### OVERHAUL

Disassembly and reassembly of generator is evident after inspection of unit and referral to exploded view in Fig. D4-1 and wiring schematic in Figs. D4-3, D4-4, D4-5, D4-6 and D4-7. Wiring should be marked to aid in reconnection. Rotor shaft and engine crankshaft end are tapered and a suitable puller is necessary to dislodge rotor from crankshaft.

**Fig. D4-1—Exploded view of generator.**
1. Cover
2. Rectifier
3. End frame
4. Brush & holder
5. Stator assy.
6. Rotor bolt
7. Lockwasher
8. Washer
9. Bearing
10. Rotor
11. Fan
12. Adapter

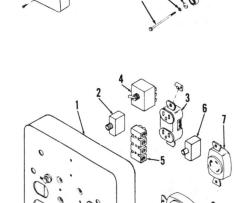

**Fig. D4-2A—Exploded view of cover assembly typical of Models 3W232, 3W232B and 3W232C.**
1. 125V-20A outlet
2. 250V-30A outlet
3. 125V-15A outlet
4. Electric start switch
5. Cover

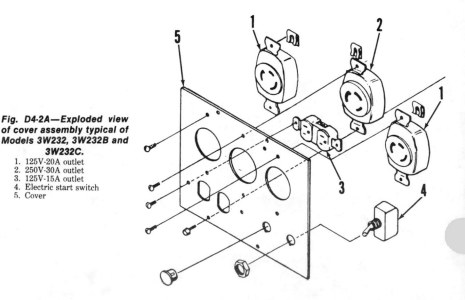

**Fig. D4-2—Exploded view of typical cover assembly. Models 3W177, 3W177A and 3W177C are similar but components (4 through 8) are not used.**

1. Cover
2. 15A circuit breaker
3. 120V outlet
4. Switch
5. Terminal strip
6. 30A circuit breaker
7. 240V twist lock
8. 120V twist lock

# ENGINE

Engine make is listed at beginning of section. Refer to Briggs and Stratton or Tecumseh engine section for engine service.

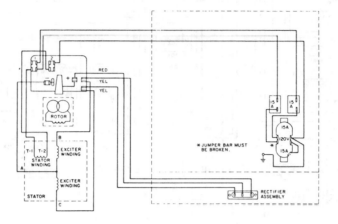

*Fig. D4-3—Wiring schematic for Models 3W177, 3W177A and 3W177C.*

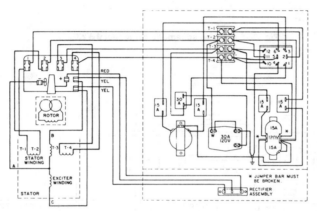

*Fig. D4-4—Wiring schematic for Models 3W178 and 3W179. Model 3W180 is similar but 20A circuit breakers are used in place of 15A circuit breakers in 240 volt circuit.*

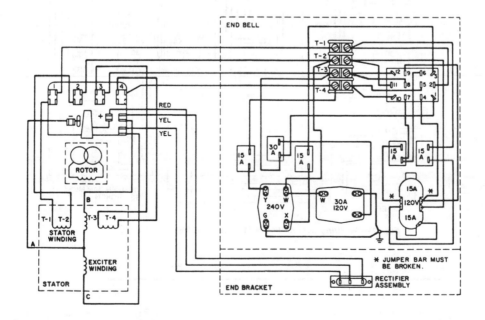

*Fig. D4-5—Wiring schematic for Models 3W178A, 3W178C, 3W179A and 3W179B. Models 3W180A and 3W180B are similar but 20A circuit breakers are used in place of 15A circuit breakers in 240 volt circuit.*

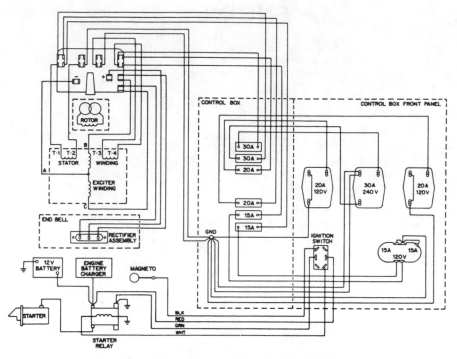

Fig. D4-6 — Wiring schematic for Model 3W232.

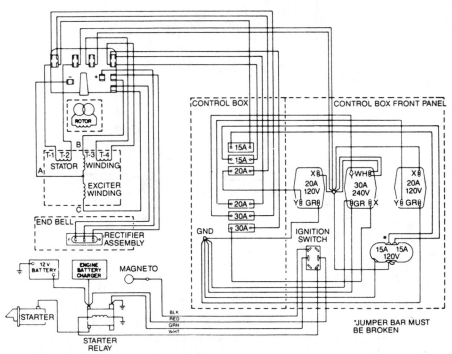

Fig. D4-7 — Wiring schematic for Models 3W232B and 3W232C.

Illustrations courtesy of Dayton Electric Mfg. Co.

# DAYTON

| Model | Output-kw | Voltage | Engine Make | Engine Model | Governed Rpm |
|---|---|---|---|---|---|
| 2W644......... | .75 | 120 | * | * | 3600 |
| 3W351......... | 1.4 | 120 | * | * | 3600 |
| 3W352......... | 1.8 | 120 | * | * | 3600 |

*Engine make and model not available.

# GENERATOR

## MAINTENANCE

Brushes are accessible after removing end cover (13–Fig. D5-1 or D5-2). Brushes should be inspected after the first 1000 hours of operation and thereafter at 500 hour intervals. Replace brushes if worn to a length of ⅜ inch (9.5 mm) or less. Inspect slip rings and polish with crocus cloth if necessary. Bearing is sealed and does not require lubrication, but should be inspected whenever cover is removed.

## TROUBLESHOOTING

If little or no current is generated, check the following: Engine must be in good condition and maintain the governed speed of 3600 rpm. Brushes and slip rings must be in good condition. All wiring connections must be clean and tight. If field magnets are too weak to generate voltage, flash the fields as follows: Run unit and using a 12 volt battery, momentarily connect negative battery terminal to generator frame and positive battery terminal to positive generator brush. Disconnect battery as soon as generator voltage rises.

Condensers (6–Fig. D5-1 and D5-2) are used to suppress radio interference. A shorted condenser will ground generator output. To check condensers, alternately disconnect condenser leads, then note generator output. If generator output improves, then condenser is faulty and must be replaced. If generator output does not improve, then condenser is not faulty and should be reconnected.

To test field coils, refer to Fig. D5-4. Remove end cover, then disconnect field coil leads from brush holders. Using an ohmmeter, check resistance between field coil leads. If resistance is infinite, then field coil is open and must be

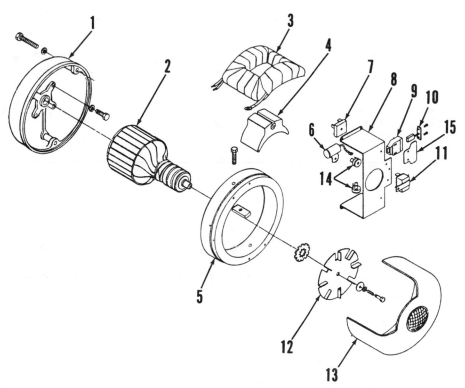

*Fig. D5-1 — Exploded view of generator typical of Models 2W644 and 3W351.*

| | | |
|---|---|---|
| 1. End housing | 6. Condenser | 11. Circuit breaker |
| 2. Armature | 7. Rectifier (full wave) | 12. Fan |
| 3. Field coil | 8. Brush mounting plate | 13. Cover |
| 4. Pole shoe | 9. Brush holder | 14. Rectifier (half wave) |
| 5. Pole shoe retainer | 10. Brush | 15. Barrier spacer |

replaced. Connect an ohmmeter lead to one of the field coil leads, then ground the other ohmmeter lead to frame as shown. If ohmmeter indicates continuity, then coil is grounded and should be replaced.

To test armature, refer to Fig. D5-5. Remove end cover and all brushes. Use an ohmmeter to check for grounded windings by connecting one ohmmeter lead to the armature shaft, then touching the other ohmmeter lead to each slip ring and each commutator segment. Use an ohmmeter to check for resistance between front and middle slip rings, then between middle and rear slip rings. Ohmmeter should indicate little or no resistance between slip rings; otherwise armature windings are open and armature must be renewed. Additional electrical tests may be performed with armature removed from generator.

Two half wave rectifiers (14 – Figs. D5-1 and D5-2) and one full wave rectifier (7 – Figs. D5-1 and D5-2) are used. Half wave rectifiers are used to excite the field coil and provide battery charging while the full wave rectifier is used in the Super Maxi-Watt boost circuit. Rectifiers may be checked with a battery operated test light with reference to Fig. D5-3. Half wave rectifiers have three terminals; two AC terminals and a positive terminal in the center which is marked by a red dot.

To test half wave rectifier, clip one lead of test light to the positive terminal and touch the other test light lead to one of the AC terminals and then the other. Test light should either light up or remain unlit when each AC terminal is touched. When test light leads are reversed, result should be opposite. If test light is lit when one AC terminal is touched and unlit when the other AC terminal is touched, then rectifier is defective and should be replaced.

The full wave rectifier has four terminals; two AC terminals, one positive terminal and one negative terminal. Test procedure for the full wave rectifier is similar to the test procedure for half wave rectifiers. Clip one test light lead to the positive terminal and touch the other test light lead to one of the AC terminals and then the other. Test light should either light up or remain unlit when each AC terminal is touched. When test light leads are reversed, result should be opposite. If test light is lit when one AC terminal is touched and unlit when the other AC terminal is touched, then rectifier is defective and should be replaced. Repeat the test sequence with one test light lead clipped to the negative terminal in place of the positive terminal. If the entire full wave rectifier is good, test light will be lit

when connected to each of the two AC terminals during only one of the two test sequences. If other results are obtained, full wave rectifier is defective and should be replaced.

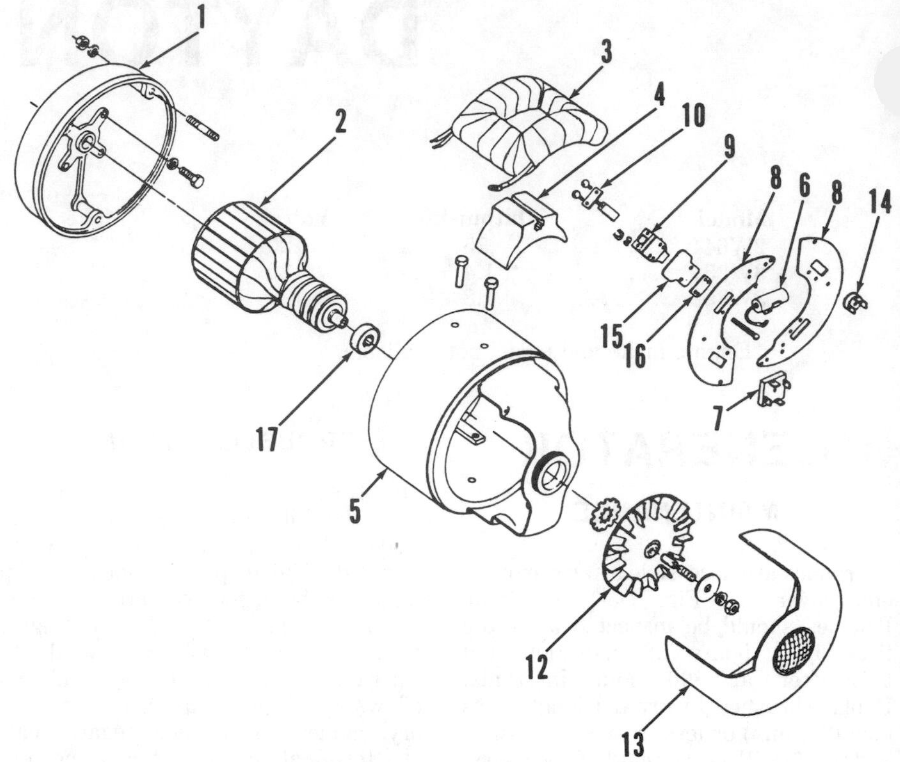

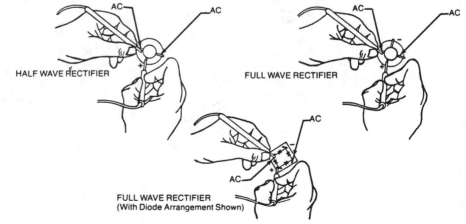

**Fig. D5-2—Exploded view of generator Model 3W352.**

| | | |
|---|---|---|
| 1. End housing | 5. Field shell | 9. Brush holder | 14. Rectifier (half wave) |
| 2. Armature | 6. Capacitor | 10. Brush | 15. Barrier spacer |
| 3. Field coil | 7. Rectifier (full wave) | 12. Fan | 16. Fiber spacer |
| 4. Pole shoe | 8. Brush mounting plate | 13. Cover | 17. Bearing |

**Fig. D5-3—View showing rectifier test sequence.**

## OVERHAUL

Disassembly and reassembly of generator is evident after inspection of unit and referral to Figs. D5-1 and D5-2.

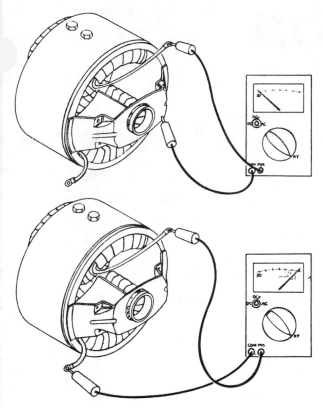

*Fig. D5-4 — Check field coil for shorted or open windings with an ohmmeter. Refer to text for procedure.*

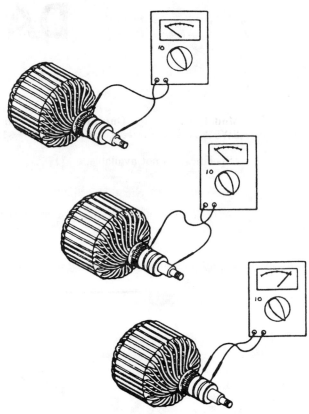

*Fig. D5-5 — Check armature for open or shorted windings with an ohmmeter. Refer to text for procedure.*

# DAYTON

| Model | Output-kw | Voltage | Engine Make | Model | Governed Rpm |
|-------|-----------|---------|-------------|-------|--------------|
| 2W512......... | .4 | 120 | Shibaura | * | * |

*Specification not available.

For service information on this unit, refer to the Kohler section and the Model Power Play 500.

Illustrations courtesy of Dayton Electric Mfg. Co.

# JOHN DEERE

**DEERE AND COMPANY**
Moline, Illinois 61265

| Model | Output-kw | Voltage | Engine Make | Engine Model | Governed Rpm |
|---|---|---|---|---|---|
| 225 . . . . . . . . . . . | 2.25 | 120 | B&S | 130212 | 3600 |
| 325 . . . . . . . . . . . | 3.25 | 120, 240 | B&S B&S | 190432 or 195432 | 3600 |
| 375 . . . . . . . . . . . | 3.75 | 120, 240 | B&S | 190432 or 195432 | 3600 |
| 500 . . . . . . . . . . . | 5.0 | 120, 240 | B&S B&S | 220432 or 221432 | 3600 |
| 650 . . . . . . . . . . . | 6.5 | 120, 240 | B&S Onan | 326435 or B43M | 3600 |

# GENERATOR

### MAINTENANCE

Generator brushes and slip rings should be inspected periodically and are accessible after removing power panel. Renew brushes if worn to a length of 5/16 inch (7.9 mm) or less as shown in Fig. JD1-10. Slip ring surfaces must be clean and bright. Use 240 grit sandpaper to smooth slip rings as necessary. When end bell (7 – Fig. JD1-2) is removed, inspect bearing (1) for smooth operation and renew if necessary.

### TROUBLESHOOTING

If little or no output is generated, check the following: Check circuit breakers and reset if needed. Engine must be in good condition and maintain a governed speed under load of 3600 rpm. All wiring connections must be clean and tight. Brushes and slip rings must be in good condition.

If field coils have lost their residual magnetism, refer to Fig. JD1-5 and unbolt power panel to gain access to brush terminals and run unit. Using a 6 or 12 volt battery, connect a lead from negative battery terminal to generator frame, then momentarily connect a lead from the positive battery terminal to the positive brush terminal. Disconnect battery as soon as generator voltage rises. Reinstall power panel. If loss of field magnetism is chronic, capacitor (5 – Fig. JD1-2) may be defective and should be renewed.

Test rotor for open windings with an ohmmeter as shown in Fig. JD1-6. Specified resistance between slip rings is as follows:

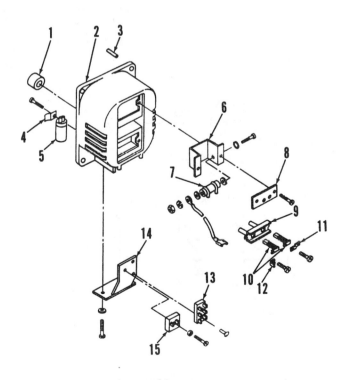

*Fig. JD1-2 – Exploded view of end bell assembly.*

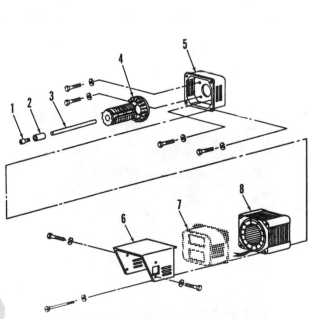

*Fig. JD1-1 – Exploded view of generator.*

| | | |
|---|---|---|
| 1. Screw | 5. Adapter | |
| 2. Sleeve | 6. Power panel shroud | |
| 3. Stud | 7. End bell | |
| 4. Rotor | 8. Stator | |

| | | |
|---|---|---|
| 1. Bearing | 6. Bracket | 11. Terminal |
| 2. Housing | 7. Resistor | 12. Terminal |
| 3. Dowel pin | 8. Guard | 13. Terminal |
| 4. Bracket | 9. Brush holder | 14. Bracket |
| 5. Capacitor | 10. Brushes | 15. Rectifier |

| MODEL | OHMS |
|-------|------|
| 225 | 21.0-22.6 |
| 325 | 21.0-25.6 |
| 375 | 22.5-27.5 |
| 500 | 26.2-32.0 |
| 650 | 31.0-37.9 |

Test rotor for shorts by placing one ohmmeter lead on rotor shaft or frame and the other lead on one of the slip rings. If there is continuity, rotor windings are grounded. Repeat procedure with the other slip ring. Renew rotor if there are opens or shorts.

Test stator for open or shorted windings with an ohmmeter as shown in Fig. JD1-7. Clip stator leads T2 and T3 together and measure between leads T1 and T4. Specified resistance values for stator windings are as follows:

| MODEL | OHMS |
|-------|------|
| 225 | 1.22-1.57 |
| 325 | 0.75-0.96 |
| 375 | 0.55-0.67 |
| 500 | 0.37-0.47 |
| 650 | 0.25-0.33 |

Specified resistance values for excitor windings, measured between leads X1 and X2, are as follows:

| MODEL | OHMS |
|-------|------|
| 225 | 1.54-2.75 |
| 325 | 1.54-2.31 |
| 375 | 1.52-1.86 |
| 500 | 1.29-2.20 |
| 650 | 1.14-2.20 |

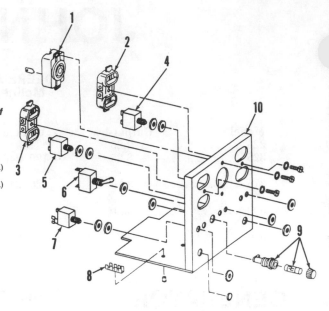

*Fig. JD1-4 — Exploded view of typical control panel.*

1. Receptacle (120/240V, 30A)
2. Receptacle (120V, 15A)
3. Receptacle (240V, 20A)
4. Circuit breaker (120V, 15A)
5. Circuit breaker (120/240V, 30A)
6. Switch
7. Circuit breaker (120/240V, 30A)
8. Terminal
9. Fuse assy. (8A)
10. Panel

Renew stator if windings are open or shorted together. Test stator for grounds by attaching one ohmmeter lead to stator frame and the other lead to one of the stator leads. Repeat for each one of the stator and excitor winding leads. If there is continuity, then windings are grounded and stator assembly must be renewed. Burnt windings or melted insulation is also an indication of a defective stator.

To test rectifier, remove exciter winding leads, X2 and X3 from AC terminals of rectifier. Refer to Fig. JD1-9. Rectifier may be tested with either an ohm-

meter or a battery operated test light. Clip one test light or ohmmeter lead to positive terminal of rectifier and touch the other lead first to one of the AC terminals of the rectifier and then to the other AC terminal. Test light or ohmmeter should either show continuity or infinity when each AC terminal is touched. When leads are reversed, the result should be opposite. If test light or ohmmeter shows continuity when one AC terminal is touched and does not show continuity when the other AC terminal is touched, then rectifier is defective and should be renewed. Repeat test se-

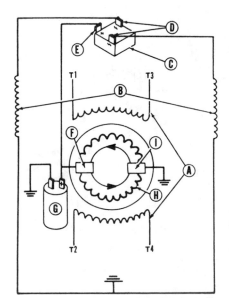

*Fig. JD1-3 — View of basic generator circuitry.*

A. Stator output windings
B. Excitor windings
C. Rectifier
D. AC terminals
E. Positive terminal
F. Positive brush
G. Capacitor
H. Rotor winding
I. Negative brush

*Fig. JD1-5 — View showing procedure for flashing fields with a 6-volt battery.*

A. Positive brush terminal
B. Positive battery terminal
C. Negative battery terminal

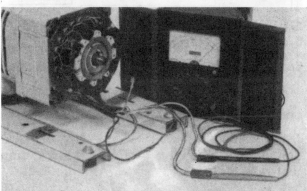

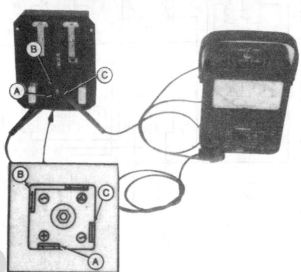

*Fig. JD1-6 — Test rotor as shown for open windings.*

*Fig. JD1-7 — Test stator for open or shorted windings with an ohmmeter as shown. Refer to text for specifications.*

*Fig. JD1-8 — View showing correct capacitor lead routing. Note that white ground lead (B) must be connected to unpainted capacitor terminal and black lead from positive brush must be connected to capacitor terminal that is painted red.*

quence with one test light or ohmmeter lead clipped to negative rectifier terminal in place of the positive terminal. If the entire rectifier is good, test light or ohmmeter will show continuity at each of the two AC terminals during only one of the two test sequences. If other results are obtained, then rectifier is defective and should be renewed.

## OVERHAUL

Refer to Fig. JD1-1 for exploded view of generator and Fig. JD1-2 for exploded view of end bell assembly. To disassemble, remove retaining screws and disconnect wiring from power panel assembly (6 – Fig. JD1-1). Remove brushes (10 – Fig. JD1-2) from brush holder (9) on end bell. Unbolt and remove end bell assembly (2). Using a soft hammer, tap stator away from engine adapter (5 – Fig. JD1-1). Remove socket head screw (1) at end of rotor (4) and tap end of rotor with a soft hammer to loosen taper. Slip rotor off stud (3).

Reassembly is reverse of disassembly procedure. When installing rotor, thread stud (3) halfway into end of crankshaft. Slide rotor over stud and torque socket head screw to 102-180 in.-lbs. (13.56-20.34 N·m). After torquing, tap end of rotor with a soft hammer to seat rotor on crankshaft taper, then retorque.

To replace end bearing (1 – Fig. JD1-2), drill out rivets which retain brush holder (9) to housing (2) with a 1/8 inch drill, then remove holder. Tap old bearing out of housing. Note that it may be necessary to support housing in the center when driving old bearing out to avoid bending housing. Tap new bearing into housing and rivet or screw brush holder back into position. Torque end bell and stator retaining cap screws to 60-72 in.-lbs. (6.78-8.14 N·m). Refer to wiring schematics when reinstalling power panels.

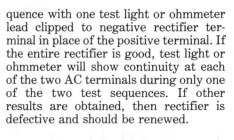

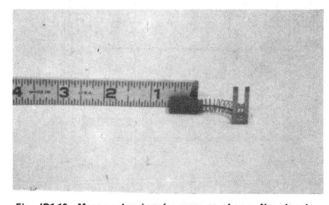

*Fig. JD1-10 — Measure brushes for wear as shown. New brushes measure 9/16 inch (14.3 mm) and minimum used brush length is 5/16 inch (7.9 mm).*

*Fig. JD1-9 — Test rectifier with an ohmmeter as shown.*
A. Positive terminal
B. AC terminal
C. AC terminal

Illustrations for Fig. JD1-6, Fig. JD1-7, Fig. JD1-8, Fig. JD1-9 and Fig. JD1-10 reproduced by permission of Deere & Company. Copyright Deere & Company.

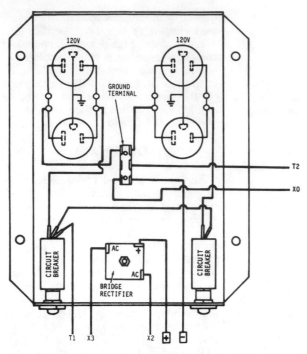

Fig. JD1-11 — *Power panel wiring schematic for Model 225 with serial number 640,281 and earlier.*

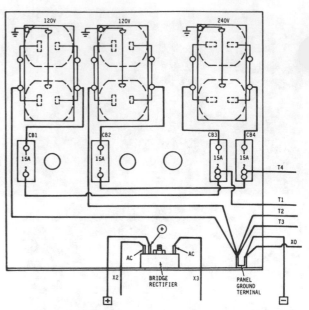

Fig. JD1-13 — *Power panel wiring schematic for Models 325 and 375 with serial number 640,281 and earlier.*

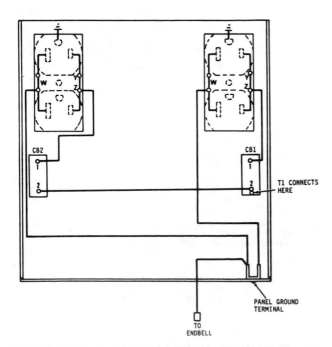

Fig. JD1-12 — *Power panel wiring schematic for Model 225 with serial number 640,282 and later.*

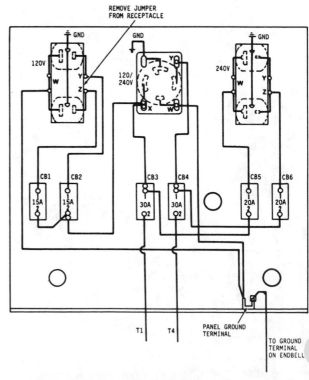

Fig. JD1-14 — *Power panel wiring schematic for Models 325, 375 and 500 with serial number 640,282 and later.*

# IDLE-MATIC

### OPERATION

The Idle-Matic system is optional on Models 325, 375, 500 and 650. The system closes the throttle when no load is imposed on the generator. A two-position switch activates the system when in "ON" mode or disables the system when in "OFF" mode. In "OFF" mode engine runs at governed speed regardless of generator load.

Wiring schematics are shown in Figs. JD1-20, JD1-21 nd JD1-22. Stator leads are routed through an electronic circuit board which controls the system. A solenoid is attached to the carburetor to raise and lower engine speed.

### TROUBLESHOOTING

If Idle-Matic system is inoperative, first check the 1 amp fuse which is mounted on the circuit board and replace if needed. Check wiring for loose connections and check solenoid for proper operation and adjustment. The electronic circuit board may only be tested by substitution.

# ENGINE

Engine make and model are listed at beginning of section. Refer to Briggs and Stratton or Onan engine section for engine service.

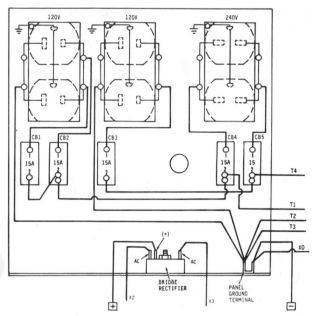

*Fig. JD1-15 — Power panel wiring schematic for Model 500 with serial number 640,281 and earlier.*

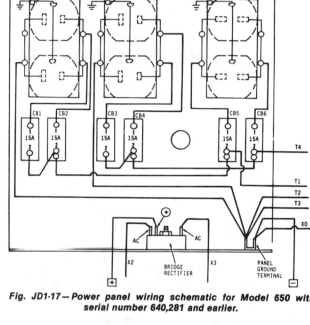

*Fig. JD1-17 — Power panel wiring schematic for Model 650 with serial number 640,281 and earlier.*

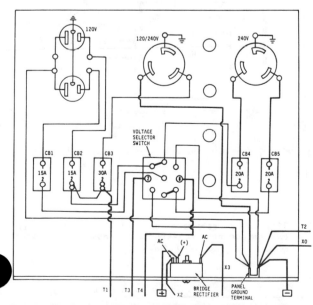

*Fig. JD1-16 — Contractor's power panel wiring schematic for Models 325, 375 and 500 with serial number 640,281 and earlier.*

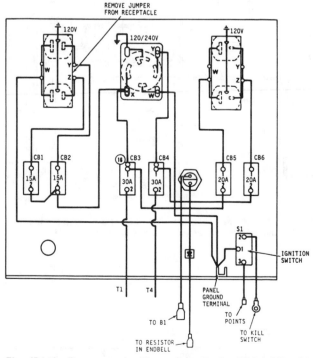

*Fig. JD1-18 — Power panel wiring schematic for Model 650 with serial number 640,282 and later.*

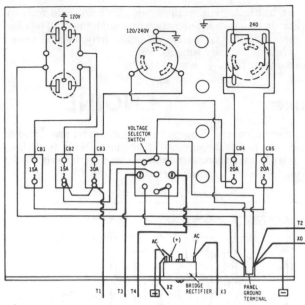

Fig. JD1-19 — Contractor's power panel wiring schematic for Model 650 with serial number 640,281 and earlier.

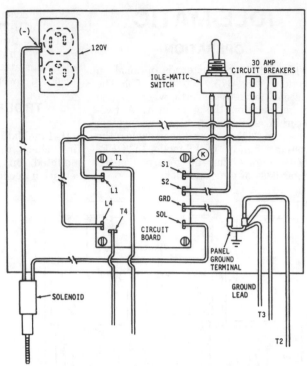

Fig. JD1-21 — Wiring schematic for Idle-Matic circuit used on generators with serial number 640,282 and later.

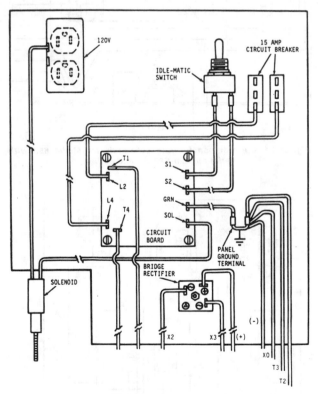

Fig. JD1-20 — Wiring schematic for Idle-Matic circuit used on generators with serial number 640,281 and earlier.

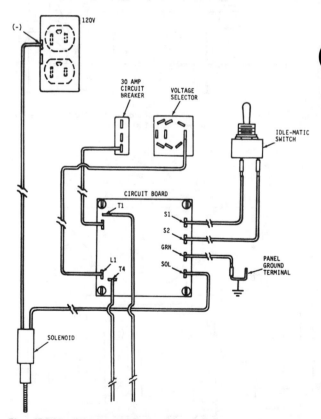

Fig. JD1-22 — Wiring schematic for Idle-Matic circuit used on generators with contractor's power panel and serial number 640,281 and earlier.

# JOHN DEERE

| Model | Output-kw | Voltage | Engine | | Governed Rpm |
|-------|-----------|---------|--------|--------|--------------|
| | | | **Make** | **Model** | |
| 1750W......... | 1.75 | 125 | Tec. | HS40 | 3600 |
| 3000W......... | 3.0 | 125, 250 | Tec. | HM80 | 3600 |
| 4000W......... | 4.0 | 125, 250 | Tec. | HH100 | 3600 |
| 5000W......... | '5.0 | 125, 250 | Tec. | HH140 | 3600 |

# GENERATOR

### MAINTENANCE

Generator brushes and slip rings should be inspected periodically and are accessible after removing panel cover (10–Fig. JD2-2). Renew brushes if worn to a length of 5/16 inch (7.9 mm) or less. Slip ring surfaces must be clean and bright. Use 240 grit sandpaper to smooth slip rings as necessry. When end bell (1–Fig. JD2-1) is removed, inspect bearing (3) for smooth operation and replace if necessary.

### TROUBLESHOOTING

If little or no output is generated, check the following: Check circuit breakers and reset if needed. Engine must be in good condition and maintain a governed speed under load of 3600 rpm. All wiring connections must be clean and tight. Brushes and slip rings must be in good condition.

If field coils have lost their residual magnetism, use the following procedure: Remove end receptacle panel or cover plate and run unit. Using a 6 or 12 volt battery, connect a lead from negative battery terminal to generator frame, then momentarily connect a lead from the positive battery terminal to the

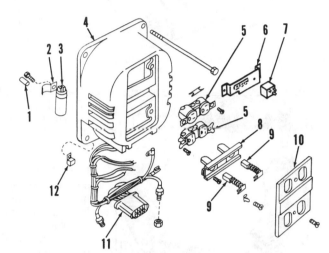

**Fig. JD2-2 – Exploded view of end bell.**
1. Dowel pin
2. Bracket
3. Capacitor
4. Housing
5. Receptacle
6. Bracket
7. Rectifier (late)
8. Brush holder
9. Brush
10. Cover
11. Harness (early)
12. Terminal plug

positive brush terminal. Disconnect battery as soon as generator voltage rises. Reinstall end receptacle panel or cover plate. If loss of field magnetism is chronic, capacitor (3–Fig. JD2-2) may be defective and should be renewed.

To check rectifier diodes (D–Fig. JD2-3) on early models, unsolder and remove diodes from end bell. Connect ohmmeter or battery operated test light leads to diode and note whether or not there is continuity. Reverse leads and again note whether or not there is continuity. If there is continuity when leads

are connected both ways or if there is no continuity when leads are connected both ways, then diode is faulty and must be renewed. The later type rectifier shown in Fig. JD2-4 may be tested with either an ohmmeter or a battery operated test light. Disconnect black positive (+) lead and X2 and X3 AC stator leads from rectifier. Clip one test light or ohmmeter lead to positive terminal of rectifier and touch the other lead to one of the AC terminals on the rectifier, then touch lead to remaining AC terminal. Test light or ohmmeter

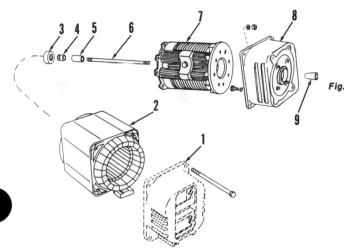

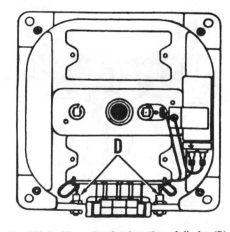

**Fig. JD2-1 – Exploded view of generator.**
1. End bell
2. Stator
3. Bearing
4. Screw
5. Sleeve
6. Stud
7. Rotor
8. Housing
9. Adapter

**Fig. JD2-3 – View showing location of diodes (D) on early models.**

Illustrations for Fig. JD2-1, Fig. JD2-2 and Fig. JD2-3 reproduced by permission of Deere & Company. Copyright Deere & Company.

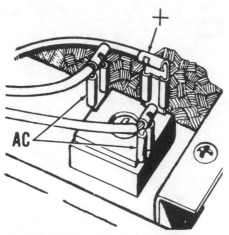

*Fig. JD2-4 — View of rectifier showing (AC) terminals and positive ( + ) terminal.*

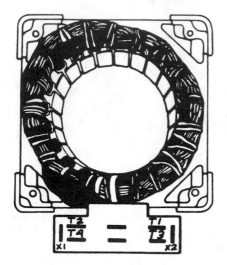

*Fig. JD2-5 — View of stator and terminal board on early models.*

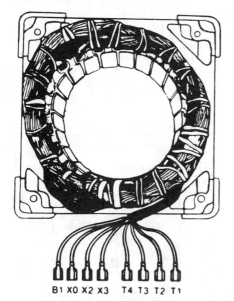

*Fig. JD2-6 — View of stator and leads on later models.*

| B1. Orange | T4. Yellow |
| T1. Blue | X0. Black |
| T2. Black | X2. Lt. Blue |
| T3. Red | X3. Lt. Blue |

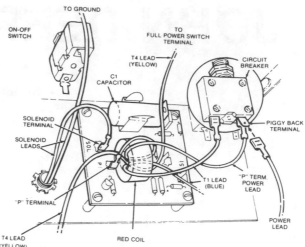

*Fig. JD2-7 — On early models, route wires to Idle-Matic control board as shown. On 120 volt only models, stator wire T3 replaces T4 wire shown and T3 wire must pass through red coil in same direction as blue T1 stator wire.*

should either show continuity or infinity when each AC terminal is touched. When leads are reversed, the result should be opposite. If test light or ohmmeter shows continuity when one AC terminal is touched and does not show continuity when the other AC terminal is touched, then rectifier is defective and should be replaced. Repeat test sequence with one test light or ohmmeter lead clipped to negative rectifier terminal in place of the positive terminal. If the entire rectifier is good, test light or ohmmeter will show continuity at each of the two AC terminals during only one of the two test sequences. If other results are obtained, then rectifier is defective and should be replaced.

To check stator, note terminal board in Fig. JD2-5 used on early models and loose wire terminals shown in Fig. JD2-6 used on later models. On all models, connect a jumper wire between terminals T2 and T3, then measure resistance between terminals T1 and T4. Note length of stator frame and desired resistance values in following table:

| Stator Frame Length | T1 to T4 Resistance |
| --- | --- |
| 1½ in. (38.1 mm) | 2.57-3.30 ohms |
| 2¼ in. (57.2 mm) | 1.22-1.57 ohms |
| 3 in. (76.2 mm) | 0.75-0.96 ohms |
| 3½ in. (88.9 mm) | 0.55-0.67 ohms |
| 4½ in. (114.3 mm) | 0.37-0.47 ohms |
| 6 in. (152.4 mm) | 0.25-0.33 ohms |

To check excitor coil in stator, connect ohmmeter leads to terminals X1 and X2 on early model terminal board (Fig. JD2-5) or to X2 and X3 stator leads of later models shown in Fig. JD2-6. Note length of stator frame and desired resistance values in following table:

| Stator Frame Length | Exciter Coil Resistance |
| --- | --- |
| 1½ in. (38.1 mm) | 2.04-3.30 ohms |
| 2¼ in. (57.2 mm) | 1.54-2.75 ohms |
| 3 in. (76.2 mm) | 1.54-2.31 ohms |
| 3½ in. (88.9 mm) | 1.52-1.86 ohms |
| 4½ in. (114.3 mm) | 1.29-2.20 ohms |
| 6 in. (152.4 mm) | 1.14-2.20 ohms |

To determine if rotor is shorted to ground, connect ohmmeter leads between slip rings and rotor shaft. Renew rotor if there is continuity between slip rings and ground. Measure length of stator frame and refer to the following table for specified resistance values in ohms when reading is taken between rotor slip rings.

| Stator Frame Length | Rotor Resistance |
| --- | --- |
| 1½ in. (38.1 mm) | 25.7-31.5 ohms |
| 2¼ in. (57.2 mm) | 18.5-22.6 ohms |
| 3 in. (76.2 mm) | 21.0-25.6 ohms |
| 3½ in. (88.9 mm) | 22.5-27.5 ohms |
| 4½ in. (114.3 mm) | 26.2-32.0 ohms |
| 6 in. (152.4 mm) | 31.0-37.9 ohms |

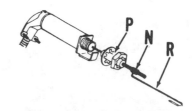

*Fig. JD2-8 — View of Idle-Matic solenoid. Refer to text for adjustment.*

### OVERHAUL

Refer to Fig. JD2-1 for exploded view of generator and Fig. JD2-2 for exploded view of end bell assembly. To disassemble, remove cover (10–Fig. JD2-2). Remove brushes (9) from brush holder (8) on end bell. Unbolt end bell assembly and gently tap away from stator with a soft hammer. Disconnect stator wiring from end bell. Using a soft hammer, tap stator (2–Fig. JD2-1) away from engine adapter (9). Remove socket head screw (4) at end of rotor (7)

and tap end of rotor with a soft hammer to loosen taper. Slip rotor off stud (6).

Reassembly is reverse of disassembly procedure. When installing rotor, thread stud (6) halfway into end of engine crankshaft. Slide rotor over stud and torque socket head screw to 102-180 in.-lbs. (13.56-20.34 N·m). After torquing, tap end of rotor with a soft hammer to seat rotor on crankshaft taper, then retorque. To replace end bearing (3), drill out rivets which retain brush holder (8–Fig. JD2-2) to housing (4) with a 1/8 inch drill, then remove holder. Tap old

bearing out of housing. Note that it may be necessary to support housing in the center when driving old bearing out to avoid bending housing. Tap new bearing into housing and rivet or screw brush holder back into position. Torque end bell and stator retaining cap screws to 60-72 in.-lbs. (6.78-8.14 N·m).

# IDLE-MATIC

Some models may be equipped with a solid state unit attached to the receptacle panel and a solenoid connected to the carburetor throttle. The system senses generator load and will reduce engine speed to idle when load is removed.

To adjust solenoid, first set engine idle speed screw so engine will idle at 1800 rpm. With engine stopped, push solenoid plunger (P–Fig. JD2-8) into solenoid until bottomed. Turn adjusting nuts (N) so linkage rod (R) pulls carburetor throttle arm against idle speed screw. Tighten adjusting nuts and check operation of solenoid.

# ENGINE

Engine model is listed at beginning of section. Refer to Tecumseh engine section for engine service.

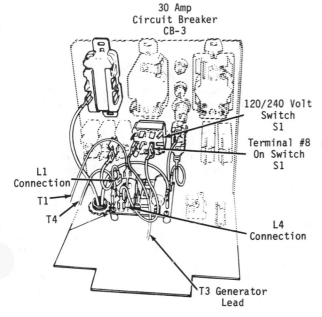

**Fig. JD2-9 — View of Idle-Matic control board on later models.**

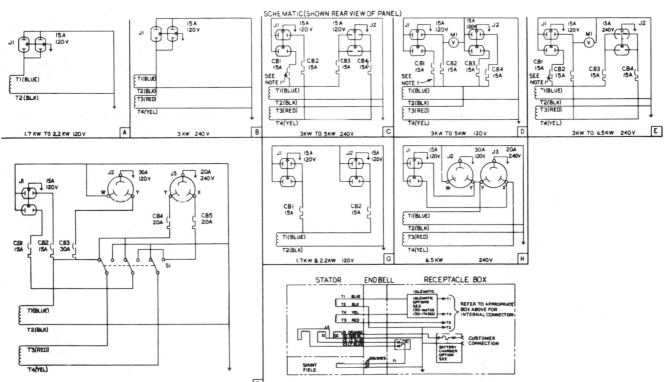

**Fig. JD2-10 — Typical wiring diagram.**

Illustrations for Fig. JD2-9 and Fig. JD2-10 reproduced by permission of Deere & Company. Copyright Deere & Company.

# GENERAC

**GENERAC CORPORATION**
**P.O. Box 8**
**Waukesha, WI 53187**

| Model | Output-watts | Voltage | Engine Make | Engine Model | Governed Rpm |
|---|---|---|---|---|---|
| G1000 | 750 | 120 | Kawasaki | FA76 | 3600 |
| G1600 | 1400 | 120 | Kawasaki | FA130 | 3600 |
| G2600 | 2400 | 120 | Kawasaki | FA210 | 3600 |
| G4000 | 3600 | 120 | Kawasaki | FG300 | 3600 |

# GENERATOR

## MAINTENANCE

All bearings are sealed and require no lubrication, but should be inspected for smooth operation when generator is disassembled. Inspect unit periodically for condition of receptacles, electrical connections and operation of all switches.

## TROUBLE-SHOOTING

If little or no output is generated, check the following: The engine must be in good condition and maintain a governed speed under load of 3600 rpm. Check condition of circuit breakers and make sure all wiring connections are clean and tight. Rotor slip rings and brushes (if used) must be in good condition.

**NOTE: Even with generator stopped the capacitor used on Model G1000 should have a retained charge if working properly. Do not touch capacitor terminals as electric shock may result. Discharge capacitor by shorting across terminals with a screwdriver or similar tool with an insulated handle.**

To test capacitor (3—Fig. GE5) on Model G1000, use an ohmmeter set to the RX1000 scale. With ohmmeter leads connected to the capacitor terminals, a meter deflection should be seen followed by a slow return to infinity. No meter deflection or continuing continuity indicates an open or shorted capacitor.

If field coil magnets have lost their residual magnetism, use the following procedure: On Model G1000, operate generator at specified speed and using a 12 volt battery as a power source, "flash" across capacitor terminals by connecting jumper wires from the battery to the capacitor terminals for not more than one second. Generator output should then build to normal level. On all models except G1000, remove end cover plate (1—Fig. GE6) to gain access to brush terminals. Operate generator at specified speed and using a 6 or 12 volt battery as a power source, connect negative battery terminal to generator frame and the positive battery terminal to the positive generator brush. Disconnect battery as soon as generator voltage rises.

All models except G1000 are equipped with a diode (3—Fig. GE6) which may be checked after removing diode from generator. Using an ohmmeter or battery powered test light, connect tester leads to diode and note reading, then reverse tester leads and again note reading. Ohmmeter should read very high for one connection and very low for the other connection. If both readings are high or low, diode is faulty and must be renewed. If using a test light, lamp should light with leads connected one way and should not light when leads are reversed.

All models except G1000 are equipped with a bridge rectifier (4—Fig. GE6) attached to generator end frame. To test rectifier, disconnect wiring leads and using an ohmmeter, connect one test lead to rectifier positive terminal and touch other test lead to first one of the AC terminals and then the other. Note ohmmeter reading, then reverse the tester connections and again note reading. Ohmmeter should show either continuity or infinity when each terminal is touched. When leads are reversed, the result should be the opposite. Repeat test sequence with one ohmmeter lead

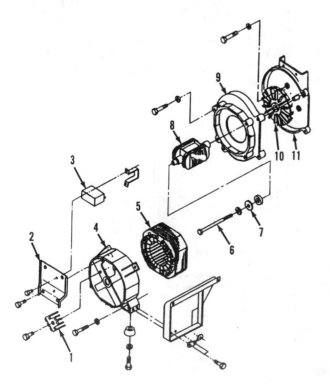

**Fig. GE5—Exploded view of Model G1000 generator**

1. Battery charge rectifier
2. End bracket
3. Capacitor
4. End housing
5. Stator assy.
6. Rotor bolt
7. Bearing
8. Rotor assy.
9. Fan housing
10. Cooling fan
11. Engine adapter housing

connected to rectifier negative terminal. If test results are not as outlined above, renew rectifier.

If the above tests do not locate malfunction, generator must be disassembled and rotor and stator tested for open and short circuits.

## OVERHAUL

Disassembly and reassembly of generator is evident after inspection of unit. Refer to Fig. GE5 or GE6 for an exploded view of typical generators. Sealed bearings should be inspected for smooth operation when generator is disassembled.

Refer to Figs. GE7 through GE9 for the wiring schematic for model being serviced.

# ENGINE

Refer to the appropriate Kawasaki engine section for engine service.

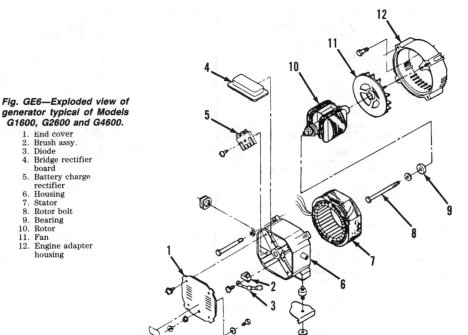

*Fig. GE6—Exploded view of generator typical of Models G1600, G2600 and G4600.*

1. End cover
2. Brush assy.
3. Diode
4. Bridge rectifier board
5. Battery charge rectifier
6. Housing
7. Stator
8. Rotor bolt
9. Bearing
10. Rotor
11. Fan
12. Engine adapter housing

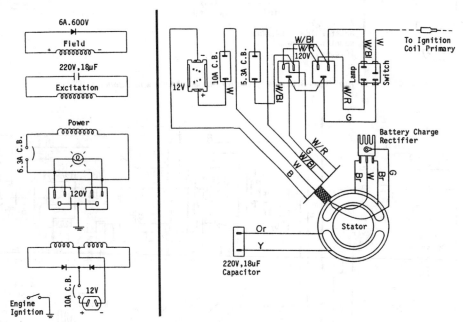

*Fig. GE7—Wiring diagram for Model G1000.*

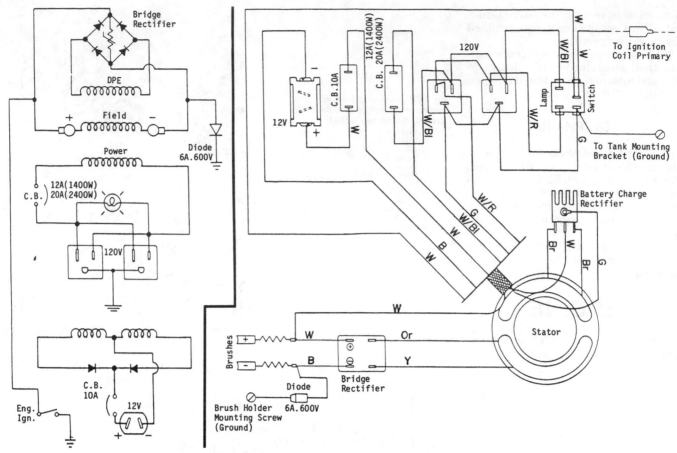

Fig. GE8—Wiring diagram for Models G1600 and G2600.

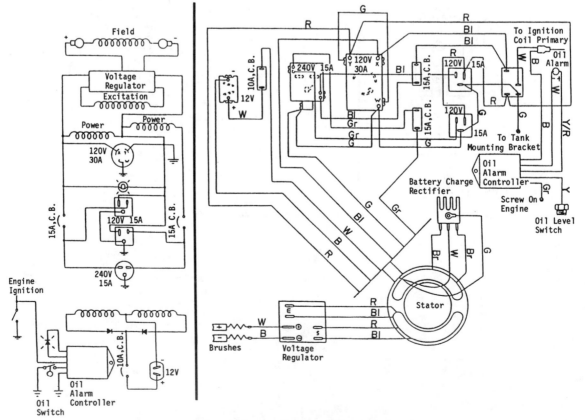

Fig. GE9—Wiring diagram for Model G4000.

# GENERAC

| Model | Output-watts | Voltage | Engine Make | Engine Model | Governed Rpm |
|-------|-------------|---------|------|-------|--------------|
| S4000 | 4000 | 120 | Tecumseh | HM80 | 3600 |
| S4002 | 4000 | 120 | Tecumseh | HM80 | 3600 |
| S5002 | 5000 | 120 | Tecumseh | HM100 | 3600 |

# GENERATOR

### MAINTENANCE

All bearings are sealed and require no lubrication, but should be inspected for smooth operation when generator is disassembled. Inspect unit periodically for condition of receptacles, electrical connections and operation of all switches.

### TROUBLE-SHOOTING

If little or no output is generated, check the following: The engine must be in good condition and maintain a governed speed under load of 3600 rpm. Check condition of circuit breakers and make sure all wiring connections are clean and tight. Rotor slip rings and brushes must be in good condition.

If field coil magnets have lost their residual magnetism, use the following procedure: Remove end cover plate (2—Fig. GE15 or GE16) to gain access to brush terminals. Operate generator at specified speed and using a 6 or 12 volt battery as a power source, connect negative battery terminal to generator frame and the positive battery terminal to the positive generator brush. Disconnect battery as soon as generator voltage rises.

All models are equipped with a bridge rectifier (7) attached to generator end frame. To test rectifier, disconnect wiring leads and using an ohmmeter, connect one test lead to rectifier positive terminal and touch other test lead to one of the AC terminals and then the other. Note ohmmeter reading, then reverse the tester connections and again note reading. Ohmmeter should show either continuity or infinity when each terminal is touched. When leads are reversed, the result should be the opposite. Repeat test sequence with one ohmmeter lead connected to rectifier negative terminal. If test results are not as outlined above, renew rectifier.

If the above tests do not locate malfunction, generator must be disassembled and rotor and stator windings tested for open and short circuits.

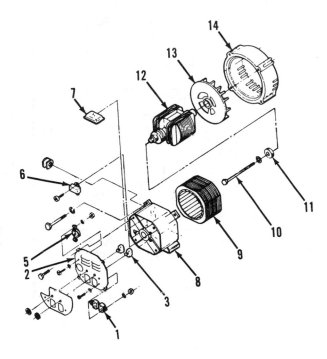

**Fig. GE15—Exploded view of Model S4000 generator.**
1. Receptacle (120V)
2. Receptacle panel
3. Circuit breaker (20A)
4. Receptacle (240V)
5. Brush holder
6. Bridge rectifier
7. Bridge rectifier
8. End housing
9. Stator assy.
10. Rotor bolt
11. Bearing
12. Rotor assy.
13. Fan
14. Engine adapter housing

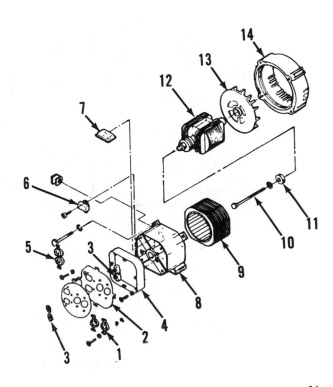

**Fig GE16—Exploded view of Model S4002 and S5002 generator.**
1. Receptacles (120/240V)
2. Receptacle panel
3. Circuit breakers (20A)
4. Panel shroud
5. Receptacle (120V)
6. Brushes
7. Bridge rectifier
8. End housing
9. Stator assy.
10. Rotor bolt
11. Bearing
12. Rotor assy.
13. Fan
14. Engine adapter housing

## OVERHAUL

Disassembly and reassembly of generator is evident after inspection of unit. Refer to Fig GE15 and GE16 for an exploded view of model being serviced.

Sealed bearings should be inspected for smooth operation when generator is disassembled.

Refer to Figs. GE17 and GE18 for the wiring schematic for model being serviced.

# ENGINE

Refer to the appropriate Tecumseh engine section for engine service.

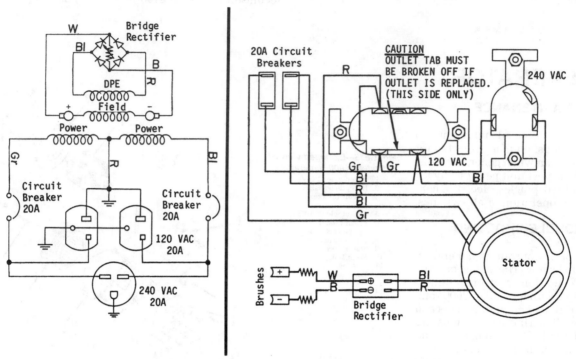

*Fig. GE17—Wiring diagram for Model S4000.*

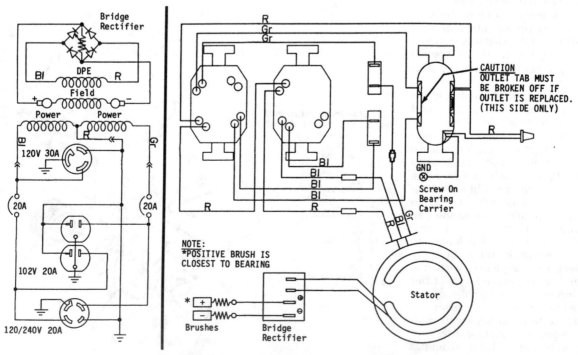

*Fig. GE18—Wiring diagram for Models S4002 and S5002.*

Illustrations courtesy of Generac Corp.

# HOMELITE

**HOMELITE DIV. OF TEXTRON**
**P.O. Box 7047**
**14401 Carowinds Blvd.**
**Charlotte, N.C. 28241**

| Model | Output-kw | Voltage | Engine Make | Engine Model | Governed Rpm |
|-------|-----------|---------|-------------|--------------|--------------|
| 9A34-1, 9A34-1A | 3.5 | 120* | Homelite | 9 | 3600 |
| 9A34-3, 9A34-3A | 3.4 | 120/240 | Homelite | 9 | 3600 |

*May be converted to 240 volt by using kit.

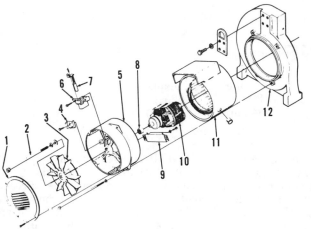

**Fig. H1-1—Exploded view of generator.**

1. Cover
2. Bolt
3. Fan
4. Rectifier
5. Brush head
6. Brush holder
7. Brush
8. Bearing
9. Resistor
10. Rotor
11. Stator assy.
12. Adapter

# GENERATOR

## MAINTENANCE

Brush length and condition should be checked after every 1000 hours of operation. Brushes can be inspected after removing brush head cover (1 – Fig. H1-1), rotor bolt (2) and fan (3).

**NOTE: Do not attempt to hold rotor with fan to unscrew rotor bolt.**

Loosen screws retaining brush holder and slide out brush holder. Brushes must be renewed if brush length is shorter than 3/0 inch (9.5 mm) as shown in Fig. H1-2. Red and black wires must be connected to brushes and routed as shown

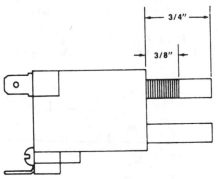

**Fig. H1-2 — Brush length should not be less than 3/8 inch (9.5 mm) when measured as shown. New brush length is 3/4 inch (19 mm).**

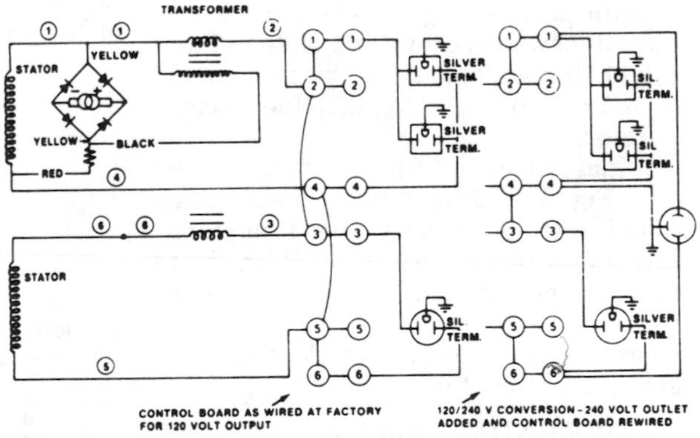

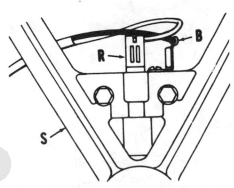

**Fig. H1-3—Red (R) and black (B) leads must be connected as shown and routed behind support (S).**

**Fig. H1-4—Wiring schematic of Model 9A34-1.**

**Fig. H1-5—Wiring schematic of Model 9A34-1A.**

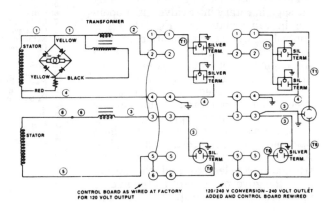

in Fig. H1-3. Wires must be routed behind rotor support to prevent interference with fan. During reassembly, fan must be mounted squarely on rotor shaft. Tighten rotor bolt to 120-140 in.-lbs.

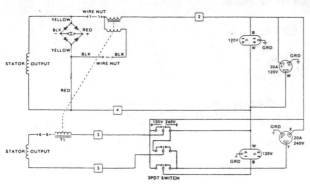

Fig. H1-6—Wiring schematic of Model 9A34-3A. Model 9A34-3 is similar.

## OVERHAUL

Refer to Fig. H1-1 for an exploded view of generator and to appropriate wiring schematic. Wires should be marked during disassembly to aid reassembly of unit. Wires should be handled carefully to prevent damage to wire, insulation or connections. Rotor (10) and engine crankshaft ends are tapered and a suitable tool is necessary to pull rotor off crankshaft. Tighten rotor bolt (2) to 120-140 in.-lbs.

# ENGINE

| Model | Cyls. | Bore | Stroke | Displ. |
|---|---|---|---|---|
| 9 | 1 | 2¾ in. (70 mm) | 2⅛ in. (54 mm) | 12.62 cu. in. (207 cc) |

## MAINTENANCE

**SPARK PLUG.** Recommended spark plug is Champion UJ11G. Spark plug electrode gap should be 0.025 inch (0.64 mm).

**CARBURETOR.** A Tillotson Model MT carburetor is used. Refer to Fig. H1-10 for an exploded view of carburetor and to Fig. H1-11 for an exploded view of fuel system. Note in Fig. H1-11 that a diaphragm type fuel pump (5, 6, 7 and 8) is used to pump fuel to carburetor.

Initial setting of idle mixture screw (23 – Fig. H1-10) is ¾ turn open while initial setting of high speed mixture screw (14) is one turn open. With float bowl cover (2) inverted, float height should be 1-13/32 inches (35.7 mm) as measured from the top of float (20) to gasket surface of cover.

**GOVERNOR.** Engine speed is regulated by varying size of inlet passage in rotary inlet valve. Inlet opening in valve plate (IV – Fig. H1-12) is controlled by the combination shutter and governor weight (GW) which when the engine is not running is held in the open position by spring (GS). When engine speed exceeds desired governed speed, centrifugal force acting on the weight overcomes opposing spring pressure and partially covers the inlet opening and thus throttles the engine. As the throttled engine slows, centrifugal force lessens and the spring pulls the governor weight so the inlet is uncovered and engine speed increased.

To adjust governor, run engine until normal operating temperature is reached then adjust carburetor for best performance. Disconnect pulse line and unscrew fitting (F – Fig. H1-13). Turn adjusting screw (AS – Fig. H1-12) to obtain desired no-load engine speed of 3750 rpm. With a load connected to alternator, a frequency meter may be used to set governed speed so output is 60 Hertz (cycles).

**IGNITION.** Breaker point gap should be 0.020 inch (0.51 mm). Ignition timing is fixed and non-adjustable. Breaker point cam (8 – Fig. H1-14) must be installed so arrow on cam faces out and notch fits pin in spacer (9). Ignition coil primary resistance is 0.95 ohms while coil secondary resistance is 5500-6000 ohms.

**LUBRICATION.** The engine is lubricated by mixing oil with the fuel. If

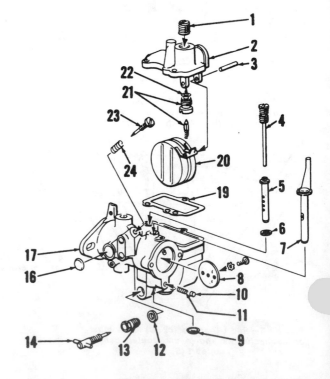

Fig. H1-10—Exploded view of Tillotson MT88A carburetor used on all models.

1. Plug
2. Float bowl cover
3. Float pin
4. By-pass tube
5. Main nozzle
6. Gasket
7. Choke shaft
8. Choke plate
9. Expansion plug
10. Choke friction pin
11. Spring
12. Gasket
13. Packing nut
14. Main fuel needle
16. Expansion plug
17. Carburetor body
19. Gasket
20. Float
21. Fuel inlet valve
22. Gasket
23. Idle mixture screw

Illustrations courtesy of Homelite Div. of Textron

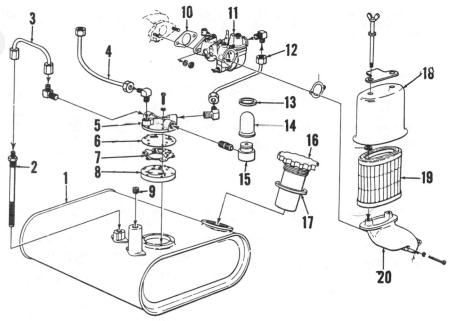

Fig. H1-11—Exploded view of fuel system. Primer bulb (14) will pump fuel into carburetor float bowl.

| | | | | | |
|---|---|---|---|---|---|
| 1. Fuel tank | 6. Pump diaphragm | 11. Carburetor | 16. Filler cap |
| 2. Fuel filter | 7. Gasket | 12. Fuel line | 17. Gasket |
| 3. Pump suction line | 8. Diaphragm cover | 13. Ring | 18. Air cleaner cover |
| 4. Pump pulse line | 9. Plug | 14. Primer bulb | 19. Element |
| 5. Pump body | 10. Gasket | 15. Adapter | 20. Adapter |

Homelite Premium SAE 40 oil is used, fuel:oil ratio should be 32:1. Fuel:oil ratio should be 16:1 if Homelite 2-Cycle SAE 30 oil or other SAE 30 oil designed for air-cooled two-stroke engine is used.

**CLEANING CARBON.** Carbon deposits should be cleaned from the exhaust ports and muffler at regular intervals. When scraping carbon be careful not to damage exhaust ports or piston.

## REPAIRS

**CYLINDER.** Cylinder bore is chrome plated. Inspect chrome bore for excessive wear and damage to chrome sur-

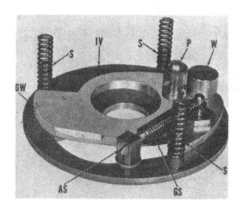

Fig. H1-12—View of inlet valve and governor assembly. Adjust governed speed by turning adjusting screw (AS).

AS. Adjusting screw
GS. Governor spring
GW. Governor weight & plate assy.
IV. Intake valve plate
P. Pivot post
S. Springs
W. Weight

face. A new cylinder must be installed if chrome is scored, cracked or worn so base metal is exposed.

**PISTON, PIN AND RINGS.** The piston is accessible after cylinder removal. Piston and pin are available only as a unit assembly. Piston skirt to

Fig. H1-14—To remove magneto rotor (3), unscrew retaining nut (2) against plate (1) which acts as puller. Notch in inner end of cam (8) must fit over pin in spacer (9).

| | |
|---|---|
| 1. Starter plate | 6. Breaker cam nut |
| 2. Rotor nut | 7. Breaker cam key |
| 3. Magneto rotor (flywheel) | 8. Breaker cam |
| 4. Wave washer | 9. Cam spacer |
| 5. Rotor key | 10. Magneto assembly |

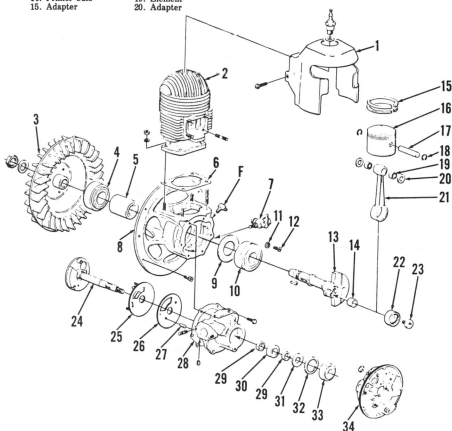

Fig. H1-13—Exploded view of engine.

| | | | |
|---|---|---|---|
| 1. Air shroud | 10. Bearing | 18. Snap rings | 26. Formica wear plate |
| 2. Cylinder | 11. Washer | 19. Bearings | 27. Pin |
| 3. Fan | 12. Bearing retaining screw | 20. Washer | 28. Timer bracket |
| 4. Bearing | 13. Crankshaft | 21. Connecting rod | 29. Spacers |
| 5. Spacer | 14. Inner race | 22. Bearing | 30. Bearing |
| 6. Gasket | 15. Piston rings | 23. Crankpin screw | 31. Seal |
| 7. Drain valve | 16. Piston | 24. Intake valve shaft | 32. Retaining ring |
| 8. Crankcase | 17. Piston pin | 25. Inlet valve & governor assy. | 33. Bearing |
| 9. Gasket | | | 34. Stator plate |

bore clearance should be 0.004-0.007 inch (0.10-0.18 mm).

Install piston so piston ring locating pin is on intake side of engine. If removed, use new piston pin retaining clips.

**CONNECTING ROD.** The connecting rod is one piece and rides on roller bearing (22–Fig. H1-13). To remove rod, remove cylinder and timer bracket (28) then unscrew crankpin screw (23). Slide rod off crankpin and remove through top of crankcase. Inspect rod, bearing and inner race (14). Two needle bearings (19) in rod small end support piston pin.

**CRANKSHAFT AND CRANK-CASE.** The crankshaft (13–Fig. H1-13) is supported in the crankcase (8) by ball bearings (4 and 10). Intake valve shaft (24) is driven by the crankshaft and is supported in timer bracket (28) by ball bearings (30 and 33).

Crankshaft runout must not exceed 0.003 inch (0.07 mm). Crankshaft journals must not be out-of-round or worn more than 0.001 inch (0.025 mm).

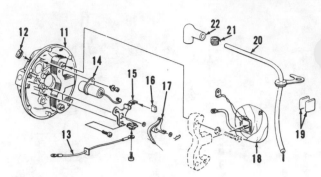

Fig. H1-15— Exploded view of typical magneto assembly (see 10—Fig. H1-14).

11. Back plate
12. Grommet
13. Ground (switch) lead
14. Condenser
15. Breaker point base
16. Cam wiper felt
17. Breaker point arm
18. Ignition coil
19. Coil wedges
20. Spark plug wire
21. Terminal
22. Spark plug boot

Inspect bearings, seals and crankcase surfaces. Be sure all crankcase passages are clean, especially the idle passage line which enters the crankcase via the intake valve. This passage is sometimes restricted by carbon which may be dislodged using a wire.

**ROTARY VALVE.** Rotary valve and governor assembly (25–Fig. H1-13) should be renewed if the following damage is noted: Sealing surfaces of valve are scored or worn enough to produce a ridge; if spring posts are loose or extend to valve seating surface; if the governor pivot point has started to wear through the surface of the valve.

Slight scoring of valve may be corrected by lapping on a lapping plate using very fine abrasive. Lapping motion should be in the pattern of a figure eight to obtain best results. Slight scoring of the Formica wear plate (26) is permissible. Homelite recommends soaking a new wear plate in oil for 24 hours prior to installation.

Illustrations courtesy of Homelite Div. of Textron

# HOMELITE

| Model | Output-kw | Voltage | Make | Model | Governed Rpm |
|-------|-----------|---------|------|-------|--------------|
| | | | — Engine — | | |
| 10A35-1L | 3.5 | 115 | Homelite | 10 | 3600 |
| 10A35-2L | 3.5 | 115/230 | Homelite | 10 | 3600 |
| 251A35-1 | 3.5 | 115 | Homelite | 251 | 3600 |
| 251A35-2 | 3.5 | 115/230 | Homelite | 251 | 3600 |

# GENERATOR

### MAINTENANCE

To remove brushes, unscrew clamp screws and slide brush cover (20–Fig. H2-1) off brush head (18). Unscrew set screws and remove brush caps (24) and brushes (25). Be sure to mark brushes so that they can be returned to their original positions. Check length of each brush as shown in Fig. H2-2. Renew brush if length is ⅝-inch (15.9 mm) or shorter. Be sure to include length of projections for brush springs when measuring brush length.

If new brushes must be installed, use the following procedure: Wrap the commutator with a medium grit (4/0 to 6/0) piece of garnet paper which will fit the commutator. Install new brushes. Remove spark plug and use engine

starter to rotate engine until brushes are seated. Remove brushes and install fine grit (8/0 to 9/0) garnet paper. Install brushes and repeat seating process. Remove brushes and garnet paper and install spark plug. Blow carbon and grit out of generator and install brushes.

### OVERHAUL

Refer to Fig. H2-1 for an exploded view of alternator and to appropriate wiring schematic. Wires should be marked to aid in reassembly. Generator rotor (5) can be removed from engine crankshaft by inserting a pin (Homelite no. 22271 cut to a length of 7¾ inches (19.7 cm) measured from tapered end) into rotor after removing retaining screw (11). Thread a jackscrew (Homelite no. S-394) into rotor, tighten jackscrew and tap with a hammer to loosen taper fit.

# LOADAMATIC

Generators 10A35-1L and 10A35-2L are equipped with an automatic idle control (Loadamatic). An electromagnet is mounted adjacent to the engine's carburetor and acts on the carburetor throttle arm. When there is no load on the generator, the electromagnet is energized and the engine governor is overridden as the electromagnet pulls the carburetor throttle arm to idle position. The governor resumes control of engine speed when a load is imposed on the generator. The electromagnet is deenergized and the throttle arm is released to be controlled by the governor.

To adjust automatic idle control (Loadamatic), proceed as follows: Refer to Fig. H2-3 and adjust height of electromagnet to place bottom of electromagnet parallel with throttle arm. Do

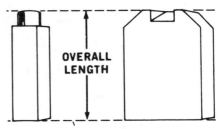

Fig. H2-2—Measure length of brushes as shown.

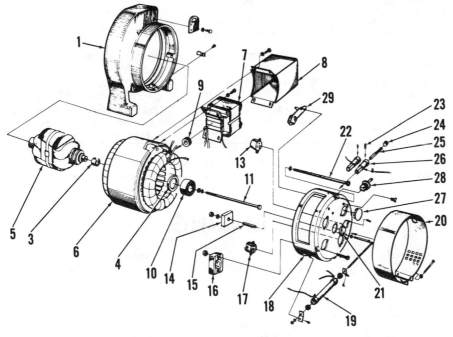

Fig. H2-1—Exploded view of generator. Switch (28) and control board assembly (29) are used on Loadamatic models only.

| | | | |
|---|---|---|---|
| 1. Fan housing | 9. Grommet | 17. Receptacle | 24. Cap |
| 3. Inner race | 10. Ball bearing | 18. Brush head | 25. Brush |
| 4. Seal | 11. Bolt | 19. Resistor | 26. Brush holder |
| 5. Rotor | 13. Receptacle | 20. Cover | 27. Plug |
| 6. Stator | 14. Rectifier | 21. Bearing cap | 28. Loadamatic switch |
| 7. Transformer | 15. Roll pin | 22. Bolt | 29. Loadamatic control board |
| 8. Cover | 16. Receptacle | 23. Set screw | |

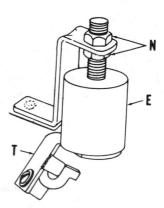

Fig. H2-3—Loadamatic is adjusted by turning nuts (N) until bottom of electromagnet (E) is parallel with throttle arm (T). Do not bend throttle arm or electromagnet bracket.

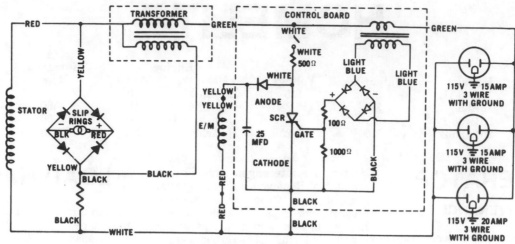

*Fig. H2-4—Wiring schematic for Model 10A35-1L. Model 251A35-1 is similar.*

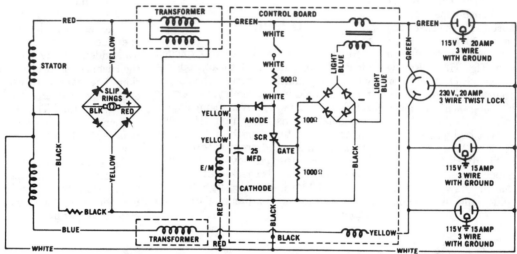

*Fig. H2-4A—Wiring schematic for Model 10A35-2L. Model 251A35-2 is similar.*

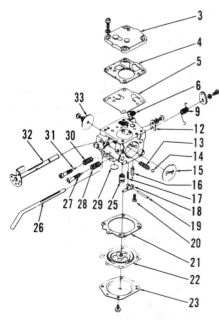

*Fig. H2-5—Exploded view of Tillotson HS carburetor.*

3. Fuel pump cover
4. Fuel pump gasket
5. Fuel pump diaphragm
6. Inlet channel screen
9. Governor spring arm
12. Throttle shaft clip

13. Choke detent spring
14. Choke detent ball
15. Choke shutter
16. Inlet needle valve
17. Inlet lever spring
18. Inlet control lever
19. Lever pinion pin
20. Pinion pin retaining screw
21. Diaphragm gasket
22. Metering diaphragm
23. Diaphragm cover

25. Nozzle check valve
26. Choke shaft
27. Main adjusting needle
28. Main needle spring
29. Welch plug
30. Idle fuel needle spring
31. Idle adjusting needle
32. Throttle shaft
33. Throttle shutter

not bend bracket or throttle arm to make this adjustment. Tighten electromagnet nuts. Position generator toggle switch to "START" to disengage Loadamatic. Start engine and allow it to reach operating temperature. If necessary, adjust carburetor for proper mixture and speed. Flip toggle switch to "AUTO" position. Engine speed should reduce to idle speed. If idle speed is not steady, adjust carburetor idle mixture. Idle speed should be 2400-2600 rpm and is adjusted by loosening electromagnet. Raising electromagnet will decrease engine speed while lowering electromagnet increases engine speed. Apply a light load to generator and then remove load. Engine speed should increase to governed speed and then return to idle after the load is removed.

# ENGINE

| Model | Cyls. | Bore | Stroke | Displ. |
|-------|-------|------|--------|--------|
| 10,251 | 1 | 2¾ in. (70 mm) | 2⅛ in. (54 mm) | 12.62 cu. in. (207 cc) |

## MAINTENANCE

**SPARK PLUG.** Either a Champion J6J, HO-8A or UJ-11G spark plug may be used. The Champion HO-8A platinum tip or UJ-11-G gold paladium tip plug will provide longer service as well as longer intervals between cleanings. Electrode gap is 0.025 inch (0.64 mm).

*Illustrations courtesy of Homelite Div. of Textron*

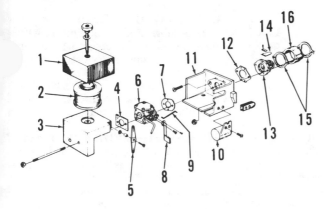

*Fig. H2-6—Exploded view of induction system. Components (8, 10 and 11) are not used on models with 251 engine.*

1. Filter cover
2. Air filter
3. Air manifold
4. Gasket
5. Bellcrank
6. Carburetor
7. Gasket
8. Throttle arm
9. Link
10. Loadamatic electromagnet
11. Housing
12. Gasket
13. Reed valve seat
14. Reed valve
15. Gasket
16. Spacer

**CARBURETOR.** Refer to Fig. H2-5. The Tillotson HS carburetor can be removed from engine by removing air manifold (3 – Fig. H2-6) and air cleaner assembly as the two manifold bolts also retain carburetor and reed valve assemblies.

When disassembling carburetor, slide diaphragm assembly towards adjustment needle side of carburetor body to disengage diaphragm from fuel inlet control lever. To remove welch plug (29 – Fig. H2-5), carefully drill through plug with a small diameter drill and pry plug out with a pin. Caution should be taken that drill bit just goes through welch plug as deeper drilling may seriously damage carburetor. Note channel screen (6) and check valve assembly (25) located in bores of carburetor body.

Inlet control metering lever (18) should be flush with metering chamber floor of carburetor body. If not, bend diaphragm end of lever up or down as required so that lever is flush.

Normal adjustment of low speed fuel mixture needle (marked "LO" on air inlet manifold or needle nearest throttle shaft) is one turn open and main mixture adjustment needle (marked "HI" on air inlet manifold or needle nearest choke shaft) should be opened ¾-turn.

*Fig. H2-7—View showing proper installation of governor bellcrank and links.*

Start engine and allow to warm up before making final carburetor adjustments. Apply load and readjust main needle so engine runs best, then remove load and adjust idle needle for smoothest operation. Refer to LOADAMATIC section for idle adjustment on models with Loadamatic.

**NOTE: The main and idle speed mixture adjustments are interdependent so that changing one needle setting often requires readjustment of other needle.**

When reinstalling carburetor, be sure governor link is connected as shown in Fig. H2-7.

**GOVERNOR.** Engine is equipped with a flyweight type governor mounted on engine crankshaft; refer to Fig. H2-8 for exploded view showing governor unit. External governor linkage is shown in Fig. H2-7.

**CAUTION: Never move governor linkage manually, or exert any pressure on lever or linkage to increase engine speed. Working governor linkage manually, even momentarily, may cause damage to governor cup and cam due to friction and burning.**

Maximum no-load speed should be about 3750 rpm; generator engine speed under load should be 3600 rpm. If necessary to readjust governor, first remove the cover plate (43 – Fig. H2-8) and slightly loosen governor shaft guide (36) retaining screws. Note that screw hole in guide at carburetor side is slotted; insert screwdriver between side of guide and shoulder machined in housing (45). Pry carburetor side of guide towards cylinder to decrease governed

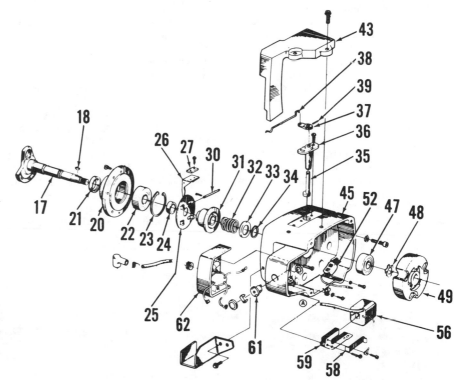

*Fig. H2-8—Exploded view showing magneto end of crankshaft (17), governor and magneto assemblies. The housing (45) retaining socket head screws can be removed by working through notches in outside of magneto rotor (49) allowing all parts to be removed as a unit after removing governor linkage and disconnecting spark plug wire.*

| | | |
|---|---|---|
| 17. Crankshaft, magneto end | 26. Governor weight arm | 37. Governor arm | 52. Magneto trigger coil |
| 18. Woodruff key | 27. Governor weight | 38. Bellcrank link | 56. Ignition coil |
| 20. Bearing housing | 30. Weight pivot pin | 39. Pin | 58. Coil spring clip |
| 21. Crankshaft seal | 31. Governor cup | 43. Governor linkage cover | 59. Armature core |
| 22. Bearing | 32. Governor spring | 45. Magneto housing | 61. Stop switch |
| 23. Snap ring | 33. Spring retainer | 47. Bearing | 62. Magneto cover assembly (includes condenser and solid state switchbox) |
| 24. Spacer | 34. Snap ring | 48. Loading spring | |
| 25. Governor back plate | 35. Governor camshaft | 49. Magneto rotor | |
| | 36. Camshaft guide | | |

speed or away from cylinder to increase speed. Adjustment provided by total length of slot will change the engine maximum governed speed about 1000 rpm.

If engine governed speed cannot be properly adjusted, check for wear on governor shaft cam and inspect governor spring connected to carburetor throttle shaft.

To renew governor spring (32), cup (31) or back plate (25), first remove starter, magneto rotor (49) and housing (45); the snap ring (34) can then be removed from crankshaft allowing removal of spring cup and back plate.

IGNITION. A breakerless solid-state ignition system is used. Refer to Fig. H2-8 for exploded view of the magneto (items 49 through 62) and to Figs. H2-9 and H2-10.

To check the solid-state magneto, disconnect spark plug wire and remove spark plug. Insert a bolt or screw in spark plug wire terminal and while holding bolt or screw about ¼-inch away from engine casting, crank engine and check for spark as with conventional magneto. If no spark occurs, refer to following inspection and test procedure.

Visually check for broken or frayed wires which would result in open circuit or short. Be sure stop switch is not permanently grounded. Inspect magneto rotor (49–Fig. H2-8), trigger coil (52) and the switch box, condenser and magneto cover assembly (62) for visible damage.

To test magneto components, remove starter assembly, magneto cover and disconnect leads as shown in Fig. H2-10, then proceed as follows:

To test ignition coil, refer to test instrument instructions; readings for Graham Model 51 and Merc-O-Tronic testers are given below:

**Graham Model 51:**

Maximum secondary . . . . . . . . . . 10,000
Maximum primary . . . . . . . . . . . . . . 1.7
Coil index . . . . . . . . . . . . . . . . . . . . . 65
Maximum coil test . . . . . . . . . . . . . . 20
Maximum gap index . . . . . . . . . . . . . 65

**Merc-O-Tronic**

Operating amperage . . . . . . . . . . . . . . 1.3
Minimum primary resistance . . . . . . . 0.6
Maximum primary resistance . . . . . . 0.7
Minimum secondary continuity . . . . . 50
Maximum secondary continuity . . . . 60

If ignition coil does not meet test specifications, renew using correct part number coil. Do not substitute a coil of other specifications with the solid-state ignition system. If coil tested ok, check switch box as follows:

With leads and condenser disconnected as shown in Fig. H2-10, connect one ohmmeter lead to one switch box test point (flag terminal or ground lead) and other ohmmeter to remaining switch box test point. The ohmmeter reading should be either between 5 to 25 ohms or from one megohm to infinity. When ohmmeter test leads are reversed, the opposite reading should be observed. If these ohmmeter readings are not observed, renew magneto cover and switch box assembly. If ignition coil and switch box both test ok, check trigger coil as follows:

Connect ohmmeter positive lead to junction of switch box and trigger coil leads and ohmmeter negative lead to magneto housing (see "Trigger Coil Test Points" in Fig. H2-10.) It is not necessary to disconnect trigger coil lead from switch box lead. The ohmmeter reading should be 22 to 24 ohms.

To check condenser, stick a pin through the condenser lead to provide a contact point, then test condenser using standard procedure to check series resistance, short and capacitance. Condenser capacitance should be 0.16-0.20 mfd.

If either the switch box or condenser tested faulty, renew the complete condenser, switch box and magneto cover assembly.

LUBRICATION. Engine on all models is lubricated by mixing oil with regular gasoline. If Homelite Premium SAE 40 chain saw oil is used, fuel:oil ratio should be 32:1. Fuel:oil ratio should be 16:1 if Homelite 2-Cycle SAE 30 oil or other SAE 30 oil designed for air-cooled two-stroke engines is used.

AIR AND FUEL FILTERS. The air cleaner element may be washed in a detergent and water solution or by sloshing it around in a container of non-oily solvent. After a number of cleanings, the filter pores may become permanently clogged, making it necessary to renew element.

The fuel filter is a part of the fuel pickup inside fuel tank and can be fished from tank filler opening using a wire hook. Under normal operations, the filter element should be changed at intervals of from one to two months. If engine is run continuously or fuel is dirty, filter may need to be changed weekly or at shorter intervals.

CARBON. The carbon should be cleaned from exhaust ports at 100 to 200 hour intervals. Remove muffler, crank piston to top dead center and use a wooden or plastic scraper to remove carbon deposits. Avoid scratching piston or damaging edges of port.

## REPAIRS

CONNECTING ROD. Connecting rod lower end is fitted with a roller bearing (82–Fig. H2-11) which rides on a renewable inner race (83). To remove piston and connecting rod assembly from crankshaft (84) crankpin, first remove cylinder and the magneto housing and crankshaft rotor end assembly. Place a block of wood between crankshaft throw and crankcase to keep crankshaft from turning, then unscrew crankpin screw (81) (counter-clockwise) with ⅜-inch Allen wrench. If renewal of inner race is indicated, remove race from crankpin. Usually, race will slide

Fig. H2-9—View of magneto assembly with rewind starter assembly removed. Armature to magneto rotor air gap and trigger coil to rotor air gap should be adjusted using a 0.0075 inch thick plastic shim.

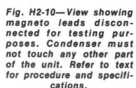

Fig. H2-10—View showing magneto leads disconnected for testing purposes. Condenser must not touch any other part of the unit. Refer to text for procedure and specifications.

Illustrations courtesy of Homelite Div. of Textron

off of pin; however, it may be necessary to pry race from crankpin with screwdrivers.

To renew crankpin needle roller bearing in the connecting rod, press old bearing out using plug (Homelite tool No. 24120-1) while supporting rod on sleeve (Homelite tool No. 24118-1). Install new bearing by supporting rod on sleeve (Homelite tool No. 24124-1) and pressing bearing in with same plug as used to remove old bearing. Shouldered face of sleeve (24124-1) will properly position bearing so that it protrudes equally from each side of rod.

To renew piston pin bearings, support rod on sleeve (Homelite tool No. 24124-1) and press bearings out with plug (24131-1). New bearings are installed separately from opposite ends of bore. Support rod on sleeve (24124-1) and using straight end of plug (24131-1) (end with recessed face), press new bearing in (press on lettered side of cage only) until shoulder of plug seats against rod. Turn rod over and press other new bearing into rod in same manner. When properly installed, recessed faces of piston pin thrust washers will clear protruding bearing races and will contact connecting rod.

When reinstalling connecting rod and bearing inner race, thoroughly lubricate all parts and tighten connecting rod cap screw to a torque of 50 ft.-lbs. (68 N·m).

**NOTE: Locate connecting rod on crankpin so that oil hole in upper end of rod will be towards intake side of engine. Piston should be assembled to connecting rod so that piston ring locating pin will be to same (intake) side of assembly as oil hole in rod.**

**PISTON, PIN AND RINGS.** Piston is accessible after removing cylinder from crankcase. Always support piston while removing or installing piston pin. Piston should be renewed if ring side clearance, measured with new ring installed in top groove, exceeds 0.004 inch (0.10 mm). Also, renew piston if piston skirt to cylinder bore clearance exceeds 0.007 inch (0.18 mm) when measured with new or unworn cylinder. Inspect piston ring locating pin and renew piston if pin has worn to half of its original thickness. Piston pin should be a snug push-fit to light press-fit in piston. Piston, pin and rings are available in standard size only. Homelite recommends that piston rings be renewed whenever engine is disassembled for service.

When reassembling piston to connecting rod, insert new snap ring in exhaust side (opposite ring locating pin) of piston. Lubricate all parts and place piston, exhaust side down, in holding fixture.

**NOTE: A used cylinder sawed in half makes a good holding fixture.**

Press pin into upper (intake) side of piston, then insert connecting rod and thrust washers into piston with oil hole in rod up and recessed sides of washers next to piston pin bearings in rod, press pin on through the assembly and secure with new snap ring. Be sure that piston and rod assembly is installed on crankpin with pinned (intake) side of piston away from exhaust port side of engine.

**CYLINDER.** Cylinder bore is chrome plated. Inspect cylinder bore for excessive wear and damage to chrome surface of bore. A new cylinder must be installed if chrome is scored, cracked or the base metal underneath exposed.

**CRANKSHAFT, BEARINGS AND SEALS.** The two-piece crankshaft can be serviced as two separate parts. Refer to following paragraphs for crankshaft service.

CRANKSHAFT, MAGNETO END. To service the shaft, bearings, seal or governor components, proceed as follows: Remove rewind starter as an assembly. Unscrew magneto rotor retaining nut and using suitable puller (Homelite tool No. AA-22560 or equivalent), remove rotor. Disconnect governor linkage and remove governor bellcrank. Using a 3/16-inch Allen wrench, remove the six socket head screws retaining magneto housing to crankcase and remove housing and shaft assembly. Remove the two screws retaining bearing housing (20 – Fig. H2-8) to magneto housing (45) and separate shaft and bearing assembly from housing (45). Bearing (47) can be renewed at this time. Remove snap ring (34), re-

tainer (33) and governor spring (32) and pry governor back plate (25), with weights, from shaft (17). Remove spacer (24), support bearing housing (20) and press shaft from housing. Remove crankshaft seal (21) and snap ring (23), then press bearing (22) from housing.

To assemble magneto end shaft and bearing assembly, proceed as follows: Install new seal (21) in housing with lip of seal towards crankcase side. Lubricate seal and insert shaft through seal and housing. Support flat inner end of shaft and press bearing (22) down over shaft and into housing until bearing inner race is seated against shoulder on shaft. Then, support housing and press bearing outer race into housing so that retaining snap ring (23) can be installed. Place spacer (24) on shaft, then drive or press governor back plate onto shaft against spacer. Install governor cup, spring, spring retainer and snap ring. Attach bearing housing to magneto housing with the two screws, then reinstall shaft, bearing housing and magneto housing to crankcase using new gasket.

CRANKSHAFT, OUTPUT END. First, remove magneto end crankshaft, bearing and magneto housing assembly as described in preceding paragraph. Remove cylinder and piston and connecting rod unit, then proceed as follows:

Remove the crankcase, output crankshaft end and blower rotor (fan) as an assembly from pump or generator; refer to exploded views of pump and generator units shown in this section. Remove fan retaining nut and washer and the two crankshaft bearing retaining screws (87 – Fig. H2-11) and washers (88). Support magneto (open) end of crankcase, then press crankshaft (84) from fan and crankcase. Remove the three corks, if so equipped, from

66. Spark plug
68. Cylinder shield
72. Cylinder gasket
73. Cylinder
75. Piston rings
76. Piston & pin assy.
77. Snap rings
78. Connecting rod
79. Needle bearings
80. Thrust washers
81. Crankpin screw
82. Roller bearing
83. Inner race
84. Crankshaft (output end)
85. Woodruff key
87. Bearing retaining screws
88. Bearing retaining washers
89. Bearing
90. Petcock
91. Crankcase gasket
94. Crankcase

95. Oil seal
96. Spacer
97. Bearing
98. Fan

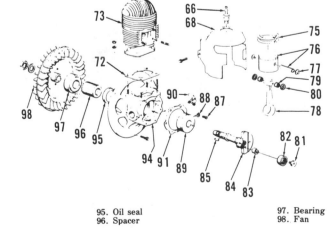

*Fig. H2-11—Exploded view showing engine crankcase, cylinder, rod and piston assembly, fan, crankshaft (output end) and related parts. Petcock (90) is to drain crankcase should it become flooded with fuel; do not attempt to start and run engine with petcock open.*

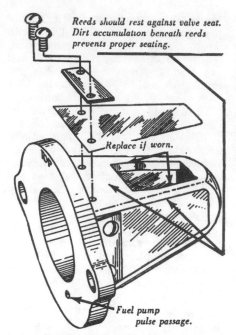

*Fig. H2-12—View showing reed valve assembly with one valve reed removed. Inspect seat and reeds as noted and be sure fuel pump pulse passage is open.*

fan, insert jackscrews into the tapped holes and push bearing (97) from fan inner hub. Remove screws and reinstall corks. Corks keep threads clean but are not necessary for operation. Support

outer race of bearing (89) and press shaft out of bearing. Remove spacer (96) and crankshaft seal (95) from crankcase.

To reassemble, proceed as follows: Press new seal into crankcase with lip towards inside (away from fan). Support outer hub of fan, then press new bearing (97) onto fan inner hub. Press new inner bearing (89) into crankcase with retaining groove in outer race properly positioned. Support bearing inner race with sleeve, then press crankshaft into bearing. Install bearing retaining screws and washers. Place spacer (96) in position on crankshaft, then carefully press fan onto shaft making sure that keys and keyways are aligned. Install fan retaining washer and nut securely. Complete reassembly by installing connecting rod and piston, cylinder and magneto end assembly.

**CRANKCASE.** To renew crankcase, follow procedures as outlined in previous paragraph "CRANKSHAFT, OUTPUT END".

**REED INTAKE VALVE.** Engine is equipped with a pyramid reed valve assembly shown in Fig. H2-6. The reed valve should be inspected whenever carburetor is removed. Refer to Fig. H2-12 for inspection points. When installing new reeds on pyramid seat, thoroughly

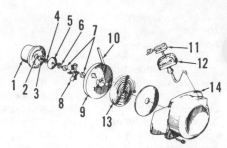

*Fig. H2-13—Exploded view of rewind starter. Washer (15) is not used on early models.*

| | |
|---|---|
| 1. Starter cup | |
| 2. Lock washer | |
| 3. Crankshaft nut | 9. Starter pulley |
| 4. Snap ring | 10. Starter rope |
| 5. Retaining washer | 11. Handle insert |
| 6. Brake spring | 12. Handle |
| 7. Brake washer | 13. Rewind spring |
| 8. Friction shoe | 14. Starter cover |
|    assembly | 15. Washer |

clean all threads and apply "Loctite" to threads on screws before installing. Be sure reeds are centered before tightening screws.

**REWIND STARTER.** Refer to Fig. H2-13 for exploded view. In an emergency in case of rewind starter failure, engine can be started by removing starter and winding a rope around starter cup (1).

# HOMELITE

| Model | Output-kw | Voltage | Engine | | Governed Rpm |
| | | | Make | Model | |
| --- | --- | --- | --- | --- | --- |
| 112A15-1 | 1.5 | 115 | B&S | 100232 | 3600 |
| 113A25-1 | 2.5 | 115 | B&S | 146432 | 3600 |
| 116A50-2 | 5.0 | 115/230 | B&S | 243431 | 3600 |
| 116A50-2L | 5.0 | 115/230 | B&S | 243431 | 3600 |
| 118A35-1 | 3.5 | 115 | Wisc. | S8D | 3600 |
| 118A35-2 | 3.5 | 115/230 | Wisc. | S8D | 3600 |
| 119A35-1 | 3.5 | 115 | B&S | 200431 | 3600 |
| 119A35-1L | 3.5 | 115 | B&S | 200431 | 3600 |
| 119A35-2 | 3.5 | 115/230 | B&S | 200431 | 3600 |
| 119A35-2L | 3.5 | 115/230 | B&S | 200431 | 3600 |

# GENERATOR

## MAINTENANCE

To remove brushes, unscrew clamp screws and slide brush cover (20–Fig. H3-1) off brush head (18). Unscrew set screws and remove brush caps (24) and brushes (25). Be sure to mark brushes so that they can be returned to their original positions. Check length of each brush as shown in Fig. H3-2. Renew brush if length is ⅝-inch (15.9 mm) or shorter. Be sure to include length of projections for brush springs when measuring brush length.

If new brushes must be installed, use the following procedure: Wrap the commutator with a medium grit (4/0 to 6/0) piece of garnet paper which will fit the commutator. Install new brushes. Remove spark plug and use engine starter to rotate engine until brushes are seated. Remove brushes and install fine grit (8/0 to 9/0) garnet paper. Install brushes and repeat seating process. Remove brushes and garnet paper and install spark plug. Blow carbon and grit out of generator and install brushes.

## TROUBLESHOOTING

If little or no output is generated, check the following: The engine must be in good condition and maintain 3600 rpm under load or a no-load governed speed of 3750 rpm. Brushes, commutator and slip rings must be in good condition. All wiring connections must be clean and tight.

If loss of field magnetism is suspected cause of low generator output, flash the field as follows: Run unit without a load connected while using a 6-volt dry cell battery, momentarily connect the negative battery terminal to the negative (black) brush and the positive battery terminal to the positive (red) brush. Disconnect battery as soon as generator voltage rises.

To test rectifier, mark and remove brushes and disconnect rectifier leads. Using an ohmmeter measure resistance between each brush and each rectifier lead then reverse ohmmeter leads and again measure resistance. Resistance should be 4-100 ohms for one reading and infinite for other reading. Renew rectifier if resistance measurements are incorrect.

Disconnect thin black transformer wire. With all load disconnected, start and run generator. If voltage rises, transformer is faulty and must be renewed. With unit stopped, measure resistance between disconnected thin black transformer wire and rectifier terminal with thin yellow wire. Renew transformer if reading is infinite. Reconnect black transformer wire.

Disconnect resistor leads and measure resistance. Resistor resistance should be 95-105 ohms for Model 112A15-1; 71.5-78.5 ohms for Model 113A25-1; 47.5-52.5 ohms for all other models. Renew resistor if measurement is incorrect.

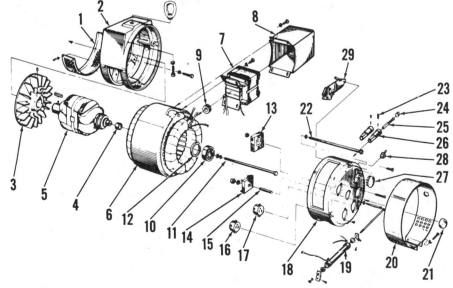

*Fig. H3-1—Exploded view of typical generator covered in this section. Switch (28) and control board assembly (29) are used on Loadamatic models only.*

1. Shield
2. End bell
3. Fan
4. Inner race
5. Rotor
6. Stator
7. Transformer
8. Cover
9. Grommet
10. Ball bearing
11. Bolt
12. Seal
13. Receptacle
14. Rectifier
15. Roll pin
16. Receptacle
17. Receptacle
18. Brush head
19. Resistor
20. Cover
21. Bearing cap
22. Bolt
23. Set screw
24. Cap
25. Brush
26. Brush holder
27. Plug
28. Loadamatic switch
29. Loadamatic control board

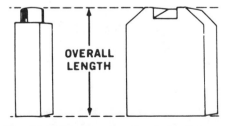

*Fig. H3-2—Measure length of brushes as shown.*

To test rotor, mark and remove brushes then measure resistance between slip rings. Resistance should be approximately 31 ohms for Model 112A15-1; 32 ohms for Model 113A25-1; 26 ohms for Models 116A50-2 and 116A50-2L; 30 ohms for all other models. Check for shorts by measuring between each slip ring and rotor shaft. Renew rotor if preceding tests are failed. It may be necessary to flash field as previously outlined after testing to restore magnetism in field magnets.

## OVERHAUL

Refer to Fig. H3-1 for an exploded view of generator and to appropriate schematic in Fig. H3-4, H3-5, H3-6 or H3-7. Be sure to mark wiring so that it may be correctly connected for reassembly. Wires should be handled carefully to prevent damage to wire, insulation and connections. Generator rotor (5—Fig. H3-1) has a taper fit with engine crankshaft and should be removed with a suitable tool.

# LOADAMATIC

Models 116A50-2L, 119A35-1L and 119A35-2L are equipped with an automatic idle control (Loadamatic). An electromagnet is mounted adjacent to the engine's carburetor and acts on the carburetor throttle arm. When there is no load on the generator, the electromagnet is energized and the engine governor is overridden as the electromagnet pulls the carburetor throttle arm to idle position. The governor

resumes control of engine speed when a load is imposed on the generator. The electromagnet is deenergized and the throttle arm is released to be controlled by the governor.

To adjust automatic idle control (Loadamatic), proceed as follows: Refer to Fig. H3-3 and adjust height of electromagnet to place bottom of electromagnet parallel with throttle arm. Do not bend bracket or throttle arm to make this adjustment. Tighten electromagnet nuts. Position generator

toggle switch to "START" to disengage Loadamatic. Start engine and allow it to reach operating temperature. If necessary, adjust carburetor for proper mixture and speed. Flip toggle switch to "AUTO" position. Engine speed should slow to idle speed. If idle speed is not steady, adjust carburetor idle mixture. Idle speed should be 2400-2600 rpm and is adjusted by loosening electromagnet. nuts and altering position of electromagnet. Raising electromagnet will decrease engine speed while lowering

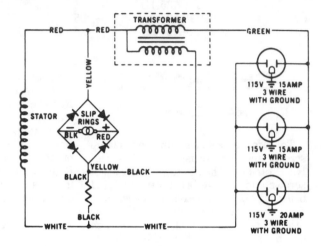

*Fig. H3-4—Schematic for generators 112A15-1, 113A25-1, 118A35-1 and 119A35-1.*

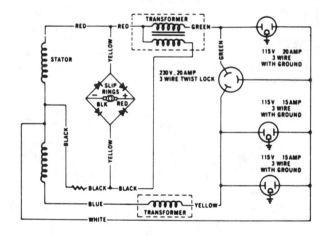

*Fig. H3-5—Schematic for generators 116A50-2, 118A35-2 and 119A35-2.*

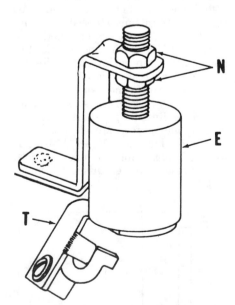

*Fig. H3-3—Loadamatic is adjusted by turning nuts (N) until bottom of electromagnet (E) is parallel with throttle arm (T). Do not bend throttle arm or electromagnet bracket.*

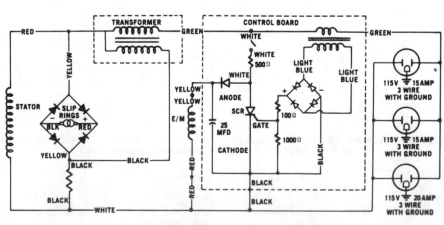

*Fig. H3-6—Schematic for generator 119A35-IL.*

electromagnet will decrease engine speed. Apply a light load to generator and then remove it. Engine speed should increase to governed speed and then return to idle after the load is removed.

# ENGINE

Engine make and model are listed at beginning of section. Refer to Briggs and Stratton or Wisconsin engine section for engine service.

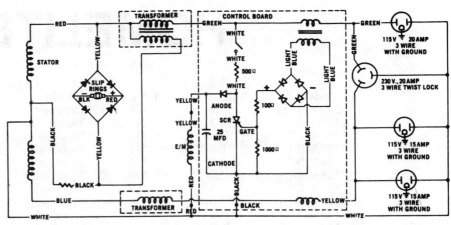

*Fig. H3-7—Schematic for generators 116A50-2L and 119A35-2L.*

# HOMELITE

| Model | Output-kw | Voltage | Engine Make | Engine Model | Governed Rpm |
|---|---|---|---|---|---|
| 124A10-1 | 1.0 | 115 | Tecumseh | H30 | 3600 |
| 125A15-1 | 1.5 | 115 | Tecumseh | H35 | 3600 |
| 126A22-1 | 2.25 | 115 | Tecumseh | H50 | 3600 |
| 127A30-1 | 3.0 | 115 | Tecumseh | H70 | 3600 |
| 128A10-1 | 1.0 | 120 | Tecumseh | H30 | 3600 |
| 128A10-1B | 1.0 | 120 | Tecumseh | H30 | 3600 |
| 129A15-1 | 1.5 | 120 | Tecumseh | H35 | 3600 |
| 129A15-1B | 1.5 | 120 | Tecumseh | H35 | 3600 |
| 130A22-1 | 2.25 | 120 | Tecumseh | H50 | 3600 |
| 130A22-1B | 2.25 | 120 | Tecumseh | H50 | 3600 |
| 130A22-1C | 2.25 | 120 | Tecumseh | HS50 | 3600 |
| 131A30-1 | 3.0 | 120 | Tecumseh | H70 | 3600 |
| 131A30-1B | 3.0 | 120 | Tecumseh | H70 | 3600 |
| 132A40-1 | 4.0 | 120 | B&S | 190412 | 3600 |
| E1350-1 | 1.35 | 120 | B&S | 80212 | 3600 |
| E1700-1 | 1.7 | 120 | B&S | 100212 | 3600 |
| E2250-1 | 2.25 | 120 | B&S | 130212 | 3600 |
| E3000-1, E3000-1A | 3.0 | 120 | B&S | 170412 | 3600 |
| E4000-1, E4000-1A | 4.0 | 120 | B&S | 190412 | 3600 |

# GENERATOR

## MAINTENANCE

**BRUSHES.** Brush length and condition should be checked after every 1000 hours of operation. Brushes can be inspected after removing brush head cover (1–Fig. H4-1), rotor bolt (2) and fan (3).

**NOTE: Do not attempt to hold rotor with fan to remove rotor bolt.**

Loosen screws retaining brush holder and slide out brush holder. Brushes must be renewed if brush length is shorter than 3/8 inch (9.5 mm) as shown in Fig. H4-2. Be sure to reinstall a used brush so that brush curvature matches collector ring. During reassembly, fan must be mounted squarely on rotor shaft. Tighten rotor bolt to 120-140 in.-lbs. (13.6-15.8 N·m).

Disassemble brush holder to install new brushes. Insert new brushes in brush holder so curvature of brush will match curvature of collector ring. Seating new brushes is not required as they are manufactured with correct curvature. Note on Models 124A10-1, 125A15-1, 126A22-1 and 127A30-1 that the red wire to the brush holder should be connected to the brush closest to the rotor bearing. On all other models, red and black wires to brushes should be connected as shown in Fig. H4-3 and wires must be routed behind rotor support to prevent interference with fan.

## OVERHAUL

**DISASSEMBLY AND REASSEMBLY.** Refer to Fig. H4-1 for an exploded view of generator and to appropriate schematic. Be sure to mark wiring so that it may be correctly connected for reassembly. Wire should be handled carefully to prevent damage to wire, insulation or connections. Generator rotor (10) has a taper fit with engine crankshaft and should be removed with a suitable tool.

**BRUSH HEAD KIT.** Brush head kit A-51450 is available as a direct replacement for the brush head assembly on

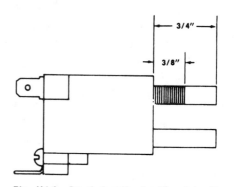

Fig. H4-2—Brush length should not be less than 3/8 inch when measured as shown above. Brush length of a new brush is 3/4 inch.

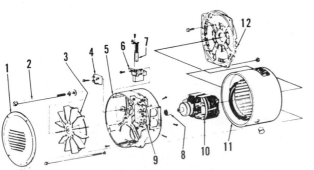

**Fig. H4-1—Exploded view of generator.**
1. Cover
2. Bolt
3. Fan
4. Rectifier
5. Brush head
6. Brush holder
7. Brush
8. Bearing
9. Receptacle
10. Rotor
11. Stator & housing
12. End bell

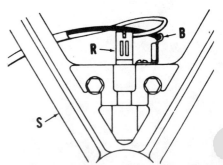

Fig. H4-3—On models indicated in text, red (R) and black (B) wires to brushes are connected and routed as shown.

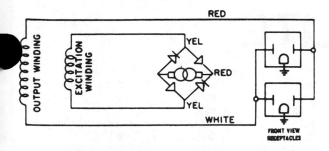

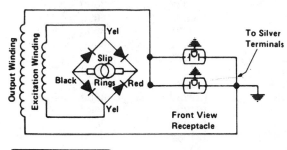

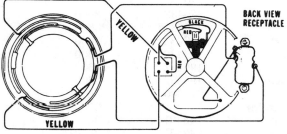

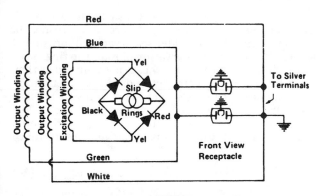

*Fig. H4-4—Schematic for generator Models 128A10-1, 129A15-1 and 130A22-1 without brush head kit A-51450 installed.*

*Fig. H4-5—Schematic and practical diagram of Models E1350-1, E1700-1 and E2250-1 and Models 128A10-1, 128A10-1A, 128A10-1B, 129A15-1, 129A15-1A, 129A15-1B, 130A22-1, 130A22-1A, 130A22-1B and 130A22-1C with brush head kit A-51450 installed.*

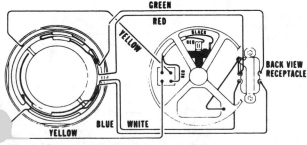

*Fig. H4-6—Schematic and practical diagram of Model E3000-1 and Models 131A30-1, 131A30-1A and 131A30-1B with brush head kit A-51450 installed.*

Models 128A10-1B, 129A15-1B, 130A22-1B, 130A22-1C and 131A30-1B. Brush head kit A-51450 also supersedes brush head #53753-2 on Models 128A10-1, 128A10-1A, 129A15-1, 129A15-1A, 130A22-1, 130A22-1A, 131A30-1 and 131A30-1A. Schematics for these models include installation of the brush head kit.

**NOTE: Installation of brush head kit must include installation of Ground Terminal and Label Kit A-51202 if not previously installed. Alternator must be grounded with #8 ground wire to a suitable ground source.**

# ENGINE

Engine make and model are listed at beginning of section. Refer to Briggs and Stratton of Tecumseh section for engine service.

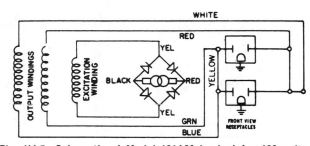

*Fig. H4-7—Schematic of Model 131A30-1 wired for 120 volts without brush head kit A-51450.*

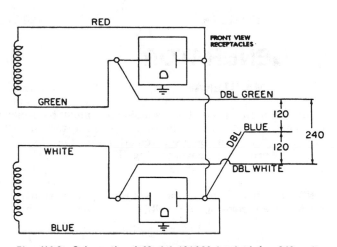

*Fig. H4-8—Schematic of Model 131A30-1 wired for 240 volts without brush head kit A-51450.*

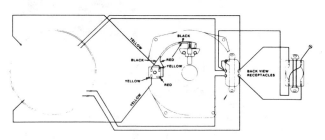

*Fig. H4-9—Wiring diagram for Models E3000-1A and E4000-1A. Models 132A40-1 and E4000-1 are similar.*

# HOMELITE

| Model | Output-kw | Voltage | Engine Make | Engine Model | Governed Rpm |
|-------|-----------|---------|------|-------|--------------|
| 151A15-1 | 1.5 | 120 | B&S | 100232 | 3600 |
| 151A15-1A | 1.5 | 120 | B&S | 100232 | 3600 |
| 151A15-1B | 1.5 | 120 | B&S | 100232 | 3600 |
| 151A25-1 | 2.5 | 120 | B&S | 170432 | 3600 |
| 152A27-1A | 2.75 | 120 | B&S | 170432 | 3600 |
| 153A35-1 | 3.5 | 120 | B&S | 190432 | 3600 |
| 153A35-1A | 3.5 | 120 | B&S | 190432 | 3600 |
| 154A20-1 | 2.0 | 120 | B&S | 130232 | 3600 |
| 155A50-1 | 5.0 | 120 | B&S | 243431 | 3600 |
| 155A50-1A | 5.0 | 120 | B&S | 243431 | 3600 |
| 170A15-1 | 1.5 | 120 | B&S | 100232 | 3600 |
| 170A15-1A | 1.5 | 120 | B&S | 100232 | 3600 |
| 170B16 | 1.5 | 120 | B&S | 131232 | 3600 |
| 172A20-1 | 2.0 | 120 | B&S | 130232 | 3600 |
| 172A20-1A | 2.0 | 120 | B&S | 131232 | 3600 |
| 172A20-1B | 2.0 | 120 | B&S | 131232 | 3600 |
| 172B26 | 2.75 | 120 | B&S | 131232 | 3600 |
| 174A27-1 | 2.75 | 120/240 | B&S | 170432 | 3600 |
| 174A27-1A | 2.75 | 120/240 | B&S | 170432 | 3600 |
| 174A27-1B | 2.75 | 120/240 | B&S | 170432 | 3600 |
| 176A35-1 | 3.5 | 120/240 | B&S | 190432 | 3600 |
| 176A35-1A | 3.5 | 120/240 | B&S | 195432 | 3600 |
| 176A35-1B | 3.5 | 120/240 | B&S | 195432 | 3600 |
| 176A35-1C | 3.5 | 120/240 | B&S | 195432 | 3600 |
| 176B40 | 3.5 | 120/240 | B&S | 170000 | 3600 |
| 177D38-1 | 3.8 | 120/240 | Lombardini | 530 | 3600 |
| 178A50-1 | 5.0 | 120/240 | B&S | 243431 | 3600 |
| 178A50-1A | 5.0 | 120/240 | B&S | 243431 | 3600 |
| 178A50-1B | 5.0 | 120/240 | B&S | 243431 | 3600 |
| 178B48 | 4.3 | 120/240 | B&S | 190000 | 3600 |
| 180A75-1 | 7.5 | 120/240 | B&S | 326431 | 3600 |
| 180A75-1A | 7.5 | 120/240 | B&S | 326431 | 3600 |

# GENERATOR

### MAINTENANCE

Brush length and condition should be checked after every 1000 hours of operation. Brushes can be inspected after removing brush head cover (1—Fig. H5-1), rotor bolt (2) and fan (3).

**NOTE: Do not attempt to hold rotor with the fan when removing rotor bolt.**

Loosen screws retaining brush holder and slide out brush holder. Brushes must be renewed if brush length is shorter than 3/8 inch (9.5 mm) as shown in Fig. H5-2. Be sure to reinstall used brush so that brush curvature matches rotor slip ring. During reassembly, fan must be mounted squarely on rotor shaft. Tighten rotor bolt to 120-140 in.-lbs. (13.5-15.8 N·m).

Disassemble brush holder to install new brushes. Insert new brushes in brush holder so curvature of brush will match curvature of rotor slip ring. Seating new brushes is not required as they are manufactured with correct curvature. Red and black wires to brushes should be connected as shown in Fig. H5-3, and wires must be routed behind rotor support to prevent interference with fan.

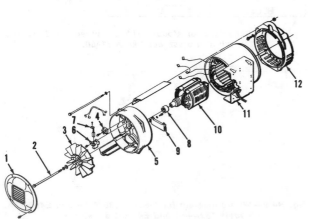

**Fig. H5-1—Exploded view of Model 172A20-1A. Other models are similar.**
1. Cover
2. Bolt
3. Fan
4. Rectifier
5. Brush head
6. Brush holder
7. Brush
8. Bearing
9. Resistor
10. Rotor
11. Stator & housing
12. Endbell

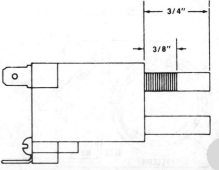

**Fig. H5-2—Brush length should not be less than 3/8 inch (9.5 mm) when measured as shown above. Brush length of a new brush is 3/4 inch (19 mm).**

### OVERHAUL

Refer to Fig. H5-1 for an exploded view of generator. Be sure to mark wiring so that it may be correctly connected for reassembly. Wires should be handled carefully to prevent damage to wire, insulation or connections. Generator rotor (10) has a taper fit with engine crankshaft and should be removed with a suitable puller.

# LOADAMATIC

Some models may be equipped with an automatic idle control (Loadamatic). An electromagnet is mounted adjacent to the engine's carburetor and acts on the carburetor throttle arm. When there is no load on the generator, the electromagnet is energized and the engine governor is overridden as the electromagnet pulls the carburetor throttle arm to idle position. The governor resumes control of engine speed when a load is imposed on the generator. Electric current to the electromagnet is cut off and the throttle arm is released to be controlled by the governor.

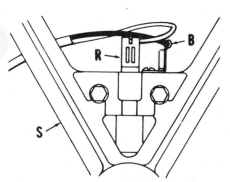

Fig. H5-3—Red (R) and black (B) wires to brushes are connected and routed as shown.

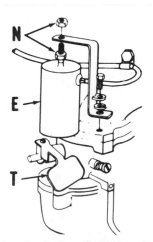

Fig. H5-4—Loadamatic is adjusted by turning nuts (N) until bottom of electromagnet (E) is parallel with throttle arm (T). Do not bend throttle arm or electromagnet bracket.

To adjust automatic idle control (Loadamatic), proceed as follows: Refer to Fig. H5-4 and adjust height of electromagnet (E) to position bottom of electromagnet parallel with throttle arm (T) by turning nuts (N). Do not bend bracket

or throttle arm to make this adjustment. Move generator toggle switch to "START" position to disengage Loadamatic. Start engine and allow it to reach operating temperature. If necessary, adjust carburetor for proper mix-

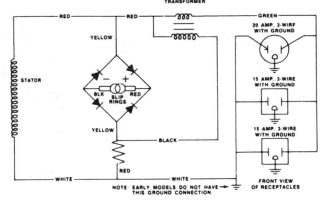

Fig. H5-5—Wiring schematic of Models 151A15-1 and 151A15-1A.

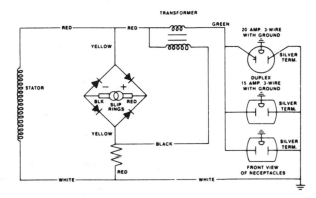

Fig. H5-6—Wiring schematic of Models 151A15-1B and 154A20-1. Models 172A20-1, 172A20-1A and 172A20-1B are similar but 20 amp outlet is not used.

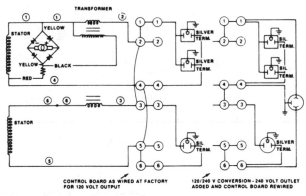

Fig. H5-7—Wiring schematic of Models 152A25-1, 153A35-1 and 155A50-1.

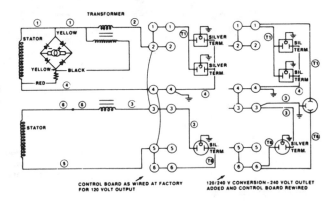

Fig. H5-8—Wiring schematic of Models 152A27-1A, 153A35-1A and 155A50-1A.

# ENGINE

ture and speed. Move toggle switch to "AUTO" position. Engine speed should reduce to idle speed. If idle speed is not steady, adjust carburetor idle mixture. Idle speed should be 2400-2600 rpm, and is adjusted by loosening electromagnet nuts and altering position of elec-tromagnet. Raising electromagnet decreases engine speed, while lowering electromagnet increases engine speed. Apply a light load to generator and then remove it. Engine speed should increase to governed speed and then return to idle after the load is removed.

Engine make and model are listed at beginning of section. Refer to appropriate Briggs & Stratton or Lombardini section for engine service.

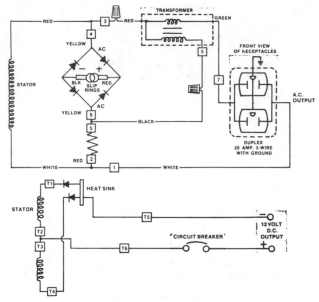

Fig. H5-8A—Wiring schematic of Model 170A15-1A.

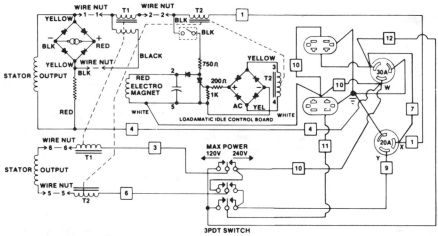

Fig. H5-9—Wiring schematic of Models 174A27-1A, 176A35-1A, 176A35-1B and 178A50-1A.

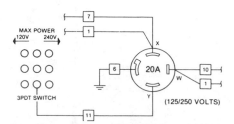

Fig. H5-10—Wiring schematic for 20 amp outlet on Models 174A27-1B, 176A35-1C and 178A50-1B. Remainder of wiring is same as shown in Fig. H5-9.

Illustrations courtesy of Homelite Div. of Textron

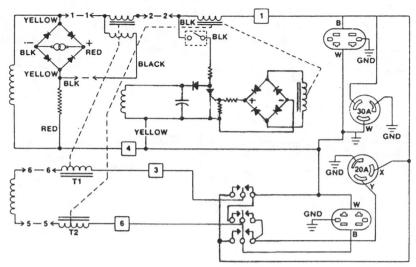

Fig. H5-11—Wiring schematic of Models 174A27-1, 176A35-1 and 178A50-1.

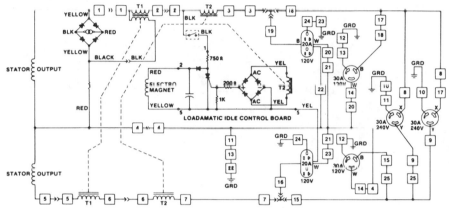

Fig. H5-12—Wiring schematic of Model 180A75-1.

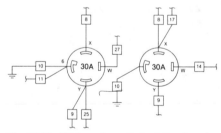

Fig. H5-13—Wiring schematic for 240 volt outlets on Model 180A75-1A. Remainder of wiring is same as shown in Fig. H5-12.

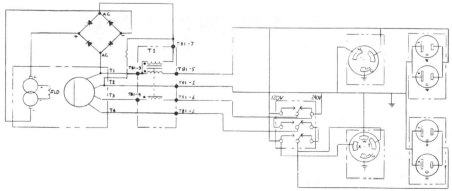

Fig. H5-14—Wiring schematic for Model 177D38-1.

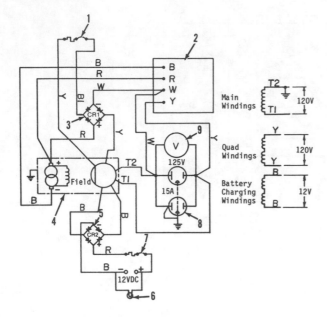

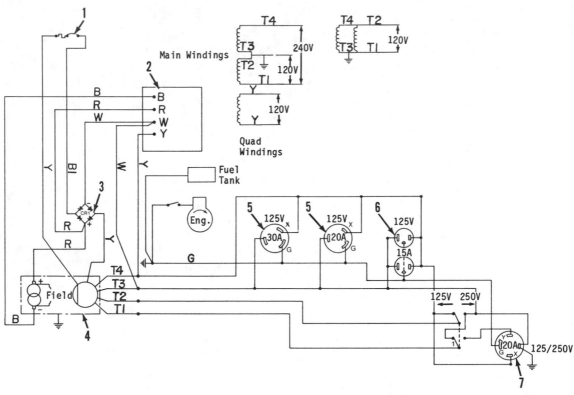

**Fig. H5-15—Wiring schematic of Models 170B16 and 172B26.**

1. Thermal circuit breaker
2. Voltage regulator
3. Bridge rectifier
4. Rotor
5. Battery charge rectifier
6. Indicator light
7. 12-volt circuit breaker
8. 125-volt receptacles
9. Voltmeter

**Fig. H5-16—Wiring schematic of Models 176B40 and 178B48.**

1. Thermal circuit breaker
2. Voltage regulator
3. Bridge rectifier
4. Rotor

5. 125-volt receptacles
6. 125-volt receptacle
7. 125/250 volt receptacle

# HOMELITE

| Model | Output-kw | Voltage | Engine Make | Engine Model | Governed Rpm |
|---|---|---|---|---|---|
| 270A20-1, 207A20-1A, 270A20-1B, 270A20-1C | 2.0 | 120 | Homelite | 270 | See text |

## GENERATOR

### MAINTENANCE

#### Model 270A20-1

To remove brushes, unscrew clamp screws and slide brush cover (20 – Fig. H6-1) off brush head (18). Unscrew set screws and remove brush caps (16) and brushes (17). Be sure to mark brushes so that they can be returned to their original positions. Check length of each brush as shown in Fig. H6-2. Renew brush if length is ⅝ inch (15.9 mm) or shorter. Be sure to include length of projections for brush springs when measuring brush length.

If new brushes must be installed, use the following procedure: Wrap the commutator with a medium grit (4/0 to 6/0) piece of garnet paper which will fit the commutator. Install new brushes. Remove spark plug and use engine starter to rotate engine until brushes are seated. Remove brushes and install fine grit (8/0 to 9/0) garnet paper. Install brushes and repeat seating process. Remove brushes and garnet paper and install spark plug. Blow carbon and grit out of generator and install brushes.

### Models 270A20-1A, 270A20-1B, 270A20-1C

Brush length and condition should be checked after every 1000 hours of operation. Brushes can be inspected after removing brush head cover (2 – Fig. H6-3), rotor bolt (1) and fan (3).

**NOTE: Do not attempt to hold rotor with fan to remove rotor bolt.**

Loosen screws retaining brush holder, and slide out brush holder. Brushes must be renewed if brush length is shorter than ⅜ inch (9.5 mm) as shown in Fig. H6-4. Be sure to reinstall a used brush so that brush curvature matches collector ring. During reassembly, fan must be mounted squarely on rotor shaft. Tighten rotor bolt to 120-140 in.-lbs. (13.6-15.8 N·m).

Disassemble brush holder to install new brushes. Insert new brushes in brush holder so curvature of brush will match curvature of collector ring. Seating new brushes is not required as they are manufactured with correct curvature. Red and black wires to brushes should be connected as shown in Fig. H6-5 and wires must be routed behind rotor support to prevent interference with fan.

### OVERHAUL

Refer to Fig. H6-1 or Fig. H6-3 for an exploded view of generator. Refer to Fig. H6-6 for wiring schematic of all models except 270A20-1.

The generator rotor is a taper fit on engine crankshaft and a suitable tool is necessary to break taper fit loose. When installing rotor, tighten rotor bolt to 120-140 in.-lbs (13.6-15.8 N·m).

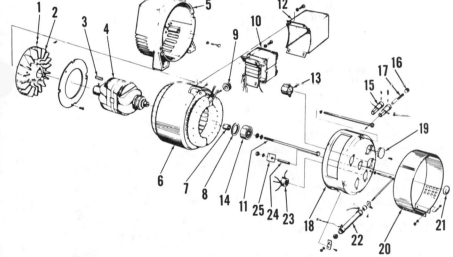

*Fig. H6-1—Exploded view of generator on Model 270A20-1. Through bolt (11) retains generator rotor (4) on tapered end of engine crankshaft. After removing rotor and fan (2) assembly from engine crankshaft, fan housing end bell (5) can be unbolted from engine crankcase.*

1. Set screw
2. Fan
3. Key
4. Generator rotor
5. End bell
6. Generator stator
7. Inner race
8. Seal
9. Grommet
10. Transformer
11. Bolt
12. Transformer box
13. Receptacle, 3-wire
14. Ball bearing
15. Brush holder
16. Brush holder cap
17. Brushes
18. Brush head
19. Plug
20. Cover
21. Bearing cap
22. Resistor
23. Receptacle, 2-wire
24. Rectifier spacer
25. Rectifier & lead assy.

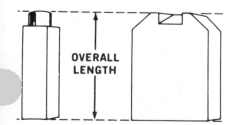

**OVERALL LENGTH**

*Fig. H6-2—Measure brush length as shown.*

*Fig. H6-3—Exploded view of generator used on Models 270A20-1A, 270A20-1B and 270A20-1C.*

1. Screw
2. Cover
3. Fan
4. Rectifier
5. Brush
6. Brush holder
7. Brush head
8. Roller bearing
9. Resistor
10. Rotor
11. Stator & housing
12. End bell

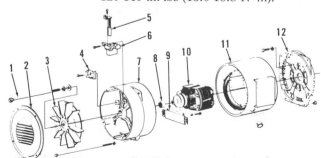

# ENGINE

| Model | Cyls. | Bore | Stroke | Displ. |
|-------|-------|------|--------|--------|
| 270 | 1 | 2½ in. (63.5 mm) | 1⅝ in. (41.3 mm) | 8.0 cu. in. (131 cc) |

## MAINTENANCE

**SPARK PLUG.** Recommended spark plug is Champion J6J or UJ11G. Spark plug electrode gap should be 0.025 inch (0.64 mm).

**CARBURETOR.** Tillotson HS-45C carburetor shown in Fig. H6-7 is used on

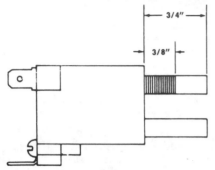

*Fig. H6-4 — Brush length should not be less than ⅜ inch (9.5 mm) when measured as shown above. Brush length of a new brush is ¾ inch (19 mm).*

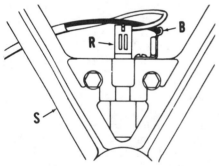

*Fig. H6-5—On models indicated in text, red (R) and black (B) wires to brushes are connected and routed as shown.*

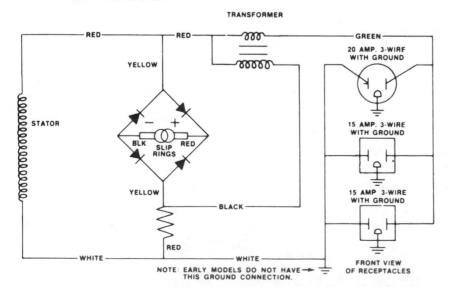

*Fig. H6-6—Schematic of generator Models 270A20-1A, 270A20-1B and 270A20-1C.*

all models. As engine speed is controlled by governor plate on rotary inlet valve, there are no governor linkage connections to throttle shaft. Throttle shaft has spring loaded detent to hold shaft in wide open position.

When disassembling, slide diaphragm assembly towards adjustment needle side of carburetor body to disengage diaphragm from fuel inlet control lever. To remove welch plugs, carefully drill through plugs with a small diameter drill and pry plugs out with a pin. Caution should be taken that drill bit just goes through welch plug as deeper drilling may seriously damage carburetor.

Inlet control metering lever (15) should be flush with metering chamber floor of carburetor body. If not, bend diaphragm end of lever up or down as required so that lever is flush.

Normal adjustment of low speed fuel mixture needle (marked "L" on carburetor body) is ¾-turn open and main mixture adjusting needle (marked "H" on carburetor body) should be opened one full turn.

Start engine and allow to warm up before making final carburetor adjustments. With carburetor throttle shaft in high speed detent position and engine running under load, adjust main (H) fuel needle for smoothest running. There is no external control for throttle, thus no need to make idle adjustments. Move throttle shaft to idle speed position and adjust idle fuel needle (L) for smoothest idle.

**GOVERNOR.** The governor is a part

of the rotary inlet valve; refer to Fig. H6-9. As engine speed increases, centrifugal force pivots governor plate on pivot pin (P) against tension of spring (S). The governor plate then closes the opening in rotary valve and thus throttles the engine. Maximum governed speed is controlled by tension of governor spring, which is adjusted by turning screw (A).

To check and adjust engine speed, proceed as follows: First, bring engine to normal operating temperature and adjust carburetor for highest speed and best performance obtainable, then check engine speed with tachometer. Governed speed at no load should be 3750 rpm.

If adjustment is necessary, stop engine and remove air filter cover and rubber plug (22—Fig. H6-8) from air filter base (13). Remove brass plug from

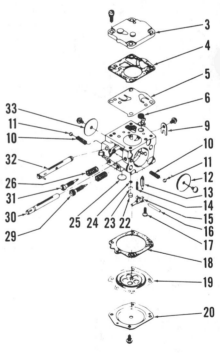

*Fig. H6-7—Exploded view of Tillotson HS carburetor.*

3. Fuel pump cover
4. Fuel pump gasket
5. Fuel pump diaphragm
6. Inlet screen
9. Throttle shaft clip
10. Shaft detent springs (2)
11. Shaft detent balls
12. Choke shutter
13. Inlet needle
14. Inlet tension spring
15. Inlet control lever
16. Inlet pinion pin
17. Inlet pinion pin retaining screw
18. Diaphragm gasket
19. Metering diaphragm
20. Diaphragm cover
22. Welch plug
23. Retaining ring for (24)
24. Body channel screen
25. Welch plug
26. Adjusting needle springs
29. Main adjusting needle
30. Choke shaft
31. Idle adjusting needle
32. Throttle shaft
33. Throttle shutter

Illustrations courtesy of Homelite Div. of Textron

engine crankcase through opening in filter base and turn engine so that adjusting screw (A – Fig. H6-9) is accessible. Then, as shown in Fig. H6-10, turn adjusting screw clockwise to increase speed or counterclockwise to decrease speed. One turn of the screw will change governed speed approximately 100 rpm. Reinstall brass plug, rubber plug and air filter cover, then recheck engine speed; readjust if necessary.

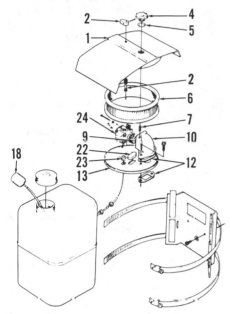

Fig. H6-8—Exploded view of air intake and fuel system. Refer to Fig. H6-7 for exploded view of carburetor (24).

| 1. Air filter cover | 12. Crankcase gasket |
| 2. Control shaft assembly | 13. Air filter mounting plate |
| 4. Cover retaining nut | 18. Fuel filter |
| 5. Nylon washer | 22. Governor adjusting hole plug |
| 6. Air filter element | 23. Grommet |
| 7. Stud | 24. Carburetor assembly |
| 9. Spacer | |
| 10. Intake manifold | |

**IGNITION AND TIMING.** Breaker points, condenser and ignition coil are accessible after removing engine flywheel. A hole is provided in fan housing and inner face of flywheel so that a pin may be inserted to hold flywheel from turning. Unscrew flywheel nut and remove flywheel using Homelite puller No. AA-22560, or equivalent.

To adjust breaker point gap, turn engine so that leading edge of breaker cam is about ⅛-inch (3.2 mm) past breaker point cam follower, then adjust point gap to 0.020 inch (0.51 mm).

**NOTE: Service crankshafts may have two breaker cam keyways for use in Model**

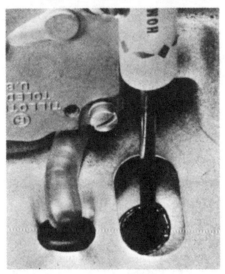

Fig. H6-10—Access to governor spring adjusting screw (A—Fig. H6-9) is gained by removing air filter cover, rubber plug (22—Fig. H6-8) and the brass plug from engine crankcase. Then, turn engine so that screw can be turned with screwdriver as shown.

270 and an earlier model engine. On Model 270 engines, install breaker cam and flywheel key in red painted keyway which is 60 degrees ahead of crankpin.

**LUBRICATION.** Engine on all models is lubricated by mixing oil with regular gasoline. If Homelite® Premium SAE 40 oil is used, fuel:oil ratio should be 32:1. Fuel:oil ratio should be 16:1 if Homelite® 2-Cycle SAE 30 oil or other SAE 30 oil designed for air-cooled two stroke engines is used.

**AIR AND FUEL FILTERS.** The air cleaner element may be washed by sloshing it around in a container of non-oily solvent. If engine is run continuously, clean air filter daily. After a number of cleanings, filter pores may become permanently clogged, making it necessary to renew element.

The fuel filter is part of the fuel pickup inside fuel tank and can be fished out with wire hook. Under normal operations, the filter element should be changed at one to two month intervals. If engine is run continuously, or fuel is dirty, the filter may need to be changed weekly, or more often if necessary.

**CARBON.** Carbon should be cleaned from exhaust ports at 100 to 200 hour intervals. Remove muffler, crank piston to top dead center and use a wooden or plastic scraper to remove carbon deposits. Avoid scratching piston or damaging edge of port.

**NOTE: For easy access to exhaust port, stand engine on recoil starter end.**

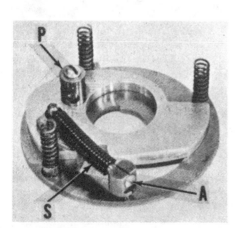

Fig. H6-9—View or rotary intake valve and governor assembly. Governor plate pivots on post (P) to close off valve opening in rotary valve plate to govern engine speed. Speed at which plate closes the opening is regulated by tension of governor spring (S) which is adjusted by turning screw (A).

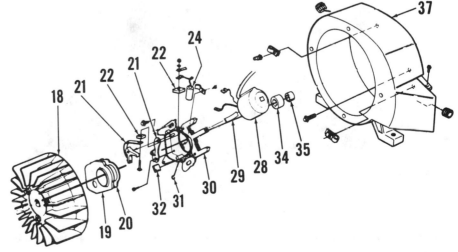

Fig. H6-11—Exploded view of magneto assembly. Magneto components are accessible after removing flywheel; refer to text.

| 18. Flywheel | 22. Terminal connection | 30. Stator & armature assy. | 32. Cam wiper felt |
| 19. Breaker box cover | 24. Condenser | 31. Breaker box cover spring | 34. Breaker cam |
| 20. Gasket | 28. Ignition coil | | 35. Felt seal |
| 21. Breaker points | 29. Coil wedge | | 37. Fan housing |

## REPAIRS

**CONNECTING ROD.** Connecting rod and piston assembly can be removed after removing cylinder from crankcase. Refer to Fig. H6-12. Be careful to remove all of the loose needle rollers when detaching rod from crankpin. There are 23 bearing rollers in rod bearing.

Renew connecting rod if bent or twisted, of if crankpin bearing surface is scored, burned or excessively worn. The caged needle roller piston pin bearing can be renewed by pressing old bearing out and pressing new bearing in with Homelite tool Nos. 24131-1 (plug) and 24124-1 (sleeve). Press on lettered end of bearing cage only.

Renew crankpin needle rollers as a set if any roller is scored, burned or has flat spots. Stick needle roller set to crankpin with heavy grease or beeswax. Using a 10-32 threaded rod or headless screw, position connecting rod cap so that mating boss on cap and connecting rod will align when pinned side of piston is on the intake (upper) side of engine; refer to Fig. H6-13. Slide connecting rod down over threaded rod or screw, then install socket head screw in opposite side of rod and cap and remove the installation tool. Install remaining socket head screw and tighten both screws to a torque of 32 inch-pounds (3.6 N·m).

**PISTON, PIN AND RINGS.** Piston assembly is accessible after removing cylinder assembly from crankcase. Always support piston when removing or installing piston pin. Piston is of aluminum alloy and is fitted with three pinned piston rings.

If piston ring locating pin is worn to half the original thickness, or if there is any visible up and down play of piston pin in piston bosses, renew piston and pin assembly. Inspect piston for cracks or holes in dome and renew if any such defect is noted. Slight scoring of piston is permissible, but if rough surfaces are accompanied by deposit of aluminum on cylinder wall, renew piston.

Always use new piston pin retaining snap rings when reassembling piston to connecting rod. Fit new piston rings in grooves, aligning ring end gaps with locating pin. Be sure locating pin side of piston is away from exhaust side of engine when installing piston and connecting rod assembly.

**CYLINDER.** Cylinder bore is chrome plated; the plating is light gray in color and does not have the appearance of polished chrome. Renew cylinder if any part of chrome plated bore is worn through; usually, the worn area is bright as the aluminum is exposed. In some instances, particles from the aluminum piston may be deposited on top of chrome plating. This condition is usually indicated by rough, flaky appearance and deposits can be removed by using a rubber compound buffing wheel on a ¼-inch electric drill. If a screwdriver can be run over the cleaned surface without leaving marks, the cylinder is suitable for further service. If screwdriver scratches surfaces, renew cylinder.

Lubricate piston, rings and cylinder bore. Compress rings, then slide cylinder down over piston. Tighten cylinder retaining nuts evenly and securely.

**CRANKSHAFT, BEARINGS AND SEALS.** To remove crankshaft, engine must first be removed from generator. Refer to Fig. H6-11 for an exploded view of engine. Remove flywheel, magneto assembly and magneto back plate. Remove and discard seal (35 – Fig. H6-11) from fan housing (37). Remove "O" ring oil slinger (not shown) from crankshaft. Remove cylinder, piston and connecting rod assembly, then separate crankcase halves (41 & 73 – Fig. H6-12). Pull governor weight away from crankshaft, then carefully remove the rotary intake valve and governor assembly (48) to avoid damaging sealing surface of crankshaft. Remove two screws (67) and washers (68) retaining ball bearing (66) in crankcase half (73) and remove shaft and bearing. Tape shaft to prevent scratching sealing surface, then remove snap ring (69) and pull bearing (66) from crankshaft. Remove intake valve plate (46) from crankcase half (41), pry seal (42) out of bore and press needle roller bearing cages (43) out of crankcase half. Remove seal (74) from opposite half.

**NOTE: Remove bearings from crankshaft and flywheel side crankcase half only if renewal is indicated.**

Renew crankshaft if it has damaged threads, enlarged keyways, or if run out exceeds 0.003 inch (0.07 mm). Flywheel end main journal must be free of pits, galling or heavy score marks. If journal is worn or out of round more than 0.001 inch (0.025 mm), renew crankshaft. Renew ball bearing at output end if bearing shows perceptible wear or feels rough when rotated. The caged needle roller bearings at flywheel end should be

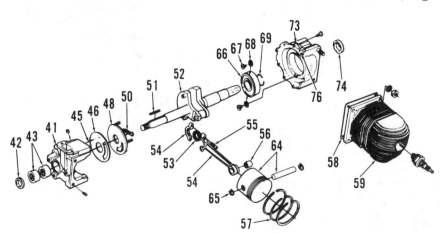

Fig. H6-12—Exploded view of engine. Crankshaft end play is controlled by ball bearing (66).

| | | | |
|---|---|---|---|
| 41. Crankcase half | 50. Governor spring | 56. Needle roller bearing | 67. Bearing retaining screws |
| 42. Crankshaft seal | 51. Cam & flywheel key | 57. Piston rings | 68. Bearing retaining washers |
| 43. Needle roller bearings | 52. Crankshaft | 58. Gasket | 69. Snap ring |
| 45. Dowel pins | 53. Needle bearing rollers (23) | 59. Cylinder | 73. Crankcase half |
| 46. Intake valve wear plate | 54. Connecting rod & cap | 64. Piston & piston pin | 74. Crankshaft oil seal |
| 48. Intake valve & governor assy. | 55. Socket head screws | 65. Snap rings | 76. Dowel pin |
| | | 66. Roller bearing | |

Fig. H6-13—A 10-32 threaded rod or headless screw is used as tool to help in assembling connecting rod to cap. Stick the 23 loose needle rollers to crankpin with beeswax or heavy grease, then carefully position cap so that when assembled, pinned side of piston is towards intake side (upper side) of engine.

Illustrations courtesy of Homelite Div. of Textron

renewed if any roller shows visible flat spot, or if rollers in either cage can be separated more than the width (diameter) of one roller.

To reassemble, support crankshaft at throw, then press new ball bearing onto shaft and secure with snap ring (69).

**NOTE: Be sure that groove in outer bearing race is towards crankshaft throw.**

Lubricate seal (74), then using suitable installation tool (Homelite No. 24120-1 or equivalent) press seal into crankcase half with lip of seal inward. Pressing against outer race of bearing (66) only, install crankshaft and bearing assembly into crankcase half (73) and secure with the two screws (67) and washers (68).

**NOTE: Use suitable seal protector (Homelite Nos. 24125-1, 24126-1 or 24127-1, or equivalent) to prevent damage to seal.**

Using Homelite tool No. 24155-1, press outer needle bearing into crankshaft half (41) with stepped end of tool, then press inner bearing into crankcase half with straight end of tool.

**NOTE: Press on lettered side of bearing cage only.**

Install seal (42) with lip in towards needle bearing with straight end of bearing installation tool.

Fit governor and rotary intake valve assembly onto crankshaft so that thrust springs fit into proper bores of crankshaft.

**NOTE: Hold governor plate away from crankshaft when installing to prevent scratching seal surface.**

Lubricate all parts thoroughly and insert intake valve plate (46) in crankcase half (41) so that it is properly positioned on dowel pins and intake opening. Using seal protector (Homelite No. 24121-1 or equivalent), assemble the crankcase half over crankshaft and governor assembly, hold assembly together against thrust spring pressure and install crankcase cap screws. Tighten cap screws to a torque of 80 inch-pounds (9.0 N·m). Tighten fan housing retaining cap screws to a torque of 80 inch-pounds (9.0 N·m). Complete reassembly by reversing disassembly procedure.

**Fig. H6-14—Exploded view of Fairbanks-Morse rewind starter.**

1. Plate
2. Cover & bushing assy.
3. Starting cord
4. Starting cord grip
5. Insert
6. Rewind spring
7. Starter pulley
8. Fiber washer
9. Friction shoe assembly
10. Brake spring
11. Brake retaining washer
12. Snap ("E") ring
13. Flywheel nut
14. Lock washer
15. Starter cup
16. Screen
17. Mounting ring

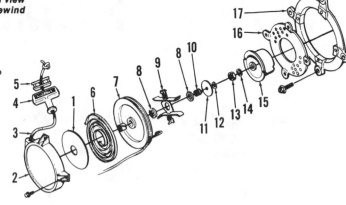

**CRANKCASE.** Be sure that all passages through crankcase are clean. The idle passage line which enters crankcase via the intake valve register may be restricted with carbon deposits.

If main bearing bore at output end has a worn appearance, bearing has been turning in bore. If so, crankcase half should be renewed. The mating surfaces of the two-piece crankcase must be free of all nicks and burrs as neither sealing compound nor gaskets are used at this joint.

**NOTE: Fuel tank bracket mounting screws are secured in engine crankcase with "Loctite". When reinstalling bracket, clean the screw threads and threads in crankcase, then apply a drop of "Loctite" to each screw. Tighten screws to a torque of 120 inch-pounds.**

**ROTARY INTAKE VALVE.** The combination rotary intake valve and governor (see Fig. H6-9) should be renewed if any of the following conditions are noted: If the sealing faces of valve or governor plate are worn or scored enough to produce a ridge; if spring post is loose or extended to valve seating surface; or, if governor pivot post has started to wear through the surface of valve. Maximum allowable

clearance between governor plate and intake valve plate is 0.006 inch. The governor spring and/or governor spring adjusting screw may be renewed separately from the assembly.

Slight scoring of valve face may be corrected by lapping on a lapping plate using a very fine abrasive. Lapping motion should be in the pattern of a figure eight to obtain best results. Slight scoring of the Formica wear plate is permissible. Homelite recommends soaking a new Formica plate in oil for 24 hours prior to installation.

**REWIND STARTER.** Refer to Fig. H6-14 for exploded view of the Fairbanks-Morse starter used on all applications. In an emergency in case of rewind starter failure, remove starter assembly and wind rope around starter cup (15) to start engine.

Refer to exploded view for proper reassembly of starter unit. Hook end of starter rope in notch of pulley and turn pulley five turns counterclockwise, then let spring wind rope into pulley for proper spring pre-tension.

To remove starter cup, insert lock pin through holes in fan housing and flywheel to hold flywheel from turning, then unscrew retaining nut.

# HOMELITE

| Model | Output-kw | Voltage | Engine Make | Engine Model | Governed Rpm |
|-------|-----------|---------|-------------|--------------|--------------|
| G3600-1, G3600-2* | 3.6 | 120/240 | Kohler | K301 | 1800 |
| G4800-1, G4800-2* | 4.8 | 120/240 | Kohler | K341 | 1800 |

*Models G3600-2 and G4800-2 are equipped with an electric starter, battery and battery charging circuit.

## GENERATOR

**NOTE: Do not start or run Models G3600-2 or G4800-2 unless battery is connected.**

### MAINTENANCE

Brush length and condition should be checked after every 1000 hours of operation. Brushes can be inspected after removing brush head cover (1–Fig. H7-1), rotor bolt (2) and fan (3).

**NOTE: Do not attempt to hold rotor with fan to remove rotor bolt.**

Loosen screws retaining brush holder and slide out brush holder. Brushes must be renewed if brush length is shorter than ⅜ inch (9.5 mm) as shown in Fig. H7-2. Be sure to reinstall a used brush so that brush curvature matches collector ring.

Disassemble brush holder to install new brushes. Insert new brushes in brush holder so curvature of brush will match curvature of collector ring. Seating new brushes is not required as they are manufactured with correct curvature. Red (No. 15) and black (No. 14) wires to brushes should be connected as shown in Fig. H7-3 and wires must be routed behind rotor support (S) to prevent interference with fan.

### OVERHAUL

Refer to Fig. H7-1 for an exploded view of generator. Refer to Fig. H7-4 for wiring schematic. When installing brushes refer to MAINTENANCE section for proper brush installation.

## ENGINE

Engine make and model are listed at beginning of section. Refer to Kohler engine section for engine service.

**NOTE: These generators are designed for operation at 1800 rpm. Adjust governor as outlined in engine section but adjust governed speed so engine runs at 1800 rpm under load.**

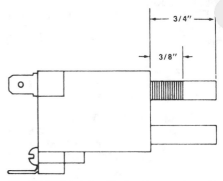

Fig. H7-2 — Brush length should not be less than ⅜ inch (9.5 mm) when measured as shown above. Brush length of a new brush is ¾ inch (19 mm).

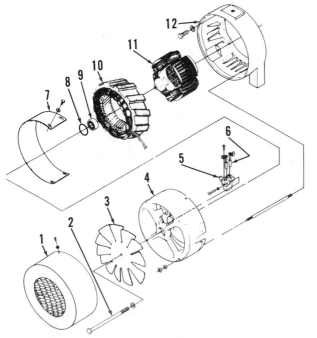

Fig. H7-1—Exploded view of generator.

1. Cover
2. Bolt
3. Fan
4. Brush head
5. Brush holder
6. Brushes
7. Cover
8. "O" ring
9. Ball bearing
10. Stator
11. Rotor
12. End bell

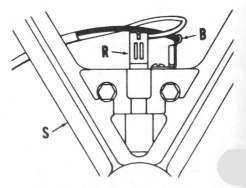

Fig. H7-3—Red (R) number "15" brush wire and black (B) number "14" brush wire are connected and routed as shown.

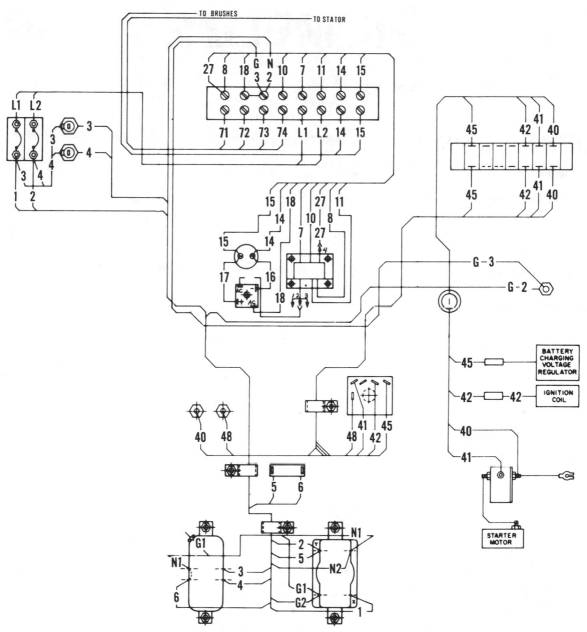

Fig. H7-4—Wiring diagram for Models G3600-2 and G4800-2. Models G3600-1 and G4800-2 are similar except for engine circuit.

# HOMELITE

| Model | Output-kw | Voltage | Engine Make | Engine Model | Governed Rpm |
|-------|-----------|---------|-------------|--------------|--------------|
| HSB50-1 | 5.0 | 120/240 | B&S | 252417 | 3600 |

# GENERATOR

## MAINTENANCE

Brush length and condition should be checked after every 1000 hours of operation. Brushes can be inspected after removing brush head cover (12–Fig. H8-1), rotor bolt (13) and fan (10).

**NOTE: Do not attempt to hold rotor with fan to remove rotor bolt.**

Loosen screws retaining brush holder and slide out brush holder. Brushes must be renewed if brush length is shorter than 3/8 inch (9.5 mm) as shown in Fig. H8-2. Be sure to reinstall a used brush so that brush curvature matches collector ring.

Disassemble brush holder to install new brushes. Insert new brushes in brush holder so curvature of brush will match curvature of collector ring. Seating new brushes is not required as they are manufactured with correct curvature. Red and black wires to brushes should be connected as shown in Fig. H8-3 and wires must be routed behind rotor support to prevent interference with fan.

## OVERHAUL

Refer to Fig. H8-1 for an exploded view of generator. Refer to Fig. H8-4 for wiring schematic. When installing brushes refer to MAINTENANCE section for proper brush installation.

# ENGINE

Engine make and model are listed at beginning of section. Refer to Briggs and Stratton section for engine service, however, note the following additional service areas.

## FUEL SYSTEM

Engine fuel is contained in the five gallon container shown in Fig. H8-5. The engine fuel pump transfers fuel from the tank to the carburetor. Filters (6 and 8) must be clean and vents (2) in tank cap

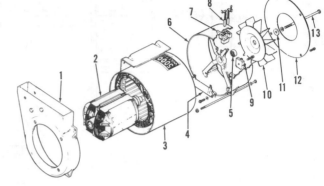

**Fig. H8-1—Exploded view of generator.**

1. End bell
2. Rotor
3. Stator assy.
4. Resistor
5. Bearing
6. Brush head
7. Brush holder
8. Brushes
9. Rectifier
10. Fan
11. Washer
12. Cover
13. Bolt

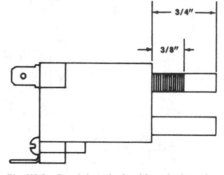

**Fig. H8-2 — Brush length should not be less than 3/8 inch (9.5 mm) when measured as shown above. Brush length of a new brush is 3/4 inch (19 mm).**

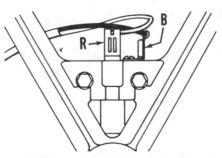

**Fig. H8-3—Red (R) and black (B) wires to brushes are connected and routed as shown.**

(3) must operate properly for adequate fuel movement. Vents (2) are one-way valves and must be installed in cap so one vent allows air in while other valve vents pressure in tank. Note quick-connect coupler (1) and valve (V) in tank cap which must also seal properly.

## LUBRICATION

Oil in the engine crankcase is replenished by oil in an auxiliary oil tank attached to the control panel. See Fig. H8-6. Oil flow into the crankcase is controlled by valve (4). With engine crankcase full, oil level in auxiliary oil tank should be even with sight glass (S). Oil amount necessary to fill oil tank from empty to sight glass is approximately one quart.

**NOTE: Always fill engine crankcase using engine oil fill tube and by reading marks on engine oil dipstick. Do not consider oil tank oil level an indication of engine crankcase oil level.**

**Fig. H8-4—Wiring schematic for HSB50-1.**

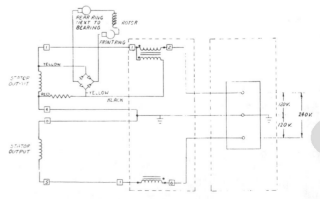

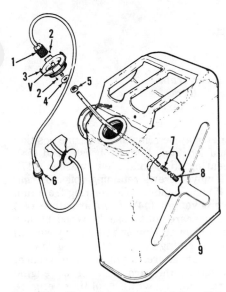

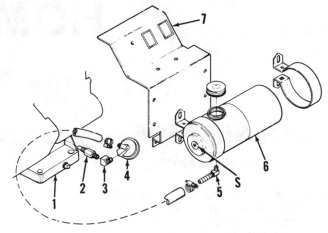

*Fig. H8-6—View of auxiliary oil tank and line. Oil level is maintained even with sight glass (S).*

1. Engine
2. Special fitting
3. Elbow
4. Valve
5. Elbow
6. Auxiliary oil tank
7. Control panel

*Fig. H8-5—View of fuel tank and line. Note quick-disconnect valve (V) in cap (3).*

1. Coupler
2. Vents
3. Cap
6. Filter
4. Washer
5. Nut
7. Spring
8. Filter
9. Fuel tank

# HOMELITE

| | | Engine | | Governor |
|---|---|---|---|---|
| **Model** | **Voltage** | **Make** | **Model** | **Rpm** |
| XLA115 | 115 | Homelite | XL-12 | See text |

## GENERATOR

### MAINTENANCE

To remove brushes, refer to Fig. H9-1 and proceed as follows: Remove two outside screws securing receptacle housing (52) to brush head (58) and move receptacle assembly aside. Remove brush holder caps (56) and brushes, taking particular care to note position and location of brushes so that they may be reinstalled in original position and location. Renew brushes if excessively worn.

Lubricate generator by removing receptacle housing (52 – Fig. H9-1) from end of generator and applying 4 or 5 drops of oil to the felt washer (47) after each 200 hours of operation.

### OVERHAUL

Refer to exploded view of the generator unit in Fig. H9-1. To gain access to the engine governor unit or engine crankshaft, proceed as follows:

Remove two outside screws holding receptacle housing (52) to brush head (58) and move receptacle assembly aside. Remove brush holder caps (56) and brushes, taking particular care to note position and location of brushes so that they may be reinstalled in same position and location. Remove brush head (58), bearing (50) and felt washer (47). Remove screws retaining collector ring (45) and remove collector ring from fan. Remove fan (44), from armature (43) taking care not to lose the Woodruff key (not shown). Unbolt generator yoke (35) from bearing housing (13) and slide yoke from armature. Using a strap wrench, unscrew the armature from engine crankshaft while holding engine flywheel. Procedure and need for further disassembly is obvious from inspection of unit. After reassembly check adjustment of governor as outlined in GOVERNOR paragraph.

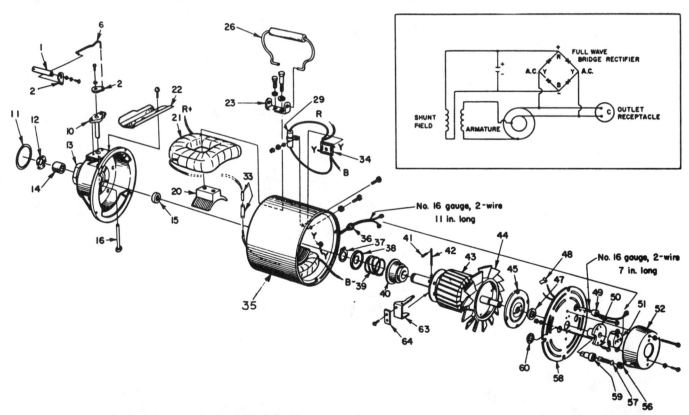

**Fig. H9-1—Exploded view of generator assembly. Wiring diagram for generator is shown in upper right corner. Crankshaft needle roller (main) bearing (14) and crankshaft seal (15) are carried in bearing housing (13). Armature (43) must be unscrewed from engine crankshaft to allow removal of bearing housing and crankshaft.**

| | | | | |
|---|---|---|---|---|
| 1. Throttle shaft extension | 14. Needle bearing | 26. Carrying handle | 38. Centering washer | 47. Felt washer |
| 2. Linkage arms | 15. Crankshaft seal | 29. Capacitor | 39. Governor spring | 48. Brush holder clip |
| 6. Governor link | 16. Governor cam & shaft | 33. Connector | 40. Governor cup | 49. Insulating bushing |
| 10. Governor shaft guide | 20. Generator field pole | 34. Rectifier | 41. Cotter pins (4) | 50. Flanged Oilite bushing |
| 11. "O" ring | 21. Generator field coil | 35. Generator yoke (frame) | 42. Pivot pins (2) | 51. Outlet receptacle |
| 12. Loading spring | 22. Linkage guard | 36. Insulating bushing | 43. Generator armature | 52. Receptacle housing |
| 13. Bearing housing | 23. Handle bracket | 37. Snap ring | 44. Fan | 56. Brush holder cap |
| | | | 45. Collector ring | |
| | | | | 57. Brush |
| | | | | 58. Brush head |
| | | | | 59. Brush holder |
| | | | | 60. Holder retaining ring |
| | | | | 63. Governor arms (2) |
| | | | | 64. Governor weights (2) |

Illustrations courtesy of Homelite Div. of Textron

# ENGINE

| Model | Cyls. | Bore | Stroke | Displ. |
|-------|-------|------|--------|--------|
| XL-12 | 1 | 1¾ in. (44.4 mm) | 1⅜ in. (34.9 mm) | 3.3 cu. in. (54.1 cc) |

## MAINTENANCE

**SPARK PLUG.** A Champion TJ-8J spark plug or equivalent is used. Set electrode gap to 0.025 inch (0.64 mm).

**CARBURETOR.** Refer to Fig. H9-2 for exploded view of Tillotson HS diaphragm type carburetor with integral fuel pump used on XL-12 engines. Carburetor is accessible after removing air box cover.

**NOTE: If early type cover gasket becomes damaged when air box cover is removed, install new gasket as follows: Carefully remove old gasket from air box and be sure that surface is free of all dirt, oil, etc. Apply "3M" or Homelite No. 22788 cement to new gasket and carefully place gasket, adhesive side down, on lip around air box chamber. On later engines, gasket is bonded to filter element; install new filter element if either gasket or filter is damaged.**

When disassembling carburetor, slide the diaphragm assembly towards adjustment needle side of carburetor body to disengage diaphragm from fuel inlet control lever. To remove welch plugs, carefully drill through large plug (27 – Fig. H9-2) with a ⅛-inch drill or through small plug (24) with a 1/16-inch drill and pry plugs out with a pin inserted through the drilled hole. Caution should be taken that the drill just goes through the welch plug as deeper drilling may seriously damage the carburetor. Note channel screen (26) and screen retaining ring (25) which are accessible after removing welch plug (24).

Inlet control lever (17) should be flush with metering chamber floor of carburetor body. If not, bend diaphragm end of lever up or down as required so that the lever is flush.

On generator engine, make initial carburetor adjustment as follows: Turn fuel adjustment needle in gently until it just contacts seat, then back needle out 1¼ turns. For final adjustment, apply load to generator to allow engine to warm up, then slowly turn needle in until engine speed starts to drop. Correct final setting is ⅛-turn open from this point.

**GOVERNOR.** Engine speed is controlled by a mechanical flyweight type governor located within the generator unit; refer to exploded view of generator in Fig. H9-1. Slotted end of extension shaft (1) fits over pinned end of carburetor throttle shaft. Centrifugal force of governor weights (64) acting through arms (63) actuate governor cup (40), cam (16) and connecting linkage against force of governor spring (39). The spring (29 – Fig. H9-3) in air box is adjusted to full tension and is used as a throttle opening spring and to keep slack out of linkage.

To adjust governed speed with tachometer or frequency meter, proceed as follows: To use an rpm indicator, remove receptacle housing (52 – Fig. H9-1) to gain access to end of generator shaft. With engine at operating temperature, governed no-load speed should be approximately 3750 rpm, or a reading of 63 cycles per second should be indicated on frequency meter. If engine speed or cycles per second is not as specified, loosen the two screws clamping governor shaft guide (10) to bearing housing (13) and move guide towards engine to increase speed, or

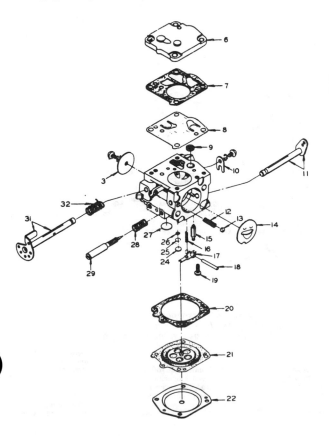

**Fig. H9-2—Exploded view of typical Tillotson HS carburetor.**

3. Throttle disc
6. Pump cover
7. Gasket
8. Pump diaphragm
9. Inlet screen
10. Throttle shaft clip
11. Choke shaft & lever
12. Detent spring
13. Choke detent ball
14. Choke disc
15. Inlet needle
16. Spring
17. Diaphragm lever
18. Lever pin
19. Pin retaining screw
20. Gasket
21. Diaphragm
22. Diaphragm cover
24. Welch plug
25. Retaining ring
26. Channel screen
27. Welch plug
28. Springs
29. Main fuel needle
31. Throttle shaft & lever
32. Throttle spring

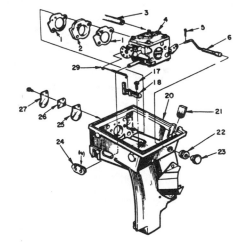

**Fig. H9-3—Exploded view of engine air box assembly. Mechanical governor parts are shown in exploded view of generator in Fig. H9-1.**

1. Gaskets
2. Spacer
3. Fuel line
4. Carburetor assy.
5. Cotter pin
6. Choke rod
17. Screws (2)
18. Adjustment plate
20. Air box
21. Felt plug
22. Grommet
23. Choke button
24. Adjustment needle grommet
25. Reed valve
26. Reed back-up
27. Reed stop
29. Throttle opening spring

towards generator to decrease speed.

**NOTE: Only one mounting hole of governor shaft guide is slotted and guide pivots on opposite screw.**

Tighten screws when proper speed is obtained.

If carburetor has been removed, reconnect throttle opening spring as in-dicated in Fig. H9-3. If plate (18) has been removed, reinstall screws loosely, push plate to apply as much spring tension as possible (to end of slot) and tighten screws. Servicing of the mechanical governor unit requires disassembly of the generator unit; refer to exploded view in Fig. H9-1 and to GENERATOR paragraph.

**MAGNETO:** A Wico flywheel type magneto with external armature and ignition coil is used. Breaker points and condenser are located behind flywheel.

Armature core and stator plate are riveted together and are serviced only as a unit. Stator plate fits firmly on shoulder of crankcase; hence, armature air gap is non-adjustable.

Magneto stator plate has slotted mounting holes, and should be rotated as far clockwise as possible before tightening mounting screws to obtain correct ignition timing of 30 degrees BTDC. Set breaker point gap to 0.015 inch (0.38 mm). Condenser capacitance should test 0.16-0.20 mfd.

**CAUTION: Be careful when installing breaker points not to bend tension spring any more than necessary; if spring is bent excessively, spring tension may be reduced causing improper breaker point operation.**

**LUBRICATION.** The engine on all models is lubricated by mixing oil with regular gasoline. Fuel:oil ratio should be 32:1 when Homelite® SAE 40 Premium Motor Oil is mixed with fuel. Fuel:oil ratio should be 16:1 if Homelite® SAE 30 2-Cycle oil or another good grade of oil

**Fig. H9-4—Exploded view of the flywheel type magneto. Connect condenser and low tension leads as indicated by letters "B" and "C".**

24. Flywheel
25. Condenser
26. Condenser clamp
28. Breaker box cover
29. Terminal block
30. Breaker point set
31. Washer
32. Pivot post clip
34. Gasket
37. Plate and armature assy.
38. Gasket
40. Felt seal
41. Coil retaining clip
42. Ignition coil assy.
43. High tension wire
44. Connector spring
45. Spark plug boot

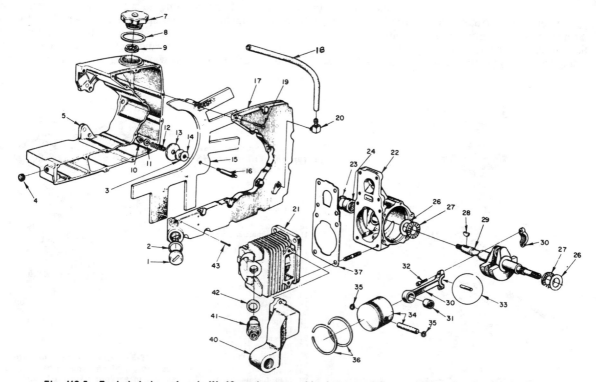

**Fig. H9-5—Exploded view of early XL-12 engine assembly. Later models are similar. Crankcase side cover, needle roller main bearing and the crankshaft seal are a part of the generator as shown in Fig. H9-1. The generator must be disassembled to gain access to crankshaft; refer to appropriate generator paragraph.**

1. Plug
2. Gasket
3. Felt washer
4. Plug
5. Fuel tank
7. Fuel cap
8. Gasket
9. Valve
10. Nut
11. Flat washer
12. Compression spring
13. Wick washer
14. Gasket
15. Fuel pick-up wick
16. Fuel pick-up stud
17. Fuel tank cover
18. Fuel line
19. Round head screws (16)
20. Fuel line elbow
21. Cylinder
22. Crankcase
23. Seal
24. Needle bearing
26. Thrust bearing race
27. Needle thrust bearing
28. Flywheel key
29. Crankshaft
30. Connecting rod
31. Needle bearing
32. Rod cap screws
33. Crankpin rollers (28)
34. Piston & pin assy.
35. Snap rings (only one used on late models)
36. Piston rings
37. Gasket
40. Manifold
41. Spark plug
42. Spark plug gasket

Illustrations courtesy of Homelite Div. of Textron

designed for air-cooled two-stroke engines is used.

**CARBON.** Muffler, manifold and cylinder exhaust ports should be cleaned periodically to prevent loss of power through carbon build up. Remove muffler and scrape free of carbon; a bent wire can be inserted through hole in a muffler to clean outer shell. With muffler or manifold removed, turn engine so that piston is at top dead center and carefully remove carbon from exhaust ports with a wooden scraper. Be careful not to damage chamfered edges of exhaust ports or to scratch piston. Do not run engine with muffler removed.

## REPAIRS

**CONNECTING ROD.** Connecting rod and piston assembly can be removed after removing cylinder from crankcase. Refer to Fig. H9-5. Be careful to remove all needle rollers when detaching rod from crankpin. Early models have 28 loose needle rollers while later models have 31 needle rollers.

Renew connecting rod if bent or twisted, or if crankpin bearing surface is scored, burned or excessively worn. The caged needle roller piston pin bearing can be renewed by pressing old bearing out and pressing new bearing in with Homelite tool No. 23756. Press on lettered end of bearing cage only.

It is recommended that the crankpin needle rollers be renewed as a set whenever engine is disassembled for service. On early models with 28 needle rollers, stick 14 needle rollers in the rod and remaining 14 needle rollers in rod cap with light grease or beeswax. On late models with 31 needle rollers, stick 16 rollers in rod and 15 rollers in rod cap. Assemble rod to cap with match marks aligned, and with open end of piston pin towards flywheel side of engine. Wiggle the rod as cap retaining screws are being tightened to align the fractured mating surfaces of rod and cap.

**PISTON, PIN AND RINGS.** The piston is fitted with two pinned compression rings. Renew piston if scored, cracked or excessively worn, or if ring side clearance in top ring groove exceeds 0.0035 inch (0.089 mm).

Recommended piston ring end gap is 0.070-0.080 inch (1.78-2.03 mm); maximum allowable ring end gap is 0.085 inch (2.16 mm). Desired ring side clearance in groove is 0.002-0.003 inch (0.05-0.07 mm).

Piston, pin and rings are available in standard size only. Piston and pin are available in a matched set, and are not available separately.

Piston pin has one open and one closed end and may be retained in piston with snap rings or a Spirol pin. A wire retaining ring is used on exhaust side of piston on some models and should not be removed.

To remove piston pin on all models, remove the snap ring at intake side of piston. On piston with Spirol pin at exhaust side, drive pin from piston and rod with slotted driver (Homelite tool No. A-23949). On all other models, insert a 3/16-inch pin through snap ring at exhaust side and drive piston pin out.

When reasembling, be sure closed end of piston pin is to exhaust side of piston (away from piston ring locating pin). Install Truarc snap ring with sharp edge out.

**CRANKSHAFT.** The crankshaft is supported in two caged needle roller bearings and crankshaft end play is controlled by a roller bearing and hardened steel thrust washer at each end of the shaft. Refer to Fig. H9-5. On generator, crankshaft end play is taken up by a loading spring (12–Fig. H9-1).

To remove crankshaft, it will be necessary to disassemble the generator. Refer to Fig. H9-1 and GENERATOR section.

**CYLINDER.** The cylinder bore is chrome plated. Renew the cylinder if chrome plating is worn away exposing the softer base metal.

**CRANKCASE, BEARING HOUSING AND SEALS.**

**CAUTION: Do not lose bearing housing-to-crankcase screws. New screws of same length must be installed in place of old screws.**

The needle roller main bearings and crankshaft seals in crankcase and bear-

*Fig. H9-6—When installing reed valve, reed back-up and reed stop, be sure reed is centered between two points indicated by black arrows.*

ing housing can be renewed using Homelite tool Nos. 23757 and 23758. Press bearings and seals from crankcase or bearing housing with large stepped end of tool No. 23757, pressing towards outside of either case.

To install new needle bearings, use the shouldered short end of tool No. 23757 and press bearings into bores from inner side of either case. Press on lettered end of bearing cage only.

To install new seals, first lubricate the seal and place seal on long end of tool No. 23758 so that lip of seal will be towards needle bearing as it is pressed into place.

To install crankshaft, lubricate thrust bearings (27–Fig. H9-5) and place on shaft as shown. Place a hardened steel thrust washer to the outside of each thrust bearing. Insert crankshaft into crankcase being careful not to damage seal in crankcase. Place a seal protector sleeve (Homelite tool No. 23759) on crankshaft and large "O" ring or gasket on shoulder of bearing housing.

**NOTE: On early production, crankcase was sealed to drivecase with an "O" ring; however, use of "O" ring has been discontinued and a gasket, rather than an "O" ring should be used on all models.**

Lubricate seal protector sleeve, seal and needle bearing and assemble bearing housing to crankshaft and crankcase. Use **NEW** bearing housing retaining screws. Clean the screw threads and apply "Loctite" to threads before installing screws. Be sure the screws are correct length; screw length is critical. Tighten the screws alternately and remove seal protector sleeve from crankshaft. Reassemble circular saw drive case, pump, generator or brush cutter upper drive housing as outlined in appropriate paragraph.

**FLAT REED INTAKE VALVE.** The reed valve is attached to the carburetor air box as shown in Fig. H9-6, and is accessible after removing air box from crankcase.

Check the reed seating surface on air box to be sure it is free of nicks, chips or burrs. Renew valve reed if rusted, pitted or cracked, or if it does not seat flatly against its seat.

The reed stop is curved so that measurement of reed lift distance is not practical. However, be sure that reed is centered over opening in air box and reed stop is aligned with reed as shown in Fig. H9-6.

**NOTE: If air box has been removed to service reed valve (25–Fig. H9-3), inspect gasket between air box and crankcase. If gasket is damaged and cylinder is not be-**

ing removed for other purposes, it is suggested that the exposed part of the old gasket be carefully removed and the new gasket be cut to fit between the air box and crankcase. Also, refer to note in CARBURETOR paragraph in MAINTENANCE section.

**REWIND STARTER.** To disassemble starter, refer to exploded view in Fig. H9-7 and proceed as follows: Pull starter rope out fully, hold pulley (7) and pry rope knot from pulley. Let pulley rewind slowly. Hold pulley while removing screw (9) and washer (8). Turn pulley counterclockwise until disengaged from spring, then carefully lift pulley off starter post. Turn open side of housing down and rap housing sharply against top of work bench to remove spring.

**CAUTION: Be careful not to dislodge spring when removing pulley as spring could cause injury if it should recoil rapidly.**

Install new spring with loop in outer end over pin in blower housing and be sure spring is coiled in direction shown in Fig. H9-7. Install pulley (7), turning pulley clockwise until it engages spring and secure with washer and screw. In-

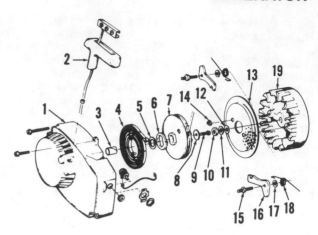

Fig. H9-7—Exploded view of recoil starter used on later models. Early models are similar.

1. Starter housing
2. Rope handle
3. Bushing
4. Rewind spring
5. Washer
6. Spring lock
7. Rope pulley
8. Washer
9. Screw
10. Nut
11. Lock washer
12. Washer
13. Screen
14. Lock nut
15. Stud
16. Pawl
17. Washer
18. Spring
19. Flywheel

sert new rope through handle and hole in blower housing. Knot both ends of the rope and harden the knots with heat or cement. Turn pulley clockwise eight turns and slide knot in rope into slot and keyhole in pulley. Let starter pulley rewind slowly.

Starter pawl spring outer ends are hooked behind air vanes on flywheel in line with starter pawls when pawls are resting against flywheel nut. Pull starter rope slowly when installing blower housing so that starter cup will engage pawls.

# HOMELITE

| | | | Engine | | Governed |
| Model | Output-kw | Voltage | Make | Model | Rpm |
| HG600 | 0.5 | 120Vac | Own | HG600 | 3600 |
| | 12Vdc | | | | |

## GENERATOR
### MAINTENANCE

The generator is equipped with a brushless, self-exciting rotor and no periodic maintenance is required. Rotor bearing (4—Fig. H10-2) is sealed and does not require lubrication, but should be inspected for smooth operation when generator is disassembled.

### TROUBLE-SHOOTING

If little or no generator output is evident, check circuit breaker (12—Fig. H10-1) and reset if necessary. Also, be sure that AC switch (10) is in the proper position. Engine must be in good enough condition to maintain the desired governed speed of 3600 rpm under load. Remove side panel and check wiring for loose connections. All wiring connectors must be clean and tight.

Test rectifier (10—Fig. H10-2) with an ohmmeter by clipping the positive ohmmeter lead to the center terminal of the rectifier. Touch negative ohmmeter lead to one of the outside terminals of rectifier and then the other outside terminal. Ohmmeter reading should show low resistance on one side and infinite resistance on the opposite side. Repeat the test by clipping negative ohmmeter lead to center terminal of rectifier, then touch positive ohmmeter lead to one of the outside terminals of rectifier and then to the other outside terminal. Readings should be opposite of those observed in the first test. Renew rectifier if other results are observed.

Test stator (7) for shorted windings with a battery operated test light or ohmmeter by touching one tester probe to stator frame and the other probe to each one of the stator winding leads at the connector (see wiring schematic). If continuity is indicated in any of the winding leads, a grounded winding is indicated and stator should be renewed. Melted insulation or a burnt appearance is a sure sign of grounded or shorted stator windings.

To test rotor windings, unsolder diode rectifiers (1) from end of rotor winding. Using a battery powered test light or an ohmmeter, check for continuity between rotor winding and rotor shaft or frame. If continuity is indicated, renew rotor. Melted insulation or windings having a burnt appearance is an indication of winding shorted together. Renew rotor if defective.

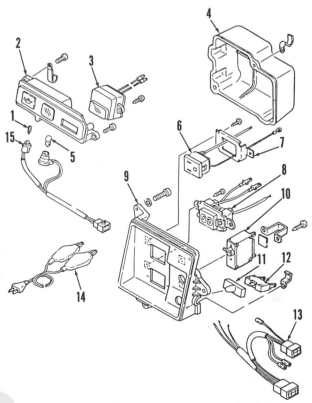

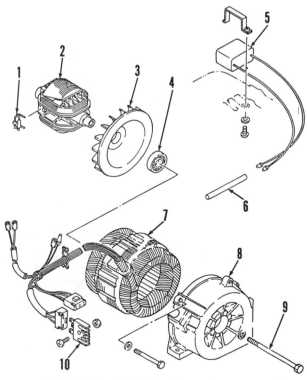

**Fig. H10-1—Exploded view of control panel assembly.**

| | | |
|---|---|---|
| 1. Bulb | 6. AC receptacle | 11. Cover |
| 2. Holder | 7. Holder | 12. Circuit breaker |
| 3. Voltmeter | 8. DC receptacle | 13. Wiring harness |
| 4. Rear cover | 9. Front cover | 14. DC charge cord |
| 5. Bulb | 10. AC switch | 15. Socket cord assy. |

**Fig. H10-2—Exploded view of generator components.**

| | |
|---|---|
| 1. Diode rectifier | 6. Tube |
| 2. Rotor | 7. Stator |
| 3. Fan | 8. Rear frame |
| 4. Bearing | 9. Through-bolt |
| 5. Condenser | 10. Rectifier |

*Illustrations courtesy of Homelite Div. of Textron*

With diodes unsoldered from rotor windings, test with battery powered test light or ohmmeter. Continuity should be indicated in only one direction through each diode. Renew diode(s) if other test results are observed.

Condenser (5) may be tested by disconnecting and then running generator to see if output returns to normal. If generator output returns to normal with condenser disconnected, then condenser is faulty and must be renewed.

## OVERHAUL

Refer to Fig. H10-1 for an exploded view of control panel assembly and to Fig. H10-2 for an exploded view of generator components. Before disconnecting wires, be sure to mark wires as necessary for correct reassembly. Rotor (2—Fig. H10-2) has a taper fit on engine crankshaft and may be removed with Homelite tools 49028-55 and 49028-57.

Inspect components and renew as necessary.

# ENGINE

| Model | Cyls. | Displacement |
|-------|-------|--------------|
| HG600 | 1 | 70.6 cc |
| | | (4.3 cu. in.) |

## MAINTENANCE

**SPARK PLUG.** Recommended spark plug is NGK BPMR6A, Champion RCJ8Y, Champion CJ8Y or equivalent. Specified electrode gap is 0.6-0.7 mm (0.024-0.027 inch).

**LUBRICATION.** Crankcase oil capacity is 350 mL (11.8 oz.). A good quality automotive type oil with API service classification of SE or SF is recommended. Use SAE 20 or SAE 30 weight oil depending on ambient temperature. Specified oil change interval is 150 hours of operation.

**AIR FILTER.** Air filter element is oil-wetted foam and should be cleaned and reoiled after every 50 hours of operation, or more often if generator is operated in extremely dusty conditions.

**FUEL FILTER.** Use the tool supplied with generator and unscrew fuel filter plug at bottom of generator. Remove cup and screen assembly, clean and reinstall. Specified service interval is 150 hours of operation.

**CARBURETOR.** Initial setting of idle mixture screw (4—Fig. H10-4) is 3/4 turn out from a lightly seated position. High speed mixture is controlled by main jet (10). Standard main jet (Homelite part 49025-95) is 56.3. Recommended main jet for altitudes above 6500 feet is 55 (part 49023-56). Recommended main jet for altitudes above 8800 feet is 52.5 (part 49025-94). Standard size of pilot jet (1) should be 37.5.

Float level should be 0.67 inch (17 mm) when measured as shown in Fig. H10-5. Measure from float to gasket surface

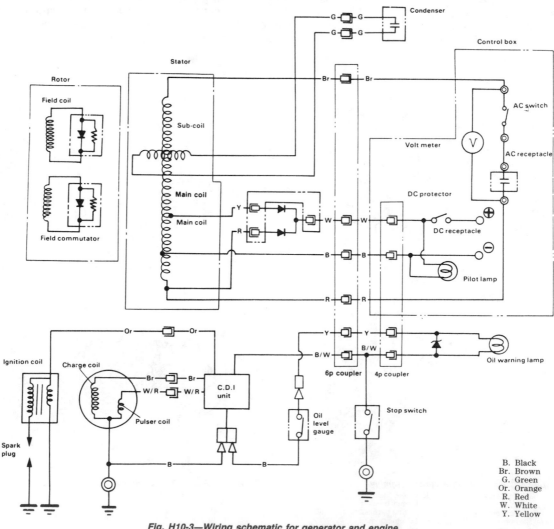

B. Black
Br. Brown
G. Green
Or. Orange
R. Red
W. White
Y. Yellow

*Fig. H10-3—Wiring schematic for generator and engine.*

Illustrations courtesy of Homelite Div. of Textron

of carburetor body. Fuel inlet valve seat size should be 0.059 inch (1.5 mm).

**FUEL PUMP.** A pressure-actuated diaphragm type fuel pump is located adjacent to the fuel tank. The pump is operated by vacuum pulsations in the intake manifold. Refer to Fig. H10-6 for a diagram of correct hose routing.

Refer to Fig. H10-7 for an exploded view of fuel pump. During overhaul, be sure to inspect diaphragm and check valves.

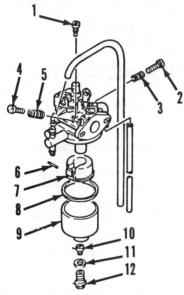

*Fig. H10-4—Exploded view of carburetor used on Model HG600 generator engine.*

1. Pilot jet
2. Idle speed screw
3. Spring
4. Idle mixture screw
5. Spring
6. Float pin
7. Float
8. Gasket
9. Float bowl
10. Main jet
11. Gasket
12. Jet holder

*Fig. H10-5—Measure float level with carburetor inverted as shown. Float level should be 0.67 inch (17 mm).*

Illustrations courtesy of Homelite Div. of Textron

**GOVERNOR.** To adjust governor, loosen nut (Fig. H10-8) and push governor arm towards the carburetor as far as it will go. Rotate governor shaft clockwise as far as possible, then retighten nut. Note that the governor arm spring must be connected to top hole in arm for 60 Hz operation.

**IGNITION SYSTEM.** Refer to Fig. H10-10 for an exploded view of the engine ignition system. To check ignition components, proceed as follows:

To test the primary coil winding, set an ohmmeter to the Rx1 scale and connect the common lead to ground and the positive lead to the positive lead wire. Specified resistance of the primary winding is 1.44-1.76 ohms.

To test the secondary coil winding, set ohmmeter to the Rx1K scale and connect the common lead to the spark plug wire and the positive lead to ground. Specified resistance of the secondary winding is 5280-7920 ohms.

To test the pulser coil winding, set ohmmeter to the Rx1 scale and connect the ohmmeter leads to the red/white wire and black wire. Specified resistance of the pulser coil winding is 33.3-40.7 ohms at 68° F (20° C).

To test the charge coil winding, set ohmmeter to Rx100 scale and connect the ohmmeter leads to the black wire and brown wire. Specified resistance of the charge coil winding is 3.15-3.85 ohms at 68° F (20° C). If test results vary significantly from the specifications, replace components as needed.

An oil level switch (7—Fig. H10-10) shuts down the ignition system when the oil level falls below a predetermined level. The oil level switch may be tested with an ohmmeter or battery operated

test light. There should be continuity through the switch when the switch is in the upright position. There should be no continuity through the switch when the switch is inverted.

## REPAIRS

**TIGHTENING TORQUE VALUES.** Special tightening torque values are listed below.

Cylinder head .........80-86 in.-lbs. (9.0-9.7 N·m)
Connecting rod ........48-57 in.-lbs. (5.4-6.4 N·m)

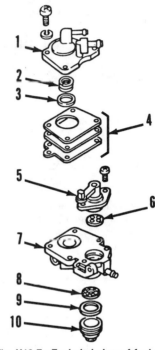

*Fig. H10-7—Exploded view of fuel pump.*

1. Cover
2. Spring
3. Washer
4. Diaphragm & gaskets
5. Control lever
6. Valve disc
7. Body
8. Filter
9. Gasket
10. Cup

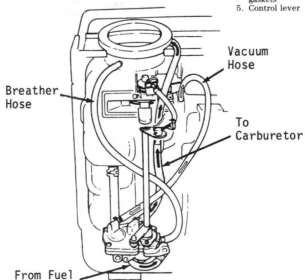

Breather Hose

Vacuum Hose

To Carburetor

From Fuel Tank

*Fig. H10-6—Diagram showing routing of fuel pump hoses.*

Crankcase cover bolt . . . 78-86 in.-lbs.
(8.8-9.7 N·m)
Charge and pulser coils. .52-70 in.-lbs.
(5.8-8.0 N·m)
Insulator mount nut . . . .52-70 in.-lbs.
(5.8-8.0 N·m)
Fan case bolt . . . . . . . . .52-70 in.-lbs.
(5.8-8.0 N·m)
Flywheel . . . . . . . . . . . . .23-26 ft.-lbs.
(31-35 N·m)

**COMPRESSION PRESSURE.** For optimum performance, cylinder compres-

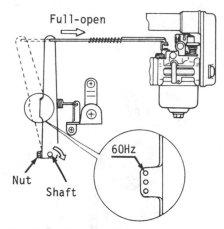

Fig. H10-8—Refer to drawing shown above and text for governor adjustment.

sion pressure should be 85 psi (586 kPa) with engine at normal operating temperature. Engine should be inspected and repaired when compression pressure is 71 psi (490 kPa) or below.

**VALVE SYSTEM.** Valve clearance with engine cold should be 0.002-0.006 inch (0.05-0.15 mm) for both intake and exhaust. Specified valve seat width for both intake and exhaust is 0.028 inch (0.07 mm) with a maximum of 0.047 inch (1.2 mm). Specified valve stem diameter is 0.217 inch (5.5 mm) for both intake and exhaust valves with a minimum allowable diameter of 0.213 inch (5.41 mm) for exhaust valve and 0.212 inch (5.38 mm) for intake valve. Specified valve spring free length is 0.906 inch (23 mm) for both intake and exhaust valves with an allowable minimum of 0.787 inch (20 mm).

**CYLINDER AND CRANKCASE.** Standard cylinder bore diameter is 1.890 inches (48 mm) with a wear limit of 1.896 inches (48.15 mm). Cylinder may be rebored for installation of oversize piston and rings if scored or excessively worn.

Specified crankcase cover journal diameter is 0.394 inch (10.0 mm) with a maximum allowable diameter of 0.398

inch (10.1 mm). When installing crankcase seal (1—Fig. H10-11), be sure that spring side of seal is facing the inside of the crankcase. Install bearings so numbered side will be toward connecting rod.

**PISTON, PIN AND RINGS.** Inspect piston skirt for excessive wear, scoring or other damage and renew as necessary. Piston and rings are available in oversizes of 0.010 and 0.020 inch (0.25 and 0.50 mm).

Specified ring groove side clearance is 0.002 inch (0.05 mm) for the top ring and 0.0015 inch (0.04 mm) for the second compression ring and the oil ring. Renew piston if ring groove side clearance (with a new ring) exceeds 0.006 inch (0.15 mm) for all rings.

Specified piston ring end gap is 0.008-0.014 inch (0.20-0.35 mm) and maximum ring end gap is 0.039 inch (1.0 mm). Install rings on piston as shown in Fig. H10-13. Stagger ring end gaps around diameter of piston. Lubricate piston, rings and cylinder wall with engine oil before installing piston in engine.

Piston pin should be a slip fit in piston with no excess play. Piston pin diameter should be 0.394 inch (10.0 mm); minimum allowable diameter is 0.392 inch (9.95 mm).

Assemble piston on connecting rod so that arrow on piston crown points toward valves.

**CRANKSHAFT AND CONNECTING ROD.** Specified crankpin diameter is

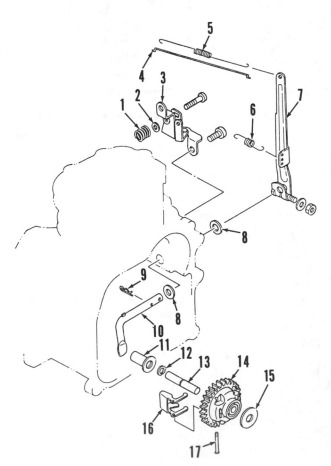

Fig. H10-9—Exploded view of governor and linkage assembly.

1. Spring
2. Washer
3. Adjuster plate
4. Rod
5. Tension spring
6. Spring
7. Arm
8. Washer
9. Clip
10. Fork
11. Collar
12. Clip
13. Shaft
14. Flyweight
15. Thrust washer
16. Flyweight
17. Pin

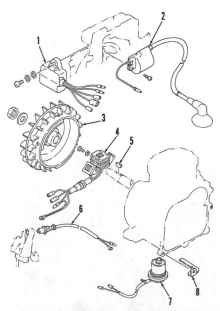

Fig. H10-10—Exploded view of engine electrical components.

1. Control module
2. Ignition coil
3. Flywheel
4. Magnetic pickup
5. Woodruff key
6. Oil stop switch
7. Oil level sensor
8. Clamp

0.709 inch (18.0 mm) with a wear limit of 0.705 inch (17.9 mm). Specified rod bearing clearance is 0.0006-0.0011 inch 0.015-0.027 mm). Side clearance between connecting rod and crankpin shoulder should be 0.004-0.016 inch (0.1-0.4 mm) with an allowable maximum clearance of 0.039 inch (1.0 mm). Maximum allowable crankshaft runout is 0.0008 inch (0.02 mm).

Connecting rod big end diameter is 0.709 inch (18.0 mm) with a maximum allowable diameter of 0.711 inch (18.05 mm). Connecting rod small end diameter is 0.394 inch (10.0 mm) with a maximum allowable diameter of 0.396 inch (10.05 mm).

Inspect crankshaft and connecting rod bearing surfaces for scoring and excessive wear.

Install connecting rod in crankcase with the "7x9" mark facing the crankshaft cover. When installing rod cap onto connecting rod, make sure that marks are aligned. Install oil slinger (8—Fig. H10-12) so dipper is towards crankshaft gear and away from oil level switch (7—Fig. H10-10). When installing camshaft and crankshaft into crankcase, be sure that the hole in the camshaft gear and the "O" mark on the crankshaft gear are aligned as shown in Fig. H10-14.

**CAMSHAFT.** Specified camshaft lobe height for both intake and exhaust is 0.772 inch (19.6 mm) with a minimum allowable height of 0.762 inch (19.35 mm). Specified camshaft bearing journal diameter is 0.394 inch (10.0 mm) with a minimum allowable diameter of 0.390 inch (9.9 mm). Replace camshaft and lifters if lobes or journals are scored or worn excessively.

When installing camshaft, be sure hole in camshaft gear aligns with "O" mark on crankshaft gear as shown in Fig. H10-14.

**REWIND STARTER.** Refer to Fig. H10-16 for an exploded view of the rewind starter. To disassemble starter, pull out rope until notch in outer edge of rope pulley (5) is adjacent to rope. Hold pulley so that it cannot turn, then pull a loop of rope back inside starter housing and engage rope in notch. Release pulley and allow pulley to unwind until spring tension is released. Remove rope handle, screw (10), plate (9), spring (8) and pulley assembly (5) while being careful not to dislodge rewind spring (2).

Use care if rewind spring must be removed; do not allow spring to uncoil uncontrolled.

When assembling starter, install rewind spring so spring coils are wrapped in a counterclockwise direction from outer end. Wind rope on pulley in a counterclockwise direction as viewed from pawl side of pulley. Preload rewind spring by turning pulley two turns against spring tension with rope engaged in pulley notch.

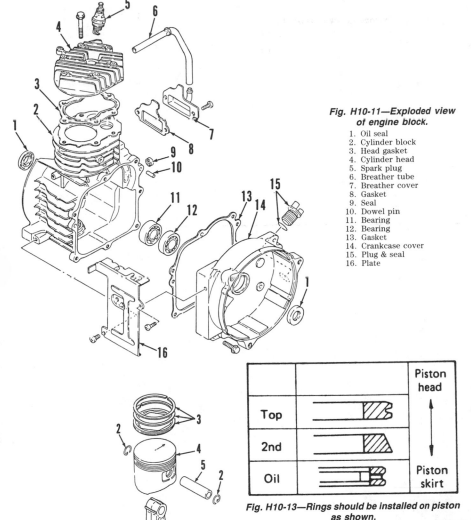

Fig. H10-11—Exploded view of engine block.

1. Oil seal
2. Cylinder block
3. Head gasket
4. Cylinder head
5. Spark plug
6. Breather tube
7. Breather cover
8. Gasket
9. Seal
10. Dowel pin
11. Bearing
12. Bearing
13. Gasket
14. Crankcase cover
15. Plug & seal
16. Plate

| | Piston head |
|---|---|
| Top | |
| 2nd | |
| Oil | Piston skirt |

Fig. H10-13—Rings should be installed on piston as shown.

Fig. H10-12—Exploded view of crankshaft and piston assembly.

1. Crankshaft
2. Retainers
3. Rings
4. Piston
5. Piston pin
6. Connecting rod
7. Rod cap
8. Oil slinger
9. Lock tab
10. Bolt

Fig. H10-14—View showing correct alignment of crankshaft timing mark "O" and timing hole in camshaft gear.

Timing hole

Timing mark

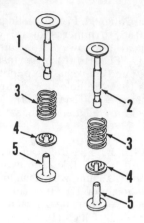

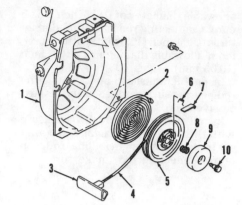

*Fig. H10-16—Exploded view of rewind starter.*

1. Starter housing
2. Rewind spring
3. Handle
4. Rope
5. Pulley
6. Spring
7. Pawl
8. Spring
9. Drive plate
10. Screw

*Fig. H10-15—Exploded view of valve components.*

1. Intake valve
2. Exhaust valve
3. Spring
4. Retainer
5. Lifters
6. Camshaft assy.

# HOMELITE

| Model | Output-kw | Voltage | Engine | | Governed Rpm |
| | | | Make | Model | |
| --- | --- | --- | --- | --- | --- |
| HG1400........ | 1.2 | 120Vac 12Vdc | Own | HG1400 | 3600 |

## GENERATOR

### MAINTENANCE

The generator is equipped with a brushless, self-exciting rotor and no maintenance is required. Rotor bearing (6 – Fig. H11-2) is sealed and does not require lubrication, but should be inspected for smooth operation when generator is disassembled.

### TROUBLESHOOTING

If little or no generator output is evident, check circuit breakers (4 and 7 – Fig. H11-1) and reset if necessary. Engine must be in good enough condition to maintain the desired governed speed under load. Remove control panel cover (11) and check wiring for loose connections. All wiring connectors must be clean and tight.

If there is little or no DC output, disconnect six-wire coupler (Fig. H11-1A) and use an ohmmeter or battery operated test light to test for continuity between terminal (C) and DC circuit breaker (7 – Fig. H11-1), between DC circuit breaker and DC receptacle (8) and between terminal (B) and DC receptacle. If there is not continuity between any of these components, repair wiring or connectors as needed. If continuity exists in all tests, disconnect three-wire coupler (Fig. H11-1B) and check resistance of DC stator coil by connecting ohmmeter leads to terminal (D) of six-wire coupler and terminal (J) of three-wire coupler. Resistance should be approximately 0.200-0.300 ohms. Renew stator if measured resistance values vary widely from specifications. If DC coil resistance value is correct, check voltage at AC receptacle (9 – Fig. H11-1). If voltage at AC receptacle is within proper range, rectifier (3 – Fig. H11-2) may be defective.

If AC output voltage is below specified values or nonexistent, disconnect six-wire coupler and use an ohmmeter or battery operated test light to test for continuity between terminal (F – Fig. H11-1A) and AC receptacle, between terminal (A) and AC circuit breaker (4 – Fig. H11-1) and between AC receptacle and AC circuit breaker. If there is no continuity between any of these components, repair wiring or connectors as needed. If continuity exists in all tests, measure resistance of main stator coil by connecting ohmmeter leads to terminals (C – Fig. H11-1A) and (D) of six-wire coupler. Resistance should be approximately 0.60-0.80 ohms. Measure resistance of stator sub-coil by connecting ohmmeter to terminals (B) and (E) of six-wire coupler. Resistance should be approximately 2-3 ohms. Renew stator if measured resistance values vary widely from specifications. If measured resistance values of stator coils are in the correct range, disconnect leads of condenser (5 – Fig. H11-1) inside control box. Touch condenser terminals together to be sure condenser is discharged. With ohmmeter set on Rx100 scale, connect ohmmeter leads to condenser terminals. Ohmmeter needle should swing rapidly across meter then return. Disconnect ohmmeter and discharge condenser by touching terminals together. Renew condenser if test is failed.

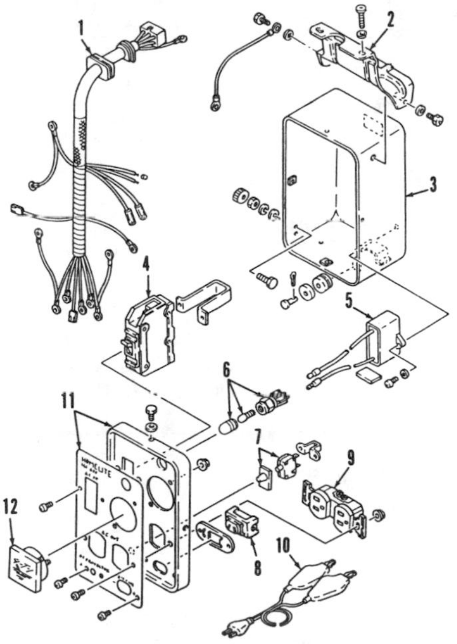

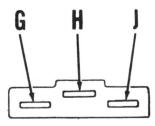

Fig. H11-1—*Exploded view of control panel assembly.*

| | |
| --- | --- |
| 1. Harness | 7. Circuit breaker assy. |
| 2. Brace | 8. DC receptacle |
| 3. Box | 9. AC receptacle |
| 4. Circuit breaker | 10. DC charge cord assy. |
| 5. Condenser | 11. Cover assy. |
| 6. Pilot light assy. | 12. Voltmeter |

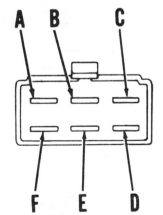

Fig. H11-1A—*Drawing of terminals in six-wire coupler.*

Fig. H11-1B—*Drawing of terminals in three-wire coupler.*

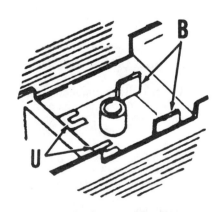

Fig. H11-1C—*View showing blade (B) and U-shaped (U) terminals of rotor. Terminals are located on both sides of rotor.*

Illustrations courtesy of Homelite Div. of Textron

If condenser tests satisfactory, remove rotor as outlined in OVERHAUL section and proceed as follows: Refer to Fig. H11-1C. Unsolder wires from blade (B) and U-shaped terminals (U) on both sides of rotor. Measure coil resistance by connecting ohmmeter leads to U-shaped terminals on each side of rotor. Field coil resistance should be approximately 3-5 ohms. Renew rotor if measured resistance values of field coils vary widely from specifications.

To check AC rectifiers (7 – Fig. H11-2), connect an ohmmeter or battery operated test light leads to blade terminals (B – Fig. H11-1C), note whether or not there is continuity, reverse leads and again note whether or not there is continuity. There should be continuity in one direction only through rectifier. Follow the same procedure using blade terminals on opposite side of rotor. Replace rectifier(s), if necessary.

## OVERHAUL

To disassemble generator, remove rear cover (1 – Fig. H11-2), then disconnect and remove control box. Unbolt and remove mounting pads under rear frame (4) legs. Unscrew four through-bolts securing rear frame and stator (5) to engine crankshaft cover. Remove rear frame and stator as an assembly. Remove rotor through-bolt (2). Rotor (8) has a taper fit on end of engine crankshaft and may be removed with Homelite tools 49028-55 and 49028-57.

Inspect components and reassemble generator by reversing disassembly procedure. Refer to Fig. H11-3 for wiring schematic. Torque rotor through-bolt to 18.6 N·m (165 in.-lbs.) and torque stator through-bolts to 7.8 N·m (69 in.-lbs.).

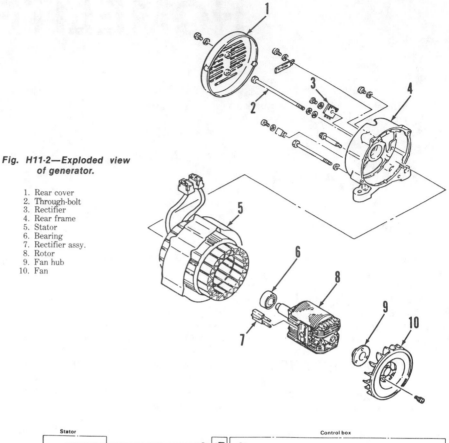

**Fig. H11-2—Exploded view of generator.**

1. Rear cover
2. Through-bolt
3. Rectifier
4. Rear frame
5. Stator
6. Bearing
7. Rectifier assy.
8. Rotor
9. Fan hub
10. Fan

# ENGINE

| Model | Cyls. | Displ. |
|-------|-------|--------|
| HG1400 | 1 | 145 cc (8.9 cu. in.) |

## MAINTENANCE

**SPARK PLUG.** Recommended spark plugs are NGK BP6HS, Champion L87Y and Champion L87YC. Specified electrode gap is 0.6-0.7 mm (0.024-0.028 in.).

**CARBURETOR.** An exploded view of the carburetor is shown in Fig. H11-5. Initial setting of idle mixture screw (4) is 1½ turns out from seated position. Specified float level is 18.5 mm (0.728 in.). Bend tangs on float (10) for adjustment.

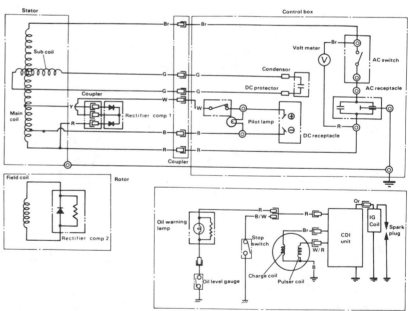

**Fig. H11-3—Wiring schematic.**

**IGNITION AND TIMING.** Refer to Fig. H11-4 for an exploded view of the electronic ignition system. Ignition timing is fixed at 20° BTDC at 3600 rpm. The pulser coil (10) and charge coil (9) are accessible after removing flywheel.

An ohmmeter may be used to check pulse, charge and ignition coils. Specified resistance reading for pulse coil with ohmmeter attached to black and red/white leads is 18-22 ohms. Specified resistance for charge coil is 378-462 ohms. Specified resistance for ignition coil primary circuit with ohmmeter attached to primary lead and coil leg is 1.6-2.0 ohms. Specified resistance for ignition coil secondary circuit with ohmmeter leads attached to secondary lead and coil leg is 8K-12K ohms.

The ignition module is equipped with an over-speed circuit which prevents excessive engine speed by disrupting ignition circuit when engine speed reaches 4050-4650 rpm.

Specified flywheel nut tightening torque is 54 N·m (40 ft.-lbs.).

Note that ignition system problems may be due to malfunction in the oil sensor (5).

**GOVERNOR.** The flyweight type governor is attached to and driven by the camshaft gear. The governor forces governor shaft (7 – Fig. H1-9) to rotate according to engine speed. A pinch bolt secures governor arm (2) to shaft. Movement of governor arm is transmitted by link (3) to carburetor throttle arm.

To adjust governed speed, loosen pinch bolt and pull upper end of governor arm away from carburetor so that throttle valve is fully open. Turn governor shaft counterclockwise until stop is contacted, then retighten pinch bolt. With engine running under load, turn adjusting screw so maximum governed speed is 4000 rpm. If available, a frequency meter may be connected to the AC receptacle and maximum governed speed adjusted so current frequency is 60 hertz.

**LUBRICATION.** Crankcase capacity is 500 mL (0.5 qts.). Use SAE 30 oil if ambient temperature is above 15°C (59°F), SAE 20 if temperature is between 0°C (32°F) and 15°C (59°F), and SAE 10 if temperature is below 0°C (32°F). Change oil after first 20 hours of operation. Thereafter change oil after every 150 hours of operation or three months, whichever occurs first.

All models are equipped with a low-oil safety system which disables the ignition system and lights a warning bulb when engine oil level is critically low. A float mounted in the engine crankcase monitors oil level. When crankcase is approximately half full, a magnet in the float closes the contacts in reed switch thereby grounding the ignition and lighting the oil warning light.

Refer to OVERHAUL section to service oil warning system.

**AIR FILTER.** Air filter element is oil-wetted foam and should be cleaned and reoiled after every 50 hours of operation at least and more often if generator is operated in dusty conditions.

**FUEL FILTER.** Turn fuel valve to "OFF" position, remove cup & screen, clean and reinstall. Specified service interval is 150 hours of operation or three months, whichever occurs first.

# OVERHAUL

**TIGHTENING TORQUES.** Recommended tightening torques are as follows:

Connecting rod . . . . . . . . . . . . .9.8 N·m
(7 ft.-lbs.)
Crankcase cover . . . . . . . . . .14.7 N·m
(11 ft.-lbs.)
Cylinder head . . . . . . . . . . . . .19.6 N·m
(14.5 ft.-lbs.)
Flywheel . . . . . . . . . . . . . . . . . .50 N·m
(40 ft.-lbs.)

**VALVE SYSTEM.** Refer to Fig. H11-8 for exploded view of valve components. Specified valve tappet gap is 0.05-0.15 mm (0.002-0.006 in.) for both intake and exhaust. Different length valves are available if tappet gap is incorrect. Valves are marked "0, 1 or 2" with "0" valve shortest and "2" valve longest.

Valve stem diameter is 6.995-6.970 mm (0.2738-0.2744 in.). Valve seat width is 0.7-0.9 mm (0.027-0.035 in.). Valve spring free length is 33.5 mm (1.319 in.).

**PISTON, PIN AND RINGS.** Refer to Fig. H11-7 for an exploded view of piston and rings. Piston (2) and con-

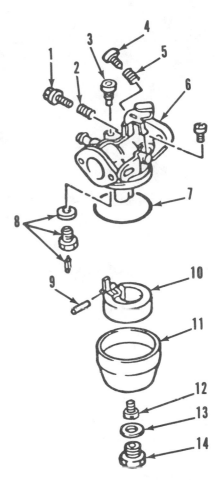

**Fig. H11-5—Exploded view of carburetor.**

1. Idle screw
2. Spring
3. Pilot jet
4. Mixture valve
5. Spring
6. Body
7. Gasket
8. Fuel inlet valve assy.
9. Float pin
10. Float
11. Float bowl
12. Main jet
13. Gasket
14. Retainer

necting rod (5) assembly are removed as a unit after cylinder head and crankcase cover are removed. Pistons are available in oversizes of 0.25 mm (0.010 in.) and 0.50 mm (0.020 in.). Specified piston clearance is 0.065-0.075 mm (0.0025-0.0030 in.). Desired ring end gap is 0.2-0.4 mm (0.008-0.016 in.). Install piston with arrow on crown pointing towards valve side of engine. Piston pin (4) is fully floating in piston and rod. Piston pin retaining clips (3) should be renewed if removed.

**CONNECTING ROD.** Refer to Fig. H11-7 for view of connecting rod. The connecting rod and piston are removed as a unit after cylinder head and crankcase cover are removed. The aluminum rod rides directly on the crankpin. Specified connecting rod-to-crankpin clearance is 0.016-0.034 mm (0.0006-0.0013 in.). Specified connecting rod side clearance is 0.2-0.8 mm (0.008-

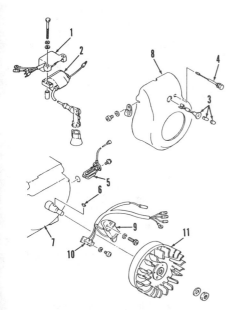

**Fig. H11-4—Exploded view of engine electrical components.**

1. Control module
2. Coil
3. Pilot light
4. Stop switch
5. Oil sensor
6. Woodruff key
7. Crankcase
8. Fan shroud
9. Charge coil
10. Pulser coil
11. Flywheel

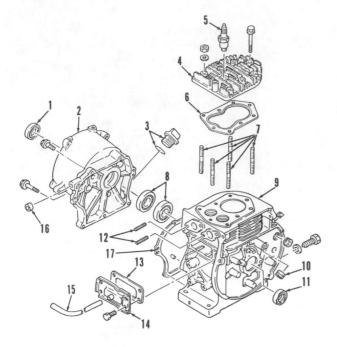

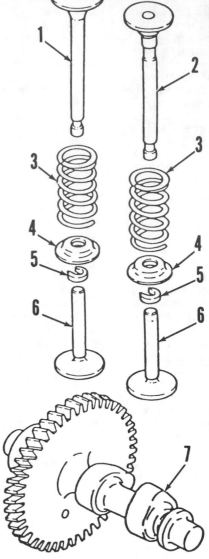

Fig. H11-6—Exploded view of engine cylinder block.

1. Seal
2. Cover
3. Plug
4. Cylinder head
5. Spark plug
6. Head gasket
7. Studs
8. Bearings
9. Block
10. Grommet
11. Seal
12. Studs
13. Gasket
14. Breather cover
15. Breather tube
16. Seal
17. Gasket

0.032 in.). Match marks on rod and cap must be aligned when installing rod. Install oil splasher (6) between rod cap and lock plate (7).

**CAMSHAFT AND GOVERNOR.** The camshaft rides directly in crankcase and crankcase cover. Inspect camshaft lobes, gear and bearing surfaces for scoring or excessive wear. Renew if needed. Specified camshaft lobe height for both intake and exhaust is 27.55 mm (1.085 in.) with a wear limit of 27.3 mm (1.075 in.). Punch marks on camshaft gear and crankshaft gear must be aligned during reassembly for proper valve timing.

Refer to Fig. H11-9 for exploded view of governor components. During reassembly, collar (5) must engage grooves on weights (6). Install governor fork (4) on shaft (7) so pads on fork ends are towards camshaft.

**CRANKSHAFT AND CRANK-CASE.** Refer to Fig. H11-6 for exploded view of crankcase. The crankshaft is supported by ball bearings (8). Specified crankpin diameter is 24.952-24.970 mm (0.9824-0.9831 in.). Inspect crankshaft for excessive wear, straightness and damage. Renew if needed.

**REWIND STARTER.** Refer to Fig. H11-10 for an exploded view of the rewind starter. With starter removed from engine, untie rope handle from rope and allow rope to wind into starter housing. Remove bolt (1) and disassem-

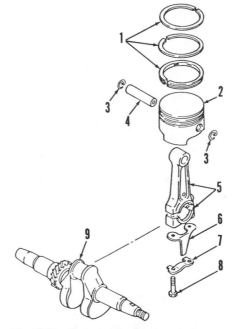

Fig. H11-7—Exploded view of engine piston and crankshaft assembly.

1. Rings
2. Piston
3. Clips
4. Pin
5. Rod
6. Splasher
7. Lock plate
8. Bolt
9. Crankshaft

ble starter. Use care when removing drum (6) so that rewind spring (7) will not be dislodged from case (8). If rewind spring must be removed, use care as uncontrolled uncoiling of spring may cause injury.

Fig. H11-8—Exploded view of engine valve system components.

1. Intake valve
2. Exhaust valve
3. Spring
4. Retainers
5. Keys
6. Lifter
7. Camshaft

When assembling starter, install rewind spring so coil direction is counterclockwise from outer end. Wind rope on drum in a counterclockwise direction as viewed with drum installed in case. Preload rewind spring by turning drum two turns against spring tension before passing rope through outlet in case.

**OIL WARNING SYSTEM.** All models are equipped with an oil warning system which disables ignition system when oil level is low. See LUBRICATION section.

To inspect oil level sensor, drain engine oil and remove sensor from side of crankcase. Handle float assembly

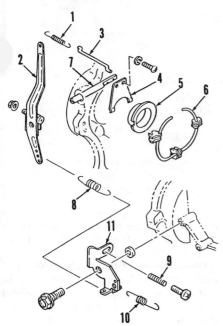

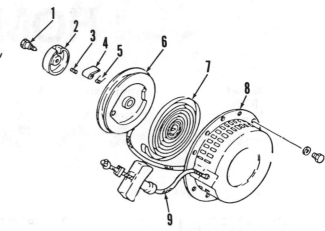

**Fig. H11-10—Exploded view of rewind starter.**

1. Bolt
2. Plate
3. Spring
4. Pawl
5. Clip
6. Drum
7. Spring
8. Case
9. Rope

**Fig. H11-9—Exploded view of engine governor components.**

1. Tension spring
2. Arm
3. Rod
4. Fork
5. Collar
6. Weights
7. Shaft
8. Spring
9. Spring
10. Spring
11. Thrust plate

carefully as reed switch may be damaged if unit is dropped. To check operation of float assembly, connect an ohmmeter to mounting flange and to wire lead. Do not use a battery operated test light as battery voltage may damage reed switch. Ohmmeter should show continuity with float assembly in normal position with float hanging down. Ohmmeter should read infinity with float assembly inverted. Oil warning light bulb should be checked and renewed if burned out.

# HOMELITE

| Model | Output-Watts | Voltage | Engine Make | Engine Model | Governed Rpm |
|---|---|---|---|---|---|
| HG1500 | 1400 | 120 | B&S | 131232 | 3600 |
| HG2100 | 1900 | 120 | B&S | | 3600 |
| HG2500 | 2300 | 120 | B&S | | 3600 |
| HG3500 | 3000 | 120/240 | B&S | | 3600 |

# GENERATOR

## MAINTENANCE

Brush length and condition should be checked after every 1000 hours of operation. Brushes can be inspected after removing brush head cover (1—Fig. H12-1), rotor bolt (2) and fan (3).

**NOTE: Do not attempt to hold rotor with fan to remove rotor bolt.**

Loosen screws retaining brush holder (6) and slide out brush holder. Brushes must be renewed if brush length is shorter than 3/8 inch (9.5 mm) as shown in Fig. H12-2. Be sure to reinstall used brushes so that brush curvature matches rotor slip ring. During reassem-

bly, fan must be mounted squarely on rotor shaft. Tighten rotor bolt to 120-140 in.-lbs. (13.5-15.8 N·m).

Disassemble brush holder to install new brushes. Insert new brushes in brush holder so curvature of brush will match curvature of rotor slip ring. Seating new brushes is not required as they are manufactured with correct curvature. Red and black wires should be con-

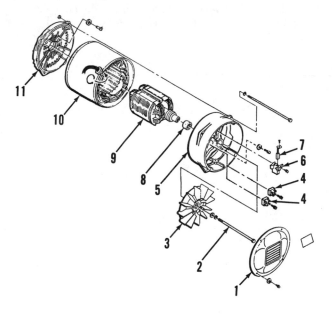

**Fig. H12-1—Exploded view of Model HG2500 generator. Other models are similar.**
1. Cover
2. Bolt
3. Fan
4. Rectifier
5. Brush head
6. Brush holder
7. Brush
8. Bearing
9. Rotor
10. Stator & housing
11. Endbell

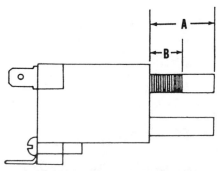

**Fig. H12-2—Worn brush length (B) should not be less than 3/8 inch (9.5 mm) when measured as shown. Brush length (A) of a new brush is 3/4 inch (19 mm).**

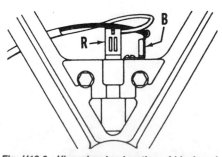

**Fig. H12-3—View showing location of black and red brush wires.**

                                                                        Illustrations courtesy of Homelite Div. of Textron

nected to brushes as shown in Fig. H12-3, and wires must be routed behind rotor support to prevent interference with —n.

## OVERHAUL

Refer to Fig. H12-1 for an exploded view of generator and Figs. H12-4 and H12-5 for wiring schematic. Be sure to mark wiring so that it may be correctly connected when reassembling. Wires should be handled carefully to prevent damage to wire, insulation or connections. Generator rotor (9—Fig. H12-1) has a taper fit with engine crankshaft and should be removed with a suitable tool.

# ENGINE

Engine make and model are listed at beginning of section. Refer to Briggs and Stratton section for engine service.

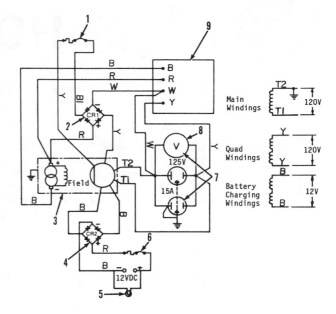

**Fig. H12-4—Wiring schematic for Models HG1500, HG2100 and HG2500.**

1. Circuit breaker
2. Bridge rectifier
3. Rotor
4. Bridge rectifier
5. Charge indicator lamp
6. Circuit breaker
7. Receptacles
8. Voltmeter
9. Voltage regulator

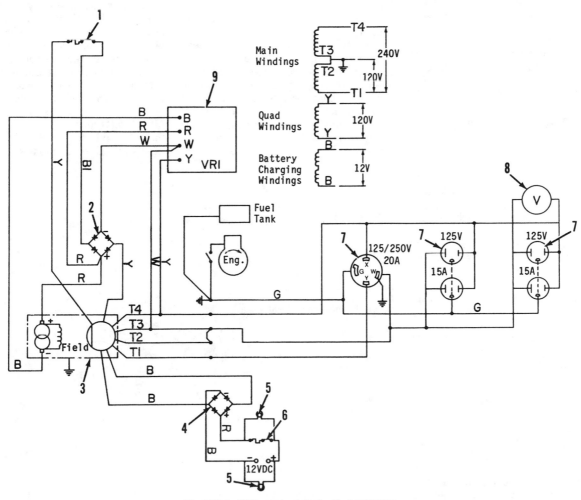

**Fig. H12-5—Wiring schematic for Model HG3500.**

| | | |
|---|---|---|
| 1. Circuit breaker | 4. Bridge rectifier | 7. Receptacles |
| 2. Bridge rectifier | 5. Indicator lamps | 8. Voltmeter |
| 3. Rotor | 6. Circuit breaker | 9. Voltage regulator |

# HONDA

AMERICAN HONDA MOTOR CO., INC.
4475 Rivergreen Parkway
Duluth, GA 30136

| Model | Output-kw | Voltage | Engine | | Governed Rpm |
| | | | Make | Model | |
|-------|-----------|---------|-------|-------|--------------|
| E300A | 0.3 | 120VAC 12VDC | Honda | .... | 3600 |
| E300C | 0.3 | 117VAC 12VDC | Honda | .... | 3600 |
| E300E | 0.3 | 220VAC 12VDC | Honda | .... | 3600 |
| E300U | 0.3 | 240VAC 12VDC | Honda | .... | 3600 |
| ER400 | 0.3 | 120VAC 13VDC | Honda | .... | 3600 |

# GENERATOR

## MAINTENANCE

The generator is equipped with a brushless, permanent magnet type rotor and no maintenance is required.

The powerplant is enclosed by louvered side panels. The side panels must be clean and unobstructed to provide adequate air movement for generator and engine cooling.

## TROUBLESHOOTING

If little or no generator output is evident, inspect fuses, switches and all wiring and connections. Engine must be in good condition and able to maintain governed speed of 3600 rpm under load. The following tests will help locate malfunction.

With Model ER400 operating at 4000 rpm under no load, AC voltage should be 115-130 volts; also, AC voltage should be at least 100 volts with a load of two amps. Replace regulator (13–Fig. HN1-1) with a new or good used regulator and recheck voltage.

To check reactor coil (10), disconnect coil leads and using an ohmmeter, measure coil resistance. Coil resistance should be 5.0-6.1 ohms on Model E300A, 4.1-5.1 ohms on Model E300C, 16.2-19.8 ohms on Model E300E, 20.3-24.8 ohms on Model E300U and 3.0-3.6 ohms on Model ER400.

To test stator coils, disconnect four-wire connector on Model ER400 and six-wire connector on other models. Connect an ohmmeter to male connector terminals as follows: On Model ER400 attach ohmmeter leads to blue and brown wire terminals and on all other models attach ohmmeter leads to blue and red/white wire terminals. Resistance should be 4.6-5.2 ohms on Model E300A, 4.0-4.6 ohms on Model E300C, 16.1-18.5 ohms on Model E300E, 21.0-24.2 ohms on Model E300U and 4.3 ohms on Model ER400.

To test rectifier diodes (7), disconnect rectifier leads. Using an ohmmeter, connect ohmmeter leads to a pink diode lead and to yellow/green rectifier lead; note reading, reverse ohmmeter leads and again note reading. Ohmmeter should read very high for one connection and very low for the other connection. If both readings are high or low, diode is faulty and must be renewed. Use same procedure to test remaining diode.

## OVERHAUL

To disassemble generator, remove side panels, fuel tank, bottom cover and frames. Remove the muffler, ignition breaker point assembly and reactor (10–Fig. HN1-1). Remove four screws retaining stator and withdraw stator. Unscrew rotor nut (3) and using a suitable puller remove rotor.

When reassembling generator, make certain wires are connected correctly. Refer to wiring schematic in Fig. HN1-2 or HN1-3. Tighten stator screws on Model ER400 to 2.9-5.9 N·m (26-52 in.-lbs.) and to 7.8-8.8 N·m (69-78 in.-lbs.) on all other models.

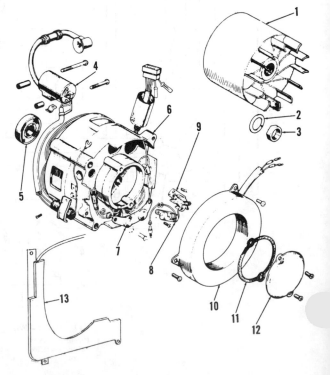

Fig. HN1-1—Exploded view of generator and ignition system. Regulator (13) is used only on ER400 models.

1. Rotor
2. Washer
3. Nut
4. Ignition coil
5. Bearing
6. Stator
7. Rectifier
8. Condenser
9. Ignition breaker points
10. Reactor
11. Gasket
12. Cover
13. Regulator

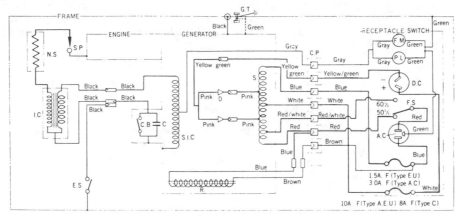

*Fig. HN1-2—Wiring diagram for all models except ER400.*

| | | | |
|---|---|---|---|
| C. Condenser | E.S. Ignition switch | G.T. Ground terminal | R. Reactor |
| C.B. Ignition breaker | F.M. Frequency meter | I.C. Ignition coil | S. Stator |
|    points | F.S. Frequency | N.S. Noise suppressor | S.I.C. Ignition stator |
| C.P. Coupler |    switch | P.L. Pilot lamp | S.P. Spark plug |
| D. Rectifier diode | | | |

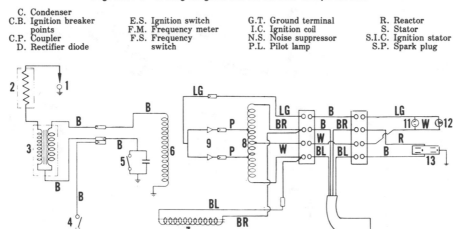

*Fig. HN1-3—Wiring diagram for Model ER400.*

| | | | |
|---|---|---|---|
| 1. Spark plug | 5. Ignition breaker | 7. Reactor | 10. Voltage regulator |
| 2. Noise suppressor |    points | 8. Generator stator | 11. DC receptacle |
| 3. Ignition coil | 6. Ignition stator coil | 9. Diodes | 12. Pilot lamp |
| 4. Ignition switch | | | 13. AC receptacle |

# ENGINE

| Bore | Stroke | Cyls. | Displ. |
|---|---|---|---|
| 42 mm | 40 mm | 1 | 55.4 cc |
| (1.65 in.) | (1.57 in.) | | (3.38 cu. in.) |

## MAINTENANCE

**SPARK PLUG:** Recommended plug for all models is NGK type CM-6. Plug gap should be set to 0.4mm (0.016 inch).

**CARBURETOR.** The float type carburetor shown in Fig. HN1-10 is used. The idle mixture needle (1–Fig. HN1-10) should be set to provide smooth running at idle speed. Initial setting is ⅞-turn open. Final setting may be 1⅜ turns open (±¼-turn). Idle speed is adjusted by turning stop screw (2). To remove carburetor, the fan cover and governor must be removed. Refer to the GOVERNOR section for removal. Refer to Fig. HN1-10 for exploded view

of carburetor. When reinstalling, gasket should be installed between cylinder and insulator. "O" ring (4) should be installed between insulator and carburetor.

Float level is 14.5mm (0.570 inch) measured with carburetor inverted from bottom of float to surface of body flange. Some engines may be equipped with manual choke, and Model ER400 has an automatic choke of type shown in Fig. HN1-12. "U"-shaped bi-metallic strip(s) acts as a thermostat to control choke valve opening. At 0°C (32°F) gap (G) at open end of "U" should be 10mm (0.394 in.). Adjustment is not recommended, however, if automatic choke does not function properly, opening or closing of

loop (L) in choke rod may restore performance so that choke will operate.

**THE FUEL VENT.** The fuel tank is located above the engine and generator and is vented through tubes in the carrying handle. When reinstalling carrying handle, note that there are two vent holes on one end and one vent hole at the other end. Be sure vent holes in carrying handle correspond to vent holes on fuel tank.

**GOVERNOR.** A flyweight type governor is used on all models. Refer to Figs. HN1-13 and HN1-14 for views of governor mechanism. Flyweight carrier (23 – Fig. HN1-13) and slider (22) ride on the engine crankshaft. As engine speed increases, weights (F) push the slider (22) out against governor shaft (10). As governor shaft rotates, governor lever (3) moves governor arm (7) through spring (4). Governor arm (7) contacts carburetor throttle shaft (5—Fig. HN1-14) to control carburetor. The force of flyweights is balanced by governor spring (12). Engine speed is varied by turning throttle control shaft (18) which has a knob on all models except ER400 which has a slot cut in end of shaft (18).

Governed speed may be adjusted using either a tachometer or frequency meter. With generator under load, turn throttle control shaft (18) to obtain a frequency of 60 hertz (cycles) or an engine speed of 3600 rpm. If turning throttle control shaft (18) does not result in engine running at governed speed, then turn governor spring adjusting nut (15). Wit

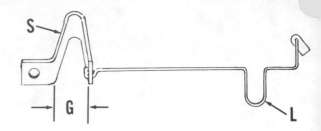

Fig. HN1-12—Diagram of choke mechanism. Refer to text for adjustment.

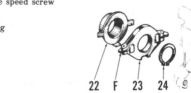

Fig. HN1-13—Exploded view of governor. Governor arm (7) contacts throttle shaft (5—Fig. HN1-14).

1. Nut
2. Washer
3. Lever
4. Spring
5. Screw
7. Arm
8. "E" ring
9. Washer
10. Shaft
11. Dowel pin
12. Spring
13. Spring stud
14. Bracket
15. Nut
16. Spring
17. Nut
18. Throttle control shaft
19. Plate
20. Spring
21. High idle speed screw
22. Slider
23. Carrier
24. Snap ring

Fig. HN1-14—View of governor mechanism. Refer to Fig. HN1-13 for parts identification except for (5) throttle shaft.

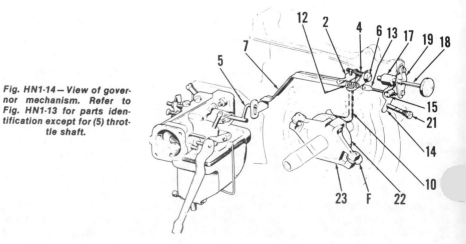

Fig. HN1-10—Exploded view of carburetor.

1. Idle mixture screw
2. Idle speed screw
3. Pilot jet
4. "O" ring
5. Throttle assy.
6. Choke assy.
7. Gasket
8. Fuel inlet valve
9. Main jet
10. Float
11. Float pin
12. Gasket
13. Float bowl
14. Drain screw

engine running under no load, turn high idle speed screw (21) so engine speed is 4200 rpm.

**IGNITION.** A magneto type ignition system is used. The low tension coil is located in the generator stator and is not serviced separately. The high tension coil is mounted on the crankcase below the cylinder. The breaker points and condenser are mounted at the rear of generator.

Breaker point gap should be 0.30-0.40 mm (0.012-0.016 in.) at maximum opening. Ignition timing should occur (points just open) when mark on generator rotor cooling fin aligns with mark on stator. If ignition timing is incorrect, adjust breaker point gap. Timing marks should align when the piston is 17-21 degrees BTDC.

**LUBRICATION.** Crankcase capacity is 0.29 liter (0.61 pint). Recommended motor oil API code may be SE or SF. Use SAE 20 or 20W oil for temperatures between 0°C (32°F) and 15°C (59°F). For temperatures below 0°C (32°F) use SAE 10W and for temperatures over 15°C (59°F), use SAE 30 oil. SAE 10W30 may be used in all seasons. Oil should be maintained above bottom of dipstick on filler plug.

**NOTE: Filler plug must not be screwed in when checking oil level.**

Engine oil should be changed every 100 hours of operation.

## REPAIRS

**CYLINDER HEAD.** When removing the cylinder head, loosen screws in reverse of sequence shown in Fig. HN1-15 to prevent distorting cylinder head. When cleaning carbon, be careful not to damage the sealing surfaces. Cylinder head stud nuts should be tightened in sequence shown in Fig. HN1-15 to 7.7-11.8 N·m (69-104 in.-lbs.) torque.

**PISTON, PIN AND RINGS.** To remove the piston, it is necessary to remove the cylinder head and generator stator. Unbolt and remove cover (23 – Fig. HN1-16) from crankcase. Unscrew 14mm oil drain plug for access to connecting rod screws. The piston and connecting rod assembly can then be withdrawn from top of cylinder. For easier access to cap of connecting rod it is advisable, though not absolutely necessary, to remove both valves and camshaft (13) after removing camshaft reservoir (14). With cap of connecting rod unbolted, piston and connecting rod can be removed from cylinder. Refer to the following specification data:

Piston pin to connecting rod bore –
Desired clearance . . . . . 0.016-0.04 mm
(0.0006-0.0016 inch)

Wear limit . . . . . . . . . . . . . . . .0.16 mm
(0.006 inch)
Piston pin to bore in piston –
Desired clearance . . . . . . .0-0.012 mm
0-0.0005 inch
Wear limit . . . . . . . . . . . . . . . .0.13 mm
(0.005 inch)
Piston diameter, top of piston at right angles to piston pin –
Desired . . . . . . . . . . .41.55-41.60 mm
(1.6358-1.6378 inch)
Minimum limit . . . . . . . . . . .41.35 mm
(1.628 inch)
Cylinder bore –
Standard
diameter . . . . . . .42.005-42.025 mm
(1.6537-1.6545 inch)
Maximum taper and/or
out-of-round . . . . . . . . . . . . .0.1 mm
(0.004 inch)
Piston ring side clearance in groove –
Top, desired . . . . . . . . . .0.01-0.04 mm
(0.0004-0.0016 inch)
Wear limit . . . . . . . . . . . . . . . .0.08 mm
(0.0032 inch)
Second (compression) and bottom (oil) –
Desired . . . . . . . . . . . .0.005-0.035 mm
(0.0002-0.0014 inch)
Wear limit . . . . . . . . . . . . . . . .0.08 mm
(0.0032 inch)
Piston ring end gap in cylinder bore –
All rings, desired . . . . . . . .0.1-0.3 mm
(0.004-0.012 inch)
Wear limit . . . . . . . . . . . . . . . .1.0 mm
(0.04 inch)

When assembling, manufacturer's marks on side of all piston rings should be toward top of piston. The top compression ring is chrome plated and second ring is unplated. Dipper on con-

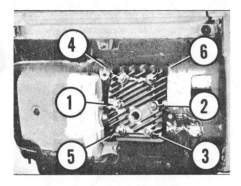

Fig. HN1-15—Cylinder head tightening sequence.

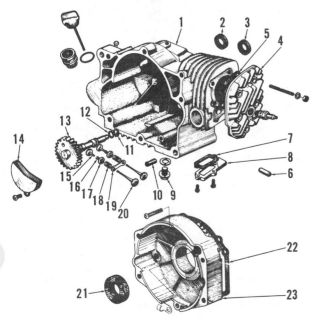

**Fig. HN1-16—Exploded view of crankcase and valve system.**

1. Crankcase/cylinder
2. Oil seal
3. Oil seal
4. Cylinder head
5. Head gasket
6. Valve guide
7. Gasket
8. Cover
9. Drain plug
10. Dowel pin
11. Snap ring
12. Washer
13. Camshaft
14. Oil reservoir
15. Valve tappet
16. Adjuster cap
17. Retainer
18. Valve spring
19. Spring seat
20. Valve
21. Oil seal
22. Gasket
23. Crankcase cover

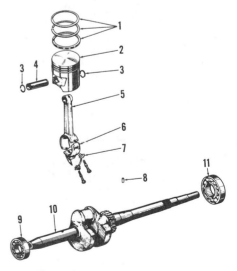

**Fig. HN1-17—Exploded view of crankshaft, rod and piston assembly.**

1. Piston rings
2. Piston
3. Snap ring
4. Piston pin
5. Connecting rod
6. Rod cap
7. Lock tabs
8. Dowel pin
9. Ball bearing
10. Crankshaft
11. Ball bearing

necting rod cap and arrow (triangle) on underside of piston crown should both be down. Tighten connecting rod screws to 2.9-3.4 N·m (26-30 in.-lbs.). Tighten cylinder head retaining nuts in sequence shown in Fig. HN1-15 to 7.8-11.8 N·m (69-104 in.-lbs.). Tighten crankcase cover screws to 2.9-5.9 N·m (26-52 in.-lbs.)

**CRANKCASE AND CRANKCASE.** The single-piece crankshaft is supported by a ball bearing at both ends. The non-removable camshaft drive gear is located on the generator end of the crankshaft.

During assembly, align crankshaft and camshaft gear timing marks. On early models, align dots on gear teeth. On later models, align dot on camshaft gear with chamfered gear tooth on crankshaft gear.

Crankshaft main bearing journal diameter on fan end should be 15.007-15.025 mm (0.5907-0.5913 inch); renew if under 14.95 mm (0.5880 inch). Main bearing journal diameter on generator end should be 17.007-17.025 mm (0.6694-0.6700 inch); renew if under 16.95 mm (0.6672 inch).

Main bearing bore diameter is 35.00 mm (1.3780 inch) in crankcase with a wear limit of 35.08 mm (1.3811 inch). Main bearing bore diameter in crankcase cover is 40.00 mm (1.5748 inch) with a wear limit of 40.02 mm (1.5756 inch). Camshaft bore diameter in crankcase and crankcase cover is 10.00 mm (0.3937 inch) with a wear limit of 10.08 mm (0.3969 inch).

**CAMSHAFT AND TAPPETS.** The camshaft is accessible after removal of generator unit and crankcase cover (23 – Fig. HN1-16). Cylinder head and valves are also removed to release valve lifters (15), and after cam gear reservoir

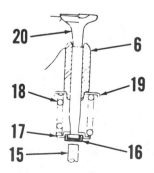

Fig. HN1-18—Cross-section of valve showing adjuster cap (16). Refer to text for valve clearance adjustment. Refer to Fig. HN1-16 for identification.

(14) is unbolted (one capscrew at bottom) camshaft assembly can be withdrawn. Both cam journals are of equal diameter – 9.972-9.987 mm (0.3924-0.3930 inch) and camshaft assembly should be renewed if either journal is worn under 9.90 mm (0.3897 inch) diameter. Cam lobe height should be 16.3-16.4 mm (0.6417-0.6457 inch) with a wear limit of 16.15 mm (0.6358 inch).

Valve lifters (tappets) are identical for all engines. Standard tappet stem diameter is 3.978-3.990 mm (0.1565-0.1570 inch). Renew if worn to less than 3.95 mm (0.1554 inch).

During assembly, align crankshaft and camshaft gear timing marks. On early models, align dots on gear teeth. On later models, align dot on camshaft gear with chamfered gear tooth on crankshaft gear.

**VALVE SYSTEM.** Valves are removable after removing cylinder head and tappet cover. Face and seat angle for both valves is 45 degrees. Use cutter to dress seats and lap valve directly to seat for proper finish. A 75 degree cutter should be used to narrow valve seats if necessary.

Valve clearance should be checked with piston at TDC on its compression stroke. Valve clearance is adjusted by installing an adjuster cap (16 – Fig. HN1-18) of different thickness. Note identifying letter codes and corresponding thicknesses in following table.

| Letter Code | Thickness |
|---|---|
| A | 3.48-3.51 mm (0.1370-0.1382 in.) |
| G | 3.42-3.45 mm (0.1347-0.1358 in.) |
| F | 3.33-3.36 mm (0.1311-0.1323 in.) |
| I | 3.24-3.27 mm (0.1276-0.1287 in.) |
| L | 3.15-3.18 mm (0.1240-0.1252 in.) |
| O | 3.06-3.09 mm (0.1205-0.1217 in.) |

Valve springs are interchangeable. Note following specifications:

Valve seat width . . . . . . . . . 0.5-0.98 mm (0.0196-0.0382 inch)
Service limit . . . . . . . . . . . . . . . . . 1.0 mm (0.0393 inch)
Tappet clearance –
Intake . . . . . . . . . . . . . . 0.04-0.06 mm (0.0016-0.0024 in.)
Exhaust . . . . . . . . . . . 0.03-0.07 mm (0.0012-0.0028 in.)

Valve spring free
length . . . . . . . . . . . . . . 21.7-21.9 mm (0.854-0.862 inch)
Service limit . . . . . . . . . . . . . . . . . 20 mm (0.787 inch)
Min. spring pressure at
13.1 mm (0.516 in.) . . . . . . . . . . 5.0 kg (11.0 lbs.)
Valve stem diameter
minimum . . . . . . . . . . . . . . . . 3.94 mm (0.155 inch)

**REWIND STARTER.** To remove starter, remove interfering covers and using a suitable puller, detach fan from crankshaft. Care should be used when removing starter spring so spring is not allowed to uncoil uncontrolled.

When assembling, the starter pulley should be rotated two turns to provide tension for pulling rope back in. Preloading the spring two turns is accomplished after rope is wound on pulley and spring is attached by leaving end of rope out of pulley notch then turning the pulley. End of rope can be guided through hole in housing after preloading the spring. The screws that attach starter housing to crankcase should be tightened to 2.9-5.9 N·m (26-52 in.-lbs.). Tighten fan retaining screw to 12-15 N·m (9-11 ft.-lbs.)

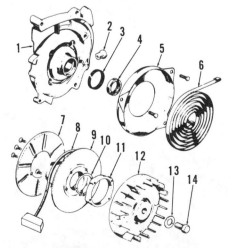

Fig. HN1-19—Exploded view of starter.

1. Starter housing
2. Bushing
3. Felt ring
4. Oil seal
5. Spring housing
6. Rewind spring
7. Pulley half
8. Rope
9. Pulley half
10. Snap ring
11. Ratchet
12. Fan
13. Washer
14. Capscrew

# HONDA

| Model | Output-kw | Voltage | Engine | | Governed |
|-------|-----------|---------|--------|--|----------|
| | | | **Make** | **Model** | **Rpm** |
| EM400 | 0.3 | 115VAC | Honda | ..... | 3600 |
| | | 13VDC | | | |

## GENERATOR

### MAINTENANCE

Brushes are accessible after removing end panel. Inspect brushes periodically and renew brushes if worn to length less than 8 mm (0.32 in.). Inspect slip rings and recondition if required.

### TROUBLESHOOTING

If little or no generator output is evident, inspect fuses, switches and all wiring and connections. Engine must be in good condition and able to maintain governed speed of 3600 rpm under load. The following tests will help locate malfunction. Refer to wiring diagram in Fig. HN2-1.

To check stator, disconnect four-wire coupler and using an ohmmeter check resistance between brown and blue terminals for stator wires. Resistance should be 3.0-4.0 ohms. Check for grounded stator coils by connecting ohmmeter leads alternately between blue and brown wire terminals and ground. Stator should be renewed if open or grounded stator windings are discovered.

To check exciter coil, disconnect red/white and green/white exciter wires from voltage regulator leads. Connect an ohmmeter to exciter wires and measure resistance. Resistance should be 4.4-5.8 ohms. Alternately connect ohmmeter leads to red/white and green/white exciter wires and ground. Exciter coil should be renewed if windings are open or grounded.

To check rotor, measure resistance between slip rings with an ohmmeter. Resistance should be 10-13 ohms. Check for grounded windings by measuring resistance between slip ring and ground. If rotor windings are open or grounded, renew rotor.

If previous tests do not locate malfunction, the voltage regulator should be replaced with a new or known to be good regulator and generator output observed.

### OVERHAUL

Disassembly and reassembly are evident after inspection and referral to wiring diagram in Fig. HN2-1. Note the following: Use a suitable puller to detach fan. Tighten rotor nut to 39-49 N·m (29-36 ft.-lbs.), stator screws to 5.9-8.8 N·m (52-78 in.-lbs.) and generator fan retaining screw to 11.8-15.7 N·m (8.7-11.5 ft.-lbs.).

## ENGINE

The engine used on Model EM400 is the same as used on Models E300 and ER400. Refer to previous section for engine service while noting the following:

Ignition timing is fixed at 23-27 degrees BTDC and should occur when mark on generator cooling fan is aligned with mark on stator frame.

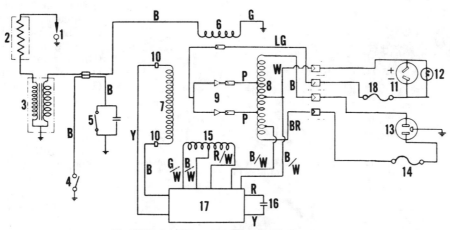

**Fig. HN2-1—Wiring diagram for Model EM400.**

1. Spark plug
2. Noise suppressor
3. Ignition coil
4. Ignition switch
5. Ignition breaker points
6. Ignition stator coil
7. Field coil
8. Generator stator
9. Diodes
10. Brushes
11. DC receptacle
12. Pilot lamp
13. AC receptacle
14. Fuse
15. Exciter coil
16. Capacitor
17. Voltage regulator
18. Fuse

# HONDA

| | | | Engine | | Governed |
| Model | Output-kw | Voltage | Make | Model | Rpm |
|---|---|---|---|---|---|
| E600A | 0.6 | 120 | Honda | G25 | 5000 |
| E600S | 0.6 | 220 | Honda | G25 | 5000 |
| E900 | 0.9 | 120VAC 12VDC | Honda | G28 | 5000 |

# GENERATOR

## OPERATION

All models are belt driven as the engine and generator are separate units. Model E900 is designed for 60 hertz (cycles) while Models E600A and E600S may be wired for 50 or 60 hertz operation. Refer to Fig. HN3-1 and wiring diagram in Fig. HN3-6 to be sure wiring is properly connected for 60 hertz use. All service procedures are directed towards units wired for 60 hertz.

## MAINTENANCE

Brushes are accessible after removing end cover. Inspect brushes periodically and renew brushes if brush length is less than 7 mm (0.28 in.). Note that negative (left) and positive (right) brush springs are different and not interchangeable. Inspect slip rings and recondition if required.

## DRIVE BELT ADJUSTMENT

Improper drive belt tension may allow drive belt to slip thereby reducing generator output. Drive belt tension should be 10 mm (³/₈ in.) with finger pressure exerted at midpoint of belt. To adjust tension on Models E600A and E600S, loosen engine clamp bolt shown in Fig. HN3-2 and turn adjust bolt shown in Fig. HN3-3. Make certain that engine clamp bolt (Fig. HN3-2) is retightened after belt tension is adjusted. On Model E900, loosen generator mounting nuts and turn adjustor bolt located in powerplant frame under generator mount pads. Retighten mounting nuts after belt adjustment.

## TROUBLESHOOTING

If little or no generator output is evident, be sure drive belt is not slipping then inspect fuses and all wiring and connections. Engine must be in good condition and able to maintain governed speed of 5000 rpm under load. The following tests will help locate malfunc-

tion. Refer to wiring diagrams in Figs. HN3-6 and HN3-7.

The exciter coil is located under the engine flywheel. To check exciter coil on Models E600A and E600S, disconnect the two blue wires shown in Fig. HN3-4 and using an ohmmeter measure resistance between wires. Resistance should be 4 ohms. With engine running at maximum governed speed, measure voltage at wires. Voltage should be at least 20VAC. Renew exciter coil if resistance or voltage measurements are incorrect. To check exciter coil on Model E900, it may be necessary to remove generator for access to coil wires—two yellow and one blue. Using an ohmmeter, check for open coil windings by checking for continuity between both yellow wires and between yellow wires and blue wire. With engine running at maximum governed speed, voltage should be 72-88VAC between yellow leads and 36-44 VAC between yellow and blue leads. Renew exciter coil if resistance or voltage measurements are incorrect.

To check rectifier diodes (10–Fig. HN3-5) on Models E600A and E600S, disconnect red wire from terminal (2–Fig. HN3-1), yellow wire from terminal (4), brown wire from point behind (BR) and both wires from brushes. With ends of wires insulated, each diode (10–Fig. HN3-5) can be checked with an ohmmeter. Diodes should have infinite resistance in one direction and low resistance with ohmmeter leads reversed.

If rectifier is suspected on Model E900, replace rectifier with a new or

Fig. HN3-1—View of terminals connected for 60 hertz operation. Refer to text for testing.

Fig. HN3-3—Belt tension is adjusted by turning the adjust bolt.

Fig. HN3-2—View showing the engine clamp bolt. Turning in direction of arrow loosens clamp.

Fig. HN3-4—Refer to text for checking the external (on engine) exciter coil.

Fig. HN3-5—View showing location of the diodes. Refer to text for testing.

known to be good rectifier and check generator output.

Model E900 is equipped with a variable resistor (18–Fig. HN3-7) in exciter circuit. With leads disconnected, resistance between fixed and center terminal should be approximately 20 ohms. Resistance is varied by moving center terminal ring. Renew resistor if desired resistance is not obtainable.

To check transformer (23–Fig. HN3-7) on Model E900, disconnect transformer leads. There should be continuity between the two yellow leads, between the light green and brown leads and between the light green and yellow/white leads. Renew transformer if open windings are found.

Test capacitor (24–Fig. HN3-7) on Model E900 or replace with a new or known to be good capacitor.

Check condition of diode (11–Fig. HN3-5) on Models E600A and E600S. Diode should have infinite resistance with ohmmeter leads connected one way and then very little resistance with ohmmeter leads reversed. If resistance readings are both high or both low, diode is faulty and must be renewed.

Check condition of DC diodes (20–Fig. HN3-7) on Model E900 using same diode testing procedure as outlined in previous paragraph.

To check stator on Models E600A and E600S, disconnect yellow/white and red/white wires from terminals (2 and 4–Fig. HN3-1). Resistance between the yellow/white wire and point (BR) between the two lower diodes should be 0.85 ohms for Model E600A or 1.40 ohms for Model E600S. Resistance between red/white wire and point (BR) should be 1.3 ohms for Model E600A or 4.5 ohms for Model E600S. A short circuit is indicated if resistance is too low: an open circuit or poor connection if resistance is too high.

To check stator on Model E900, connect an ohmmeter lead to lead end of brown stator wire then alternately connect remaining ohmmeter lead to all other stator wires. There should be continuity between brown stator wire and

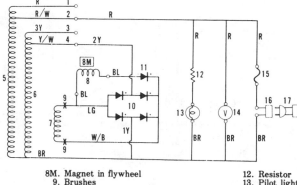

Fig. HN3-6—Wiring diagram for Models E600A and E600S. Magnet (8M) is in engine flywheel. Terminals (1 & 3) are used for 50 hertz units.

- BL. Blue
- BR. Brown
- LG. Light green
- R. Red
- R/W. Red/White stripe
- W/B. White/Blue stripe
- Y. Yellow
- Y/W. Yellow/White stripe

- 1,2,3 & 4. Terminals
- 5. AC generator coil
- 6. Internal exciter coil (stator)
- 7. Armature (field) coil
- 8. External exciter coil (engine)

- 8M. Magnet in flywheel
- 9. Brushes
- 10. Bridge rectifier (4 diodes)
- 11. External exciter diode

- 12. Resistor
- 13. Pilot light
- 14. Voltmeter
- 15. Fuse
- 16. Connector
- 17. Load

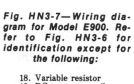

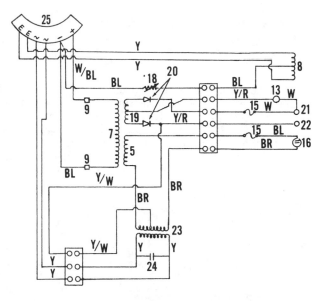

Fig. HN3-7—Wiring diagram for Model E900. Refer to Fig. HN3-6 for identification except for the following:

- 18. Variable resistor
- 19. DC generator coil
- 20. DC diodes
- 21. Neg. DC terminal
- 22. Pos. DC terminal
- 23. Transformer
- 24. Capacitor
- 25. Rectifier

all other stator wires or there is an open winding in stator and it must be renewed. However, if any of the resistance readings is below 1.6 ohms, then a short exists and the stator must be renewed.

To check field coils, measure resistance between slip rings. Resistance should be approximately 55 ohms but not less than 45 ohms. If open or shorted windings are found, renew rotor.

## OVERHAUL

Disassembly and reassembly are evident after inspection of unit and referral to wiring diagram. Use a suitable puller to remove drive pulley. Note flanged washers located between bearings and rotor. At drive end of rotor, flange side of washer is towards rotor. At slip ring end of rotor, flange side of washer is towards bearing.

# ENGINE

| Model | Bore | Stroke | Cyls. | Displ. |
|---|---|---|---|---|
| G25 | 46 mm (1.811 in.) | 35.6 mm (1.402 in.) | 1 | 59 cc (3.61 cu. in.) |
| G28 | 49 mm (1.929 in.) | 35.6 mm (1.402 in.) | 1 | 67 cc (4.10 cu. in.) |

## MAINTENANCE

**SPARK PLUG.** For Model G25, NGK spark plug type C-6H is used. C-6HB is recommended for Model G28. Plug elec-

trode gap is set to 0.6-0.7 mm (0.024-0.028 inch).

**CARBURETOR.** A Keihin float type carburetor is used on Model G25. Refer

to Fig. HN3-10 for cross-sectional view. Idle fuel mixture is adjusted by turning needle (8). Initial setting is ¾-1 turn open for the idle mixture needle. Main fuel mixture is controlled by size of main

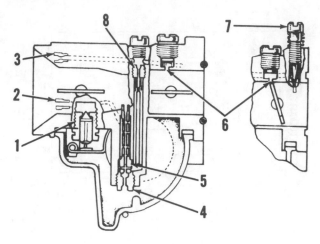

*Fig. HN3-10—Cross-sectional view of carburetor used on G25 engines. Fuel inlet needle valve (1) is provided with a spring inside needle.*

1. Fuel inlet needle valve
2. Air jet
3. Pilot air jet
4. Main jet
5. Main nozzle
6. Bypass
7. Idle mixture needle
8. Pilot jet

jet (4). Standard main jet size is #70; however, a smaller main jet size (number) may provide better operation at high altitudes. Turn idle speed stop screw (S – Fig. HN3-11) until idle speed is approximately 1500 rpm. With engine idling, the clutch should be disengaged and output shaft should not be turning.

To check for proper float level, refer to Fig. HN3-12. Measure distance (H) from bottom side of float to carburetor body when holding float so that float lever just contacts needle valve tip.

*Fig. HN3-11—The idle speed stop screw (S) and fuel bowl drain (D) are accessible through holes in cover as shown.*

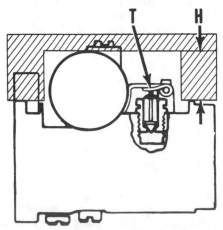

*Fig. HN3-12—Float height (H) should be set as described in text by bending tang (T).*

Float level is correct if height (H) is 16-17 mm (⅝ to 21/32-inch).

**NOTE: Needle valve is spring loaded and any pressure such as weight of float resting against needle valve may result in incorrect measurement of float level.**

Model G28 is furnished with a carburetor of style shown in Fig. HN3-13. Initial setting of idle mixture adjustment screw (20) is ¾ to 1¼ turns open. With engine running, mixture screw is adjusted alternately with throttle stop (idle speed) screw (26) to provide smooth idle at 1200-1500 rpm or so that automatic clutch is disengaged and output shaft is not turning. Mixture adjustment screw (20) of this carburetor also serves as an air bleed at no-load rated rpm, and final adjustment should be a balance between idle speed and open throttle to rated rpm (no load on output shaft) with no surge or hesitation as throttle/governor knob is changed from idle to operating speed setting.

Main jet (11) is fixed, providing proper fuel to air ratio when engine operates under load. Standard main jet size is No. 68, with substitute sizes ranging from No. 60 to 72 to meet operational requirements at higher or lower elevations.

Float valve setting of this carburetor calls for use of a template gage which straddles float when carburetor body is inverted. Correct space between float and flange of carburetor body is 2 mm (0.080 inch) which is approximate diameter of 5/64-inch drill which might be used for lack of another gage. Because float needle valve contains a small plunger which is loaded by a very light action spring, contact with float arm must be very slight or a small clearance (0.01 mm or 0.0004 inch) allowed between tip of valve and float arm to insure correct fuel level.

Model E900 generator is fitted with an automatic choke as shown in Fig. HN3-14. This choke operates by expansion and contraction of U-shaped bi-

metallic strip (1) and may be checked for proper performance by reference to the following:

| AIR TEMPERATURE | CHOKE VALVE POSITION |
|---|---|
| 0°C/32°F. | Fully closed |
| 10°C/50°F. | ⅛ Open |
| 20°C/69°F. | ¼ Open |
| *80°C/176°F. | Wide open |

*Air temperature raised by heat radiated from running engine.

Correct choke opening for a given temperature is adjusted by releasing rod clamp (2) so that rod length can be changed to set choke valve to proper position. Take care not to bend or distort bi-metallic control strip. If engine is sluggish, stalls or lacks power,

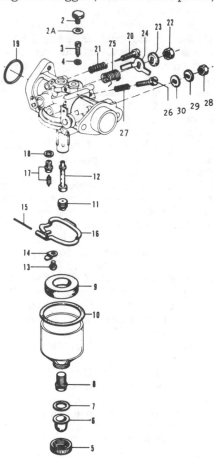

*Fig. HN3-13—Exploded view of carburetor used on Model G28 engine. Refer to text for service.*

| | |
|---|---|
| 2. Cap plug | 16. Float arm |
| 2A. Flat washer | 17. Float valve assembly |
| 3. Throttle shaft guide screw | 18. Gasket |
| 4. Lock washer | 19. "O" ring |
| 5. Sediment bulb retainer | 20. Idle mixture screw |
| 6. Sediment bulb | 21. Spring |
| 7. Gasket | 22. Throttle shaft nut |
| 8. Float boal retainer | 23. Lock washer |
| 9. Float | 24. Throttle lever |
| 10. Float bowl gasket | 25. Throttle shaft spring |
| 11. Main jet | 26. Idle speed screw |
| 12. Main nozzle | 27. Spring |
| 13. Pin set plate screw | 28. Choke shaft nut |
| 14. Pin set plate | 29. Lock washer |
| 15. Float pin | 30. Spacer |

especially during warm-up, operation of automatic choke should be checked.

**GOVERNOR.** A mechanical yweight type governor is used. Governor weights (1–Fig. HN3-15) are mounted on the clutch center (Fig. HN3-20) at end of crankshaft. Engine governed speed is adjusted by turning control knob (2–Fig. HN3-15) to increase or decrease governor spring tension. Length of rod (5) should be so that throttle valve is wide open when engine is not running and the speed control knob (2) is set at the high speed position. If rod length is incorrect, bend loop (L) in rod as necessary. If rod length is incorrect, engine speed may vary excessively (hunt).

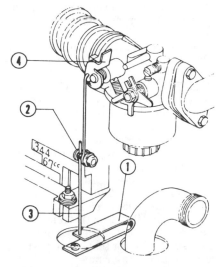

**Fig. HN3-14—Thermostatic controlled automatic choke used on G28 engine when fitted to Model E900 generator. Note that bi-metallic strip is mounted below engine deck and near heat source for stable temperatures. See text.**

1. Bi-metallic sensor
2. Adjusting clamp
3. Control rod
4. Choke lever

**IGNITION.** The ignition system is the energy transfer type with the low tension coil and breaker points located under the flywheel. High tension coil and condenser are mounted on engine below fuel tank.

Breaker point gap should be approximately 0.3-0.4 mm (0.012-0.016 inch); however, gap should be adjusted to provide correct ignition timing. To adjust ignition timing, remove the recoil starter and cooling shroud from side of engine. Remove starter pulley and screens from flywheel and turn flywheel in normal direction of rotation (counter-clockwise as viewed from flywheel end) until "F" mark on flywheel is aligned with pointer on crankcase as shown in Fig. HN3-16. Breaker points should just open when "F" mark aligns with pointer. If points open too soon, decrease the breaker point gap. Breaker point gap can be adjusted through holes in flywheel, but flywheel must be removed to remove points.

**LUBRICATION.** Engines use a splash lubrication system with a dipper on connecting rod. The camshaft, valves and rocker arms are lubricated by oil picked up by the cam chain and tension roller.

Crankcase capacity is approximately 1½ pints. Oil level should be maintained between marks on the oil level dipstick. The filler plug (dipstick) should NOT be screwed into crankcase when checking oil level. Recommended motor oil is API classification SE or SF.

SAE 10W/30 oil is recommended for all season use, however, if single viscosity oils are used, SAE 30 is called for in temperatures over 15°C (59°F), SAE 10W is required below 0°C (32°F) and SAE 20W should be used in mid-range temperatures 0°C to 15°C (32°F to 59°F). Engine oil should be checked for

level every 20 hours of operation and changed after every 100 hours.

**VALVE CLEARANCE.** The valves are actuated by a chain driven camshaft located in the cylinder head and rocker arms (cam followers). Valve clearance should be set when engine is cold and piston at TDC on compression stroke. The flywheel has a mark "T" indicating top dead center when aligned with crankcase mark (P–Fig. HN3-16). Clearance should be 0.05 mm (0.002 inch) for both valves.

## REPAIRS

**VALVES AND CYLINDER HEAD.** Remove the cylinder head cover (with rocker arms). Remove dowel pin (1–Fig. HN3-17), withdraw the camshaft center shaft from opposite end of head and remove camshaft. Remove the four stud nuts and one screw then lift off the cylinder head. Cylinder head gasket surfaces must be perfectly flat.

Six hollow dowels are used on these engines for alignment of cylinder, cylinder head and cylinder head cover. Two of these dowels are installed between cylinder and crankcase over

**Fig. HN3-16—Ignition breaker points should just open as mark "F" aligns with pointer (P). Piston is at TDC when mark (T) is aligned with pointer.**

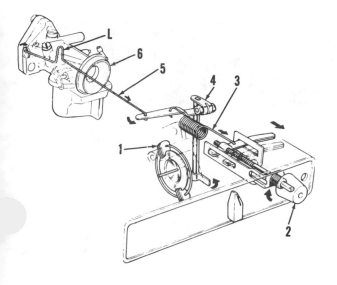

**Fig. HN3-15—Drawing of speed control system. Arrows show direction of movement to increase engine speed.**

1. Governor weights
2. Speed control knob
3. Governor spring
4. Governor lever and shaft
5. Link rod
6. Carburetor

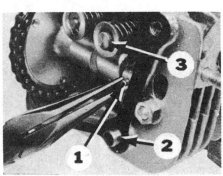

**Fig. HN3-17—Camshaft center shaft is retained in head by dowel pin (1). After center shaft is pulled from opposite end, camshaft and sprocket can be removed.**

1. Shaft retainer dowel
2. Hollow dowel
3. Valve rotator (Exh. only)

diagonally opposite stud bolts. Two more are fitted at joint between cylinder and cylinder head, and two are also used to line up cylinder head cover. During assembly, be sure that these dowels are fitted into correct bolt holes and that assembly gaskets are correctly placed.

When assembling, hollow dowels should be installed around two of the studs between the cylinder and cylinder head. Tighten the four cylinder head retaining nuts diagonally to 7.9-8.8 N·m (70-78 in.-lbs.) torque. Make certain that all are tightened evenly. Turn the flywheel until "T" mark is aligned with mark (P – Fig. HN3-16) on crankcase and install camshaft with holes in sprocket parallel to gasket surface as shown in Fig. HN3-18. When the camshaft center shaft is installed, make certain that oil guide is correctly positioned between cylinder head and camshaft sprocket. Align hole in center shaft and install dowel pin (1 – Fig. HN3-17). Valve clearance must be adjusted after assembling, and after head bolts are tightened to specified torque value.

**IMPORTANT: Do not tighten cylinder head after adjusting valve tappet clearance.**

Cylinder head cover is also tightened to 7.9-8.8 N·m (70-78 in.-lbs.) torque.

**PISTON, RINGS AND CYLINDER.** To remove the cylinder, first remove the cylinder head as in the previous paragraphs, then withdraw the cylinder. Rings should be installed with marked side toward closed end of piston. Arrow on top of piston should point down. Piston and rings are available in standard size only for Model G25 engine. For Model G28, oversizes of 0.25, 0.50 and 0.75 mm (0.010, 0.020 and 0.030 inch) are available with corresponding oversizes in renewal cylinders.

If cylinder bore is out-of-round or tapers more than 0.05 mm (0.002 inch) cylinder should be renewed. Piston ring side clearance is 0.02-0.06 mm (0.0008-0.0024 inch) for top compression ring. Side clearance for second and oil control rings is 0.01-0.05 mm (0.0004-0.002 inch). Service limit of side clearance for all rings is 0.15 mm (0.006 inch).

Assembly details are as follows: Stagger piston ring end gaps equally in ring grooves. Manufacturer's marks are imprinted on top side of ring. Install with these marks toward top of piston. Top compression ring is chromed and easily identifiable. Second compression ring and oil ring are parkerized. Arrow mark on head of piston should point down. Piston pin retaining snap rings should be installed with open part of snap ring NOT aligned with cut-out in piston. Hollow dowels are installed between cylinder and crankcase on two of the studs.

**CRANKSHAFT AND CONNECTING ROD.** The crankshaft and connecting rod are available only as an assembly and should not be disassembled. Removal is accomplished after removing the clutch and reduction assembly, cylinder, piston, flywheel and separating the crankcase halves.

When assembling, make certain that crankcase halves are clean and free from nicks and burrs.

**CLUTCH AND REDUCTION UNIT.** The automatic clutch (3 – Fig. HN3-19) is located on the crankshaft and drives the output shaft (4) via a chain reduction drive. The clutch should be disengaged when engine is idling. The clutch and reduction unit can be removed after removing the crankcase side cover. Refer to Fig. HN3-20 for cross-sectional view of the clutch assembly. Clutch shown in Fig. HN3-20 is that installed in Model G25 engine. Clutch used in Model G28 is similar except that only two shoes are fitted to clutch drum.

**REWIND STARTER.** Refer to Figs.

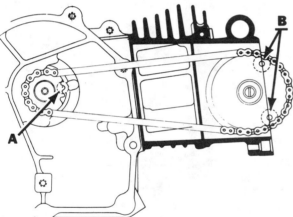

Fig. HN3-18—Valve timing. Align mark on crankshaft sprocket (A) with crankcase cutout and two holes (B) of camshaft sprocket on gasket surface of cylinder head as shown. "T" mark on flywheel should be in alignment with timing rib on left crankcase.

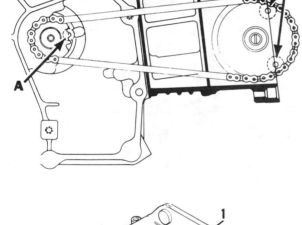

Fig. HN3-19—View of engine showing layout of parts. The automatic clutch (3) disengages at idle speed.

1. Crankshaft
2. Camshaft
3. Clutch
4. Output shaft
5. Chain tension roller
6. Oil guide
7. Oil separator

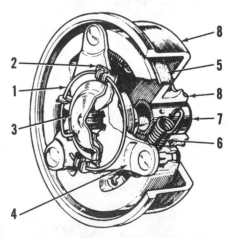

Fig. HN3-20—Cross-sectional view of clutch and governor weights. Assembly is mounted on end of crankshaft.

1. Weight retaining ring
2. Governor weights
3. Governor slider
4. Drive plate
5. Clutch shoes (weights)
6. Springs
7. Bushing
8. Clutch drum and sprocket

HN3-21 and HN3-22 for views of rewind starter used. Starter can be disassembled as follows: Pull rope out slightly and untie knot at handle end while holding the rope pulley (6) from turning, remove handle from rope and allow spring to unwind slowly. Remove snap ring (1), washer, friction plate (2), springs (3 & 9), cup (10), and washer. Remove ratchets (4) from pulley and remove pulley and spring from housing. Reassemble by reversing disassembly procedure and check action of starter before reinstalling on engine.

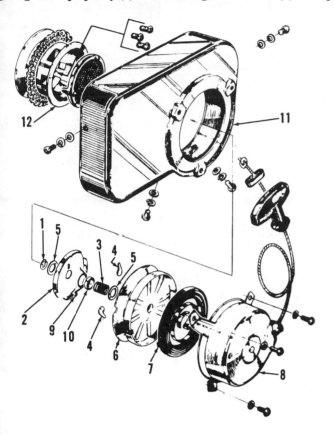

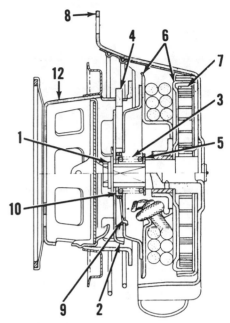

*Fig. HN3-21—Exploded view of starter assembly. Refer also to Fig. HN3-22.*

*Fig. HN3-22—Cross-sectional view of starter. Refer to Fig. HN3-21 for legend.*

| | |
|---|---|
| 1. Snap ring | 7. Rewind spring |
| 2. Friction plate | 8. Housing |
| 3. Friction spring | 9. Set spring |
| 4. Ratchet pawls | 10. Spring cup |
| 5. Washers | 11. Cooling shroud |
| 6. Pulley | 12. Starter cup |

# HONDA

| Model | Output-kw | Voltage | Engine Make | Engine Model | Governed Rpm |
|-------|-----------|---------|------|-------|----------|
| E1000A | 1.0 | 120VAC 12VDC 24VDC | Honda | G40 | 3600 |
| E1000S | 1.0 | 220VAC 12VDC 24VDC | Honda | G40 | 3600 |
| E1500A | 1.25 | 120VAC 12VDC 24VDC | Honda | G40 | 3600 |
| E1500S | 1.25 | 250VAC 12VDC 24VDC | Honda | G40 | 3600 |
| E2000A | 2.0 | 120VAC 12VDC 24VDC | Honda | G65 | 3600 |
| E2000S | 2.0 | 220VAC 12VDC 24VDC | Honda | G65 | 3600 |
| E2500A | 2.0 | 120VAC 12VDC 24VDC | Honda | G65 | 3600 |
| E2500S | 2.0 | 220VAC 12VDC 24VDC | Honda | G65 | 3600 |

# GENERATOR

## OPERATION

All models are belt driven as the engine and generator are separate units. The generator may be wired for 50 or 60 hertz (cycles) operation. Refer to wiring diagram in Fig. HN4-4, HN4-5 or HN4-6 and note marked terminals in generator and control box. All service procedures are directed towards 60 hertz units.

Fig. HN4-1—View of diodes on Models E1000A and E1000S.

Fig. HN4-2—View showing location of diodes (11 & 20) in excitor circuits of Figs. HN4-5 and HN4-6.

Fig. HN4-3—Exploded view of Model E1500A generator. Other models are similar.

1. Fan cover
2. Nut
3. Lockwasher
4. Washer
5. Generator pulley
6. Engine pulley
7. Retainer
8. Snap ring
9. Front cover
10. Mount bracket
11. Bearing
12. Retainer
13. Rotor
14. Washer
15. Bearing
16. Stator
17. Frame
18. DC diode assy.
19. Brushes
20. Terminal
21. Brush holder
22. End frame
23. End cover

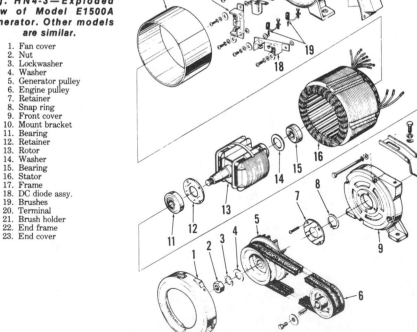

## MAINTENANCE

Brushes are accessible after removing end cover. Inspect brushes periodically and renew brushes if brush is worn excessively. Inspect slip rings and recondition if required.

## DRIVE BELT ADJUSTMENT

Improper drive belt tension may allow drive belt to slip thereby reducing generator output. Drive belt tension should be 10 mm (³/₈ in.) with finger pressure exerted at midpoint of belt. To adjust belt tension, loosen generator mounting nuts and turn adjuster bolt located beneath side of generator. Retighten generator mounting nuts after adjustment.

## TROUBLESHOOTING

If little or no generator output is produced, be sure drive belts are not slipping then inspect fuses and all wiring and connections. Engine must be in good condition and able to maintain governed speed of 3600 rpm under load. The following tests will help locate malfunctions. Refer to Figs. HN4-4, HN4-5 and HN4-6.

Remove generator end cover and with engine running at maximum governed speed and no load connected to generator, measure DC voltage present at brush terminals. Voltage should be 26-32 volts if exciter circuit is functioning correctly. If voltage reading is incorrect, disconnect wire leads to isolate diodes and test diodes shown in Fig. HN4-1 and HN4-2. Check resistance between ends of each diode with an ohmmeter, then reverse the ohmmeter leads. Diode should have infinite resistance with one connection and little resistance with ohmmeter leads reversed. If resistance is either high or low for both readings then diode is faulty.

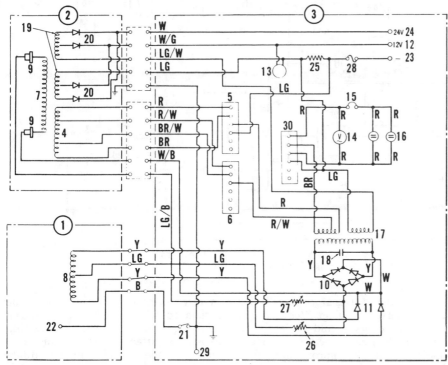

*Fig. HN4-4—Wiring diagram for E1000A and E1000S generators. External exciter coil (8) is under engine flywheel. Plug (30) is attached to connection (5) for 50 hertz operation or connection (6) for 60 hertz operation.*

1. Engine
2. Generator
3. Control box
4. AC generator coil
5. Connector (for 50 Hz)
6. Connector (for 60 Hz)
7. Armature
8. External exciter coil (engine)
9. Brushes
10. Bridge rectifier (4 diodes)
11. External exciter diodes
12. DC (12 Volt) positive connection
13. Pilot light
14. Voltmeter
15. AC circuit breaker
16. AC outlet connections
17. Transformer
18. Capacitor
19. DC generator coils
20. Diodes for DC output
21. Engine stop (kill) switch
22. Ignition breaker points
23. DC ground connection
24. DC (24 volt) positive connection
25. Resistor
26. Variable resistor
27. Voltage regulator (resistor)
28. DC fuse
29. Ground terminal
30. Plug

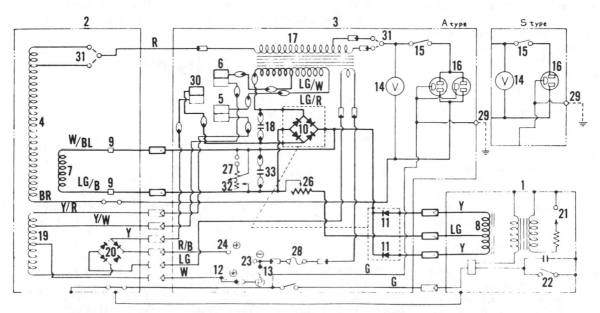

*Fig. HN4-5—Wiring diagram for Models E1500A and E1500S; Models E2000A and E2000S are similar. Plug (30) is attached to connector (5) for 50 hertz operation or connector (6) for 60 hertz operation. Wires at terminals (31) must be properly connected for 50 or 60 hertz. Refer to Fig. HN4-4 for identification except for: 31. Frequency terminals; 32. Variable resistor; 33. Capacitor.*

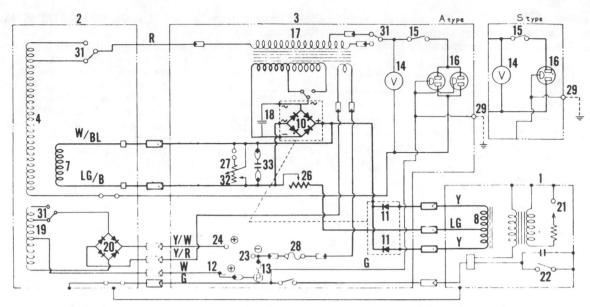

Fig. HN4-6 — Wiring diagram for Models E2500A and E2500S. Wires at terminals (31) must be properly connected for 50 or 60 hertz. Refer to Fig. HN4-4 for identification.

Check capacitors by shorting across terminals to discharge capacitor, then disconnect one capacitor lead. If capacitor is good, the ohmmeter needle will swing sharply then return when ohmmeter leads are connected to capacitor terminals. Check resistance of resistors (26 and 27 – Fig. HN4-4) on Models E1000A and E1000S and resistors (26, 27 and 32 – Fig. HN4-5 and HN4-6) on all other models. Resistor (27) is adjustable by turning knob on control panel. Maximum reference of resistor (27) should be 600 ohms on Models E1000A and E1000S or 1200 ohms on all other models. Resistor (26) on all models and resistor (32) on all models except E1000A and E1000S is variable. Resistor (26) resistance should be 100 ohms on Models E1000A, E1000S, E2000A and E2000S; 60 ohms on Models E1500A and E1500S; 30 ohms on Models E2500A and E2500S. Resistor (32) resistance should be 150 ohms on Models E1500A, E1500S, E2500A and E2500S. To check excitor coil located under engine flywheel, disconnect two yellow wires and one light green wire between engine and control box. Resistance between light green wire and either yellow wire should be 9-11 ohms. If resistance is correct, check AC voltage output of coil by running engine at maximum governed speed and measure voltage between light green wire and alternately each yellow wire. Voltage should be 58-72 volts on Models E1000A and E1000S, 52-64 volts on Models E2000A and E2000S and 63-77 volts on all other models. If resistance and/or voltage readings are incorrect, remove engine flywheel and renew exciter coil.

Check current transformer (17) for open or shorted windings. Resistance in primary windings should be less than one ohm while resistance in secondary windings should be approximately four ohms.

Check for open or shorted windings in AC stator. Resistance should be less than one ohm.

Measure resistance between rotor slip rings to check condition of rotor windings. Resistance should be approximately 90 ohms. Check for grounded windings by measuring resistance between slip ring and ground. If rotor windings are open or grounded, renew rotor.

Check DC generator coils for open or shorted windings. Check DC diodes (20) by disconnecting leads then measuring resistance with an ohmmeter, noting reading, then again measuring resistance after reversing ohmmeter leads. Diode should have infinite resistance with one connection and little resistance with leads reversed. If resistance readings are both high or both low then diode is faulty.

### OVERHAUL

Disassembly and reassembly of generator is evident after inspection of unit and referral to exploded view in Fig. HN4-3 and wiring diagrams in Figs. HN4-4, HN4-5 and HN4-6.

# ENGINE

| Model | Bore | Stroke | Cyls. | Displ. |
|---|---|---|---|---|
| G40 | 66 mm | 50 mm | 1 | 170 cc |
| | (2.60 in.) | (1.97 in.) | | (10.4 cu. in.) |
| G65 | 72 mm | 59 mm | 1 | 240 cc |
| | (2.83 in.) | (2.32 in.) | | (14.6 cu. in.) |

### MAINTENANCE

**SPARK PLUG.** Recommended spark plug for Model G40 is NGK type B-6H. G65 uses B-7H spark plug. Set electrode gap at 0.7 mm (0.028 inch).

**CARBURETOR.** A float-type, vacuum-controlled variable venturi carburetor is used on early engines. Carburetor is identified as CV (constant vacuum) type. Refer to Figs. HN4-10 and HN4-11.

Idle mixture is adjusted at needle valve (4 – Fig. HN4-12).

Higher operating range speeds are controlled by varying size of main jet (1 – Fig. HN4-11) by lifting of valve needle (9) as venturi slide (6) opens. Operating rpm limits are determined by setting of governor control knob. When overrich mixture condition is indicated, as by plug fouling, a main jet with smaller orifice should be installed. Main jet size should also be reduced for operation at high elevations.

Carburetor float level setting calls for use of a special gage which bridges over

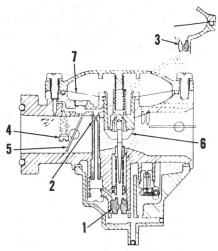

*Fig. HN4-10—Cross-sectional view of carburetor at low operating speed. Refer also to Fig. HN4-11.*

1. Main jet
2. Idle jet
3. Air jet
4. Idle mixture needle
5. Throttle plate
6. Venturi slide
7. Diaphragm

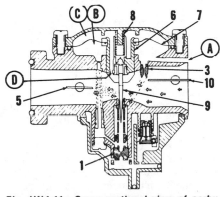

*Fig. HN4-11—Cross-sectional view of carburetor at high operating speed. Air enters port (A) into chamber (B) and vacuum at venturi enters chamber (C) through port (D). The higher pressure in chamber (B) under diaphragm (7) raises venturi slide (6) against pressure of spring (8). Tapered needle (9) changes size of main nozzle opening as venturi and needle move up.*

1. Main jet
3. Air jet
5. Throttle plate
6. Venturi slide
7. Diaphragm
8. Spring
9. Valve needle
10. Choke plate

*Fig. HN4-12—Idle mixture is adjusted at needle (4) and idle speed at screw (S). Fuel can be drained from float bowl at screw (D). Fuel filter and shut off valve is shown at (F).*

floats to measure height of float pontoons above float bowl flange. Bend float arm so that it barely touches seated needle valve. This needle valve contains a spring-loaded plunger, and if float arm exerts any pressure, a high fuel level and resultant flooding condition will result.

In later production of G40 engine (after serial number 1112157) another style carburetor has been used. Refer to Fig. HN4-13. This carburetor does not have vacuum operated variable venturi or double pontoon float as in the CV type but maintenance services are essentially similar. Adjust idle speed and mixture screws by the same procedure except that idle mixture adjustment screw (5 – Fig. HN4-13) is also used to smooth out engine operation when throttle/governor is opened to rated rpm with no load. This screw performs an air bleed function and should be adjusted for smooth performance whether engine is idling or with open throttle. Best in-

dication of a good adjustment is a uniform transition from idle to running speed with no surge or flutter.

Float level adjustment requires use of a template type gage. To adjust, invert carburetor body and lift float arm (20) so that very slight clearance 0.01 mm (0.004 in.) is evident between tip of float valve (23) and contact point on float arm with float and gage in place. Bend float arm close to pin (19) to adjust. Float level is critical to performance of this engine so great care should be taken with this adjustment. Number 78 main jet (21) is standard size with numbers 68 thru 85 available for substitution as required.

On later Model G65 engines a standard float type carburetor with throttle and choke plates is used. Refer to Fig. HN4-13A for exploded view of carburetor. Note that main jet (13) secures float bowl to carburetor body. Standard size of main jet (13) is number 92 while

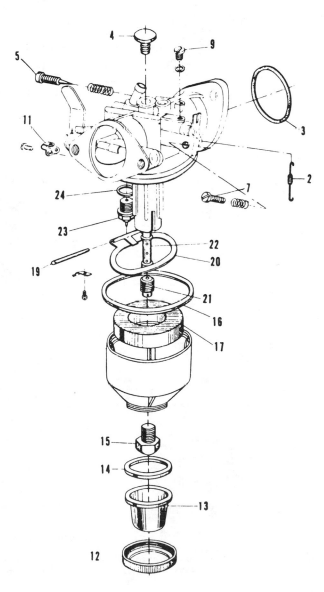

*Fig. HN4-13—Exploded view of carburetor fitted to G40 engines serial number 1112158 and after.*

2. Throttle return spring
3. Manifold "O" ring
4. Rubber cap
5. Idle mixture needle
7. Idle speed screw
9. Throttle shaft retainer
11. Clip plate
12. Bulb retainer ring
13. Sediment bulb
14. Gasket
15. Assembly bolt
16. Float chamber gasket
17. Float
19. Float pin
20. Float arm
21. Main jet
22. Main nozzle
23. Float valve assembly
24. Fiber washer

standard size of pilot jet (5) is number 48.

**GOVERNOR.** A mechanical, fly-weight governor is used. Governor weights are mounted on the camshaft gear as shown in Fig. HN4-14. The camshaft must be removed to service governor weights or slider. Engine speed is adjusted by turning control knob (K–Fig. HN4-15) to increase or decrease governor spring tension. Length of governor rod (R) should be so that movement of the lever (L) will completely open and close the carburetor throttle.

Engine governed speed should be 3600 rpm for all models. High speed limiting nuts (N) should be adjusted so that engine speed is 3600 rpm with full load and controls set at maximum speed position.

**IGNITION.** The magneto breaker points, coil and condenser are located under the flywheel. On generator units, the external exciter coil is also located under flywheel.

Breaker point gap should be approximately 0.3-0.4 mm (0.012-0.016 inch); however, gap should be adjusted to provide correct ignition timing. To set timing, remove recoil starter, cooling shroud, pulley cup and screen from flywheel side of engine. Turn flywheel until "F" mark on side of flywheel is aligned with crankcase joint on muffler side as shown in Fig. HN4-16. Move the small breaker point cover out of the way and adjust breaker points to just open when "F" mark aligns with crankcase joint. If breaker points open too soon, decrease gap. Breaker point gap (timing) can be adjusted through hole in flywheel, but flywheel must be removed to remove breaker points or condenser. Timing marks are aligned when piston is 20 degrees BTDC on Model G40. Model G65 is correctly aligned at 25 degrees BTDC.

Correct breaker point spring tension is 700-900 gr. (25-32 oz.). Condenser capacitance is 0.24 microfarads ($\pm$10%).

Coil output is adequate if spark length is 8 mm (0.32 in.) at 300 rpm. Renewal of coil should be considered if spark length falls below 7 mm (0.27 in.).

**LUBRICATION.** Crankcase capacity of Model G40 is 1.2 pints. Model G65 requires 1.7 US pints. Recommended motor oil is API classification SF. Use SAE 20 or 20W for temperatures between 0°C (32°F) and 15°C (59°F). For colder temperatures, use SAE 10W and for temperatures over 15°C (59°F), use SAE 30 or SAE 40. SAE 10W-30 weight may be used in all seasons. Oil should be maintained between marks on filler plug dipstick if engine has this optional item.

**NOTE: Do not screw filler plug in when checking oil level.**

Engine oil should be changed after first 20 hours operation of a new or rebuilt engine and at 100 hour intervals thereafter unless extremely dusty conditions are encountered.

### REPAIRS

**CYLINDER HEAD.** The cylinder head can be removed after removing muffler, fuel tank, air cleaner, fan cover and the cylinder and head cooling shroud. When installing head, tighten cylinder head retaining screws in sequence shown in Fig. HN4-17 to torque of 21.5-23.5 N·m (16-17 ft.-lbs.).

**PISTON, PIN AND RINGS.** To remove the piston, the crankcase halves must be separated, then the piston together with connecting rod and crankshaft can be withdrawn from bottom. Refer to CRANKSHAFT section for removal procedure. The following specification data applies:

Piston pin bore in piston:
G40 Standard . . . . . 18.000-18.006 mm
(0.7086-0.7089 inch)
G65 Standard . . . . . 19.000-19.006 mm
(0.7480-0.7483 inch)

Fig. HN4-13A—Exploded view of carburetor used on later Model G65 engines.

1. Idle speed screw
2. "O" ring
3. Main air jet
4. Pilot air jet
5. Pilot jet
6. Fuel inlet valve
7. Nozzle
8. Float
9. Gasket
10. Float bowl
11. Screw
12. Gasket
13. Main jet

Fig. HN4-15—View of governor external parts. Governor control link rod is shown at (R).

Fig. HN4-14—As engine speed increases, governor weights (W) push the slider (S) out against governor rod.

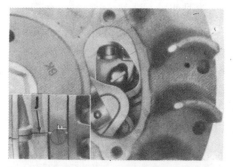

Fig. HN4-16—Ignition breaker point gap (timing) can be set through hole in flywheel. Inset shows timing "F" mark on side of flywheel aligned with joint (J) between halves of crankcase (muffler side).

Fig. HN4-17—Cylinder head retaining screws should be tightened in sequence shown. Make certain that screws (6, 7 & 8) are special type.

Pin clearance in bore:
  G40 & G65 .........0.012-0.070 mm
              (0.0005-0.0028 inch)
Piston pin bore in connecting rod:
  G40 Standard .....18.010-18.033 mm
              (0.7090-0.7100 inch)
  G65 Standard ....19.007-19.028 mm
              (0.7483-0.7491 inch)
Maximum bore I.D.
  G40 ....................18.08 mm
              (0.7118 inch)
  G65 ....................19.08 mm
              (0.7512 inch)
Cylinder bore diameter:
  G40 Standard .......66.00-66.01 mm
              (2.5984-2.5988 inch)
  G65 Standard ......72.00-72.01 mm
              (2.8346-2.835 inch)
Maximum bore I.D.,
  G40 ....................66.15 mm
              (2.6043 inch)
  G65 ....................72.15 mm
              (2.8405 inch)
Piston to cylinder,
  Clearance ...........0.07-0.10 mm
              (0.0028-0.0039 inch)
  Wear limit................0.25 mm
              (0.0098 inch)
Piston ring clearance in ring groove:
  Top ring ............0.02-0.06 mm
              (0.0008-0.0024 inch)
  Wear limit................0.15 mm
              (0.006 inch)
  Second ring .........0.01-0.05 mm
              (0.0004-0.0019 inch)
  Wear limit................0.15 mm
              (0.006 inch)
  Third (oil) ring ........0.01-0.05 mm
              (0.0004-0.0019 inch)
  Wear limit................0.15 mm
              (0.006 inch)
End gap (all rings) .........0.2-0.4 mm
              (0.008-0.016 in.)
May not exceed ..............1.5 mm
              (0.06 in.)

When assembling, manufacturer's marks on side of all piston rings should be toward top of piston. Pistons and rings are available in standard size and oversizes of 0.25, 0.50 and 0.75 mm (0.010, 0.020 and 0.030 in.). The top compression ring is chrome plated and the second compression ring is parkerized. Refer to the CRANKSHAFT section for installing crankshaft, rod and piston assembly and the lower half of crankcase. Tighten the cylinder head retaining screws in sequence shown in Fig. HN4-17 to torque of 21.5-23.5 N·m (16-17 ft.-lbs.).

## CRANKSHAFT AND CONNECTING ROD.
The connecting rod, crankshaft and piston can be removed after separating the crankcase halves. Connecting rod and crankpin bearing are removed by pressing crankshaft apart.

**NOTE: The crankshaft should be disassembled ONLY if required tools are available to correctly check and align the reassembled crankshaft.**

Crankshaft and connecting rod specifications:

Inside diameter, small end:
  G40 .............18.010-18.033 mm
              (0.709-0.710 inch)
  Max. I.D.................18.08 mm
              (0.718 inch)
  G65 .............19.007-19.028 mm
              (0.7483-0.7491 inch)
  Max. I.D.................19.08 mm
              (0.7512 inch)
Inside diameter, large end:
  G40 .............30.999-31.014 mm
              (1.2204-1.2210 inch)
  Max. I.D.................31.05 mm
              (1.222 inch)
  G65 .............35.569-35.589 mm
              (1.4003-1.4011 inch)
  Max. I.D.................35.64 mm
              (1.403 inch)
Crankpin O.D. specifications:
  G40 .............26.035-26.045 mm
              (1.0249-1.0253 inch)
  Min. O.D. ...............25.99 mm
              (1.0232 inch)
  G65 .............29.565-29.575 mm
              (1.1639-1.1643 inch)
  Min. O.D. ...............29.52 mm
              (1.162 inch)
Connecting rod end play:
  G40 ...............0.004-0.006 mm
              (0.00015-0.00024 inch)
  G65 ...............0.008-0.016 mm
              (0.0003-0.0006 inch)
Connecting rod side (axial) play:
  All ..................0.10-0.35 mm
              (0.004-0.014 inch)

Eccentricity at ends with crankshaft supported at main bearings should not exceed 0.2 mm (0.008 in.). The connecting rod, bearing and/or crankpin should be renewed if the top (piston pin) end of connecting rod rocks (shakes) back and forth more than 3.0 mm (0.12 inch).

To separate the crankcase halves, remove the muffler, carburetor, cooling shrouds, fan (flywheel), magneto and cylinder head. Remove the eight cap screws that attach the lower half of crankcase to the upper half and cylinder. Turn engine upside down and lift lower half of crankcase off locating dowels.

**NOTE: Be careful not to damage sealing surface between crankcase halves.**

When reassembling, install crankshaft with "O" marks on camshaft and crankshaft gears aligned. The set ring in top half of crankcase should correctly engage groove (G–Fig. HN4-18). Oil seals should be positioned on crankshaft before seating crankshaft in the top half of crankcase. No gasket is used between

crankcase halves and sealing surfaces must be smooth and flat. Coat mating surfaces of lower half of crankcase with a non-hardening sealer. The four 6 mm crankcase retaining screws should be torqued to 7.8-8.8 N·m (69-78 in.-lbs.) and the four 8 mm screws should be torqued to 19.6-20.6 N·m (14.5-15.2 ft.-lbs.); recheck several times to make certain that final torque is correct.

## CAMSHAFT AND TAPPETS.
The camshaft and cam followers (tappets) can be removed after separating the crankcase halves as outlined in the CRANKSHAFT section. The camshaft holder (center shaft) can be withdrawn from magneto end after removing the retaining screw. Refer to the following specification data:

Camshaft bore to holder clearance—
  Both ends, desired ...0.016-0.061 mm
              (0.0006-0.0024 inch)
  Wear limit.................0.1 mm
              (0.004 inch)
Camshaft end clearance in crankcase—
  Desired ..............0.1-0.5 mm
              (0.004-0.020 inch)
  Wear limit.................1.0 mm
              (0.04 inch)
Maximum lobe diameter—
  Wear limit................29.0 mm
              (1.14 inch)

Align "O" marks (Fig. HN4-18) on camshaft and crankshaft gears when reassembling. Refer to CRANKSHAFT section when installing lower crankcase half. Refer to VALVE SYSTEM section for adjusting valve clearance.

## VALVE SYSTEM.
Valves are removable after removing cylinder head and tappet cover. Valve face and seat angle for both valves is 45 degrees. Desired seat width is 0.7-1.0 mm (0.028-0.039 inch). Narrow seats using a 75 degree cutter if width exceeds 2.0 mm (0.08 inch). Valve clearance should be 0.07-0.1 mm (0.003-0.004 inch) for both inlet and exhaust valves.

Valve clearance should be checked when piston is at TDC on compression

*Fig. HN4-18—The "O" marks on camshaft timing gears must be aligned when assembling. Set ring in crankcase top half should completely engage groove (G) in bearing.*

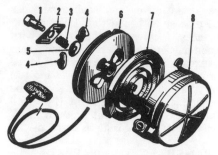

Fig. HN4-19—Exploded view of the rewind starter assembly. Starter cup (attached to engine flywheel) is not shown.

1. Special bolt
2. Friction plate
3. Friction spring
4. Ratchet pawls
5. Washer
6. Rope pulley
7. Rewind spring
8. Housing

Fig. HN4-20—Exploded view of rewind starter used on some models. Starter cup is attached to flywheel and not shown.

1. "E" ring
2. Washer
3. Pawls
4. Friction plate
5. Set spring
6. Friction spring
7. Washer
8. Pulley
9. Rewind spring
10. Housing
11. Starter rope

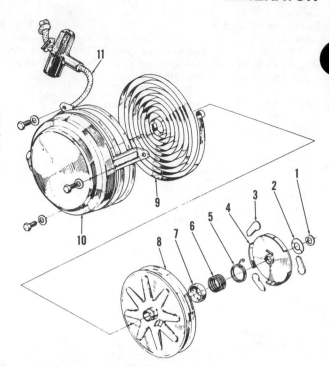

stroke. Clearance is increased by carefully grinding end of valve stem. To decrease clearance, reseat valve further in cylinder. Inlet and exhaust valve springs are interchangeable. Model G40 valve spring should be renewed if free length is less than 26 mm (1.02 in.). G65 valve springs should have a minimum free length of 30 mm (1.2 in.). Minimum pressure tests of valve springs are as follows.

G40, more than 5.9 kg at 22.5 mm
　　　(13 lbs. at 0.886 in.)
G65, more than 4.8 kg at 27.6 mm
　　　(10.6 lbs. at 1.087 in.)
G40, more than 12.5 kg at 16.7 mm
　　　(27.6 lbs. at 0.657 in.)
G65, more than 12.5 kg at 20.6 mm
　　　(27.6 lbs. at 0.812 in.)

Inlet valve heads are larger than those of exhaust valves by design. Valve stem to valve guide clearance of all engines should be 0.02 to 0.05 mm (0.0008-0.002 in.). Valve and/or valve guide should be renewed if clearance exceeds 0.12 mm (0.005 inch).

**REWIND STARTER.** Refer to Fig. HN4-19 or HN4-20 for an exploded view of rewind starter. To disassemble starter, remove rope handle and allow rope to wind into starter. Unscrew bolt (1—Fig. HN4-19) or detach "E" ring (1—Fig. HN4-20) and remove starter components while noting position of pawls. If removal of rewind spring is necessary, care should be used not to allow spring to uncoil uncontrolled.

Install rewind spring with coils wound in counter-clockwise direction from outer end. Wind starter rope on pulley in counter-clockwise direction as viewed with pulley in housing. Check operation of starter after assembly.

# KAWASAKI

**KAWASAKI MOTORS CORP., USA**
**9950 Jeronimo Road**
**Irvine, CA 92718-2016**

| Model | Output-kw | Voltage | Engine Make | Engine Model | Governed Rpm |
|---|---|---|---|---|---|
| GA1800A ...... | 1.6 | 110, 220 | Own | FG200 | 3600 |
| GA2300A ...... | 2.0 | 110, 220 | Own | FG200 | 3600 |
| GA3200A/AS ... | 2.9 | 110, 220 | Own | FG300 | 3600 |
| KG4000A/AS ... | 3.2 | 120, 240 | Own | KF82 | 3600 |
| KG5000A/AS ... | 4.0 | 120, 240 | Own | KF100 | 3600 |

# GENERATOR

## OPERATION

All models may be easily adjusted for 50 or 60 Hz operation by changing governed engine rpm. 50 Hz operation uses 3000 rpm and 60 Hz operation uses 3600 rpm. For Models GA1800, GA2300 and GA3200, refer to Fig. KA2-1A and position governor spring (2) as required. For Models KG4000 and KG5000, refer to Fig. KA2-1B and position adjustor plate so that desired frequency will appear in display window (3).

## MAINTENANCE

Brushes and brush holder (8—Fig. KA2-1) are accessible after removing generator end cover (3). Renew brushes if they are worn to a length of 6.5 mm (¼ inch) or less. Slip ring surfaces must be clean and bright. Polish slip rings with fine sandpaper, if necessary. When generator end frame (5) is removed, inspect bearing for smooth operation and renew, if needed.

**Fig. KA2-1A — Reposition governor spring as shown to adjust frequency to either 50 or 60 Hz on Models GA1800, GA2300 and GA3200.**
1. Lower hole (50 Hz)
2. Governor spring
3. Upper hole (60 Hz)

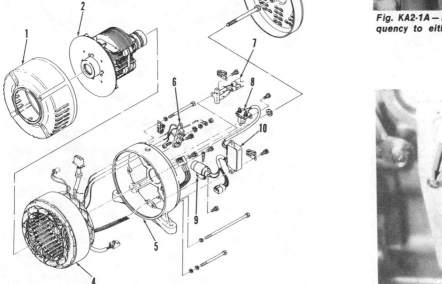

**Fig. KA2-1 — Exploded view of generator.**

| | |
|---|---|
| 1. Engine adapter | 6. Terminal block |
| 2. Rotor | 7. Diode assy. |
| 3. End cover | 8. Brush holder |
| 4. Stator | 9. Capacitor |
| 5. End frame | 10. Voltage regulator |

**Fig. KA2-1B — View of sliding plate used to adjust frequency to either 50 or 60 Hz on Models KG4000 and KG5000.**
1. Frequency adjuster nut
2. Lock screw
3. Display window
4. Fine tune adjuster

## TROUBLESHOOTING

If little or no output is generated, check the following: Check circuit breakers and reset if needed. Engine must be in good condition and maintain a governed speed under load of 3600 rpm for 60 Hz operation. All wiring connections must be clean and tight. Brushes and slip rings must be in good condition.

It is not necessary to completely disassemble these generators in order to perform most of the electrical tests. Removing end cover (3–Fig. KA2-1) will gain access to most of the stator wiring, brushes and voltage regulator.

To check stator main coils, disconnect the three wires from terminal block as shown in Fig. KA2-2. Refer to appropriate chart shown in Fig. KA2-3 or Fig. KA2-4 and measure resistance between terminals as shown. Check for continuity between each of the stator main coil leads and generator frame. Ohmmeter reading should be infinite when checking for grounds. If stator coil is grounded to generator frame or if inter-terminal resistance readings vary widely from specified resistance values shown in the charts, stator is defective and must be renewed. Refer to Fig. KA2-5 and measure resistance of stator sub coil as shown. Specified stator sub coil resistance in ohms is as follows:

| MODEL | OHMS |
|---|---|
| GA1800 | 2.4 |
| GA2300 | 1.8 |
| GA3200 | 1.9 |
| KG4000 | 1.7 |
| KG5000 | 1.5 |

**Fig. KA2-2 – Test main stator coil as shown. Refer to charts for specified resistance values.**

A. Test leads
B. Black
C. White
D. Red

### Main Coil Resistance

| Model | Circuit | Wire Color Connections | Meter Reading (Ω) |
|---|---|---|---|
| KG4000 | 120 V | Black-White | 0.5 |
| | 240 V | Black-Red | 1.5 |
| KG5000 | 120 V | Black-White | 0.35 |
| | 240 V | Black-Red | 1.2 |

**Fig. KA2-4 – Specified resistance values for stator main coil on Models KG4000 and KG5000.**

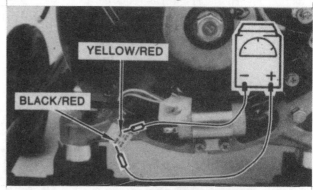

**Fig. KA2-5 – Test stator sub coil (exciter coil) as shown. Refer to text for resistance values.**

| Model | Circuit | | Wire color connections | Meter Reading ★ |
|---|---|---|---|---|
| GA1800 | Single voltage | 110, 120V | Blue-Red | 1 Ω |
| | | 220, 240V | Blue-Red | 3.8 Ω |
| | Dual voltage | 115V | Black-White | 1.6 Ω |
| | | 230V | Black-Red | 4.9 Ω |
| GA2300 | Single voltage | 110, 120V | Blue-Red | 0.6 Ω |
| | | 220, 240V | Blue- Red | 2.4 Ω |
| | Dual voltage | 115V | Black-White | 1 Ω |
| | | 230V | Black-Red | 3 Ω |
| GA3200 | Single voltage | 110, 120V | Blue-Red | 0.4 Ω |
| | | 220, 240V | Blue-Red | 1.6 Ω |
| | Dual voltage | 115, 120V | Black-White | 0.6 Ω |
| | | 230, 240V | Black-Red | 1.8 Ω |

★ Typical resistance within normal range.

**Fig. KA2-3 – Specified resistance values for stator main coil on Models GA1800, GA2300 and GA3200.**

**Fig. KA2-6 – Test rotor as shown. Refer to text for specified resistance values.**

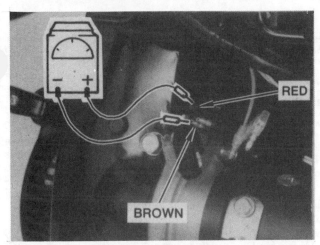

Fig. KA2-7 — View showing exciter coil testing procedure on Models KG4000 and KG5000. Approximate resistance values in ohms are 370 for Model KG4000 and 260 for Model KG5000.

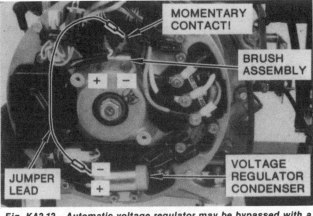

Fig. KA2-12 — Automatic voltage regulator may be bypassed with a jumper lead as shown. Refer to text for test procedure.

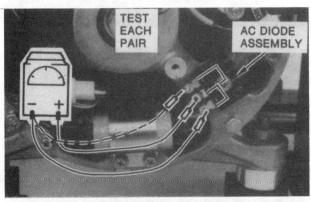

Fig. KA2-8 — View showing AC voltage to diode test procedure. Refer to text.

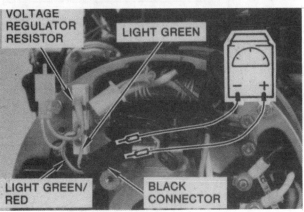

Fig. KA2-9 — View showing hookup procedure for testing voltage regulator resistor. Refer to charts for resistance values.

| Model | Meter Reading (Ω) | | | | | |
|---|---|---|---|---|---|---|
| | Single Voltage | | | | Dual Voltage | |
| | 110V | 120V | 220V | 240V | 115/230V | 120/240V |
| GA1800 | 108 | 184 | | 108 | 108 | 184 |
| GA2300 | to | to | None | to | to | to |
| GA3200 | 162 | 276 | | 162 | 162 | 276 |

Fig. KA2-10 — Chart showing voltage regulator resistor resistance values for Models GA1800, GA2300 and GA3200.

| Model | Meter Reading (Ω) |
|---|---|
| KG4000 | 230 |
| KG5000 | 230 |

Fig. KA2-11 — Chart showing voltage regulator resistor resistance values for Models KG4000 and KG5000.

Resistance values taken between either of the brush holder terminals and the generator or rotor frame should be infinite. Rotor winding resistance readings may also be taken with rotor removed by placing ohmmeter leads directly on slip rings. If ohmmeter readings vary widely from those specified, then shorted or open field windings are indicated and rotor must be renewed. Refer to Fig. KA2-7 and test exciter coil on Models KG4000 and KG5000. Approximate resistance values in ohms are 370 for Model KG4000 and 260 for Model KG5000.

AC output voltage to diodes may be measured as shown in Fig. KA2-8. Run engine at 3600 rpm. Connect an AC voltmeter between center wire and each outside wire as shown. Specified voltage is 19-34V AC for models GA1800, GA2300 and GA3200. Specified voltage is 29-44V AC for models KG4000 and KG5000.

To test diode assembly, unbolt diode assembly from end frame and unsolder leads. Connect one ohmmeter lead to center terminal of diode and the other lead to one of the outside terminals. Note ohmmeter reading. Ohmmeter should either indicate the specified resistance value or infinity. Reverse leads and reading should be opposite. Repeat the procedure for the other outside diode terminal. If entire diode is good, ohmmeter should show the specified resistance value at each of the two outside terminals during only one of the two test sequences. If other results are obtained, diode is defective and should be renewed. Specified diode resistance is 5.6-8.4 ohms for Models GA1800, GA2300 and GA3200. Specified diode resistance is 7 ohms for Models KG4000 and KG5000.

To test voltage regulator resistor, refer to Fig. KA2-9 and connect ohmmeter leads as shown. Specified resistance values are shown in Fig.

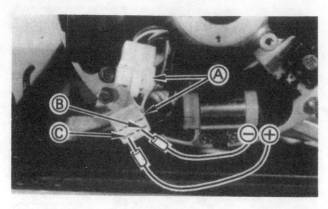

*Fig. KA2-13—View showing test lead hookup for voltage regulator sampling circuit of stator. Refer to chart for specified resistance values for Models GA1800, GA2300 and GA3200. Specified resistance value for Models KG4000 and KG5000 is 0.1 ohm.*

A. Voltage regulator connector
B. Blue
C. Lt. Green

KA2-10 for Models GA1800, GA2300 and GA3200. Specified resistance values are shown in Fig. KA2-11 for Models KG4000 and KG5000. Renew resistor if resistance values vary widely from those specified.

Automatic voltage regulator may be bypassed momentarily to check for proper operation. Refer to Fig. KA2-12. With an AC voltmeter connected to receptacle and with generator running at 3600 rpm, the following problems are indicated when the voltage regulator bypass procedure is used: If voltage jumps to 157-199V AC when jumper is attached and falls to 119-135V AC when jumper is removed, weakened magnets in rotor is indicated. If voltage drops to 95V AC or below when jumper is attached, a layer short in main stator coil is indicated. If voltage falls to near zero when jumper is removed, a faulty voltage regulator is indicated. If AC voltage stays very low when jumper wire is attached, a possible open or shorted winding in the rotor may be indicated.

| Model | | | Meter Reading ★ |
|---|---|---|---|
| GA1800 | Single voltage | 110V 120V | 0.2 Ω |
| | | 220V 240V | 0.3 Ω |
| | Dual voltage | 115/230V | 0.3 Ω |
| GA2300 | Single voltage | 110V 120V | 0.1 Ω |
| | | 220V 240V | 0.2 Ω |
| | Dual voltage | 115/230V | 0.2 Ω |
| GA3200 | Single voltage | 110V 120V | 0.1 Ω |
| | | 220V 240V | 0.1 Ω |
| | Dual voltage | 115/230V 120/240V | 0.1 Ω |

★ Typical resistance within normal range.

*Fig. KA2-14—Chart showing specified resistance values for voltage regulator sampling circuit of stator for Models GA1800, GA2300 and GA3200.*

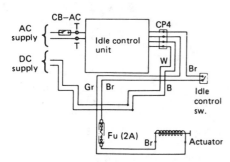

*Fig. KA2-14A—Wiring schematic for idle control circuit used on Models GA1800, GA2300 and GA3200.*

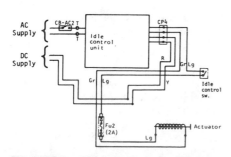

*Fig. KA2-14B—Wiring schematic for idle control circuit used on Models KG4000 and KG5000.*

**Wiring Code**

| Symbol | Name |
|---|---|
| AC (1P) out | AC outlet, single phase |
| Ⓑ | Brush |
| $C_1$ | Condenser |
| CB-AC1, 2 | AC circuit breaker |
| CB | DC circuit breaker |
| CP1 | 1 Pin coupler |
| CP2 | 2 Pin coupler |
| CP4 | 4 Pin coupler |
| CP6 | 6 Pin coupler |
| $D_1 \sim D_4$ | DC diodes |
| DC out | DC outlet |
| ENG.SW. | Engine switch |
| Fc | Field coil |
| Fu 1, 2 | Fuse |
| M | Starter motor |
| Mc (DC) | DC coil |
| Mc (1P) | Main coil, single phase |
| P | Pilot lamp |
| Ⓡ | Resistor |
| Ⓡ₁ | Resistor (variable) |
| Sc | Sub coil |
| Sp | Spark plug |
| T | Terminal (Screw type) |
| V | Voltage meter |

**Color Code**

| Symbol | Color |
|---|---|
| B | Black |
| Br | Brown |
| BR | Black/Red |
| BW | Black/White |
| BY | Black/Yellow |
| G | Green |
| Gr | Grey |
| L | Blue |
| Lg | Light green |
| Lg/B | Light green/Black |
| Lg/R | Light green/Red |
| Lg/W | Light green/White |
| R | Red |
| RW | Red/White |
| W | White |
| WL | White/Blue |
| WR | White/Red |
| WY | White/Yellow |
| Y | Yellow |
| YR | Yellow/Red |

*Fig. KA2-15—Wiring and color code charts for Models GA1800, GA2300, GA3200, KG4000 and KG5000.*

Test voltage regulator sampling circuit of stator as shown in Fig. KA2-13. If ohmmeter reading is not as specified, then stator must be renewed.

## OVERHAUL

Refer to Fig. KA2-1 for exploded view of generator. Remove battery and battery carrier, if so equipped. Remove muffler. Unbolt and remove end cover (3 – Fig. KA2-1). Unplug generator output connector and remove plug from wiring on generator side. On Models GA1800, GA2300 and GA3200, shut off fuel at fuel valve and disconnect fuel line at carburetor. Unbolt and remove fuel tank and control panel. Remove engine-to-frame and generator-to-frame mounting bolts. Remove generator/engine assembly from frame. On all models, pull output wiring through grommet on end frame (5). Disconnect wiring from terminal block as shown in Fig. KA2-2. Disconnect ground wire from end frame (5 – Fig. KA2-1). Disconnect and remove

brush holder assembly (8). Unplug voltage regulator and unbolt diode assembly (7) from end frame. On Models KG4000 and KG5000, remove end frame-to-generator frame mounting bolts and spacers. Support engine/generator assembly at rear of engine on all models. Remove stator through-bolts and end frame-to-stator mounting bolts. Carefully work end frame away from stator and rotor bearing. Using a soft hammer, tap stator away from engine adapter (1). Rotor through-bolt has right-hand threads; therefore, turn bolt counterclockwise for removal. Tap rotor with a soft hammer to unseat from crankshaft taper and remove.

# IDLE CONTROL CIRCUIT

Some models may be equipped with an idle control circuit which throttles engine back to idle speed when no AC

load is applied. The system used a two-position switch on the control panel. When switch is placed in "Auto" or "I" position, system is in operation and engine will idle until a load is applied. When switch is placed in "Off" or "O" position, engine will run at rated rpm constantly. The system also consists of a solid state idle control module and a throttle solenoid. If system is inoperative, first check the 2 amp fuse and renew, if needed. The solenoid may be tested for proper operation by connecting directly to a 12V battery. If fuse, switch, wiring and solenoid are all in good condition, it may be assumed that solid state control unit is at fault if unit is inoperative. Refer to wiring schematics shown in Figs. KA2-14A and KA2-14B.

# ENGINE

Engine model is listed at beginning of section. Refer to Kawasaki engine section for service procedures.

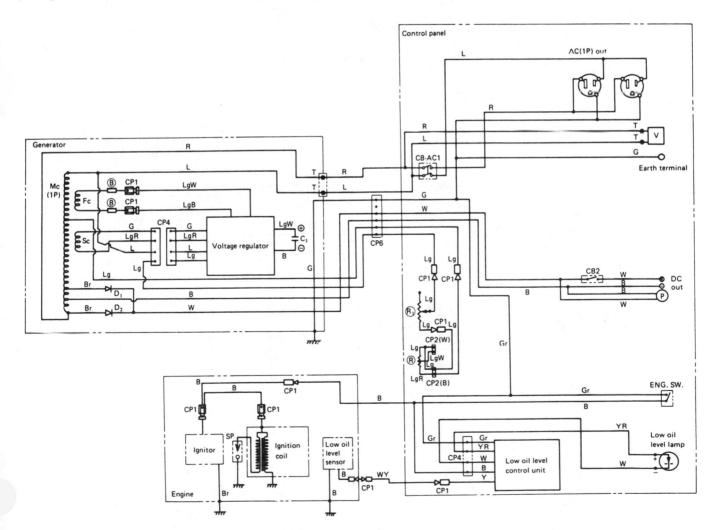

*Fig. KA2-16 — Wiring schematic for Model GA1800.*

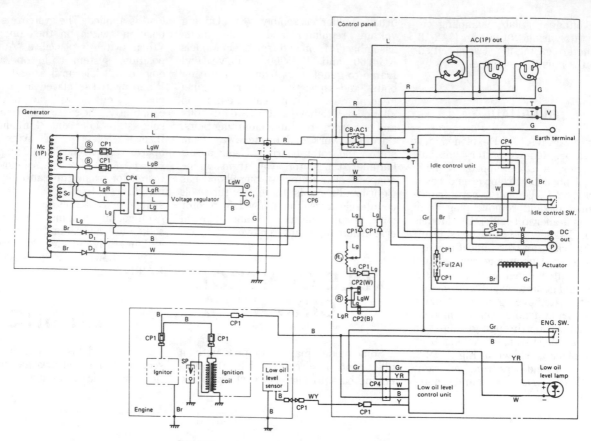

*Fig. KA2-17 — Wiring schematic for Model GA2300A.*

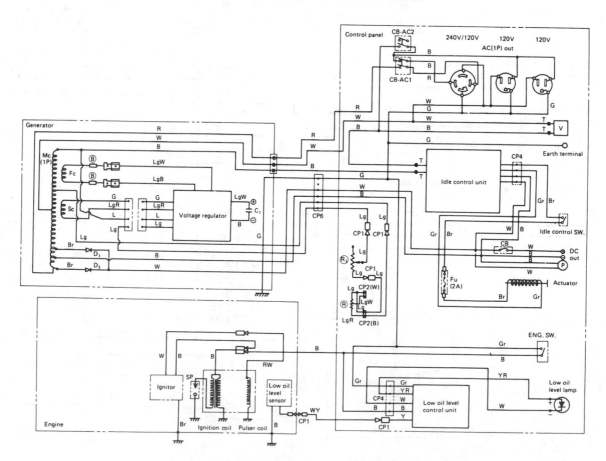

*Fig. KA2-18 — Wiring schematic for Model GA3200A (manual start).*

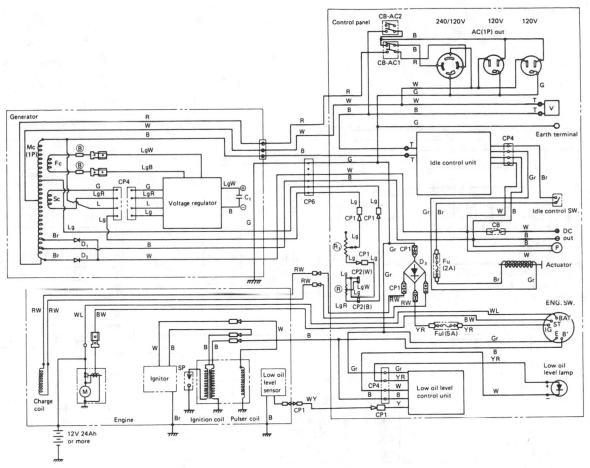

*Fig. KA2-19 — Wiring schematic for Model GA3200-S (electric start).*

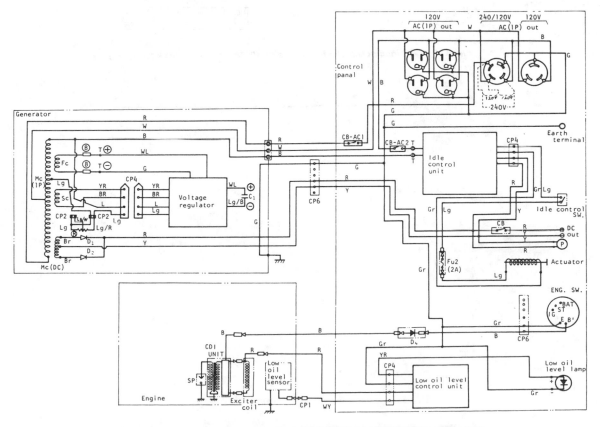

*Fig. KA2-20 — Wiring schematic for Models KG4000A and KG5000A (manual start).*

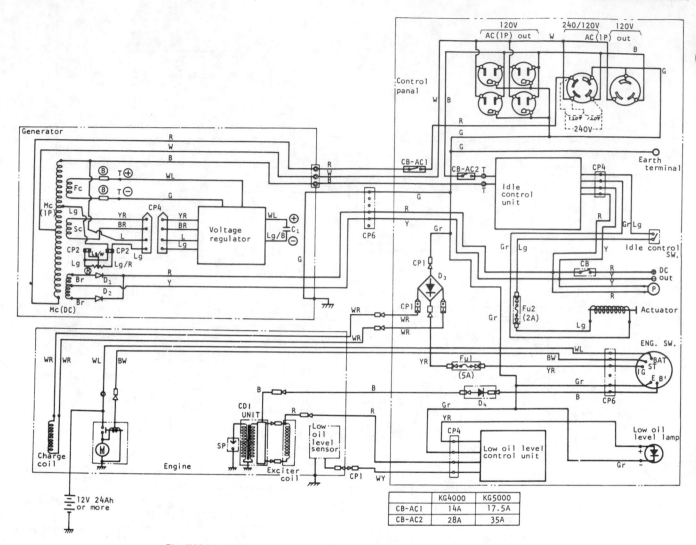

**Fig. KA2-21 — Wiring schematic for Models KG4000AS and KG5000AS (electric start).**

| | KG4000 | KG5000 |
|---|---|---|
| CB-AC1 | 14A | 17.5A |
| CB-AC2 | 28A | 35A |

# KAWASAKI

| Model | Output-watts | Voltage | Engine | | Governed Rpm |
| --- | --- | --- | --- | --- | --- |
| | | | **Make** | **Model** | |
| KG700A | 600 | 120 | Kawasaki | KF24 | 3600 |
| KG1000B | 900 | 120 | Kawasaki | KF34 | 3600 |
| KG1500B | 1300 | 120 | Kawasaki | KF53 | 3600 |
| KG2600B | 2300 | 120 | Kawasaki | KF64 | 3600 |

# GENERATOR

### MAINTENANCE

All bearings are sealed and require no lubrication, but should be inspected for smooth operation when generator is disassembled. Inspect unit periodically for condition of receptacles, electrical connections and operation of all switches.

### TROUBLE-SHOOTING

### All Models

**AC VOLTAGE SIDE.** Correct voltage reading at the AC receptacle with engine operating at specified speed, unloaded is 119-133 volts AC. Refer to Fig KA2-40, KA2-41 or KA2-42 for appropriate wiring diagram for model being serviced. To test area indicated during trouble-shooting, refer to the appropriate TEST section following TROUBLE-SHOOTING section.

No voltage output at the AC receptacle with engine running at specified speed and pilot lamp illuminated indicates a poor receptacle connection, blown fuse/faulty circuit breaker or an open main coil winding.

No or low voltage output at the AC receptacle with engine running at specified speed and pilot lamp not illuminated indicates an open rotor coil, worn brushes or poor contact, open circuit in the voltage regulator sensor coil or a faulty voltage regulator.

High voltage output at the AC receptacle with engine running at specified speed and continual burning out of pilot lamp indicates a faulty voltage regulator or a faulty sensor coil circuit.

Low voltage output at the AC receptacle with engine running at specified speed indicates a short in stator coil or the sensor coil.

Varying voltage output at the AC receptacle with engine running steadily at specified speed indicates an open capacitor in the voltage regulator or poor connector lead connection.

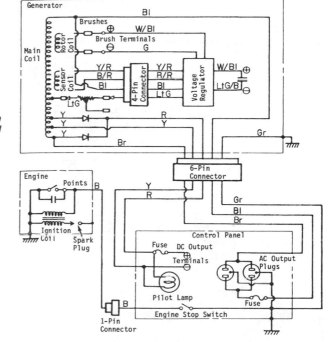

Fig KA2-40-Wiring diagram for Models KG700B and KG1000B.

B. Black
G. Green
R. Red
W. White
Y. Yellow
Bl. Blue
Br. Brown
Gr. Gray
Lg. Light green

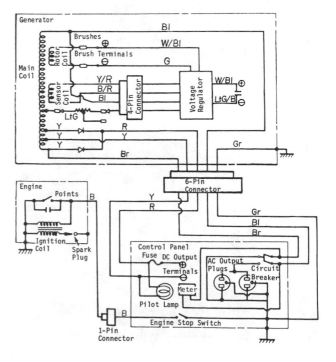

Fig. KA2-41-Wiring diagram for Models KG1500B. Refer to Fig. KA2-40 for color code.

Circuit breaker repeatedly tripping indicates either too heavy a load or a faulty circuit breaker.

**DC VOLTAGE SIDE.** Correct voltage reading at the DC receptacle with engine running at specified speed should be 12-18 volts DC. Refer to Fig. KA2-40, KA2-41 or KA2-42 for appropriate wiring diagram for model being serviced. To test area indicated during trouble-shooting, refer to the appropriate TEST section following TROUBLE-SHOOTING section.

No DC output at the DC receptacle with engine running at specified speed and pilot lamp illuminated indicates a tripped circuit breaker.

If AC output at the AC receptacle is normal but no DC output is at the DC receptacle and pilot lamp is not illuminated indicates a faulty diode assembly or an open lead (one diode defective).

If no output is registered at the AC or DC receptacles, an open stator coil is indicated.

Fig. KA2-42—Wiring diagram for Model KG2600B. Refer to Fig. KA2-40 for color code.

## TEST

### All Models

Voltage regulator resistor resistance should be 260 ohms for all models. Resistance is checked at the leads from each end of the resistor as shown in Fig. KA2-43.

Voltage regulator may be bypassed momentarily to check for proper operation. With an AC voltmeter connected to a receptacle and with generator running with no load, momentarily connect a jumper wire to negative side of capacitor (green/black lead) and to negative brush terminal as shown in Fig. KA2-44.

**CAUTION: Do not connect jumper wire longer than 10 seconds, otherwise rotor windings will be damaged.**

The following problems are indicated when the voltage regulator bypass procedure is used: If voltage jumps to 157-199 volts AC when jumper is attached and falls to 119-135 volts AC when jumper is removed, weak magnets in rotor is indicated. If voltage drops to 95 volts AC or below when jumper is attached, a short in main stator coil is indicated. If voltage jumps to 157-199 volts AC when jumper is attached and falls to near zero when jumper is removed, a faulty voltage regulator is indicated. If AC voltage stays very low when jumper wire is attached, a possible open or shorted winding in the rotor may be indicated.

Rotor coil resistance should be 48-53 ohms for Model KG700B, 53-58 ohms for Model KG1000B, 60-65 ohms for Model KG1500B and 67-72 ohms for Model KG2600B. Resistance is measured between the two brush holder terminals with brush leads disconnected (Fig. KA2-45. If coil resistance is higher than specified but not infinite, brushes may

Fig. KA2-43—View showing ohmmeter test lead connections for testing voltage regulator resistor. Resistance should be 260 ohm for all models.

Fig. KA2-44—Voltage regulator may be bypassed with a jumper lead as shown. Refer to text for test procedure.

not be making good contact with slip rings. Clean slip rings and inspect brushes, then recheck. If resistance is either infinite or less than specified, rotor is faulty.

Stator pick-up coil resistance should be 265 ohms for all models. Disconnect voltage regulator connector and measure resistance between the light green and blue wires at the stator side of the connector.

Voltage regulator sensor coil resistance should be 4.8 ohms for Model KG700B, 3.9 ohms for Model KG1000B, 3.1 ohms for Model KG1500B and 1.8 ohms for Model KG2600B. Resistance is measured between the black/red and yellow/red wires at the stator side of connector (Fig. KA2-46).

Diodes (Fig. KA2-47) should have continuity in one direction only. To check the diodes, unsolder yellow lead from one of the diodes. Be careful not to use excessive heat as diode will be damaged. Check resistance through each diode lead to ground using an ohmmeter, then reverse ohmmeter test leads and again check resistance. The resistance should be very high in one direction and very low in the other. If either diode has the same resistance in both directions, renew diode assembly.

### OVERHAUL

Remove control panel mounting bolts and disconnect electrical connectors. Remove control panel, fuel tank and muffler. Remove governor end cover. Remove brush holder mounting screws (the wires may remain attached). Remove four through-bolts that hold generator together. Remove generator-to-frame mounting bolts, then tap generator housing mounting feet with a soft hammer to separate generator main housing from fan housing. Resistor, voltage regulator, capacitor and diode assembly may now be removed from generator housing. To separate stator housing from main housing, remove output connector cover and separate wires from the connector. Remove two screws attaching stator housing to main housing and separate the housings while feeding output wires through the grommet. To remove rotor from engine crankshaft, remove retaining bolt from end of rotor (bolt has left-hand threads). Thread a short bolt into end of rotor and use a suitable slide hammer puller to grip the bolt and pull rotor off the crankshaft. Unbolt and remove generator fan housing from engine if necessary.

Sealed bearings should be inspected for smooth operation when generator is disassembled. Use an ohmmeter to check diodes and coil windings for open, shorted or grounded circuits. Brushes should be renewed if worn shorter than 6.5 mm (1/4 inch). Make sure that brush holder springs hold brushes firmly in place against rotor slip rings.

To reassemble, reverse the disassembly procedure.

# ENGINE

Refer to the appropriate Kawasaki engine section for engine service.

Fig. KA2-45—Rotor coil resistance is checked at brush holder terminals with wire leads disconnected.

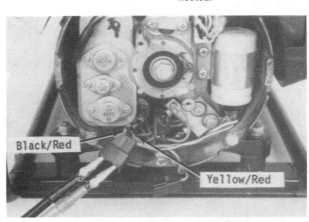

Fig. KA2-46—Voltage regulator sensor coil resistance is checked between black/red and yellow/red wires at stator side of voltage regulator connector.

Fig. KA2-47—Diodes should show continuity in one direction only. Diodes may be checked using ohmmeter or battery powered test light.

# KAWASAKI

| Model | Output-watts | Voltage | Engine | | Governed Rpm |
| --- | --- | --- | --- | --- | --- |
| | | | Make | Model | |
| KG550A | 450 | 120 | Kawasaki | FA76 | 3600 |
| KG550B | 450 | 120 | Kawasaki | FA76 | 3600 |
| KG750A | 600 | 120 | Kawasaki | KF24 | 3600 |
| KG750B | 600 | 120 | Kawasaki | KF24 | 3600 |
| KG1100A | 900 | 120 | Kawasaki | KF34 | 3600 |
| KG1100B | 900 | 120 | Kawasaki | KF34 | 3600 |
| KG1600A | 1300 | 120 | Kawasaki | KF53 | 3600 |
| KG1600B | 1300 | 120 | Kawasaki | KF53 | 3600 |
| KG2900A | 2300 | 120/240 | Kawasaki | KF68 | 3600 |
| KG2900B | 2300 | 120/240 | Kawasaki | KF68 | 3600 |

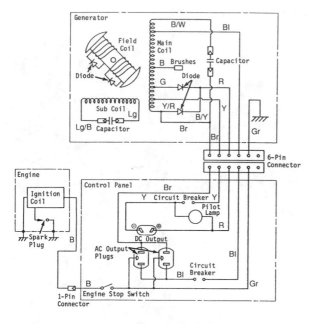

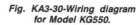

**Fig. KA3-30-Wiring diagram for Model KG550.**

- B. Black
- G. Green
- R. Red
- W. White
- Y. Yellow
- Bl. Blue
- Br. Brown
- Gr. Gray
- LtG. Light green

**Fig. KA3-31-Wiring diagram for Models KG750 and KG1100. Refer to Fig. KA3-30 for wiring color codes.**

# GENERATOR

### MAINTENANCE

All bearings are sealed and require no lubrication, but should be inspected for smooth operation when generator is disassembled. Inspect unit periodically for condition of receptacles, electrical connections and operation of all switches.

### TROUBLE-SHOOTING

### All Models

**AC VOLTAGE SIDE.** Correct voltage reading at the AC receptacle with engine operating at specified speed, unloaded is 119-133 volts AC. Refer to Fig KA3-30 through KA3-33 for appropriate wiring diagram for model being serviced. To test area indicated during trouble-shooting, refer to the appropriate TEST section following TROUBLE-SHOOTING section.

No voltage output at the AC receptacle with engine running at specified speed and pilot lamp illuminated indicates a poor receptacle connection, tripped or faulty circuit breaker or an open main coil winding.

No or low voltage output at the AC receptacle with engine running at specified speed and pilot lamp not illuminated indicates an open rotor coil, worn brushes or poor contact, open sub coil or automatic voltage regulator (AVR) sensor coil or a faulty AVR.

High voltage output at the AC receptacle with engine running at specified speed and continual burning out of pilot lamp indicates a faulty AVR or a faulty AVR sensor coil.

Low voltage output at the AC receptacle with engine running at specified speed indicates a short in stator coil (including sub coil) or an open rotor field on Model KG550.

Varying voltage output at the AC receptacle with engine running steadily at specified speed indicates an open AVR capacitor or poor connector connection.

Circuit breaker repeatedly tripping indicates either too heavy a load or a faulty circuit breaker.

**DC VOLTAGE SIDE.** Correct voltage reading at the DC receptacle with engine running at specified speed should be 12-18 volts DC. Refer to Fig. KA3-30 through KA3-33 for appropriate wiring diagram for model being serviced. To test area indicated during troubleshooting, refer to the appropriate TEST section following TROUBLE-SHOOTING section.

No DC output at the DC receptacle with engine running at specified speed with pilot lamp illuminated indicates a tripped circuit breaker.

If AC output at the AC receptacle is normal but no DC output is at the DC receptacle and pilot lamp is not illuminated indicates a faulty diode assembly or an open lead (one diode defective).

If no output is registered at the AC or DC receptacles, an open stator coil is indicated.

## TEST

### Model KG550

Main coil resistance should be 2.4 ohms. To check, disconnect output connector and connect ohmmeter test leads to the blue and brown leads of male connector (Fig. KA3-34).

**IMPORTANT: Discharge the condensers before and after testing to avoid possible electrical shock and to ensure accurate ohmmeter test results.**

To check condensers, first disconnect wires from condensers (Fig. KA3-35). Connect ohmmeter test leads to condenser wires. Meter needle should swing into range of 20K ohms to 60K ohms, then quickly drop back to infinity. Reverse tester leads and repeat test. The same results should be obtained as first test. Renew condenser if meter needle does not move to specified range, or if needle moves up but does not drop back to infinity.

Sub coil resistance reading should be 10.4 ohms with ohmmeter test leads connected at the light green/black and light green leads (Fig. KA3-35).

With ohmmeter test leads connected to ends of each rotor diode (Fig. KA3-36), field coil resistance reading should be 14 ohms with ohmmeter leads connected one way and 5 ohms with leads reversed.

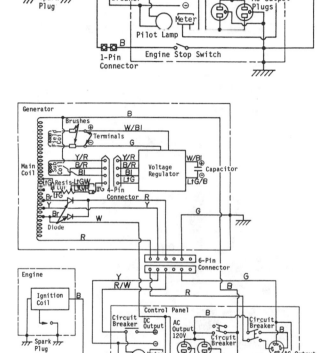

*Fig. KA3-32-Wiring diagram for Model KG1600. Refer to Fig. KA3-30 for wiring color codes.*

*Fig. KA3-33-Wiring diagram for Model KG2900. Refer to Fig. KA3-30 for wiring color codes.*

*Fig. KA3-34—Check main coil resistance at blue and brown wires in male connector.*

With field coil lead disconnected from one end of rotor diode, resistance of diode should be 5 ohms in one direction and infinite with ohmmeter leads reversed. If rotor diode requires renewal, be sure to note position of yellow band around one end of ceramic portion of diode and install new diode with yellow band facing in the same direction. Be careful when soldering field coil leads to diodes as excessive heat will damage diode.

To check DC diode assembly, unsolder the green, red and yellow/red wires from diode. Connect ohmmeter test leads to center post of diode and to one outside terminal on diode assembly. Note meter reading then reverse test leads. DC diode resistance should be 7 ohms in one direction and infinite with ohmmeter leads reversed. Repeat test with ohmmeter test leads connected to center terminal and to terminal on other end of diode assembly. Renew diode assembly if either diode tests faulty.

### All Models Except Model KG550

If AC voltage is lower or higher than specified, check resistor in voltage regulator circuit as follows: Disconnect leads from each end of resistor (Fig. KA3-3[ ]) and check resistance between the tw[ ] end leads of resistor. Resistance should be 265 ohms.

Voltage regulator may be bypassed momentarily to check for proper operation. With an AC voltmeter connected to output receptacle and with generator running with no load, momentarily connect a jumper wire to negative side of condenser (green/black lead) and to negative brush terminal as shown in Fig. KA3-38.

**CAUTION: Do not connect jumper wire longer than 10 seconds, otherwise rotor windings will be damaged.**

The following problems are indicated when the voltage regulator bypass procedure is used: If voltage jumps to 157-199 volts AC when jumper is attached and falls to 119-135 volts AC when jumper is removed, weak magnets in rotor is indicated. If voltage drops to 95 volts AC or below when jumper is attached, a short in main stator coil is indicated. If voltage jumps to 157-199 volts AC when jumper is attached and falls to near zero when jumper is removed, a faulty voltage regulator is indicated. If AC voltage stays very lo[ ] when jumper wire is attached, a possible open or shorted winding in the rotor may be indicated.

To check main coil resistance, connect ohmmeter test leads to brown and blue wires in output connector. Main coil resistance reading should be 2.5 ohms for Model KG750, 1.3 ohm for Model KG1100, 0.7 ohm for Model KG1600 and 0.6 ohm for Model KG2900.

Rotor coil resistance may be checked by connecting ohmmeter test leads to brush holder terminal posts (with brush post wires disconnected) as shown in Fig. KA3-39. Resistance reading should be 48-53 ohms for Model KG750, 53-58

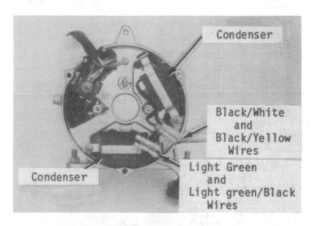

*Fig. KA3-36—Field coil resistance may be checked at ends of rotor diodes.*

*Fig. KA3-35—View of Model KG550 generator showing correct wiring connections to condensers.*

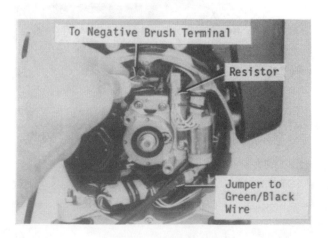

*Fig. KA3-38—On all models except KG550, operation of voltage regulator can be checked by momentarily connecting jumper wire between green/black wire and negative brush terminal while measuring AC voltage output. Refer to text.*

*Fig. KA3-39—Rotor coil resistance can be measured between the two brush terminals.*

ohms for Model KG1100, 60-65 ohms for Model KG1600 and 67-72 ohms for Model KG2900. If rotor coil resistance is higher than specified but not infinite, check condition of rotor slip rings and brushes. Renew brushes if shorter than 6.5 mm (1/4 inch). Be sure that springs hold brushes firmly against slip rings.

Stator pick up coil resistance, measured at the light green wire and blue wire at the stator side (male end) of the voltage regulator connector, should be 265 ohms for all models.

Voltage regulator sensor coil resistance is measured at black/red wire and yellow/red wire at stator side of voltage regulator connector. Resistance should be 4.8 ohms for Model KG750, 3.9 ohms for Model KG1100, 3.1 ohms for Model KG1600 and 1.8 ohms for Model KG2900.

To check DC diode assembly, disconnect the three wires from diode (Fig. KA3 40). Connect ohmmeter test leads to center post of diode and to one outside terminal on diode assembly. Note meter reading then reverse test leads. DC diode resistance should be 7 ohms in one direction and infinite with ohmmeter leads reversed. Repeat test with ohmmeter test leads connected to center terminal and to terminal on other end of diode assembly. Renew diode assembly if either diode tests faulty.

## OVERHAUL

To disassemble generator unit, first remove generator control panel assembly. On all models except KG550, remove governor control plate. Remove muffler cover, governor arm (Models KG1600 and KG2900) and muffler. Remove the fuel tank. Remove generator end cover. On all models except KG550, remove the brush holder retaining screws (the wires may remain attached). On all models, remove through-bolts that hold generator together. On Models KG1600 and KG2900, remove bolts retaining generator to the frame. Remove the stator assembly and end housing as an assembly. Do not separate end housing from stator assembly until components are removed from end housing.

On Model KG550, remove mounting screws from condensers (4—Fig. KA3-41) and diode assembly (8). Separate stator assembly (9) from end housing (7). On Models KG750, KG1100, KG1600 and KG2900, remove resistor (7—Fig. KA3-42), voltage regulator (8) and condenser (4) mounting screws. Disconnect output connector and remove the wiring pins from the male connector. Remove two screws retaining stator (12) and end housing (11) together, then separate stator from end housing while feeding output wires through grommet in end housing. To remove rotor (2—Fig. KA3-41 or

KA3-42) on all models, remove retaining bolt (3). Note that bolt has left-hand threads on all models except KG550. Thread a short 10 mm bolt into end of rotor, then attach suitable puller to head of bolt and pull rotor off engine crankshaft. Unbolt and remove generator fan housing (1) from engine if necessary.

Sealed bearings should be inspected for smooth operation when generator is disassembled.

To reassemble generator, reverse the disassembly procedure while noting the following special instructions: To install rotor on crankshaft, first turn crankshaft until piston is at top dead center (either compression or exhaust stroke). Position rotor on crankshaft so that an imaginary line (Fig. KA3-43) drawn through rounded sides (poles) of rotor is aligned with marks cast in the fan housing. Rotor retaining bolt should be

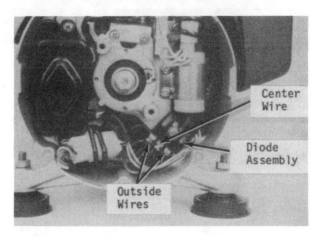

Fig. KA3-40—View of DC diode assembly. Refer to text for test procedure.

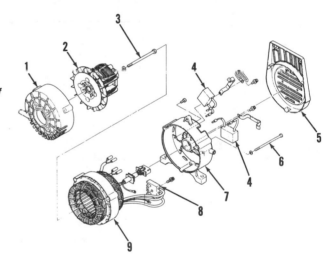

**Fig. KA3-41-Exploded view of Model KG550 generator.**

1. Fan housing
2. Rotor assy.
3. Rotor bolt
4. Condensers
5. End cover
6. Through-bolt
7. Main housing
8. Diode assy.
9. Stator assy.

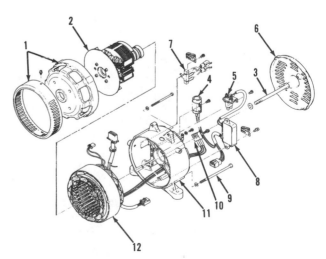

**Fig. KA3-42-Exploded view of Models KG750, KG1100, KG1600 and KG2900 generator.**

1. Fan housing assy.
2. Rotor assy.
3. Rotor bolt
4. Condenser
5. Brushes
6. End cover
7. Resistor
8. Voltage regulator
9. Through-bolt
10. Diode assy.
11. Main housing
12. Stator assy.

tightened to specified torque after governor housing is installed. Specified rotor bolt torque is 8 N·m (6 ft.-lbs.) on Model KG550; 19 N·m (14 ft.-lbs.) on Models KG750, KG1100 and KG1600; 38 N·m (28 ft.-lbs.) on Model KG2900. Refer to appropriate Fig. KA3-44 or KA3-45 for correct installation of output wiring in connector plug. Be sure that the pins snap into place in connector plug.

On Model KG550, be sure to connect condensers to correct wires as shown in Fig. KA3-35.

# ENGINE

Refer to the appropriate Kawasaki engine section for engine service.

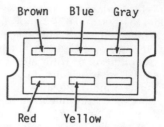

Fig. KA3-44—On all models except KG2900, wires should be positioned in output connector as shown.

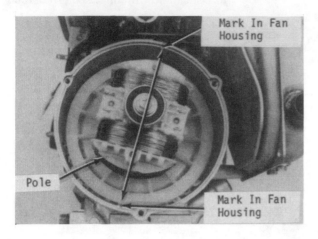

Fig. KA3-43—To "time" rotor to engine, position piston at TDC and align rotor poles with marks in fan housing.

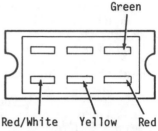

Fig. KA3-45—On Model KG2900, wires should be positioned in output connector as shown.

# KOHLER

**Kohler, Wisconsin 53044**

| Model | Output-kw | Voltage | Engine Make | Engine Model | Governed Rpm |
|---|---|---|---|---|---|
| 1.75MBM25 | 1.75 | 120 | B&S | 100232 | 3600 |
| 1.75MM26 | 1.75 | 120 | Kohler | K91 | 3600 |
| 2.25MBM25 | 2.25 | 120 | B&S | 130232 | 3600 |
| 3.5CM65 | 3.5 | 120/240 | Kohler | K181 | 3600 |
| 3.5MBM25 | 3.5 | 120 | B&S | 190412 | 3600 |
| 3.5MBM65 | 3.5 | 120/240 | B&S | 190412 | 3600 |
| 3.5MM25 | 3.5 | 120 | Kohler | K181 | 3600 |
| 3.5MM65 | 3.5 | 120/240 | Kohler | K181 | 3600 |
| 5CM65 | 5.0 | 120/240 | Kohler | K301 | 3600 |
| 5MBM25 | 5.0 | 120/240 | B&S | 252412 | 3600 |
| 5MBM65 | 5.0 | 120/240 | B&S | 252412 | 3600 |
| 5MM25 | 5.0 | 120/240 | Kohler | K301 | 3600 |
| 5MM65 | 5.0 | 120/240 | Kohler | K301 | 3600 |

# GENERATOR

**WARNING: Prior to any servicing involving powerplant wiring, depress Power Monitor button on generator power panel to discharge capacitor in generator circuit. Failure to discharge capacitor may produce an electrical shock.**

## MAINTENANCE

Manufacturer recommends inspecting and servicing, if necessary, generator brushes and slip rings after every 100 hours of operation. Refer to OVERHAUL section and remove end frame for access to brushes and slip rings. Be sure outer brush lead is connected to positive (+) regulator terminal and inner brush lead is connected to negative (-) regulator terminal.

The rotor shaft is supported in end frame by a ball bearing which does not normally require attention. However, bearing should be inspected and repacked with good quality high temperature grease whenever end frame is removed.

## TROUBLESHOOTING

Use the following troubleshooting sequence when an electrical malfunction occurs: Be sure fuse(s) and circuit breaker(s) are good. Check operation of Power Monitor switch. When depressed to start generator, Power Monitor switch discharges a capacitor through slip rings into rotor to excite field coils thereby activating generator. Power Switch button light should glow during generator operation. Refer to Figs. KO1-1 and KO1-2.

Detach power panel from end frame with all leads attached then place a rubber sheet or other insulating material between panel and end frame. Connect a jumper wire between panel and end frame. Reach inside end frame and disconnect brush leads from voltage regulator terminals. Refer to Fig. KO1-3 and connect a 12 volt battery and DC ammeter to brush leads as shown with

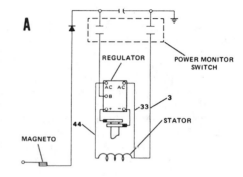

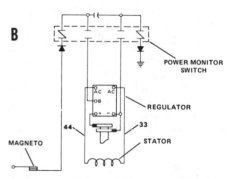

Fig. KO1-2—Wiring schematic for six-terminal power monitor switch (View A) and eight-terminal power monitor switch (View B).

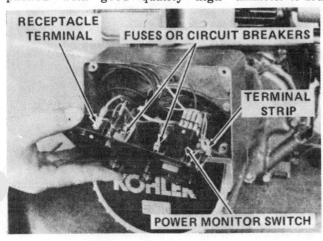

Fig. KO1-1—View of power panel detached from end frame.

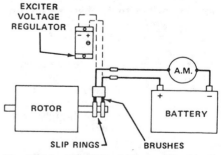

Fig. KO1-3—Connect a battery and ammeter (A.M.) as shown and refer to text for troubleshooting.

outer brush lead connected to positive battery terminal. Ammeter should read 1.3-2.0 amps on 1.75 and 2.25 kw units, 1.1-1.3 amps on 3.5 kw units or 1.0-1.1 amps on 5 kw units. If ammeter reading is incorrect, check rotor as outlined in subsequent paragraph.

With auxiliary battery connected as outlined in previous paragraph, run powerplant without load and check current frequency at receptacles. Current frequency should be 60 hertz (cycles). Measure voltage at receptacle. On units with only a 240 volt receptacle, measure between ground and one output slot. If voltage reading is 100 volts or more, malfunction is located in excitation circuit or exciter voltage regulator. Refer to following paragraph. If voltage reading is less than 100 volts, inspect rotor and stator as outlined in subsequent paragraph. Disconnect auxiliary battery and reconnect brush leads to regulator.

If not presently removed, refer to previous paragraph and detach power panel. Disconnect red wire lead from terminal strip (Fig. KO1-1) and connect an AC voltmeter to wire and engine or generator ground. Red wire supplies voltage from engine magneto to charge Power Monitor capacitor. With engine running, voltmeter should indicate approximately two volts. If little or no voltage is present, check engine magneto circuit. Reconnect red wire lead. If sufficient voltage is present at red wire, depress Power Monitor switch to discharge capacitor and connect ohmmeter leads to red and blue wire terminals on terminal strip (Fig. KO1-1) to check diode. Note ohmmeter reading, reverse leads and observe second reading. Diode should conduct in one direction only. If diode conducts in both or neither direction, renew diode being sure leads are connected properly. If diode is not faulty, proceed to following paragraph.

Using a DC voltmeter, ground black lead to end frame and connect red meter lead to white wire lead of capacitor. With engine running, voltmeter should indicate at least 40 volts. If less than 40 volts is indicated, depress Power Monitor switch to discharge capacitor then renew terminal strip. If 40 volts or more is present, proceed to following paragraph.

Refer to OVERHAUL section, detach end frame and remove regulator from end frame. To test regulator, refer to Fig. KO1-4 and construct circuit shown using two 100 watt lightbulbs with sockets, a 20 amp switch and two amp fuse with holder. Connect a DC voltmeter as shown in schematic and obtain current from a source of 110/120 VAC. With switch closed, lightbulbs

should glow immediately and voltmeter should read 10 to 50 volts. However, bulbs may flicker or glow dimly depending on strength of test current. Momentarily touch a jumper wire between "B" and either "AC" terminal. Bulbs should glow brighter and voltmeter reading should be 50 to 75 volts. If desired voltage readings are obtained, regulator is good. Otherwise, regulator must be renewed. Note wiring diagram in Fig. KO1-2.

To check rotor or stator, refer to OVERHAUL section and detach end frame. Be sure slip rings are clean then use an ohmmeter to check resistance between slip rings. Resistance should be 6-9 ohms on 1.75 and 2.25 kw units, 9-10 ohms on 3.5 kw units or 11-12 ohms on 5 kw units. Low or infinite resistance indicates windings are shorted or broken and rotor must be renewed. Check for continuity between either slip ring and rotor shaft or pole. If continuity exists rotor must be renewed as coils are grounded.

Stator condition may be checked using an ohmmeter as follows: Note that stator leads are numbered. Resistance between 1 and 2 and leads 3 and 4 should be approximately 0.25 ohm. There should be continuity between leads 3, 4, 33 and 44. There should not be continuity from lead 1 to lead 3, 4, 33 or 44. There should not be continuity from lead 2 to lead 3, 4, 33 or 44. There should not be continuity from any stator lead to ground, stator frame or housing. If stator does not pass all of the above tests then stator is defective and must be renewed.

## OVERHAUL

To disassemble generator, detach power panel, disconnect interfering wires and set panel aside. Insert a wire through holes in brush holder (3–Fig. KO1-5) to hold brushes up and away from rotor slip rings. Place a support under drive end housing (12) then unscrew four through bolts in end frame (2). Remove end frame by pulling or tapping end frame off rotor shaft bearing (8) then disconnect stator leads numbered 33 and 44 connected to regulator (1). Remove stator assembly (6) being careful not to damage wiring or rotor. Rotor is threaded on end of engine crankshaft. To dislodge rotor, place a wood block against trailing edge of rotor pole as shown in Fig. KO1-6 then strike against block. Do not attempt to turn rotor using a wrench or other device.

Inspect components for mechanical failure and use testing procedure in TROUBLESHOOTING section to locate electrical faults. Rotor shaft ball bearing (8–Fig. KO1-5) should be repacked with high temperature grease. Turn rotor slip rings in a lathe if burnt or pitted.

To reassemble generator, clean threads of engine crankshaft then apply an anti-seize compound to threads. Screw rotor (9) onto crankshaft and tighten by striking with moderate blows against a wood block held against a rotor pole. Be sure bearing seat (7) is properly installed in end frame bearing bore. Remainder of assembly is reverse of disassembly procedure. A thin coating of thermal compound (Kohler

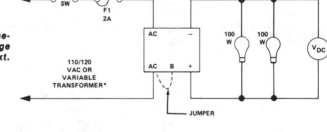

*Fig. KO1-4—Wiring schematic for testing voltage regulator. Refer to text.*

*Fig. KO1-5—Exploded view of generator. Spacer (13) and washer (14) are only used on 3.5 and 5 kw units with Briggs & Stratton engines.*

1. Voltage regulator
2. End frame
3. Brush holder
4. Spring
5. Brush
6. Stator assy.
7. Bearing seat
8. Bearing
9. Rotor
10. Rotor bolt
11. Fan
12. Housing
13. Spacer
14. Washer

Illustrations courtesy of Kohler Co.

287945) should be applied to regulator seating surface in end frame (2). Shims are available to adjust brush holder position so brushes are centered on slip rings. On models equipped with Econo-Throttle, be sure stator lead numbered 3 on 1.75 kw units or stator leads numbered 1 and 3 on 3.5 and 5 kw units are routed through Econo-Throttle transformer coil as shown in Fig. KO1-7. Refer to wiring diagrams in this section for proper connections. Tighten through bolts (10 – Fig. KO1-5) to 10 ft.-lbs. Be sure wire retaining brushes in brush holder is removed or regulator will be damaged if unit is run.

# ECONO-THROTTLE

## OPERATION

Units rated 1.75 kw with Kohler engines and all 3.5 kw and 5 kw units may be equipped with Econo-Throttle system. The Econo-Throttle system lowers engine speed to idle speed when electrical load is 40 watts or less. Elec-

trical load is sensed by a transformer coil which surrounds stator leads numbered 1 and 3 on 3.5 and 5 kw units or stator lead numbered 3 on 1.75 kw units. Transformer coil leads attach to a circuit board mounted on generator end frame. Electrical current is supplied to circuit board from the "AC" terminals of the voltage regulator. The circuit board then controls a solenoid which operates the carburetor throttle. Refer to Fig. KO1-9 for a typical wiring diagram of Econo-Throttle circuit.

## ADJUSTMENT

### Models With Briggs & Stratton Engine

Start and run engine until normal operating temperature is reached then switch on generator. Loosen screws (S – Fig. KO1-10) securing solenoid bracket then rotate throttle shaft counter-clockwise so throttle is against idle speed screw. Measure voltage at outlet and adjust idle speed so voltage is 60-75 volts. Power Monitor light should flicker or just go out.

**NOTE Do not run engine for a prolonged**

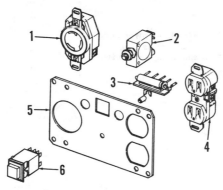

Fig. KO1-8—Exploded view of typical power panel.

1. Twist outlet
2. Circuit breaker
3. Terminal strip
4. Outlet
5. Panel
6. Power monitor switch

period at 2000 to 2700 rpm or overheating may result.

Hold governor arm in idle position then move solenoid bracket until it contacts governor control plate and tighten screws (S).

### Models With Kohler Engine

Start and run engine until normal operating temperature is reached then switch on generator. Loosen screws (S – Fig. KO1-11) securing plate to governor arm then push arm towards carburetor until throttle crank contacts idle speed screw. Note that plate is closer to solenoid than slotted flange on

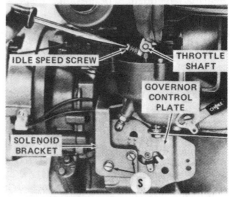

Fig. KO1-10—View of Econo-Throttle on models equipped with Briggs & Stratton engines.

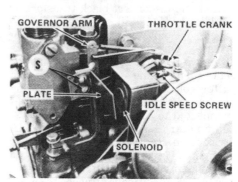

Fig. KO1-11—View of Econo-Throttle on models equipped with Kohler engines.

Fig. KO1-6—Hold a wood block against rotor pole (P) and strike block to loosen rotor.

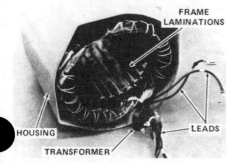

Fig. KO1-7—View of stator assembly. Note Econo-throttle transformer coil around stator leads.

Fig. KO1-9—Wiring schematic of Econo-Throttle circuit. Stator lead 3 on 1.75 kw units or stator leads 1 and 3 on 3.5 and 5 kw units are routed through Econo-Throttle transformer coil as shown in Fig. KO1-7.

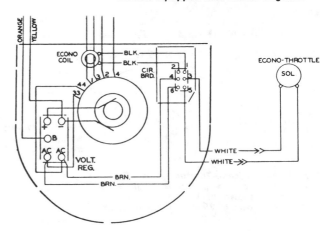

3.5 and 5 kw units while slotted flange is closer to solenoid than plate on 1.75 kw units. See Fig. KO1-12. Measure voltage at outlet and adjust idle speed so voltage is 55 volts on 1.75 kw units or 60-75 volts on 3.5 and 5 kw units. Power monitor light should flicker or just go out.

**NOTE: Do not run engine for a prolonged period at 2000 or 2700 rpm or overheating may result.**

Hold governor arm and throttle crank in idle position, push plate against solenoid and tighten screws (S).

## TROUBLESHOOTING

If Econo-Throttle does not operate properly, be sure powerplant is not overloaded during operation. Inspect governor and throttle linkage and all electrical connections. Adjust Econo-Throttle as outlined in previous section. Test solenoid, circuit board and transformer coil as follows:

Disconnect leads to solenoid and with a voltmeter, measure DC voltage in wires from circuit board. Voltage should be 35-50 volts with engine idling. If voltage is correct, renew solenoid. To check sensing transformer coil, remove end frame as outlined in OVERHAUL section. Disconnect transformer coil leads and using an ohmmeter measure coil resistance. Renew coil if resistance is not 53-63 ohms.

To test circuit board, detach power panel and disconnect black transformer coil leads from coil terminals. Disconnect solenoid leads in solenoid wiring harness. On some models, in-line connectors are not present and wires must be cut then in-line connectors installed after concluding tests. Using an ohmmeter check resistance of solenoid wires connected to circuit board then reverse ohmmeter leads and again check resistance. Resistance should be 150-330 ohms and 1900-2100 ohms. Using an ohmmeter check resistance of black transformer coil leads attached to circuit board. See Fig. KO1-13. Resistance

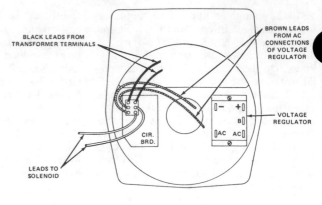

Fig. KO1-13—View of generator end frame showing wire leads to Econo-Throttle circuit board.

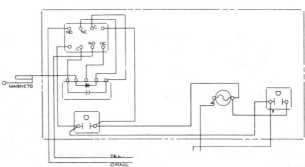

Fig. KO1-14—Power panel wiring diagram for early 1.75 and 2.25 kw models.

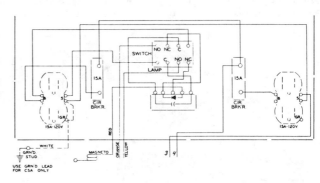

Fig. KO1-15—Power panel wiring diagrams for late 1.75 and 2.25 kw models.

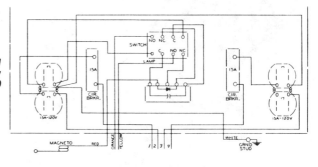

Fig. KO1-16—Power panel wiring diagrams for early 3.5 kw models with 120 volt output.

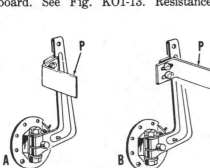

Fig. KO1-12—Econo-Throttle plate (P) on models equipped with Kohler engines is installed as shown in View A on 1.75 kw units or as shown in View B on 3.5 and 5 kw units.

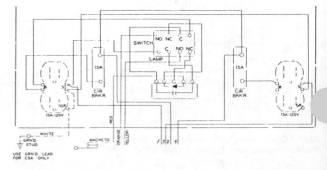

Fig. KO1-17—Power panel wiring diagram for late 3.5 kw models with 120 volt output.

Illustrations courtesy of Kohler Co.

should be 10,000-15,000 ohms. Renew circuit board if resistance tests are not correct. Remove end frame as outlined in OVERHAUL section. Disconnect brown leads from "AC" terminals of voltage regulator and connect a 120 volt AC power source to brown leads. Measure DC voltage at ends of solenoid wires attached to circuit board. Voltage should be 35-45 volts. With 120 volt AC power source still connected to brown leads, connect a 12 volt DC power source to black transformer coil leads attached to circuit board. Voltage at ends of solenoid wires attached to circuit board should now drop to zero. Disconnect 12 volt DC power source and voltage at solenoid wires should return to 35-45 volts DC. If voltage tests are not correct, renew circuit board.

# ENGINE

Engine make and model are listed at beginning of section. Refer to Briggs and Stratton or Kohler engine section for engine service.

# WIRING DIAGRAMS

Refer to Fig. K01-9 for a typical wiring diagram of generator stator and rotor leads. On 1.75 kw units, stator leads 1 and 2 are not used. Also refer to Fig. K01-2 for wiring diagrams of six and eight-terminal Power Monitor switches. Wiring diagrams are typical, and where applicable, include wiring for the outlet ground wire required by Canadian Standards Association (CSA).

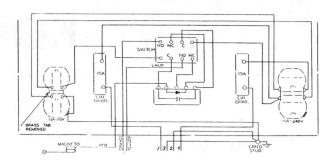

Fig. KO1-18—Power panel wiring diagram for early 3.5 kw models with 120/240 volt output.

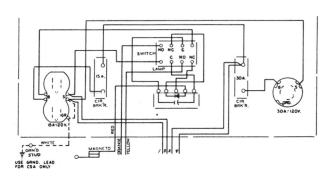

Fig. KO1-19—Power panel wiring diagram for 5 kw models with 120 volt output.

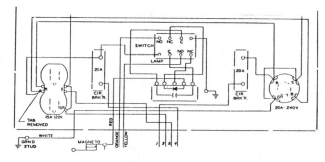

Fig. KO1-20—Power panel wiring diagram for 5 and late 3.5 kw models with 120/240 volt output.

# KOHLER

| Model | Output-kw | Voltage | Engine Make | Engine Model | Governed Rpm |
|---|---|---|---|---|---|
| Power Play 600 (Dayton 2W512) | .45 | 120 | Shibaura | W1-145 | 3600 |

# GENERATOR

## OPERATION

The generator is equipped with a brushless self exciting rotor (6–Fig. KO2-1). Adjustments for engine speed may be made as shown in Figs. KO2-5 and KO2-6. Specified engine speed is 3000 rpm for 50 Hz operation and 3600 rpm for 60 Hz operation. The generator is belt driven.

## MAINTENANCE

Rotor bearings (5–Fig. KO2-1) are sealed and do not require lubrication but should be inspected for smooth operation when generator is disassembled. Renew as needed.

Oil level in governor chamber should be checked every 100 hours as shown in Fig. KO2-2. Specified lubricant is SAE 10W-40 engine oil and level should be even with fill hole.

## DRIVE BELT ADJUSTMENT

Improper drive belt tension may allow drive belt to slip thereby reducing generator output. Inspect belt for fraying or cracking and renew as needed. Refer to Figs. KO2-3 and KO2-4. Specified drive belt tension is 7-8 mm (0.27-0.31 inch) while exerting a pressure of approximately 8 kg (17.6 pounds) at midpoint of belt. To adjust belt, remove cover and loosen the four generator mounting bolts. Turn bolt tension adjustment bolt until tension is correct. Tighten mounting bolts and reinstall cover.

## TROUBLESHOOTING

If little or no generator output is evident, check circuit breaker and reset if necessary. Check belt for slippage and readjust or renew as needed. Engine must be in good enough condition to maintain desired speed under load and

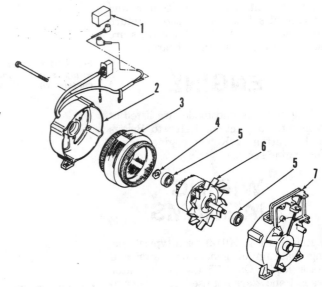

Fig. KO2-1—Exploded view of generator.
1. Condenser
2. Rear frame
3. Stator
4. Wave washer
5. Bearing
6. Rotor
7. Front frame

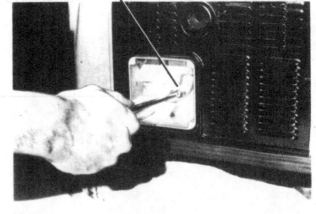

Fig. KO2-2—Check governor chamber oil level as shown.

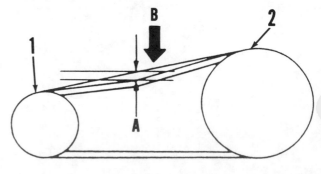

Fig. KO2-3—Specified belt deflection between engine pulley (1) and generator pulley (2) should be 7-8 mm (0.27-0.31 inch) (A) when a force of 8 kg (17.6 pounds) (B) is applied.

Illustrations courtesy of Kohler Co.

Fig. KO2-4 — View showing belt tension adjusting nuts.

Fig. KO2-5 — View showing governor adjusting screw.

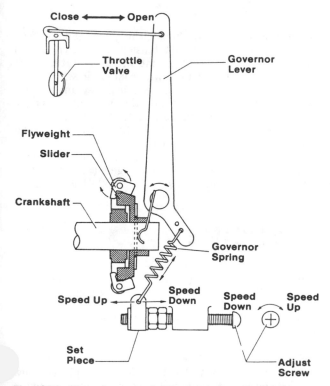

Fig. KO2-6 — View of governor linking showing speed adjusting procedure.

Fig. KO2-6A — Internal view of governor assembly showing slider (1), thrust washer (2) and flyweight (3).

Fig. KO2-6B — View of governor housing rear cover showing internal components.

Fig. KO2-7 — View of rotor showing diode and resistor location.

governor must be properly adjusted for desired frequency. Remove enclosure cover and check wiring for loose connections. All wiring connections must be clean and tight.

The rotor has a diode/resistor soldered to field windings as shown in Fig. KO2-7. To check the diode, it must be unsoldered from rotor terminals and checked for continuity with an ohmmeter or battery operated test light. Current should flow through diode in one direction only. Specified field resistance is 9.5-10 ohms and specified resistance of field resistor is 15000

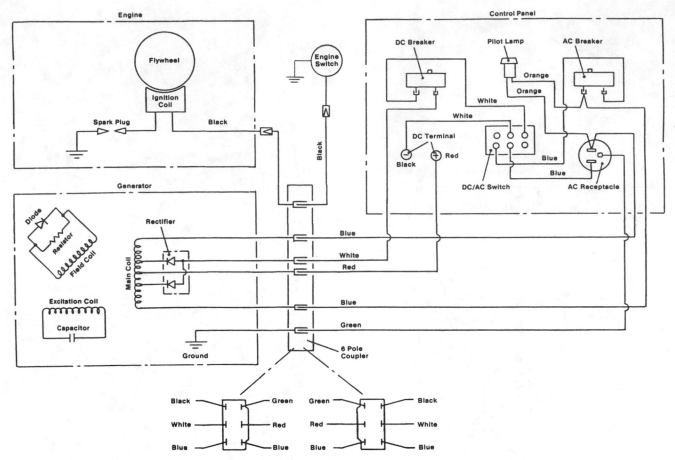

**Fig. KO2-8 — Wiring schematic.**

ohms. Check rotor for grounds by placing one ohmmeter lead on terminal of winding and other lead on rotor shaft or frame. There should be no continuity. Renew rotor if windings are grounded or if readings vary widely from specified resistance values.

Specified stator excitation coil resistance is 11 ohms when checked between pink leads of connector. Specified stator main coil resistance should be approximately 5 ohms when checked between blue leads of connector. Connector red lead is the center tap of the main coil and resistance should be ½ of the total coil resistance or 2.5 ohms when checked between red lead and either of the blue leads. There should be no continuity between frame of stator and any one of the three coils. Renew stator if any of the coils are grounded or if resistance values vary widely from specifications.

Refer to Fig. KO2-8 for wiring schematic.

### OVERHAUL

To disassemble generator, first unbolt and remove receptacle panel. Unplug wiring connector. Remove battery cable storage cover and choke lever assembly. Turn generator on its side and remove the four mounting screws which secure enclosure cover to fuel tank base. Turn generator assembly right side up and remove fuel cap and grommet. Lift enclosure cover away from generator assembly. Disconnect fuel line from carburetor. Remove fuel tank from engine/generator sub base and remove the four rubber vibration mounts.

Remove top baffle plate and muffler. Loosen belt tension adjuster nuts and generator mounting bolts. Lift generator away from engine.

Remove generator pulley and remove the four stator through-bolts. Using a soft hammer, carefully tap rear frame (2 – Fig. KO2-1), front frame (7) and stator (3) apart.

Reassembly is reverse of disassembly procedure. Torque generator mounting bolts to 5.9-6.9 N·m (52-61 in.-lbs.).

# ENGINE

| Cyls. | Bore | Stroke | Displ. |
|---|---|---|---|
| 1 | 40 mm | 30 mm | 41.5 cc |
| | (1.57 in.) | (1.3 in.) | (2.53 cu. in.) |

The engine is a two cycle design and specified fuel and oil mix ratio is 50 to 1.

### MAINTENANCE

**SPARK PLUG.** Recommended spark plug is either NGK BMR6A or Champion RCJ-8. Specified spark plug electrode gap is 0.6-0.7 mm (0.025-0.028 inch).

**CARBURETOR.** An exploded view of the float type carburetor is shown in Fig. KO2-10. Adjust float hinge height (A – Fig. KO2-10A) to 4mm (0.16 inch).

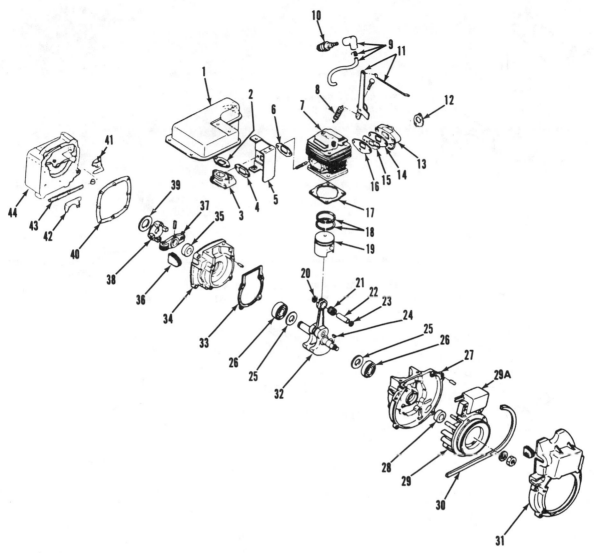

**Fig. KO2-9 — Exploded view of engine.**

| | | | |
|---|---|---|---|
| 1. Muffler | 10. Spark plug | 19. Piston | 28. Bearing | 36. Rubber |
| 2. Gasket | 11. Lever & link | 20. Clip | 29. Flywheel | 37. Flyweight |
| 3. Manifold | 12. Gasket | 21. Needle bearing | 29A. Coil | 38. Governor slider |
| 4. Gasket | 13. Insulator | 22. Wrist pin | 30. Belt | 39. Thrust washer |
| 5. Cowling | 14. Gasket | 23. Clip | 31. Shroud | 40. Gasket |
| 6. Gasket | 15. Gasket | 24. Key | 32. Crankshaft assy. | 41. Set piece |
| 7. Cylinder | 16. Gasket | 25. Shim | 33. Gasket | 42. Follower |
| 8. Spring | 17. Gasket | 26. Bearing | 34. Crankcase, rear | 43. Governor shaft |
| 9. Spark plug wire | 18. Rings | 27. Crankcase, front | 35. Seal | 44. Housing |

**IGNITION AND TIMING.** An electronic ignition system is used with a coil and control module molded together. Specified air gap between coil assembly and flywheel magnet is 0.35-0.5 mm (0.0138-0.0197 inch). Ignition timing is preset at 30° BTDC and no adjustments are possible. Specified torque for flywheel nut is 24.5-29.4 N·m (18-21 ft.-lbs.).

**GOVERNOR.** A flywheel type governor is attached to the end of the crankshaft. To adjust governor, turn adjustment screw clockwise to increase engine speed and counterclockwise to decrease engine speed. Internal views of governor assembly are shown in Figs. KO2-6A and KO2-6B. If governor shaft (43 – Fig. KO2-9) is removed from housing, (44), readjust follower (42) mounting screws as follows: Install gasket (40) on housing rear cover and place a straightedge across housing above the follower contacts. Rotate shaft so that follower contacts are up against straightedge. Contacts should be parallel with one another and clearance between straightedge and contacts should be less than 0.015 mm (0.0059 inch). Use a thread locking compound on

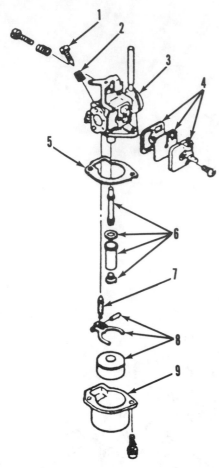

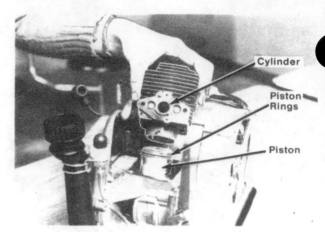

Governor specifications are as follows:

Follower contact width . . . . . . . 2.5 mm
(0.1 in.)
Thrust washer width . . . . . . . . . . 8 mm
(0.315 in.)
Governor shaft diameter . . . 5.9-6.0 mm
(0.234-0.236 in.)
Governor shaft bore
diameter . . . . . . . . . . . . . 6.0-6.012 mm
(0.2362-0.2367 in.)

## OVERHAUL

**TIGHTENING TORQUES.** Recommended tightening torques are as follows:

Crankcase assembly screws 5.9-6.9 N·m
(52-61 in.-lbs.)
Cylinder assembly screws . . 5.9-6.9 N·m
(52-61 in.-lbs.)
Flywheel . . . . . . . . . . . . . . 24.5-29.4 N·m
(18-21 ft.-lbs.)

**CYLINDER.** The cylinder may be removed as an assembly as shown in Fig. KO2-11A. Cylinder head should be removed and cleaned of carbon at 100 hour intervals. Pulse hole (2) provides pressure for fuel pump operation.

**PISTON, PIN AND RINGS.** Piston (19 – Fig. KO2-9) and pin assembly (22) may be serviced with cylinder removed. Piston and pin specifications are as follows:

Bore diameter . . . . . . . . . . 39.9-40.0 mm
(1.574-1.5742 in.)
Piston diameter . . . . . . . 39.96-39.98 mm
(1.5735-1.574 in.)
Piston pin outside
diameter . . . . . . . . . . . 9.994-10.0 mm
(0.3935-0.3937 in.)

**CONNECTING ROD AND CRANKSHAFT.** Connecting rod is permanently assembled to crankshaft which is pressed together at the connecting rod journal. Connecting rod and crankshaft

**Fig. KO2-10 — Exploded view of carburetor.**

1. Needle valve
2. Spring
3. Carburetor body
4. Diaphragm & cover assy.
5. Gasket
6. Jet assy.
7. Needle valve
8. Float assy.
9. Float bowl

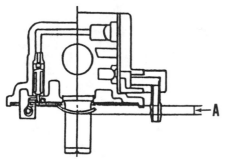

**Fig. KO2-10A — Adjust float hinge height (A) to 4mm (0.16-inch).**

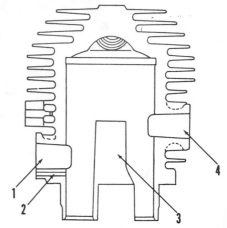

**Fig. KO2-11A — Cross-sectional view of cylinder assembly.**

1. Intake port
2. Pulse hole
3. Scavenging port
4. Exhaust port

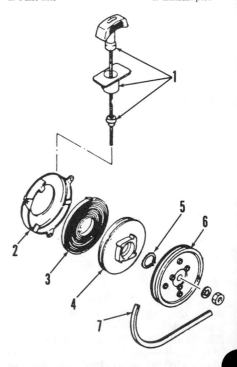

**Fig. KO2-12 — Exploded view of recoil starter assembly.**

1. Rope & handle assy.
2. Housing
3. Spring
4. Reel
5. Snap ring
6. Pulley, generator
7. Belt

follower mounting screws after final adjustment is made. Torque governor cover mounting screws to 6.0-6.9 N·m (52.4-60.8 in.-lbs.). When installing governor lever on governor shaft, rotate shaft clockwise as far as possible. With lever (11) connected to carburetor, open throttle as far as possible, then tighten set screw.

Illustrations courtesy of Kohler Co.

must be renewed as an assembly. Specifications for connecting rod/crankshaft assembly are as follows:

Connecting rod small end bore
   diameter . . . . . . . . . . 14.00-14.01 mm
               (0.5512-0.5515 in.)
Crankshaft end play . . . . . . . 0.1-0.2 mm
               (0.0039-0.0078 in.)

**REWIND STARTER.** Refer to Fig. KO2-12 for exploded view of rewind starter assembly. Remove enclosure cover as outlined in overhaul paragraph to gain access to starter. Remove drive belt. Remove pulley nut and using a puller, remove belt pulley (6–Fig. KO2-12) from tapered end of crank-shaft. Remove snap ring (5) and remove the three starter assembly mounting screws. With starter removed, service rope and spring as needed. When reas-sembling, torque starter assembly mounting screws to 3.4-4.4 N·m (30-39 in.-lbs.) and torque pulley nut to 24.5-29.4 N·m (18-21 ft.-lbs.).

# KOHLER

| Model | Output-kw | Voltage | Engine | | Governed |
| --- | --- | --- | --- | --- | --- |
| | | | Make | Model | Rpm |
| Power Play | 0.9 | 120 | Shibaura | * | 3600 |
| 800/950 | | | | | |
| *Specification not available. | | | | | |

# GENERATOR

## OPERATION

The generator is equipped with a brushless, self-exciting rotor (6–Fig. KO3-1). Refer to Fig. KO3-3 and adjust governor to change engine speed and frequency. Specified engine speed is 3000 rpm for 50 Hz operation and 3600 rpm for 60 Hz operation.

## MAINTENANCE

Rotor bearing (3–Fig. KO3-1) is sealed and does not require lubrication, but should be inspected for smooth operation when generator is disassembled. Renew as needed.

## TROUBLESHOOTING

If little or no generator output is evident, check circuit breakers and reset if necessary. Engine must be in good enough condition to maintain desired speed under load and governor must be properly adjusted for desired frequency. Remove control panel (Fig. KO3-2) and check wiring for loose connections. All wiring connections must be clean and tight.

The rotor has a diode/resistor soldered to the field windings. To check diode, unsolder from rotor winding terminals and check for continuity with an ohmmeter or battery operated test light. There should be continuity through diode in one direction only. Specified resistance value of field windings is 11 ohms and specified resistance value of field resistor is 15000 ohms. Check rotor for grounds by placing one ohmmeter lead on terminal of field winding and the other lead on rotor shaft or frame. There should be no continuity. Renew rotor if windings are grounded or if resistance readings vary widely from specified resistance values.

Stator excitation coil resistance value should be 6.8-7.8 ohms when measured between black wire terminals of connector. Stator main coil resistance value should be 1.8-8.4 ohms when measured between blue wire and black wire terminals of connector. Stator DC coil resistance should be 1.8-8.4 ohms when checked between blue & white wire and green & white wire terminals of the connector. There should be no continuity between frame of stator and any of the three coils. Renew stator, if any of the coils are grounded or if resistance values vary widely from specifications.

Refer to Fig. KO3-3A for wiring schematic.

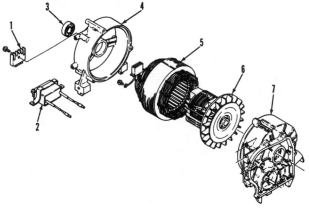

**Fig. KO3-1 – Exploded view of generator.**

1. Rectifier
2. Capacitor
3. Bearing
4. End bracket
5. Stator
6. Rotor
7. Crankcase cover

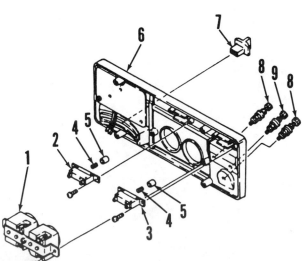

**Fig. KO3-2 – Exploded view of control panel.**

1. Receptacle
2. AC circuit breaker
3. DC circuit breaker
4. Spring
5. Cover
6. Panel assy.
7. Knob
8. DC terminal (black)
9. DC terminal (red)

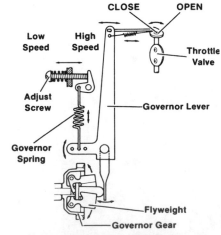

**Fig. KO3-3 – View of governor linkage showing speed adjustment procedure.**

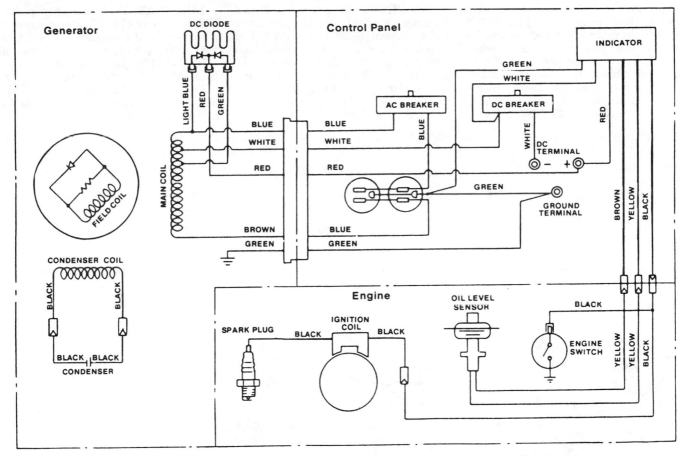

Fig. KO3-3A — Wiring schematic for generator.

## OVERHAUL

To disassemble generator, drain fuel and oil, then remove rear and side covers. Remove knob (7 – Fig. KO3-2) from control lever. Disconnect fuel hose and wire from fuel valve, then disconnect choke cable from carburetor. Remove muffler cover (29 – Fig. KO3-4), air filter assembly and carburetor. Remove muffler. Disconnect choke cable. Unplug and remove control panel (Fig. KO3-2) along with front cover. Remove air intake elbow. Remove spark plug and cylinder head top cover. Disconnect capacitor wires from generator and disconnect float switch wiring from side of engine crankcase. Remove bolts securing engine/generator assembly to rubber dampers and base assembly. Lift engine/generator assembly away from base. Remove rotor through-bolt (12 – Fig. KO3-13). Note that rotor through-bolt has left-hand threads and must be turned clockwise for removal. Remove the three starter, stator and end bracket (4 – Fig. KO3-1) bolts. Lift rewind starter away from end bracket.

With a soft hammer, carefully tap stator and end bracket away from crankcase cover (7). Rotor (6) is keyed to a taper on engine crankshaft and may sometimes be removed by tapping on rotor housing with a soft hammer. A rotor puller, part number 222630 is available from the manufacturer.

Reassemble by reversing removal procedure. Torque rotor through-bolt to 9.82-13.74 N·m (7.25-10.14 ft.-lbs.) and torque the three starter, end bracket and stator bolts to 4.92-7.86 N·m (3.63-5.80 ft.-lbs.).

# ENGINE

| Cyls. | Bore | Stroke | Displ. |
|-------|------|--------|--------|
| 1 | * | * | 98 cc (6.0 cu. in.) |

*Specification not available.

## MAINTENANCE

**SPARK PLUG.** Recommended spark plug is either NGK BPR6HS or Champion RL-12Y. Specified spark plug gap is 0.6-0.7 mm (0.024-0.028 inch).

**CARBURETOR.** An exploded view of the float type carburetor is shown in Fig. KO3-5. Initial setting for mixture needle valve (20) is 1¼ turns out from its fully closed position.

**IGNITION AND TIMING.** An electronic ignition system is used with a coil and control module (42–Fig. KO3-4) molded together. Specified air gap between coil assembly and flywheel magnet is 0.3-0.5 mm (0.012-0.020 inch). Ignition timing is preset at 22-24° BTDC and no adjustments are possible. Specified torque for flywheel nut is 29.50-34.25 N·m (21.47-34.25 ft.-lbs.).

**GOVERNOR.** Refer to Fig. KO3-6 for governor adjustment. With governor rod and springs installed, push top of governor lever as shown to fully open throttle. With throttle held open, turn governor shaft fully clockwise and tighten clamp. Refer to Fig. KO3-3 for a view of governor linkage and speed adjustment procedure.

**LUBRICATION.** Crankcase capacity is 0.45 liter (0.475 qt.). Recommended motor oil is API classification SF. Use SAE 10W-30 for temperatures between −10°C (12.2°F) and 45°C (115°F). Use SAE 20 for temperatures between −30°C (3°F) and 4°C (40°F). Use SAE 30 for temperatures between 4°C (40°F) and 45°C (115°F).

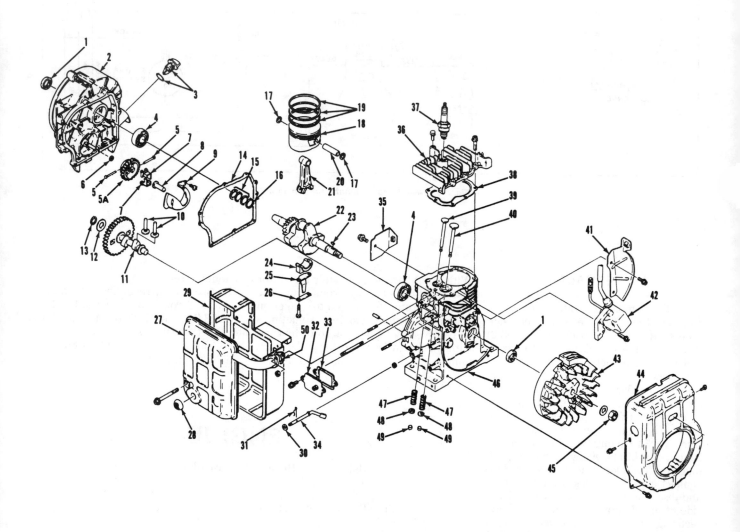

*Fig. KO3-4 — Exploded view of engine.*

| | | | |
|---|---|---|---|
| 1. Seal | 10. Tappets | 20. Pin | 30. Washer | 40. Intake valve |
| 2. Crankcase cover | 11. Camshaft | 21. Rod | 31. Pin | 41. Support |
| 3. Oil plug & seal | 12. Thrust washer | 22. Crankshaft | 32. Breather | 42. Ignition module |
| 4. Bearing | 13. Shim | 23. Woodruff key | 33. Gasket | 43. Flywheel |
| 5. Pin | 14. Gasket | 24. Rod cap | 34. Shaft, governor | 44. Housing, blower |
| 5A. Gear, governor | 15. Shims | 25. Splasher | 35. Cowling | 45. Nut, flywheel |
| 6. Shim | 16. Thrust washer | 26. Lock tab | 36. Head | 46. Block |
| 7. Weight, governor | 17. Clip | 27. Muffler | 37. Spark plug | 47. Valve spring |
| 8. Slider, governor | 18. Piston | 28. Spark arrestor | 38. Head gasket | 48. Retainer |
| 9. Cover | 19. Rings | 29. Cover | 39. Exhaust valve | 49. Cap |

**OIL LEVEL SENSOR.** An oil level sensor float switch is mounted in crankcase cover. The system shuts down the ignition system when oil level falls

Fig. KO3-6 — View showing governor adjustment procedure.

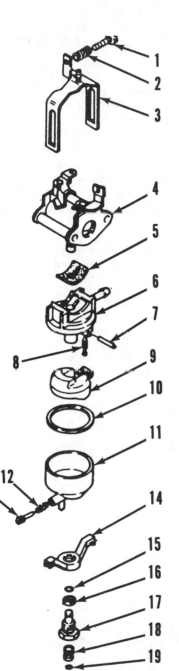

Fig. KO3-5 — Exploded view of carburetor.

1. Throttle stop screw
2. Spring
3. Bracket
4. Carburetor body
5. Gasket
6. Float chamber cover
7. Float pin
8. Float valve
9. Float
10. Gasket
11. Float chamber
12. Spring
13. Drain screw
14. Arm
15. "O" ring
16. Gasket
17. Main jet
18. Spring
19. "O" ring
20. Main mixture needle

Illustrations courtesy of Kohler Co.

below a certain level. Refer to Fig. KO3-12 for an exploded view of float switch. Use a thread locking compound and torque float switch nut (7) to 0.49-0.98 N·m (4.3-8.7 in.-lbs.).

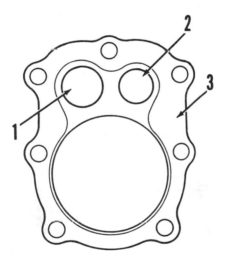

Fig. KO3-7 — View of head gasket (3) showing intake port (1) and exhaust port (2).

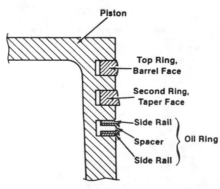

Fig. KO3-8 — Cross-sectional view showing proper piston ring installation.

## OVERHAUL

**TIGHTENING TORQUES.** Recommended tightening torques are as follows:

Connecting rod . . . . . . . . . 5.9-8.8 N·m (52.2-78.2 in.-lbs.)
Cylinder head. . . . . . . . . 7.86-10.81 N·m (69.6-95.7 in.-lbs.)
Crankcase cover . . . . . . . 4.91-7.86 N·m (43.5-69.6 in.-lbs.)
Flywheel nut . . . . . . . . 29.50-46.54 N·m (21.47-34.25 ft.-lb.)

**CONNECTING ROD.** Connecting rod and piston are removed from cylinder head end of block as an assembly. The aluminum alloy connecting rod rides directly on the crankpin. Splash lubrication is used. Specified connecting rod-to-crankpin running clearance is 0.03-0.05 mm (0.001-0.002 inch). When assembling, be sure to match marks on rod and rod cap.

**PISTON, PIN AND RINGS.** An aluminum alloy piston carries three cast iron rings. Refer to Fig. KO3-8 for cross-sectional view of rings. Specified running clearance between piston and cylinder is 0.04-0.09 mm (0.0016-0.0036 inch). Specified piston ring gap is as follows:

Top & second ring . . . . . . . 0.15-0.30 mm (0.006-0.012 in.)
Oil ring . . . . . . . . . . . . . . . . 0.2-0.8 mm (0.008-0.032 in.)

Renew pistons showing visible signs of wear, scoring or scuffing. When installing piston and rod assembly, be sure oil hole on rod faces toward flywheel side of engine.

**CYLINDER.** If cylinder taper or out-of-round exceeds 0.15 mm (0.006 inch), block must be renewed. Minor scuffing or scoring of the cylinder may be removed with a hone or fine grain sandpaper.

**CRANKSHAFT AND MAIN BEARINGS.** Crankshaft is supported by ball bearings (4 – Fig. KO3-4) on each end. Renew bearings, if roughness is evident. Inspect crankpin for out-of-round or scoring and renew crankshaft, if needed. Specified crankshaft end play is 0.05-0.2 mm (0.002-0.008 inch) and is adjusted with shims (15 – Fig. KO3-4). Install any combination of available 0.1 mm (0.004 inch) or 0.2 mm (0.008 inch) shims to obtain desired crankshaft end play.

**CAMSHAFT AND TAPPETS.** Inspect camshaft (11 – Fig. KO3-4) and tappets (10) for scoring on lobe contact surfaces. Renew as needed. Refer to Fig. KO3-11 for a view of valve timing mark location.

**VALVE SYSTEM.** Specified tappet clearance for both intake and exhaust valves is 0.15-0.51 mm (0.006-0.020 inch) when engine is cold. Tappet clearance is adjusted by installing different thicknesses of adjuster caps (5 – Fig. KO3-9). Adjuster caps are color coded for the various thicknesses. Refer to Fig. KO3-10 for a color code chart showing adjuster cap thicknesses. Specified clearance between intake valve stem and valve guide is 0.031-0.071 mm (0.0012-0.0028 inch) with a maximum allowable clearance of 0.15 mm (0.006 inch). Specified clearance between exhaust valve stem and valve guide is 0.05-0.095 mm (0.002-0.0038 inch) with a maximum allowable clearance of 0.2 mm (0.008 inch).

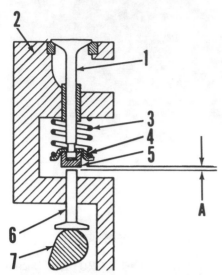

Fig. KO3-9 – Cross-sectional view of valve assembly. Specified valve clearance (A) is 0.15-0.51 mm (0.006-0.020 inch).

1. Valve
2. Block
3. Spring
4. Retainer
5. Cap
6. Tappet
7. Camshaft

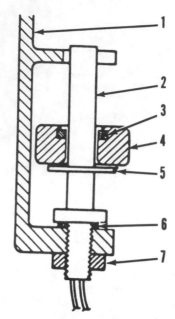

Fig. KO3-12 – Cross-sectional view of oil level sensor float switch.

1. Crankcase cover
2. Body
3. Magnet
4. Float
5. Clip
6. Seal
7. Nut

| Range of Adjustment Inches | (Millimeters) | Color of Cap | Part Number |
|---|---|---|---|
| .071 | Over 1.775 | Not adjustable | |
| .069–.071 | 1.725–1.775 | RED | 222489 |
| .067–.069 | 1.675–1.725 | YELLOW | 222490 |
| .065–.067 | 1.625–1.675 | PINK | 222491 |
| .063–.065 | 1.575–1.625 | No color | 222492 |
| .061–.063 | 1.525–1.575 | GREEN | 222493 |
| .059–.061 | 1.475–1.525 | WHITE | 222494 |
| .057–.059 | 1.425–1.475 | BLUE | 222495 |
| .054–.057 | 1.375–1.425 | BLACK | 222634 |
| Under .053 | Under 1.325 | Not adjustable | |

Fig. KO3-10 – Color code chart for valve clearance adjuster caps.

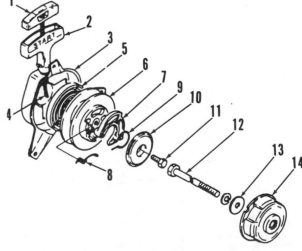

Fig. KO3-13 – Exploded view of rewind starter.

1. Cap
2. Knob
3. Housing
4. Rope
5. Spring
6. Reel
7. Ratchet
8. Spring, return
9. Spring, friction
10. Plate, friction
11. Bolt
12. Rotor through-bolt
13. Washer
14. Pulley

Fig. KO3-11 – View showing proper alignment of valve timing marks.

# KUBOTA

**550 W. Artesia Blvd.**
**Compton, California 90220**

| Model | Output-kva* | Voltage | Engine | | Governed Rpm |
|-------|-------------|---------|--------|--|--------------|
| | | | **Make** | **Model** | |
| A500 | .40 | 120 | Kubota | GS90 | 3600 |
| A650 | .55 | 120 | Kubota | GS90 | 3600 |

*1 kva equals 1000 watts.

# GENERATOR

## MAINTENANCE

All bearings are sealed and require no lubrication, but should be inspected for smooth operation when generator is disassembled. Inspect unit periodically for condition of brushes (Model A650), receptacles, electrical connections and operation of all switches.

## TROUBLE-SHOOTING

### Model A500

**AC VOLTAGE SIDE.** If generator output at the AC receptacle is zero, check the following: The engine must be in good condition and maintain desired governed speed of 3600 rpm. All wiring connections must be clean and tight.

If all wiring connections are in good condition, run generator at specified speed and measure voltage at the AC receptacle. Correct generator output at the AC receptacle is 113-127 volts. If no voltage is present, check circuit breaker (Fig. KU4). If circuit breaker is in good condition and voltage is still not present, remove the control panel cover and disconnect the connector between the cover and the stator assembly. Use an ohmmeter to check the main coil resistance by connecting ohmmeter leads to the red and blue wires at the generator output connector (Fig. KU6). If resistance is not within the range of

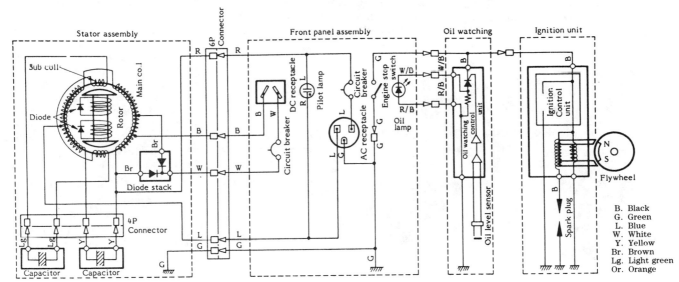

*Fig. KU4-Wiring schematic for Kubota A500 generator.*

B. Black
G. Green
L. Blue
W. White
Y. Yellow
Br. Brown
Lg. Light green
Or. Orange

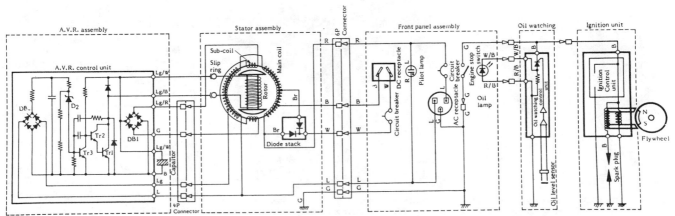

*Fig. KU5-Wiring schematic for Kubota A650 generator. Refer to Fig. KU4 for wiring color codes.*

2.42-2.71 ohms, the main coil is faulty. If resistance is as specified, there is a faulty receptacle contact, a faulty switch contact or a short in wiring.

If AC voltage is present at AC receptacle but is 96 volts or less, use an ohmmeter and check the resistance of the main coil by connecting ohmmeter leads to the red and blue wires (Fig. KU6). If resistance at main coil is below 2 ohms, there is a short in the layers of the main coil. If main coil resistance is within specified range, remove end cover and disconnect wiring connectors between

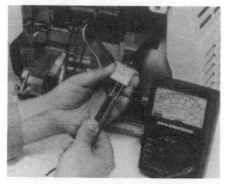

Fig. KU6—To measure main coil resistance, disconnect connector between stator and control panel and connect ohmmeter leads between red and blue terminals.

Fig. KU8—To measure rotor coil resistance on Model A500, connect ohmmeter leads between coil terminals as shown.

stator and capacitors. Measure sub-coil winding resistance by connecting ohmmeter leads to the two light green wire terminals (Fig. KU7). If resistance is not within range of 10.55-12.08 ohms, renew stator assembly. If resistance at sub-coil is within specified range, remove rotor and check resistance at the rotor coil by connecting ohmmeter leads to coil windings as shown in Fig. KU8. Measure the resistance of rotor coils with ohmmeter leads connected one way and note reading, then reverse ohmmeter leads and check resistance again. Resistance should be 13.11-14.7 ohms. If resistance is 2-5 ohms in normal direction and below 9.8 ohms in the reverse direction, there is a short in the layers of the rotor coil. If resistance is 2-5 ohms in normal direction and infinite in the reverse direction, there is a open circuit in wiring on one side. If resistance is below 5 ohms in each direction, there is a defective diode.

If a direct current of 3-15 volts is measured at the AC receptacle with generator running at specified engine speed, use an ohmmeter to measure the resistance of the sub-coil by connecting ohmmeter leads to the two light green wire terminals as shown in Fig. KU7 at the connector. If resistance is less than

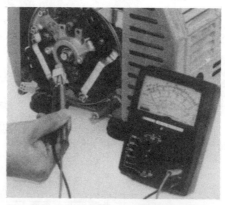

Fig. KU7—To measure sub-coil resistance on Model A500, disconnect connector between stator and capacitors and connect ohmmeter leads between two light green terminals.

Fig. KU9—To measure capacitor resistance on Model A500, connect ohmmeter leads to positive and negative leads of capacitor.

10.55-12.08 ohms or infinite, there is a short in one of the connections or in the layers of the sub-coil. If resistance is as specified, remove rotor and use an ohmmeter to measure the rotor coil resistance as shown in Fig. KU8. Measure the resistance of rotor coils with ohmmeter leads connected one way and note reading, then reverse ohmmeter leads and check resistance again. Resistance should be 13.11-14.7 ohms. If resistance is 2 to 5 ohms in the normal direction and infinite in the reverse direction, the rotor coil is shorted. If resistance is less than specified or 5 ohms in both normal and reverse direction, the rotor diode is defective. If resistance is as specified, check the capacitor by disconnecting the connector at the capacitor and connecting ohmmeter leads to positive and negative sides of capacitor (Fig. KU9). Note ohmmeter reading, then reverse the leads. Resistance should be infinite with leads connected in either direction. If resistance is not as specified, renew capacitor.

**DC VOLTAGE SIDE.** Correct DC voltage measured at the DC receptacle is 10.5 to 12.0 volts. If generator output at the DC receptacle is zero, check for blown circuit breaker on the DC side. If circuit breaker is in good condition, check DC voltage at the white wire and black or black-white wire at the generator output connector. If voltage is 10-16 volts, the contact at the receptacle is bad or there is a short in wiring. If voltage is zero, measure the AC voltage at the AC receptacle. If AC voltage is zero, check AC system as previously outlined. If voltage is 30-40 volts, there is a defective diode.

## Model A650

**AC VOLTAGE SIDE.** Correct AC voltage measured at the AC receptacle with generator running at specified speed should be 113-127 volts. If no voltage is present, check for blown circuit breaker (Fig. KU5). If circuit breaker is in good condition, remove the control panel cover and disconnect the connector between the cover and the stator assembly. Measure resistance of the main coil by connecting ohmmeter leads to the red and blue lead terminals as shown in Fig. KU6. Resistance should be 3.21-3.62 ohms. If resistance is not within specified range, main coil is faulty. If resistance is as specified, there is a faulty contact at receptacle, short in wiring or the permanent magnet is demagnetized.

If 3-6 volts AC is present at AC receptacle with generator running at specified speed, stop engine and remove the brush assembly. Use an ohmmeter and measure the resistance of the rotor coil by connecting ohmmeter leads to the slip rings (Fig. KU10). If resistance is infinite, there is an open circuit in the rotor windings. If resistance is 26.22-28.98 ohms, remove the end cover and disconnect the connector between the stator assembly and the automatic voltage regulator (AVR). Use an ohmmeter and measure the resistance between the light green and green lead terminals as shown in Fig. KU11. If resistance is infinite, there is open circuit in the sub-coil. If resistance is 3.10-3.55 ohms, there is a faulty brush contact or a defective diode in diode stack.

**DC VOLTAGE SIDE.** Correct DC voltage at the DC receptacle with generator running at specified speed should be 10.5-12.0 volts. If voltage is less than specified and AC voltage output is normal, there is a defective diode. If no voltage is present, check for a blown circuit on the DC side. If circuit breaker is in good condition, measure the DC voltage between the white wire and black or black-white wire terminals at the generator output connector. If 10-16 volts are measured, there is a faulty contact at receptacle or a short in wiring. If no voltage is measured, check voltage at the AC receptacle. If voltage at AC receptacle is 30-40 volts, there is a defective diode. If voltage at the AC receptacle is 0 volts, check AC side of generator as previously outlined.

### OVERHAUL

To disassemble generator, remove control panel, side cover and end cover. Remove the fuel tank, engine base plate, muffler heat shield, muffler and oil watching unit. Remove brush holder on Model A650. On all models, remove stator through-bolts and withdraw stator housing as an assembly. Remove rotor mounting bolt and pull rotor off engine crankshaft.

Refer to Figs. KU12 or KU13 for exploded view of generator components. Sealed bearings should be inspected for smooth operation when generator is disassembled. Minimum brush length for Model A650 is 3.5 mm (0.138 inch), measured as shown in Fig. KU14. Make certain all parts are clean before reassembling.

To reassemble, reverse the disassembly procedure. Tighten rotor retaining bolt to torque of 12-16 N·m (9-11 ft.-lbs.), and tighten stator through-bolts to 3.5-4.9 N·m (30-43 in.-lbs.).

# ENGINE

Refer to the appropriate Kubota engine section for engine service.

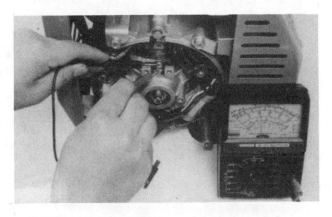

Fig. KU10—Rotor coil resistance can be measured on Model A650 by connecting ohmmeter leads between rotor slip rings.

Fig. KU11—To measure sub-coil resistance on Model A650, disconnect connector between stator and voltage regulator (AVR). Connect ohmmeter leads between light green and green terminals.

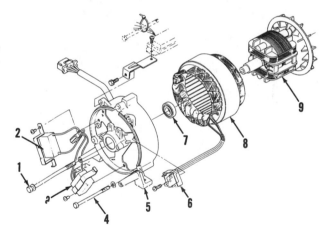

Fig. KU12—Exploded view of Kubota A500 generator assembly.

1. Rotor bolt
2. Main coil capacitor
3. Sub-coil capacitor
4. Through-bolt
5. Stator housing
6. Diode assy.
7. Bearing
8. Stator
9. Rotor

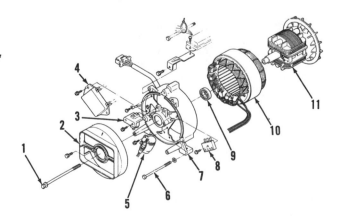

Fig. KU13—Exploded view of Kubota A650 generator.

1. Rotor bolt
2. End cover
3. Brush assy.
4. Voltage regulator
5. Capacitor
6. Through-bolt
7. Stator housing
8. Diode assy.
9. Bearing
10. Stator
11. Rotor

Fig. KU14—On Model A650, minimum allowable brush length is 3.5 mm (0.140 inch).

# KUBOTA

| Model | Output-kva* | Voltage | Engine | | Governed Rpm |
| --- | --- | --- | --- | --- | --- |
| | | | Make | Model | |
| A450 | .40 | 120/220 | Kubota | GN550 | 3600 |
| A1000 | .80 | 120/220 | Kubota | GS130 | 3600 |
| A1400 | 1.2 | 120/220 | Kubota | GS130 | 3600 |
| A2100 | 1.8 | 120/220 | Kubota | GN1850 | 3600 |
| A3000 | 2.5 | 120/220 | Kubota | GN2500 | 3600 |
| A3500 | 3.0 | 120/220 | Kubota | GS280 | 3600 |

*1 kva equals 1000 watts.

# GENERATOR

### MAINTENANCE

All bearings are sealed and require no lubrication, but should be inspected for smooth operation when generator is disassembled. Check brushes (if used) after every 500 hours of operation. Renew brushes if length beyond brush holder is less than 3.5 mm (0.140 inch). Inspect unit periodically for condition of receptacles, electrical connections and operation of all switches.

### TROUBLE-SHOOTING

#### Models A450-A1000

**AC VOLTAGE SIDE.** The engine must be in good condition and maintain desired governed speed of 3600 rpm. All wiring connections must be clean and tight.

If all wiring connections are in good condition, run generator at specified speed and measure voltage at the AC receptacle. Correct generator output at the AC receptacle is 108-144 volts. If no voltage is present, check for blown circuit breaker. If circuit breaker is in good condition and voltage output is 0 volts, stop generator, remove control panel and disconnect the generator output connector. Measure the main coil resistance by connecting ohmmeter leads to the red and blue wire terminals of the generator output connector (Fig. KU20). Resistance should be 2.43 ohms for Model A450 or 1.30 ohms for Model A1000. If reading is as specified but 0 volts are measured at receptacle, a faulty receptacle contact, faulty switch contact or disconnected wiring is indicated. A resistance reading less than specified or an infinite reading indicates a short or open main coil winding.

If generator output at specified speed is 96 volts or less when measured at the 120 volt AC receptacle, check the main coil resistance as previously outlined. If main coil resistance is 1.8 ohm or less for Model A450 or 0.97 ohm or less for

Model A1000, a short in the main coil is indicated. If voltage at the 120 volt AC receptacle is 96 volts or less and main coil resistance is as specified, use an ohmmeter and check the sub-coil resistance as follows: Disconnect capacitor wiring connector and connect ohmmeter leads to the two light green wire terminals of male connector (Fig. KU21). An ohmmeter reading of 7.9 ohms or less for Model A450 or 5.6 ohms or less for Model A1000, indicates a short in the sub-coil. If resistance is 10.5 ohms for Model A450 or 7.4 ohms for Model A1000, a faulty rotor coil is indicated. Capacitor main coil resistance is measured between the two yellow lead wires

of capacitor connector. Specified main coil resistance is 14.6 ohms for Model A450 or 10.2 ohms for Model A1000.

If generator output is 3-15 volts DC measured at the 120 volt AC receptacle with engine at specified speed, use an ohmmeter to measure the sub-coil resistance. Disconnect wiring connector at capacitors and connect ohmmeter leads to the two light green wire leads at the capacitor (Fig. KU21). If resistance is less than 10.5 ohms for Model A450 or less than 7.4 ohms for Model A1000 or infinite for either model, a short in sub coil is indicated. A resistance reading of 10.5 ohms for Model A450 or 7.4 ohms for Model A1000 may indicate a defec-

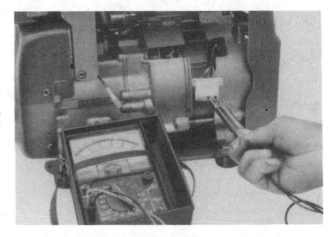

Fig. KU20—On Models A450 and A1000, stator main coil winding resistance may be measured using an ohmmeter connected between red and blue lead wires in output connector.

Fig. KU21—On Models A450 and A1000, sub-coil winding resistance is measured between the two light green lead wires at capacitor connector. Main capacitor coil resistance is measured between the two yellow lead wires.

tive sub-coil capacitor. If sub-coil capacitor is known to be good and output is still 3-15 volts DC at the AC receptacle, a faulty rotor is indicated.

**DC VOLTAGE SIDE.** Correct DC voltage measured at the DC receptacle is 10.5 to 12.0 volts. If generator output at the DC receptacle is zero, check for

blown circuit breaker (fuse) on the DC side. If circuit breaker is in good condition, measure DC voltage between the white wire and black or black-white wire at the generator output connector. If voltage is 10-16 volts, the contact at the receptacle is bad or there is a short in wiring. If voltage is zero, measure the AC voltage at the AC receptacle. If AC voltage is normal, a defective diode is indicated.

Fig. KU22—On Models A1400, A2100, A3000 and A3500, stator main coil winding resistance may be measured using an ohmmeter connected between red and green wires at output terminal as shown.

## Models A1400-A2100-A3000-A3500

**AC VOLTAGE SIDE.** Correct AC voltage measured at the AC receptacle with generator running at specified speed should be 108-132 volts. If no voltage is present, check for blown circuit breaker (fuse). If circuit breaker is in good condition, remove generator end cover and use an ohmmeter to measure the stator coil resistance at the generator output terminals (Fig. KU22). Specified resistance is 1.7 ohm for Model A1400, 0.94 ohm for Model A2100, 0.6 ohm for Model A3000 or 0.58 ohm for Model A3500. If resistance is less than specified or infinity, a short or open circuit is indicated in stator coil. If specified resistance is measured, check for a faulty contact at receptacle, disconnected wire, or demagnetization of permanent magnet.

If AC voltage measured at AC receptacle is 3-6 volts, stop the engine and remove brush holder. Use an ohmmeter to measure rotor coil resistance by touching tester probes to rotor slip rings. Rotor coil resistance should be 45 ohms for Model A1400, 54 ohms for Model A2100, 68 ohms for Model A3000 and 70 ohms for Model A3500. If resistance is less than specified or infinity, a short or open circuit is indicated in rotor coil. If specified resistance is measured, check sub-coil resistance as follows: Disconnect four wire connector and measure resistance between terminals (1) and (2) as shown in Fig. KU23. Sub-coil resistance should be 3.1 ohms for Model A1400, 2.6 ohms for Model A2100, 2.4 ohms for Model A3000 and 2.2 ohms for Model A3500. If resistance is infinite, there is open circuit in sub-coil. If specified resistance is measured, faulty brush contact or voltage regulator is indicated.

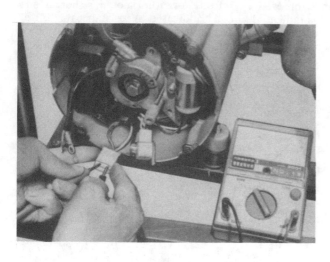

Fig. KU23—To check sub-coil resistance on Models A1400, A2100, A3000 and A3500, disconnect four wire connector and measure resistance between terminals (1) and (2).

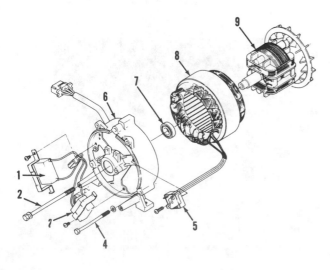

Fig. KU24—Exploded view of Model A450 and A1000 generator components.

1. Capacitor
2. Rotor bolt
3. Capacitor
4. Through-bolt
5. Diode assy.
6. Stator housing
7. Bearing
8. Stator assy.
9. Rotor assy.

## OVERHAUL

### Models A450-A1000

To disassemble generator, remove control panel, side cover and end cover.

Illustrations courtesy of Kubota Tractor Corp.

Close fuel shut-off valve and remove fuel tank. Unbolt and remove engine base plate. Remove the muffler. Remove through-bolts from generator and withdraw stator assembly. Remove rotor retaining bolt, then use a suitable puller to remove rotor from engine crankshaft.

Refer to Fig. KU24 for exploded view of generator assembly. Inspect for burned wiring or other damage and renew parts as necessary. Sealed bearings should be inspected for smooth operation. Make certain all parts are clean during reassembly.

To reassemble, reverse the disassembly procedure while noting the following special instructions: Tighten rotor retaining bolt to 11.8-15.7 N·m (105-139 in.-lbs.). Tighten stator housing through-bolts to 5.9-8.8 N·m (52-78 in.-lbs.). After assembly, pull engine recoil starter rope to make sure rotor rotates smoothly.

## Models
## A1400-A2100-A3000-A3500

To disassemble generator, remove end cover. Disconnect generator ground wire and output wires. Tag wires if necessary to ensure correct reassembly. Remove retaining screws from brush holder. Disconnect brush lead wires and automatic voltage regulator connector. Remove stator housing through-bolts and withdraw stator assembly. Remove rotor retaining bolt and use a suitable puller to remove rotor from engine crankshaft.

Refer to Fig. KU25 for exploded view of generator assembly. Inspect for burned wiring or other damage and renew parts as necessary. Sealed bearings should be inspected for smooth operation. Brushes should be renewed if brushes extend less than 3.5 mm (0.140 inch) from end of brush holder. Make certain all parts are clean during reassembly.

To reassemble, reverse the disassembly procedure while noting the following special instructions: Tighten rotor retaining bolt to 16-20 N·m (12-15 ft.-lbs.) on Models A1400 and A2100, or 32-43 N·m (23-31 ft.-lbs.) on Models A3000 and A3500. Tighten stator housing through-bolts to 5.9-8.8 N·m (52-78 in.-lbs.). Be sure that positive lead wire is connected to positive brush terminal (there is a "+" sign on positive wire and on brush holder). Make sure that red and blue output wires are matched according to color.

# ENGINE

Refer to the appropriate Kubota engine section for engine service.

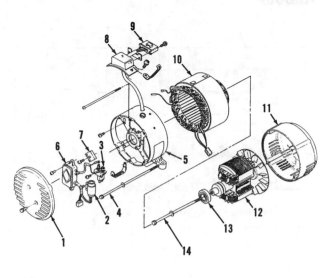

*Fig. KU25—Exploded view of generator components typical of Models A1400, A2100, A3000 and A3500.*

1. End cover
2. Condenser
3. Brushes
4. Through-bolt
5. Stator housing
6. Voltage regulator
7. Diode assy.
8. Cover
9. Output connector
10. Stator assy.
11. Fan housing
12. Rotor assy.
13. Bearing
14. Rotor bolt

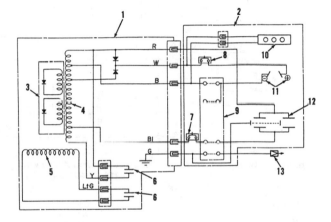

*Fig. KU26—Wiring schematic of Models A450 and A1000 produced for the U.S.A.*

1. Generator
2. Control panel
3. Rotor
4. Main coil
5. Sub-coil
6. Capacitors
7. Circuit breaker, AC
8. Circuit breaker, DC
9. Selector switch
10. Voltage indicator
11. Receptacle, DC
12. Receptacle, AC
13. Engine stop
B. Black
G. Green
R. Red
W. White
Y. Yellow
Bl. Blue
LtG. Light green

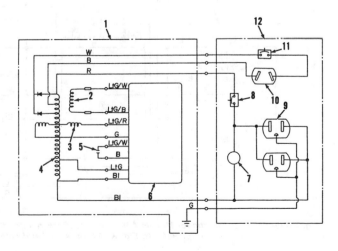

*Fig. KU27—Wiring schematic for Models A1400 and A2100 produced for the U.S.A.*

1. Generator
2. Field coil
3. Sub-coil
4. Main coil
5. Capacitor
6. Voltage regulator
7. Indicator lamp
8. Circuit breaker, AC
9. Receptacle, AC
10. Receptacle, DC
11. Circuit breaker, DC
12. Control panel
B. Black
G. Green
R. Red
W. White
Bl. Blue
LtG. Light green

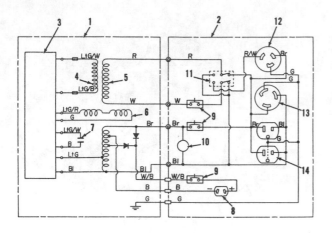

Fig. KU28—Wiring schematic for Models A3000 and A3500 with serial number up to 701010 produced for U.S.A.

1. Generator
2. Control panel
3. Voltage regulator
4. Field coil
5. Main coil
6. Sub-coil
7. Capacitor
8. Receptacle, DC
9. Circuit breakers
10. Indicator lamp
11. Switch
12. Receptacle
13. Receptacle
14. Receptacle
B. Black
G. Green
R. Red
W. White
Bl. Blue
Br. Black
LtG. Light green

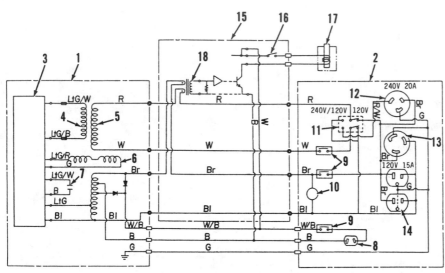

Fig. KU29—Wiring schematic for Models A3000 and A3500 with auto idler between serial numbers 919441-999924.

1. Generator
2. Control panel
3. Voltage regulator
4. Field coil
5. Main coil
6. Sub-coil
7. Capacitor
8. Receptacle, DC
9. Circuit breakers
10. Indicator lamp
11. Switch
12. Receptacle
13. Receptacle
14. Receptacle
15. Auto idler assy.
16. Switch
17. Solenoid
18. Transformer
B. Black
G. Green
R. Red
W. White
Bl. Blue
Br. Brown
LtG. Light green

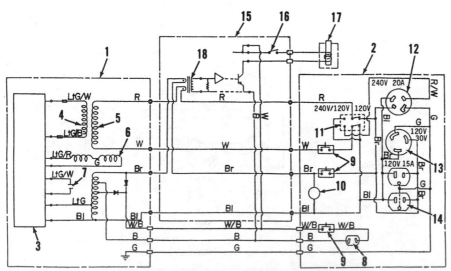

Fig. KU30-Wiring schematic for Models A3000 and A3500 with auto idler from serial number 138598 and up. Refer to Fig. KU29 for legend and color code.

# McCULLOCH

6085 S. McCulloch Drive
Tucson, AZ 85706

| Model | Output-kw | Voltage | Engine Make | Engine Model | Governed Rpm |
|-------|-----------|---------|------|-------|-----|
| HP 1200 | 1.2 | 120 | B&S | * | 3600 |
| HP 2000 | 2.0 | 120 | B&S | * | 3600 |
| HP 3000 | 3.0 | 120 | B&S | * | 3600 |
| HP 3500 | 3.5 | 120/240 | B&S | * | 3600 |
| HP 3500 DLX | 3.5 | 120/240 | B&S | * | 3600 |
| RA 120 | 1.2 | 110 | B&S | * | 3600 |
| RA 121 | 1.2 | 110 | B&S | * | 3600 |
| RA 122 | 1.2 | 110/220 | B&S | * | 3600 |
| RA 150 | 1.5 | 110 | B&S | * | 3600 |
| RA 151 | 1.5 | 110 | B&S | * | 3600 |
| RA 152 | 1.5 | 110/220 | B&S | * | 3600 |
| RA 200 | 2.0 | 110 | B&S | * | 3600 |
| RA 201 | 2.0 | 110 | B&S | * | 3600 |
| RA 202 | 2.0 | 110/220 | B&S | * | 3600 |
| RA 330 | 3.3 | 110/220 | B&S | * | 3600 |
| RA 330A | 3.3 | 110 | B&S | * | 3600 |
| RA 330ES | 3.3 | 110/220 | B&S | * | 3600 |
| RA 331 | 3.3 | 110/220 | B&S | * | 3600 |
| RA 332 | 3.3 | 110/220 | B&S | * | 3600 |

*Note model number on engine ID plate and refer to appropriate Briggs and Stratton section.

# GENERATOR

## MAINTENANCE

These models are not equipped with a commutator, slip rings or brushes due to permanent magnet type design. Maintenance is accomplished by checking for loose fasteners and electrical connections.

## TROUBLESHOOTING

If little or no output is generated, check the following: The engine must be in good condition and maintain desired governed speed under load. Inspect fuse(s) and all wiring.

Refer to OVERHAUL section and measure rotor air gap and side clearance

between rotor and stator. Disconnect wiring to stator terminals. Using an ohmmeter measure stator resistance and compare measurement with a known good stator to determine if stator is shorted or open. If preceding tests fail to locate problem, renew rotor. Be sure correct components are used as pieces may appear identical but differ electrically.

## OVERHAUL

Refer to Fig. M1-1 for an exploded view of generator and to appropriate wiring diagrams.

To disassemble generator, unscrew cover (7) screws. On Models RA 120 and RA 150 disconnect receptacle wires and on Models RA 121 and RA 151 unscrew

receptacle screws, then on all models remove cover (7). Disconnect stator wires and remove stators (4). Unscrew rotor screw (6). Install McCulloch rotor puller screw 21883 on Models HP 3000, HP 3500, HP 3500 DLX, RA 330, RA 330A, RA 330 ES, RA 331 and RA 332 or rotor puller screw 21882 on all other models and pull rotor free from engine crankshaft. DO NOT attempt to remove rotor by pulling on outer edge of rotor or by striking rotor.

Check air gap between rotor halves. Note that this measurement may be done with rotor installed. Rotor air gap should not be less than 0.550 inch (14.0 mm) at any point around rotor or it must be renewed. Install rotor and tighten rotor

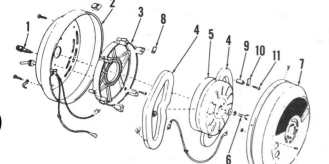

Fig. M1-1—Typical exploded view of generator.

1. Fuse holder
2. Housing
3. Stator mount
4. Stator halves
5. Rotor
6. Screw
7. Cover
8. Shims
9. Bushing
10. Stator retainer
11. Screw

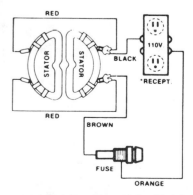

Fig. M1-2—Wiring schematic for Models RA 120 and RA 150. Models RA 121 and RA 151 are similar but receptacle is grounded.

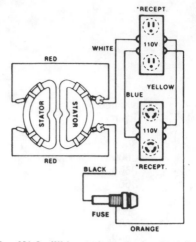

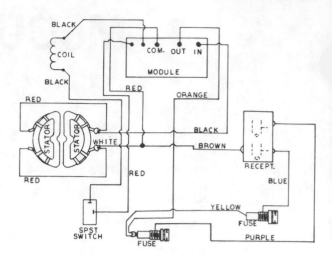

*Fig. M1-6—Wiring schematic for Model RA 330A.*

*Fig. M1-3—Wiring schematic for Model RA 200.*

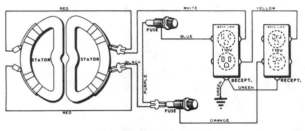

*Fig. M1-4—Wiring schematic for Model RA 201.*

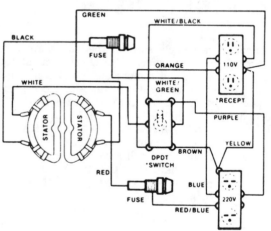

*Fig. M1-5—Wiring schematic for Models RA 122, RA 152, RA 202, RA 330, RA 330ES and RA 332.*

screw (6) to 240-260 in.-lbs. (27.1-29.3 N·m). When installing stator half (4), first tighten retaining screws finger tight. Grasp stator half near center screw and pull windings away from rotor until inner windings are tight against upper and lower stator bushings (9) then position center bushing against windings and tighten center retaining screw to 55-60 in.-lbs. (6.2-6.8 N·m). Use same procedure to install other stator half.

Check side clearance of stator halves in rotor gap. There should be 0.030 inch (0.76 mm) clearance on each side of stator half at any point in rotor gap. Stator location is determined by shims (8) which are special adhesive backed insulators. Shims are available in thicknesses of 1/32 and 1/16 inch.

# ENGINE

A Briggs & Stratton engine is used on all models. Refer to appropriate Briggs and Stratton section for engine service.

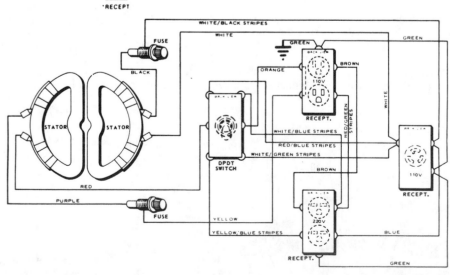

*Fig. M1-7—Wiring schematic for Model RA 331.*

# McCULLOCH

| Model | Output-kw | Voltage | Engine Make | Model | Governed Rpm |
|-------|-----------|---------|------|-------|--------------|
| HP 5000 | 5.0 | 120/240 | B&S | * | 3600 |
| HP 5000 DLX | 5.0 | 120/240 | B&S | * | 3600 |

*Note model number on engine ID plate and refer to appropriate Briggs and Stratton engine section.

# GENERATOR

### MAINTENANCE

Brushes and slip rings are accessible after removing control panel. Minimum brush length is 5/16 inch (7.9 mm). Brushes should move freely in holders. Clean slip rings using 240 grit sandpaper.

### TROUBLESHOOTING

If little or no output is generated, check the following: The engine must be in good condition and maintain a governed speed under load of 3600 rpm. Circuit breakers must operate properly and all wiring connections must be clean and tight. Brushes and slip rings must be in good condition.

If field coil magnets have lost their residual magnetism, use the following procedure: Remove control panel and run unit. Using a 6 volt dry cell battery, momentarily connect the negative battery terminal to the generator frame and the positive battery terminal to the positive brush terminal. Disconnect battery as soon as generator voltage rises. If loss of field magnetism is chronic, capacitor located inside end frame may be defective and should be renewed.

To test rectifier attached to control panel, disconnect leads to rectifier. Using an ohmmeter, connect ohmmeter leads to an (AC) terminal and positive (+) terminal, note reading, reverse ohmmeter leads and again note reading. One reading should be high and the other low. If both readings are high or both are low the rectifier is defective and must be renewed. Repeat test by connecting ohmmeter leads between remaining (AC) terminal and positive (+) terminal. If test readings using remaining (AC) terminal are both high or both low, rectifier is faulty and must be renewed.

If above tests do not locate malfunction, generator must be disassembled and rotor and stator tested for open and short circuits.

### OVERHAUL

Disassembly and reassembly of generator is evident after inspection of unit, however, note that rotor shaft end is tapered and rotor is dislodged by tapping with a soft hammer.

# ENGINE

All models are powered by Briggs and Stratton engines. Refer to appropriate Briggs and Stratton section for engine service.

# ONAN

1400 73rd Avenue Northeast
Minneapolis, Minnesota 55432

| Model | Output-kw | Voltage | Engine Make | Model | Governed Rpm |
|---|---|---|---|---|---|
| 0.5AK-1M | 0.5 | 120 | Onan | AK | 1800 |
| 0.5AK-2M | 0.5 | 240 | Onan | AK | 1800 |
| 0.5AK-1R | 0.5 | 120 | Onan | AK | 1800 |
| 0.5AK-2R | 0.5 | 240 | Onan | AK | 1800 |
| 0.7AK-1M | 0.75 | 120 | Onan | AK | 1800 |
| 0.7AK-2M | 0.75 | 240 | Onan | AK | 1800 |
| 0.7AK-1R | 0.75 | 120 | Onan | AK | 1800 |
| 0.7AK-2R | 0.75 | 240 | Onan | AK | 1800 |
| 1.0AJ-1M | 1.0 | 120 | Onan | AJ | 1800 |
| 1.0AJ-2M | 1.0 | 240 | Onan | AJ | 1800 |
| 1.0AJ-1R | 1.0 | 120 | Onan | AJ | 1800 |
| 1.0AJ-2R | 1.0 | 240 | Onan | AJ | 1800 |
| 1.5AK-1M | 1.5 | 120 | Onan | AK | 3600 |
| 1.5AK-2M | 1.5 | 240 | Onan | AK | 3600 |
| 1.5AK-1P | 1.5 | 120 | Onan | AK | 3600 |
| 1.5AK-2P | 1.5 | 240 | Onan | AK | 3600 |
| 2.5AJ-1M | 2.5 | 120 | Onan | AJ | 3600 |
| 2.5AJ-2M | 2.5 | 240 | Onan | AJ | 3600 |
| 2.5AJ-3M | 2.5 | 120/240 | Onan | AJ | 3600 |
| 2.5AJ-1P | 2.5 | 120 | Onan | AJ | 3600 |
| 2.5AJ-2P | 2.5 | 240 | Onan | AJ | 3600 |
| 2.5AJ-3P | 2.5 | 120/240 | Onan | AJ | 3600 |
| 2.5AJ-1R | 2.5 | 120 | Onan | AJ | 3600 |
| 2.5AJ-2R | 2.5 | 240 | Onan | AJ | 3600 |
| 2.5AJ-3R | 2.5 | 120/240 | Onan | AJ | 3600 |

# GENERATOR

## MAINTENANCE

Generator brushes should be removed and inspected after every 500 hours of operation. Brushes are accessible after removing end cover (12–Fig. O1-1). Two types of brushes are used. Rectangular brushes should be renewed if worn to 5/8 inch (15.9 mm) or less while square brushes should be renewed if worn to 5/16 inch (7.9 mm) or less. On some models, the position of brush plate (13) is adjustable. An alignment mark is provided on some models so plate may be properly positioned, however, if mark is not found on plate, install plate so minimum brush arcing is observed and maximum generator output is obtained.

Note condition of commutator and slip rings and recondition if necessary.

## TROUBLESHOOTING

If little or no generator output is evident, be sure brushes, slip rings, and if so equipped, commutator are in good condition. The engine must be capable of maintaining governed rpm under load. Check wiring and be sure connections are clean and tight.

If field coil magnets have lost their residual magnetism, use the following procedure to flash the field; Remove end cover and run unit. Using a 6 volt dry cell battery, momentarily connect the negative battery terminal to the negative brush and the positive battery terminal to the positive brush. Disconnect battery as soon as generator voltage rises.

DC field current is derived either by use of a commutator on early models or by a diode on later models. To check diode, disconnect diode lead. Connect ohmmeter leads to diode terminals; note reading, then reverse ohmmeter leads and again note reading. If both readings are high or both are low, then diode is faulty and must be renewed.

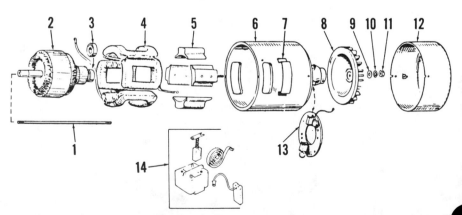

Fig. O1-1—Exploded view of generator. Brushes and brush holder (14) are used on model having a commutator on armature (2).

1. Stud
2. Armature
3. Bearing
4. Field coils
5. Field magnets
6. Frame
7. Cover
8. Fan
9. Washer
10. Lockwasher
11. Nut
12. End cover
13. Brush plate
14. Brush assy.

To check stator windings, disconnect all stator leads and test for continuity between each stator lead and a clean, paint free area of frame. Renew stator if any windings are grounded. Measure field coil resistance by connecting ohmmeter leads between negative and positive brush terminals with all brushes raised off commutator. Refer to following table for desired resistance readings which may be slightly higher if coils are warm from running.

| Model | Field Resistance (Ohms) |
|---|---|
| 0.5AK-1M | 4.38 |
| 0.5AK-2M | 4.38 |
| 0.5AK-1R | 4.38 |
| 0.5AK-2R | 4.38 |
| 0.7AK-1M | 15.4 |
| 0.7AK-2M | 15.4 |
| 0.7AK-1R | 15.4 |
| 0.7AK-2R | 15.4 |
| 1.0AJ-1M | 2.7 |
| 1.0AJ-2M | 2.7 |
| 1.0AJ-1R | 2.7 |
| 1.0AJ-2R | 2.7 |
| 1.5AK-1M | 12.0 |
| 1.5AK-2M | 12.0 |
| 1.5AK-1P | 12.0 |
| 1.5AK-2P | 12.0 |
| 2.5AJ-1M | 7.72 |
| 2.5AJ-2M | 7.72 |
| 2.5AJ-3M | 7.72 |
| 2.5AJ-1P | 7.72 |
| 2.5AJ-2P | 7.72 |
| 2.5AJ-3P | 7.72 |
| 2.5AJ-1R | 7.72 |
| 2.5AJ-2R | 7.72 |
| 2.5AJ-3R | 7.72 |

To test armature, use an ohmmeter or continuity tester and check for continuity between slip rings and between each slip ring and armature shaft. There should be continuity between slip rings except on models with four slip rings where continuity should not exist between the two center rings. No continuity should exist between any slip ring and shaft. Check for continuity between ad-

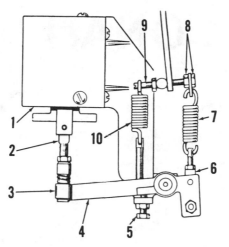

**Fig. O1-2—Diagram of Idle-Matic mechanism. Refer to text for adjustment.**

1. Solenoid
2. Plunger
3. Rod end
4. Lever
5. Screw
6. Rod end
7. Idle-Matic spring
8. Nuts
9. Governor sensitivity screw
10. Governor spring

jacent bars of commutator on models so equipped. If continuity does not exist, then there is an open armature winding. As a final test, armature may be removed and tested with a growler.

### OVERHAUL

Disassembly and reassembly is evident after inspection of unit and referral to Fig. 01-1. Note that armature shaft end is tapered to fit engine crankshaft and may have to be tapped to break taper loose. Armature shaft runout must not exceed 0.012 inch (0.30 mm) measured at bearing end with armature removed and supported at center. With armature mounted on engine crankshaft, runout at bearing must not exceed 0.002 inch (0.05 mm). Note that improper runout with armature mounted on crankshaft may be due to poorly mated tapers on armature shaft and engine crankshaft. Tighten armature shaft retaining nut to 40 ft.-lbs. (54.2 N·m).

# IDLE-MATIC

Some models may be equipped with a load sensing mechanism (Idle-Matic) which reduces engine speed when generator load is removed. A transformer in the control box senses generator load and trips a relay which actuates solenoid (1 – Fig. O1-2) when generator load is removed. As the solenoid plunger retracts, lever (4) tightens spring (7) which overcomes governor spring (10) tension to reduce governed speed to approximately 1800 rpm.

To adjust Idle-Matic, first check adjustment of carburetor and governor and adjust if necessary.

**NOTE: On Idle-Matic models, rod end (6 – Fig. O1-2) must be detached from lever (4) during governor adjustment.**

With idle control switch in "off" position, turn nuts (8 – Fig. O1-2) so they are positioned at outer end of sensitivity screw (9) then detach rod end (6) from lever (4). Back out screw (5) until solenoid plunger (2) is fully extended. Reconnect rod end (6) and run generator under no load with idle control switch "on". Adjust rod ends (3 and 6) so engine idles at 1800 rpm. Apply load to generator and turn screw (5) in so spring (7) is taut but not stretched. Be sure all locknuts are tight.

# ENGINE

## OVERHAUL

Engine make and model are listed at beginning of section. Refer to Onan engine section for engine service.

### ENGINE CONTROLS

On some models, the generator also serves as the engine starting motor. Refer to SERVICING ONAN ACCESSORIES in engine section. Note that on some models a low oil pressure switch and high air temperature switch may be used to protect engine.

# ONAN

| Model | Output-kw | Voltage | Engine Make | Engine Model | Governed Rpm |
|-------|-----------|---------|-------------|--------------|--------------|
| 2.5LK-1M | 2.5 | 120 | Onan | LK | 1800 |
| 2.5LK-1R | 2.5 | 120 | Onan | LK | 1800 |
| 2.5LK-3CR | 2.5 | 120/240 | Onan | LK | 1800 |
| 2.5LK-3M | 2.5 | 120/240 | Onan | LK | 1800 |
| 2.5LK-3R | 2.5 | 120/240 | Onan | LK | 1800 |
| 4.0BFA-1R | 4.0 | 120 | Onan | B43M | 1800 |
| 4.0CCK-3CE | 4.0 | 120/240 | Onan | CCK | 1800 |
| 4.0CCK-3CP | 4.0 | 120/240 | Onan | CCK | 1800 |
| 4.0CCK-3CR | 4.0 | 120/240 | Onan | CCK | 1800 |
| 5.0BGA-3CR | 5.0 | 120/240 | Onan | B48M | 1800 |
| 5.0CCK-3CE | 5.0 | 120/240 | Onan | CCK | 1800 |
| 5.0CCK-3CP | 5.0 | 120/240 | Onan | CCK | 1800 |
| 5.0CCK-3CR | 5.0 | 120/240 | Onan | CCK | 1800 |
| 6.5NH-3CE | 6.5 | 120/240 | Onan | NH | 1800 |
| 6.5NH-3CR | 6.5 | 120/240 | Onan | NH | 1800 |

# GENERATOR

### MAINTENANCE

Generator brushes, commutator and slip rings should be inspected after every 500 hours of operation.

To determine brush length on early CCK and LK models, remove end cover and measure brush length. If length is ⅝ inch (15.9 mm) or less, renew brush. The position of brush plate (12 – Fig. O2-1) is adjustable. Mark (M – Fig. O2-2) on brush plate (12) should be aligned with edge of end bell (13). If mark is absent from brush plate, adjust position of plate to obtain maximum voltage and minimum brush arcing with brushes seated and generator hot.

On later CCK and LK models and all other models, brush wear may be determined by measuring depth of brush in brush holder as shown in Fig. O2-3. Measure from top of brush holder to top of brush. Renew AC brushes if depth is 1-1/16 inches (27.0 mm) or more and renew DC brushes if depth is 1 inch (25.4 mm) or more.

Note condition of commutator and slip rings and recondition if necessary.

### TROUBLESHOOTING

If little or no generator output is evident, be sure brushes, commutator and slip rings are in good condition. The engine must be capable of maintaining governed rpm under load. Check wiring and be sure connections are clean and tight.

#### Early Models CCK And LK

If field coil magnets have lost their residual magnetism, use the following procedure to flash the field: Remove end cover and run unit. Using a 6 volt dry cell battery, momentarily connect the negative battery terminal to the negative brush and the positive battery terminal to the positive brush. Disconnect battery as soon as generator voltage rises.

To check stator windings, disconnect all stator leads and test for continuity between each stator lead and a clean, paint free area of frame. Renew stator if any windings are grounded. Measure field coil resistance by connecting ohmmeter leads between negative and positive brush terminals with all brushes raised off commutator. Refer to following table for desired resistance readings which may be slightly higher if coils are warm from running.

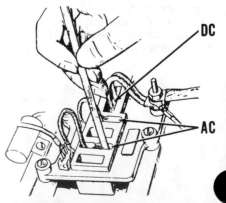

Fig. O2-2—Mark (M) on brush plate (12) should align with edge of end bell (13) on early CCK and LK models.

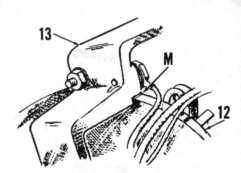

Fig. O2-3—View showing location of AC and DC brushes. To determine brush wear measure depth from top of brush holder to brush top as shown and refer to text.

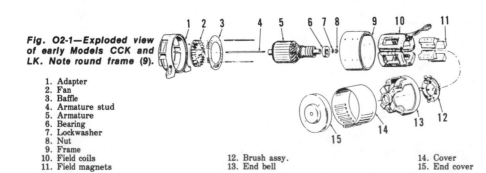

Fig. O2-1—Exploded view of early Models CCK and LK. Note round frame (9).

1. Adapter
2. Fan
3. Baffle
4. Armature stud
5. Armature
6. Bearing
7. Lockwasher
8. Nut
9. Frame
10. Field coils
11. Field magnets
12. Brush assy.
13. End bell
14. Cover
15. End cover

| Model | Field Resistance (Ohms) |
|---|---|
| 2.5LK-1M, 2.5LK-1R, 2.5LK-3M | 1.00 |
| 3.5CCK | 0.80 |
| 4.0CCK | 0.73 |
| 5.0CCK | 0.73 |

To test armature, use an ohmmeter or continuity tester and check for continuity between slip rings and between each slip ring and armature shaft. There should be continuity between slip rings but no continuity between either slip ring and shaft. Check for continuity be-tween adjacent bars of commutator. If continuity does not exist, then there is an open armature winding. As a final test, armature may be removed and tested with a growler.

## All Other Models

To check field coils, disconnect field lead wires shown in Fig. 02-5. Measure resistance of shunt field coil by connecting ohmmeter leads to field coil leads F2 and S1. Measure resistance of series field coil by connecting ohmmeter leads to field coil leads S1 and S2, then measure resistance between field coil leads S1 and S3. Refer to following table for desired resistance readings which may be slightly higher if coils are warm from running.

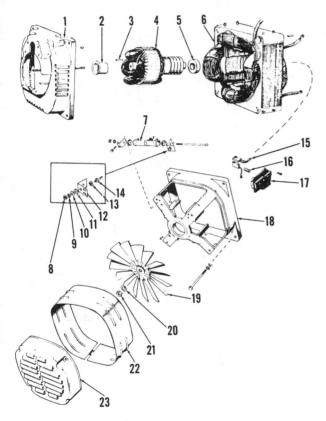

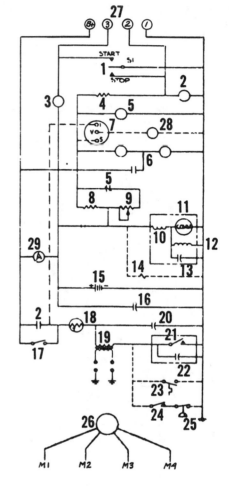

*Fig. 02-4—Typical view of later CCK and LK models and all other models. Note square frame (6). A fan is used in place of hub (2) on some models.*

1. Adapter
2. Hub
3. Key
4. Armature
5. Bearing
6. Stator & frame
7. Resistor
8. Nut
9. Lockwasher
10. Washer
11. Insulator washer
12. Bracket
13. Shouldered insulator washer
14. Diode
15. Brushes
16. Brush spring
17. Brush holder
18. End bell
19. Fan
20. Lockwasher
21. Nut
22. Cover
23. End cover

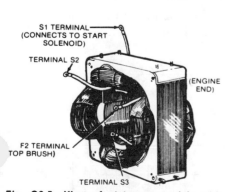

*Fig. 02-5—View of stator on models with square frame.*

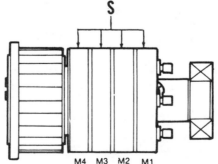

*Fig. 02-6—Slip rings (S) are designated as shown. M3 and M4 rings are not used on Model 4.0BFA-1R.*

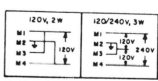

*Fig. 02-7—Wiring schematic for early Models CCK and LK.*

1. Start switch
2. Stop relay
3. Start solenoid
4. Resistor
5. Voltage regulator
6. Start disconnect relay assy.
7. Vacuum switch
8. Resistor
9. Resistor
10. Series field
11. Commutator
12. Shunt field
13. Capacitor
14. Electric choke
15. Battery
16. Capacitor
17. Manual start switch
18. Resistor
19. Ignition coil
20. Capacitor
21. Ignition breaker points
22. Capacitor
23. High air termperature switch
24. Oil pressure defeat switch
25. Oil pressure switch
26. Generator output
27. Remote control terminals
28. Fuel solenoid
29. Ammeter

| Model | Shunt Field Resistance (Ohms) | Series Field Resistance (Ohms) |
|---|---|---|
| 2.5L-3CR | 0.89 | 0.015 |
| 4.0BFA-1R | 1.00 | 0.019 |
| 4.0CCK-3CE | 1.48 | 0.014 |
| 4.0CCK-3CP | 1.48 | 0.014 |
| 4.0CCK-3CR | 1.48 | 0.014 |
| 5.0BGA-3CR | 1.88 | 0.014 |
| 5.0CCK-3CE | 1.48 | 0.014 |
| 5.0CCK-3CP | 1.48 | 0.014 |
| 5.0CCK-3CR | 1.48 | 0.014 |
| 6.5NH-3CE | 0.94 | 0.010 |
| 6.5NH-3CR | 0.94 | 0.010 |

To check armature, refer to Fig. 02-6 and using a continuity tester or ohmmeter determine if continuity exists between slip rings M1 and M2 on all models and between M3 and M4 on all models except 4.0BFA-1R. Continuity should exist. However, continuity should NOT exist between slip rings M2 and M3 on all models except 4.0BFA-1R. Check for continuity between adjacent bars of commutator. If continuity does not exist, then there is an open armature winding. Continuity should NOT exist between any slip ring and commutator, between commutator and armature shaft or between any slip ring and armature shaft. As a final test, armature may be removed and tested with a growler.

## OVERHAUL

Disassembly and reassembly is evident after inspection of unit and referral to Fig. O2-1 or O2-4. Note that armature shaft end is tapered to fit engine crankshaft and may have to be tapped to break taper loose. Be sure armature shaft is properly mounted on engine crankshaft taper so bearing end of armature runs true in end bell. Do not tighten armature shaft nut until end bell retaining nuts are tightened. Tighten armature shaft nut to 30-40 ft.-lbs. (40.6-54.2 N·m) on early Models CCK and LK or to 45-50 ft.-lbs. (60.9-67.7 N·m) on all other models.

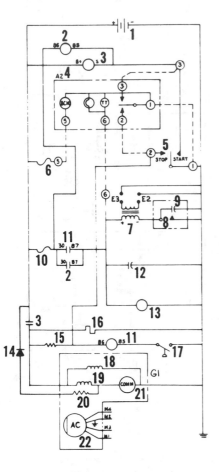

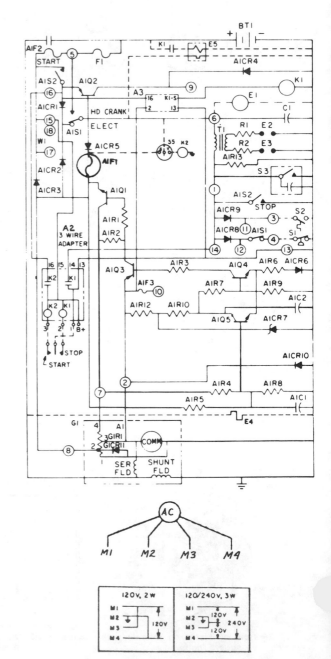

*Fig. O2-9—Wiring schematic for Models CCK and NH equipped with a printed circuit board in control box. Note that fuse (A151) was used on early models while later models are equipped with fuse (F1). Fuse (F1) is a 9 amp in-line fuse and should be installed on early models.*

BT1. Battery
E1. Fuel pump or gas valve
E4. Electric choke
K1. Start solenoid
K2. Fuel solenoid
S1. Oil pressure switch
S2. High air temperature switch
S3. Ignition breaker points
S5. Vacuum switch
T1. Ignition coil
A1S1. Oil pressure defeat switch
A1S2. Start switch

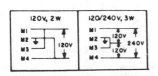

*Fig. O2-8—Wiring schematic for BFA, SGA and NH recreational vehicle generators.*

1. Battery
2. Ignition start relay
3. Start solenoid
4. Remote control board
5. Start switch
6. Fuse
7. Ignition coil
8. Ignition breaker points
9. Capacitor
10. Fuse
11. Ignition run relay
12. Capacitor
13. Fuel pump
14. Diode
15. Resistor
16. Electric choke
17. High air temperature switch
18. Shunt field
19. Series field
20. Resistor
21. Commutator
22. Generator output

Illustrations courtesy of Onan Corp.

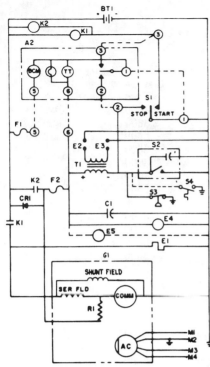

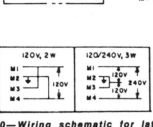

# ENGINE

## OVERHAUL

Engine make and model are listed at beginning of section. Refer to Onan engine section for engine service.

## ENGINE CONTROLS

On some models, the generator also serves as the engine starting motor. Refer to SERVICING ONAN ACCESSORIES in engine section. Note that on some models a low oil pressure switch and high air temperature switch may be used to protect engine.

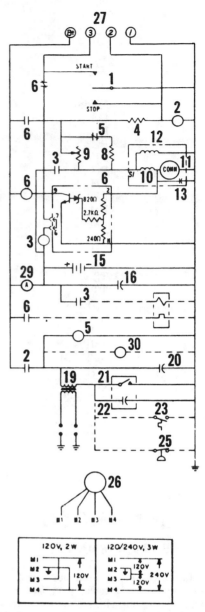

**Fig. O2-10—Wiring schematic for later Models CCK, LK and NH.**

| | |
|---|---|
| A2. Remote control | S1. Start switch |
| BT1. Battery | S2. Ignition breaker points |
| E1. Electric choke | S3. Oil pressure switch |
| E4. Fuel pump | S4. High air temperature switch |
| E5. Fuel solenoid | |
| K1. Start solenoid | T1. Ignition coil |
| K2. Ignition start relay | |

**Fig. O2-11—Wiring schematic for early NH models. Refer to Fig. O2-7 for component identification except for: 30. Fuel pump or gas valve.**

# ONAN

| Model | Output-kw | Voltage | Make | Model | Governed Rpm |
|---|---|---|---|---|---|
| | | | — Engine — | | |
| PC | 1.75 | 120 | B&S | * | 3600 |
| PD | 1.2 | 120 | B&S | * | 3600 |
| PE | 2.0 | 120 | B&S | * | 3600 |
| PF | 3.0 | 120/240 | B&S | * | 3600 |
| PG | 4.5 | 120/240 | B&S | * | 3600 |
| PH | 5.5 | 120/240 | B&S | * | 3600 |
| PK | 4.5 | 120/240 | B&S | * | 3600 |
| PM | 5.5 | 120/240 | B&S | * | 3600 |
| TM | 1.2 | 120 | Tecumseh | * | 3600 |
| TN | 1.75 | 120 | Tecumseh | * | 3600 |
| TP | 2.0 | 120 | Tecumseh | * | 3600 |
| TR | 3.0 | 120/240 | Tecumseh | * | 3600 |
| TS | 4.5 | 120/240 | Tecumseh | * | 3600 |
| TT | 5.5 | 120/240 | Tecumseh | * | 3600 |
| 1.7PC-1P | 1.75 | 120 | B&S | 100212 | 3600 |
| 2.2PE-1P | 2.25 | 120 | B&S | 130212 | 3600 |
| 3.0PL-1P | 3.0 | 120 | B&S | 170412 | 3600 |
| 3.0PL-3P | 3.0 | 120/240 | B&S | 170412 | 3600 |
| 3.2PF-3P | 3.25 | 120/240 | B&S | 190432 | 3600 |
| 3.7PF-3P | 3.75 | 120/240 | B&S | 190432 | 3600 |
| 5.0PK-3E | 5.0 | 120/240 | B&S | 220435 | 3600 |
| 5.0PK-3P | 5.0 | 120/240 | B&S | 220432 | 3600 |
| 5.0PN-3E | 5.0 | 120/240 | B&S | 252415 | 3600 |
| 5.0PN-3P | 5.0 | 120/240 | B&S | 252412 | 3600 |
| 6.5PM-3E | 6.5 | 120/240 | B&S | 326435 | 3600 |
| 6.5PM-3P | 6.5 | 120/240 | B&S | 326432 | 3600 |

*Engine model number was not available.

# GENERATOR

## MAINTENANCE

Manufacturer recommends inspecting and servicing, if necessary, generator brushes and slip rings once every year. Brushes are accessible after removing receptacle end panel or cover plate while the end frame must be removed to service the slip rings. Renew brushes if worn to a length of 5/16 inch (7.9 mm) or less.

## TROUBLESHOOTING

If little or no output is generated, check the following: The engine must be in good condition and maintain a governed speed under load of 3600 rpm. Check condition of circuit breakers and/or fuses. All wiring connections must be clean and tight. Brushes and slip rings must be in good condition.

If field coil magnets have lost their residual magnetism, use the following procedure: Remove end receptacle panel or cover plate and run unit. Using a 6 volt dry cell battery, momentarily connect the negative battery terminal to the generator frame and the positive battery terminal to the positive generator brush. Disconnect battery as soon as generator voltage rises. If loss of field magnetism is chronic, capacitor (11–Fig. O3-2) may be defective and should be renewed.

To check rectifier diodes (D–Fig. O3-3) on early models, unsolder and remove diodes from end bell. Connect

Fig. O3-1—Exploded view of generator. Refer to Fig. O3-2 for end frame components.

1. Adapter
2. Rotor
3. Stud
4. Sleeve
5. Stator
6. Nut
7. End frame

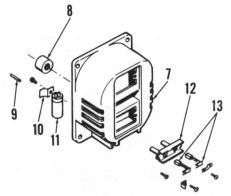

Fig. O3-2—Exploded view of end frame components.

7. End frame
8. Bearing
9. Dowel pin
10. Retainer
11. Capacitor
12. Brush holder
13. Brushes

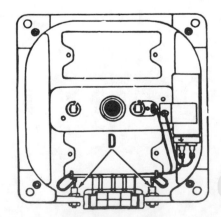

Fig. O3-3—View showing location of diodes (D) on early models.

         Illustrations courtesy of Onan Corp.

ohmmeter leads to diode, note reading, then reverse ohmmeter leads and again note reading. If both readings are high or both are low, then diode is faulty and must be renewed.

To check modular type rectifier on intermediate and later models, disconnect black positive (+) lead and "X2" and "X3" AC stator leads from rectifier. See Figs. O3-4 and O3-5. Connect ohmmeter leads to an AC terminal and positive (+) terminal, note reading, reverse ohmmeter leads and again note reading. One reading should be high and the other low. If both readings are high or both

are low, the rectifier is defective and must be renewed. Repeat test by connecting ohmmeter leads between re-

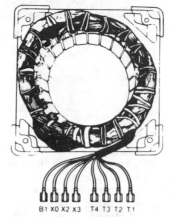

Fig. O3-7—View of stator and leads on later models.

| | |
|---|---|
| B1. Orange | T4. Yellow |
| T1. Blue | X0. Black |
| T2. Black | X2. Lt. Blue |
| T3. Red | X3. Lt. Blue |

maining AC terminal and positive (+) terminal. If test readings using remaining AC terminal are both high or both low, rectifier is faulty and must be renewed.

To check stator, note terminal board in Fig. O3-6 used on early models and loose stator wire terminals as shown in Fig. O3-7. On all models, connect a jumper wire between terminals T2 and T3 then measure resistance between terminals T1 and T4. Note length of stator frame and desired resistance reading in following table:

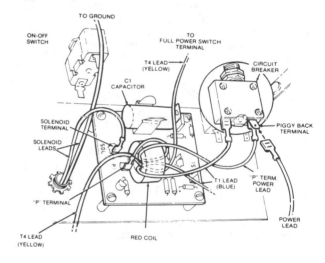

Fig. O3-8—View of Idle-matic solenoid. Refer to text for adjustment.

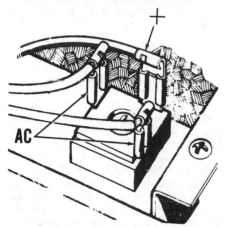

Fig. O3-4—View of rectifier on intermediate models showing (AC) terminals and positive (+) terminal.

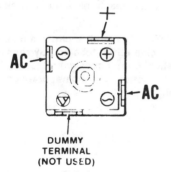

Fig. O3-5—View of rectifier on later models showing (AC) terminals and positive (+) terminal.

Fig. O3-9—On early models, route wires to Idle-matic control board as shown. On 120 volt only models, stator wire T3 replaces T4 wire shown and T3 wire must pass through red coil in same direction as blue T1 stator wire.

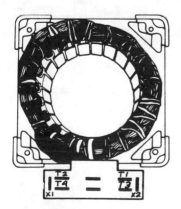

Fig. O3-6—View of stator and terminal board on early models.

Fig. O3-10—View of Idle-matic control board on later models.

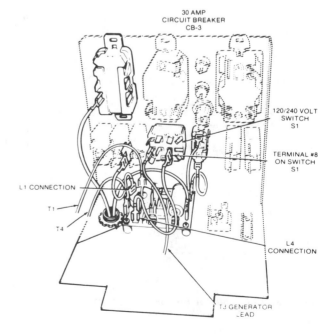

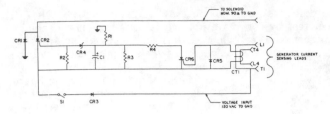

*Fig. O3-11—Wiring diagram of Idle-matic control board.*

# IDLE-MATIC

Some models may be equipped with a solid-state unit attached to the receptacle panel and a solenoid connected to the carburetor throttle which senses generator load and will reduce engine speed to idle when the load is removed.

To adjust solenoid, first set engine idle speed screw so engine will idle at 1800 rpm. With engine stopped, push solenoid plunger (P – Fig. O3-8) into solenoid until bottomed. Turn adjusting nuts (N) so linkage rod (R) pulls carburetor throttle arm against idle speed screw. Tighten adjusting nuts and check operation of solenoid.

# ENGINE

Engine make and model are listed at beginning of section. Refer to Briggs and Stratton or Tecumseh engine section for engine service except for the following:

**EXTERNAL OIL SYSTEM.** On models designed for extended use, an external oil system is added to prevent depletion of engine crankcase oil. Probe (S – Fig. O3-12) senses crankcase oil level and valve (V) allows oil from oil tank (T) to flow into crankcase. With engine crankcase full, oil level in oil tank should just reach mark on oil tank sight glass.

**NOTE: Always fill engine crankcase using engine oil fill tube and by reading oil dipstick. Do not consider oil tank oil level an indication of engine crankcase oil level.**

| Stator Frame Length | T1 to T4 Resistance |
|---|---|
| 1½ in. (38.1 mm) | 2.57-3.30 ohms |
| 2¼ in. (57.2 mm) | 1.22-1.57 ohms |
| 3 in. (76.2 mm) | 0.75-0.96 ohms |
| 3½ in. (89.9 mm) | 0.55-0.67 ohms |
| 4½ in. (114.3 mm) | 0.37-0.47 ohms |
| 6 in. (152.4 mm) | 0.25-0.33 ohms |

To check exciter coil in stator, connect ohmmeter leads to terminals X1 and X2 on early model terminal board (Fig. O3-6) or to X2 and X3 stator leads of later models shown in Fig. O3-7. Measure length of stator frame and refer to following table of desired resistance readings:

| Stator Frame Length | Exciter Coil Resistance |
|---|---|
| 1½ in. (38.1 mm) | 2.04-3.30 ohms |
| 2¼ in. (57.2 mm) | 1.54-2.75 ohms |
| 3 in. (76.2 mm) | 1.54-2.31 ohms |
| 3½ in. (89.9 mm) | 1.52-1.86 ohms |
| 4½ in. (114.3 mm) | 1.29-2.20 ohms |
| 6 in. (152.4 mm) | 1.14-2.20 ohms |

To determine if rotor is shorted to ground, connect ohmmeter leads between slip rings and rotor shaft. Renew rotor if there is continuity between slip rings and ground. Measure resistance between slip rings and ground. Measure length of stator frame. Measure resistance between rotor slip rings and renew rotor if resistance is not as listed in following table:

| Stator Frame Length | Rotor Resistance |
|---|---|
| 1½ in. (38.1 mm) | 25.7-31.5 ohms |
| 2¼ in. (57.2 mm) | 18.5-22.6 ohms |
| 3 in. (76.2 mm) | 21.0-25.6 ohms |
| 3½ in. (89.9 mm) | 22.5-27.5 ohms |
| 4½ in. (114.3 mm) | 26.2-32.0 ohms |
| 6 in. (152.4 mm) | 31.0-37.9 ohms |

## OVERHAUL

Disassembly and reassembly of generator is evident after inspection of unit and referral to Fig. O3-1 and O3-2. Note that rotor shaft is tapered and rotor must be tapped with a soft hammer to break taper loose. Tighten rotor nut (6 – Fig. O3-1) to 10-15 ft.-lbs. (13.5-20.3 N·m).

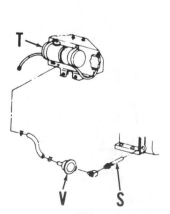

*Fig. O3-12—Diagram of external oil system used on some models.*

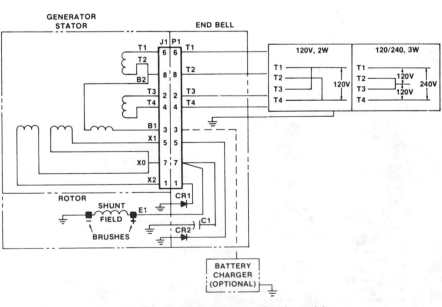

*Fig. O3-13—Wiring diagram of early models.*

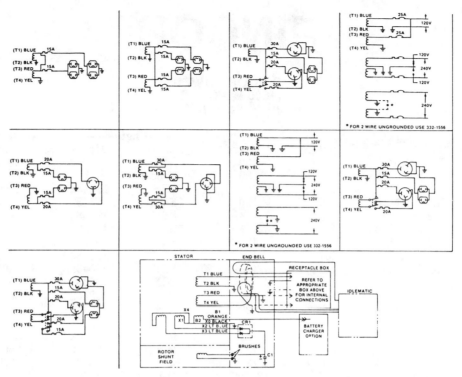

**Fig. O3-14—Wiring diagrams for intermediate models.**

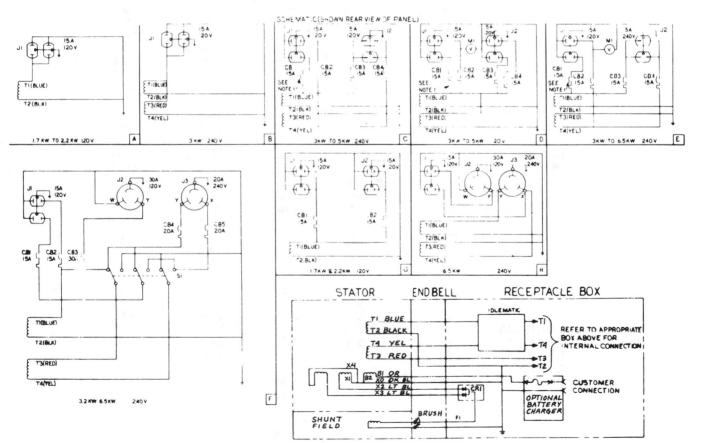

**Fig. O3-15—Wiring diagrams for late models.**

# PINCOR

**PINCOR POWER**
**11600 W. King St.**
**Franklin Park, IL 60131**

| Model | Output-kw | Voltage | Engine Make | Engine Model | Governed Rpm |
|---|---|---|---|---|---|
| CRF20HM1 | 2.0 | 120 | *B&S | 131232 | 3600 |
| CRF20HM1A | 2.0 | 120 | Wisconsin | Robin | 3600 |
| CRF30HM1 | 3.0 | 120 | *B&S | 195432 | 3600 |
| CRF30HM1A | 3.0 | 120 | *B&S | 195432 | 3600 |
| CRF30HM5 | 3.0 | 120/240 | *B&S | 195432 | 3600 |
| CRF30HM5A | 3.0 | 120/240 | *B&S | 195432 | 3600 |
| CRF37HM1 | 3.7 | 120 | *B&S | 195432 | 3600 |
| CRF37HM1A | 3.7 | 120 | *B&S | 195432 | 3600 |
| CRF37HM5 | 3.7 | 120/240 | *B&S | 195432 | 3600 |
| CRF37HM5A | 3.7 | 120/240 | *B&S | 195432 | 3600 |
| CRF45HM1 | 4.5 | 120 | B&S | 243432 | 3600 |
| CRF45HM1A | 4.5 | 120 | B&S | 243432 | 3600 |
| CRF45HM5 | 4.5 | 120/240 | B&S | 243432 | 3600 |
| CRF45HM5A | 4.5 | 120/240 | B&S | 243432 | 3600 |
| CRF55HM1 | 5.5 | 120 | B&S | 326432 | 3600 |
| CRF55HM1A | 5.5 | 120 | B&S | 326432 | 3600 |
| CRF55HM5 | 5.5 | 120/240 | B&S | 326432 | 3600 |
| CRF55HM5A | 5.5 | 120/240 | B&S | 326432 | 3600 |
| FRF30HE5 | 3.0 | 120/240 | B&S | 195437 | 3600 |
| FRF30HM5 | 3.0 | 120/240 | B&S | 190412 | 3600 |
| FRF45HE5 | 4.5 | 120/240 | B&S | 252417 | 3600 |
| FRF45HM5 | 4.5 | 120/240 | B&S | 252412 | 3600 |
| FRF55HE5 | 5.5 | 120/240 | B&S | 302437 | 3600 |
| FRF55HM5 | 5.5 | 120/240 | B&S | 302432 | 3600 |
| RF10 | 1.0 | 115VAC 12VDC | ** | ** | 3600 |
| RF12HM1 | 1.25 | 120 | B&S | 80232 | 3600 |
| RF20HM1 | 2.0 | 120 | B&S | 130232 | 3600 |
| RF30HE1 | 3.0 | 120 | B&S | 195437 | 3600 |
| RF30HM1 | 3.0 | 120 | B&S | 190412 | 3600 |
| RF30HM1A | 3.0 | 120 | B&S | 190412 | 3600 |
| RF30HE2 | 3.0 | 120/240 | B&S | 195437 | 3600 |
| RF30HM2 | 3.0 | 120/240 | B&S | 190412 | 3600 |
| RF30HM2A | 3.0 | 120/240 | B&S | 190412 | 3600 |
| RF30HE5 | 3.0 | 120/240 | B&S | 195437 | 3600 |
| RF30HM5 | 3.0 | 120/240 | B&S | 190412 | 3600 |
| RF30HM5A | 3.0 | 120/240 | B&S | 190412 | 3600 |
| RF37HE1 | 3.7 | 120 | B&S | 195437 | 3600 |
| RF37HM1 | 3.7 | 120 | B&S | 190412 | 3600 |
| RF37MH1A | 3.7 | 120 | B&S | 190412 | 3600 |
| RF37HE2 | 3.7 | 120/240 | B&S | 195437 | 3600 |
| RF37HM2 | 3.7 | 120/240 | B&S | 190412 | 3600 |
| RF37HM2A | 3.7 | 120/240 | B&S | 190412 | 3600 |
| RF37HE5 | 3.7 | 120/240 | B&S | 195437 | 3600 |
| RF37HM5 | 3.7 | 120/240 | B&S | 190412 | 3600 |
| RF37HM5A | 3.7 | 120/240 | B&S | 190412 | 3600 |
| RF50HE1 | 5.0 | 120 | B&S | 252417 | 3600 |
| RF50HM1 | 5.0 | 120 | B&S | 252412 | 3600 |
| RF50HM1A | 5.0 | 120 | B&S | 252412 | 3600 |
| RF50HE2 | 5.0 | 120/240 | B&S | 252417 | 3600 |
| RF50HM2 | 5.0 | 120/240 | B&S | 252412 | 3600 |
| RF50HM2A | 5.0 | 120/240 | B&S | 252412 | 3600 |
| RF50HE5 | 5.0 | 120/240 | B&S | 252417 | 3600 |
| RF50HM5 | 5.0 | 120/240 | B&S | 252412 | 3600 |
| RF50HM5A | 5.0 | 120/240 | B&S | 252412 | 3600 |
| RF60HE1 | 6.0 | 120 | B&S | 401417 | 3600 |
| RF60HM1 | 6.0 | 120 | B&S | 401000 | 3600 |
| RF60HM1A | 6.0 | 120 | B&S | 401000 | 3600 |

Illustrations courtesy of Pioneer Gen-E-Motor Corp.

| Model | Output-kw | Voltage | Engine Make | Engine Model | Governed Rpm |
|-------|-----------|---------|------|-------|--------------|
| RF60HE2 | 6.0 | 120/240 | B&S | 401417 | 3600 |
| RF60HM2 | 6.0 | 120/240 | B&S | 401000 | 3600 |
| RF60HM2A | 6.0 | 120/240 | B&S | 401000 | 3600 |
| RF65HE1 | 6.5 | 120 | B&S | ** | 3600 |
| RF65HM1 | 6.5 | 120 | B&S | ** | 3600 |
| RF65HM1A | 6.5 | 120 | B&S | ** | 3600 |
| RF65HE2 | 6.5 | 120/240 | B&S | ** | 3600 |
| RF65HM2 | 6.5 | 120/240 | B&S | ** | 3600 |
| RF65HM2A | 6.5 | 120/240 | B&S | ** | 3600 |
| RF70HE1 | 7.0 | 120 | B&S | 401417 | 3600 |
| RF70HM1 | 7.0 | 120 | B&S | 401000 | 3600 |
| RF70HM1A | 7.0 | 120 | B&S | 401000 | 3600 |
| RF70HE2 | 7.0 | 120/240 | B&S | 401417 | 3600 |
| RF70HM2 | 7.0 | 120/240 | B&S | 401000 | 3600 |
| RF70HM2A | 7.0 | 120/240 | B&S | 401000 | 3600 |
| RF1250H | 1.25 | 115 | ** | ** | 3600 |
| RF1500H | 1.5 | 115 | ** | ** | 3600 |
| RF1600H | 1.6 | 115VAC 12VDC | ** | ** | 3600 |
| RF2000H | 2.0 | 115 | ** | ** | 3600 |
| RF2500H | 2.5 | 115 | ** | ** | 3600 |
| RF2500HDV | 2.5 | 115/230 | ** | ** | 3600 |
| RF3000H | 3.0 | 115 | ** | ** | 3600 |
| RF3000HDV | 3.0 | 115/230 | ** | ** | 3600 |
| RF3000HFP | 3.0 | 115/230 | ** | ** | 3600 |
| RF3500H | 3.5 | 115 | ** | ** | 3600 |
| RF3500HDV | 3.5 | 115/230 | ** | ** | 3600 |
| RF3800HDV | 3.8 | 115/230 | ** | ** | 3600 |
| RF4000H | 4.0 | 115 | ** | ** | 3600 |
| RF4000HDV | 4.0 | 115/230 | ** | ** | 3600 |
| RF4000HFP | 4.0 | 115/230 | ** | ** | 3600 |

*May also be equipped with Wisconsin Robin engine.
**Note engine make and model on ID plate and refer to appropriate engine section.

# GENERATOR

## MAINTENANCE

Brushes are accessible after removing end bell (3—Fig. P1-1 or P1-2). Renew brushes if worn to 5/16 inch (7.9 mm) or less. Note condition of rotor slip rings and recondition if necessary.

Bearing (10—Fig. P1-1 or 9—P1-2) is sealed and does not require maintenance, however, manufacturer recommends renewing bearing after 1000 hours of operation.

## TROUBLESHOOTING

If little or no output is generated, check the following: On models so equipped, check fuses and circuit breakers. The engine must be in good condition and maintain desired governed speed of 3600 rpm under load. Brushes and slip rings must be in good condition. All wiring connections must be clean and tight. Be sure wire leads are marked before disconnecting.

If residual magnetism is insufficient to induce generator output, it may be necessary to flash field as follows: Run unit at maximum governed speed. Manufacturer recommends momentarily inserting probes connected to a 115 VAC source into the slots of one of the generator AC receptacles. An alternate method is to momentarily connect a 12 VDC battery so the negative battery terminal is connected to the negative brush terminal and the positive battery and brush terminals are connected. Using either method, connect external voltage source only until generator output rises. If loss of residual magnetism is a chronic problem, then rotor should be renewed. Rotor magnet strength may be measured using a spring scale and Pincor test fixture 22916-000 as shown in

*Fig. P1-1—Exploded view of generator used on models prior to 1977. Note that negative brush holder (14) and rectifier diodes (15) are mounted on backside of plate (5). On Model RF10, a diode for battery charging circuit is also mounted on plate.*

1. Receptacle cover
2. Receptacle
3. End bell
4. Positive brush holder
5. Plate
6. Stator assy.
7. Bolt
8. Lockwasher
9. Washer
10. Bearing
11. Rotor
12. Fan
13. End frmae
14. Negative brush holder
15. Diodes

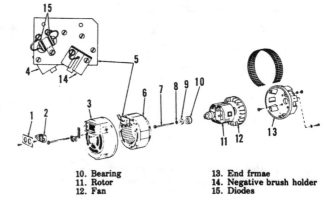

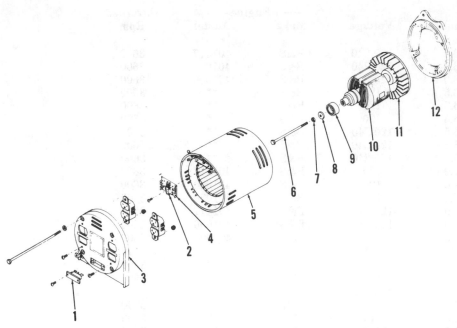

Fig. P1-2—Exploded view of generator used on models after 1976.

1. Rectifier
2. Brushes
3. End bell
4. Plate
5. Stator assy.
6. Bolt
7. Lockwasher
8. Washer
9. Bearing
10. Rotor
11. Fan
12. Adapter

Fig. P1-3. Each rotor magnet should exert 3½ pounds (1.6 Kg) pull. An alternate method may be used by comparing magnet strength of new and suspect rotor.

To test exciter rectifier diodes (15 – Fig. P1-1) on models prior to 1977, use the following procedure: Remove end bell and place insulating material under brushes so brushes are isolated from slip rings. Connect an ohmmeter lead to positive brush terminal (BT – Fig. P1-4) and the other ohmmeter lead to one of the rectifier terminals (RT). Note ohmmeter reading, reverse ohmmeter leads using same (RT) terminal and again note ohmmeter reading. One reading should be high and the other reading should be low. If both readings are either high or low, then diode connected to (RT) terminal tested faulty and must be renewed. Test remaining diode using same procedure with ohmmeter connected to positive brush terminal (BT) and opposite rectifier terminal (RT).

To test rectifier diode module (1 – Fig. P1-2) on models after 1976, remove and disconnect diode module from end bell (3). Using an ohmmeter, connect an ohmmeter lead to center module terminal (C – Fig. P1-5) and to one of the outer terminals (O). Note ohmmeter reading then reverse ohmmeter leads and again note reading. One reading should be high while the other reading should be low. If both readings are high or both low, then a rectifier diode is faulty and module must be renewed. Repeat test by connecting ohmmeter to center terminal (C) and remaining outer terminal (O). If readings are both high or both low, then remaining rectifier diode is faulty and diode module must be renewed.

To check stator windings, disassemble generator sufficiently to disconnect stator leads. Wire leads are marked on some models, if not, refer to Fig. P1-6. Using a sensitive ohmmeter, connect ohmmeter to leads indicated in table shown in Fig. P1-7 and note desired resistance. Renew stator if high resistance should be present between stator leads T1 and T3, T1 and T4, T2 and T3, and T2 and T4. Infinite resistance should also exist between any AC lead (T1, T2, T3 or T4) and any exciter lead (A, B or C). On Model RF10, infinite resistance should exist between battery charger coil leads (AA and BB) and any AC lead (T1, T2, T3 or T4) or exciter lead (A, B or C). To check for grounded stator windings, manufacturer recommends fabricating a test bulb circuit as shown in Fig. P1-8. Plug test wire into a 115 VAC source, ground one test wire to stator frame and alternately connect remaining test wire to each stator wire lead. If bulb glows at all, then a stator winding is grounded and stator must be renewed.

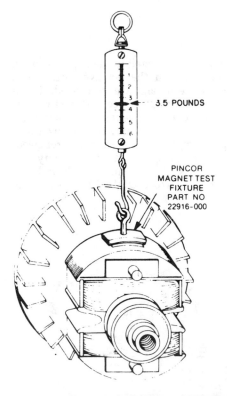

Fig. P1-3—Use Pincor test fixture 22916-000 and a spring scale to test magnet strength as outlined in text.

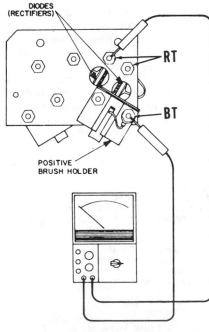

Fig. P1-4—To check rectifier diodes on models prior to 1977, connect ohmmeter to positive brush terminal (BT) and diode terminals (RT). Refer to text.

Fig. P1-5—To check rectifier module on models after 1976, connect ohmmeter to center (C) and outer (O) terminals as outlined in text.

Illustrations courtesy of Pioneer Gen-E-Motor Corp.

To check rotor, use ohmmeter and measure resistance between slip rings.

If ohmmeter reading is not as listed in following table, renew rotor.

To determine if rotor windings are grounded, manufacturer recommends fabricating a test bulb circuit as shown

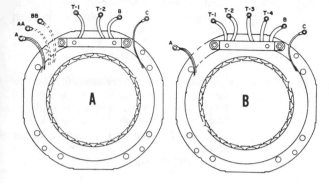

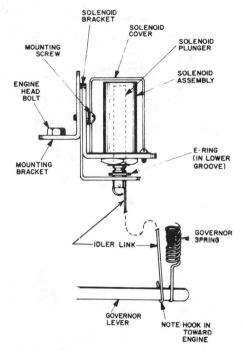

*Fig. P1-6—Diagram of stator wire leads. Stator (A) is used on generators with less than 3 kw output. Leads (AA and BB) are used on Model RF1600H; refer to schematic in Fig. P1-34. Stator (B) is used on generators with 3 kw or more output.*

| Output-kw | Excitor Windings | | | AC Windings | Battery Charger Windings | |
|---|---|---|---|---|---|---|
| | A to B | B to C | A to C | T1 to T2 T3 to T4 | A to AA A to BB | AA to BB |
| 1.0 | 0.80 | 1.6 | 0.80 | 0.99 | | |
| 1.25 | 0.28 | 0.57 | 0.28 | 0.66 | | |
| 1.5 | 0.27 | 0.55 | 0.27 | 0.42 | | |
| 1.6 | 0.27 | 0.55 | 0.27 | 0.42 | 0.18 | 0.36 |
| 2.0 | 0.30 | 0.60 | 0.30 | 0.43 | | |
| 2.5 | 0.22 | 0.44 | 0.22 | 0.45 | | |
| 3.0 | 0.23 | 0.46 | 0.23 | 0.45 | | |
| 3.5 | 0.20 | 0.40 | 0.20 | 0.22 | | |
| 3.7 | 0.16 | 0.32 | 0.16 | 0.22 | | |
| 3.8 | 0.20 | 0.40 | 0.20 | 0.22 | | |
| 4.0 | 0.20 | 0.40 | 0.20 | 0.22 | | |
| 4.5 | 0.16 | 0.32 | 0.16 | 0.22 | | |
| 5.0 | 0.16 | 0.32 | 0.16 | 0.22 | | |
| 5.5 | 0.16 | 0.32 | 0.16 | 0.22 | | |
| 6.0 | 0.23 | 0.45 | 0.23 | 0.13 | | |
| 7.0 | 0.23 | 0.45 | 0.23 | 0.13 | | |

*Fig. P1-7—Stator resistance table.*

*Fig. P1-10—View of Auto-Idle solenoid used on later models. Note position of idler link hook around governor lever.*

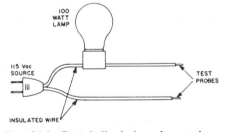

*Fig. P1-8—Test bulb device shown above may be used to test stator and rotor as outlined in text.*

| Output-kw | Rotor Resistance (ohms) |
|---|---|
| 1.0 | 5.70 |
| 1.25 | 3.45 |
| 1.5, 1.6, 2.0 | 4.20 |
| 2.5, 3.0 | 5.00 |
| 3.5, 3.7, 3.8, 4.0, 4.5, 5.0, 5.5 | 6.20 |
| 6.0, 6.5, 7.0 | 7.50 |

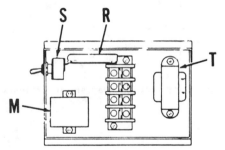

*Fig. P1-11—Diagram showing Auto-Idle control box components on early models.*

M. Sensor module
R. Resistor
S. Switch
T. Transformer

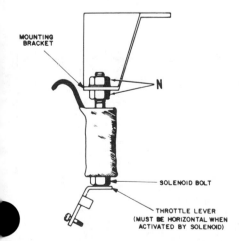

*Fig. P1-9—View of Auto-Idle solenoid used on early models. Refer to text for adjustment.*

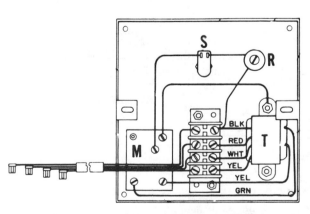

*Fig. P1-12—Diagram showing Auto-Idle control box components on later models. Refer to Fig. P1-11 for parts identification.*

Illustrations courtesy of Pioneer Gen-E-Motor Corp.

in Fig. P1-8. Plug test wire into a 115 VAC source, ground one test wire to rotor shaft and alternately connect remaining test wire to each slip ring. I bulb glows at all, then rotor is grounded and must be renewed.

## OVERHAUL

Disassembly and reassembly is evident after inspection of unit and referral to exploded views in Fig. P1-1 and P1-2. Note that rotor shaft and crankshaft ends are tapered and a suitable puller should be used to break tapers loose. Be sure rotor shaft is centered on

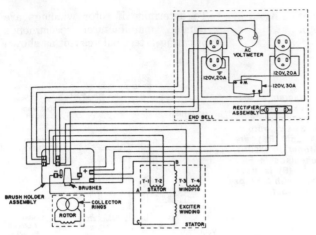

Fig. P1-15—*Wiring diagram for Models CRF45HM1 and CRF55HM1.*

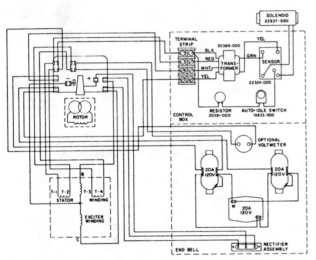

Fig. P1-16—*Wiring diagram for Models CRFHM1A and CRF55HM1A.*

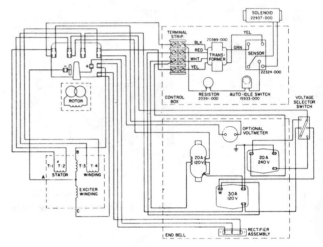

Fig. P1-18—*Wiring diagram for Models CRF45HM5A and CRF55HM5A.*

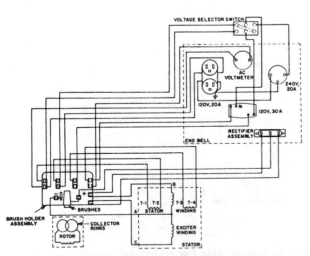

Fig. P1-17—*Wiring diagram for Models CRF45HM5 and CRF55HM5.*

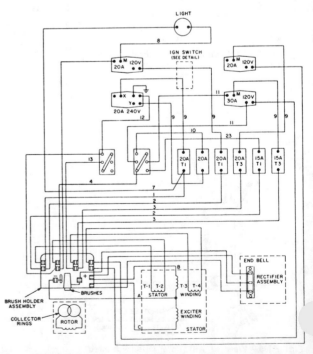

Fig. P1-19—*Wiring diagram for Models FRF30H, FRF45H and FRF55H.*

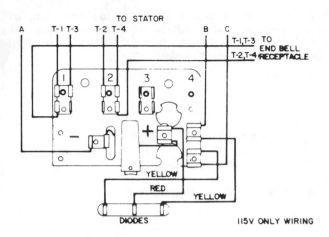

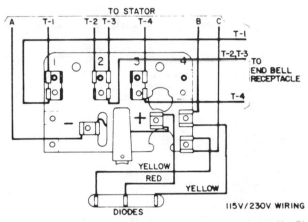

*Fig. P1-20—Typical wiring diagrams of brush plate (4—Fig. P1-2) on models after 1976. Refer also to stator wiring in Fig. P1-6 and receptacle wiring in Figs. P1-21 and P1-22 as well as wiring diagrams for specific models.*

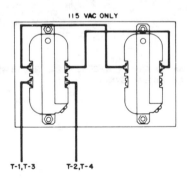

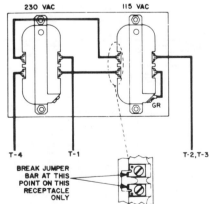

*Fig. P1-21—Typical wiring diagrams of receptacles on non-CSA (Canadian) models.*

crankshaft during assembly and tighten rotor bolt to 12 ft.-lbs. (16.3 N·m).

# AUTO-IDLE

Some models are equipped with the Auto-Idle system which decreases engine speed to approximately 1800 rpm when generator load is removed. A sensor circuit monitors the load imposed on

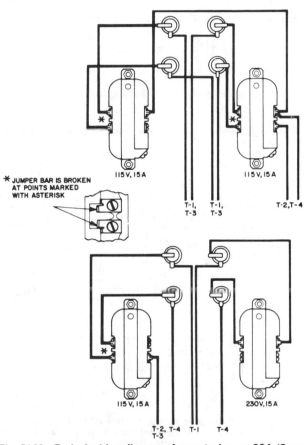

*Fig. P1-22—Typical wiring diagram of receptacles on CSA (Canadian) models. Fuses are used in generator output to comply with Canadian Standards Association.*

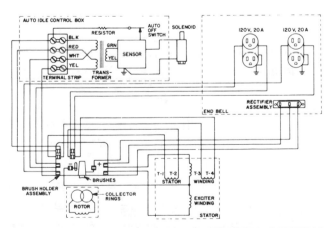

*Fig. P1-23—Wiring diagram of Models RF30HM1A, RF37HM1A and RF50HM1A. CSA (Canadian) models have fuses in receptacle circuit.*

the generator and will activate a solenoid to idle down the engine when the load is removed or deactivate the solenoid so engine speed increases.

When a load of from 40 to 60 watts is imposed, the engine should accelerate to full governed speed. On early models, the activated solenoid pushes the car-

buretor throttle lever to idle position. On later models, refer to Fig. P1-10 and note that solenoid pulls the governor lever to idle down engine.

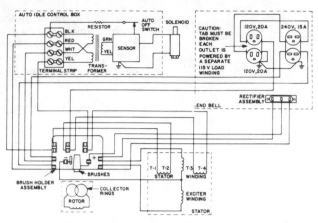

Fig. P1-24—Wiring diagram of Models RF30HM2A, RF37HM2A and RF50HM2A. CSA (Canadian) models have fuses in receptacle circuit.

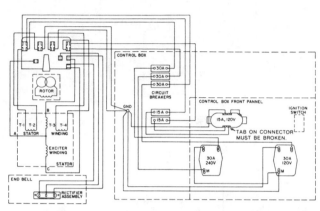

Fig. P1-27—Wiring diagram of RF65HE2 and RF65HM2 models prior to September 1978.

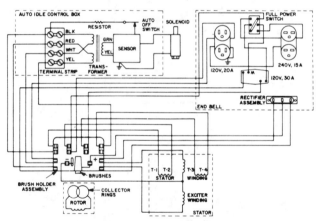

Fig. P1-25—Wiring diagrams of Models RF30HM5A, RF37HM5A and RF50HM5A. CSA (Canadian) models have fuses in receptacle circuit.

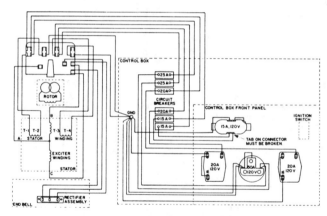

Fig. P1-28—Wiring diagram of Models RF65HE1 and RF65HM1 after August 1978.

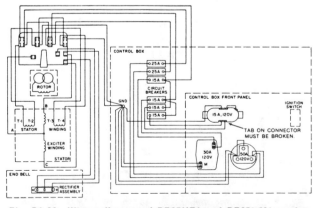

Fig. P1-26—Wiring diagram of RF65HE1 and RF65HM1 models prior to September 1978.

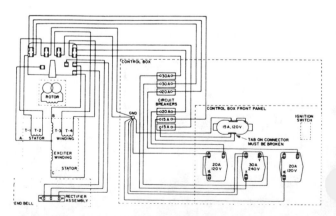

Fig. P1-29—Wiring diagram of Models RF65HE2 and RF65HM2 after August 1978.

Illustrations courtesy of Pioneer Gen-E-Motor Corp.

### ADJUSTMENT

To adjust Auto-Idle solenoid on early models, first switch off Auto-Idle and be sure engine runs at 3600 rpm with generator under load. If not, refer to engine section and adjust carburetor and governor as necessary. Remove load from generator and switch on Auto-Idle. Engine speed should decrease to 2000 rpm. To adjust idle speed, adjust height of solenoid by turning adjusting nuts (N – Fig. P1-9). Throttle lever should be horizontal at idle speed as shown in Fig. P1-9 for proper operation of Auto-Idle. Cycle generator unit several times with Auto-Idle on by adding and removing load to be sure engine accelerates smoothly and returns to desired idle.

To adjust Auto-Idle solenoid on later

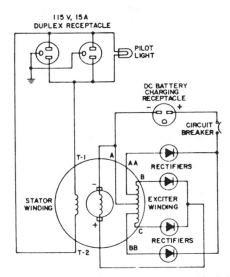

Fig. P1-34—*Wiring schematic of Model RF1600H.*

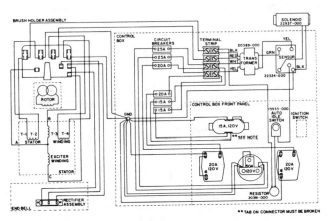

*Fig. P1-30—Wiring diagram of Model RF65HM1A.*

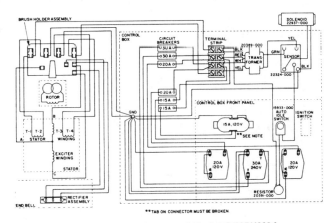

*Fig. P1-31—Wiring diagram of Model RF65HM2A.*

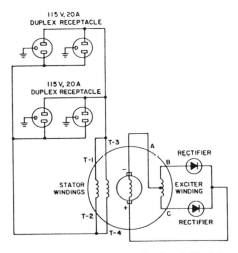

*Fig. P1-35—Wiring schematic of Models RF2500H, RF3000H, RF3500H and RF4000H. Refer to Fig. P1-38 for CSA (Canadian) models.*

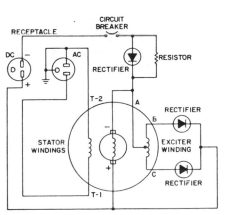

*Fig. P1-32—Wiring schematic of Model RF10.*

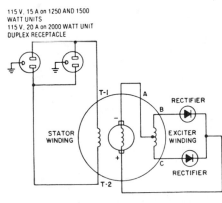

*Fig. P1-33—Wiring schematic of Models RF1250H, RF1500H and RF2000H. CSA (Canadian) models have receptacle ground connected to a common ground lug.*

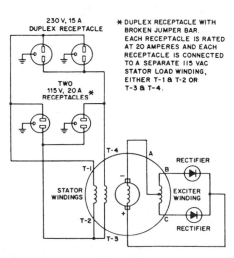

*Fig. P1-36—Wiring schematic of Models RF2500HDV, RF3000HDV, RF3500HDV and RF4000HDV. Refer to Fig. P1-39 for CSA (Canadian) models.*

Illustrations courtesy of Pioneer Gen-E-Motor Corp.

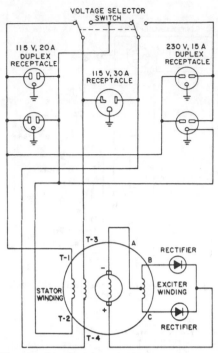

**Fig. P1-37—Wiring schematic of Models RF3000HFP and RF4000HFP.**

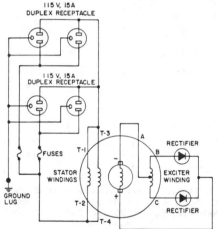

**Fig. P1-38—Wiring schematic of CSA (Canadian) Models RF2500H, RF3000H, RF3500H and RF4000H.**

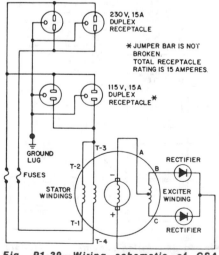

**Fig. P1-39—Wiring schematic of CSA (Canadian) Models RF2500HDV, RF3000HDV and RF4000HDV.**

models, first switch off Auto-Idle and be sure engine runs at 3600 rpm with generator under load. If not, refer to engine section and adjust carburetor and governor as necessary. Remove load and switch Auto-Idle on. Engine speed should decrease to 1800-1900 rpm. Turn carburetor idle speed screw to adjust idle speed. Be sure idle speed screw is snug against stop during idle. If not, loosen solenoid mounting screws and relocate solenoid higher on bracket to exert more tension on governor lever.

## TROUBLESHOOTING

If Auto-Idle does not function, proceed as follows: Run generator unit under no load and actuate Auto-Idle switch. If solenoid is not activated, check Auto-Idle switch and if good, stop engine and measure resistance of resistor (R – Fig. P1-11 or P1-12). Renew

resistor if resistance is not 60-90 ohms. If resistor is good, check solenoid for open or shorted windings and renew solenoid if found faulty. If solenoid is good, renew sensor module (M) and recheck operation. If solenoid does not deactivate when a 100 watt or greater load is imposed on generator, measure voltage between yellow and green transformer (T) leads. If voltage is 3.6-4.4 volts, renew sensor module (M) and check operation. If voltage is not 3.6-4.4 volts, renew transformer (T).

# ENGINE

Engine make and model are listed at beginning of section. Refer to Briggs and Stratton or Wisconsin engine section for engine service.

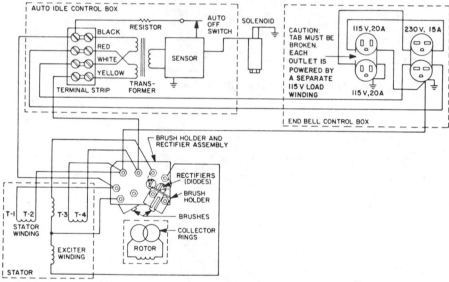

**Fig. P1-40—Wiring diagram of Auto-Idle models prior to 1977 with 115 volt output. CSA (Canadian) models have fuses in receptacle circuit.**

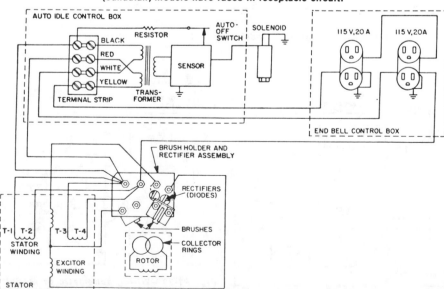

**Fig. P1-41—Wiring diagram of Auto-Idle models prior to 1977 with 115/230 volt output. CSA (Canadian) models have fuses in receptacle circuit.**

# PINCOR

| Model | Output-kw | Voltage | Engine Make | Engine Model | Governed Rpm |
|-------|-----------|---------|------|-------|-----|
| CAG30SM1 | 3.0 | 120 | B&S | 243432 | 1800 |
| CAG30SM1 | 3.0 | 120 | Wisconsin | AENL | 1800 |
| CAG30SM2 | 3.0 | 120/240 | B&S | 243432 | 1800 |
| CAG30SM2 | 3.0 | 120/240 | Wisconsin | AENL | 1800 |
| CAG40SE1 | 4.0 | 120 | B&S | 326437 | 1800 |
| CAG40SE2 | 4.0 | 120/240 | B&S | 326437 | 1800 |
| CAG40SM1 | 4.0 | 120 | B&S | 326432 | 1800 |
| CAG40SM2 | 4.0 | 120/240 | B&S | 326432 | 1800 |

# GENERATOR

## MAINTENANCE

Brushes are accessible after removing end cover (1–Fig. P2-1) and should be inspected after ever 1000 hours of operation. Renew brushes if worn to length of ¼ inch (6.4 mm) or less. Note condition of slip rings and recondition if necessary.

## TROUBLESHOOTING

If little or no output is generated, check the following: Be sure all circuit breakers function properly. The engine must be in good condition and able to maintain desired governed speed of 1800 rpm under load. Brushes and slip rings must be in good condition. All wiring connections must be clean and tight.

If residual magnetism is insufficient to induce generator output, it may be necessary to flash field as follows: Run unit at maximum governed speed. Manufacturer recommends momentarily inserting probes connected to a 115 VAC source into the slots of one of the generator AC receptacles. An alternate method is to momentarily connect a 12 VDC battery so the negative battery teminal is connected to the negative brush terminal and the positive battery and brush terminals are connected. Using either method, connect external voltage source only until generator output rises. If loss of residual magnetism is a chronic problem, then rotor should be renewed.

Test remainder of generator components by referring to wiring schematics in Fig. P2-2 or P2-3.

## OVERHAUL

Generator overhaul is evident after inspection of unit and referral to exploded view in Fig. P2-1 and wiring schematic in Fig. P2-2 or P2-3.

# ENGINE

Engine make and model are listed at beginning of section. Refer to Briggs and Stratton or Wisconsin section for engine service.

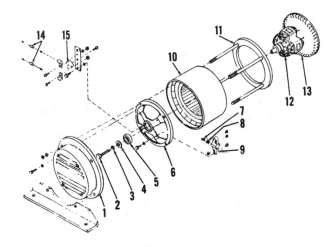

**Fig. P2-1—Exploded view of generator.**

1. End cover
2. Rotor bolt
3. Lockwasher
4. Washer
5. Bearing
6. Bearing carrier
7. Brush
8. Brush spring
9. Brush holder
10. Stator
11. Bracket
12. Rotor
13. Fan
14. Suppressor diodes
15. Bridge rectifier.

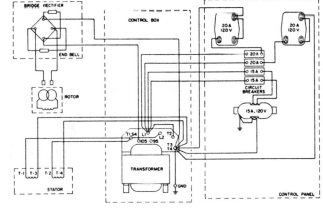

**Fig. P2-2—Wiring diagram of Models CAG30SM1, CAG40SE1 and CAG40SM1.**

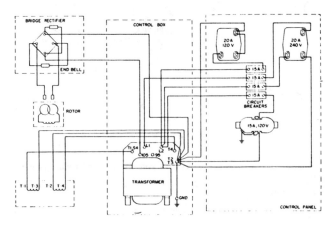

**Fig. P2-3—Wiring diagram of Models CAG30SM2, CAG40SE2 and CAG40SM2.**

# ROBIN

**NAGATA AMERICA CORP.**
**400 Lively Blvd.**
**Elk Grove Village, IL 60007**

| Model | Rated Output | Voltage | Engine | | Governed Rpm |
| | | | Make | Model | |
|---|---|---|---|---|---|
| R600, R650 | 550 watts | 120 | Robin | EY08D | 3600 |

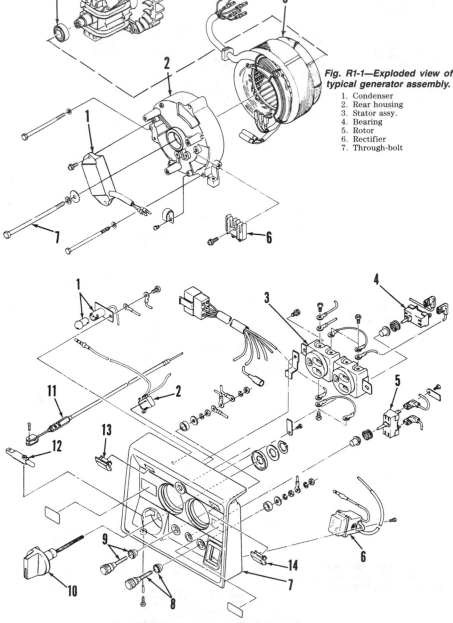

**Fig. R1-1—Exploded view of typical generator assembly.**

1. Condenser
2. Rear housing
3. Stator assy.
4. Bearing
5. Rotor
6. Rectifier
7. Through-bolt

**Fig. R1-2—Exploded view of typical power panel.**

1. Indicator lamp
2. Micro switch
3. AC receptacle
4. Circuit breaker (AC)
5. Circuit breaker (DC)
6. Frequency meter (R600) or volt meter (R650)
7. Control panel
8. Terminal bolt (red)
9. Terminal bolt (black)
10. Engine control knob
11. Choke cable
12. Lever
13. Green cap
14. Black cap

# GENERATOR

## MAINTENANCE

The generator is equipped with a brushless, self-exciting rotor and no periodic maintenance is required. Rotor bearing (4—Fig. R1-1) is sealed and does not require lubrication, but should be inspected for smooth operation when generator is disassembled.

## TROUBLESHOOTING

If little or no generator output is evident, check circuit breaker and reset if necessary. Engine must be in good enough condition to maintain desired governed speed (3600 rpm) under load. Remove side panel and check wiring for loose connections. All wiring connections must be clean and tight.

To check stator windings, remove left side cover and disconnect wiring coupler connecting generator to control panel. To check AC winding, connect ohmmeter test leads to brown and blue wire leads on Model R600 (Fig. R1-7) or to brown and white wire leads on Model R650 (Fig. R1-8). Specified resistance is 2.7 ohms for both models. To check DC winding, disconnect wire leads from rectifier (6—Fig. R1-3) and connect ohmmeter test leads between green and white wire leads. Specified resistance is 0.9 ohm for Model R600 and 0.7 ohm for Model R650. Check stator for shorted windings by connecting one test lead of ohmmeter to each of the stator winding leads and the other test lead to stator core. If there is continuity, winding is shorted and stator must be renewed.

To check condenser coil, disconnect wires connecting condenser coil to condenser (1—Fig. R1-3). Use an ohmmeter to measure resistance between the condenser coil lead wires. Specified resistance is 7.6 ohms for Model R600 and 7.1 ohms for Model R650. The simplest method to determine if condenser is faulty, is to install a condenser that is known to be good and run generator to see if output returns to normal.

To check rotor, generator must be disassembled and rotor removed. Unsolder one wire of diode (D—Fig. R1-4) and one wire of resistor (R). Use an ohmmeter to check resistance between wire ends of rotor winding. Specified resistance is 11.5 ohms for all models. Check for shorted winding by connecting ohmmeter test lead to one of the soldered terminals and other lead to rotor core. If there is continuity, winding is shorted and rotor must be renewed.

To check diode, connect ohmmeter test leads to diode terminals as shown in Fig. R1-4. Note meter reading, then reverse the tester leads. Resistance should be about 16 ohms in one test and infinite resistance when leads are reversed. If both readings are the same, renew diode. When renewing diode, note position of cathode mark (CM—R1-4) on diode body and be be sure to install new diode with mark in same direction.

To check rectifier, refer to Fig. R1-5 or R1-6 and use an ohmmeter to check resistance between red and green terminals and red and white terminals. Rectifier is normal if resistance is 16 ohms in one test and infinite when tester leads are reversed as indicated in chart.

## OVERHAUL

Refer to Fig. R1-1 for an exploded view of typical generator and to Fig. R1-2 for view of typical power panel.

To disassemble, remove side covers and disconnect wiring couplers between generator and power panel. Disconnect choke cable and fuel line. Remove rear cover, fuel tank and front cover and power panel assembly. Unbolt and remove muffler cover, muffler and base plate. Remove recoil starter and starter pulley. Remove through-bolts retaining stator housing to front housing and tap stator housing with a soft hammer to separate the housings. Remove capacitor and rectifier from stator housing and separate stator assembly from rear housing. Use a suitable puller to remove rotor from engine crankshaft. If necessary, unbolt and remove generator front housing from engine.

To reassemble, install generator front housing and tighten retaining screws to 8-10 N·m (6-7 ft.-lbs.). Position rotor and flywheel on engine crankshaft making sure that keyway is aligned. If ignition unit was removed, reinstall on generator housing and set air gap between ignition coil and flywheel magnet to 0.4-0.5 mm (0.016-0.020 inch). Assemble stator, rear housing, condenser and rectifier and reconnect wiring leads. Tighten housing mounting screws to 6-7 N·m (4-5 ft.-lbs.). Install starter pulley and tighten rotor through-bolt to 10-15 N·m (7-11 ft.-lbs.). Complete reassembly by reversing disassembly procedure.

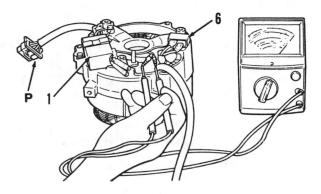

*Fig. R1-3—Procedure for checking condenser coil.*
  P. 6-pin coupler
  1. Condenser
  6. Rectifier

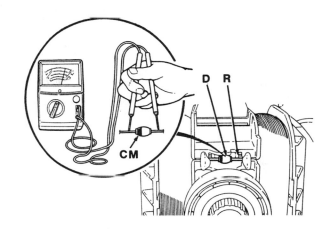

*Fig. R1-4—Diode (D) and resistor (R) must be unsoldered from rotor winding leads before checking resistance. Be sure to note position of diode cathode mark (CM) and reinstall in same position.*

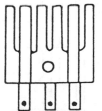

| Green | Red | White ⓒ | |
|---|---|---|---|
| | (+) | (−) | ... 16Ω |
| (−) | (+) | | ... 16Ω |
| | (−) | (+) | ... ∞Ω |
| (+) | (−) | | ... ∞Ω |

The polarity of the circuit tester

*Fig. R1-5—Table showing specified resistance of rectifier used on Model R600.*

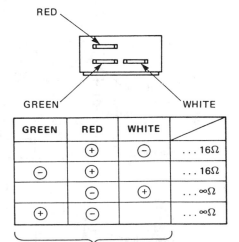

| GREEN | RED | WHITE | |
|---|---|---|---|
| | (+) | (−) | ... 16Ω |
| (−) | (+) | | ... 16Ω |
| | (−) | (+) | ... ∞Ω |
| (+) | (−) | | ... ∞Ω |

The polarity of the circuit tester

*Fig. R1-6—Table showing specified resistance of rectifier used on Model R650.*

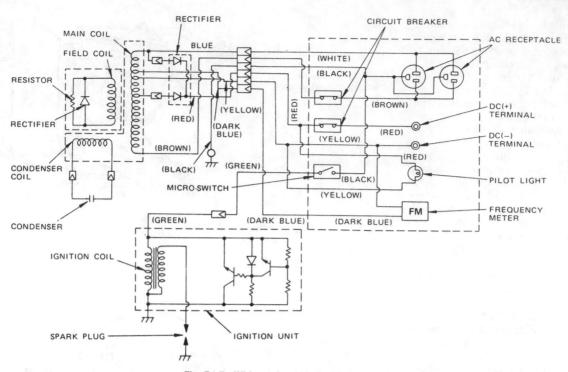

Fig. R1-7—Wiring schematic for Model R600.

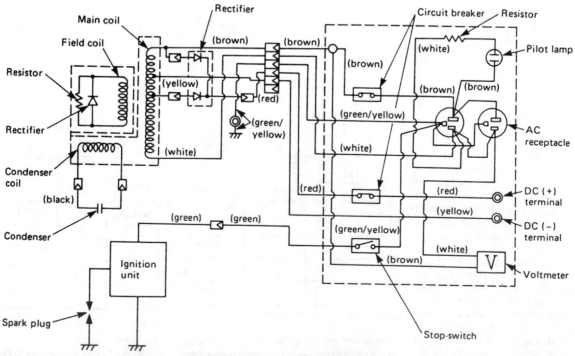

Fig. R1-8—Wiring schematic for Model R650.

# ENGINE

| Model | No. Cyls. | Bore | Stroke | Displacement |
|-------|-----------|------|--------|--------------|
| EY08D | 1 | 51.00 mm (2.0 in.) | 38.48 mm (1.515 in.) | 78 cc (4.76 cu. in.) |

## MAINTENANCE

**SPARK PLUG.** Spark plug should be removed and cleaned after every 50 hours of operation. Renew spark plug if excessively worn, burned or damaged.

Recommended spark plug is NGK BMR4A or Champion RCJ8. Spark plug electrode gap should be set to 0.6-0.7 mm (0.024-0.028 inch) on all models.

**AIR FILTER.** The engine is equipped with a foam type air filter element. The air filter element should be removed and cleaned after every 50 hours of operation, or more often if operating in extremely dusty conditions. To clean element, wash in a nonflammable cleaning solvent and gently squeeze element dry. Oil element with clean engine oil and gently squeeze out the excess oil.

**FUEL FILTER.** Some models may be equipped with an inline fuel filter located in the fuel line between tank and carburetor. Inline filter should be renewed after every 200 hours of operation or every six months, whichever comes first.

**CARBURETOR.** Refer to Fig. R1-10 for exploded view of float type carburetor. Carburetor is accessible after removing generator side cover and the air filter. The pilot jet (7) controls fuel supply at low engine speed. Main fuel mixture is controlled by a fixed main jet (8) for middle and high engine speeds. There is no provision for adjustment of pilot jet or main jet on this carburetor. Engine idle speed is adjusted by turning throttle stop screw (6) as necessary.

If carburetor related fuel system problems are encountered, disassemble carburetor and clean using suitable solvent and compressed air. Do not use drills or wire to clean fuel passages as calibration of carburetor fuel:air mixture will be affected if orifices of main jet or pilot jet are enlarged.

**GOVERNOR.** A mechanical flyweight governor is used. The governor unit (17 through 21—Fig. R1-14) is enclosed within the engine crankcase and is driven by the camshaft gear (16). Refer to Fig. R1-17 for view of governor external linkage.

Adjustment is made with carburetor and governor linkage installed. Loosen governor lever clamp bolt (3—Fig. R1-

11). With a screwdriver, rotate governor shaft (2) fully clockwise while pushing governor lever (1) to fully open throttle valve. While holding parts in this position, tighten governor lever clamp bolt.

Maximum engine speed is adjusted by turning adjusting screw (4). Specified engine speed at no load is 3700-3750 rpm for 60 Hz operation.

**IGNITION.** An electronic, breakerless ignition system is used. No periodic maintenance or adjustments are required. Ignition unit is mounted on the flywheel/generator adapter housing. Specified air gap between flywheel and ignition coil is 0.4-0.5 mm (0.016-0.020 inch).

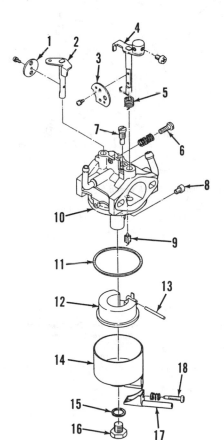

**Fig. R1-10—Exploded view of float type carburetor used on all engines.**

1. Throttle valve
2. Throttle shaft
3. Choke valve
4. Choke shaft
5. Return spring
6. Idle speed adjusting screw
7. Pilot jet
8. Main jet
9. Inlet needle valve
10. Carburetor body
11. Gasket
12. Float
13. Pin
14. Float bowl
15. Gasket
16. Screw
17. Cap
18. Drain screw

**LUBRICATION.** Manufacturer recommends changing engine oil after first 20 hours of operation and every 100 hours thereafter. To drain engine oil, remove oil filler cap/dipstick and tilt engine to drain oil from filler opening. To ensure quick and complete draining, it is recommended that oil be drained while engine is warm.

Use engine oil with API service classification SE or SF. SAE 10W-30 or SAE 10W-40 oil is recommended for use in all temperatures.

**CLEANING CARBON.** Cylinder head should be removed and carbon cleaned from cylinder head and piston after every 500 hours of operation. Refer to CYLINDER HEAD paragraph for removal and installation procedure.

**VALVE ADJUSTMENT.** Clearance between tappet and end of valve stem should be checked after every 500 hours of operation. To check clearance, remove tappet chamber cover (18—Fig. R1-13) and breather plate (17). Rotate crankshaft to position piston at top dead center of compression stroke. Use a feeler gage to measure clearance between tappet and valve stem (Fig. R1-12). Specified clearance with engine cold is 0.08-0.12 mm (0.003-0.005 inch) for intake and exhaust.

If clearance is smaller than specified, increase clearance by carefully grinding

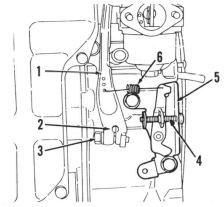

**Fig. R1-11—View of governor external control linkage.**

1. Governor lever
2. Governor shaft
3. Clamp bolt
4. High speed adjusting screw
5. Speed control lever
6. Governor spring

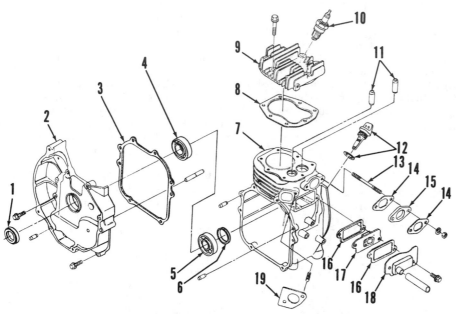

Fig. R1-12—Valve clearance can be checked using a feeler gage. Refer to text for procedure.

end of valve stem. To decrease clearance, renew valve and/or tappet. Refer to VALVE SYSTEM paragraph for valve removal and installation procedure.

## REPAIRS

**CYLINDER HEAD.** To remove cylinder head, remove generator side covers, end cover, fuel tank, head cover and cylinder air baffle. Remove cylinder head retaining bolts and remove cylinder head and gasket.

Clean all carbon deposits from cylinder head. Inspect cylinder head for distortion or other damage. If cylinder head is warped more than 0.15 mm (0.006 inch), resurface head or renew as necessary.

Reinstall cylinder head using a new head gasket. Tighten head bolts evenly in a crisscross pattern to 9-11 N·m (7-8 ft.-lbs.).

**PISTON, PIN AND RINGS.** To remove piston and connecting rod assembly, engine must be separated from generator unit as outlined in GENERATOR section. Remove cylinder head as previously outlined. Unbolt and remove crankcase cover (2—Fig. R1-13) from cylinder block (7). Remove connecting rod cap (4—Fig. R1-14) and withdraw piston and connecting rod assembly from top of cylinder block. Remove retaining ring (8), push out piston pin (9) and separate piston from connecting rod if necessary. Heating piston in warm water or oil will make removal and installation of piston pin easier.

After removing piston rings and separating piston from connecting rod, carefully clean carbon and other deposits from piston surface and ring lands.

**CAUTION: Extreme care should be exercised when cleaning ring lands. Do not damage squared edges or widen ring grooves. If ring lands are damaged, piston must be renewed.**

Inspect piston for scoring or excessive wear and renew as necessary. Piston pin-to-piston bore clearance is 0.009 mm (0.00035 inch) interference to 0.010 mm (0.00040 inch) loose. Refer to the following specification data for piston and rings:

Fig. R1-13—Exploded view of engine cylinder block assembly.

| | | | |
|---|---|---|---|
| 1. Oil seal | 6. Rubber plug | 11. Valve guides | 16. Gasket |
| 2. Crankcase cover | 7. Cylinder block | 12. Dipstick | 17. Breather plate |
| 3. Gasket | 8. Gasket | 13. Stud | 18. Tappet chamber cover |
| 4. Main bearing | 9. Cylinder head | 14. Gasket | |
| 5. Main bearing | 10. Spark plug | 15. Insulator | 19. Gasket |

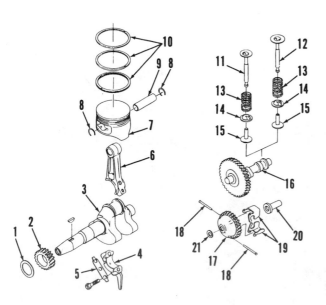

Fig. R1-14—Exploded view of engine internal components.

1. Spacer
2. Gear
3. Crankshaft
4. Rod cap & oil dipper
5. Lock plate
6. Connecting rod
7. Piston
8. Retaining rings
9. Piston pin
10. Rings
11. Exhaust valve
12. Intake valve
13. Spring
14. Retainer
15. Tappet
16. Camshaft
17. Governor gear
18. Pins
19. Governor weights
20. Sleeve
21. Thrust washer

Piston-to-cylinder wall clearance—
Desired . . . . . . . . . . .0.008-0.047 mm
(0.0004-0.002 in.)
Wear limit . . . . . . . . . . . . . .0.14 mm
(0.0055 in.)
Piston ring end gap—
Top & second . . . . . . . .0.20-0.40 mm
(0.008-0.016 in.)
Oil ring . . . . . . . . . . . . .0.05-0.25 mm
(0.002-0.010 in.)
Side clearance between ring and ring groove—
Top . . . . . . . . . . . . .0.090-0.135 mm
(0.0035-0.0053 in.)
Wear limit . . . . . . . . . . . . . .0.26 mm
(0.010 in.)
Second . . . . . . . . . .0.060-0.105 mm
(0.0025-0.0040 in.)
Wear limit . . . . . . . . . . . . . .0.23 mm
(0.009 in.)
Oil . . . . . . . . . . . . . .0.010-0.065 mm
(0.0004-0.0025 in.)

Wear limit . . . . . . . . . . . . . 0.19 mm
(0.008 in.)

When assembling piston and rings, marked side of top and second rings must face toward top of piston. It is recommended that piston pin retaining rings (8—Fig. R1-14) be renewed whenever they are removed. Lubricate piston pin, piston and cylinder prior to assembly. Stagger piston ring end gaps 90 degrees apart around piston. Install piston and connecting rod so connecting rod end cap and oil dipper will be positioned as shown in Fig. R1-15.

**CRANKSHAFT AND CONNECTING ROD.** The crankshaft is supported at each end by ball type main bearings (4 and 5—Fig. R1-13). To remove crankshaft, separate engine from generator. Remove crankcase cover and cylinder head. Remove connecting rod cap and push piston and connecting rod through top of cylinder. Slide camshaft out of cylinder block and remove tappets. Withdraw crankshaft from block.

Inspect crankshaft and connecting rod for excessive wear or damage and renew as necessary. Make certain that ball type main bearings turn smoothly. Bearings should be a light press fit on crankshaft journals. Renew oil seal (1) in crankcase cover if necessary. Refer to the following crankshaft and connecting rod specifications:

Connecting rod small
end inside diameter—
　Standard . . . . . . . 11.010-11.021 mm
(0.4335-0.4339 in.)
　Wear limit . . . . . . . . . . . . 11.08 mm
(0.4362 in.)
Clearance between piston
pin and connecting rod
small end . . . . . . . . . 0.010-0.029 mm
(0.0004-0.0011 in.)
　Wear limit . . . . . . . . . . . . . 0.12 mm
(0.0047 in.)
Connecting rod large
end inside diameter—
　Standard . . . . . . . 20.000-20.013 mm
(0.7874-0.7879 in.)
　Wear limit . . . . . . . . . . . 20.050 mm
(0.7894 in.)
Crankpin outside diameter—
　Standard . . . . . . . 19.950-19.963 mm
(0.7855-0.7860 in.)
　Wear limit . . . . . . . . . . . . 19.92 mm
(0.7842 in.)
Clearance between connecting
rod big end and crankpin—

Standard . . . . . . . . . 0.037-0.063 mm
(0.0015-0.0025 in.)
Wear limit . . . . . . . . . . . . . 0.13 mm
(0.005 in.)
Side clearance between connecting
rod and crankshaft—
　Standard . . . . . . . . . . . 0.10-0.70 mm
(0.004-0.027 in.)
　Wear limit . . . . . . . . . . . . . 1.0 mm
(0.040 in.)

To reassemble, reverse the disassembly procedure while noting the following special instructions: Be sure that connecting rod is installed so connecting rod cap and oil dipper is positioned as shown in Fig. R1-15. Use a new lockplate (5—Fig. R1-14) on connecting rod cap screws and tighten screws to 6-8 N·m (4.5-6.0 ft.-lbs.). Align timing marks on camshaft gear with timing mark on crankshaft gear (Fig. R1-16). Be careful not to damage lip of oil seal when installing crankcase cover. Tighten crankcase cover retaining screws to 8-10 N·m

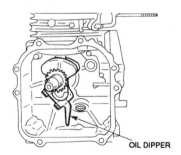

**Fig. R1-15—Piston and connecting rod must be installed in engine so oil dipper is positioned as shown.**

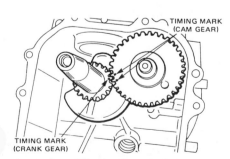

**Fig. R1-16—Timing marks on camshaft gear and crankshaft gear must be aligned as shown to provide correct valve timing.**

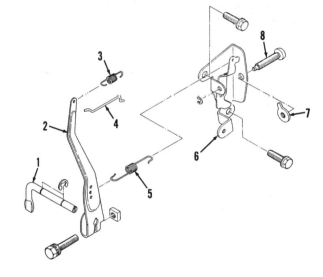

**Fig. R1-17—Exploded view of governor external control linkage.**

1. Governor shaft
2. Governor lever
3. Rod spring
4. Governor rod
5. Governor spring
6. Control bracket
7. Adjusting nut
8. High speed adjusting screw

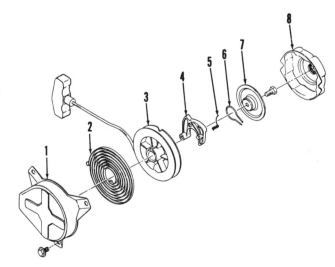

**Fig. R1-18—Exploded view of recoil starter assembly.**

1. Housing
2. Recoil spring
3. Rope & pulley
4. Ratchet
5. Spring
6. Friction spring
7. Friction plate
8. Starter hub

(6-7 ft.-lbs.). Check crankshaft end play using a dial indicator. Specified end play is 0.0-0.2 mm (0.0-0.008 inch). If necessary, remove crankcase cover and install different thickness spacer (1—Fig. R1-14) to obtain recommended end play.

**CAMSHAFT AND TAPPETS.** The camshaft (16—Fig. R1-14) and tappets (15) can be removed from cylinder block after removing crankcase cover as outlined in CRANKSHAFT section. Inspect camshaft and tappets for excessive wear, scoring, pitting or other damage and renew as necessary. It is recommended that camshaft and tappets be renewed as a set.

When installing camshaft, be sure to align timing marks on camshaft gear with timing mark on crankshaft gear (Fig. R1-16).

**GOVERNOR.** The internal centrifugal flyweight governor assembly (17 through 21—Fig. R1-14) is mounted on a stub shaft in the crankcase cover. Movement of governor weights is transmitted through external linkage (Fig. R1-17) to the carburetor throttle valve to maintain a relatively constant engine speed under various loads.

Governor assembly should be checked for free movement of governor weights. Governor should be renewed if weights (19), pivot pins (18), gear (17) or stub shaft is worn or damaged. Adjust governor external linkage as outlined in MAINTENANCE section.

**VALVE SYSTEM.** Valves can be removed after removing cylinder head and tappet chamber cover (18—Fig. R1-13). Use a screwdriver or other suitable tool to compress valve spring, then slip spring retainer (14—Fig. R1-14) off end of valve stem and remove valve and spring.

Valve face and seat angle for both intake and exhaust valves is 45 degrees. Valve seating contact width should be 0.5-1.0 mm (0.020-0.040 inch). The following specifications apply to valve guides and valves:

Valve stem OD—
Intake . . . . . . . . . . . 5.468-5.480 mm
(0.2153-0.2158 in.)
Exhaust . . . . . . . . . 5.426-5.444 mm
(0.2136-0.2143 in.)

Valve guide ID . . . . . . 5.500-5.518 mm
(0.2165-0.2172 in.)
Valve stem-to-guide clearance—
Intake . . . . . . . . . . . . 0.020-0.050 mm
(0.0008-0.0020 in.)
Exhaust . . . . . . . . . 0.056-0.072 mm
(0.0022-0.0028 in.)
Wear limit—intake &
exhaust . . . . . . . . . . . . . . . . 0.20 mm
(0.008 in.)

Specified valve clearance with engine cold is 0.08-0.12 mm (0.003-0.005 inch) for intake and exhaust valves. To check clearance, remove tappet chamber cover and breather plate. Turn crankshaft to position piston at top dead center of compression stroke. Use a feeler gage to measure clearance between end of valve stem and tappet. Valve clearance may be increased by carefully grinding end of valve stem. To decrease clearance, renew valve and/or tappet.

**RECOIL STARTER.** Recoil starter is mounted on end of generator opposite engine. Refer to Fig. R1-18 for exploded view of starter assembly.

# ROBIN

| Model | Output-kw | Voltage | Engine | | Governed Rpm |
| | | | Make | Model | |
| --- | --- | --- | --- | --- | --- |
| R1200 | 1.0 | 110,120 | Robin | W1-145 | 3600 |

# GENERATOR

## MAINTENANCE

Brushes (3 – Fig. R2-1) are accessible after removing end cover (1). Renew brushes, if they are worn to a length of 5 mm (0.20 inch) or less as shown in Fig. R2-3. Slip ring surfaces must be clean and bright. Polish slip rings with fine sandpaper, if necessary. When generator is disassembled, inspect bearing (11) for smooth operation and renew as needed.

## TROUBLESHOOTING

If little or no output is generated, check the following: Engine must be in good condition and maintain desired governed speed under load. Check circuit breakers and reset, if needed. Brushes and slip rings must be clean and tight.

To test stator, refer to Fig. R2-4 and measure resistance of windings as shown. Specified resistances in ohms between stator leads are as follows:

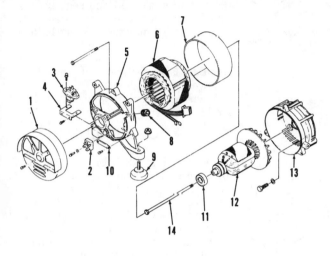

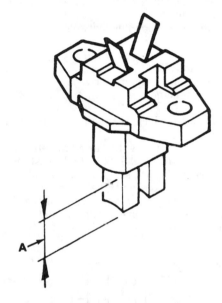

**Fig. R2-1 — Exploded view of generator.**

1. End cover
2. Rectifier
3. Brush holder
4. Brush bracket
5. Rear bracket
6. Stator
7. Stator cover
8. Grommet
9. Mount
10. Bracket
11. Bearing
12. Rotor
13. Front bracket
14. Through-bolt

**Fig. R2-3 — View of typical brush holder showing usable brush length (A) of 5-15 mm (0.20-0.59 inch).**

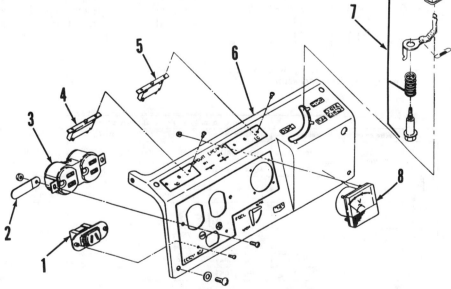

**Fig. R2-2 — Exploded view of control panel.**

1. DC receptacle
2. Clamp
3. AC receptacle
4. AC circuit breaker
5. DC circuit breaker
6. Panel
7. Throttle lever assy.
8. Voltmeter

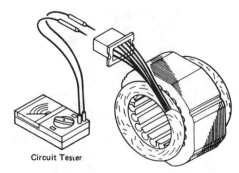

Circuit Tester

**Fig. R2-4 — Test stator with an ohmmeter as shown. Refer to text for specifications.**

100V Model:
Main coil (Coupler, between
white and red)..............0.6
DC coil (Between white
and yellow) ................0.23
Sub coil (Coupler, between
white and green) ............3.6
110V Model:
Main coil (Coupler, between
white and red)..............0.9
DC coil (Between white
and yellow) ................0.32
Sub coil (Coupler, between
white and green)............3.6
120V Model:
Main coil (Coupler, between
white and red)..............0.9
DC coil (Between white
and yellow) ................0.32
Sub coil (Coupler, between
white and green)............3.6
220V Model:
Main coil (Coupler, between
white and red)..............3.5
DC coil (Between white
and yellow) ................0.42
Sub coil (Coupler, between
white and green)............3.6

Any lead which is inactive indicates a faulty stator.

Check rotor with an ohmmeter as shown in Fig. R2-6. Specified resistance between the two slip rings is 10.7 ohms. If there is continuity between either of the slip rings and rotor frame or shaft, rotor windings are grounded and rotor must be renewed.

Red ① — ④ White
Orange ② — ⑤ Brown
Green ③ — ⑥ Black

*Fig. R2-5 — Color code chart showing proper placement of stator wiring in connector.*

Automatic voltage regulator should be visually inspected for a burnt or melted appearance which is a sure sign of trouble. Use an ohmmeter to check inter-lead resistance as shown in Fig. R2-8. A color code for the automatic voltage regulator connector is shown in Fig. R2-9 and specified inter-terminal resis-

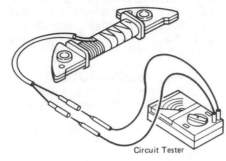

*Fig. R2-7 — Test exciting coil as shown for a specified resistance of 1-3 ohms. Specified voltage at rated rpm is 10-30V AC.*

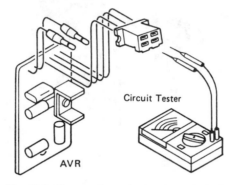

Circuit Tester

AVR

*Fig. R2-8 — Use an ohmmeter as shown to test automatic voltage regulator. Refer to Fig. R2-10 for specifications.*

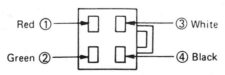

Red ① — ③ White
Green ② — ④ Black

*Fig. R2-9 — Color code for wiring on automatic voltage regulator connector.*

tances are shown in Fig. R2-10. If test results are other than those specified, renew regulator.

The exciting coil eliminates the need for flashing fields when field poles have lost their residual magnetism. Coil is mounted behind engine flywheel and utilizes magnets mounted in flywheel for generation of voltage. Test exciting coil winding with an ohmmeter as shown in Fig. R2-7 for a specified resistance of 1-3 ohms. Specified voltage output of exciting coil at rated rpm is 10-30V AC. Renew exciting coil, if test results vary from specifications.

Test rectifier by referring to circuit diagram shown in Fig. R2-11 and chart shown in Fig. R2-12. The rectifier has four terminals; two AC terminals, one positive terminal and one negative terminal. Rectifier may be tested with either an ohmmeter or a battery operated test light. If results other than those specified in the chart are obtained, rectifier is defective and should be renewed.

A wiring schematic is shown in Fig. R2-13.

## OVERHAUL

Refer to Fig. R2-1 for exploded view of generator. To disassemble, shut off fuel at strainer valve and remove side and rear sheet metal covers. Pull knob off throttle lever (7 – Fig. R2-2) and disconnect control panel coupler. Disconnect automatic voltage regulator coupler. Remove control panel mounting screws and remove panel. Remove air cleaner assembly. Disconnect fuel line running from fuel strainer to carburetor, then disconnect strainer from frame. Remove fuel tank, handle and strainer as an assembly. Remove sheet metal bracket and generator end cover (1 – Fig. R2-1). Disconnect wiring from brushes and rectifier. Remove brush

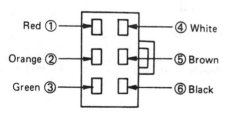

*Fig. R2-6 — Check rotor with an ohmmeter as shown. Refer to text for specifications.*

| | Wire color | Connect to the ⊖ terminal of the circuit tester | | | | |
| | | Yellow | Red | White | Green | Black |
|---|---|---|---|---|---|---|
| Connect to the ⊕ terminal of the current tester | Yellow | ∞ | 600K~1MΩ | 75K~120KΩ | 0Ω or ∞ according to the porality | 7K~10KΩ |
| | | ↑ | 400K~500KΩ | ↑ | ↑ | ↑ |
| | Red | | | 250K~300KΩ | ∞ | 400K~500KΩ |
| | | | | 120K~130KΩ | ↑ | 200K~220KΩ |
| | White | 250K~300KΩ | | | ∞ | 45K~50KΩ |
| | | 120K~130KΩ | | | ↑ | |
| | Green | 500K~1MΩ | 75K~110KΩ | | | 7K~9KΩ |
| | | 400K~500KΩ | ↑ | | | |
| | Black | 400K~500KΩ | 40K~46KΩ | ∞ | | |
| | | 200K~250KΩ | ↑ | ↑ | | |

*Fig. R2-10 — Chart showing specified inter-terminal resistances for automatic voltage regulator. Note that upper box in each space pertains to 220, 230, 240V models and lower box pertains to 100, 110, 120V models.*

assembly (3) and bracket (4). Remove rectifier (2). Unbolt and remove rear bracket (5) by gently tapping away from stator (6) with a plastic hammer. Pull off stator cover (7) and gently tap stator away from front bracket (13) with a plastic hammer. Loosen through-bolt (14) and hit bolt on head with a plastic

hammer to loosen rotor (12) from taper on end of crankshaft. Remove through-bolt and rotor. Remove front bracket (13) from engine crankcase cover.

Reassemble generator by reversing disassembly procedure. When reassembling, torque rotor through-bolt to 9.8-14.7 N·m (7.2-10.8 ft.-lbs.). Torque

rear bracket bolts to 5.4-7.4 N·m (4.0-5.5 ft.-lbs.).

# ENGINE

Engine make and model are listed at beginning of section. Refer to Wisconsin Robin engine section for engine service.

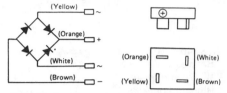

Fig. R2-11 — View showing rectifier circuitry along with color code for wiring.

|  |  | Connect black ⊖ terminal of the circuit tester | | | |
|---|---|---|---|---|---|
|  |  | Yellow | White | Orange | Brown |
| Connect red ⊕ terminal of the circuit tester | Yellow |  | No continuity | No continuity | Continuity |
|  | White | No continuity |  | No continuity | Continuity |
|  | Orange | Continuity | Continuity |  | Continuity |
|  | Brown | No continuity | No continuity | No continuity |  |

Fig. R2-12 — Continuity chart for rectifier.

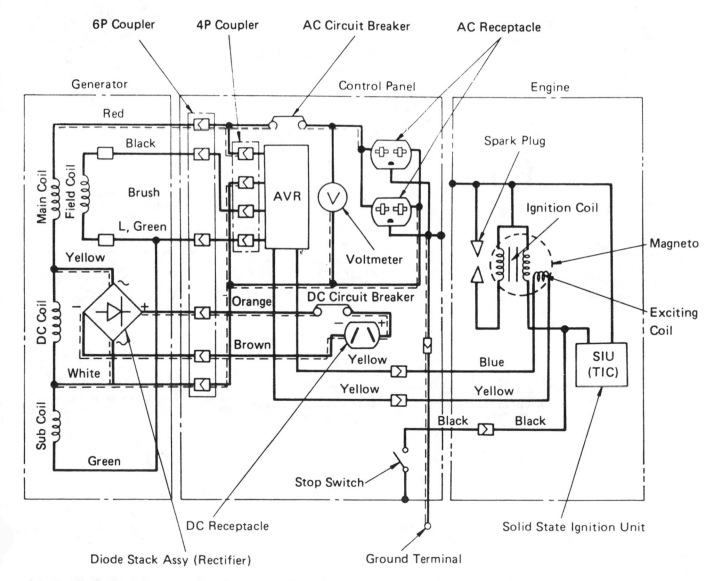

Fig. R2-13 — Wiring schematic.

# ROBIN

| Model | Output-kw | Voltage | Engine Make | Engine Model | Governed Rpm |
|-------|-----------|---------|-------------|--------------|--------------|
| RGX180 | 1.5 | 110, 220 | Robin | W1-185 | 3600 |
| RGX240 | 2.0 | 110, 220 | Robin | W1-185 | 3600 |
| RGX305 | 2.9 | 110, 220 | Robin | EY-25W | 3600 |
| RGD351 | 3.3 | 110, 220 | Robin | DY30D | 3600 |
| RGD500 | 4.5 | 110, 220 | Robin | DY41D | 3600 |

# GENERATOR

## MAINTENANCE

Brushes (1–Fig. R3-1 or 11–Fig. R3-2) are accessible after removing brush cover. Renew brushes, if they are worn to a length of 5 mm (0.20 inch) or less as shown in Fig. R3-3. Slip ring surfaces must be clean and bright. Polish slip rings with fine sandpaper, if necessary. When generator rear cover (11–Fig. R3-1 or 7–Fig. R3-2) is removed, inspect bearing for smooth operation and renew if needed.

## TROUBLESHOOTING

If little or no output is generated, check the following: Engine must be in good condition and maintain desired governed speed under load. Brushes and slip rings must be clean and bright.

To test stator, refer to Fig. R3-4 and measure resistance of windings as shown. Specified resistances in ohms between stator leads are as follows:

RGX180 (110V, 120V):
Between: White and red . . . . . . . . .2.3
White and green . . . . . . . . . . . . . .3.3

RGX240 (110V, 120V):
Between: White and red . . . . . . . . .0.4
White and green . . . . . . . . . . . . . .3.0
Blue and blue . . . . . . . . . . . . . . . .0.35

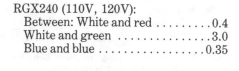

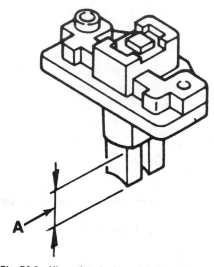

**Fig. R3-3 – View of typical brush holder showing usable brush length (A) of 5-15 mm (0.20-0.59 inch).**

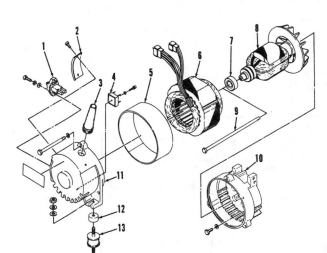

**Fig. R3-1 – Exploded view of generator typical of Models RGX180, RGX240 and RGX305.**

1. Brush holder
2. Brush cover
3. Grommet
4. Rectifier
5. Cover
6. Stator
7. Bearing
8. Rotor
9. Bolt
10. Front cover
11. Rear cover
12. Spacer
13. Damper

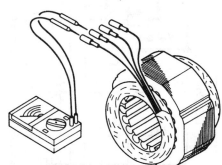

**Fig. R3-4 – Test stator with an ohmmeter as shown. Refer to text for specified resistances.**

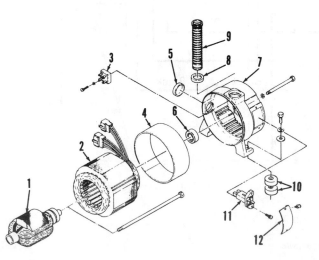

**Fig. R3-2 – Exploded view of generator typical of Models RGD351 and RGD500.**

1. Rotor
2. Stator
3. Rectifier
4. Cover
5. Cap
6. Bearing
7. Rear cover
8. Cap
9. Grommet
10. Spacer
11. Brush assy.
12. Brush cover

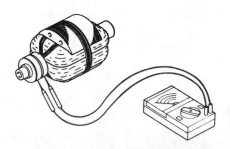

**Fig. R3-5 – Check rotor with an ohmmeter as shown. Refer to text for specifications.**

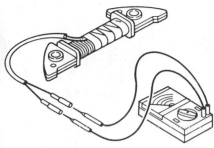

*Fig. R3-6 — Test exciting coil on Models RGX180, RGX240 and RGX305 with an ohmmeter as shown for a specified resistance of 10-30 ohms. Specified voltage at rated rpm is 10-30V AC.*

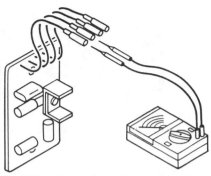

*Fig. R3-9 — Use an ohmmeter as shown to test automatic voltage regulator. Refer to charts for specifications.*

RGX240 (110V/220V, 120V/240V):
Between: White and red . . . . . . . . .0.8
White and black . . . . . . . . . . . . .0.8
White and green . . . . . . . . . . . . .3.0
Blue and blue . . . . . . . . . . . . . . .0.35

RGX305 (110V, 120V,
110V/220V, 120V/240V):
Between: White and red . . . . .0.4
White and green . . . . . . . . . . . . .1.7
Black and yellow . . . . . . . . . . . .0.7
Blue and blue . . . . . . . . . . . . . . .0.17

RGD351 (110V, 120V,
110V/220V, 120/240V):
Between: White and red . . . . . . . .0.5
White and green . . . . . . . . . . . . .2.2
Black and yellow . . . . . . . . . . . .0.5
Blue and blue . . . . . . . . . . . . . . .0.2

RGD351 (220V):
Between: White and red . . . . . . . .1.0
White and green . . . . . . . . . . . . .2.2
Blue and blue . . . . . . . . . . . . . . .0.2

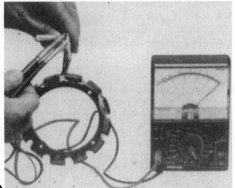

*Fig. R3-7 — Test exciting coil on Models RGD351 and RGD500 with an ohmmeter as shown for a specified resistance of 2-3 ohms. Specified voltage at rated rpm is 10-20V AC.*

| Tester polarity (−) / (+) | Yellow | Red | White | Green | Brown |
|---|---|---|---|---|---|
| Yellow | ∞ | 300K ~ 500KΩ | 10K ~ 70KΩ | One wire: 0Ω / Another: ∞ | lead resistance existence |
| | ∞ | 150K ~ 250KΩ | 10K ~ 70KΩ | One wire: 0Ω / Another: ∞ | lead resistance existence |
| Red | ∞ | − | 220K ~ 250KΩ | ∞ | 300KΩ |
| | ∞ | − | 110K ~ 125KΩ | ∞ | 150KΩ |
| White | ∞ | 220K ~ 250KΩ | − | ∞ | lead resistance |
| | ∞ | 110K ~ 125KΩ | − | ∞ | existence |
| Green | ∞ | 330K ~ 500KΩ | 10K ~ 70KΩ | − | lead resistance |
| | ∞ | 150K ~ 250KΩ | 10K ~ 70KΩ | − | existence |
| Brown | ∞ | 300K ~ 500KΩ | 10K ~ 40KΩ | ∞ | − |
| | ∞ | 150K ~ 250KΩ | 10K ~ 40KΩ | ∞ | − |

*Upper lines are for the 220, 230, 240V specifications; lower lines for the 110, 120V specifications and the dual voltage type.*

*Fig. R3-10 — Chart showing specified inter-terminal resistances for automatic voltage regulator used on Models RGX180, RGX240 and RGD500.*

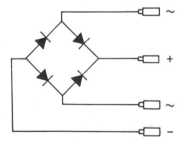

*Fig. R3-8 — View showing rectifier circuitry along with polarity chart.*

| TERMINAL | CURRENT | TERMINAL |
|---|---|---|
| ~ | →○→ / ✕ | + |
| ~ | ✕→ / ○→ | − |

| Tester polarity (−) / (+) | Yellow | Red | White | Green | Brown |
|---|---|---|---|---|---|
| Yellow | ∞ | 700K ~ 1MΩ | 72K ~ 120KΩ | One wire: 0Ω / Another: ∞ | 65K ~ 10KΩ |
| | ∞ | 400K ~ 500KΩ | 72K ~ 120KΩ | One wire: 0Ω / Another: ∞ | 65K ~ 10KΩ |
| Red | − | − | 250K ~ 300KΩ | ∞ | 400K ~ 500KΩ |
| | − | − | 130K ~ 140KΩ | ∞ | 220K ~ 250KΩ |
| White | − | 250K ~ 300KΩ | − | ∞ | 45K ~ 50KΩ |
| | − | 130K ~ 140KΩ | − | ∞ | 45K ~ 50KΩ |
| Green | − | 600K ~ 1MΩ | 70K ~ 110KΩ | − | 7K ~ 9.5KΩ |
| | − | 400K ~ 500KΩ | 70K ~ 110KΩ | − | 6.5K ~ 8.5KΩ |
| Brown | − | 400K ~ 500KΩ | 40K ~ 46KΩ | ∞ | − |
| | − | 250K ~ 300KΩ | 40K ~ 46KΩ | ∞ | − |

*Upper lines are for the 220, 230, 240V specifications; lower lines for the 110, 120V specifications and the dual voltage type.*

*Fig. R3-11 — Chart showing specified inter-terminal resistances for automatic voltage regulator used on Models RGX305 and RGD351.*

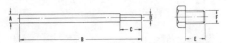

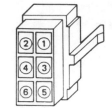

**Fig. R3-12 – A rotor puller for Model RGX305 may be fabricated using the following dimensions:**
A. 10 mm (0.39 in.)   D. 8 mm (0.31 in.)
B. 250 mm (9.8 in.)   E. 20 mm (0.79 in.)
C. 33 mm (1.3 in.)    F. 12 mm (0.47 in.)

RGD500 (110V, 120V,
    110V/220V, 120V/240V):
    Between: White and red . . . . . . . . 0.2
    White and green . . . . . . . . . . . . . . 1.3
    Black and yellow . . . . . . . . . . . . . . 0.2
    Blue and blue . . . . . . . . . . . . . . . . 0.12

RGD500 (220V):
    Between: White and red . . . . . . . . 0.4
    White and green . . . . . . . . . . . . . . 1.3
    Blue and blue . . . . . . . . . . . . . . . . 0.12

Any lead which is inactive indicates a faulty stator.

Refer to Fig. R3-5 and check rotor with an ohmmeter as shown. Specified resistance in ohms between the two slip rings is 6-10 for Models RGX180 and RGX240, 5-7.5 for Model RGX305 and 5-10.5 for Models RGD351 and RGD500. If there is continuity between either of the slip rings and rotor frame or shaft, rotor windings are grounded and rotor must be renewed.

Automatic voltage regulator should be visually inspected for a burnt or melted appearance which is a sure sign of trouble. Use an ohmmeter to check inter-lead resistance as shown in Fig. R3-9. Specified resistances are shown in Fig. R3-10 for Models RGX180, RGX240 and RGD500. Refer to Fig. R3-11 for Models RGX305 and RGD351. If test results are

|        | ① | ② | ③ | ④ | ⑤ | ⑥ | ⑦ | ⑧ | ⑨ | ⑩ | ⑪ | ⑫ |
|--------|---|---|---|---|---|---|---|---|---|---|---|---|
| RGX180 | — | — | — | — | — | — | WHITE | RED | BROWN | GREEN | WHITE | RED |
| RGX240 | — | — | WHITE | RED | BROWN / WHITE | ORANGE | WHITE | RED | BROWN | GREEN | — | — |
| RGX240D | — | BLACK | WHITE | RED | BROWN / WHITE | ORANGE | WHITE | RED | BROWN | GREEN | | |

**Fig. R3-14 – Wiring color code for connectors on Models RGX180 and RGX240.**

other than those specified, renew regulator.

The exciting coil eliminates the need for flashing fields when field poles have lost their residual magnetism. Coil is mounted behind engine flywheel and utilizes magnets mounted in flywheel for generation of voltage. On gasoline models, test exciting coil windings as shown in Fig. R3-6 for a specified resistance of 10-30 ohms and a specified voltage of 10-30V AC at rated rpm. On diesel models, test exciting coil windings as shown in Fig. R3-7 for a specified resistance of 2-3 ohms and a specified voltage of 10-20V AC at rated rpm. The field excitation circuit may also be tested as follows: Disconnect control box and

operate generator at rated rpm. Disconnect the two yellow wires that run from exciting coil to automatic voltage regulator (AVR), and momentarily connect the positive and negative leads of a 12 volt battery in their place. If no voltage is generated, reconnect battery leads the other way. If voltage is generated, the exciting coil is defective If no voltage is generated by performing the previous test, operate generator at rated rpm and momentarily connect a 12 volt battery to brushes. Note that green lead is positive and black or brown lead is negative. If voltage is generated, the primary exciting circuit in automatic voltage regulator is defective and regulator must be renewed.

**Fig. R3-13 – View of puller set up to remove rotor.**

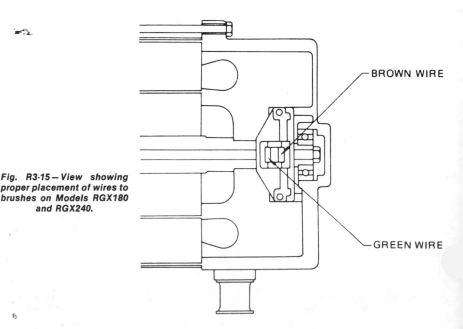

**Fig. R3-15 – View showing proper placement of wires to brushes on Models RGX180 and RGX240.**

BROWN WIRE

GREEN WIRE

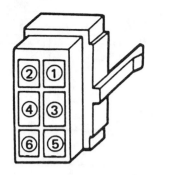
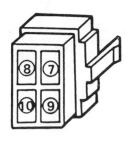
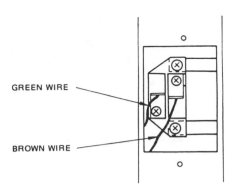

Fig. R3-18 — View showing proper placement of wires to brushes on Models RGX305, RGD351 and RGD500.

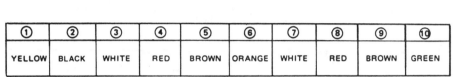

| ① | ② | ③ | ④ | ⑤ | ⑥ | ⑦ | ⑧ | ⑨ | ⑩ |
|---|---|---|---|---|---|---|---|---|---|
| YELLOW | BLACK | WHITE | RED | BROWN | ORANGE | WHITE | RED | BROWN | GREEN |

Fig. R3-16 — Wiring color code for connectors on Model RGX305.

Test rectifier by referring to the circuit diagram and chart shown in Fig. R3-8. The rectifier has four terminals; two AC terminals, one positive terminal and one negative terminal. Rectifier may be tested with either an ohmmeter or a battery operated test light. Clip one test light lead to positive terminal of rectifier and touch the other test light lead to one of the AC terminals of rectifier and then the other. Test light should either light up or remain unlit when each AC terminal is touched. When test light lead is reversed, the result should be opposite. If test light is lit when one AC terminal is touched and unlit when the other AC terminal is touched, rectifier is defective and should be renewed. Repeat test sequence with one test light lead clipped to negative rectifier terminal in place of positive terminal. If entire rectifier is good, test light will be lit at each of the two AC terminals during only one of the two test sequences. If other results are obtained, rectifier is defective and should be renewed.

## OVERHAUL

Refer to Fig. R3-1 for exploded view of generator. To disassemble, shut off fuel valve at strainer and disconnect fuel line at carburetor or injector pump. Unbolt fuel strainer bracket and fuel tank, then remove as an assembly. Remove battery and frame on electric start models. Disconnect couplers at rear of control box. Unbolt control box from frame and remove. Remove front and rear side plates. Unbolt and remove generator/engine assembly from frame. Remove brush cover (2 – Fig. R3-1 or 12 – Fig. R3-2). Disconnect and remove brush assembly (1 – Fig. R3-1 or 11 – Fig. R3-2). Disconnect stator wiring from couplers. Remove rear cover bolts and remove rear cover by tapping gently with a soft hammer. Slide stator wiring out through grommet in rear cover. With a screwdriver, gently pry stator assembly away from front cover. Remove rotor through-bolt. Rotor is taper fitted to rear of engine crankshaft and may sometimes be removed by striking outside of core with a soft hammer. If needed, a puller with part number 367 5433 08 is available from manufacturer for Model RGX305. A puller may also be fabricated. Refer to Fig. R3-12 for dimensions.

Reassemble generator by reversing disassembly procedure. Color codes for reassembling stator wiring to connectors are shown in Figs. R3-14, R3-16 and R3-17. Brush wiring is shown in Figs. R3-15 and R3-18. Refer to Figs. R3-19, R3-20 or R3-21 for wiring schematics.

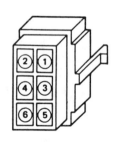
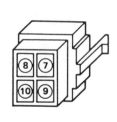

| | SPEC. | ① | ② | ③ | ④ | ⑤ | ⑥ | ⑦ | ⑧ | ⑨ |
|---|---|---|---|---|---|---|---|---|---|---|
| Hand start | 220, 230, 240V | YELLOW | YELLOW | GREEN | BROWN | ORANGE | BROWN/WHITE | RED | WHITE | — |
| | D type, 110, 120V | YELLOW | YELLOW | GREEN | BROWN | ORANGE | BROWN/WHITE | RED | WHITE | BLACK |
| Electric start | 220, 230, 240V | — | — | GREEN | BROWN | ORANGE | BROWN/WHITE | RED | WHITE | — |
| | D type, 110, 120V | — | — | GREEN | BROWN | ORANGE | BROWN/WHITE | RED | WHITE | BLACK |

| | SPEC. | ⑩ | ⑪ | ⑫ | ⑬ | ⑭ | ⑮ | ⑯ | ⑰ | ⑱ |
|---|---|---|---|---|---|---|---|---|---|---|
| Hand start | 220, 230, 240V | — | — | — | — | — | — | — | — | — |
| | D type, 110, 120V | YELLOW | — | — | — | — | — | — | — | — |
| Electric start | 220, 230, 240V | — | YELLOW | YEELOW | — | — | GRAY | PINK | (WHITE) | — |
| | D type, 110, 120V | YELLOW | YELLOW | YELLOW | — | — | GRAY | PINK | (WHITE) | — |

Fig. R3-17 — Wiring color code for connectors on Models RGD351 and RGD500.

# ENGINE

Engine make and model are listed at beginning of section. Refer to Wisconsin Robin engine section for engine service.

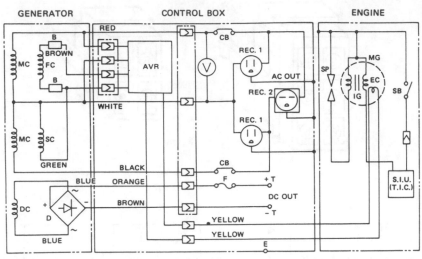

Fig. R3-19—View of typical wiring schematic along with code chart for Models RGX180 and RGX240.

| Symbols | Name of Parts |
|---------|---------------|
| MC | AC Winding |
| SC | Auxiliary Winding |
| DC | DC Winding |
| FC | Field Winding |
| B | Brush |
| AVR | Automatic Voltage Regulator |
| V | Voltmeter |
| PL | Pilot Lamp |
| D | Diode Stack Assy |
| E | Earth (Ground) Terminal |
| F | Fuse |

| Symbols | Name of Parts |
|---------|---------------|
| REC₁ | Receptacle (110V or 120V) |
| REC₂ | Receptacle (220V or 240V) |
| T | Terminal |
| CB | Circuit Breaker |
| SP | Spark Plug |
| MG | Magneto |
| IG | Ignition Coil |
| EC | Exciting Coil |
| SB | Stop Button |
| S.I.U. | Solid State Ignition Unit |

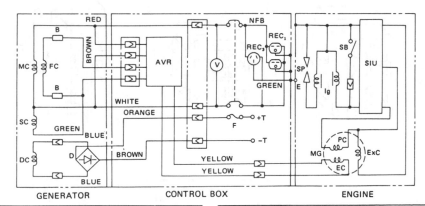

Fig. R3-20—View of typical wiring schematic along with code chart for Model RGX305.

| Symbols | Name of Parts |
|---------|---------------|
| MC | AC Winding |
| SC | Auxiliary Winding |
| DC | DC Winding |
| FC | Field Winding |
| B | Brush |
| AVR | Auto Voltage Regulator |
| V | Voltage Meter |
| EPSw | Full Power Switch |
| D | Diode Stack Assy |
| F | Fuse |
| REC₁ | AC Output Receptacle (Total 15A Max) |
| REC₂ | AC Output Receptacle (110V/120V) |

| Symbols | Name of Parts |
|---------|---------------|
| REC₃ | AC Output Receptacle (220V/230V/240V) |
| T | DC Output Terminal |
| NFB | No-Fuse Breaker |
| SP | Spark Plug |
| Ig | Ignition Coil |
| EC | Exciting Coil (Charge Coil) |
| ExC | Excitor Coil |
| PC | Pulser Coil |
| MG | Magneto |
| SB | Stop Button |
| SIU | Solid State Ignition Unit |
| E | Earth (Ground) Terminal |

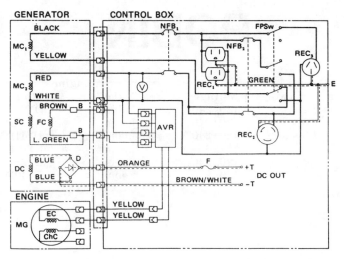

REC₃ for the U.S.A. and Canada is of the shape

| Symbols | Name of Parts |
|---------|---------------|
| MC | AC Winding |
| SC | Auxiliary Winding |
| DC | DC Winding |
| FC | Field Winding |
| B | Brush |
| AVR | Auto Voltage Regulator |
| V | Voltage Meter |
| FPSw | Full Power Switch |
| D | Diode Stack Assy |
| F | Fuse |
| REC₁ | AC Output Receptacle (Total 15A Max) |
| REC₂ | AC Output Receptacle (110/120V) |
| REC₃ | AC Output Receptacle (220/230/240V) |

| Symbols | Name of Parts |
|---------|---------------|
| T | DC Output Terminal |
| NFB | No-Fuse Breaker |
| E | Earth Terminal (Ground) |
| MG | Magneto |
| EC | Exciting Coil |
| ChC | Charge Coil |
| SM | Starting Motor |
| BAT | Battery |
| KS | Key Switch |

The wire symbols in the circuit diagrams signify as follows. Use the wire sizes specified.

| | |
|---|---|
| ——— | 0.75 mm² |
| ------- | 1.25 mm² |
| ——— | 2.0 mm² |
| === | 3.5 mm² |
| ===== | 5.0 mm² |

*Fig. R3-21 — View of typical wiring schemtic along with code chart for Models RGD351 and RGD500.*

# ROBIN

| Model | Rated Output-kw | Voltage | Engine Make | Engine Model | Governed Rpm |
|-------|-----------------|---------|-------------|--------------|--------------|
| RGD2500 | 2.2 | 120, 240 | Robin | DY23D | 3600 |
| RGD3300 | 3.0 | 120, 240 | Robin | DY27D | 3600 |
| RGD5000 | 4.5 | 120, 240 | Robin | DY41D | 3600 |
| RGX2400 | 2.0 | 120 | Robin | EY20D | 3600 |
| RGX3500 | 3.0 | 120, 240 | Robin | EY28D | 3600 |
| RGX5500 | 4.8 | 120, 240 | Robin | EY40D | 3600 |

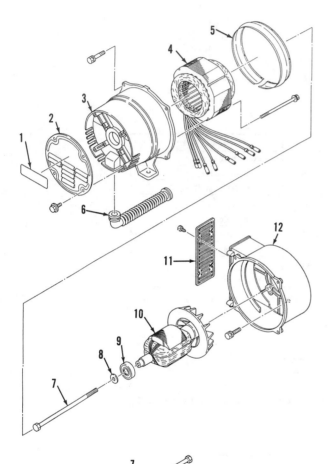

Fig. R4-1—Exploded view of generator assembly typical of all models except Model RGD5000.

1. Specification plate
2. End cover
3. Rear cover
4. Stator assy.
5. Support ring
6. Wiring harness cover
7. Rotor through-bolt
8. Washer
9. Bearing
10. Rotor assy.
11. Plate
12. Adapter housing

R4-1A—Exploded view of Model RGD5000 generator assembly. Refer to Fig. R4-1 for legend except for spacer (13) and stater pulley (14).

# GENERATOR

## MAINTENANCE

The generator is equipped with a brushless, self-exciting rotor and no periodic maintenance is required. Rotor bearing (9—Fig. R4-1 or R4-1A) is sealed and does not require lubrication, but should be inspected for smooth operation when generator is disassembled.

## TROUBLESHOOTING

If little or no generator output is evident, check circuit breaker and reset if necessary. Engine must be in good enough condition to maintain desired governed speed (3600 rpm) under load. Check wiring for loose connections. All wiring connections must be clean and tight.

To check stator windings, remove power panel and disconnect wires at the couplers. Use an ohmmeter to measure resistance between stator lead wires. Refer to chart shown in Fig. R4-5 for normal resistance. Renew stator if there is a wide variation from specified values. Check for continuity between each of the stator leads and stator frame. Renew stator if there is continuity.

To check rotor field coil, generator must be disassembled and rotor removed. Unsolder coil wire ends from terminals on rotor. Use ohmmeter to measure resistance between the two wire leads. Refer to following chart for normal field coil resistance values.

### Field Coil Resistance

RGD2500 . . . . . . . . . . . . . . . . .3.7 ohms
RGD3300 . . . . . . . . . . . . . . . . .3.3 ohms
RGD5000 . . . . . . . . . . . . . . . . .1.6 ohms
RGX2400 . . . . . . . . . . . . . . . . .2.7 ohms
RGX3500 . . . . . . . . . . . . . . . . .2.2 ohms
RGX5500 . . . . . . . . . . . . . . . . .1.6 ohms

To check rotor diode, unsolder and remove diode from rotor. Connect ohmmeter test leads to diode terminals and note gage reading, then reverse test leads and again note gage reading. There should be continuity in one test and infinite resistance when test leads are reversed. If both readings are the same, diode is faulty and must be renewed. Measure resistance of surge absorber connected to diode holder. Normal resistance is infinite. Renew rotor if surge absorber is faulty.

The simplest method to determine if condenser (8—Fig. R4-2 or R4-3) is faulty, is to install a condenser that is known to be good and run generator to see if output returns to normal.

To check diode rectifier, use an ohmmeter to check for continuity between each terminal as shown in Fig. R4-6 or R4-7. Renew rectifier if any of the tests indicate a failure of rectifier.

## OVERHAUL

Refer to Fig. R4-1 or R4-1A for exploded view of generator. To disassemble, first disconnect battery cables and remove starting motor battery on models so equipped. Drain fuel and remove the fuel tank. Remove power panel and disconnect generator wiring couplers. Remove mounting bolts and remove engine and generator assembly from frame.

On models equipped with recoil starter mounted on generator end cover, unbolt and remove recoil starter from generator. On all other models, remove rear cover plate (2—Fig. R4-1) from generator. On all models, remove through-bolt (7) from rotor shaft. Remove four cap screws that retain rear cover (3) to engine adapter housing (12). Use a suitable puller to remove rear cover as shown in Fig. R4-8. If a puller is not available, rear cover may be separated from adapter housing by carefully tapping against generator mounting legs and the boss at the top of rear cover using a soft mallet. Remove stator retaining bolts, then pull stator support ring (5—Fig. R4-1 or R4-1A) from rear cover and withdraw stator (4) from cover. Use suitable puller bolt to remove rotor (10) from engine crankshaft, or lightly tap rotor core with plastic hammer while pulling on rotor to remove rotor from shaft taper. Do not strike rotor windings or plastic insulators. If necessary, unbolt and remove adapter housing from engine.

To reassemble, proceed as follows: On Model RGD5000, if driving shaft (Fig. R4-9) was removed from flywheel, assemble shaft to flywheel and tighten cap screws to 55-70 N·m (40-50 ft.-lbs.). Install adapter housing, making sure that flat side of housing faces downward and mounting boss for fuel filter is on air cleaner side of engine. Tighten housing retaining cap screws to 20-23 N·m

(15-17 ft.-lbs.). Install rotor and temporarily install rotor through-bolt.

On all other models, if louver plate (11—Fig. R4-1) was removed from adapter housing (12), reinstall plate with louvers facing inward and upward. Tighten adapter housing retaining cap screws to 12-14 N·m (9-10 ft.-lbs.). Clean tapered portion of crankshaft and rotor shaft, then install rotor and tighten through-

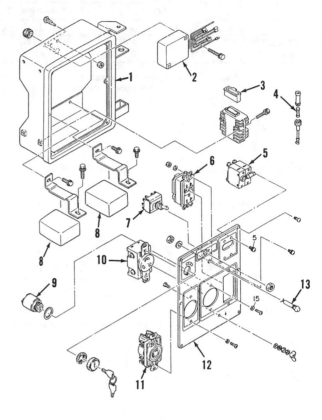

**Fig. R4-2—Exploded view of power panel used on Model RGD3300 equipped with electric starter and engine low oil level sensor. Model RGD2500 is similar.**

1. Control box
2. Oil sensor unit
3. Voltage regulator
4. Fuse
5. Circuit breaker
6. AC receptacle, 120V
7. Full power switch
8. Condensers
9. Ignition switch
10. AC receptacle, 120V
11. AC receptacle, 240V
12. Control panel
13. Indicator light

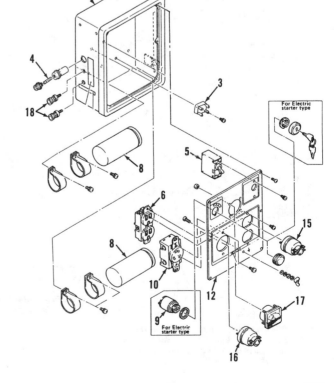

**Fig. R4-3—Exploded view of power panel used on Model RGX3500. Models RGX2400, RGX5500 and RGD5000 are similar.**

1. Control box
3. Diode rectifier
4. Fuse
5. Circuit breaker
6. AC receptacle, 120V
8. Condensers
9. Ignition switch
10. AC receptacle, 240V
12. Control panel
15. AC plug, 120V
16. AC plug, twist lock
17. Voltmeter
18. DC terminals

| Model | AC Winding | | DC Winding | Condenser Winding |
|---|---|---|---|---|
| | White/Red | Black/Blue | Brown/Brown | Yellow/Yellow |
| RGD5000 | 0.26Ω | 0.26Ω | 0.13Ω | 0.57Ω |
| RGX2400 | 0.84Ω | 0.84Ω | 0.27Ω | 0.51Ω |
| RGX3500 | 0.69Ω | 0.69Ω | 0.22Ω | 0.52Ω |
| RGX5500 | 0.24Ω | 0.26Ω | 0.14Ω | 0.58Ω |
| Model | AC Winding | | DC Winding | Condenser Winding |
| | Brown/White | Blue/LtBlue | Yellow/Yellow | Black/Orange |
| RGD2500 | 0.66Ω | 0.66Ω | 0.12Ω | 1.9Ω |
| RGD3300 | 0.44Ω | 0.44Ω | 0.11Ω | 1.6Ω |

Fig. R4-5—Chart showing normal resistance values for generator stator windings when tested at ambient temperature of 20° C (68° F).

bolt to 11-13 N·m (8-9 ft.-lbs.) on Models RGD2500 and RGX2400 or 23-25 N·m (17-18 ft.-lbs.) on Models RGD3300, RGX3500 and RGX5500.

On all models, install support ring (5—Fig. R4-1 or R4-1A) around stator (4) making sure that hooking holes in ring are positioned at flat sides of stator. Assemble stator and rear cover, positioning stator leads through opening in cover. Tighten stator retaining screws evenly in several steps to 8-10 N·m (6-7 ft.-lbs.). Install stator and rear cover over the rotor and use a plastic hammer

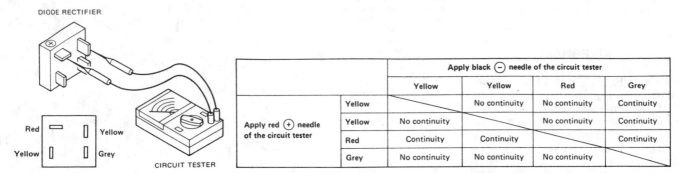

| | | Apply black (−) needle of the circuit tester | | | |
|---|---|---|---|---|---|
| | | Yellow | Yellow | Red | Grey |
| Apply red (+) needle of the circuit tester | Yellow | | No continuity | No continuity | Continuity |
| | Yellow | No continuity | | No continuity | Continuity |
| | Red | Continuity | Continuity | | Continuity |
| | Grey | No continuity | No continuity | No continuity | |

Fig. R4-6—An ohmmeter may be used to test continuity of diode rectifier. Continuity chart shown is for Models RGD2500 and RGD3300. Refer to Fig. R4-7 for all other models.

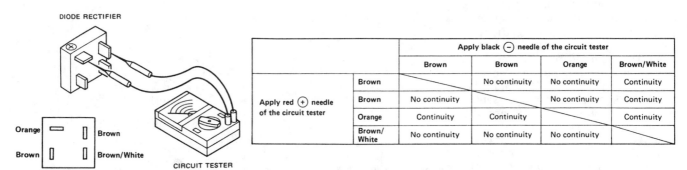

| | | Apply black (−) needle of the circuit tester | | | |
|---|---|---|---|---|---|
| | | Brown | Brown | Orange | Brown/White |
| Apply red (+) needle of the circuit tester | Brown | | No continuity | No continuity | Continuity |
| | Brown | No continuity | | No continuity | Continuity |
| | Orange | Continuity | Continuity | | Continuity |
| | Brown/White | No continuity | No continuity | No continuity | |

Fig. R4-7—Diode rectifier continuity chart for Models RDG5000, RGX2400, RGX3500 and RGX5500.

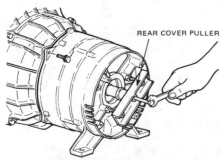

Fig. R4-8—Use a suitable puller to remove generator rear cover. If puller is not available, carefully tap legs and boss of rear cover with a plastic hammer to separate cover from adapter housing.

Fig. R4-9—View of generator adapter housing and engine flywheel used on Model RGD5000. Refer to text for installation procedure.

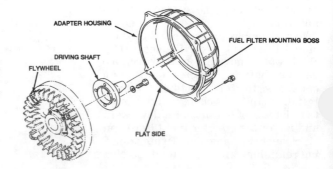

to tap rear cover onto rotor bearing. Tighten rear cover cap screws to 5-6 N·m (3.5-4.5 ft.-lbs.).

On Model RGD5000, remove rotor through-bolt and install spacer (13—Fig. R4-1A) and starter pulley (14). Tighten

rotor through-bolt to 24-30 N·m (18-22 ft.-lbs.). Install recoil stater and tighten retaining screws to 4-6 N·m (3-4 ft.-lbs.).

On all models, complete reassembly by reversing disassembly procedure.

# ENGINE

Engine make and model are listed at beginning of section. Refer to Wisconsin Robin engine section for engine service.

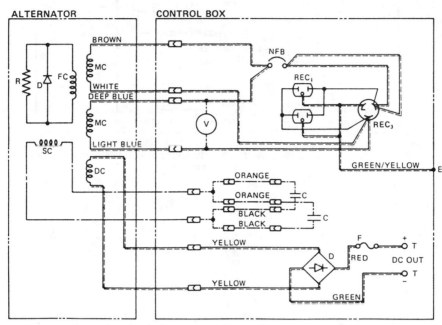

***Fig. R4-12—Wiring diagram for Models RGD2500 and RGD3300, 120V generators.***

| | | | |
|---|---|---|---|
| C. Condenser | R. Resistor | | |
| D. Diode | V. Voltmeter | FC. Field coil | SC. Sub coil |
| F. Fuse | DC. DC coil | MC. Main coil | NFB. Circuit breaker |

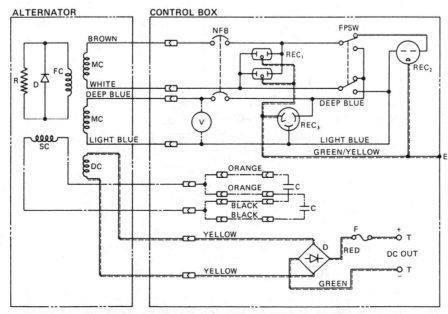

***Fig. R4-13—Wiring diagram for Models RGD2500 and RGD3300, 120V/240V generators.***

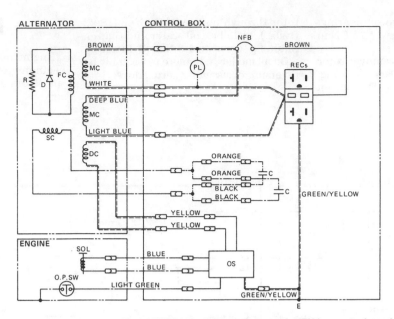

Fig. R4-14—Wiring diagram for Model RGD2500, 120V generator with NEMA receptacle and engine oil sensor.

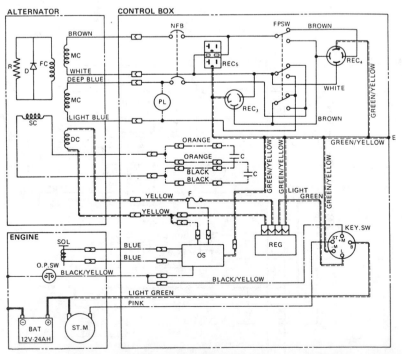

Fig. R4-15—Wiring diagram for Model RGD3300, 120V/240V generator with NEMA receptacle, engine oil sensor and electric starter.

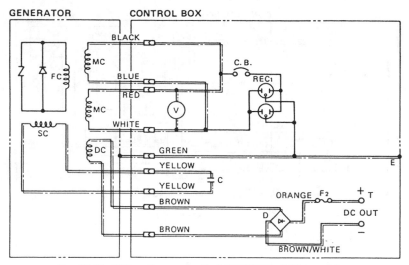

Fig. R4-16—Wiring diagram for Model RGX2400, 120V generator.

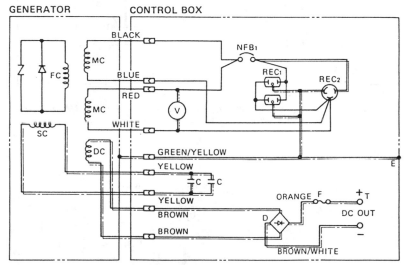

Fig. R4-17—Wiring diagram for Models RGD5000, RGX3500 and RGX5500, 120V generators.

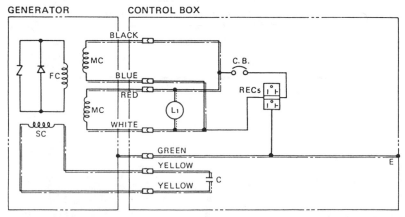

Fig. R4-18—Wiring diagram for Model RGX2400, 120V generator with NEMA receptacle.

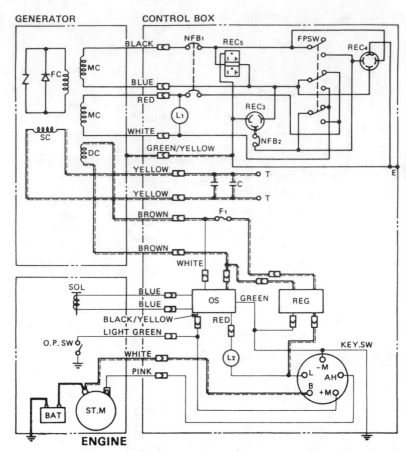

*Fig. R4-19—Wiring diagram for Model RGD5000, 120V/240V generator with NEMA receptacle, engine oil sensor and electric starter.*

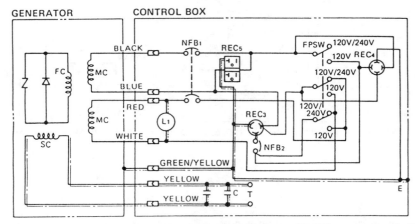

*Fig. R4-20—Wiring diagram for Models RGX3500 and RGX5500, 120V/240V generators with NEMA receptacle.*

# SUZUKI

**AMERICAN SUZUKI MOTOR CORP.**
**3251 E. Imperial Highway**
**Brea, CA 92621-6722**

| Model | Output-kva* | Voltage | Engine Make | Model | Governed Rpm |
|-------|-------------|---------|-------------|-------|--------------|
| SE300 | .31 | 120 | Suzuki | ** | 3600 |

*1 kva equals 1000 watts.
**Engine model number was not available.

# GENERATOR

## MAINTENANCE

All bearings are sealed and require no lubrication. Inspect unit periodically for condition of receptacles and operation of all switches. Inspect fuses (F1 and F2—Fig. SU1). To remove fuses, turn the fuse holder counterclockwise using a screwdriver. Use only AC 4A fuse at F1 or DC 10A fuse at F2.

## TROUBLE-SHOOTING

If generator output is zero, check the following: The engine must be in good condition and maintain desired governed speed of 3600 rpm. All wiring connections must be clean and tight. Check condition of fuse (F1—Fig. SU1). If the fuse is good, stop engine and disconnect thermistor. Check main coil resistance at the brown and blue wire leads (Fig. SU2) with an ohmmeter set at the Rx1 scale. Specified resistance is 3.0-5.0 ohms. If reading is infinite or less than 3.0 ohms, there is an open or short circuit in main coil. If the reading is within specified limits, outlet receptacle is faulty, fuse is defective or the permanent magnet in rotor is weak.

If generator output is low, check the following: The engine must be in good condition and maintain desired governed speed of 3600 rpm. All wiring connections must be clean and tight. Replace capacitor and check output voltage. If output voltage is still low, disconnect wires to capacitor and use an ohmmeter set at the Rx1 scale (Fig. SU3) to measure exciting coil resistance at the two pink leads. Specified resistance is 11.4 ohms. If reading is 11.4 ohms, wiring is disconnected or outlet receptacle is faulty. If reading is infinite or less than 11.4 ohms, exciting coil is shorted or a connection is poor. If exciting coil reading is 11.4 ohms and all contacts are good, check rotor coil resistance using an ohmmeter set at the Rx1 scale (Fig. SU4). If reading is infinite or less than 9.4 ohms, the rotor coil is shorted or a connection is poor. If specified resistance is obtained, rotor diode is faulty.

## OVERHAUL

Refer to Fig. SU5 for an exploded view of generator assembly. To disassemble generator, disconnect wiring connectors and remove control panel, side cover, end cover and muffler. Remove bolts retaining generator housing to base plate. Remove rotor retaining bolt. Remove stator housing through-bolts and withdraw stator assembly. Use suitable slide hammer puller to remove rotor from engine crankshaft.

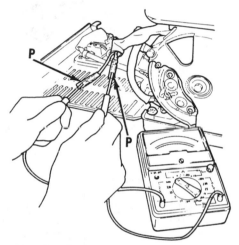

**Fig. SU3-Connect ohmmeter set at the Rx1 scale between two pink leads (P) as shown to check exciting coil resistance.**

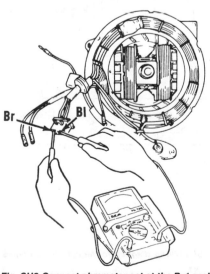

**Fig. SU1-View showing locations of fuse holders on Model SE300.**

| | |
|---|---|
| M. Frequency meter | F2. DC 10A fuse |
| S. Engine switch | R1. AC receptacle |
| F1. AC 4A fuse | R2. DC receptacle |

**Fig. SU2-Connect ohmmeter set at the Rx1 scale between brown lead (Br) and blue lead (Bl) as shown to check main coil resistance.**

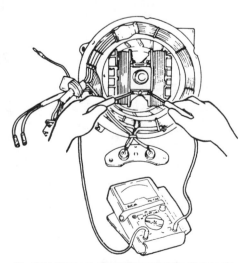

**Fig. SU4-Connect ohmmeter set at the Rx1 scale between rotor leads as shown to check rotor coil resistance.**

To reassemble, reverse the disassembly procedure while noting the following special instructions: Tighten rotor retaining bolt to 16-22 N·m (12-16 ft.-lbs.). Tighten stator housing through-bolts to 9-11 N·m (7-8 ft.-lbs.).

# ENGINE

Refer to the appropriate Suzuki engine section for engine service.

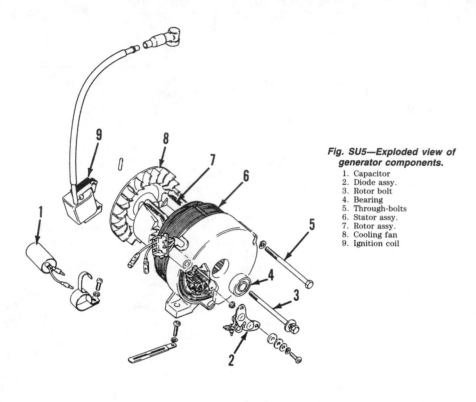

**Fig. SU5—Exploded view of generator components.**

1. Capacitor
2. Diode assy.
3. Rotor bolt
4. Bearing
5. Through-bolts
6. Stator assy.
7. Rotor assy.
8. Cooling fan
9. Ignition coil

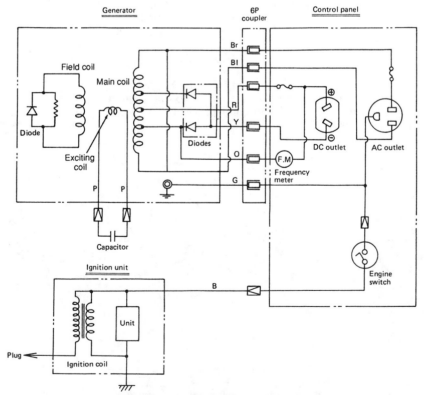

**Fig. SU6-Wiring diagram for Model SE300 generator.**

| | |
|---|---|
| B. Black | O. Orange |
| Bl. Blue | P. Pink |
| Br. Brown | R. Red |
| G. Green | Y. Yellow |

Illustrations courtesy of Suzuki Motor Corp.

# SUZUKI

|  Model | Output-kva* | Voltage | Engine Make | Engine Model | Governed Rpm |
|--------|-------------|---------|-------------|--------------|--------------|
| SE500  | .40         | 120     | Suzuki      | **           | 3600         |
| SE500A | .40         | 120     | Suzuki      | **           | 3600         |

*1 kva equals 1000 watts.

**Engine model number was not available.

# GENERATOR

## MAINTENANCE

All bearings are sealed and require no lubrication. Inspect unit periodically for condition of receptacles and operation of all switches. Inspect fuses (F1 and F2—Fig. SU11). To remove fuses, turn the fuse holder counterclockwise using a screwdriver. Use an AC 4A fuse (F1) for 120 volt AC 60 Hz circuit and use a DC 10A fuse (F2) for DC current circuit.

## TROUBLE-SHOOTING

### Model SE500

If generator AC output is zero, check the following: The engine must be in good condition and maintain desired governed speed of 3600 rpm. All wiring connections must be clean and tight. Check condition of fuse (F1—Fig. SU11). If the fuse is good, stop engine and disconnect generator output connector. Check main coil resistance using an ohmmeter set at the Rx1 scale and connected between the brown lead and light blue lead as shown in Fig. SU12. Resistance should be 3.6 ohms. If reading is less than specified, main coil is shorted or a connection is poor (open circuit). If the main coil resistance reading is as specified, outlet receptacle is faulty, fuse is defective or the permanent magnet in rotor is weak.

If generator AC output is low (0.5-2.0 volts), check the following: The engine must be in good condition and maintain desired governed speed of 3600 rpm. All wiring connections must be clean and tight. Replace capacitor and recheck output voltage. If output voltage is still low, disconnect wires to capacitor and measure exciting coil resistance at the two pink leads (Fig. SU13) using an ohmmeter set at the Rx1 scale. Specified resistance is 7.5 ohms. If specified resistance is obtained, there is a contact failure or wiring is disconnected. If reading is less than specified or infinite, exciting coil is shorted or a connection is poor. If exciting coil reading is as specified and all contacts are good, check rotor coil using an ohmmeter set at the Rx1 scale as shown in Fig. SU14. Specified rotor coil resistance is 9.5 ohms. If specified resistance is obtained, the rotor diode is defective. If the reading is less than 9.5 ohms or infinite, the rotor coil is shorted or a connection is poor.

### Model SE500A

If generator output is zero, check the following: The engine must be in good condition and maintain desired governed speed of 3600 rpm. All wiring connections must be clean and tight. Check condition of fuse (F1—Fig. SU11). If the fuse is good, stop engine and disconnect generator output connector. Measure main coil resistance at the brown and light blue leads (Fig. SU16) with an ohmmeter set at the Rx1 scale. Specified resistance is 3.2 ohms. If reading is infinite, there is a broken wire in main coil; if reading is less than 3.2 ohms, there is a layer short in main coil winding. If main coil resistance is as specified, outlet receptacle is faulty, fuse is defective or the permanent magnet in rotor is weak.

If generator output is low, check the following: The engine must be in good condition and maintain desired governed speed of 3600 rpm. All wiring connections must be clean and tight. Connect 12-volt battery positive terminal to brush holder positive terminal (red) and connect battery negative ter-

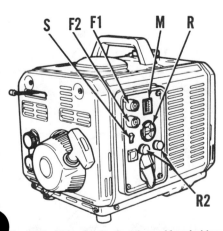

Fig. SU11—View showing locations of fuse holders on Models SE500 and SE500A.

| | |
|---|---|
| M. Frequency meter | F2. DC 10A fuse |
| S. Engine switch | R1. AC receptacle |
| F1. AC 4A fuse | R2. DC receptacle |

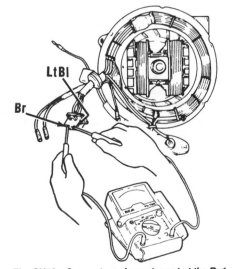

Fig. SU12—Connect an ohmmeter set at the Rx1 scale between brown and light blue leads as shown to measure main coil resistance on Model SE500.

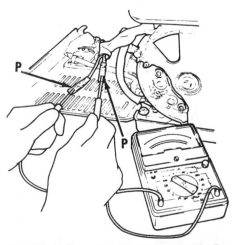

Fig. SU13—Connect an ohmmeter set at the Rx1 scale between two pink leads (P) as shown to check exciting coil resistance on Model SE500.

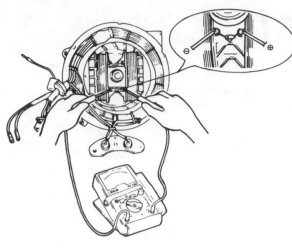

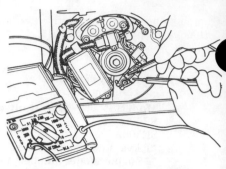

Fig. SU14—Connect an ohm-meter set at the Rx1 scale to rotor leads as shown to check the rotor coil resistance on Model SE500.

Fig. SU18—Connect an ohmmeter set at the Rx1 scale between the brushes as shown to check the rotor coil resistance on Model SE500A.

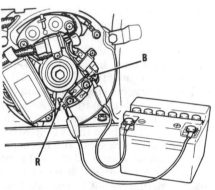

Fig. SU15—If AC output is low (1-3 volts) on Model SE500A generator, connect a 6 or 12-volt battery to brush holder terminals as shown and recheck output voltage. Refer to text.

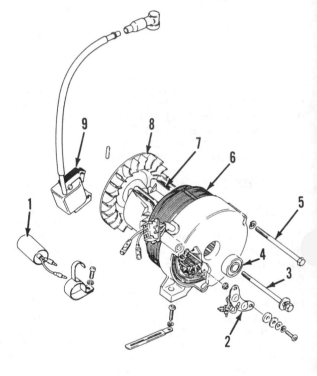

Fig. SU19—Exploded view of Model SE500 generator components.

1. Capacitor
2. Diode assy.
3. Rotor bolt
4. Bearing
5. Through-bolt
6. Stator assy.
7. Rotor assy.
8. Cooling fan
9. Ignition coil

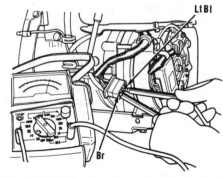

Fig. SU16—Connect an ohmmeter set at the Rx1 scale to brown lead (Br) and light blue lead (LtBl) in output connector as shown to check the main coil resistance on Model SE500A.

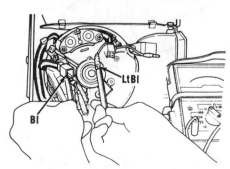

Fig. SU17—Connect an ohmmeter set at the Rx1 scale between blue lead (Bl) and light blue lead (LtBl) as shown to measure the exciting coil resistance on Model SE500A.

Fig. SU20—Exploded view of Model SE500A generator components.

1. Capacitor
2. Voltage regulator
3. Bearing
4. Rotor bolt
5. Diode assy.
6. Through-bolt
7. Brush holder assy.
8. Stator assy.
9. Rotor assy.
10. Cooling fan
11. Ignition coil

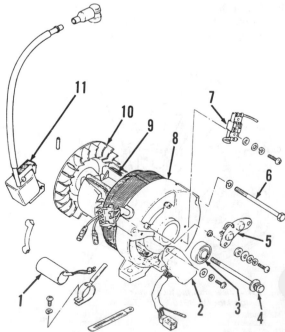

Illustrations courtesy of Suzuki Motor Corp.

minal to brush holder negative terminal (black) as shown in Fig. SU15.

**CAUTION: If battery terminal connections are reversed at brush holder, automatic voltage regulator will be damaged.**

If AC output voltage increases with battery connected to brush terminals, disconnect battery wires. If output voltage remains at rated level, either permanent magnet in rotor is demagnetized or automatic voltage regulator is faulty. If no output voltage is produced when battery leads are disconnected, stop engine and measure exciting coil resistance at the blue and light blue leads (Fig. SU17) using an ohmmeter set at the Rx1 scale. Specified resistance is 4.1 ohms. If exciter coil resistance is normal, voltage regulator is faulty or there is poor contact at brushes. If resistance is infinite, there is an open circuit in coil winding. If resistance is lower than specified, there is a layer short in coil winding.

If there is no change in AC output voltage with battery connected to brush terminals, stop engine and measure rotor coil resistance by connecting ohmmeter leads to brushes as shown in Fig. SU18. Specified rotor coil resistance is 12.9 ohms. If the reading is less than 12.9 ohms or infinite, the rotor coil is shorted or has a broken wire. If resistance is normal, measure excitor coil resistance as outlined above.

## OVERHAUL

Refer to appropriate Fig. SU19 or Fig. SU20 for an exploded view of generator assembly. To disassemble generator, disconnect wiring connectors and remove control panel, side cover, end cover and muffler. Remove bolts retaining generator housing to base plate. Remove rotor retaining bolt. Remove stator housing through-bolts and withdraw stator assembly. Use suitable slide hammer puller to remove rotor from engine crankshaft.

To reassemble, reverse the disassembly procedure while noting the following special instructions: Tighten rotor retaining bolt and stator housing through-bolts to 10 N·m (7 ft.-lbs.).

# ENGINE

Refer to the appropriate Suzuki engine section for engine service.

Illustrations courtesy of Suzuki Motor Corp.

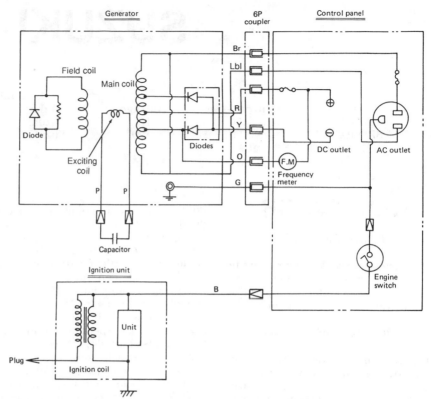

*Fig. SU21—Wiring diagram for Model SE500.*

| | | |
|---|---|---|
| B. Black | | P. Pink |
| Bl. Blue | G. Green | R. Red |
| Br. Brown | O. Orange | Y. Yellow |

*Fig. SU22—Wiring diagram for Model SE500A.*

| | | |
|---|---|---|
| B. Black | | P. Pink |
| Bl. Blue | G. Green | R. Red |
| Lbl. Light blue | Gr. Gray | W. White |
| Br. Brown | O. Orange | Y. Yellow |

# SUZUKI

| Model | Output-kva* | Voltage | Engine Make | Model | Governed Rpm |
|-------|-------------|---------|-------------|-------|--------------|
| SE600A | .50 | 120 | Suzuki | GY15D | 3600 |
| SE800A | .70 | 120 | Suzuki | GY20D | 3600 |

*1 kva equals 1000 watts.

## GENERATOR

### MAINTENANCE

All bearings are sealed and require no lubrication. Inspect unit periodically for condition of receptacles and operation of all switches and breaker units.

### TROUBLE-SHOOTING

If there is no AC voltage output, check the following: The engine must be in good condition and maintain desired governed speed of 3600 rpm. All wiring connections must be clean and tight. Make certain that breaker switch is not in "OFF" position. Make certain that all couplers and wiring plugs are making good connection.

Stop engine and disconnect generator output connector. Using an ohmmeter set at the Rx1 scale, measure main coil resistance between yellow and green wires (Fig. SU30). Main coil resistance should be 2.6 ohms for Model SE600A or 1.8 ohms for Model SE800A. If reading is less than specified or infinite, main coil is shorted or there is a poor connection or a broken wire in main coil. If specified resistance is obtained, outlet receptacle is faulty, circuit breaker is defective or the permanent magnet in rotor is weak.

If generator AC output is low (0.5-2.0 volts), check the following: The engine must be in good condition and maintain desired governed speed of 3600 rpm. All wiring connections must be clean and tight. Install a new capacitor and recheck output voltage. If output voltage is still low, stop engine and disconnect capacitor wire leads. Measure capacitor coil resistance using an ohmmeter set at the Rx1 scale between the capacitor coil leads (black wires) as shown in Fig. SU31. Capacitor coil resistance should be 6.6 ohms for Model SE600A and 5.7 ohms for Model SE800A. If reading is as specified, there is a contact failure or a short in wiring. If reading is less than specified or infinite, capacitor coil is shorted or a connection is poor. If capacitor coil reading is as specified and all contacts are good, measure rotor coil resistance using an ohmmeter set at the Rx1 scale. Rotor coil resistance should be 13 ohms. If reading is as specified, the rotor diode or the resistor is defective. If the reading is infinite or less than specified, the rotor coil is shorted or a connection is poor.

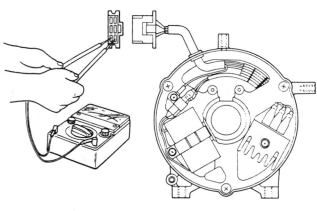

Fig. SU30—Connect ohmmeter leads between yellow and green wires in output connector as shown to measure main coil resistance. Refer to text.

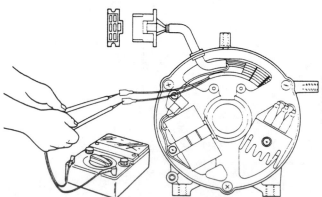

Fig. SU31—To check capacitor coil resistance, disconnect the two black wires from capacitor and connect ohmmeter leads to the wires as shown.

### OVERHAUL

To disassemble generator, disconnect wiring between generator and control panel and remove control panel, side cover and end cover. Remove fuel tank and muffler. Unbolt and remove stator assembly. Remove rotor retaining bolt, then remove rotor using suitable puller.

To reassemble, reverse the disassembly procedure. Tighten rotor retaining bolt to 5.0-7.3 N·m (44-64 in.-lbs.) on Model SE600A or 9.8-14.7 N·m (87-130 in.-lbs.) on Model SE800A.

## ENGINE

Refer to the appropriate Suzuki engine section for engine service.

Illustrations courtesy of Suzuki Motor Corp.

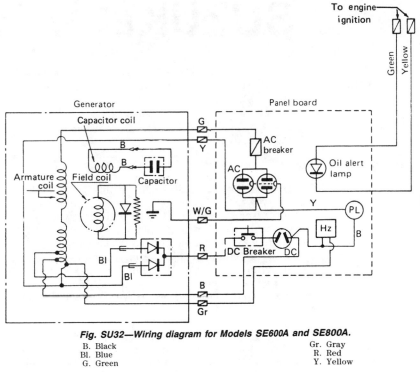

**Fig. SU32—Wiring diagram for Models SE600A and SE800A.**

| | |
|---|---|
| B. Black | Gr. Gray |
| Bl. Blue | R. Red |
| G. Green | Y. Yellow |

# SUZUKI

| Model | Output-kva* | Voltage | Engine | | Governed Rpm |
| | | | Make | Model | |
| SE700 | .66 | 120/220 | Suzuki | ** | 3600 |
| SE700A | .66 | 120/220 | Suzuki | ** | 3600 |

*1 kva equals 1000 watts.
**Engine model number was not available.

# GENERATOR

## MAINTENANCE

All bearings are sealed and require no lubrication. Inspect unit periodically for condition of receptacles and operation of all switches.

## TROUBLE-SHOOTING

### Model SE700

If generator AC output is zero, check the following: The engine must be in good condition and maintain desired governed speed of 3600 rpm. All wiring connections must be clean and tight. Check condition of circuit breaker (6—Fig. SU40). If circuit breaker is good, stop engine and disconnect generator output connector. Measure main coil resistance using an ohmmeter set at the Rx1 scale and connected between brown and light blue wires on 120 volt model or between brown and pink wires on 220 volt model as shown in Fig. SU41. Resistance should be 1.85 ohms for 120 volt model or 7.35 ohms for 220 volt model. If reading is less than specified or infinite, main coil is shorted or a connection is poor (open circuit). If the main coil resistance reading is as specified, outlet receptacle is faulty, fuse is defective or the permanent magnet in rotor is weak.

If generator AC output is low (0.5-2.0 volts), check the following: The engine must be in good condition and maintain desired governed speed of 3600 rpm. All wiring connections must be clean and tight. Install a new capacitor and recheck output voltage. If output voltage is still low, stop engine and disconnect wires to capacitor. Measure exciting coil resistance at the pink and green leads (Fig. SU42) using an ohmmeter set at the Rx1 scale. Specified resistance is 7.1 ohms. If specified resistance is obtained, there is a contact failure or wiring is disconnected. If reading is less than specified or infinite, exciting coil is shorted or a connection is poor. If exciting coil reading is as specified and all contacts are good, check rotor coil using an ohmmeter set at the Rx1 scale as shown in Fig. SU43. Specified rotor coil resistance is 10.5 ohms. If specified resistance is obtained, the rotor diode is defective. If the reading is less than 9.5 ohms or infinite, the rotor coil is shorted or a connection is poor.

### Model SE700A

If generator AC output is zero, check the following: The engine must be in good condition and maintain desired governed speed of 3600 rpm. All wiring connections must be clean and tight. Check condition of circuit breaker (6—Fig. SU40). If the circuit breaker is good, stop engine and disconnect generator output connector. Measure main coil resistance at the brown and light blue leads on 120 volt model or between brown and gray leads on 220 volt model (Fig. SU41) with an ohmmeter set at the Rx1 scale. Specified resistance is 2.05 ohms for 120 volt model or 8.32 ohms for 220 volt model. If reading is infinite, there is a broken wire in main coil; if reading is less than specified, there is a layer short in main coil winding. If main coil resistance is as specified, outlet receptacle is faulty, fuse is defective or the permanent magnet in rotor is weak.

If generator AC output is low (1-3 volts), check the following: The engine must be in good condition and maintain desired governed speed of 3600 rpm. All wiring connections must be clean and tight. Connect 12-volt battery positive terminal to brush holder positive terminal (red) and connect battery negative terminal to brush holder negative terminal (black) as shown in Fig. SU44.

**CAUTION: If battery terminal connections are reversed at brush holder, automatic voltage regulator will be damaged.**

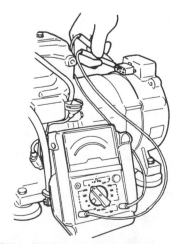

*Fig. SU40—View of control panel for Model SE700A generator. Model SE700 is similar.*

1. Indicator light
2. Oil level alarm lamp (SE700A)
3. Engine switch
4. DC receptacles
5. AC receptacles
6. AC circuit breaker
7. DC circuit breaker

*Fig. SU41—Disconnect generator output connector and use an ohmmeter set at the Rx1 scale to measure main coil resistance. Refer to text.*

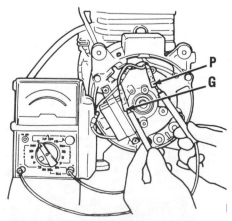

*Fig. SU42—Connect ohmmeter leads between pink and green wires as shown to check exciting coil resistance on Model SE700.*

If AC output voltage increases with battery connected to brush terminals, disconnect battery wires. If output voltage remains at rated level, either permanent magnet in rotor is demagnetized or automatic voltage regulator is faulty. If no output voltage is produced when battery leads are disconnected, stop engine and measure exciting coil resistance at the blue and green leads (Fig. SU45) using an ohmmeter set at the Rx1 scale. Specified resistance is 5.92 ohms. If exciting coil resistance is normal, voltage regulator is faulty or there is poor contact at brushes. If resistance is infinite, there is an open circuit in coil winding. If resistance is lower than specified, there is a layer short in coil winding.

If there is no change in AC output voltage with battery connected to brush terminals, stop engine and measure rotor coil resistance by connecting ohmmeter leads to brushes as shown in Fig. SU46. Specified rotor coil resistance is 14.6 ohms. If the reading is less than 14.6 ohms or infinite, the rotor coil is shorted or has a broken wire. If resistance is normal, measure excitor coil resistance as outlined above.

## OVERHAUL

Refer to appropriate Fig. SU48 or Fig. SU49 for an exploded view of generator assembly. To disassemble generator, disconnect wiring connectors and remove control panel, fuel tank, muffler cover, muffler, heat shield and generator end cover. Remove brush holder and stator housing through-bolts, then withdraw stator assembly. Remove rotor retaining bolt. Use suitable slide hammer puller to remove rotor from engine crankshaft.

To reassemble, reverse the disassembly procedure while noting the following special instructions: Tighten rotor retaining bolt to 16-23 N·m (12-16 ft.-lbs.) and tighten stator housing through-bolts to 4-7 N·m (3-5 ft.-lbs.).

# ENGINE

Refer to the appropriate Suzuki engine section for engine service.

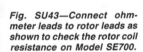

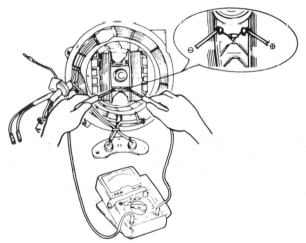

Fig. SU43—Connect ohmmeter leads to rotor leads as shown to check the rotor coil resistance on Model SE700.

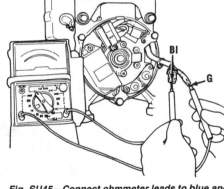

Fig. SU45—Connect ohmmeter leads to blue and green leads as shown to check the exciting coil resistance on Model SE700A.

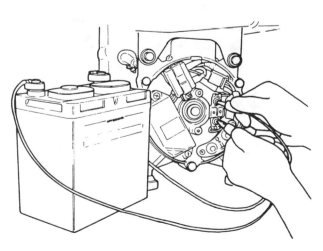

Fig. SU44—If AC output is low (1-3 volts) on Model SE700A generator, connect a 6 or 12-volt battery to brush holder terminals as shown and recheck output voltage. Refer to text.

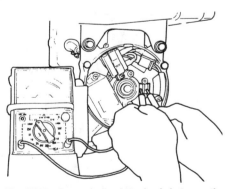

Fig. SU46—Connect ohmmeter leads between the brushes as shown to check the rotor coil resistance on Model SE700A.

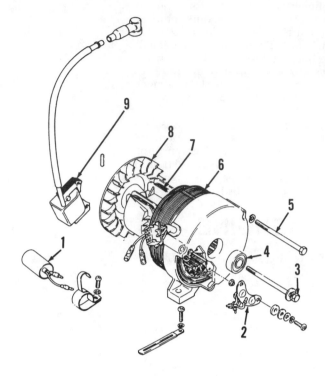

**Fig. SU48—Exploded view of Model SE700 generator components.**

1. Capacitor
2. Diode assy.
3. Rotor bolt
4. Bearing
5. Through-bolt
6. Stator assy.
7. Rotor assy.
8. Cooling fan
9. Ignition coil

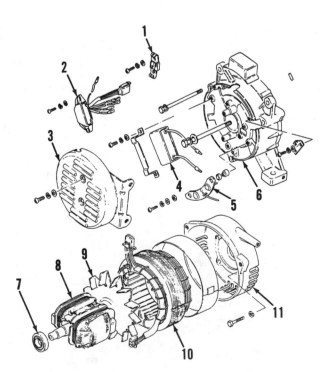

**Fig. SU49—Exploded view of Model SE700A generator components.**

1. Brush holder
2. Voltage regulator
3. End cover
4. Capacitor
5. Diode assy.
6. Stator housing
7. Bearing
8. Rotor assy.
9. Fan
10. Stator assy.
11. Fan housing

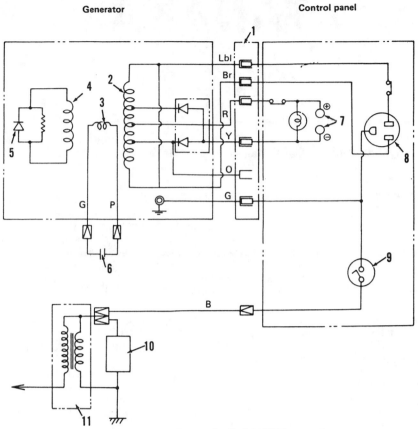

Fig. SU50—*Wiring diagram for Model SE700 generator.*

1. Connector
2. Main coil
3. Exciting coil
4. Rotor coil
5. Diode
6. Capacitor
7. DC receptacle

8. AC receptacle
9. Engine switch
10. Ignition unit
11. Ignition coil
B. Black
Br. Brown

G. Green
Lbl. Light blue
O. Orange
P. Pink
R. Red
Y. Yellow

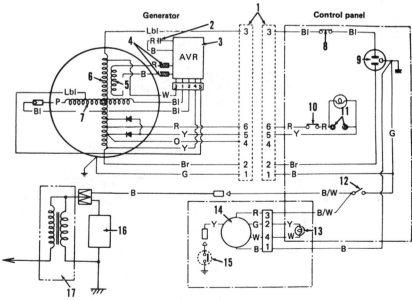

Fig. SU51—*Wiring diagram for Model SE700A generator.*

1. Connector
2. Capacitor
3. Voltage regulator
4. Brushes
5. Rotor coil
6. Main coil
7. Exciting coil
8. AC circuit breaker
9. AC receptacle
10. DC circuit breaker

11. DC receptacle
12. Engine switch
13. Oil level alarm lamp
14. Oil level unit
15. Oil level sensor
16. Ignition unit
17. Ignition coil
B. Black

Bl. Blue
Br. Brown
G. Green
Lbl. Light blue
O. Orange
P. Pink
R. Red
W. White
Y. Yellow

# TANAKA

**TANAKA KOGYO (USA) LTD.**
**7509 S. 228th St.**
**Kent, WA 98031**

| Model | Output-kw | Voltage | Engine Make | Engine Model | Governed Rpm |
|-------|-----------|---------|------|-------|--------------|
| REG-800 . . . . . . . | 0.8 | 115/240 | Own | RE-11 | 3600 |

## GENERATOR

### MAINTENANCE

Generator brushes and brush holder 8–Fig. TA1-1 are accessible after removing generator end cover (10). Renew brushes if they are worn to a length of 7 mm (0.275 inch) or less. Slip ring surfaces must be clean and bright. Polish slip rings with fine sandpaper if necessary. When generator is disassembled, inspect bearings (5 and 7) for smooth operation and renew as needed. Inspect drive belt (1) every 30 hours of operation for fraying or cracking and renew as needed. Specified belt deflection when measured midway between pulleys is 10 mm (4 inches).

### TROUBLESHOOTING

If little or no output is generated, check the following: Check circuit breaker and reset if necessary. Engine must be in good condition and maintain a governed speed under load of 3600 rpm for 60 Hz operation. Check drive belt for slippage and renew or tighten as needed. All wiring connections must be clean and tight. Brushes and slip rings must be in good condition. Removing end cover (10–Fig. TA1-1) will gain access to most of the stator wiring and brushes.

Visually inspect stator for burnt windings or melted insulation and renew as needed. Measure resistance in stator windings with an ohmmeter for open circuit or grounds. Ohmmeter reading should be infinite when measuring be-tween any of the stator leads and stator frame.

Check rotor with an ohmmeter by measuring resistance between the slip rings. There should be continuity between the slip rings and no continuity between slip rings and rotor shaft or frame.

### OVERHAUL

Refer to Fig. TA1-1 for exploded view of generator. Drain fuel and disconnect fuel line from carburetor. Unplug control panel from wiring harness. Remove the four screws and lift frame, fuel tank and control panel assembly away from base. Remove air filter. Remove the four generator-to-base mounting bolts and lift generator from base after removing drive belt. Unbolt and remove fan pulley

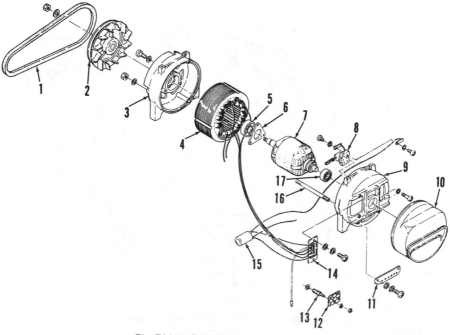

*Fig. TA1-1 — Exploded view of generator.*

| | | | |
|---|---|---|---|
| 1. Drive belt | 6. Retainer | 10. Rubber cover | 14. Diode |
| 2. Pulley | 7. Rotor | 11. Terminal plate | 15. Condenser |
| 3. Rear case | 8. Brush assy. | 12. Regulator diode | 16. Case bolt |
| 4. Stator | 9. Front case | 13. Diode bolt | 17. Bearing |
| 5. Bearing | | | |

(2), then remove the three nuts which hold generator case halves (3 and 9) and stator (4) together. Remove brush and holder assembly (8). Using a soft hammer, tap loose and remove rear case half (3) and rotor (7) away from stator (4) and front case (9). Remove retainer (6) and separate rotor and rear case. Inspect components and renew as needed. Reassemble by reversing disassembly procedure.

## OPERATION

Engine is a two-stroke design with a recommended fuel/oil mix ratio of 25:1. An exploded view of the engine is shown in Fig. TA1-3.

# ENGINE

| Model | Cyls. | Bore | Stroke | Displ. |
|-------|-------|------|--------|--------|
| RE-11 . . . . . . . . . . . . | 1 | 45 mm<br>(1.772 inch) | 38 mm<br>(1.496 inch) | 60 cc<br>(3.66 cu. in.) |

## MAINTENANCE

**SPARK PLUG.** Recommended spark plug is NGK B-7S or equivalent. Specified spark plug gap is 0.6 mm (0.024 inch).

**CARBURETOR.** An exploded view of the float type carburetor is shown in Fig. TA1-4. Initial setting for needle valve (5) is 1¾-2¼ turns out from its fully closed position.

**IGNITION AND TIMING.** Refer to Fig. TA1-5 for view of engine ignition components. Timing is preset at 30°

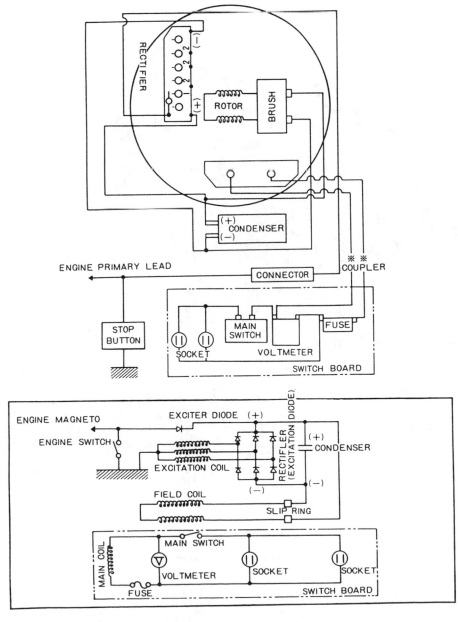

*Fig. TA1-2 — Wiring schematic for Model REG-800.*

BTDC. Specified breaker point gap is 0.35 mm (0.013-0.014 inch). Specified air gap between coil arms and flywheel is 0.3 mm (0.011-0.012 inch).

**GOVERNOR.** Governor is enclosed in a separate oil filled chamber. Refill capacity of governor chamber is 30 cc (0.06 pint).

## OVERHAUL

**CONNECTING ROD AND CRANKSHAFT.** Connecting rod and crankshaft assembly (18 – Fig. TA1-3) is removable after removing cylinder (11) and splitting crankcase. Specified maximum clearance between crankshaft and connecting rod journal is 0.08 mm (0.003 inch). If clearance is exceeded or other damage is evident, renew crankshaft.

**PISTON, PIN AND RINGS.** The piston (13A – Fig. TA1-3) carries two rings (13). Specified running clearance between piston and cylinder is 0.01-0.05 mm (0.0004-0.002 inch) with a maximum allowable clearance of 0.20 mm (0.008 inch). Specified clearance between ring and ring groove is 0.04-0.08 mm (0.002-0.003 inch). Specified end gap for both rings is 0.20-0.35 mm (0.008-0.014 inch). Specified maximum clearance between piston pin (15) and piston pin bore is 0.07 mm (0.003 inch).

**CYLINDER.** If cylinder taper or out-of-round exceeds 0.12 mm (0.005 inch), cylinder must be renewed. Minor scuffing or scoring of cylinder may be removed with a hone or fine grain sandpaper.

**RECOIL STARTER.** Refer to Fig. TA1-6 for an exploded view of recoil starter.

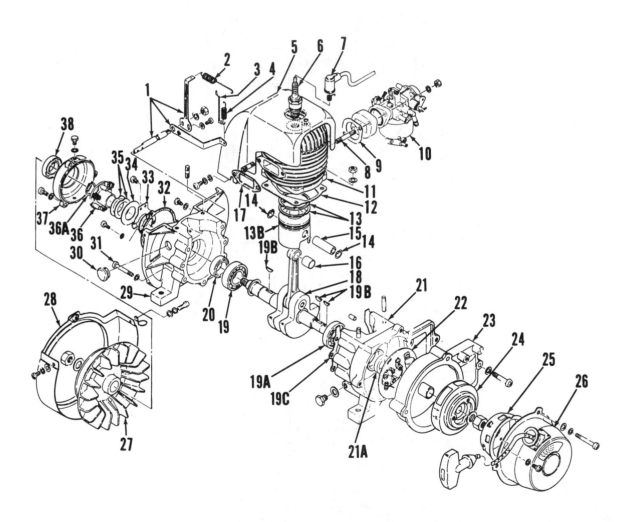

**Fig. TA1-3 — Exploded view of engine.**

| | | | |
|---|---|---|---|
| 1. Governor linkage | 9. Gasket | 15. Piston pin | 19C. Gasket | 25. Hub | 32. Gasket |
| 2. Governor spring | 10. Carburetor | 16. Bushing | 20. Seal | 26. Starter | 33. Governor arm |
| 3. Governor rod | 11. Cylinder | 17. Gasket | 21. Crankcase half | 27. Fan | 34. Governor plate |
| 4. Governor rod spring | 12. Head gasket | 18. Crankshaft assy. | 21A. Seal | 28. Fan cover | 35. Governor rings |
| 5. Cylinder cover | 13. Rings | 19. Bearing | 22. Contact plate assy. | 29. Crankcase half | 36. Governor |
| 6. Spark plug | 13A. Piston | 19A. Bearing | 23. Starter base | 30. Oil plug | 36A. Snap ring |
| 7. Plug wire | 14. Retainer clips | 19B. Key | 24. Magneto | 31. Crankcase bolt | 37. Cover |
| 8. Stud | | | | | 38. Seal |

Illustrations courtesy of Tanaka Kogyo

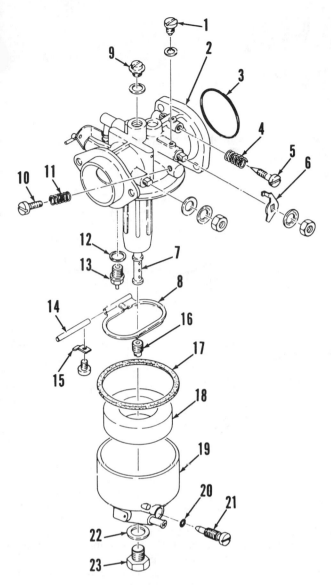

**Fig. TA1-4 — Exploded view of carburetor.**

1. Screw
2. Body
3. "O" ring
4. Spring
5. Needle valve
6. Lever
7. Main nozzle
8. Float arm
9. Screw
10. Screw
11. Spring
12. Gasket
13. Float valve
14. Float pin
15. Retainer
16. Main jet
17. Gasket
18. Float
19. Float bowl
20. Packing
21. Drain screw
22. Gasket
23. Bolt

**Fig. TA1-5 — View of engine ignition components.**

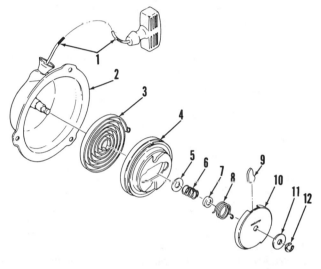

**Fig. TA1-6 — Exploded view of recoil starter assembly.**

1. Rope & handle
2. Housing
3. Recoil spring
4. Pulley
5. Washer
6. Friction spring
7. Spring case
8. Pawl spring
9. Pawl
10. Friction plate
11. Washer
12. Clip

# YAMAHA

**YAMAHA MOTOR CORPORATION**
**6555 Katella Avenue**
**Cypress, California 90630**

| Model | Output-kw | Voltage | Engine Make | Engine Model | Governed Rpm |
|-------|-----------|---------|-------------|--------------|--------------|
| EF1200 | 1.2 | 120VAC 12VDC | Yamaha | MF150 | 4000 |
| EF1400 | 1.2 | 120VAC 12VDC | Yamaha | MF150 | 4000 |

# GENERATOR

### MAINTENANCE

The generator is equipped with a brushless, permanent magnet type rotor and no maintenance is required.

### TROUBLESHOOTING

Prior to using the following troubleshooting procedure, be sure circuit breaker, switch, indicator bulb and outlets operate correctly and all wiring connections are tight and clean. Governed engine speed must be 4000 rpm.

If DC output is faulty, disconnect six-wire coupler and using an ohmmeter check for continuity between terminal (C–Fig. Y1-1) and DC circuit breaker, between DC circuit breaker box and DC outlet and between terminal (B) and DC outlet. If continuity does not exist in all checks, wiring harness is faulty and must be renewed. If continuity exists in all checks, disconnect three-wire coupler

and using an ohmmeter check resistance of DC stator coil by connecting ohmmeter leads to terminal (D) of six-wire coupler and terminal (J–Fig. Y1-2) of three-wire coupler. Resistance should be 0.222-0.272 ohms; if not, DC stator coil is faulty and stator must be renewed. If resistance is correct, check AC output voltage at outlet. If AC voltage is normal, the rectifier may be faulty and must be renewed. If AC voltage is lower than normal, proceed to following AC troubleshooting paragraph.

If AC output is faulty, disconnect six-wire coupler and using an ohmmeter check for continuity between terminal (F–Fig. Y1-1) and AC outlet, between terminal (A) and AC switch/circuit breaker and between AC outlet and AC

switch/circuit breaker. If continuity does not exist in all checks, wiring harness is faulty and must be renewed. If continuity exists in all checks, measure resistance of main stator coil by connecting ohmmeter leads to terminals (C and D) of six-wire coupler. Resistance should be 0.63-0.76 ohms. If reading is incorrect, main coil is faulty and stator must be renewed. If stator main coil

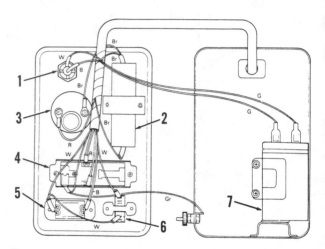

Fig. Y1-3—View of control box showing wiring and components of Model EF 1200.

B. Black
Br. Brown
B/W. Black/White
G. Green
Gr. Gray
R. Red
W. White
1. Indicator light
2. AC switch/circuit breaker
3. Voltmeter
4. AC outlet
5. DC circuit breaker
6. DC outlet
7. Condenser

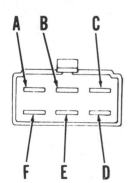

Fig. Y1-1—Drawing of terminals in six-wire coupler.

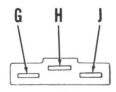

Fig. Y1-2—Drawing of terminals in three-wire coupler.

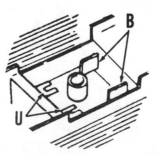

Fig. Y1-4—View showing location of U-shaped (U) and blade (B) terminals on rotor. Terminals are located on both sides of rotor.

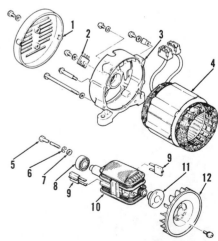

Fig. Y1-5—Exploded view of generator.

| | |
|---|---|
| 1. End cover | 7. Washer |
| 2. DC rectifier | 8. Bearing |
| 3. Stator housing | 9. AC rectifiers |
| 4. Stator | 10. Rotor |
| 5. Bolt | 11. Flange coupling |
| 6. Lockwasher | 12. Flywheel |

# ENGINE

| Model | Cyls. | Bore | Stroke | Displ. |
|-------|-------|------|--------|--------|
| MF150 | 1 | 62 mm<br>(2.44 in.) | 48 mm<br>(1.89 in.) | 145 cc<br>(8.84 cu. in.) |

resistance is correct, measure resistance of stator sub-coil by connecting ohmmeter to terminals (B and E) of six-wire coupler. Resistance should be 2.35-2.87 ohms. If reading is incorrect, sub-coil is defective and stator must be renewed. If reading is correct, open control box and disconnect leads to condenser. Place a jumper wire between condenser terminals to be sure condenser is discharged then remove jumper wire. With ohmmeter set on Rx100 scale, connect ohmmeter leads to condenser terminals. Ohmmeter needle should swing rapidly across meter then return. Disconnect ohmmeter and use a jumper wire to discharge condenser. Renew condenser if test is failed. If condenser tests satisfactory, remove rotor as outlined in OVERHAUL section and proceed as follows: Unsolder wires from blade (B – Fig. Y1-4) and U-shaped (U) terminals on both sides of rotor. Measure coil resistance by connecting ohmmeter leads to U-shaped terminals on other rotor side. Resistance should be 3.82-4.68 ohms or field coils are defective and rotor must be renewed. To check AC rectifiers, connect ohmmeter leads to blade terminals (B), note reading, reverse ohmmeter leads and again note reading. Follow same procedure using blade terminals on opposite side of rotor. Ohmmeter reading should be infinite for one reading and approximately 8.3 ohms for other reading. Renew rectifier (9 – Fig. Y1-5) if readings are incorrect.

A faulty rectifier may decrease magnetic field of rotor magnet. To strengthen magnetic field, assemble and run powerplant at rated speed then momentarily connect a 12 volt battery to terminals of condenser in control box being sure negative terminals of battery and condenser are connected together as well as positive terminals.

## OVERHAUL

To disassemble generator, remove end cover (1 – Fig. Y1-5) then disconnect and remove control box. Unbolt and remove mounting pads under stator housing legs. Unscrew four through bolts securing stator housing (3) and remove stator housing. Unscrew rotor bolt, then using a suitable puller separate rotor from engine crankshaft.

Inspect components and reassemble generator by reversing disassembly procedure. Refer to wiring schematic for proper wiring connections. Tighten rotor bolt to 18.6 N·m (165 in.-lbs.) and stator through bolts to 7.8 N·m (69 in.-lbs.).

## MAINTENANCE

**SPARK PLUG.** Recommended spark plug is Champion L87Y or NGK BP-6HS. Spark plug electrode gap should be 0.6-0.7 mm (0.024-0.028 in.).

**CARBURETOR.** A Mikuni BV18-14 float type carburetor is used. Initial setting of idle mixture screw is 1½ turns out. Float level should be 18.5 mm (0.728 in.) and float drop should be 27 mm (1.063 in.). Bend tangs on float for adjustment. Note following specifications:

| | |
|---|---|
| Fuel inlet valve | 2.0 |
| Main jet | 92.5 |
| Pilot air jet | 1.2 |
| Pilot jet | 50 |

**IGNITION AND TIMING.** All MF150 units are equipped with a capacitor discharge (CD) ignition system. Refer to Fig. Y1-6 for schematic of ignition. Ignition timing is fixed at 20° BTDC at 3600 rpm. The pulse coil (P – Fig. Y1-7) and charge coil (C) are accessible after flywheel removal.

An ohmmeter may be used to check pulse, charge and ignition coils. Resistance reading for pulse coil with ohmmeter attached to black and red/white pulse coil leads should be 18-22 ohms. Resistance reading for charge coil leads should be 378-462 ohms. With ohmmeter leads connected to ignition coil primary lead and coil leg, ignition coil primary resistance reading should be 1.6-2.0 ohms. Resistance reading for ignition coil secondary winding with ohmmeter leads connected to secondary lead and coil leg should be 8K-12K ohms.

The ignition module is equipped with an over-speed circuit which prevents excessive engine speed by disrupting the ignition circuit when engine speed

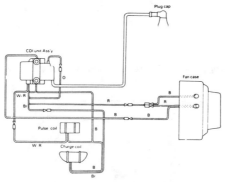

**Fig. Y1-6—Wiring diagram of ignition system.**

B. Black
Br. Brown
O. Orange
R. Red
W/R. White/Red

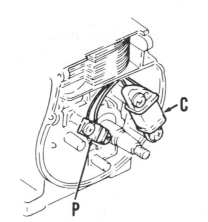

**Fig. Y1-7—View showing location of ignition pulse coil (P) and charge coil (C).**

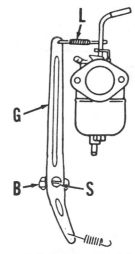

**Fig. Y1-8—Drawing of governor linkage. Refer to text for operation.**

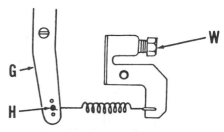

**Fig. Y1-9—Adjust governed engine speed as outlined in text. Spring end is attached to middle hole (H).**

reaches 4050-4650 rpm.

Flywheel nut tightening torque is 54 N·m (40 ft.-lbs.).

Note that an erratic or faulty ignition system may be due to an oil warning system malfunction. Refer to LUBRICATION section.

**GOVERNOR.** All models are equipped with a flyweight type governor attached to and driven by the camshaft gear. The governor forces governor shaft (S–Fig. Y1-8) to rotate according to engine speed. Pinch bolt (B) secures governor arm (G) to shaft (S). Movement of governor arm (G) is transmitted by link (L) to carburetor throttle arm.

Note in Fig. Y1-9 that governor spring end is connected to middle hole (H) of governor arm (G).

To adjust governed speed, loosen pinch bolt (B–Fig. Y1-8). Pull upper end of governor arm (G) to left (away from carburetor) so carburetor throttle is fully open. Turn governor shaft (S) counterclockwise until stop is contacted then retighten pinch bolt (B). With engine running under load, turn adjusting screw (W–Fig. Y1-9) so maximum governed speed is 4000 rpm. If available, a frequency meter may be connected to an outlet and maximum governed speed adjusted so current frequency is 60 hertz (cycles).

**LUBRICATION.** Crankcase capacity 500 cc (0.5 qts.). Use SAE 30 oil if ambient temperature is above 15°C (60°F),

SAE 20 if temperature is between 0°C (32°F) and 15°C (60°F), and SAE 10 if temperature is below 0°C (32°F).

All models are equipped with a low-oil safety system which disables the ignition system and lights a warning bulb when engine oil level is critically low. A float (F–Fig. Y1-10) mounted in the engine crankcase monitors oil level and when crankcase is approximately half full a magnet in the float closes the contacts in reed switch (R) thereby grounding the ignition and lighting the oil warning light.

Refer to OVERHAUL section to service oil warning system.

## OVERHAUL

**TIGHTENING TORQUES.** Recommended tightening torques are as follows:

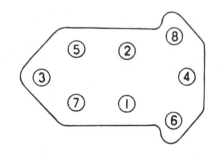

Fig. Y1-12—Tightening sequence for cylinder head screws.

Connecting rod . . . . . . . . . . . . . 9.8 N·m
(7 ft.-lbs.)
Crankcase cover . . . . . . . . . . 14.7 N·m
(11 ft.-lbs.)
Cylinder head . . . . . . . . . . . . . 19.6 N·m
(14.5 ft.-lbs.)
Flywheel . . . . . . . . . . . . . . . . . . 50 N·m
(40 ft.-lbs.)

**CYLINDER HEAD.** The cylinder head is removable. Be sure head gasket is installed as shown in Fig. Y1-11. Tighten cylinder head screws to 19.6 N·m (14.5 ft.-lbs.) in sequence shown in Fig. Y1-12.

**VALVE SYSTEM.** Valve tappet gap is 0.05-0.15 mm (0.002-0.006 in.). Different length valves are available if valve tappet gap is incorrect. Valves are marked "0, 1 or 2" with "0" valve shortest and "2" valve longest.

Valve stem diameter is 6.955-6.970 mm (0.2738-0.2744 in.). Valve seat width should be 0.7-0.9 mm (0.027-0.035 in.). Valve spring free length is 33.5 mm (1.319 in.).

**PISTON, PIN AND RINGS.** Piston and connecting rod are removed as a unit after cylinder head and crankcase cover are removed. Piston oversizes available are 0.25 mm (0.010 in.) and 0.50 mm (0.020 in.). Piston clearance should be 0.065-0.075 mm (0.0025-0.0030 in.). Desired ring end gap is 0.2-0.4 mm (0.008-0.016 in.). Install piston so arrow on crown points towards valve side of engine as shown in Fig. Y1-13. Piston pin is fully floating in piston and rod. Piston pin retaining clips should be renewed if removed.

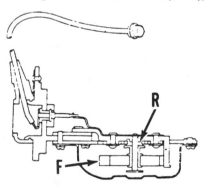

Fig. Y1-10—Cross-sectional view of oil level warning system.

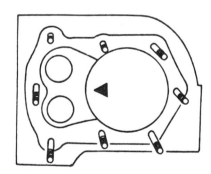

Fig. Y1-13—Install piston so arrow on crown points towards valves.

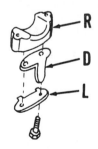

Fig. Y1-15—A separate oil dipper (D) is used between rod cap (R) and lock plate (L).

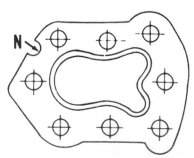

Fig. Y1-11—Install head gasket so notch (N) is towards flywheel side of engine.

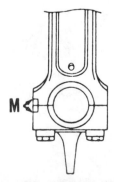

Fig. Y1-14—Match marks (M) on side of rod and cap must be aligned during assembly.

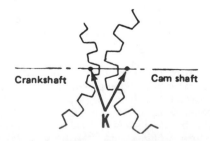

Fig. Y1-16—Align crankshaft and camshaft timing marks (K) during assembly.

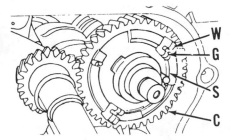

Fig. Y1-17—View of governor weights (W) and spring (S) installed on camshaft gear (C). Governor collar is not shown but must engage groove (G) in weights.

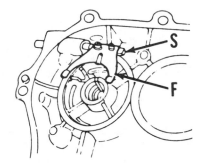

Fig. Y1-18—Install governor fork (F) on governor shaft (S) with pads on fork ends towards camshaft.

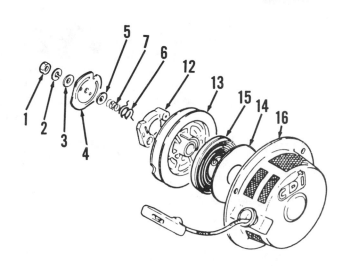

Fig. Y1-19—Exploded view of rewind starter.

1. Nut
2. Lockwasher
3. Washer
4. Plate
5. Washer
6. Return spring
7. Spring
12. Pawl assy.
13. Rope pulley
14. Plate
15. Rewind starter
16. Starter housing

**CONNECTING ROD.** The connecting rod and piston are removed as a unit after cylinder head and crankcase cover are removed. The aluminum rod rides directly on the crankpin. Connecting rod to crankpin clearance is 0.016-0.034 mm (0.0006-0.0013 in.). Connecting rod side clearance is 0.2-0.8 mm (0.008-0.032 in.). Match marks (M–Fig. Y1-14) must be aligned when installing rod. Be sure oil dipper (D–Fig. Y1-15) is clamped between rod cap (R) and lock plate (L). Tighten connecting rod screws to 9.8 N·m (7 ft.-lbs.).

**CAMSHAFT AND GOVERNOR.** The camshaft rides directly in crankcase and crankcase cover. Inspect camshaft lobes, gear and bearing surfaces. Desired camshaft lobe height is 27.55 mm (1.085 in.) with a minimum height of 27.3 mm (1.075 in.). Crankshaft and camshaft timing marks are aligned as shown in Fig. Y1-16 during assembly.

The governor weights, spring and actuating collar are mounted on camshaft as shown in Fig. Y1-17. Be sure collar properly engages groove (G) in weights (W). Install governor fork (F–Fig. Y1-18) on governor shaft (S) so pads on fork ends are towards camshaft.

**CRANKSHAFT AND CRANKCASE.** The crankshaft is supported by ball bearings. Inspect crankshaft for excessive wear, straightness and damage. Crankpin diameter is 24.952-24.970 mm (0.9824-0.9831 in.). Tighten crankcase cover screws to 14.7 N·m (11 ft.-lbs.).

**REWIND STARTER.** Refer to Fig. Y1-19 for an exploded view of rewind starter. With starter removed from engine, untie rope handle from rope and allow rope to wind into starter housing. Unscrew nut (1) to disassemble starter. Use care when removing rope pulley (13) so rewind spring (15) is not dislodged from housing (16). If spring (15) must be removed, use care as uncontrolled uncoiling of spring may cause injury.

When assembling starter, install rewind spring (15) so coil direction is counter-clockwise from outer end. Wind rope on rope pulley in counter-clockwise direction as viewed with pulley (13) installed in housing (16). Preload rewind spring by turning rope pulley two turns against spring tension before passing rope through rope outlet in housing.

**OIL WARNING SYSTEM.** All models are equipped with an oil warning system which disables ignition system when low oil level may damage engine. See LUBRICATION section.

To inspect oil level assembly, drain engine oil and remove unit from side of engine. Handle float assembly carefully as reed switch may be damaged if unit is dropped. To check operation of float assembly, connect an ohmmeter to mounting flange and to wire lead. Do not use a continuity tester as battery may damage reed switch. Ohmmeter should show continuity with float assembly in normal position and float hanging down. Ohmmeter should read infinity with float assembly inverted. Oil warning light should be checked and renewed if defective.

# YAMAHA

| Model | Output-kw | Voltage | Engine Make | Engine Model | Governed Rpm |
|---|---|---|---|---|---|
| 799 | 1.5 | 120VAC | Yamaha | MF180 | 3600 |
|  |  | 12VDC |  |  |  |
| 7A4 | 1.5 | 220VAC | Yamaha | MF180 | 3600 |
|  |  | 12VDC |  |  |  |
| 7E1 | 2.2 | 120VAC | Yamaha | MF260 | 3600 |
|  |  | 12VDC |  |  |  |
| 7E3 | 2.2 | 220VAC | Yamaha | MF260 | 3600 |
|  |  | 12VDC |  |  |  |

# GENERATOR

## MAINTENANCE

Brushes are accessible after detaching end cover. Minimum brush length is 12 mm (0.472 in.). Inspect rotor slip rings. Stator assembly must be removed as outlined in OVERHAUL section to service slip rings. Use 600 sandpaper to polish slip rings or a suitable lathe to remove deep imperfections. Minimum slip ring diameter is 36 mm (1.417 in.).

## TROUBLESHOOTING

Prior to using the following troubleshooting procedure, be sure fuse, indicator bulb, power switch and outlets operate correctly and all wiring connections are tight and clean.

If DC output is low or zero, connect a DC voltmeter to DC outlet and measure DC voltage. If voltage is abnormal, connect an AC voltmeter to AC outlet and

measure AC voltage. If AC voltage is abnormal refer to following paragraph and locate fault. If AC voltage is satisfactory, remove end cover and connect AC voltmeter leads to a diode lead and to diode plate as shown in Fig. Y2-1. If AC voltage is zero, check wiring to diodes. If AC voltage is approximately 30 volts, diode is defective and diode plate assembly must be renewed.

If AC output is low or zero, connect an AC voltmeter to AC outlet and measure AC voltage. If AC output is low, check field coil resistance with engine stopped by connecting an ohmmeter to terminals (T – Fig. Y2-2) on brush holder.

**CAUTION: Positive lead of ohmmeter must be connected to negative brush terminal and negative tester must be connected to positive brush terminal. Reversed tester leads may reduce magneto force in field coil magnets.**

Resistance should be 53.1-64.9 ohms on Models 799 and 7A4 or 63.0-77.0 ohms on Models 7E1 and 7E3. If resistance is incorrect, brushes are making poor contact with slip rings and should be inspected or rotor is defective and should be renewed. If resistance is correct, disconnect coupler shown in Fig. Y2-3, connect jumper wires (J) to terminals (T) and using an ohmmeter connected to jumper wires measure resistance of subcoil. Resistance should be 2.3-2.9 ohms on Models 799 and 7A4 or 1.4-1.8 ohms on Models 7E1 and 7E3.

If resistance is incorrect, renew subcoil. If resistance is correct, renew voltage regulator.

If AC voltage at outlet is zero, check resistance of stator coils with engine stopped as follows: On Models 799 and 7A4, disconnect wire coupler on top of alternator leading to control box. Refer to Fig. Y2-4 and attach ohmmeter leads to lower left and middle terminals (T) of coupler. Resistance should be 0.68-0.84 ohms for 799 models or 2.43-2.97 ohms for 7A4 models. On 7E1 and 7E3 models, remove end cover and detach red and brown control box wires from terminals (T – Fig. Y2-5) and attach

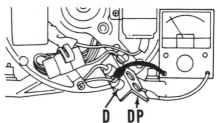

*Fig. Y2-1—Attach voltmeter leads to diode (D) and diode plate (DP).*

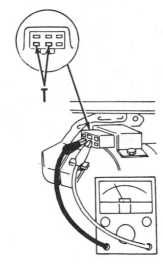

*Fig. Y2-4—On MF180 models, connect ohmmeter leads to coupler terminals (T).*

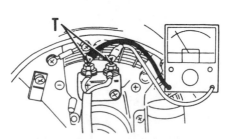

*Fig. Y2-2—Attach ohmmeter leads to brush holder terminals (T).*

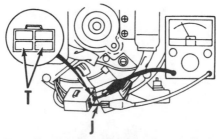

*Fig. Y2-3—Connect jumper wires (J) to coupler terminals (T) then attach ohmmeter leads to jumper wires.*

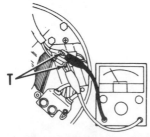

*Fig. Y2-5—On MF260 models, disconnect red and brown control box wires from terminals (T) and attach ohmmeter leads.*

ohmmeter leads to terminals. Resistance should be 0.41-0.51 ohms for 7E1 models or 1.49-1.82 ohms for 7E3 models. Renew stator if desired resistance readings are not obtained.

### OVERHAUL

To disassemble generator, remove end cover then disconnect and remove control box. Disconnect and remove brush holder assembly. Unbolt and remove mounting cushions from stator housing. Unscrew four through bolts securing stator housing and remove stator housing. Unscrew rotor through bolt, then using a slide hammer, break rotor loose from crankshaft taper and remove rotor.

Inspect components and reassemble generator by reversing disassembly procedure. Refer to wiring schematic for proper wiring connections. Tighten rotor bolt to 18.6 N·m (14 ft.-lbs.) and stator through bolts to 7.8 N·m (6 ft.-lbs.).

# ENGINE

| Model | Cyls. | Bore | Stroke | Displ. |
|-------|-------|------|--------|--------|
| MF180 | 1 | 65 mm (2.559 in.) | 54 mm (2.126 in.) | 179 cc (10.9 cu. in.) |
| MF260 | 1 | 73 mm (2.874 in.) | 61 mm (2.402 in.) | 256 cc (15.6 cu. in.) |

### MAINTENANCE

**SPARK PLUG.** Recommended spark plug is a Champion L87Y or NGK BP-6HS. Spark plug electrode gap is 0.6-0.7 mm (0.024-0.028 in.).

**CARBURETOR.** MF180 models are equipped with a Mikuni BV18-15 carburetor while MF260 models are equipped with a Keihin BB24-17 carburetor. Initial setting of idle mixture screw is 1¾ turns on BV18-15 carburetor or 1⅞ turns on BB24-17 carburetor. Refer to following tables for carburetor specifications:

**Mikuni BV18-15**
Fuel inlet valve . . . . . . . . . . . . . . . . . . .2.0
Main jet. . . . . . . . . . . . . . . . . . . . . . . . .80
Pilot air jet . . . . . . . . . . . . . . . . . . . .3.0
Pilot jet . . . . . . . . . . . . . . . . . . . . . .57.5

**Keihin BB24-17**
Fuel inlet valve . . . . . . . . . . . . . . . . . . .2.0
Main air jet . . . . . . . . . . . . . . . . . . . .130
Main jet. . . . . . . . . . . . . . . . . . . . . . . .85
Pilot jet . . . . . . . . . . . . . . . . . . . . . . .40

**IGNITION AND TIMING.** All models are equipped with a capacitor discharge ignition system. The trigger coil and charge coil are attached to a stator plate behind the flywheel while the ignition module is attached to the fan housing. Ignition timing is not adjustable.

To check trigger coil and charge coil, remove fan housing, disconnect coil leads and remove flywheel. Connect an ohmmeter to trigger coil lead and coil base. Ohmmeter should read 603-737 ohms; if not, renew trigger coil. Connect ohmmeter to charge coil lead and coil base. Ohmmeter should read 342-418 ohms; if not, renew charge coil. When reinstalling flywheel, tighten flywheel nut to 54 N·m (40 ft.-lbs.) on MF180 models or 74 N·m (54 ft.-lbs.) on MF260 models.

Note that an erratic or faulty ignition system may be due to an oil warning system malfunction. Refer to LUBRICATION section.

**GOVERNOR.** All models are equipped with a flyweight type governor attached to and driven by the camshaft gear. The governor forces governor shaft (S – Fig. Y2-6) to rotate according to engine speed. Pinch bolt (B) secures governor arm (G) to shaft (S). Movement of governor arm (G) is transmitted by link (L) to carburetor throttle arm.

Note in Fig. Y2-7 that governor spring end is connected to middle hole (H) of MF180 governor arm (G) or to lower hole (H) of MF260 governor arm (G).

To adjust governed speed, loosen pinch bolt (B – Fig. Y2-6). Pull upper end of governor arm (G) to left (away from carburetor) so carburetor throttle is fully open. Turn governor shaft (S) counterclockwise until stop is contacted then retighten pinch bolt (B). With engine running under no load, turn adjusting screw (W – Fig. Y2-7) so maximum governed speed is 3800 rpm. If available, a frequency meter may be connected to an outlet and maximum governed speed adjusted so desired current frequency is obtained.

**LUBRICATION.** Crankcase capacity is 680 cc (0.7 qt.) on MF180 models or 950 cc (1.0 qt.) on MF260 models. Use SAE

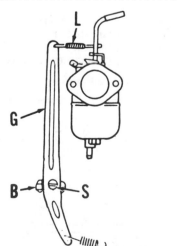

Fig. Y2-6—Drawing of governor linkage. Refer to text for operation.

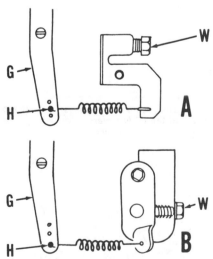

Fig. Y2-7—Adjust governed engine speed as outlined in text. Note position of spring end in hole (H) for MF180 models in View A or for MF260 models in View B.

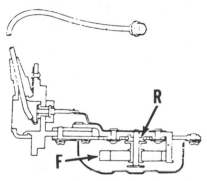

Fig. Y2-8—Cross-sectional view of oil level warning system.

30 oil if ambient temperature is above 15°C (60°F), SAE 20 if temperature is between 0°C (32°F) and 15°C (60°F), and SAE 10 if temperature is below 0°C (32°F).

All models are equipped with a low-oil safety system which disables the ignition system and lights a warning bulb when engine oil level is critically low. A float (F – Fig. Y2-8) mounted in the engine crankcase monitors oil level and when the oil level lowers to less than 325 cc (0.34 qt.) on MF180 models or to less than 450 cc (0.48 qt.) on MF260 models, a magnet in the float closes the contacts in reed switch (R) thereby grounding the ignition and lighting the oil warning light.

Refer to OVERHAUL section to service oil warning system.

## OVERHAUL

**TIGHTENING TORQUES.** Recommended tightening torques are as follows:

Connecting rod
MF180 . . . . . . . . . . . . . . . . .11.8 N·m
                       (104 in.-lbs.)
MF260 . . . . . . . . . . . . . . . . .15.7 N·m
                       (139 in.-lbs.)
Crankcase cover
MF180 . . . . . . . . . . . . . . . . . .9.8 N·m
                       (7 ft.-lbs)
MF260 . . . . . . . . . . . . . . . . . . .17 N·m
                       (12 ft.-lbs.)
Cylinder head . . . . . . . . . . . . . .19.6 N·m
                       (14.5 ft.-lbs)
Flywheel
MF180 . . . . . . . . . . . . . . . . . .54 N·m
                       (40 ft.-lbs)
MF260 . . . . . . . . . . . . . . . . . .74 N·m
                       (54 ft.-lbs.)

**CYLINDER HEAD.** The cylinder head is removable. Be sure head gasket is installed as shown in Fig. Y2-9 or Y2-10. Tighten cylinder head screws to 19.6 N·m (14.5 ft.-lbs.) in sequence shown in Fig. Y2-11.

**VALVE SYSTEM.** Valve tappet gap is 0.05-0.15 mm (0.002-0.006 in.) on MF180 models or 0.1-0.3 mm (0.004-0.012 in.) on MF260 models. Different length valves are available if valve tappet gap is incorrect. Valves are marked "0, 1 or 2". Note valve lengths in following tables:

| Valve Marking | Length | |
| --- | --- | --- |
| | **MF180** | **MF260** |
| 0 | 91.70-91.80 mm (3.610-3.614 in.) | 104.55-104.65 mm (4.116-4.120 in.) |
| 1 | 91.80-91.90 mm (3.614-3.618 in.) | 104.65-104.75 mm (4.120-4.234 in.) |
| 2 | 91.90-92.00 mm (3.618-3.622 in.) | 104.75-104.85 mm (4.124-4.128 in.) |

Intake valve stem diameter for all models is 6.955-6.970 mm (0.2738-0.2744 in.). Exhaust valve stem diameter is 6.950-6.970 mm (0.2736-0.2744 in.) on MF180 models or 6.915-6.930 mm (0.2722-0.2728 in.) on MF260 models. Valve seat width should be 0.7-0.9 mm (0.027-0.035 in.) for all valves. Valve spring free length is 33.5 mm (1.319 in.) for MF180 models or 37.0 mm (1.457 in.) for MF260 models.

**PISTON, PIN AND RINGS.** Piston and connecting rod are removed as a unit after cylinder head and crankcase cover are removed. Piston oversizes available are 0.25 mm (0.010 in.) and 0.05 mm

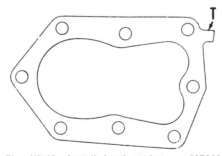

**Fig. Y2-10—Install head gasket on MF260 models so tab (T) is towards flywheel side of engine.**

(0.020 in.). Piston clearance should be 0.085 mm (0.0033 in.) for MF180 models or 0.070 mm (0.0027 in.) for MF260 models. Piston ring end gap should be 0.2-0.4 mm (0.008-0.016 in.). Install piston so arrow on crown points towards valve side of engine as shown in Fig. Y2-12. Piston pin is fully floating in piston and rod. Piston pin retaining clips should be renewed if removed.

**CONNECTING ROD.** The connecting rod and piston are removed as a unit after cylinder head and crankcase cover are removed. The aluminum rod rides directly on crankpin. Connecting rod to crankpin clearance is 0.016-0.034 mm (0.0006-0.0013 in.). Connecting rod side clearance is 0.2-0.8 mm (0.008-0.032 in.). Match marks (M – Fig. Y2-13) must be aligned when installing rod. Note that oil dipper on MF260 models is cast into rod cap while oil dipper on MF180 models is separate and fits between rod cap and lock plate as shown in Fig. Y2-14. Tighten rod screws to 11.8 N·m (104 in.-lbs.) on MF180 models or to 15.7 N·m (139 in.-lbs.) on MF260 models.

**CAMSHAFT AND GOVERNOR.** The camshaft rides directly in crankcase and crankcase cover. Inspect camshaft lobes, gear and bearing surfaces for damage and excessive wear. Crankshaft and camshaft timing marks must be aligned as shown in Fig. Y2-15 during assembly.

The governor weights, spring and actuating collar are mounted on camshaft as shown in Fig. Y2-16. Nine governor weights are held by governor spring on MF260 models while only four weights are used on MF180 models. Be sure collar properly engages groove (G) in weights (W). Install governor fork (F – Fig. Y2-17) on governor shaft (S) of MF260 models so pads on fork ends are towards camshaft.

**CRANKSHAFT AND CRANKCASE.** The crankshaft is supported by ball bearings. Inspect crankshaft for excessive wear, straightness and damage.

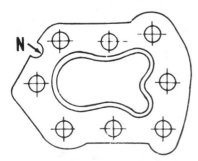

**Fig. Y2-9—Install head gasket on MF180 models so notch (N) is towards flywheel side of engine.**

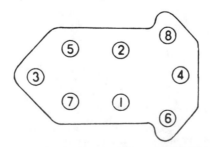

**Fig. Y2-11—Tightening sequence for cylinder head screws.**

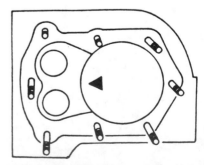

**Fig. Y2-12—Install piston so arrow on crown points towards valves.**

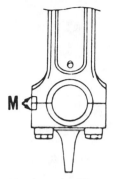

Fig. Y2-13—Match marks (M) on side of rod and cap must be aligned during assembly.

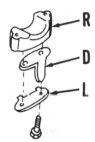

Fig. Y2-14—A separate oil dipper (D) is used on MF180 models between rod cap (R) and lock plate (L).

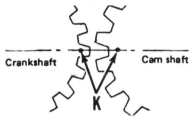

Fig. Y2-15—Align crankshaft and camshaft timing marks (K) during assembly.

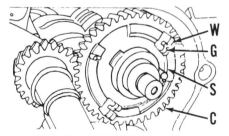

Fig. Y2-16—View of MF180 governor weights (W) and spring (S) installed on camshaft gear (C). MF260 assembly is similar but more weights are used. Governor collar is not shown but must engage groove (G) in weights.

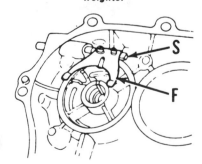

Fig. Y2-17—Install governor fork (F) on governor shaft (S) of MF260 models with pads on fork ends towards camshaft.

Crankpin diameter is 24.952-24.970 mm (0.9824-0.9831 in.) on MF180 models or 28.952-28.970 mm (1.1398-1.1405 in.) on MF260 models. Tighten crankcase cover

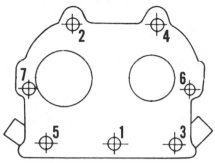

Fig. Y2-18—Tightening sequence for crankcase cover screws on MF180.

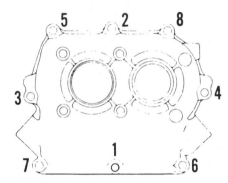

Fig. Y2-19—Tightening sequence for crankcase cover screws on MF260.

Fig. Y2-20—Exploded view of rewind starter on MF180.

1. Nut
2. Lockwasher
3. Washer
4. Plate
5. Washer
6. Return spring
7. Spring
12. Pawl assy.
13. Rope pulley
14. Plate
15. Rewind spring
16. Starter housing

Fig. Y2-21—Exploded view of rewind starter on MF260.

1. Nut
2. Lockwasher
3. Washer
4. Plate
6. Return spring
7. Spring
8. Washer
9. Pawl
10. Spring
11. Bushing
13. Rope pulley
14. Plate
15. Rewind spring
16. Starter housing

screws to 9.8 N·m (7 ft.-lbs.) on MF180 models or to 17 N·m (12 ft.-lbs.) on MF260 models. Refer to Fig. Y2-18 or Y2-19 for tightening sequence.

**REWIND STARTER.** Refer to Fig. Y2-20 or Y2-21 for an exploded view of rewind starter. With starter removed from engine, untie rope handle from rope and allow rope to wind into starter housing. Unscrew nut (1) to disassemble starter. Use care when removing rope pulley (13) so rewind spring (15) is not dislodged from housing (16). If spring (15) must be removed, use care as uncontrolled uncoiling of spring may cause injury.

When assembling starter, install rewind spring (15) so coil direction is counterclockwise from outer end. Wind rope on rope pulley in counter-clockwise direction as viewed with pulley (13) installed in housing (16). Preload rewind spring by turning rope pulley two turns against spring tension before passing rope through rope outlet in housing.

**OIL WARNING SYSTEM.** All models are equipped with an oil warning system which disables ignition system when low oil level may damage engine. See LUBRICATION section.

To inspect oil level assembly, drain engine oil and remove unit from side of

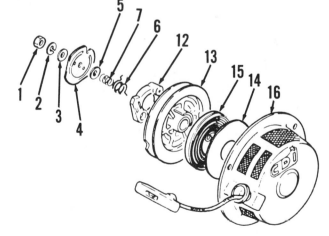

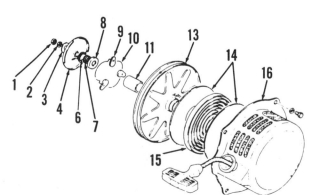

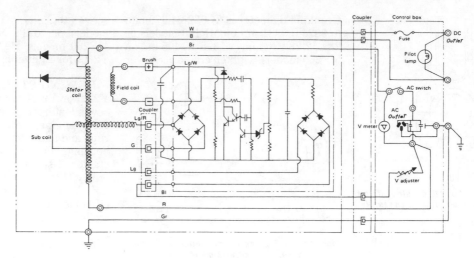

*Fig. Y2-22—Generator wiring schematic.*

engine. Handle float assembly carefully as reed switch may be damaged if unit is dropped. To check operation of float assembly, connect an ohmmeter to mounting flange and to wire lead. Do not use a continuity tester as battery may damage reed switch. Ohmmeter should show continuity with float assembly in normal position and float hanging down. Ohmmeter should read infinity with float assembly inverted. Oil warning light should be checked.

# METRIC
# CONVERSION

| | | | |
|---|---|---|---|
| Cubic meters | = .02832 | x | Cubic Feet |
| Cubic meters | x 1.308 | = | Cubic Yards |
| Cubic meters | = .765 | x | Cubic Yards |
| Liters | x 61.023 | = | Cubic Inches |
| Liters | = .01639 | x | Cubic Inches |
| Liters | x .26418 | = | U.S. Gallons |
| Liters | = 3.7854 | x | U.S. Gallons |
| Grams | x 15.4324 | = | Grains |
| Grams | = .0648 | x | Grains |
| Grams | x .03527 | = | Ounces, avoirdupois |
| Grams | = 28.3495 | x | Ounces, avoirdupois |
| Kilograms | x 2.2046 | = | Pounds |
| Kilograms | = .4536 | x | Pounds |
| Kilograms per square centimeter | x 14.2231 | = | Pounds per square Inch |
| Kilograms per square centimeter | = .0703 | x | Pounds per square Inch |
| Kilograms per cubic meter | x .06243 | = | Pounds per cubic Foot |
| Kilograms per cubic meter | = 16.01890 | x | Pounds per cubic Foot |

| | | | |
|---|---|---|---|
| Metric tons (1,000 kilograms) | x 1.1023 | = | Tons (2,000 Pounds) |
| Metric tons (1,000 kilograms) | = .9072 | x | Tons (2,000 Pounds) |
| Kilowatts | = 1.3405 | x | Horsepower |
| Kilowatts | x .746 | = | Horsepower |
| Millimeters | x .03937 | = | Inches |
| Millimeters | = 25.400 | x | Inches |
| Meters | x 3.2809 | = | Feet |
| Meters | = .3048 | x | Feet |
| Kilometers | x .621377 | = | Miles |
| Kilometers | = 1.6093 | x | Miles |
| Square centimeters | x .15500 | = | Square Inches |
| Square centimeters | = 6.4515 | x | Square Inches |
| Square meters | x 10.76410 | = | Square Feet |
| Square meters | = .09290 | x | Square Feet |
| Cubic centimeters | x .061025 | = | Cubic Inches |
| Cubic centimeters | = 16.3866 | x | Cubic Inches |
| Cubic meters | x 35.3156 | = | Cubic Feet |

# FUNDAMENTALS SECTION

## ENGINE FUNDAMENTALS

### OPERATING PRINCIPLES

The small engines used to power lawn mowers, garden tractors and many other items of power equipment in use today are basically similar. All are technically known as ''Internal Combustion Reciprocating Engines.''

The source of power is heat formed by the burning of a combustible mixture of petroleum products and air. In a reciprocating engine, this burning takes place in a closed cylinder containing a piston. Expansion resulting from the heat of combustion applies pressure on the piston to turn a shaft by means of a crank and connecting rod.

The fuel-air mixture may be ignited by means of an electric spark (Otto Cycle Engine) or by heat formed from compression of air in the engine cylinder (Diesel Cycle Engine). The complete series of events which must take place in order for the engine to run may occur in one revolution of the crankshaft (two strokes of the piston in cylinder) which is referred to as a ''Two Stroke Cycle Engine,'' or in two revolutions of the crankshaft (four strokes of the piston in cylinder) which is referred to as a ''Four-Stroke Cycle Engine.''

**OTTO CYCLE.** In a spark ignited engine, a series of five events is required in order for the engine to provide power. This series of events is called the ''Cycle'' (or ''Work Cycle'') and is repeated in each cylinder of the engine as long as work is being done. This series of events which comprise the ''Cycle'' are as follows:

1. The mixture of fuel and air is pushed into the cylinder by atmospheric pressure when the pressure within the engine cylinder is reduced by the piston moving downward in the cylinder (or by applying pressure to the fuel-air mixture as by crankcase compression in the crankcase of a ''Two-Stroke Cycle Engine'' which is described in a later paragraph).

2. The mixture of fuel and air is compressed by the piston moving upward in the cylinder.

3. The compressed fuel-air mixture is ignited by a timed electric spark.

4. The burning fuel-air mixture expands, forcing the piston downward in the cylinder thus converting the chemical energy generated by combustion into mechanical power.

5. The gaseous products formed by the burned fuel-air mixture are exhausted from the cylinder so that a new ''Cycle'' can begin.

The above described five events which comprise the work cycle of an engine are commonly referred to as (1), INTAKE; (2), COMPRESSION; (3), IGNITION; (4), EXPANSION (POWER); and (5), EXHAUST.

**DIESEL CYCLE.** The Diesel Cycle differs from the Otto Cycle in that air alone if drawn into the cylinder during the intake period. The air is heated from being compressed by the piston moving upward in the cylinder, then a finely atomized charge of fuel is injected into the cylinder where it mixes with the air and is ignited by the heat of the compressed air. In order to create sufficient heat to ignite the injected fuel, an engine operating on the Diesel Cycle must compress the air to a much greater degree than an engine operating on the Otto Cycle where the fuel-air mixture is ignited by an electric spark. The power and exhaust events of the Diesel Cycle are similar to the power and exhaust events of the Otto Cycle.

**TWO-STROKE CYCLE ENGINES.** Two stroke cycle engines may be of the Otto Cycle (spark ignition) or Diesel Cycle (compression ignition) type. However, since the two-stroke cycle engines listed in the repair section of this manual are all of the Otto Cycle type, operation of two-stroke Diesel Cycle engines will not be discussed in this section.

In two-stroke cycle engines, the piston is used as a sliding valve for the cylinder intake and exhaust ports. The intake and exhaust ports are both open when the piston is at the bottom of its downward stroke (bottom dead center or ''BDC''). The exhaust port is open to atmospheric pressure; therefore, the fuel-air mixture must be elevated to a higher than atmospheric pressure in order for the mixture to enter the cylinder. As the crankshaft is turned from BDC and the piston starts on its upward stroke, the intake and exhaust ports are closed and the fuel-air mixture in the cylinder is compressed. When piston is at or near the top of its upward stroke (top dead center or ''TDC''), an electric spark across the electrode gap

of the spark plug ignites the fuel-air mixture. As the crankshaft turns past TDC and the piston starts on its downward stroke, the rapidly burning fuel-air mixture expands and forces the piston downward. As the piston nears bottom of its downward stroke, the cylinder exhaust port is opened and the burned gaseous products from combustion of the fuel-air mixture flows out the open port. Slightly further downward travel of the piston opens the cylinder intake port and a fresh charge of fuel-air mixture is forced into the cylinder. Since the exhaust port remains open, the incoming flow of fuel-air mixture helps clean (scavenge) any remaining burned gaseous products from the cylinder. As the crankshaft turns past BDC and the piston starts on its upward stroke, the cylinder intake and exhaust ports are closed and a new cycle begins.

Since the fuel-air mixture must be elevated to a higher than atmospheric pressure to enter the cylinder of a two-stroke cycle engine, a compressor pump must be used. Coincidentally, downward movement of the piston decreases the volume of the engine crankcase. Thus, a compressor pump is made available by sealing the engine crankcase and connecting the carburetor to a port in the crankcase. When the piston moves upward, volume of the crankcase is increased which lowers pressure within the crankcase to below atmospheric. Air will then be forced through the carburetor, where fuel is mixed with the air, and on into the engine crankcase. In order for downward movement of the piston to compress the fuel-air mixture in the crankcase, a valve must be provided to close the carburetor to crankcase port. Three different types of valves are used. In Fig. 1-1, a reed type inlet valve is shown in the schematic diagram of the two-stroke cycle engine. Spring steel reeds (R) are forced open by atmospheric pressure as shown in view ''B'' when the piston is on its upward stroke and pressure in the crankcase is below atmospheric. When piston reaches TDC, the reeds close as shown in view ''A'' and fuel-air mixture is trapped in the crankcase to be compressed by downward movement of the piston. In Fig 1-2, a schematic diagram of a two-stroke cycle engine is shown in which the piston is utilized as a sliding carburetor-crankcase port (third port) valve. In Fig. 1-3, a schematic diagram of a two-stroke cycle engine is shown in which a slotted disc (rotary valve) attached to the engine crankshaft opens the carburetor-crankcase port when the piston is on its upward stroke. In each of the three basic designs shown, a transfer port (TP—Fig. 1-2) connects the crankcase compression chamber to the cylinder; the transfer port is the cylinder intake port through which the compressed fuel-air mixture in the crankcase is transferred to the cylinder when the piston is at bottom of stroke as shown in view "A."

Due to rapid movement of the fuel-air mixture through the crankcase, the crankcase cannot be used as a lubricating oil sump because the oil would be carried into the cylinder. Lubrication is accomplished by mixing a small amount of oil with the fuel; thus, lubricating oil for the engine moving parts is carried into the crankcase with the fuel-air mixture. Normal lubricating oil to fuel mixture ratios vary from one part of oil mixed with 16 to 20 parts of fuel by volume. In all instances, manufacturer's recommendations for fuel-oil mixture ratio should be observed.

**FOUR-STROKE CYCLE.** In a four-stroke engine operating on the Otto Cycle (spark ignition), the five events of the cycle take place in four strokes of the piston, or in two revolutions of the engine crankshaft. Thus, a power stroke occurs only on alternate downward

strokes of the piston.

In view "A" of Fig. 1-4, the piston is on the first downward stroke of the cycle. The mechanically operated intake valve has opened the intake port and, as the downward movement of the piston has reduced the air pressure in the cylinder to below atmospheric pressure, air is forced through the carburetor, where fuel is mixed with the air, and into the cylinder through the open intake port. The intake valve remains open and the fuel-air mixture continues to flow into the cylinder until the piston reaches the bottom of its downward stroke. As the piston starts on its first upward stroke, the mechanically operated intake valve

closes and, since the exhaust valve is closed, the fuel-air mixture is compressed as in view "B."

Just before the piston reaches the top of its first upward stroke, a spark at the spark plug electrodes ignites the compressed fuel-air mixture. As the engine crankshaft turns past top center, the burning fuel-air mixture expands rapidly and forces the piston downward on its power stroke as shown in view "C." As the piston reaches the bottom of the power stroke, the mechanically operated exhaust valve starts to open and as the pressure of the burned fuel-air mixture is higher than atmospheric pressure, it starts to flow out the open exhaust port. As the engine crankshaft

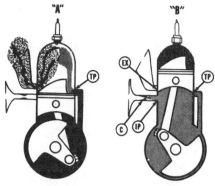

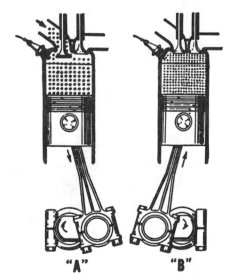

Fig. 1-2—Schematic diagram of two-stroke cycle engine operating on Otto Cycle. Engine differs from that shown in Fig. 1-1 in that piston is utilized as a sliding valve to open and close intake (carburetor to crankcase) port (IP) instead of using reed valve (R—Fig. 1-1).

C. Carburetor
EX. Exhaust port
IP. Intake port
  (carburetor to
  crankcase)

TP. Transfer port
  (crankcase to
  cylinder)

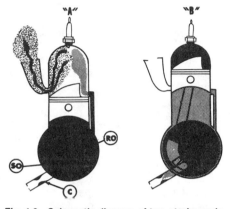

Fig. 1-3—Schematic diagram of two-stroke cycle engine similar to those shown in Figs. 1-1 and 1-2 except that a rotary carburetor to crankcase port valve is used. Disc driven by crankshaft has rotating opening (RO) which uncovers stationary opening (SO) in crankcase when piston is on upward stroke. Carburetor is (C).

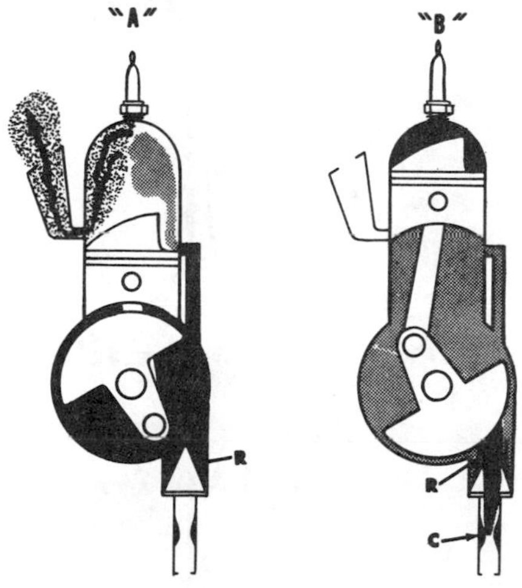

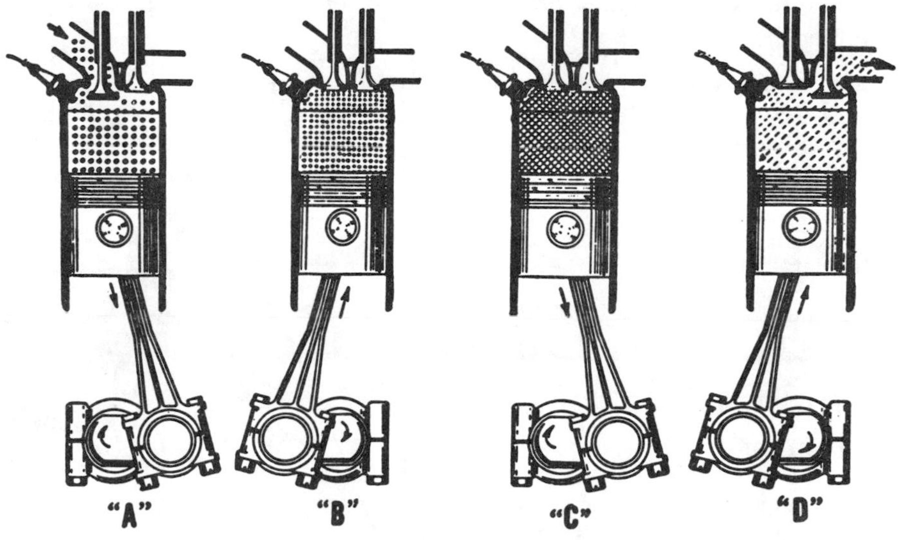

Fig. 1-1—Schematic diagram of a two-stroke engine operating on the Otto Cycle (spark ignition). View "B" shows piston near top of upward stroke and atmospheric pressure is forcing air through carburetor (C), where fuel is mixed with the air, and the fuel-air mixture enters crankcase through open reed valve (R). In view "A", piston is near bottom of downward stroke and has opened the cylinder exhaust and intake ports; fuel-air mixture in crankcase has been compressed by downward stroke of engine and flows into cylinder through open port. Incoming mixture helps clean burned exhaust gases from cylinder.

"A"    "B"    "C"    "D"

Fig. 1-4—Schematic diagram of four-stroke cycle engine operating on the Otto (spark ignition) cycle. In view "A", piston is on first downward (intake) stroke and atmospheric pressure is forcing fuel-air mixture from carburetor into cylinder through the open intake valve. In view "B", both valves are closed and piston is on its first upward stroke compressing the fuel-air mixture in cylinder. In view "C", spark across electrodes of spark plug has ignited fuel-air mixture and heat of combustion rapidly expands the burning gaseous mixture forcing the piston on its second downward (expansion or power) stroke. In view "D", exhaust valve is open and piston on its second upward (exhaust) stroke forces the burned mixture from cylinder. A new cycle then starts as in view "A."

turns past bottom center, the exhaust valve is almost completely open and remains open during the upward stroke of the piston as shown in view "D." Upward movement of the piston pushes the remaining burned fuel-air mixture out of the exhaust port. Just before the piston reaches the top of its second upward or exhaust stroke, the intake valve opens and the exhaust valve closes. The cycle is completed as the crankshaft turns past top center and a new cycle begins as the piston starts downward as shown in view "A."

In a four-stroke cycle engine operating on the Diesel Cycle, the sequence of events of the cycle is similar to that described for operation on the Otto Cycle, but with the following exceptions: On the intake stroke, air only is taken into the cylinder. On the compression stroke, the air is highly compressed which raises the temperature of the air. Just before the piston reaches top dead center, fuel is injected into the cylinder and is ignited by the heated, compressed air. The remainder of the cycle is similar to that of the Otto Cycle.

plate is closed as shown by dotted line in Fig. 2-2, air cannot enter into the carburetor and pressure in the carburetor decreases greatly as the engine is turned at cranking speed. Fuel can then flow from the fuel nozzle. In manufacturing the carburetor choke plate or disc, a small hole or notch is cut in the plate so that some air can flow through the plate when it is in closed position to provide air for the starting fuel-air mixture. In some instances after starting a cold engine, it is advantageous to leave the choke plate in a partly closed position as the restriction of air flow will decrease the air pressure in carburetor venturi, thus causing more flow fuel to flow from the nozzle, resulting in a richer fuel-air mixture. The choke plate or disc should be in full open position for normal engine operation.

If, after the engine has been started the throttle plate is in the wide-open position as shown by the solid line in Fig. 2-2, the engine can obtain enough fuel and air to run at dangerously high speeds. Thus, the throttle plate or disc must be partly closed as shown by the dotted lines to control engine speed. At no load, the engine requires very little air and fuel to run at its rated speed and the throttle must be moved on toward the closed position as shown by the dash lines. As more load is placed on the engine, more fuel and air are required for the engine to operate at its rated speed and the throttle must by moved

# CARBURETOR FUNDAMENTALS

### OPERATING PRINCIPLES

Function of the carburetor on a spark-ignition engine is to atomize the fuel and mix the atomized fuel in proper proportions with air flowing to the engine intake port or intake manifold. Carburetors used on engines that are to be operated at constant speeds and under even loads are of simple design since they only have to mix fuel and air in a relatively constant ratio. On engines operating at varying speeds and loads, the carburetors must be more complex because different fuel-air mixtures are required to meet the varying demands of the engine.

**FUEL-AIR MIXTURE RATIO REQUIREMENTS.** To meet the demands of an engine being operated at varying speeds and loads, the carburetor must mix fuel and air at different mixture ratios. Fuel-air mixture ratios required for different operating conditions are approximately as follows:

|  | Fuel | Air |
| --- | --- | --- |
| Starting, cold weather . . . . . . | 1 lb. | 7 lbs. |
| Accelerating . . . . | 1 lb. | 9 lbs. |
| Idling (no load) . | 1 lb. | 11 lbs. |
| Part open throttle . . . . . . | 1 lb. | 15 lbs. |
| Full load, open throttle . . | 1 lb. | 13 lbs. |

**BASIC DESIGN.** Carburetor design is base on the venturi principle which simply means that a gas or liquid flowing through a necked-down section (venturi) in a passage undergoes an increase velocity (speed) and a decrease in pressure as compared to the velocity and pressure in full size sections of the passage. The principle is illustrated in Fig. 2-1, which shows air passing through carburetor venturi. The figures

given for air speeds and vacuum are approximate for a typical wide-open throttle operating condition. Due to low pressure (high vacuum) in the venturi, fuel is forced out through the fuel nozzle by the atmospheric pressure (0 vacuum) on the fuel; as fuel is emitted from the nozzle, it is atomized by the high velocity air flow and mixes with the air.

In Fig. 2-2, the carburetor choke plate and throttle plate are shown in relation in the venturi. Downward pointing arrows indicate air flow through carburetor.

At cranking speeds, air flows through the carburetor venturi at a slow speed; thus, the pressure in the venturi does not usually decrease to the extent that atmospheric pressure on the fuel will force fuel from the nozzle. If the choke

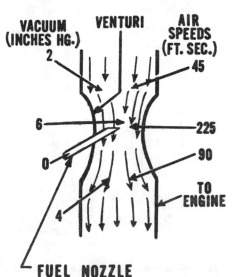

Fig. 2-1—Drawing illustrating the venturi principle upon which carburetor design is based. Figures at left are inches of mercury vacuum and those at right are air speeds in feet per second that are typical of conditions found in a carburetor operating at wide open throttle. Zero vacuum in fuel nozzle corresponds to atmospheric pressure.

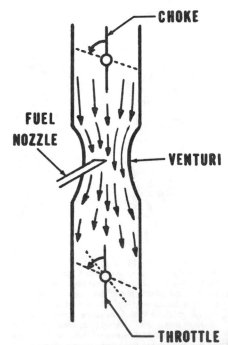

Fig. 2-2—Drawing showing basic carburetor design. Text explains operation of the choke and throttle valves. In some carburetors, a primer pump may be used instead of the choke valve to provide fuel for the starting fuel-air mixture.

closer to the wide open position as shown by the solid line. When the engine is required to develop maximum power or speed, the throttle must be in the wide open position.

Although some carburetors may be as simple as the basic design just described, most engines require more complex design features to provide variable fuel-air mixture ratios for different operating conditions. These design features will be described in the following paragraphs which outline the different carburetor types.

## CARBURETOR TYPES

Carburetors used on small engines are usually classified by types as to method of delivery of fuel to the carburetor fuel nozzle. The following paragraphs describe the features and operating principles of the different type carburetors from the most simple suction lift type to the more complex float and diaphragm types.

**SUCTION LIFT CARBURETOR.** A cross-sectional drawing of a typical suction lift carburetor is shown in Fig. 2-4. Due to the low pressure at the orifice (O) of the fuel nozzle and to atmospheric pressure on the fuel in fuel supply tank, fuel is forced up through the fuel pipe and out of the nozzle into the carburetor venturi where it is mixed with the air flowing through the venturi. A check ball is located in the lower end of the fuel pipe to prevent pulsations of air pressure in the venturi from forcing fuel back down through the fuel pipe. The lower end of the full pipe has a fine mesh screen to prevent foreign material or dirt in fuel from entering the fuel nozzle. Fuel-air ratio can be adjusted by opening or closing the adjusting needle (N) slightly; turning the needle in will decrease flow of fuel out of nozzle orifice (O).

In Fig. 2-3, a cut-away view is shown of a suction type carburetor used on several models of a popular make small engine. This carburetor features an idle fuel passage, jet and adjustment screw. When carburetor throttle is nearly closed (engine is at low idle speed), air pressure is low (vacuum is high) at inner side of throttle plate. Therefore, atmospheric pressure in fuel tank will force fuel through the idle jet and adjusting screw orifice where it is emitted into the carburetor throat and mixes with air passing the throttle plate. The adjustment screw is turned in or out until an optimum fuel-air mixture is obtained and engine runs smoothly at idle speed. When the throttle is opened to increase engine speed, air velocity through the venturi increases, air

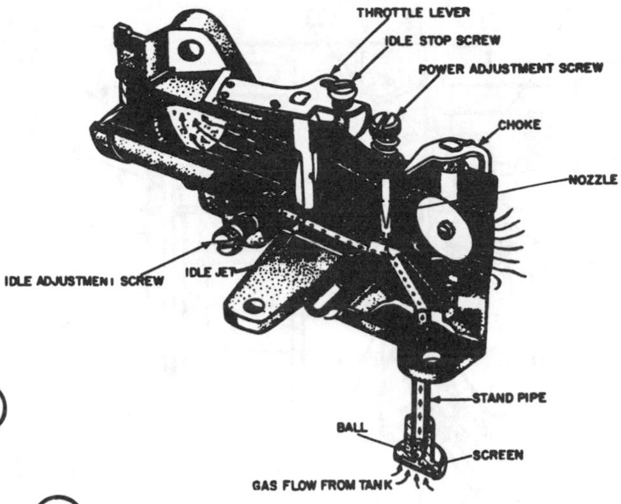

Fig. 2-3—Cut-away drawing of a suction lift carburetor used on one well known make engine. Ball in stand pipe prevents fuel in carburetor from flowing back into fuel tank.

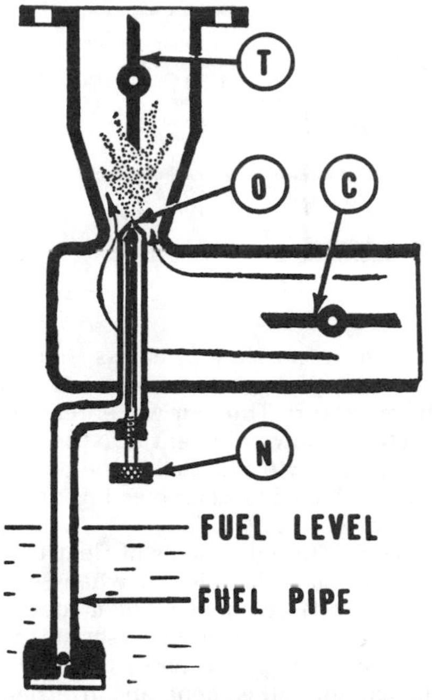

Fig. 2-4—Principle of suction lift carburetor is illustrated in above drawing. Atmospheric pressure on fuel forces fuel up through pipe and out nozzle orifice (O). Needle (N) is used to adjust amount of fuel flowing from nozzle to provide correct fuel-air mixture for engine operation. Choke (C) and throttle (T) valves are shown in wide open position.

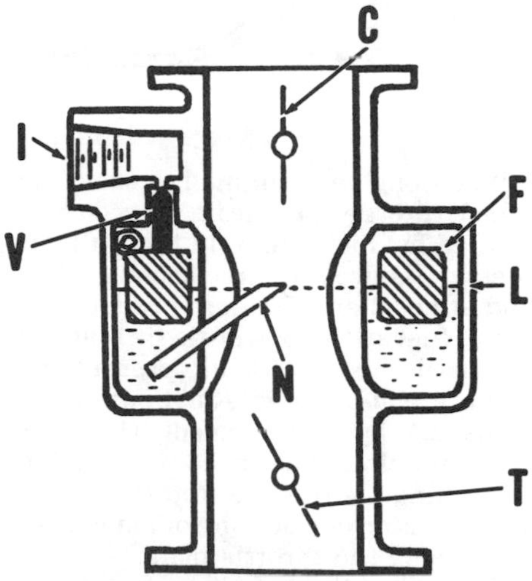

Fig. 2-5—Drawing showing basic float type carburetor design. Fuel must be delivered under pressure either by gravity or by use of fuel pump, to the carburetor fuel inlet (I). Fuel level (L) operates float (F) to open and close inlet valve (V) to control amount of fuel entering carburetor. Also shown are the fuel nozzle (N), throttle (T) and choke (C).

pressure in the venturi decreases and fuel is emitted from the nozzle. Power adjustment screw (high speed fuel needle) is turned in or out to obtain proper fuel-air mixture for engine running under operating speed and load.

**FLOAT TYPE CARBURETOR.** The principle of float type carburetor operation is illustrated in Fig. 2-5. Fuel is delivered at inlet (I) by gravity with fuel tank placed above carburetor, or by a fuel lift pump when tank is located below carburetor inlet. Fuel flows into the open inlet valve (V) until fuel level (L) in bowl lifts float against fuel valve needle and closes the valve. As fuel is emitted from the nozzle (N) when engine is running, fuel level will drop, lowering the float and allowing valve to open so that fuel will enter the carburetor to meet the requirements of the engine.

In Fig. 2-6, a cut-away view of a well known make of small engine float type carburetor is shown. Atmospheric pressure is maintained in fuel bowl through passage (20) which opens into carburetor air horn ahead of the choke plate (21). Fuel level is maintained at just below level of opening (O) in nozzle (22) by float (19) actuating inlet valve needle (8). Float height can be adjusted by bending float tang (5).

When starting a cold engine, it is necessary to close the choke plate (21) as shown by dotted lines so as to lower the air pressure in carburetor venturi (18) as engine is cranked. Then, fuel will flow up through nozzle (22) and will be emitted from openings (O) in nozzle. When an engine is hot, it will start on a leaner fuel-air mixture than when cold and may start without the choke plate being closed.

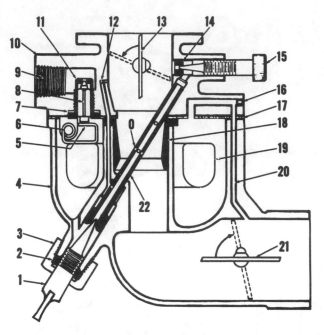

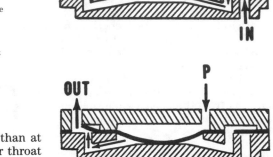

Fig. 2-6—Cross-sectional drawing of float type carburetor used on a popular small engine.

0. Orifice
1. Main fuel needle
2. Packing
3. Packing nut
4. Carburetor bowl
5. Float tang
6. Float hinge pin
7. Gasket
8. Inlet valve
9. Fuel inlet
10. Carburetor body
11. Inlet valve seat
12. Vent
13. Throttle plate
14. Idle orifice
15. Idle fuel needle
16. Plug
17. Gasket
18. Venturi
19. Float
20. Fuel bowl vent
21. Choke
22. Fuel nozzle

Fig. 2-9—Operating principle of diaphragm type fuel pump is illustrated in above drawings. Pump valves (A & B) are usually a part of diaphragm (D). Pump inlet is (I) and outlet is (O). Chamber above diaphragm is connected to engine crankcase by passage (C). When piston is on upward stroke, vacuum (V) at crankcase passage allows atmospheric pressure to force fuel into pump fuel chamber as shown in middle drawing. When piston is on downward stroke, pressure (P) expands diaphragm downward forcing fuel out of pump as shown in lower drawing.

When engine is running as slow idle speed (throttle plate nearly closed as indicated by dotted lines in Fig. 2-6), air pressure above the throttle plate is low and atmospheric pressure in fuel bowl forces fuel up through orifice in seat (14) where it mixes with air passing the throttle plate. The idle fuel mixture is adjustable by turning needle (15) in or out as required. Idle speed is adjustable by turning the throttle stop screw (not shown) in or out of control amount of air passing the throttle plate.

When throttle plate is opened to increase engine speed, velocity of air flow through venturi (18) increases, air pressure at venturi decreases and fuel will flow from openings (O) in nozzle instead of through orifice in idle seat (14). When engine is running at high speed, pressure in nozzle (22) is less than at vent (12) opening in carburetor throat above venturi. Thus, air will enter vent and travel down the vent into the nozzle and mix with the fuel in the nozzle. This is referred to as air bleeding and is illustrated in Fig. 2-7.

Many different designs of float type carburetors will be found when servicing the different makes and models of small engines. Reference should be made to the engine repair section of this manual for adjustment and overhaul specifications. Refer to carburetor servicing paragraphs in fundamentals sections for service hints.

**DIAPHRAGM TYPE CARBURETOR.** Refer to Fig. 2-8 for cross-sectional drawing showing basic design of a diaphragm type carburetor. Fuel is delivered to inlet (I) by gravity with fuel tank above carburetor, or under pressure from a fuel pump. Atmospheric pressure is maintained on lower side of diaphragm (D) through vent hole (V). When choke plate (C) is closed and engine is cranked, or when engine is running, pressure at orifice (O) is less than atmospheric pressure; this low

Fig. 2-7—Illustration of air bleed principle explained in text.

Fig. 2-8—Cross-section drawing of basic design diaphragm type carburetor. Atmospheric pressure actuates diaphragm (D).

C. Choke
D. Diaphragm
F. Fuel chamber
I. Fuel inlet
IV. Inlet valve needle
L. Lever
N. Nozzle
O. Orifice
P. Pivot pin
S. Spring
T. Throttle
V. Vent
VS. Valve seat

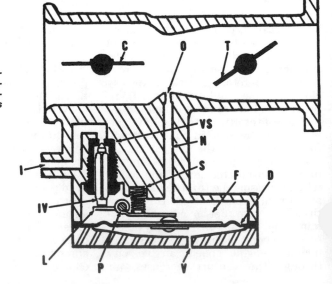

pressure, or vacuum, is transmitted to fuel chamber (F) above diaphragm through nozzle channel (N). The higher (atmospheric) pressure at lower side of diaphragm will then push the diaphragm upward compressing spring (S) and allowing inlet valve (IV) to open and fuel will flow into the fuel chamber.

Some diaphragm type carburetors are equipped with an integral fuel pump. Although design of the pump may vary as to type of check valves, etc., all operate on the principle shown in Fig. 2-9. A channel (C) (or pulsation passage) connects one side of the diaphragm to the engine crankcase. When engine piston is on upward stroke, vacuum (V) (lower than atmospheric pressure) is present in channel; thus atmospheric pressure on fuel forces inlet valve (B) open and fuel flows into chamber below the diaphragm as shown in middle view. When piston is on downward stroke, pressure (P) (higher than atmospheric pressure) is present in channel (C); thus, the pressure forces the diaphragm downward closing the inlet valve (B)

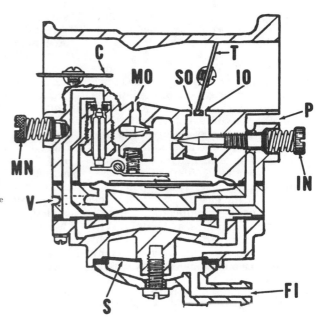

Fig. 2-10—Cross-sectional view of a popular make diaphragm type carburetor with integral fuel pump. Refer to Fig. 2-8 for view of basic diaphragm carburetor and to Fig. 2-9 for views showing operation of the fuel pump.

C. Choke
FI. Fuel inlet
IN. Idle fuel adjusting needle
IO. Idle orifice
MN. Main fuel adjusting needle
MO. Main orifice
P. Pulsation channel (fuel pump)
S. Screen
SO. Secondary orifice
T. Throttle
V. Vent (atmosphere to carburetor diaphragm)

and causes the fuel to flow out by the outlet valve (A) as shown in lower view.

In Fig. 2-10, a cross-sectional view of a popular make diaphragm type carburetor, with integral diaphragm type pump, is shown.

# IGNITION SYSTEM FUNDAMENTALS

The ignition system provides a properly timed surge of extremely high voltage electrical energy which flows across the spark plug electrode gap to create the ignition spark. Small engines may be equipped with either a magneto or battery ignition system. A magneto ignition system generates electrical energy, intensifies (transforms) this electrical energy to the extremely high voltage required and delivers this electrical energy at the proper time for the ignition spark. In a battery ignition system, a storage battery is used as a source of electrical energy and the system transforms the relatively low electrical voltage from the battery into the high voltage required and delivers the high voltage at proper time for the ignition spark. Thus, the function of the two systems is somewhat similar except for the basic source of electrical energy. The fundamental operating principles of ignition systems are explained in the following paragraphs.

## MAGNETISM AND ELECTRICITY

The fundamental principles upon which ignition systems are designed are presented in this section. As the study of magnetism and electricity is an entire scientific field, it is beyond the scope of this manual to fully explore these subjects. However, the following information will impart a working knowledge of basic principles which should be of value in servicing small engines.

**MAGNETISM.** The effects of magnetism can be shown easily while the theory of magnetism is too complex to be presented here. The effects of magnetism were discovered many years ago when fragments of iron ore were found to attract each other and also attract other pieces of iron. Further, it was found that when suspended in air, one end of the iron ore fragment would always point in the direction of the North Star. The end of the iron ore fragment pointing north was called the "north pole" and the opposite end the "south pole." By stroking a piece of steel with a "natural magnet," as these iron ore fragments were called, it was found that the magnetic properties of the natural magnet could be transferred or "induced" into the steel.

Steel which will retain magnetic properties for an extended period of time after being subjected to a strong magnetic field are called "permanent magnets;" iron or steel that loses such magnetic properties soon after being subjected to a magnetic field are called "temporary magnets." Soft iron will lose magnetic properties almost immediately after being removed from a magnetic field, and so is used where this property is desirable.

The area affected by a magnet is called a "field of force." The extent of this field of force is related to the strength of the magnet and can be determined by use of a compass. In practice, it is common to illustrate the field of force surrounding a magnet by lines as shown in Fig. 3-1 and field of force is usually called "lines of force" or "flux." Actually, there are no "lines," however, this is a convenient method of illustrating the presence of the invisible magnetic forces and if a certain magnetic force is defined as a "line of force," then all magnetic forces may be measured by comparison. The number of "lines of force" making up a strong magnetic field is enormous.

Most materials when placed in a magnetic field are not attracted by the magnet, do not change the magnitude or direction of the magnetic field, and so are called "non-magnetic materials." Materials such as iron, cobalt, nickel or their alloys, when placed in a magnetic field will concentrate the field of force and hence are magnetic conductors or "magnetic materials." There are no materials known in which magnetic fields will not penetrate and magnetic lines of force can be deflected only by magnetic materials or by another magnetic field.

Alnico, an alloy containing aluminum, nickel and cobalt, retains magnetic properties for a very long period of time after being subjected to a strong magnetic field and is extensively used as a permanent magnet. Soft iron, which loses magnetic properties quickly, is used to concentrate magnetic fields as in Fig 3-1.

**ELECTRICITY.** Electricity, like magnetism, is an invisible physical force whose effects may be more readily explained than the theory of what electricity consists of. All of us are familiar with the property of electricity to produce light, heat and mechanical power. What must be explained for the purpose of understanding ignition system operation is the inter-relationship of magnetism and electricity and how the ignition spark is produced.

Electrical current may be defined as a flow of energy in a conductor which, in some ways, may be compared to flow of water in a pipe. For electricity to flow, there must be a pressure (voltage) and a complete circuit (closed path) through which the electrical energy may return, a comparison being a water pump and a pipe that receives water from the outlet (pressure) side of the pump and returns the water to the inlet side of the pump. An electrical circuit may be completed by electricity flowing through the earth (ground), or through the metal framework of an engine or other equipment ("grounded" or "ground" connections). Usually, air is an insulator through which electrical energy will not flow. However, if the force (voltage) becomes great, the resistance of air to the flow of electricity is broken down and a current will flow, releasing energy in the form of a spark. By high voltage electricity breaking down the resistance of the air gap between the spark plug electrodes, the ignition spark is formed.

**ELECTROMAGNETIC INDUCTION.** The principle of electro-magnetic induc-tion is as follows: When a wire (conductor) is moved through a field of magnetic force so as to cut across the lines of force (flux), a potential voltage or electromotive force (emf) is induced in the wire. If the wire is part of a completed electrical circuit, current will flow through the circuit as illustrated in Fig. 3-2. It should be noted that the movement of the wire through the lines of magnetic force is a relative motion; that is, if the lines of force of a moving magnetic field cut across a wire, this will also induce an emf to the wire.

The direction of an induced current is related to the direction of magnetic force and also to the direction of movement of the wire through the lines of force, or flux. The voltage of an induced current is related to the strength, or concentration of lines of force, of the magnetic field and to the rate of speed at which the wire is moved through the flux. If a length of wire is wound into a coil and a section of the coil is moved

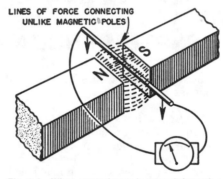

Fig. 3-2—When a conductor is moved through a magnetic field so as to cut across lines of force, a potential voltage will be induced in the conductor. If the conductor is a part of a completed electrical circuit, current will flow through the circuit as indicated by the gage.

through magnetic lines of force, the voltage induced will be proportional to the number of turns of wire in the coil.

**ELECTRICAL MAGNETIC FIELDS.** When current is flowing in a wire, a magnetic field is present around the wire as illustrated in Fig. 3-3. The direction of lines of force of this magnetic field is related to the direction of current in the wire. This is known as the left hand rule and is stated as follows: If a wire carrying a current is grasped in the left hand with thumb pointing in direction current is flowing, the curved fingers will point the direction of lines of magnetic force (flux) encircling the wire.

If a current is flowing in a wire that is wound into a coil, the magnetic flux surrounding the wire converge to form a stronger magnetic field as shown in Fig. 3-4. If the coils of wire are very close together, there is little tendency for magnetic flux to surround individual loops of the coil and a strong magnetic field will surround the entire coil. The strength of this field will vary with the current flowing through the coil.

**STEP-UP TRANSFORMERS (IGNITION COILS).** In both battery and magneto ignition systems, it is necessary to step-up, or transform, a relatively low primary voltage to the 15,000 to 20,000 volts required for the ignition spark. This is done by means of an ignition coil which utilizes the inter-relationship of magnetism and electricity as explained in preceding paragraphs.

Fig. 3-3—A magnetic field surrounds a wire carrying an electrical current. The direction of magnetic force is indicated by the "left hand rule"; that is, if thumb of left hand points in direction that electrical current is flowing in conductor, fingers of left hand will indicate direction of magnetic force.

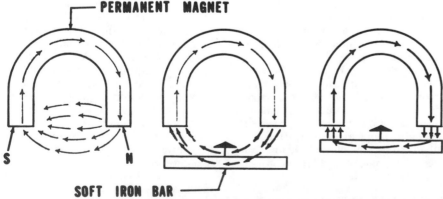

Fig. 3-1—In left view, field of force of permanent magnet is illustrated by arrows showing direction of magnetic force from north pole (N) to south pole (S). In center view, lines of magnetic force are being attracted by soft iron bar that is being moved into the magnetic field. In right view, the soft iron bar has been moved close to the magnet and the field of magnetic force is concentrated within the bar.

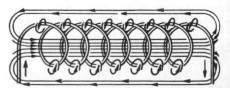

Fig. 3-4—When a wire is wound in a coil, the magnetic force created by a current in the wire wi tend to converge in a single strong magnetic field as illustrated. If the loops of the coil are wound closely together, there is little tendency for lines of force to surround individual loops of the coil.

Basic ignition coil design is shown in Fig. 3-5. The coil consists of two separate coils of wire which are called the primary coil winding and the secondary coil winding, or simply the primary winding and secondary winding. The primary winding as indicated by the heavy, black line is of larger diameter wire and has a smaller number of turns when compared to the secondary winding indicated by the light line.

A current passing through the primary winding creates a magnetic field (as indicated by the "lines of force") and this field, concentrated by the soft iron core, surrounds both the primary and secondary windings. If the primary winding current is suddenly interrupted, the magnetic field will collapse and the lines of force will cut through the coil windings. The resulting induced voltage in the secondary winding is greater than the voltage of the current that was flowing in the primary winding and is related to the number of turns of wire in each winding. Thus:

Induced secondary voltage = primary voltage ×

$$\frac{\text{No. of turns in secondary winding}}{\text{No. of turns in primary winding}}$$

For example, if the primary winding of an ignition coil contained 100 turns of wire and the secondary winding contained 10,000 turns of wire, a current having an emf of 200 volts flowing in the primary winding, when suddenly interrupted, would result in an emf of:

$$200 \text{ Volts} \times \frac{10{,}000 \text{ turns of wire}}{100 \text{ turns of wire}}$$

$$= 20{,}000 \text{ volts}$$

SELF-INDUCTANCE. It should be noted that the collapsing magnetic field resulting from the interrupted current in the primary winding will also induce a current in the primary winding. This effect is termed "self-inductance." This self-induced current is such as to oppose any interruption of current in the primary winding, slowing the collapse of the magnetic field and reducing the efficiency of the coil. The self-induced primary current flowing across the slightly open breaker switch, or contact points, will damage the contact surfaces due to the resulting spark.

To momentarily absorb, then stop the flow of current across the contact points, a capacitor or, as commonly called, a condenser is connected in parallel with the contact points. A sim-

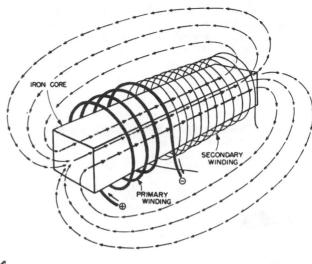

*Fig. 3-5—Drawing showing principles of ignition coil operation. A current in primary winding will establish a magnetic field surrounding both the primary and secondary windings and the field will be concentrated by the iron core. When primary current is interrupted, the magnetic field will "collapse" and the lines of force will cut the coil windings inducing a very high voltage in the secondary winding.*

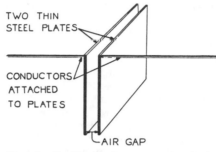

*Fig. 3-6a—Drawing showing construction of a simple condenser. Capacity of such a condenser to absorb current is limited due to the relatively small surface area. Also, there is a tendency for current to arc across the air gap. Refer to Fig. 3-7 for construction of typical ignition system condenser.*

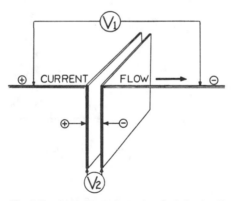

*Fig. 3-6b—A condenser in an electrical circuit will absorb flow of current until an opposing voltage (V2) is built up across condenser plates which is equal to the voltage (V1) of the electrical current.*

ple condenser is shown in Fig. 3-6a; however, the capacity of such a condenser to absorb current (capacitance) is limited by the small surface area of the plates. To increase capacity to absorb current, the condenser used in ignition systems is constructed as shown in Fig. 3-7.

EDDY CURRENTS. It has been found that when a solid soft iron bar is used as a core for an ignition coil, stray electrical currents are formed in the core.

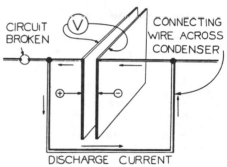

*Fig. 3-6c—When flow of current is interrupted in circuit containing condenser (circuit broken), the condenser will retain a potential voltage (V). If a wire is connected across the condenser, a current will flow in reverse direction of charging current until condenser is discharged (voltage across condenser plates is zero).*

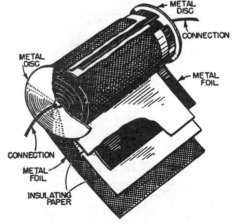

*Fig. 3-7—Drawing showing construction of typical ignition system condenser. Two layers of metal foil, insulated from each other with paper, are rolled tightly together and a metal disc contacts each layer, or strip, of foil. Usually, one disc is grounded through the condenser shell.*

These stray, or "eddy currents," create opposing magnetic forces causing the core to become hot and also decrease the efficiency of the coil. As a means of preventing excessive formation of eddy currents within the core, or other magnetic field carrying parts of a magneto, a laminated plate construction as shown in Fig. 3-8 is used instead of solid material. The plates, or laminations, are insulated from each other by a natural oxide coating formed on the plate surfaces or by coating the plates with varnish. The cores of some ignition coils are constructed of soft iron wire instead of plates and each wire is in-

sulated by a varnish coating. This type construction serves the same purpose as laminated plates.

## BATTERY IGNITION SYSTEMS

Some small engines are equipped with a battery ignition system. A schematic diagram of a typical battery ignition system for a single cylinder engine is shown in Fig. 3-9. Designs of battery ignition systems may vary, especially as to location of breaker points and method for actuating the points; however, all operate on the same basic principles.

**BATTERY IGNITION SYSTEM PRINCIPLES.** Refer to the schematic diagram on Fig. 3-9. When the timer cam is turned so that the contact points are closed, a current is established in the primary circuit by the emf of the battery. This current flowing through the primary winding of the ignition coil establishes a magnetic field concentrated in the core laminations and surrounding the windings. A cutaway view of a typical ignition coil is shown in Fig. 3-10. At the proper time for the ignition spark, the contact points are opened by

the timer cam and the primary ignition circuit is interrupted. The condenser, wired in parallel with the breaker contact points between the timer terminal and ground, absorbs the self-induced current in the primary circuit for an instant and brings the flow of current to a quick, controlled stop. The magnetic field surrounding the coil rapidly cuts the primary and secondary windings creating an emf as high as 250 volts in the primary winding and up to 25,000 volts in the secondary winding. Current absorbed by the condenser is discharged as the cam closes the breaker points, grounding the condenser lead wire.

Due to resistance of the primary winding, a certain period of time is required for maximum primary current flow after the breaker contact points are closed. At high engine speeds, the points remain closed for a smaller interval of time, hence the primary current does not build up to maximum and secondary voltage is somewhat less than at low engine speed. However, coil design is such that the minimum voltage available at high engine speed exceeds the normal maximum voltage required for the ignition spark.

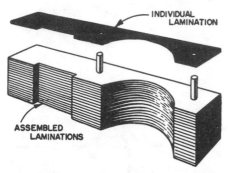

Fig. 3-8—To prevent formation of "eddy currents" within soft iron cores used to concentrate magnetic fields, core is assembled of plates or "laminations" that are insulated from each other. In a solid core, there is a tendency for counteracting magnetic forces to build up from stray currents induced in the core.

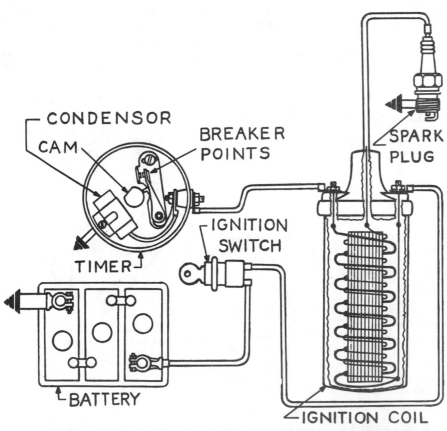

Fig. 3-9—Schematic diagram of typical battery ignition system. On unit shown, breaker points are actuated by timer cam; on some units, the points may be actuated by cam on engine camshaft. Refer to Fig. 3-10 for cutaway view of typical battery ignition coil. In view above, primary coil winding is shown as heavy black line (outside coil loops) and secondary winding is shown by lighter line (inside coil loops).

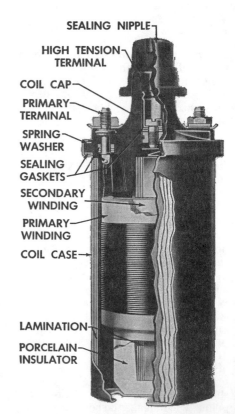

Fig. 3-10—Cutaway view of typical battery ignition system coil. Primary winding consists of approximately 200-250 turns (loops) of heavier wire; secondary winding consists of several thousand turns of fine wire. Laminations concentrate the magnetic lines of force and increase efficiency of the coil.

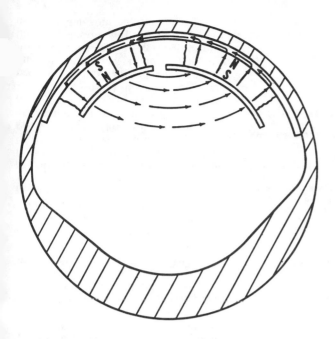

*Fig. 3-11—Cutaway view of typical engine flywheel used with flywheel magneto type ignition system. The permanent magnets are usually cast into the flywheel. For flywheel type magnetos having the ignition coil and core mounted to outside of flywheel, magnets would be flush with outer diameter of flywheel.*

## MAGNETO IGNITION SYSTEMS

By utilizing the principles of magnetism and electricity as outlined in previous paragraphs, a magneto generates an electrical current of relatively low voltage, then transforms this voltage into the extremely high voltage necessary to produce the ignition spark. This surge of high voltage is timed to create the ignition spark and ignite the compressed fuel-air mixture in the engine cylinder at the proper time in the Otto cycle as described in the paragraphs on fundamentals of engine operation principles.

Two different types of magnetos are used on small engines and, for discussion in this section of the manual, will be classified as "flywheel type magnetos" and "self-contained unit type magnetos." The most common type of ignition system found on small engines is the flywheel type magneto.

### Flywheel Type Magnetos

The term "flywheel type magneto" is derived from the fact that the engine flywheel carries the permanent magnets and is the magneto rotor. In some similar systems, the magneto rotor is mounted on the engine crankshaft as is the flywheel, but is a part separate from the flywheel.

**FLYWHEEL MAGNETO OPERATING PRINCIPLES.** In Fig. 3-11, a cross-sectional view of a typical engine flywheel (magneto rotor) is shown. The arrows indicate lines of force (flux) of the permanent magnets carried by the flywheel. As indicated by the arrows, direction of force of the magnetic field

is from the north pole (N) of the left magnet to the south pole (S) of the right magnet.

Figs. 3-12, 3-13, 3-14, and 3-15 illustrate the operational cycle of the flywheel type magneto. In Fig. 3-12, the flywheel magnets have moved to a position over the left and center legs of the armature (ignition coil) core. As the magnets moved into this position, their magnetic field was attracted by the armature core as illustrated in Fig. 3-1 and a potential voltage (emf) was induced in the coil windings. However, this emf was not sufficient to cause current to flow across the spark plug electrode gap in the high tension circuit and the points were open in the primary circuit.

In Fig. 3-13, the flywheel magnets have moved to a new position to where

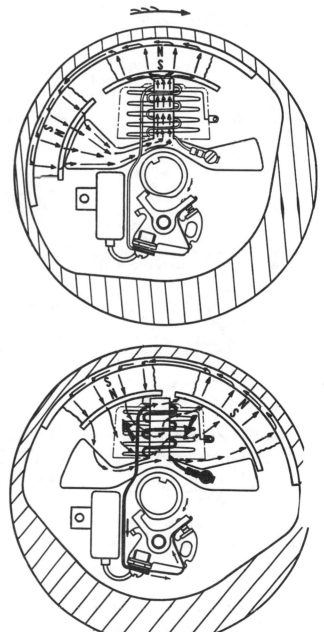

*Fig. 3-12—View showing flywheel turned to a position so that lines of force of the permanent magnets are concentrated in the left and center core legs and are interlocking the coil windings.*

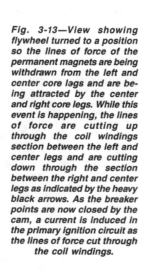

*Fig. 3-13—View showing flywheel turned to a position so the lines of force of the permanent magnets are being withdrawn from the left and center core legs and are being attracted by the center and right core legs. While this event is happening, the lines of force are cutting up through the coil windings section between the left and center legs and are cutting down through the section between the right and center legs as indicated by the heavy black arrows. As the breaker points are now closed by the cam, a current is induced in the primary ignition circuit as the lines of force cut through the coil windings.*

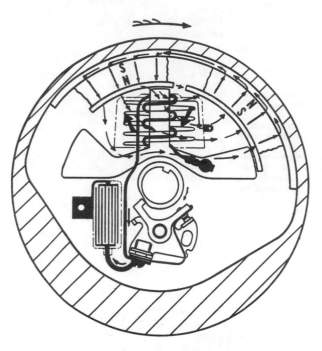

Fig. 3-14—The flywheel magnets have now turned slightly past the position shown in Fig. 3-13 and the rate of movement of lines of magnetic force cutting through the coil windings is at the maximum. At this instant, the breaker points are opened by the cam and flow of current in the primary circuit is being absorbed by the condenser, bringing the flow of current to a quick, controlled stop. Refer now to Fig. 3-15.

their magnetic field is being attracted by the center and right legs of the armature core, and is being withdrawn from the left and center legs. As indicated by the heavy black arrows, the lines of force are cutting up through the section of coil windings between the left and center legs of the armature and are cutting down through the coil windings section between the center and right legs. If the left hand rule, as explained in a previous paragraph, is applied to the lines of force cutting through the coil sections, it is seen that the resulting emf induced in the primary circuit will cause a current to flow through the primary coil windings and the breaker points which have now been closed by action of the cam.

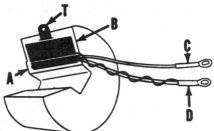

Fig. 3-16—Drawing showing construction of a typical flywheel magneto ignition coil. Primary winding (A) consists of about 200 turns of wire. Secondary winding (B) consists of several thousand turns of fine wire. Coil primary and secondary ground connection is (D); primary connection to breaker point and condenser terminal is (C); and coil secondary (high tension) terminal is (T).

At the instant the movement of the lines of force cutting through the coil winding sections is at the maximum rate, the maximum flow of current is obtained in the primary circuit. At this time, the cam opens the breaker points interrupting the primary circuit and for an instant, the flow of current is absorbed by the condenser as illustrated in Fig. 3-14. An emf is also induced in the secondary coil windings, but the voltage is not sufficient to cause current to flow across the spark plug gap.

The flow of current in the primary windings created a strong electromagnetic field surrounding the coil windings and up through the center leg of the armature core as shown in Fig. 3-15. As the breaker points were opened by the cam, interrupting the primary circuit, this magnetic field starts to collapse cutting the coil windings as indicated by the heavy black arrows. The emf induced in the primary circuit would be sufficient to cause a flow of current across the opening breaker points were it not for the condenser absorbing the flow of current and bringing it to a controlled stop. This allows the electro-magnetic field to collapse at such a rapid rate to induce a very high voltage in the coil high tension or secondary windings. This voltage, in the order of 15,000 to 25,000 volts, is sufficient to break down the resistance of the air gap between the spark plug electrodes and a current will flow across the gap. This creates the ignition spark which ignites the compressed fuel-air mixture in the engine cylinder.

## Self-Contained Unit Type Magnetos

Some four-stroke cycle engines are equipped with a magneto which is a self-contained unit as shown in Fig. 3-20. This type magneto is driven from the engine timing gears via a gear or coupling. All components of the

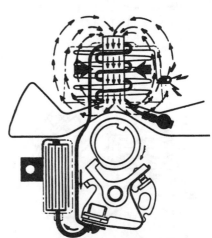

Fig. 3-15—View showing magneto ignition coil, condenser and breaker points at same instant as illustrated in Fig. 3-14; however, arrows shown above illustrate lines of force of the electromagnetic field established by current in primary coil windings rather than the lines of force of the permanent magnets. As the current in the primary circuit ceases to flow, the electro-magnetic field collapses rapidly, cutting the coil windings as indicated by heavy arrows and inducing a very high voltage in the secondary coil winding resulting in the ignition spark.

Fig. 3-17—Exploded view of a flywheel type magneto in which the breaker points (14) are actuated by a cam on engine camshaft. Push rod (9) rides against cam to open and close points. In this type unit, an ignition spark is produced only on alternate revolutions of the flywheel as the camshaft turns at one-half engine speed.

1. Flywheel
2. Ignition coil
3. Coil clamps
4. Coil ground lead
5. Breaker-point lead
6. Armature core (laminations)
7. Crankshaft bearing retainer
8. High tension lead
9. Push rod

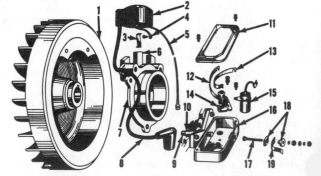

10. Bushing
11. Breaker box cover
12. Point lead strap
13. Breaker-point spring
14. Breaker-point assy.

15. Condenser
16. Breaker box
17. Terminal bolt
18. Insulators
19. Ground (stop) spring

magneto are enclosed in one housing and the magneto can be removed from the engine as a unit.

**UNIT TYPE MAGNETO OPERATING PRINCIPLES.** In Fig. 3-21, a schematic diagram of a unit type magneto is shown. The magneto rotor is driven through an impulse coupling (shown at right side of illustration). The function of the impulse coupling is to increase the rotating speed of the rotor, thereby increasing magneto efficiency, at engine cranking speeds.

A typical impulse coupling for a single cylinder engine magneto is shown in Fig. 3-22. When the engine is turned at cranking speed, the coupling hub pawl engages a stop pin in the magneto housing as the engine piston is coming up on compression stroke. This stops rotation of the coupling hub assembly and magneto rotor. A spring within the coupling shell (See Fig. 3-23) connects the shell and coupling hub; as the engine continues to turn, the spring winds up until the pawl kickoff contacts the pawl and disengages it from the stop pin. This occurs at the time an ignition spark is required to ignite the compressed fuel-air mixture in the engine cylinder. As the pawl is released, the spring connecting the coupling shell and hub unwinds and rapidly spins the magneto rotor.

The magneto rotor (See Fig. 3-21) carries permanent magnets. As the rotor turns, alternating the position of the magnets, the lines of force of the magnets are attracted, then withdrawn from the laminations. In Fig. 3-21, arrows show the magnetic field concentrated within the laminations, or armature core. Slightly further rotation of the magnetic rotor will place the magnets to where the laminations will have greater attraction for opposite poles of the magnets. At this instant, the lines of force as indicated by the arrows

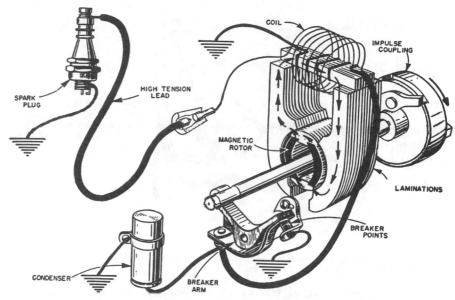

Fig. 3-21—Schematic diagram of typical unit type magneto for single cylinder engine. Refer to Figs. 3-22, 3-23 and 3-23A for views showing construction of impulse couplings.

Fig. 3-22—Views of typical impulse coupling for magneto driven by engine shaft with slotted drive connection. Coupling drive spring is shown in Fig. 3-23. Refer to Fig. 3-23A for view of combination magneto drive gear and impulse coupling used on some magnetos.

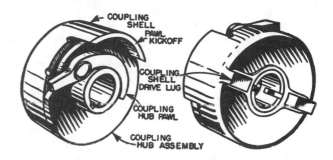

Fig. 3-20—Some engines are equipped with a unit type magneto having all components enclosed in a single housing (H). Magneto is removable as a unit after removing retaining nuts (N). Stop button (B) grounds out primary magneto circuit to stop engine. Timing window is (W).

will suddenly be withdrawn and an opposing field of force will be established in the laminations. Due to this rapid movement of the lines of force, a current will be induced in the primary magneto circuit as the coil windings are cut by the lines of force. At the instant the maximum current is induced in the primary windings, the breaker points are opened by a cam on the magnetic rotor shaft interrupting the primary circuit. The lines of magnetic force established by the primary current (Refer to Fig. 3-5) will cut through the secondary windings at such a rapid rate to induce a very high voltage in the secondary (or high tension) circuit. This voltage will break down the resistance of the spark plug electrode gap and a spark across the electrodes will result.

At engine operating speeds, centrifugal force will hold the impulse coupling hub pawl (See Fig. 3-22) in a position so that it cannot engage the stop pin in magneto housing and the magnetic rotor will be driven through the spring (Fig. 3-23) connecting the coupling shell to coupling hub. The im-

pulse coupling retards the ignition spark, at cranking speeds, as the engine piston travels closer to top dead center while the magnetic rotor is held stationary by the pawl and stop pin. The difference in degrees of impulse coupling shell rotation between the position of retarded spark and normal running spark is known as the impulse coupling lag angle.

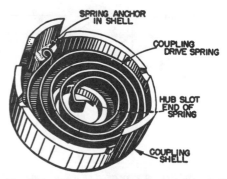

Fig. 3-23—View showing impulse coupling shell and drive spring removed from coupling hub assembly. Refer to Fig. 3-22 for views of assembled unit.

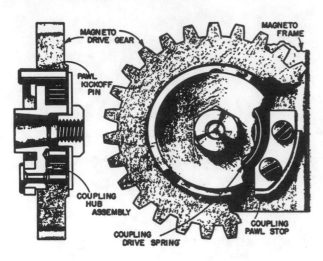

Fig. 3-23A—Views of combination magneto drive gear and impulse coupling used on some magnetos.

## SOLID-STATE IGNITION SYSTEM

**BREAKERLESS MAGNETO SYSTEM.** The solid-state (breakerless) magneto ignition system operates somewhat on the same basic principles as the conventional type flywheel magneto previously described. The main difference is that the breaker contact points are replaced by a solid-state electronic Gate Controlled Switch (GCS) which has no moving parts. Since, in a conventional system breaker points are closed over a longer period of crankshaft rotation than is the "GCS", a diode has been added to the circuit to provide the same characteristics as closed breaker points.

BREAKERLESS MAGNETO OPERATING PRINCIPLES. The same basic principles for electro-magnetic induction of electricity and formation of magnetic fields by electrical current as outlined for the conventional flywheel type magneto also apply to the solid-state magneto. Therefore the principles of the different components (diode and GCS) will complete the operating principles of the solid-state magneto.

The diode is represented in wiring diagrams by the symbol shown in Fig. 3-24. The diode is an electronic device that will permit passage of electrical current in one direction only. In electrical schematic diagrams, current flow is opposite direction the arrow part of symbol is pointing.

The symbol shown in Fig. 3-24A is used to represent the gate controlled switch (GCS) in wiring diagrams. The GCS acts as a switch to permit passage of current from cathode (C) terminal to anode (A) terminal when in "ON" state and will not permit electric current to flow when in "OFF" state. The GCS can be turned "ON" by a positive surge of electricity at the gate (G) terminal and will remain "ON" as long as current remains positive at the gate terminal or as long as current is flowing through the GCS from the cathode (C) terminal to anode (A) terminal.

The basic components and wiring diagram for the solid-state breakerless magneto are shown schematically in Fig. 3-24B. In Fig. 3-24C, the magneto rotor (flywheel) is turning and the ignition coil magnets have just moved into position so that their lines of force are cutting the ignition coil windings and producing a negative surge of current in the primary windings. The diode allows current to flow opposite to the direction of diode symbol arrow and action is same as conventional magneto with breaker points closed. As rotor (flywheel) continues to turn as shown in Fig. 3-24D, direction of magnetic flux lines will reverse in the armature center leg. Direction of current will change in the primary coil circuit and the previously conducting diode will be shut off. At this point, neither diode is conducting. As voltage begins to build up as rotor continues to turn, the condenser acts as

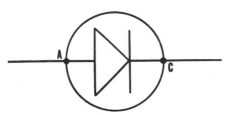

Fig. 3-24—In a diagram of an electrical circuit, the diode is represented by the symbol shown above. The diode will allow current to flow in one direction only, from cathode (C) to anode (A).

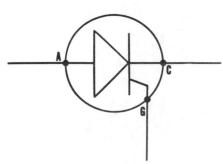

Fig. 3-24A—The symbol used for a Gate Controlled Switch (GCS) in an electrical diagram is shown above. The GCS will permit current to flow from cathode (C) to anode (A) when "turned on" by a positive electrical charge at gate (G) terminal.

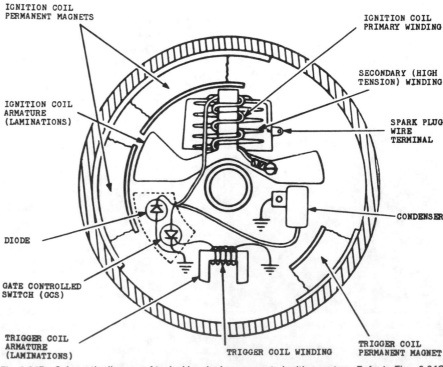

Fig. 3-24B—Schematic diagram of typical breakerless magneto ignition system. Refer to Figs. 3-24C, 3-24D and 3-24E for schematic views of operating cycle.

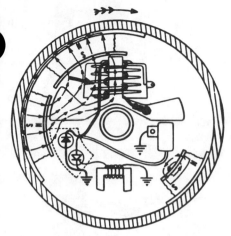

*Fig. 3-24C—View showing flywheel of breakerless magneto system at instant of rotation where lines of force of ignition coil magnets are being drawn into left and center legs of magneto armature. The diode (See Fig. 3-24) acts as a closed set of breaker points in completing the primary ignition circuit at this time.*

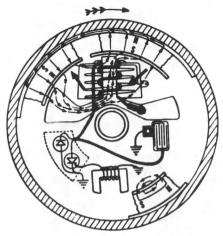

*Fig. 3-24D—Flywheel is turning to point where magnetic flux lines through armature center leg will reverse direction and current through primary coil circuit will reverse. As current reverses, diode which was previously conducting will shut off and there will be no current. When magnetic flux lines have reversed in armature center leg, voltage potential will again build up, but since GCS is in "OFF" state, no current will flow. To prevent excessive voltage build up, the condenser acts as a buffer.*

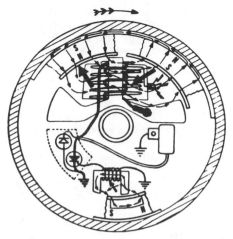

*Fig. 3-24E—With flywheel in the approximate position shown, maximum voltage potential is present in windings of primary coil. At this time the triggering coil armature has moved into the field of a permanent magnet and a positive voltage is induced on the gate of the GCS. The GCS is triggered and primary coil current flows resulting in the formation of an electromagnetic field around the primary coil which inducts a voltage of sufficient potential in the secondary windings to "fire" the spark plug.*

a buffer to prevent excessive voltage build up at the GCS before it is triggered.

When the rotor reaches the approximate position shown in Fig. 3-24E, maximum flux density has been achieved in the center leg of the armature. At this time the GCS is triggered. Triggering is accomplished by the triggering coil armature moving into the field of a permanent magnet which induces a positive voltage on the gate of the GCS. Primary coil current flow results in the formation of an electromagnetic field around primary coil which inducts a voltage of sufficient potential in the secondary coil windings to "fire" the spark plug.

When the rotor (flywheel) has moved the magnets past the armature, the GCS will cease to conduct and revert to the "OFF" state until it is triggered. The condenser will discharge during the time that the GCS was conducting.

**CAPACITOR DISCHARGE SYSTEM.** The capacitor discharge (CD) ignition system uses a permanent magnet rotor (flywheel) to induce a current in a coil, but unlike the conventional flywheel magneto and solid-state breakerless magneto described previously, the current is stored in a capacitor (condenser). Then the stored current is discharged through a transformer coil to create the ignition spark. Refer to Fig. 3-24F for a schematic of a typical capacitor discharge ignition system.

CAPACITOR DISCHARGE OPERATING PRINCIPLES. As the permanent flywheel magnets pass by the input generating coil (1—Fig. 3-24F), the current produced charges capacitor (6). On-

*Fig. 3-24F—Schematic diagram of a typical capacitor discharge ignition system.*
1. Generating coil
2. Zener diode
3. Diode
4. Trigger coil
5. Gate controlled switch
6. Capacitor
7. Pulse transformer (coil)
8. Spark plug

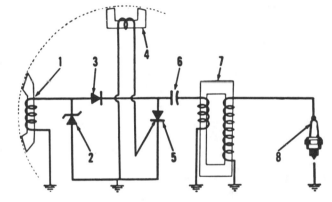

ly half of the generated current passes through diode (3) to charge the capacitor. Reverse current is blocked by diode (3) but passes through Zener diode (2) to complete the reverse circuit. Zener diode (2) also limits maximum voltage of the forward current. As the flywheel continues to turn and magnets pass the trigger coil (4), a small amount of electrical current is generated. This current opens the gate controlled switch (5) allowing the capacitor to discharge through the pulse transformer (7). The rapid voltage rise in the transformer primary coil induces a high voltage secondary current which forms the ignition spark when it jumps the spark plug gap.

## THE SPARK PLUG

In any spark ignition engine, the spark plug (See Fig. 3-25) provides the means for igniting the compressed fuel-air mixture in the cylinder. Before an electric

charge can move across an air gap, the intervening air must be charged with electricity, or ionized. If the spark plug is properly gapped and the system is not shorted, not more than 7,000 volts may be required to initiate a spark. Higher

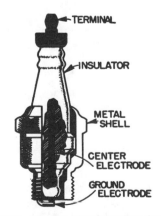

*Fig. 3-25—Cross-sectional drawing of spark plug showing construction and nomenclature.*

voltage is required as the spark plug warms up, or if compression pressures or the distance of the air gap is increased. Compression pressures are highest at full throttle and relatively slow engine speeds, therefore, high voltage requirements or a lack of available secondary voltage most often shows up as a miss during maximum acceleration from a slow engine speed.

There are many different types and sizes of spark plugs which are designed for a number of specific requirements.

**THREAD SIZE.** The threaded, shell portion of the spark plug and the attaching hole in the cylinder are manufactured to meet certain industry established standards. The diameter is referred to as "Thread Size." Those commonly used are: 10 mm, 14 mm, 18 mm, 7/8 inch and 1/2 inch pipe. The 14 mm plug is almost universal for small engine use.

**REACH.** The length of thread, and the thread depth in cylinder head or wall are also standardized throughout the industry. This dimension is measured from gasket seat of plug to cylinder end of thread. See Fig. 3-26. Four different reach plugs commonly used are 3/8 inch, 7/16 inch, 1/2 inch and 3/4 inch. The first two mentioned are the ones commonly used in small engines.

**HEAT RANGE.** During engine operation, part of the heat generated during combustion is transferred to the spark plug, and from the plug to the cylinder through the shell threads and gasket. The operating temperature of the spark plug plays an important part in engine operation. If too much heat is retained by the plug, the fuel-air mixture may be ignited by contact with heated surface before the ignition spark occurs. If not enough heat is retained, partially burned combustion products (soot, carbon and oil) may build up on the plug tip resulting in "fouling" or shorting out of the plug. If this happens, the secondary current is dissipated uselessly as it is generated instead of bridging the plug gap as a useful spark, and the engine will misfire.

The operating temperature of the plug tip can be controlled, within limits, by altering the length of the path the heat

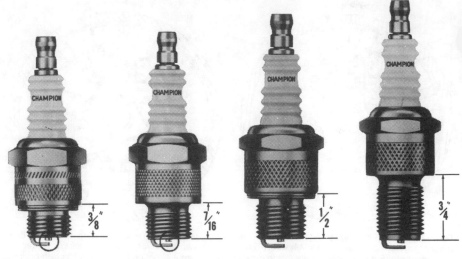

Fig. 3-26—Views showing spark plugs with various "reaches" available; small engines are usually equipped with a spark plug having a 3/8 inch reach. A 3/8 inch reach spark plug measures 3/8 inch from firing end of shell to gasket surface of shell. The two plugs at left side illustrate the difference in plugs normally used in two-stroke cycle and four-stroke cycle engines; refer to the circled electrodes. Spark plug at left has a shortened ground electrode and is specifically designed for two-stroke cycle engines. Second spark plug from left is normally used in four-stroke cycle engines although some two-stroke cycle engines may use this type plug.

Fig. 3-27—Spark plug tip temperature is controlled by the length of the path heat must travel to reach cooling surface of the engine cylinder head.

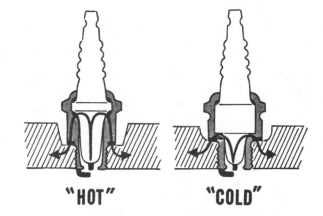

must follow to reach the threads and gasket of the plug. Thus, a plug with a short, stubby insulator around the center electrode will run cooler than one with a long, slim insulator. Refer to Fig. 3-27. Most plugs in the more popular sizes are available in a number of heat ranges which are interchangeable within the group. The proper heat range is determined by engine design and type of service. Refer to SPARK PLUG SERVICING FUNDAMENTALS for additional information on spark plug selection.

**SPECIAL TYPES.** Sometimes, engine design features or operating conditions call for special plug types designed for a particular purpose. Of special interest when dealing with two-cycle engines is the spark plug shown in the left hand view, Fig. 3-26. In the design of this plug, the ground electrode is shortened so that its end aligns with center of insulated electrode rather than completely overlapping as with the conventional plug. This feature reduces the possibility of the gap bridging over by carbon formations.

# ENGINE POWER AND TORQUE RATINGS

The following paragraphs discuss the terms used in expressing engine horsepower and torque ratings and explains the methods for determining the different ratings. Some small engine repair shops are now equipped with a dynamometer for measuring engine torque and/or horsepower and the mechanic should be familiar with terms, methods of measurement and how actual power developed by an engine can vary under different conditions.

Force →

Fig. 4-1—A force, measured in pounds, is defined as an action tending to move an object or to accelerate movement of an object.

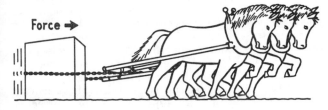

Force →

Fig. 4-2—If a force moves an object from a state of rest or accelerates movement of an object, then work is done.

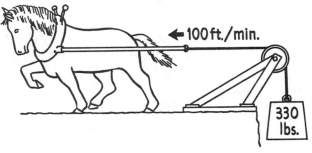

← 100 ft./min.

330 lbs.

Fig. 4-3—This horse is doing 33,000 foot-pounds of work in one minute, or one horsepower.

## GLOSSARY OF TERMS

**FORCE.** Force is an action against an object that tends to move the object from a state of rest, or to accelerate the movement of an object. For use in calculating torque or horsepower, force is measured in pounds.

**WORK.** When a force moves an object from a state of rest, or accelerates the movement of an object, work is done. Work is measured by multiplying the force applied by the distance the force moves the object, or:

work = force x distance.

Thus, if a force of 50 pounds moved an object 50 feet, work done would equal 50 pounds times 50 feet, or 2500 pounds-feet(or as it is usually expressed, 2500 foot-pounds).

**POWER.** Power is the rate at which work is done; thus, if:

then:  work=force × distance,

$$power = \frac{force \times distance}{time}$$

From the above formula, it is seen that power must increase if the time in which work is done decreases.

**HORSEPOWER.** Horsepower is a unit of measurement of power. Many years ago, James Watt, a Scotsman noted as the inventor of the steam engine, evaluated one horsepower as being equal to doing 33,000 foot-pounds of work in one minute. This evaluation has been universally accepted since that time. Thus, the formula for determining horsepower is:

$$horsepower = \frac{pounds \times feet}{33,000 \times minutes}$$

When referring to engine horsepower ratings, one usually finds the rating expressed as brake horsepower or rated horsepower, or sometimes as both.

**BRAKE HORSEPOWER.** Brake horsepower is the maximum horsepower available from an engine as determined by use of a dynamometer, and is usually stated as maximum observed brake horsepower or as corrected brake horsepower. As will be noted in a later paragraph, observed brake horsepower of a specific engine will vary under different conditions of temperature and atmospheric pressure. Corrected brake horsepower is a rating calculated from observed brake horsepower and is a means of comparing engines tested at varying conditions. The method for calculating corrected brake horsepower will be explained in a later paragraph.

**RATED HORSEPOWER.** An engine being operated under a load equal to the maximum horsepower available (brake horsepower) will not have reserve power for overloads and is subject to damage from overheating and rapid wear. Therefore, when an engine is being selected for a particular load, the engine's brake horsepower rating should be in excess of the expected normal operating load. Usually, it is recommended that the engine not be operated in excess of 80 percent of the engine maximum brake horsepower rating; thus, the "rated horsepower" of an engine is usually equal to 80 percent of maximum horsepower that the engine will develop.

**TORQUE.** In many engine specifications, a "torque rating" is given. Engine torque can be defined simply as the turning effort exerted by the engine output shaft when under load. Thus, it is possible to calculate engine horsepower being developed by measuring torque being developed and engine output speed. Refer to the following paragraphs.

## MEASURING ENGINE TORQUE AND HORSEPOWER

**THE PRONY BRAKE.** The prony brake is the most simple means of testing engine performance. Refer to diagram in Fig. 4-4. A torque arm is at-

Fig. 4-4—Diagram showing a prony brake on which the torque being developed by an engine can be measured. By also knowing the RPM of the engine output shaft, engine horsepower can be calculated.

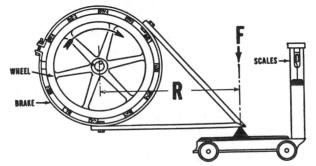

WHEEL

BRAKE

F

SCALES

R

tached to a brake on wheel mounted on engine output shaft. The torque arm, as the brake is applied, exerts a force (F) on scales. Engine torque is computed by multiplying the force (F) times the length of the torque arm radius (R), or:

$$\text{engine torque} = F \times R.$$

If, for example, the torque arm radius (R) is 2 feet and the force (F) being exerted by the torque arm on the scales is 6 pounds, engine torque would be 2 feet x 6 pounds or 12 foot-pounds.

To calculate engine horsepower being developed by use of the prony brake, we must also count revolutions of the engine output shaft for a specific length of time. In the formula for calculating horsepower:

$$\text{horsepower} = \frac{\text{feet x pounds}}{33,000 \times \text{minutes}}$$

feet will equal the circumference transcribed by the torque arm radius multiplied by the number of engine output shaft revolutions. Thus:

$$\text{feet} = 2 \times 3.14 \times \text{radius} \times \text{revolutions.}$$

Pounds in the formula will equal the force (F) of the torque arm. If, for example, the force (F) is 6 pounds, torque arm radius is 2 feet and engine output shaft speed is 3300 revolutions per minute, then:

$$\text{horsepower} = \frac{2 \times 3.14 \times 2 \times 3300 \times 6}{33,000 \times 1}$$

or,

$$\text{horsepower} = 7.54$$

**DYNAMOMETERS.** Some commercial dynamometers for testing small engines are now available, although the cost may be prohibitive for all but the larger small engine repair shops. Usually, these dynamometers have a hydraulic loading device and scales indicating engine speed and load; horsepower is then calculated by use of a slide rule type instrument. For further information on commercial dynamometers, refer to manufacturers listed in special service tool section of this manual.

## HOW ENGINE HORSEPOWER OUTPUT VARIES

Engine efficiency will vary with the amount of air taken into the cylinder on each intake stroke. Thus, air density has a considerable effect on the horsepower output of a specific engine. As air density varies with both temperature and atmospheric pressure, any change in air temperature, barometric pressure, or elevation will cause a variance in observed engine horsepower. As a general rule, engine horsepower will:

A. Decrease approximately 3 percent for each 1000 foot increase above 1000 ft. elevation;

B. Decrease approximately 3 percent for each 1 inch drop in barometric pressure; or,

C. Decrease approximately 1 percent for each 10° rise in temperature (Farenheit).

Thus, to fairly compare observed horsepower readings, the observed readings should be corrected to standard temperature and atmospheric pressure conditions of 60° F., and 29.92 inches of mercury. The correction formula specified by the Society of Automotive Engineers is somewhat involved; however, for practical purposes, the general rules stated above can be used to approximate the corrected brake horsepower of an engine when the observed maximum brake horsepower is known.

For example, suppose the engine horsepower of 7.54 as found by use of the prony brake was observed at an altitude of 3000 feet and at a temperature of 100 degrees. At standard atmospheric pressure and temperature conditions, we could expect an increase of 4 percent due to temperature (100°-60° x 1% per 10°) and an increase of 6 percent due to altitude (3000 ft.—1000 ft. x 3% per 1000 ft.) or a total increase of 10 percent. Thus, the corrected maximum horsepower from this engine would be approximately 7.54 + .75, or approximately 8.25 horsepower.

# SERVICE SECTION
## TROUBLE-SHOOTING

When servicing an engine to correct a specific complaint, such as engine will not start, is hard to start, etc., a logical step-by-step procedure should be followed to determine cause of trouble before performing any service work. This procedure is "TROUBLE-SHOOTING."

Of course, if an engine is received in your shop for a normal tune up or specific repair work is requested, trouble-shooting procedure is not required and the work should be performed as requested. It is wise, however, to fully check the engine before repairs are made and recommend any additional repairs or adjustments necessary to ensure proper engine performance.

The following procedures, as related to a specific complaint or trouble, have proven to be a satisfactory method for quickly determining cause of trouble in a number of small engine repair shops.

**NOTE: It is not suggested that the trouble-shooting procedure as outlined in following paragraphs be strictly adhered to at all times. In many instances, customer's comments on when trouble was encountered will indicate cause of trouble. Also, the mechanic will soon develop a diagnostic technique that can only come with experience. In addition to the general trouble-shooting procedure, reader should also refer to special notes**

**following this section and to the information included in the engine, carburetor and magneto servicing fundamentals sections.**

### If Engine Will Not Start— Or Is Hard To Start

1. If engine is equipped with a rope or crank starter, turn engine slowly. As the engine piston is coming up on compression stroke, a definite resistance to

*Fig. 5-1—Diagnosing cause of trouble, or "trouble-shooting" is an important factor in servicing small engines.*

turning should be felt on rope or crank. See Fig. 5-2. This resistance should be noted every other crankshaft revolution on a single cylinder four-stroke cycle engine and on every revolution of a two-stroke cycle engine crankshaft. If alternate hard and easy turning is noted, the engine compression can be considered as not the cause of trouble at this time.

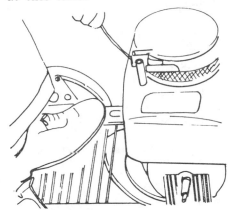

Fig. 5-2—Checking engine compression by slowly cranking engine; a definite resistance should be felt on starter rope each time piston comes up on compression stroke.

NOTE: Compression gages for small gasoline engines are available and are of value in trouble-shooting engine service problems.

Where available from engine manufacturer, specifications will be given for engine compression pressure in the engine service sections of this manual. On engines having electric or impulse starters, remove spark plug and check engine compression with gage; if gage is not available, hold thumb so that spark plug hole is partly covered. An alternating blowing and suction action should be noted as the engine is cranked.

If very little or no compression is noted, refer to appropriate engine repair section for repair of engine. If check indicates engine is developing compression, proceed to step 2.

2. Remove spark plug wire and hold wire terminal about ⅛ inch (3.18 mm) away from cylinder. (On wires having rubber spark plug boot, insert a small screw or bolt in terminal.

NOTE: If available, use of a test plug is recommended. See Fig. 5-3.

While cranking engine, a bright blue spark should snap across the ⅛ inch (3.18 mm) gap. If spark is weak or yellow, or if no spark occurs while cranking engine, refer to following IGNITION SYSTEM SERVICE section for information on appropriate type system.

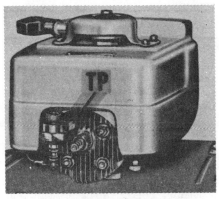

Fig. 5-3—Checking ignition spark across gap of special test plug (TP).

NOTE: A test plug with ⅛ inch (3.18 mm) gap is available or a test plug can be made by adjusting the electrode gap of a new spark plug to 0.125 inch (3.18 mm).

If spark is satisfactory, remove and inspect spark plug. Refer to SPARK PLUG SERVICING under following IGNITION SYSTEM SERVICE section. If in doubt about spark plug condition install a new plug.

NOTE: Before installing plug, be sure to check electrode gap with proper gage and, if necessary, adjust to value given in engine repair section of this manual. DO NOT guess or check gap with a "thin dime"; a few thousandths variation from correct spark plug electrode gap can make an engine run unsatisfactorily, or under some conditions, not start at all. See Fig. 5-4.

If ignition spark is satisfactory and engine will not start with new plug, proceed with step 3.

3. If engine compression and ignition spark seem to be OK, trouble within the fuel system should be suspected. Remove and clean or renew air cleaner or cleaner element. Check fuel tank (See Fig. 5-5) and be sure it is full of fresh gasoline (four-stroke cycle engines) or fresh gasoline and lubricating oil mixture (two-stroke cycle engines)

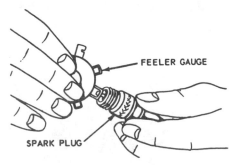

Fig. 5-4—Be sure to check spark plug electrode gap with proper size feeler gage and adjust gap to specification recommended by manufacturer.

as prescribed by engine manufacturer. Refer to LUBRICATION paragraph in each two-stroke cycle engine service (engine repair) section for proper fuel:oil mixture for each make and model. If equipped with a fuel shut-off valve, be sure valve is open.

If engine is equipped with remote throttle controls that also operate carburetor choke plate, check to be sure that when controls are placed in choke position, carburetor choke plate is fully closed. If not, adjust control linkage so that choke will fully close; then, try to start engine. If engine does not start after several turns, remove air cleaner assembly; carburetor throat should be wet with gasoline. If not, check for reason fuel is not getting to carburetor. On models with gravity feed from fuel tank to carburetor (fuel tank above carburetor), disconnect fuel line at carburetor to see that fuel is flowing through the line. If no fuel is flowing, remove and clean fuel tank, fuel line and any fuel filters or shut-off valve.

On models having a fuel pump separate from carburetor, remove fuel line at carburetor and crank engine through several turns; fuel should spurt from open line. If not, disconnect fuel line from tank to fuel pump at pump connection. If fuel will not run from open line, remove and clean fuel tank, line and if so equipped, fuel filter and/or shut-off valve. If fuel runs from open line, remove and overhaul or renew the fuel pump.

After making sure that clean, fresh fuel is available at carburetor, again try to start engine. If engine will not start, refer to recommended initial adjustments for carburetor in appropriate engine repair section of this manual and adjust carburetor idle and/or main fuel needles.

If engine will not start when compression and ignition test OK and clean, fresh fuel is available to carburetor, remove and clean or overhaul carburetor as outlined in following CARBURETOR SERVICE section.

4. The preceding trouble-shooting techniques are based on the fact that to run, an engine must develop compression, have an ignition spark and receive the proper fuel:air mixture. In some instances, there are other factors involved. Refer to the special notes following this section for service hints on finding common causes of engine trouble that may not be discovered in normal trouble-shooting procedure.

### If Engine Starts, Then Stops

This complaint is usually due to fuel starvation, but may be caused by a faulty ignition system. Recommended trouble-shooting procedure is as follows:

1. Remove and inspect fuel tank cap; on all except a few early two-stroke cycle engines, fuel tank is vented through breather in fuel tank cap so that air can enter the tank as fuel is used. If engine stops after running several minutes, a clogged breather should be suspected. On some engines, it is possible to let the engine run with fuel tank cap removed and if this permits engine to run without stopping, clean or renew the cap.

CAUTION: Be sure to observe safety precautions before attempting to run engine without fuel tank cap in place. If there is any danger of fuel being spilled on engine or spark entering open tank, DO NOT attempt to run engine without fuel tank cap in place. If in doubt, try a new cap.

2. If clogged breather in fuel tank cap is eliminated as cause of trouble, a partially clogged fuel filter or fuel line should be suspected. Remove and clean fuel tank and line and if so equipped, clean fuel shut-off valve and/or fuel tank filter. On some engines, a screen or felt type fuel filter is located in the carburetor fuel inlet; refer to engine repair section for appropriate engine make and model for carburetor construction.

3. After cleaning fuel tank, line, filters, etc., if trouble is still encountered, a sticking or faulty carburetor inlet needle valve, float or diaphragm may be cause of trouble. Remove, disassemble and clean carburetor using data in engine repair section and in following CARBURETOR SERVICE section as a guide.

4. If fuel system is eliminated as cause of trouble by performing procedure outlined in steps 1, 2 and 3, check magneto or battery ignition coil on tester if such equipment is available. If not, check for ignition spark immediately after engine stops. Renew coil, condenser and breaker points if no spark is noted. Also, on four-stroke cycle engines, check for engine compression immediately after engine stops; trouble may be caused by sticking intake or exhaust valve or cam followers (tappets). If no or little compression is noted immediately after engine stops, refer to ENGINE SERVICE section and to engine repair data in the appropriate engine repair section of this manual.

## Engine Overheats

When air cooled engines overheat, check for:

1. "Winterized" engine operated in warm temperatures.

2. Remove blower housing and shields and check for dirt or debris accumulated on or between cooling fins on cylinder.

3. Missing or bent shields or blower housing. (Never attempt to operate an air cooled engine without all shields and blower housing in place.)

4. A too lean main fuel-air adjustment of carburetor.

5. Improper ignition spark timing. Check breaker-point gap, and on engine with unit type magneto, check magneto to engine timing. On battery ignition units with timer, check for breaker points opening at proper time.

6. Engines being operated under loads in excess of rated engine horsepower or at extremely high ambient (surrounding) air temperatures may overheat.

7. Two-stroke cycle engines being operated with an improper fuel-lubricating oil mixture may overheat due to lack of lubrication; refer to appropriate engine service section in this manual for recommended fuel-lubricating oil mixture.

## Engine Surges When Running

Trouble with an engine surging is usually caused by improper carburetor adjustment or improper governor adjustment.

1. Refer to CARBURETOR paragraphs in the appropriate engine repair section and adjust carburetor as outlined.

2. If adjusting carburetor did not correct the surging condition, refer to GOVERNOR paragraph and adjust governor linkage.

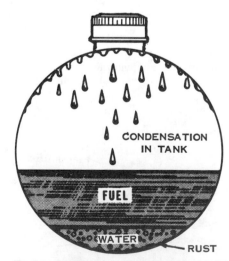

Fig. 5-5—Condensation can cause water and rust to form in fuel tank even though only clean fuel has been poured into tank.

3. If any wear is noted in governor linkage and adjusting linkage did not correct problem, renew worn linkage parts.

4. If trouble is still not corrected, remove and clean or overhaul carburetor as necessary. Also check for any possible air leaks between the carburetor to engine gaskets or air inlet elbow gaskets.

## Engine Misses Out When Running (Two-Stroke Cycle Engines)

1. If engine misses out only at no load high idle speed, first be sure that engine is not equipped with an ignition cut-out governor. (If so equipped, engine will miss out at high speed due to cut-out action.) If not so equipped, refer to appropriate engine repair section and adjust carburetor as outlined in CARBURETOR paragraph. Some two-stroke cycle engines will miss out (four-cycle) when not under load, even though carburetor is adjusted properly. If a two-cycle engine fires evenly under normal load, it can usually be considered OK.

## Special Notes on Engine Trouble-shooting

SPECIAL APPLICATION ENGINES. An engine may be manufactured or modified to function as a powerplant for a unique type of equipment. Trouble shooting and servicing must be performed while noting any out-of-the-ordinary features of the engine.

TWO-STROKE CYCLE ENGINES WITH REED VALVE. On two-stroke cycle engines, the incoming fuel-air mixture must be compressed in engine crankcase in order for the mixture to properly reach the engine cylinder. On engines utilizing reed type carburetor to crankcase intake valve, a bent or broken reed will not allow compression build up in the crankcase. Thus, if such an engine seems otherwise OK, remove and inspect the reed valve unit. Refer to appropriate engine repair section in this manual for information on individual two-stroke cycle engine models.

TWO-STROKE CYCLE ENGINE EXHAUST PORTS. Two-stroke cycle engines, and especially those being operated on an overly rich fuel-air mixture or with too much lubricating oil mixed with the fuel, will tend to build up carbon in the cylinder exhaust ports. It is recommended that the muffler be removed on two-stroke cycle engines periodically and the carbon removed from the exhaust ports. Recommended procedure varies somewhat with different makes of engines; therefore, refer to CARBON paragraph of maintenance instructions in appropriate engine repair section of this manual.

On two-stroke cycle engines that are hard to start, or where complaint is loss of power, it is wise to remove the muffler and inspect the exhaust ports for carbon build up.

**FOUR-STROKE CYCLE ENGINES WITH COMPRESSION RELEASE.** Several different makes of four-stroke cycle engines now have a compression release that reduces compression pressure at cranking speeds, thus making it easier to crank the engine. Most models having this feature will develop full compression when turned in a reverse direction. Refer to the appropriate engine repair section in this manual for detailed information concerning the compression release used on different makes and models.

# IGNITION SYSTEM SERVICE

The fundamentals of servicing ignition systems are outlined in the following paragraphs. Refer to appropriate heading for type ignition system being inspected or overhauled.

## BATTERY IGNITION SERVICE FUNDAMENTALS

Service of battery ignition systems used on small engines is somewhat simplified due to the fact that no distribution system is required as on automotive type ignition systems. Usually all components are readily accessible and while use of test instruments is sometimes desirable, condition of the system can be determined by simple checks. Refer to following paragraphs.

**GENERAL CONDITION CHECK.** Remove spark plug wire and if terminal is rubber covered, insert small screw or bolt in terminal. Hold uncovered end of terminal, or bolt inserted in terminal about ⅛ inch (3.18 mm) away from engine or connect spark plug wire to test plug as shown in Fig. 5-3. Crank engine while observing gap between spark plug wire terminal and engine; if a bright blue spark snaps across the gap, condition of the system can be considered satisfactory. However, ignition timing may have to be adjusted. Refer to timing procedure in appropriate engine repair section.

**VOLTAGE, WIRING AND SWITCH CHECK.** If no spark, or a weak yellow-orange spark occurred when checking system as outlined in preceding paragraph, proceed with following checks:

Test battery condition with hydrometer or voltmeter. If check indicates a dead cell, renew the battery; recharge battery if a discharged condition is indicated.

NOTE: On models with electric starter or starter-generator unit, battery can be assumed in satisfactory condition if the starter cranks the engine freely.

If battery checks OK, but starter unit will not turn engine, a faulty starter unit is indicated and ignition trouble may be caused by excessive current draw of such a unit. If battery and starting unit, if so equipped, are in satisfactory condition, proceed as follows:

Remove battery lead wire from ignition coil and connect a test light of same voltage as the battery between the disconnected lead wire and engine ground. Light should go on when ignition switch is in "on" position and go off when switch is in "off" position. If not, renew switch and/or wiring and recheck for satisfactory spark. If switch and wiring check OK, but no spark is obtained, proceed as follows:

**BREAKER POINTS AND CONDENSER.** Remove breaker box cover and, using small screwdriver, separate and inspect breaker points. If burned or deeply pitted, renew breaker points and condenser. If point contacts are clean to grayish in color and are only slightly pitted, proceed as follows: Disconnect condenser and ignition coil lead wires from breaker point terminal and connect a test light and battery between terminal and engine ground. Light should go on when points are closed and should go out when points are open. If light fails to go out when

points are open, breaker arm insulation is defective and breaker points must be renewed. If light does not go on when points are in closed position, clean or renew the breaker points. In some instances, new breaker point contact surfaces may have an oily or wax coating or have foreign material between the surfaces so that proper contact is prevented. Check ignition timing and breaker point gap as outlined in appropriate engine repair section of this manual.

Connect test light and battery between condenser lead and engine ground; if light goes on, condenser is shorted out and should be renewed. Capacity of condenser can be checked if test instrument is available. It is usually good practice to renew the condenser whenever new breaker points are being installed if tester is not available.

**IGNITION COIL.** If a coil tester is available, condition of coil can be checked. However, if tester is not available, a reasonably satisfactory performance test can be made as follows:

Disconnect high tension wire from spark plug. Turn engine so that cam has allowed breaker points to close. With ignition switch on, open and close points with small screwdriver while holding high tension lead about ⅛ to ¼ inch (3.18 to 6.35 mm) away from engine ground. A bright blue spark should snap across the gap between spark plug wire and ground each time the points are opened. If no spark occurs, or spark is weak and yellow-orange, renewal of the ignition coil is indicated.

Sometimes, an ignition coil may perform satisfactorily when cold, but fail after engine has run for some time and coil is hot. Check coil when hot if this condition is indicated.

## FLYWHEEL MAGNETO SERVICE FUNDAMENTALS

In servicing a flywheel magneto ignition system, the mechanic is concerned with trouble shooting, service adjustments and testing magneto components. The following paragraphs out-

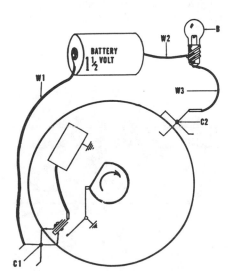

Fig. 5-6—Drawing showing a simple test lamp for checking ignition timing and/or breaker-point opening.

| | |
|---|---|
| B. 1-1/2 volt bulb | W1. Wire |
| C1. Spring clamp | W2. Wire |
| C2. Spring clamp | W3. Wire |

line the basic steps in servicing a flywheel type magneto. Refer to the appropriate engine section for adjustment and test specifications for a particular engine.

## Trouble-shooting

If the engine will not start and malfunction of the ignition system is suspected, make the following checks to find cause of trouble.

Check to be sure that the ignition switch (if so equipped) is in the "On" or "Run" position and that the insulation on the wire leading to the ignition switch is in good condition. The switch can be checked with the timing and test light as shown in Fig. 5-6. Disconnect the lead from the switch and attach one clip of the test light to the switch terminal and the other clip to the engine. The light should go on when the switch is in the "Off" or "Stop" position, and should go off when the switch is in the "On" or "Run" position.

Inspect the high tension (spark plug) wire for worn spots in the insulation or breaks in the wire. Frayed or worn insulation can be repaired temporarily with plastic electrician's tape.

If no defects are noted in the ignition switch or ignition wires, remove and inspect the spark plug as outlined under the SPARK PLUG SERVICING. If the spark plug is fouled or is in questionable condition, connect a spark plug of known quality to the high tension wire, ground the base of the spark plug to engine and turn engine rapidly with the starter. If the spark across the electrode gap of the spark plug is a bright blue, the magneto can be considered in satisfactory condition.

NOTE: Some engine manufacturers specify a certain type spark plug and a specific test gap. Refer to appropriate engine service section; if no specific spark plug type or electrode gap is recommended for test purposes, use spark plug type and electrode gap recommended for engine make and model.

If spark across the gap of the test plug is weak or orange colored, or no spark occurs as engine is cranked, magneto should be serviced as outlined in the following paragraphs.

## Magneto Adjustments

BREAKER CONTACT POINTS. Adjustment of the breaker contact points affects both ignition timing and magneto edge gap. Therefore, the breaker contact point gap should be carefully adjusted according to engine manufacturer's specifications. Before adjusting the breaker contact gap, inspect contact points and renew if condition of contact surfaces is questionable. It is sometimes desirable to check the condition of points as follows: Disconnect the condenser and primary coil leads from the breaker-point terminal. Attach one clip of a test light (See Fig. 5-6) to the breaker-point terminal and the other clip of the test light to magneto ground. The light should be out when contact points are open and should go on when the engine is turned to close the breaker contact points. If the light stays on when points are open, insulation of breaker contact arm is defective. If light does not go on when points are closed, contact surfaces are dirty, oily or are burned.

Adjust breaker-point gap as follows unless manufacturer specifies adjusting breaker gap to obtain correct ignition timing. First, turn engine so that points are closed to be sure that the contact surfaces are in alignment and seat squarely. Then, turn engine so that breaker point opening is maximum and adjust breaker gap to manufacturer's specification. A wire type feeler gage is recommended for checking and adjusting the breaker contact gap. Be sure to recheck gap after tightening breaker point base retaining screws.

IGNITION TIMING. On some engines, ignition timing is nonadjustable and a certain breaker-point gap is specified. On other engines, timing is adjustable by changing the position of the magneto stator plate. (See Fig. 5-7) with a specified breaker-point gap or by simply varying the breaker-point gap to obtain correct timing, Ignition timing is usually specified either in degrees of engine (crankshaft) rotation or in piston travel before the piston reaches top dead center position. In some instances, a specification is given for ignition timing even though the timing may be nonadjustable; if a check reveals timing is incorrect on these engines, it is an indication of incorrect breaker-point adjustment or excessive wear of breaker cam. Also, on some engines, it may indicate that a wrong breaker cam has been installed or that the cam has been installed in a reversed position on engine crankshaft.

Some engines may have a timing mark or flywheel locating pin to locate the flywheel at proper position for the ignition spark to occur (breaker points begin to open). If not, it will be necessary to measure piston travel as illustrated in Fig. 5-8 or install a degree indicating device on the engine crankshaft.

A timing light as shown in Fig. 5-6 is a valuable aid in checking or adjusting engine timing. After disconnecting the ignition coil lead from the breaker-point terminal, connect the leads of the timing light as shown. If timing is adjustable by moving the magneto stator plate, be sure that the breaker point gap is adjusted as specified. Then, to check timing, slowly turn engine in normal direction of rotation past the point at which ignition spark should occur. The timing light should be on, then go out (breaker points open) just as the correct timing location is passed. If not, turn engine to proper timing location and adjust timing by relocating the magneto stator plate or varying the breaker contact gap as specified by engine manufacturer. Loosen the screws retaining the stator plate or breaker points and adjust position of stator plate or points so that points are closed (timing light is on). Then, slowly move adjustment until timing light goes out (points open) and tighten the retaining screws. Recheck timing to be sure adjustment is correct.

*Fig. 5-7—On some engines, timing is adjustable by moving magneto stator plate in slotted mounting holes. Marks should be applied to stator plate and engine after engine is properly timed; marks usually appear on factory assembled stator plate and engine cylinder block.*

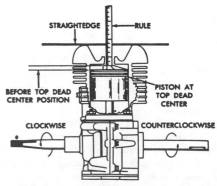

*Fig. 5-8—On some engines, it will be necessary to measure piston travel with rule, dial indicator or special timing gage when adjusting or checking ignition timing.*

**ARMATURE AIR GAP.** To fully concentrate the magnetic field of the flywheel magnets within the armature core, it is necessary that the flywheel magnets pass as closely to the armature core as possible without danger of metal to metal contact. The clearance between the flywheel magnets and the legs of the armature core is called the armature air gap.

On magnetos where the armature and high tension coil are located outside of the flywheel rim, adjustment of the armature air gap is made as follows: Turn the engine so that the flywheel magnets are located directly under the legs of the armature core and check the clearance between the armature core and flywheel magnets. If the measured clearance is not within manufacturers specifications, loosen the armature core mounting screws and place shims of thickness equal to minimum air gap specification between the magnets and armature core (Fig. 5-9). The magnets will pull the armature core against the shim stocks. Tighten the armature core mounting screws, remove the shim stock and turn the engine through several revolutions to

be sure the flywheel does not contact the armature core.

Where the armature core is located under or behind the flywheel, the following methods may be used to check and adjust armature air gap: On some engines, slots or openings are provided in the flywheel through which the armature air gap can be checked. Some engine manufacturers provide a cutaway flywheel that can be installed temporarily for checking the armature air gap. A test flywheel can be made out of a discarded flywheel (See Fig. 5-10), or out of a new flywheel if service volume on a particular engine warrants such expenditure. Another method of checking the armature air gap is to remove the flywheel and place a layer of plastic tape equal to the minimum specified air gap over the legs of the armature core. Reinstall flywheel and turn engine through several revolutions and remove flywheel; no evidence of contact between the flywheel magnets and plastic tape should be noticed. Then cover the legs of the armature core with a layer of tape of thickness equal to the maximum specified air gap; then, reinstall flywheel and turn engine through several revolutions. Indication of the flywheel magnets contacting the plastic tape should be noticed after the flywheel is again removed. If the magnets contact the first thin layer of tape applied to the armature core legs, or if they do not contact the second thicker layer of tape, armature air gap is not within specifications and should be adjusted.

NOTE: Before loosening armature core mounting screws, scribe a mark on mounting plate against edge of armature core so that adjustment of air gap can be gaged.

In some instances, it may be necessary to slightly enlarge the armature core mounting holes before proper air gap adjustment can be made.

**MAGNETO EDGE GAP.** The point of maximum acceleration of the movement of the flywheel magnetic field through the high tension coil (and therefore, the point of maximum current induced in the primary coil windings) occurs when the trailing edge of the flywheel magnet is slightly past the left hand leg of the armature core. The exact point of maximum primary current is determined by using electrical measuring devices, the distance between the trailing edge of the flywheel magnet and the leg of the armature core at this point is measured and becomes a service specification. This distance, which is stated either in thousandths of an inch or in degrees of flywheel rotation, is called the Edge Gap or "E" Gap.

For maximum strength of the ignition spark, the breaker points should just start to open when the flywheel magnets are at the specified edge gap position. Usually, edge gap is non-adjustable and will be maintained at the proper dimension if the contact breaker points are adjusted to the recommended gap and the correct breaker cam is installed. However magneto edge gap can change (and spark intensity thereby reduced) due to the following:

a. Flywheel drive key sheared
b. Flywheel drive key worn (loose)
c. Keyway in flywheel or crankshaft worn (oversized)
d. Loose flywheel retaining nut which can also cause any above listed difficulty
e. Excessive wear on breaker cam
f. Breaker cam loose on crankshaft
g. Excessive wear on breaker point rubbing block or push rod so that points cannot be properly adjusted.

### Unit Type Magneto Service Fundamentals

Improper functioning of the carburetor, spark plug or other components often causes difficulties that are thought to be an improperly functioning magneto. Since a brief inspection will often locate other causes for engine malfunction, it is recommended that one be certain the magneto is at fault before opening the magneto housing. Magneto malfunction can easily be determined by simple tests as outlined in following paragraph.

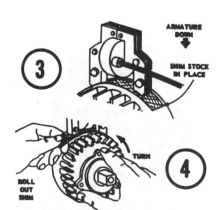

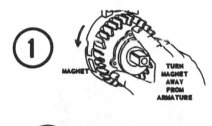

Fig. 5-9—Views showing adjustment of armature air gap when armature is located outside flywheel. Refer to Fig. 5-10 for engines having armature located inside flywheel.

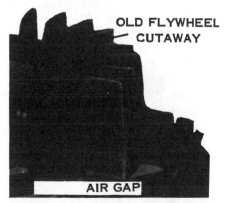

Fig. 5-10—Where armature core is located inside flywheel, check armature gap by using a cutaway flywheel unless other method is provided by manufacturer; refer to appropriate engine repair section. Where possible, an old discarded flywheel should be used to cutaway section for checking armature air gap.

## Trouble-shooting

With a properly adjusted spark plug in good condition, the ignition spark should be strong enough to bridge a short gap in addition to the actual spark plug gap. With engine running, hold end of spark plug wire not more than 1/16 inch (1.59 mm) away from spark plug terminal. Engine should not misfire.

To test the magneto spark if engine will not start, remove ignition wire from magneto end cap socket. Bend a short piece of wire so that when it is inserted in the end cap socket, other end is about ⅛ inch (3.18 mm) from engine casting. Crank engine slowly and observe gap between wire and engine; a strong blue spark should jump the gap the instant that the impulse coupling trips. If a strong spark is observed, it is recommended that the magneto be eliminated as the source of engine difficulty and that the spark plug, ignition wire and terminals be thoroughly inspected.

If, when cranking the engine, the impulse coupling does not trip, the magneto must be removed from the engine and the coupling overhauled or renewed. It should be noted that if the impulse coupling will not trip, a weak spark will occur.

## Magneto Adjustments and Service

**BREAKER POINTS.** Breaker points are accessible for service after removing the magneto housing end cap. Examine point contact surfaces for pitting or pyramiding (transfer of metal from one surface to the other); a small tungsten file or fine stone may be used to resurface the points. Badly worn or badly pitted points should be renewed. After points are resurfaced or renewed, check breaker point gap with rotor turned so that points are opened maximum distance. Refer to MAGNETO paragraph in appropriate engine repair section for point gap specifications.

When replacing the magneto end cap, both the end cap and housing mating surfaces should be thoroughly cleaned and a new gasket be installed.

**CONDENSER.** Condenser used in unit type magneto is similar to that used in other ignition systems. Refer to MAGNETO paragraph in appropriate engine repair section for condenser test specifications. Usually, a new condenser should be installed whenever the breaker points are being renewed.

**COIL.** The ignition coil can be tested without removing the coil from the housing. The instructions provided

with coil tester should have coil test specifications listed.

**ROTOR.** Usually, service on the magneto rotor is limited to renewal of bushings or bearings, if damaged. Check to be sure rotor turns freely and does not drag or have excessive end play.

**MAGNETO INSTALLATION.** When installing a unit type magneto on an engine, refer to MAGNETO paragraph in appropriate engine repair section for magneto to engine timing information.

## SOLID-STATE IGNITION SERVICE FUNDAMENTALS

Because of differences in solid-state ignition construction, it is impractical to outline a general procedure for solid-state ignition service. Refer to the specific engine section for testing, overhaul notes and timing of solid-state ignition systems.

## SPARK PLUG SERVICING

**ELECTRODE GAP.** The spark plug electrode gap should be adjusted by bending the ground electrode. Refer to Fig. 5-11. The recommended gap is listed in the SPARK PLUG paragraph in appropriate engine repair section of this manual.

**CLEANING AND ELECTRODE CONDITIONING.** Spark plugs are usually cleaned by an abrasive action commonly referred to as "sand blasting." Actually, ordinary sand is not used, but a special abrasive which is nonconductive to electricity even when melted, thus the abrasive cannot short out the plug current. Extreme care should be used in cleaning the plugs after sand blasting, however, as any particles of abrasive left on the plug may cause damage to piston rings, piston or cylinder walls. Some engine

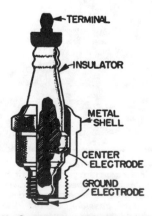

Fig. 5-11—Cross-sectional drawing of spark plug showing construction and nomenclature.

Fig. 5-12—Normal plug appearance in four-stroke cycle engine. Insulator is light tan to gray in color and electrodes are not burned. Renew plug at regular intervals as recommended by engine manufacturer.

Fig. 5-13—Appearance of four-stroke cycle spark plug indicating cold fouling. Cause of cold fouling may be use of a too-cold plug, excessive idling or light loads, carburetor choke out of adjustment, defective spark plug wire or boot, carburetor adjusted too "rich" or low engine compression.

Fig. 5-14—Appearance of four-stroke cycle spark plug indicating wet fouling; a wet, black oily film is over entire firing end of plug. Cause may be oil getting by worn valve guides, worn oil rings or plugged breather or breather valve in tappet chamber.

Fig. 5-15—Appearance of four-stroke cycle spark plug indicating overheating. Check for plugged cooling fins, bent or damaged blower housing, engine being operated without all shields in place or other causes of engine overheating. Also can be caused by too lean a fuel-air mixture or spark plug not tightened properly.

*Fig. 5-16—Normal appearance of plug removed from a two-stroke cycle engine. Insulator is light tan to gray in color, few deposits are present and electrodes not burned.*

*Fig. 5-17—Appearance of plug from two-stroke cycle engine indicating wet fouling. A damp or wet black carbon coating is formed over entire firing end. Could be caused by a too-cold plug, excessive idling, improper fuel-lubricating oil mixture or carburetor adjustment too rich.*

manufacturers recommend that the spark plug be renewed rather than cleaned because of possible engine damage from cleaning abrasives.

After plug is cleaned by abrasive, and before gap is set, the electrode surfaces between the grounded and insulated electrodes should be cleaned and returned as nearly as possible to

*Fig. 5-18—Appearance of plug from two-stroke cycle engine indicating overheating. Insulator has gray or white blistered appearance and electrodes may be burned. Could be caused by use of a too-hot plug, carburetor adjustment too lean, "sticky" piston rings, engine overloaded, or cooling fins plugged causing engine to run too hot.*

original shape by filing with a point file. Failure to properly dress the electrodes can result in high secondary voltage requirements, and misfire of the plug.

**PLUG APPEARANCE DIAGNOSIS.** The appearance of a spark plug will be altered by use, and an examination of the plug tip can contribute useful information which may assist in obtaining better spark plug life. It must be remembered that the contributing factors differ in two-stroke cycle and four-stroke cycle engine operations and although the appearance of two spark plugs may be similar, the corrective measures may depend on whether the engine is of two-stroke cycle or four-stroke cycle design. Figs. 5-12 through 5-18 are provided by Champion Spark Plug Company to illustrate typical observed conditions. Refer to Figs. 5-12 through 5-15 for four-stroke cycle engines and to Figs. 5-16 through 5-18 for two-stroke cycle engines. Listed in captions are the probable causes and suggested corrective measures.

# CARBURETOR SERVICE

The bulk of carburetor service consists of cleaning, inspection and adjustment. After considerable service it may become necessary to overhaul the carburetor and renew worn parts to restore original operating efficiency. Although carburetor condition affects engine operating economy and power, ignition and engine compression must also be considered to determine and correct causes of poor performance.

Before dismantling carburetor for cleaning or overhaul, clean all external surfaces and remove accumulated dirt and grease. Refer to appropriate engine repair section for carburetor exploded or cross-sectional views. Dismantle carburetor and note any discrepancies to assure correction during overhaul. Thoroughly clean all parts and inspect for damage or wear. Wash jets and passages and blow clear with clean, dry compressed air. Do not use a drill or wire to clean jets as the possible enlargement of calibrated holes will disturb operating balance. The measurement of jets to determine the extent of wear is difficult and new parts are usually installed to assure satisfactory results.

Carburetor manufacturers provide for many of their models an assortment of gaskets and other parts usually needed to do a correct job of cleaning

and overhaul. These assortments are usually catalogued as Gasket Kits and Overhaul Kits respectively.

On float type carburetors, inspect float pin and needle valve for wear and renew if necessary. Check metal floats for leaks and where a dual type float is installed, check alignment of float sections. Check cork floats for loss of protective coating and absorption of fuel.

**NOTE: Do not attempt to recoat cork floats with shellac or varnish or to resolder leaky metal floats. Renew part if defective.**

Check the fit of throttle and choke valve shafts. Excessive clearance will cause improper valve plate seating and will permit dust or grit to be drawn into the engine. Air leaks at throttle shaft bores due to wear will upset carburetor calibration and contribute to uneven engine operation. Rebush valve shaft holes where necessary and renew dust seals. If rebushing is not possible, renew the body part supporting the shaft. Inspect throttle and choke valve plates for proper installation and condition.

Power or idle adjustment needles must not be worn or grooved. Check

condition of needle seal packing or "O" ring and renew packing or "O" ring if necessary.

Reinstall or renew jets, using correct size listed for specific model. Adjust power and idle settings as described for specific carburetors in engine service section of manual.

It is important that the carburetor bore at the idle discharge ports and in the vicinity of the throttle valve be free of deposits. A partially restricted idle port will produce a "flat spot" between idle and mid-range rpm. This is because the restriction makes it necessary to open the throttle wider than the designed opening to obtain proper idle speed. Opening the throttle wider than the design specified amount will uncover more of the port than was intended in the calibration of the carburetor. As a result an insufficient amount of the port will be available as a reserve to cover the transition period (idle to the mid-range rpm) when the high speed system begins to function.

Refer to Fig. 5-20 for service hints on diaphragm type carburetors.

When reassembling float type carburetors, be sure float position is properly adjusted. Refer to CARBURETOR paragraph in appropriate engine repair section for float level adjustment specifications.

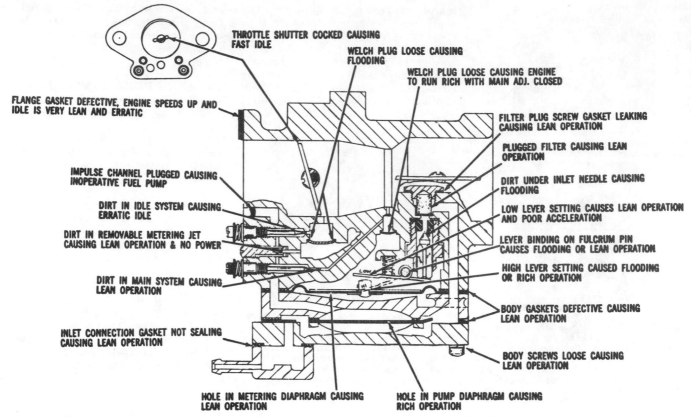

THROTTLE SHUTTER COCKED CAUSING
FAST IDLE

WELCH PLUG LOOSE CAUSING
FLOODING

WELCH PLUG LOOSE CAUSING ENGINE
TO RUN RICH WITH MAIN ADJ. CLOSED

FLANGE GASKET DEFECTIVE, ENGINE SPEEDS UP AND
IDLE IS VERY LEAN AND ERRATIC

FILTER PLUG SCREW GASKET LEAKING
CAUSING LEAN OPERATION

PLUGGED FILTER CAUSING LEAN
OPERATION

IMPULSE CHANNEL PLUGGED CAUSING
INOPERATIVE FUEL PUMP

DIRT UNDER INLET NEEDLE CAUSING
FLOODING

DIRT IN IDLE SYSTEM CAUSING
ERRATIC IDLE

LOW LEVER SETTING CAUSES LEAN OPERATION
AND POOR ACCELERATION

DIRT IN REMOVABLE METERING JET
CAUSING LEAN OPERATION & NO POWER

LEVER BINDING ON FULCRUM PIN
CAUSES FLOODING OR LEAN OPERATION

HIGH LEVER SETTING CAUSED FLOODING
OR RICH OPERATION

DIRT IN MAIN SYSTEM CAUSING
LEAN OPERATION

BODY GASKETS DEFECTIVE CAUSING
LEAN OPERATION

INLET CONNECTION GASKET NOT SEALING
CAUSING LEAN OPERATION

BODY SCREWS LOOSE CAUSING
LEAN OPERATION

HOLE IN METERING DIAPHRAGM CAUSING
LEAN OPERATION

HOLE IN PUMP DIAPHRAGM CAUSING
RICH OPERATION

Fig. 5-20—Schematic cross-sectional view of a diaphragm type carburetor illustrating possible causes of malfunction. Refer to appropriate engine repair section for adjustment information and for exploded and/or cross-sectional view of actual carburetors used.

# ENGINE SERVICE

## DISASSEMBLY AND ASSEMBLY

Special techniques must be developed in repair of engines of aluminum alloy or magnesium alloy construction. Soft threads in aluminum or magnesium castings are often damaged by carelessness in overtightening fasteners or in attempting to loosen or remove seized fasteners. Manufacturer's recommended torque values for tightening screw fasteners should be followed closely.

**NOTE: If damaged threads are encountered, refer to following paragraph, "REPAIRING DAMAGED THREADS."**

A given amount of heat applied to aluminum or magnesium will cause it to expand a greater amount than will steel under similar conditions. Because of the different expansion characteristics, heat is usually recommended for easy installation of bearings, pins, etc., in aluminum or magnesium castings. Sometimes, heat can be used to free parts that are seized or where an inter-

ference fit is used. Heat, therefore, becomes a service tool and the application of heat is one of the required service techniques. An open flame is not usually advised because it destroys the paint and other protective coatings and because a uniform and controlled temperature with open flame is difficult to obtain. Methods commonly used are heating in oil or water, with a heat lamp, electric hot plate, or in an oven or kiln. See Fig. 5-21. The use of water or oil gives a fairly accurate temperature control but is somewhat limited as to the size and type of part than can be handled. Thermal crayons are available which can be used to determine the temperature of a heated part. These crayons melt when the part reaches a specified temperature, and a number of crayons for different temperatures are available. Temperature indicating crayons are usually available at welding equipment supply houses.

The crankcase and combustion chambers of a two-stroke cycle engine must be sealed against pressure and vacuum. To assure a perfect seal, nicks,

scratches and warpage are to be avoided. Slight imperfections can be removed by using a fine-grit sandpaper. Flat surfaces can be lapped by using a surface plate or a smooth piece of plate glass, and à sheet of 120-grit sandpaper or lapping compound. Use a figure-eight motion with minimum pressure, and remove only enough

Fig. 5-21—In small engine repair, heat can be used efficiently as a disassembly and assembly tool. Heating crankcase halves on electric hot plate (above) will allow bearings to be easily removed.

metal to eliminate the imperfection. Bearing clearances, if any, must not be lessened by removing metal from the joint.

Use only the specified gaskets when re-assembling, and use an approved gasket cement or sealing compound unless the contrary is stated. Seal all exposed threads and repaint or retouch with an approved paint.

## REPAIRING DAMAGED THREADS

Damaged threads in castings can be renewed by use of thread repair kits which are recommended by a number of equipment and engine manufacturers. Use of thread repair kits is not difficult, but instructions must be carefully followed. Refer to Figs. 5-22 through 5-25 which illustrate the use

Fig. 5-22—Damaged threads in casting before repair. Refer to Figs. 5-23, 5-24 and 5-25 for steps in installing thread insert. (Series of photos provided by Heli-Coil Corp., Danbury, Conn.)

Fig. 5-23—First step in repairing damaged threads is to drill out old threads using exact size drill recommended in instructions provided with thread repair kit. Drill all the way through an open hole or all the way to bottom of blind hole, making sure hole is straight and that centerline of hole is not moved in drilling process.

of Heli-Coil thread repair kits that are manufactured by the Heli-Coil Corporation, Danbury, Connecticut.

Heli-Coil thread repair kits are available through the parts departments of most engine and equipment manufacturers; the thread inserts are available in all National Coarse (USS) sizes from #8 to 1½ inch, National Fine (SAE) sizes from #10 to 1½ inch, Metric Coarse sizes from M3 to M20 and Metric

Fig. 5-24—Special drill taps are provided in thread repair kit for threading drilled hole to correct size for outside of thread insert. A standard tap cannot be used.

Fig. 5-25—A thread insert and a completed repair are shown above. Special tools are provided in thread repair kit for installation of thread insert.

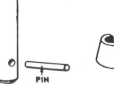
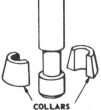

Fig. 5-27—Drawing showing three types of valve spring keepers used.

PIN     COLLARS     RETAINER

Fine sizes form M8 to M20. Also, sizes for repairing M10, M12, M14, M18 and ⅞ inch spark plug ports are available.

## VALVE SERVICE FUNDAMENTALS

### (Four-Stroke Cycle Engines)

When overhauling small four-stroke cycle engines, obtaining proper valve sealing is of primary importance. The following paragraphs cover the fundamentals of servicing the intake and exhaust valves, valve seats and valve guides.

**REMOVING AND INSTALLING VALVES.** A valve spring compressor, one type of which is shown in Fig. 5-26, is a valuable aid in removing and installing the intake and exhaust valves. This tool is used to hold the spring compressed while removing or installing the pin, collars or retainer from the valve stem. Refer to Fig. 5-27 for views showing some of the different methods of retaining valve spring to valve stem.

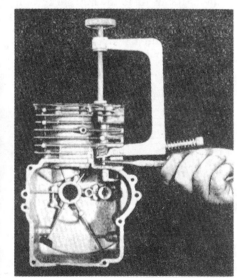

Fig. 5-26—View showing one type of valve spring compressor being used to remove keeper. (Block is cutaway to show valve spring.)

**VALVE REFACING.** If the valve face (See Fig. 5-28) is slightly worn, burned or pitted, the valve can usually be refaced providing proper equipment is available. Many small engine shops will usually renew the valves, however, rather to invest in somewhat costly valve refacing tools.

Before attempting to reface a valve, refer to specifications in appropriate engine repair section for valve face angle. On some engines, manufacturer recommends grinding the valve face to an angle of ½ to 1 degree less than that of the valve seat. Refer to Fig. 5-29. Also, nominal valve face angle may be either 30 or 45 degrees.

After valve is refaced, check thickness of valve "margin" (See Fig. 5-28). If margin is less than manufacturer's minimum specification (refer to specifications in appropriate engine repair section), or is less than one-half the margin of a new valve, renew the valve. Valves having excessive material removed in refacing operation will not give satisfactory service.

When refacing or renewing a valve, the seat should also be reconditioned, or in engines where valve seat is renewable, a new seat should be installed. Refer to following paragraph "RESEATING OR RENEWING VALVE SEATS." Then, the seating surfaces should be lapped in using a fine valve grinding compound.

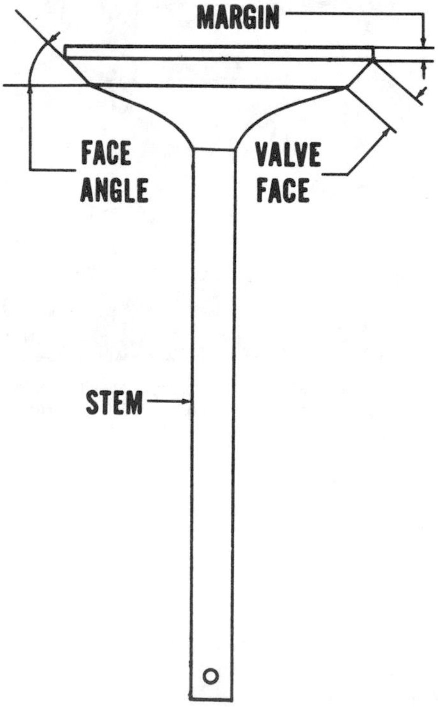

**RESEATING OR RENEWING VALVE SEATS.** On engines having the valve seat machined in the cylinder block casting, the seat can be reconditioned by using a correct angle seat grinding stone or valve seat cutter. When reconditioning valve seat, care should be taken that only enough material is removed to provide a good seating or valve contact surface. The width of the seat should then be measured (See Fig. 5-30) and if width exceeds manufacturer's maximum specifications, the seat should be narrowed by using one stone or cutter with an angle 15 degrees greater than valve seat angle and a second stone or cutter with an angle 15 degrees less than seat angle. When narrowing the seat, coat seat lightly with Prussian blue and check where seat contacts valve face by inserting valve in guide and rotating valve lightly against seat. Seat should contact approximate center of valve face. By using only the narrow angle seat narrowing stone or cutter, seat contact will be moved toward outer edge of valve face.

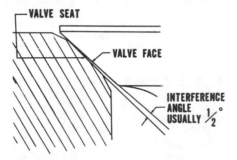

Fig. 5-29—Drawing showing line contact of valve face with valve seat when valve face is ground at smaller angle than valve seat; this is specified on some engines.

On engines having renewable valve seats, refer to appropriate engine repair section in this manual for recommended method of removing old seat and installing new seat. Refer to Fig. 5-31 for one method of installing new valve seats. Seats are retained in cylinder block bore by an interference fit; that is, seat is slightly larger than the bore in block. It sometimes occurs that the valve seat will become loose in the bore, especially on engines with aluminum crankcase. Some manufacturers provide oversize valve seat inserts (insert O.D. larger than standard part) so that if standard size insert fits loosely, the bore can be cut oversize and a new insert be tightly installed. After installing valve seat insert in engines of aluminum construction, the metal around the seat should be peened as shown in Fig. 5-32. Where a loose insert is encountered and an oversize insert is not available, the loose insert can usually be tightened by center-punching the cylinder block material at three equally spaced points around the insert, then peening completely around the insert as shown in Fig. 5-32.

For some engines with cast iron cylinder blocks, a service valve seat insert is available for reconditioning the valve seat, and is installed by counter-boring the cylinder block to specified dimensions, then driving insert into place. Refer to appropriate engine repair section in this manual for information on availability and installation of service valve seat inserts for cast iron engines.

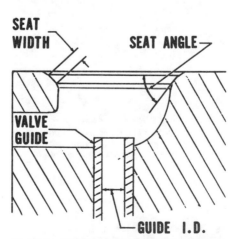

Fig. 5-30—Cross-sectional drawing of typical valve seat and valve guide as used on small engines. Valve guide may be integral part of cylinder block; on some models so constructed, valve guide ID may be reamed out and an oversize valve stem installed. On other models, a service valve guide may be installed after counter-boring cylinder block.

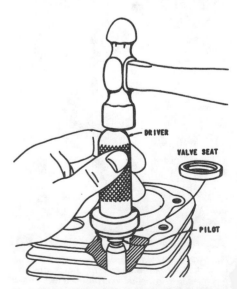

Fig. 5-31—View showing one method used to install valve seat insert. Refer to appropriate engine repair section for manufacturer's recommended method.

Fig. 5-28—Drawing showing typical four-stroke cycle engine valve. Face angle is usually 30 or 45 degrees. On some engines, valve face is ground to an angle of ½ or 1 degree less than seat angle.

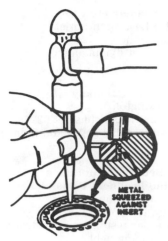

Fig. 5-32—It is usually recommended that on aluminum block engines, metal be peened around valve seat insert after insert is installed.

## INSTALLING OVERSIZE PISTON AND RINGS

Some small engine manufacturers have oversize piston and ring sets available for use in repairing engines in which the cylinder bore is excessively worn and standard size piston and rings cannot be used. If care and approved procedure are used in oversizing the cylinder bore, installation of an oversize piston and ring set should result in a highly satisfactory overhaul.

The cylinder bore may be oversized by using either a boring bar or a hone; however, if a boring bar is used, it is usually recommended the cylinder bore be finished with a hone. Refer to Fig. 5-33.

Where oversize piston and rings are available, it will be noted in the appropriate engine repair section of this manual. Also, the standard bore diameter will be given. Before attempting to rebore or hone the cylinder to oversize, carefully measure the cylinder bore to be sure that standard size piston and rings will not fit within tolerance. Also, it may be possible that the cylinder is excessively worn or damaged and that reboring or honing to largest oversize will not clean up the worn or scored surface.

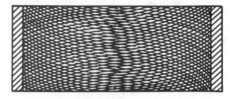

Fig. 5-33—A cross-hatch pattern as shown should be obtained when honing cylinder. Pattern is obtained by moving hone up and down cylinder bore as it is being turned by slow speed electric drill.

# SERVICE SHOP TOOL BUYER'S GUIDE

This listing of Service Shop Tools is solely for the convenience of users of this manual and does not imply endorsement or approval by Intertec Publishing Corporation of the tools and equipment listed. The listing is in response to many requests for information on sources for purchasing special tools and equipment. Every attempt has been made to make the listing as complete as possible at time of publication and each entry is made from the latest material available.

Special engine service tools such as seal drivers, bearing drivers, etc., which are available from the engine manufacturer are not listed in this section of the manual. Where a special service tool is listed in the engine service section of this manual, the tool is available from the central parts or service distributors listed at the end of most engine service sections, or from the manufacturer.

**NOTE TO MANUFACTURERS AND NATIONAL SALES DISTRIBUTORS OF ENGINE SERVICE TOOLS AND RELATED SERVICE EQUIPMENT. To obtain either a new listing for your products, or to change or add to an existing listing, write to Intertec**

**Publishing Corporation, Book Division, P.O. Box 12901, Overland Park, Kansas 66212.**

## Engine Service Tools

**Ammco Tools, Inc.**
Wacker Park
North Chicago, Illinois 60064
Valve spring compressor, torque wrenches, cylinder hones, ridge reamers, piston ring compressors, piston ring expanders.

**Black & Decker Mfg. Co.**
701 East Joppa Road
Towson, Maryland 21204
Valve grinding equipment.

**Bloom, Inc.**
Route Four, Hiway 20 West
Independence, Iowa 50644
Engine repair stand with crankshaft straightening attachment.

**Brush Research Mfg. Inc.**
4642 East Floral Drive
Los Angeles, California 90022
Cylinder hones.

**E-Z Lok**
P.O. Box 2069
Gardena, California 90247
Thread repair insert kits for metal, wood and plastic.

**Foley-Belsaw Company**
**Outdoor Power Equipment Parts Division**
6301 Equitable Road
P.O. Box 419593
Kansas City, Missouri 64141
Crankshaft straightener and repair stand, valve refacers, valve seat grinders, parts washers, cylinder hones, gages and ridge reamers, piston ring expanders and compressor, flywheel pullers, torque wrenches, rotary mower blade balancers.

**Frederick Mfg. Co., Inc.**
1400 C., Agnes Avenue
Kansas City, Missouri 64127
Crankshaft straightener.

**Heli-Coil Products Division**
**Heli-Coil Corporation**
Shelter Rock Lane
Danbury, Connecticut 06010

Thread repair kits, thread inserts, installation tools.

## K-D Tools
3575 Hempland Road
Lancaster, Pennsylvania 17604
Thread repair kits, valve spring compressors, reamers, micrometers, dial indicators, calipers.

## Keystone Reamer & Tool Co.
Post Office Box 310
Millersburg, Pennsylvania 17061
Valve seat cutter and pilots, adjustable reamers.

## Ki-Sol Corporation
100 Larkin Williams Ind. Court
Fenton, Missouri 63026
Cylinder hone, ridge reamer, ring compressor, ring expander, ring groove cleaner, torque wrenches, valve spring compressor, valve refacing equipment.

## K-Line Industries, Inc.
315 Garden Avenue
Holland, Michigan 49423
Cylinder hone, ridge reamer, ring compressor, valve guide tools, valve spring compressor, reamers.

## K.O. Lee Company
101 South Congress
Aberdeen, South Dakota 57401
Valve refacing, valve seat grinding, valve reseating and valve guide reaming tools.

## Kwik-Way Mfg. Co.
500 57th Street
Marion, Iowa 52302
Cylinder boring equipment, valve facing equipment, valve seat grinding equipment.

## Lisle Corporation
807 East Main
Clarinda, Iowa 51632
Cylinder hones, ridge reamers, ring compressors, valve spring compressors.

## Microdot, Inc.
P.O. Box 3001
800 So. State College Blvd.
Fullerton, California 92631
Thread repair insert kits.

## Mighty Midget Mfg. Co.,
## Div. of Kansas City Screw Thread Co.
2908 E. Truman Road
Kansas City, Missouri 64127
Crankshaft straightener.

## Neway Manufacturing, Inc.
1013 No. Shiawassee
Corunna, Michigan 48817
Valve seat cutters.

## Owatonna Tool Company
436 Eisenhower Drive
Owatonna, Minnesota 55060
Valve tools, spark plug tools, piston ring tools, cylinder hones.

## Power Lawnmower Parts Inc.
1920 Lyell Avenue
P.O. Box 60860
Rochester, NY 14606-0860
Gasket cutter tool, gasket scraper tool, crankshaft cleaning tool, ridge reamer, valve spring compressor, valve seat cutters, thread repair kits, valve lifter, piston ring expander.

## Precision Manufacturing
## & Sales Co., Inc.
2140 Range Rd., P.O. Box 149
Clearwater, Florida 33517
Cylinder boring equipment, measuring instruments, valve equipment, hones, porting, hand tools, test equipment, threading, presses, parts washers, milling machines, lathes, drill presses, glass beading machines, dynos, safety equipment.

## Sunnen Product Company
7910 Manchester Avenue
Saint Louis, Missouri 63143
Cylinder hones, rod reconditioning, valve guide reconditioning.

## Rexnord Specialty Fastener Division
3000 W. Lomita Blvd.
Torrance, California 90505
Thread repair insert kits (Keenserts) and installation tools.

## Vulcan Tools
## Div. of TRW, Inc.
2300 Kenmore Avenue
Buffalo, New York 14207
Cylinder hones, reamers, ridge removers, valve spring compressors, valve spring testers, ring compressor, ring groove cleaner.

## Test Equipment and Gages

## Allen Test Products
2101 North Pitcher Street
Kalamazoo, Michigan 49007
Coil and condenser testers, compression gages.

## AW Dynamometer, Inc.
131-1/2 East Main Street
Colfax, Illinois 61728
Engine test dynamometer.

## B.C. Ames Company
131 Lexington
Waltham, Massachusetts 02254
Micrometer dial gages and indicators.

## The Bendix Corporation
## Engine Products Division
Delaware Avenue
Sidney, New York 13838
Condenser tester, magneto test equipment, timing light.

## Burco
208 Delaware Avenue
Delmar, New York 12054
Coil and condenser tester, compression gage, carburetor tester, grinders, Pow-R-Arms, chain breakers, rivet spinners.

## Dixson, Inc.
Post Office Box 1449
Grand Junction, Colorado 81501
Tachometer, compression gage, timing light.

## Foley-Belsaw Company
## Outdoor Power Equipment
## Parts Division
6301 Equitable Road
P.O. Box 419593
Kansas City, Missouri 64141
Cylinder gage, amp/volt testers, condenser and coil tester, magneto tester, ignition testers for conventional and solid-state systems, tachometers, spark testers, compression gages, timing lights and gages, micrometers and calipers, torque wrenches, carburetor testers, vacuum gages.

## Graham-Lee Electronics, Inc.
4200 Center Avenue NE
Minneapolis, Minnesota 55421
Coil and condenser tester.

## K-D Tools
3575 Hempland Road
Lancaster, Pennsylvania 17604
Diode tester and installation tools, compression gage, timing light, timing gages.

## Ki-Sol Corporation
100 Larkin Williams Ind. Court
Fenton, Missouri 63026
Micrometers, telescoping gages, compression gages, cylinder gages.

**K-Line Industries, Inc.**
315 Garden Avenue
Holland, Michigan 49423
Compression gage, tachometers.

**Merc-O-Tronic Instruments Corporation**
215 Branch Street
Almont, Michigan 48003
Ignition analyzers for conventional solid-state and magneto systems, electric tachometers, electronic tachometer and dwell meter, power timing lights, ohmmeters, compression gages, mechanical timing devices.

**Owatonna Tool Company**
436 Eisenhower Drive
Owatonna, Minnesota 55060
Feeler gages, hydraulic test gages.

**Power Lawnmower Parts Inc.**
1920 Lyell Avenue
P.O. Box 60860
Rochester, NY 14606-0860
Condenser and coil tester, compression gage, flywheel magneto tester, ignition tester.

**Prestolite Electronics Div.**
**An Allied Company**
Post Office Box 931
Toledo, Ohio 43694
Magneto test plug.

**Simpson Electric Company**
853 Dundee Avenue
Elgin, Illinois 60120
Electrical and electronic test equipment.

**L.S. Starrett Company**
1-165 Crescent Street
Athol, Massachusetts 01331
Micrometers, dial gages, bore gages, feeler gages.

**Stevens Instrument Company, Inc.**
Post Office Box 193
Waukegan, Illinois 60085
Ignition analyzers, timing lights, tachometers, volt-ohmmeters, spark checkers, CD ignition testers.

**Stewart-Warner Corporation**
1826 Diversey Parkway
Chicago, Illinois 60614
Compression gage, ignition tachometer, timing light ignition analyzer.

**P.A. Sturtevant Co.,**
**Division Dresser Ind.**
3201 North Wolf Road
Franklin Park, Illinois 60131
Torque wrenches, torque multipliers, torque analyzers.

**Sun Electric Corporation**
**Instrument Products Div.**
1560 Trimble Road
San Jose, California 95131
Compression gage, hydraulic test gages, coil and condenser tester, ignition tachometer, ignition analyzer.

**Westberg Mfg. Inc.**
3400 Westach Way
Sonoma, California 95476
Ignition tachometer, magneto test equipment, ignition system analyzer.

## Shop Tools and Equipment

**Applied Power, Inc.,**
**Auto. Division**
Box 27207
Milwaukee, Wisconsin 53227
Arc and gas welding equipment, welding rods, accessories, battery service equipment.

**Black & Decker Mfg. Co.**
701 East Joppa Road
Towson, Maryland 21204
Air and electric powered tools.

**Bloom, Inc.**
Route 4, Hiway 20 West
Independence, Iowa 50644
Lawn mower repair bench with built-in engine cranking mechanism.

**Campbell Chain Division**
**McGraw-Edison Co.**
Post Office Box 3056
York, Pennsylvania 17402
Chain type engine and utility slings.

**Champion Pneumatic Machinery Co.**
1301 No. Euclid Avenue
Princeton, Illinois 61356
Air compressors.

**Champion Spark Plug Co.**
Post Office Box 910
Toledo, Ohio 43661
Spark plug cleaning and testing equipment, gap tools and wrenches.

**Chicago Pneumatic Tool Co.**
2200 Bleecker Street
Utica, New York 13503
Air impact wrenches, air hammers, air drills and grinders, nut runners, speed ratchets.

**Clayton Manufacturing Company**
415 North Temple City Boulevard
El Monte, California 91731
Steam cleaning equipment.

**E-Z Lok**
P.O. Box 2069
240 E. Rosecrans Avenue
Gardena, California 90247
Thread repair insert kits for metal, wood and plastic.

**Foley-Belsaw Company**
**Outdoor Power Equipment**
**Parts Division**
6301 Equitable Road
P.O. Box 419593
Kansas City, Missouri 64141
Torque wrenches, parts washers, micrometers and calipers.

**G & H Products, Inc.**
Post Office Box 770
St. Paris, Ohio 43027
Motorized lawn mower stand.

**General Scientific Equipment Company**
Limekiln Pike & Williams Avenue
Box 27309
Philadelphia, Pennsylvania 19118-0255
Safety equipment.

**Graymills Corporation**
3705 North Lincoln Avenue
Chicago, Illinois 60613
Parts washing stand.

**Heli-Coil Products Div.,**
**Heli-Coil Corp.**
Shelter Rock Lane
Danbury, Connecticut 06010
Thread repair kits, thread inserts and installation tools.

**Ingersoll-Rand**
P. O. Box 1776
Liberty Corner, NJ 07938
Air and electric impact wrenches, electric drills and screwdrivers.

**Jaw Manufacturing Co.**
39 Mulberry Street, P.O. Box 213
Reading, Pennsylvania 19603
Files for renewal of damaged threads, hex nut size rethreader dies, flexible shaft drivers and extensions, screw extractors, impact drivers.

**Jenny Division
of Homestead Ind., Inc.**
Box 348
Carapolis, Pennsylvania 15108-0348
Steam cleaning equipment, pressure
washing equipment.

**Keystone Reamer & Tool Co.**
Post Office Box 310
Millersburg, Pennsylvania 17061
Adjustable reamers, twist drills, tape,
dies, etc.

**K-Line Industries, Inc.**
315 Garden Avenue
Holland, Michigan 49423
Air and electric impact wrenches.

**Microdot, Inc.**
P.O. Box 3001
800 So. State College Blvd.
Fullerton, California 92631
Thread repair insert kits.

**Owatonna Tool Company**
436 Eisenhower Drive
Owatonna, Minnesota 55060
Bearing and gear pullers, hydraulic shop
presses.

**Power Lawnmower Parts Inc.**
1920 Lyell Avenue
P.O. Box 60860
Rochester, NY 14606-0860
Flywheel puller, starter wrench and
flywheel holder, blade sharpener, chain
breakers, rivet spinners, gear pullers.

**Precision Mfg. & Sales Co.**
2140 Range Road
Clearwater, FL 34625
Boring bar, test equipment,
hand tools.

**Shure Manufacturing Corp.**
1601 South Hanley Road
Saint Louis, Missouri 63144
Steel shop benches, desks, engine
overhaul stand.

**Sioux Tools, Inc.**
2801-2999 Floyd Blvd.
Sioux City, Iowa 51102
Portable air and electric tools.

**Rexnord Specialty
Fastener Division**
3000 W. Lomita Blvd.
Torrance, California 90505
Thread repair insert kits (Keenserts) and
installation tools.

**Vulcan Tools**
2300 Kenmore Avenue
Buffalo, New York 14207
Air and electric impact wrenches.

## Sharpening and Maintenance Equipment for Small Engine Powered Implements

**Bell Industries,
Saw & Machine Division**
Post Office Box 2510
Eugene, Oregon 97402
Saw chain grinder.

**Desa Industries, Inc.**
25000 South Western Ave.
Park Forest, Illinois 60466
Saw chain breakers and rivet spinners,
chain files, file holder and filing guide.

**Foley-Belsaw Company
Outdoor Power Equipment
Parts Division**
6301 Equitable Road
P.O. Box 419593
Kansas City, Missouri 64141
Circular, band and hand saw filers,
heavy duty grinders, saw setters,
retoothers, circular saw vises, lawn
mower sharpener, saw chain sharpening
and repair equipment.

**Granberg Industries**
200 S. Garrard Blvd.
Richmond, California 94804
Saw chain grinder, file guides, chain
breakers and rivet spinners, chain saw
lumber cutting attachments.

**Ki-Sol Corporation**
100 Larkin Williams Ind. Court
Fenton, Missouri 63026
Mower blade balancer.

**Magna-Matic Div.,
A.J. Karrels Co.**
Box 348
Port Washington, Wisconsin 53074
Rotary mower blade balancer, "track"
checking tool.

**Omark Industries, Inc.**
4909 International Way
Portland, Oregon 97222
Saw chain and saw bars, maintenance
equipment, to include file holders, filing vises, rivet spinners, chain breakers,
filing depth gages, bar groove gages,
electric chain saw sharpeners.

**Power Lawnmower Parts Inc.**
1920 Lyell Avenue
P.O. Box 60860
Rochester, NY 14606-0860
Grinding wheels, parts cleaning brush,
tube repair kits, blade balancer.

**S.I.P. Grinding Machines
American Marsh Pumps**
P.O. Box 23038
722 Porter Street
Lansing, Michigan 48909

Lawn mower sharpeners, saw chain
grinders, lapping machine.

**Specialty Motors Mfg.**
641 California Way,
P.O. Box 157
Longview, Washington 98632
Chain saw bar rebuilding equipment.

## Mechanic's Hand Tools

**Channellock, Inc.**
1306 South Main Street
Meadville, Pennsylvania 16335

**John H. Graham & Company, Inc.**
617 Oradell Avenue
Oradell, New Jersey 07649

**Jaw Manufacturing Company**
39 Mulberry Street
Reading, Pennsylvania 19603

**K-D Tools**
3575 Hempland Road
Lancaster, Pennsylvania 17604

**K-Line Industries, Inc.**
315 Garden Avenue
Holland, Michigan 49423

**New Britain Tool Company
Division of Litton Industrial
Products**
P.O. Box 12198
Research Triangle Park, N.C. 27709

**Owatonna Tool Company**
436 Eisenhower Drive
Owatonna, Minnesota 55060

**Power Lawnmower Parts Inc.**
1920 Lyell Avenue
P.O. Box 60860
Rochester, NY 14606-0860
Crankshaft wrench, spark plug wrench,
hose clamp pliers.

**Snap-On Tools**
2801 80th Street
Kenosha, Wisconsin 53140

**Triangle Corporation—Tool
Division**
Cameron Road
Orangeburg, South Carolina 29115

**Shop Supplies (Chemicals, Metallurgy Products, Seals, Sealers, Common Parts Items, etc.)**

**ABEX Corp.**
**Amsco Welding Products**
Fulton Industrial Park
3-9610-14
P.O. Box 258
Wauseon, Ohio 43567
Hardfacing alloys.

**Atlas Tool & Manufacturing Co.**
7100 S. Grand Avenue
Saint Louis, Missouri 63111
Rotary mower blades.

**Bendix Automotive Aftermarket**
1094 Bendix Drive, Box 1632
Jackson, Tennessee 38301
Cleaning chemicals.

**CR Industries**
900 North State Street
Elgin, Illinois 60120
Shaft and bearing seals.

**Clayton Manufacturing Company**
415 North Temple City Blvd.
El Monte, California 91731
Steam cleaning compounds and solvents.

**E-Z Lok**
P.O. Box 2069,
240 E. Rosecrans Ave.
Gardena, California 90247
Thread repair insert kits for metal, wood and plastic.

**Eutectic Welding Alloys Corp.**
40-40 172nd Street
Flushing, New York 11358
Specialized repair and maintenance welding alloys.

**Foley-Belsaw Company**
**Outdoor Power Equipment**
**Parts Division**
6301 Equitable Road
P.O. Box 419593
Kansas City, Missouri 64141
Parts washers, cylinder head rethreaders, micrometers, calipers, gasket material, nylon rope, universal lawnmower blades, oil and grease products, bolt, nut, washer and spring assortments.

**Frederick Manufacturing Co., Inc.**
1400 C Agnes Street
Kansas City, Missouri 64127
Throttle controls and parts.

**Heli-Coil Products Division**
**Heli-Coil Corporation**

Shelter Rock Lane
Danbury, Connecticut 06810
Thread repair kits, thread inserts and installation tools.

**King Cotton Cordage**
617 Oradell Avenue
Oradell, New Jersey 07649
Starter rope, nylon rod.

**Loctite Corporation**
705 North Mountain Road
Newington, Connecticut 06111
Threading locking compounds, bearing mounting compounds, retaining compounds and sealants, instant and structural adhesives.

**McCord Gasket Division**
**Ex-Cell-O Corporation**
2850 West Grand Boulevard
Detroit, Michigan 48202
Gaskets, seals.

**Microdot, Inc.**
P.O. Box 3001
800 So. State College Blvd.
Fullerton, California 92631
Thread repair insert kits.

**Permatex Industrial**
705 N. Mountain Road
Newington, Connecticut 06111
Cleaning chemicals, gasket sealers, pipe sealants, adhesives, lubricants.

**Power Lawnmower Parts Inc.**
1920 Lyell Avenue
P.O. Box 60860
Rochester, NY 14606-0860
Grinding compound paste, gas tank sealer stick, shop aprons, rotary mower blades, oil seals, gaskets, solder, throttle controls and parts, nylon starter rope.

**Radiator Specialty Co.**
Box 34689
Charlotte, North Carolina 28234
Cleaning chemicals (Gunk), and solder seal.

**Rexnord Specialty**
**Fastener Div.**
3000 W. Lomita Blvd.
Torrance, California 90506
Thread repair insert kits (Keenserts) and installation tools.

**Union Carbide Corporation**
**Home & Auto Products Division**
Old Ridgebury Road
Danbury, CT 06817
Cleaning chemicals, engine starting fluid.

# ACME

**ACME NORTH AMERICA CORP.**
**5209 W. 73rd Street**
**Minneapolis, MN 55435**

| Model | No. Cyls. | Bore | Stroke | Displacement |
|---|---|---|---|---|
| ALN 290W | 1 | 75 mm | 65 mm | 290 cc |
| | | (2.95 in.) | (2.56 in.) | (17.5 cu. in.) |
| ALN 330W | 1 | 80 mm | 65 mm | 327 cc |
| | | (3.15 in.) | (2.56 in.) | (20.0 cu. in.) |

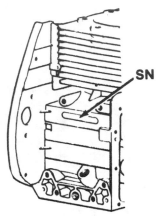

Fig. AC50—Engine serial number (SN) is stamped on side of the crankcase.

All models are four-stroke, air-cooled, horizontal crankshaft, single cylinder engines. Engine model number is on a plate mounted on cooling shroud on the right side of engine as viewed from flywheel side. The engine serial number (SN—Fig. AC50) is stamped on the engine block approximately 75 mm (3 inches) below model number plate.

## MAINTENANCE

**SPARK PLUG.** Recommended spark plug is Champion L88, AC C47 or equivalent. Spark plug should be removed, cleaned and inspected at 100 hour intervals. Spark plug electrode gap should be 0.6-0.8 mm (0.024-0.032 inch).

**AIR FILTER.** An oil bath type air filter is standard equipment on all engines. Air filter element should be removed and cleaned and old oil discarded at 50 hour intervals, or more frequently if operating in extremely dirty conditions. Replenish with new engine oil to level indicated on oil reservoir housing.

**CARBURETOR.** All models are equipped with a float type carburetor. Carburetor shown in Fig. AC51 is used on early production engines (Model ALN 290W prior to serial number 373729 and Model ALN 330W prior to serial number 368921). Carburetor shown in Fig. AC52 is used on late production engines.

Initial adjustment of idle mixture screw (3) from a lightly seated position is 1-1/4 turns open. Main fuel mixture is controlled by a fixed main jet (13).

Final adjustments should be made with engine at operating temperature and running. Adjust engine idle speed to 1100-1200 rpm by turning throttle stop screw (5). Adjust idle mixture screw (3) to obtain smoothest engine idle and smooth acceleration.

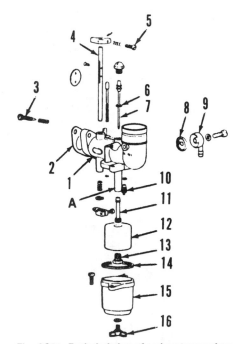

Fig. AC51—Exploded view of carburetor used on early production engines.

1. Carburetor body
2. Gasket
3. Idle mixture screw
4. Throttle shaft
5. Throttle stop screw
6. Gasket
7. Idle jet
8. Screen (filter)
9. Filter housing
10. Fuel inlet needle valve
11. Nozzle
12. Float
13. Main jet
14. Gasket
15. Float bowl
16. Wing bolt

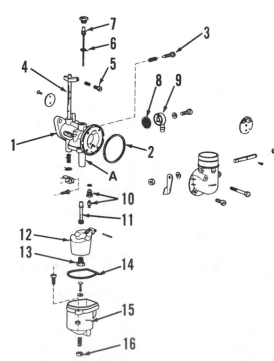

Fig. AC52—Exploded view of carburetor used on late production engines.

1. Carburetor body
2. Gasket
3. Idle mixture screw
4. Throttle shaft
5. Throttle stop screw
6. Gasket
7. Idle jet
8. Filter screen
9. Fuel inlet housing
10. Fuel inlet needle & seat
11. Nozzle
12. Float
13. Main jet
14. Gasket
15. Float bowl
16. Nut

Illustrations courtesy of Acme North America Corp.

When installing new fuel inlet needle and seat (10), install seat and measure from point (A) to edge of fuel inlet needle seat. Measurement should be 34.8-35.2 mm (1.370-1.386 inches) for early model carburetor and 36.8-37.2 mm (1.449-1.464 inches) for late model carburetor. Vary number of fiber washers between fuel inlet seat and carburetor body to obtain correct measurement.

To check float level, carefully remove float bowl and measure distance from top of bowl to top of fuel level in bowl (Fig. AC53). Measurement should be 29-31 mm (1.142-1.220 inches) for early model carburetor and 25-27 mm (0.985-1.063 inches) for late model carburetor. If float level is incorrect, fuel inlet needle seat must be shimmed as outlined in previous paragraph.

**GOVERNOR.** All engines are equipped with a flyball type centrifugal governor. Governor assembly is mounted on camshaft gear. The governor regulates engine speed via external linkage.

To adjust external linkage, first make certain all linkage moves freely with no binding or loose connections. Lock throttle control lever (5—Fig. AC54) in mid-position with throttle locknut (6). Be sure that tension spring (3) is hooked in hole "B" for 3600 rpm engine speed. Place throttle lever (5) in full speed position, loosen carburetor-to-governor lever adjustment lock (7) and push throttle fully open. Tighten rod adjustment lock (7) in this position. Governor and throttle linkage should work freely, and easily return to normal set position.

**IGNITION SYSTEM.** Early production engines are equipped with breaker-point ignition system. Late production engines (starting at serial number A/425001) are equipped with electronic ignition system.

On breaker-point ignition system, the point set and condenser are located on left side of crankcase as viewed from flywheel side and ignition coil is located behind the flywheel. Breaker points are actuated by a plunger (1—Fig. AC55) which is driven by a cam on the engine camshaft.

Ignition timing should be set at 21 degrees before top dead center (piston on compression stroke) by aligning the "1A" mark on flywheel with "PMS" mark on crankcase. Points should just begin to open at this position. To adjust point opening, shift breaker-point plate (3) as necessary. Rotate flywheel until points are fully open and measure point gap using a feeler gage. Maximum point gap should not exceed 0.5 mm (0.020 inch).

Ignition coil, which is located behind the flywheel, should have 0.6-0.8 mm (0.024-0.031 inch) clearance between coil poles and flywheel magnets.

Late production engines are equipped with electronic ignition system, which has no moving parts except the flywheel. No adjustment is required on this ignition system. Correct engine timing will be obtained when ignition coil unit is installed so that air gap between coil laminations and outer surface of flywheel is 0.4-0.5 mm (0.016-0.020 inch).

**LUBRICATION.** The engine is splash lubricated by an oil slinger mounted on bottom of the connecting rod cap. Engine oil level should be checked prior to each operating interval.

Engine oil should be changed after every 50 hours of operation. Crankcase oil capacity is approximately 0.75 L (0.8 qt.).

Manufacturer recommends using oil with API service classification of SE or SF. Use SAE 10W oil when ambient temperature is -10° C (14° F) or below. Use SAE 20W/20 oil when temperature is between -10° C (14° F) and 10° C (50° F). Use SAE 30 oil when temperature is between 10° C (50° F) and 30° C (86° F). Use SAE 40 oil when temperature is above 30° C (86° F).

**VALVE ADJUSTMENT.** Valve clearance should be checked and adjusted, if necessary, at 200 hour intervals. To check clearance, remove valve tappet chamber cover and breather assembly and turn crankshaft to position piston at TDC on compression stroke. Measure clearance between valve stem end and tappet using a feeler gage. Specified clearance is 0.010-0.15 mm (0.004-0.006 inch) for intake and exhaust.

If clearance is not as specified, remove or install shims (7 and 8—Fig. AC56) in shim holder (9) as necessary.

**CRANKCASE BREATHER.** Crankcase breather must be removed and cleaned at 50 hour intervals. Breather is located on tube mounted on valve chamber cover. Rubber valve must be installed as shown in Fig. AC57 to ensure crankcase vacuum.

**GENERAL MAINTENANCE.** Check and tighten all loose bolts, nuts or clamps daily. Check for fuel or oil leakage and repair as necessary.

Clean dust, dirt, grease or any foreign material from cylinder head and cylinder block cooling fins at 100 hour intervals, or more often in extremely dirty operating conditions. Inspect fins for damage and repair as necessary.

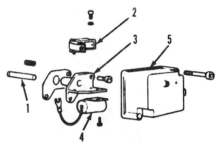

**Fig. AC55—Ignition breaker points and condenser are located on left side crankcase behind cover (5).**

1. Plunger
2. Breaker points
3. Point plate
4. Condenser
5. Cover

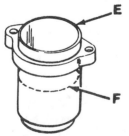

**Fig. AC53—Measure distance from fuel level (F) to top of bowl (E). Distance should be 32-34 mm (1.26-1.34 in.). Fuel inlet valve seat must be shimmed if fuel level is incorrect. Refer to text.**

*Fig. AC54—View showing relative position of governor external linkage and throttle parts. Spring (4) should be installed in hole "B" of governor lever for 3600 rpm engine speed.*

1. Governor lever-to-carburetor rod
2. Governor lever
3. Tension spring
4. Spring
5. Throttle control lever
6. Throttle lever locknut
7. Set screw

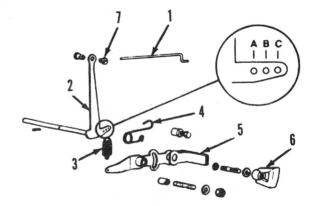

## REPAIRS

**TIGHTENING TORQUES.** Recommended tightening torque specifications are as follows:

Cylinder head . . . . . . . . . . . . . .30 N·m
(22 ft.-lbs.)
Connecting rod . . . . . . . . . . . .17 N·m
(13 ft.-lbs.)
Crankcase cover . . . . . . . . . . .30 N·m
(22 ft.-lbs.)
Flywheel . . . . . . . . . . . . . . . .157 N·m
(116 ft.-lbs.)

**CYLINDER HEAD.** To remove cylinder head, remove cooling shrouds. Loosen head bolts evenly and remove cylinder head and gasket.

**CAUTION: Cylinder head bolts should not be loosened while engine is hot as cylinder head warpage may result.**

Clean carbon and other deposits from cylinder head. Check flatness of cylinder head mounting surface using straightedge and feeler gage. Cylinder head should be resurfaced or renewed if warpage exceeds 0.5 mm (0.020 inch).

Reinstall cylinder head using a new head gasket. Do not use sealer or gasket cement on head gasket. Note that the two longest head bolts should be installed on exhaust side of engine (positions 6 and 3—Fig. AC58). Tighten head bolts gradually to specified torque following sequence shown in Fig. AC58.

**CONNECTING ROD.** To remove connecting rod, remove cylinder head and crankcase cover. Use care when crankcase cover is removed as governor flyweight balls will fall out of ramps in camshaft gear. Remove carbon or ring ridge, if present, from top of cylinder prior to removing connecting rod and piston. Remove connecting rod cap and push connecting rod and piston assembly out of cylinder head end of cylinder block. Remove piston pin retaining rings and separate piston from connecting rod.

Aluminum alloy connecting rod rides directly on crankpin journal. Clearance between crankpin and connecting rod big end bore should be 0.040-0.064 mm (0.0016-0.0025 inch). Connecting rod side play on crankpin should be 0.15-0.25 mm (0.006-0.010 inch). Renew connecting rod if bore is excessively worn or out-of-round more than 0.10 mm (0.004 inch).

Clearance between piston pin and connecting rod pin bore should be 0.016-0.039 mm (0.0006-0.0015 inch).

Piston can be installed on connecting rod either way, however, when installing connecting rod in cylinder block, triangle match marks on connecting rod and cap must be aligned and facing crankcase cover side of engine. Lubricate piston, cylinder and connecting rod bearing bore with engine oil prior to installation. Tighten connecting rod bolts to specified torque. Use grease to retain governor flyweight balls in camshaft ramps to aid crankcase cover installation. Tighten crankcase cover to specified torque.

**PISTON, PIN AND RINGS.** Refer to CONNECTING ROD paragraphs for piston removal and installation procedure.

Inspect piston for wear, deep scratches or scoring and renew as necessary.

Compression ring end gap should be 0.30-0.50 mm (0.012-0.020 inch). Oil control ring end gap should be 0.25-0.50 mm (0.010-0.020 inch).

Piston pin should be 0.005 mm (0.0002 inch) interference to 0.005 mm (0.0002 inch) loose fit in piston pin bore. It may be necessary to heat piston slightly to aid in pin removal and installation.

When installing rings on piston, be sure that notched edge of second compression ring is toward bottom of piston. Stagger ring end gaps at 120 degree intervals around diameter of piston. Lubricate piston, rings and cylinder with engine oil prior to reassembly.

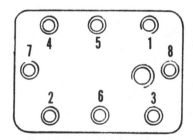

Fig. AC58—Tighten cylinder head bolts following the sequence shown.

Fig. AC56—View of valve system component parts. Shims (7 and 8) are used to adjust valve clearance. Refer to text.

1. Valve
2. Guide
3. Seal
4. Seat
5. Spring
6. Retainer
7. Shim
8. Shim
9. Shim holder
10. Tappet

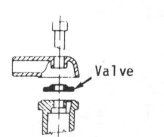

Fig. AC57—Breather valve must be installed as shown to ensure crankcase vacuum.

**Fig. AC59—Exploded view of crankshaft, connecting rod and piston assembly.**
1. Retaining ring
2. Piston pin
3. Compression rings
4. Oil control ring
5. Piston
6. Connecting rod
7. Connecting rod cap & oil slinger
8. Lock plate
9. Bolts
10. Seal
11. Main bearing
12. Key
13. Crankshaft
14. Crankshaft gear key
15. Crankshaft gear
16. Main bearing
17. Seal

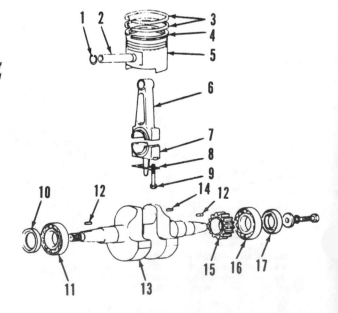

**CYLINDER AND CRANKCASE.** Cylinder and crankcase are an integral casting of aluminum alloy with a high density perlite cylinder sleeve cast as an integral part of the cylinder block.

Standard cylinder bore diameter is 75.000-75.013 mm (2.9528-2.9533 inches) for Model ALN 290W and 80.000-80.013 mm (3.1496-3.1501 inches) for Model ALN 330W. If cylinder bore is 0.06 mm (0.0024 inch) or more out-of-round or tapered, cylinder should be bored to nearest oversize for which piston and rings are available.

Crankshaft ball type main bearings should be a slight press fit in crankcase and crankcase cover bores. Renew bearings if they are loose, rough or damaged.

An oil slinger trough (2—Fig. AC60) is installed in the crankcase. If trough is removed, it must be securely repositioned prior to engine reassembly.

**CRANKSHAFT.** The crankshaft is supported at each end by ball bearing type main bearings (11 and 16—Fig. AC59). Crankshaft timing gear (15) is a press fit on crankshaft (13).

To remove crankshaft, first remove cylinder head, piston and connecting rod assembly, camshaft assembly and flywheel. Withdraw crankshaft from cylinder block.

Standard clearance between crankpin journal and connecting rod crankpin bore is 0.040-0.064 mm (0.0016-0.0025 inch). Standard crankpin journal diameter is 29.985-30.000 mm (1.1805-1.1811 inch). If crankpin journal is worn or out-of-round 0.05 mm (0.002 inch) or more, regrind crankpin journal or renew crankshaft.

Standard main bearing journal diameter is 30.002-30.015 mm (1.1812-1.1817 inches). Main journals cannot be reground. Ball type main bearings should be a slight press fit on crankshaft journal. Renew bearings if they are rough, loose or damaged. Note that crankshaft should be supported on counterweights when pressing bearings or crankshaft timing gear onto or off of crankshaft to prevent bending crankshaft.

When installing crankshaft, make certain that crankshaft and camshaft gear timing marks are aligned.

**CAMSHAFT.** Camshaft and camshaft gear are an integral casting which rides directly in bearing bores in crankcase and crankcase cover. A compression release mechanism is mounted on back side of camshaft gear and governor flyballs are located in ramps machined into face of gear.

Spring-loaded compression release mechanism should snap back against camshaft when weighted lever is pulled against spring tension and released.

Inspect camshaft journals and lobes and renew camshaft if worn, scored or damaged. Camshaft journal standard diameter is 15.973-15.984 mm (0.6289-0.6293 inch). Exhaust cam lobe height should be 26.540-26.570 mm (1.045-1.046 inch) and intake cam lobe height should be 19.975-20.025 mm (0.786-0.788 inch). Camshaft and tappets should be renewed as a set.

When installing camshaft, retain governor flyballs in ramps with grease and make certain that camshaft and crankshaft gear timing marks are aligned.

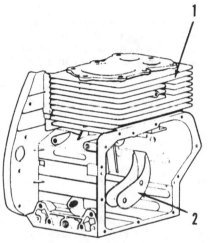

*Fig. AC60—View of cylinder block (1) and oil slinger trough (2). If trough is removed, it must be securely repositioned prior to assembly of engine.*

**VALVE SYSTEM.** Refer to VALVE ADJUSTMENT paragraph in MAINTENANCE section for valve clearance adjustment procedure.

Valve face and seat angles are 45 degrees for intake and exhaust. Standard valve seat contact width is 1.2-1.3 mm (0.047-0.051 inch). If valve head margin is 0.5 mm (0.020 inch) or less after grinding valve face, renew the valve.

Standard valve stem diameter is 6.965-6.987 mm (0.2742-0.2751 inch) for intake and exhaust.

Valve guide inside diameter for cast iron intake valve guide or bronze exhaust valve guide should be 7.015-7.025 mm (0.2762-0.2766 inch). Guides are renewable using suitable removal and installation tools.

Standard valve spring free length is 35 mm (1.380 inch). If valve spring free length is 32 mm (1.260 inch) or less, renew valve spring.

# ACME

| Model | No. Cyls. | Bore | Stroke | Displacement |
|---|---|---|---|---|
| AT 330 | 1 | 80 mm | 65 mm | 327 cc |
| | | (3.15 in.) | (2.56 in.) | (19.95 cu. in.) |
| VT 88W | 1 | 88 mm | 79 mm | 480 cc |
| | | (3.46 in.) | (3.11 in.) | (29.28 cu. in.) |
| VT 94W | 1 | 94 mm | 79 mm | 548 cc |
| | | (3.70 in.) | (3.11 in.) | (33.43 cu. in.) |

All models are four-stroke, air-cooled, single cylinder, overhead valve engines. Cylinder and crankcase are cast as a single unit for Model AT 330. Cylinder and crankcase are separate assemblies for Models VT 88 and VT 94.

Engine model number for Model AT 330 is on a plate mounted on cooling shroud on the right side of engine as viewed from flywheel side. The engine serial number (SN—Fig. AC100) is stamped on the engine block about 76 mm (3 inches) below model number plate.

Engine model and serial numbers for Models VT 88 and VT 94 are stamped on crankcase just above oil filler cap and cylinder-to-crankcase mating surface (Fig. AC101).

## MAINTENANCE

**SPARK PLUG.** Recommended spark plug is Champion L86, Bosch W95T1 or equivalent.

Spark plug should be removed, cleaned and inspected at 100 hour intervals. Spark plug electrode gap for all models should be 0.6-0.8 mm (0.024-0.032 inch).

**CARBURETOR.** All models are equipped with a float type carburetor. Refer to Fig. AC102 for an exploded view of carburetor used on Model AT 330 and to Fig. AC103 for exploded view of carburetor used on Models VT 88 and VT 94.

On Model AT 330, initial adjustment of idle mixture screw (3—Fig. AC102) is 1-1/2 turns out from a lightly seated position. Main fuel mixture is controlled by a fixed main jet (23). Final adjustments should be made with engine running at normal operating temperature. Adjust engine idle speed screw (5) to obtain desired low idle speed. Adjust idle mixture screw (3) to obtain smoothest engine idle and smooth acceleration.

To check float level on Model AT 330, invert carburetor body (1) with fuel inlet valve (14 and 15) and float (18) installed. Float should be parallel with carburetor body gasket surface.

On Models VT 88 and VT 94, initial adjustment of idle mixture screw (3—Fig. AC103) is 1-1/4 turns out from a lightly seated position. Main fuel mixture is controlled by a fixed main jet (13).

Final adjustments should be made with engine running at normal operating temperature. Adjust throttle stop screw (5) to obtain desired low idle speed. Adjust idle mixture screw (3) to obtain smoothest engine idle and smooth acceleration.

When installing new fuel inlet needle and seat, install seat and measure from point (A) to edge of fuel inlet needle seat. Measurement should be 33-37 mm (1.30-1.46 inch). If necessary, vary number of fiber washers between fuel inlet seat and carburetor body to obtain correct measurement.

To check float level, carefully remove float bowl (15) and measure distance from top of float bowl to top of fuel level in the bowl (Fig. AC104). Measurement should be 32-34 mm (1.26-1.34 inch). If float level is incorrect, fuel inlet needle seat must be shimmed as outlined in previous paragraph.

Fuel filter screen (8—Fig. AC103) should be removed and cleaned at 50 hour intervals.

**GOVERNOR.** All models are equipped with a flyball type centrifugal governor.

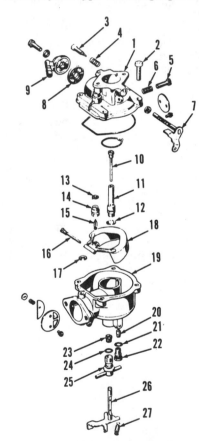

**Fig. AC102—Exploded view of float type carburetor used on Model AT 330.**

1. Carburetor body
2. Screw
3. Idle mixture screw
4. Spring
5. Throttle stop screw
6. Spring
7. Throttle shaft
8. Screen (filter)
9. Filter housing
10. Nozzle
11. Nozzle holder
12. Gasket
13. Gasket
14. Seat
15. Fuel inlet needle
16. Float pin
17. Retainer
18. Float
19. Carburetor body/float bowl
20. Idle jet
21. Gasket
22. Idle jet holder
23. Main jet
24. Gasket
25. Main jet holder
26. Choke shaft
27. Spring

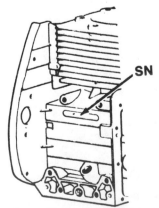

**Fig. AC100—On Model AT 330, engine serial number (SN) is stamped on the crankcase.**

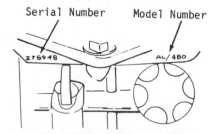

Serial Number　　Model Number

276948　　　AL/480

**Fig. AC101—On Models VT 88 and VT 94, model and serial numbers are stamped on crankcase as shown.**

On Model AT 330, governor assembly is located on the camshaft gear (Fig. AC105). On Models VT 88 and VT 94, the governor assembly is incorporated on the magneto gear and drive shaft (Fig. AC106). On all models, governor lever should be adjusted to hold carburetor throttle plate in wide open position when engine is not running.

**IGNITION SYSTEM.** Model AT 330 is equipped with an electronic ignition system (Fig. AC107). Models VT 88 and VT 94 are equipped with a self-contained magneto unit (Fig. AC 108).

The electronic ignition unit used on Model AT 330 requires no regular maintenance. Air gap between ignition unit and flywheel should be 0.45-0.50 mm (0.018-0.020 inch).

The magneto unit on Models VT 88 and VT 94 is equipped with a breaker point set, condenser and ignition coil which are located within the magneto unit. Point gap should be checked and adjusted at 100 hour intervals. To check point gap, remove cover (3—Fig. AC108) and turn crankshaft until points are fully open. Measure gap using a suitable feeler gage. Specified point gap is 0.4-0.5 mm (0.016-0.020 inch).

Magneto should be installed so the points just begin to open when flywheel is at the 30 degrees BTDC mark. This mark should be located on the flywheel 62.8 mm (2.472 inch) before top dead center position.

**LUBRICATION.** On all models, check engine oil level prior to each operating interval. Maintain oil level at lower edge of fill plug opening. Change oil at 50 hour intervals.

Manufacturer recommends oil with an API service classification of SE or SF. Use SAE 40 oil for ambient temperatures above 30° C (86° F); SAE 30 oil for temperatures between 10° C (50° F) and 30° C (86° F); SAE 20W-20 oil for temperatures between 10° C (50° F) and −10° C (14° F) and SAE 10W oil for temperatures below −10° C (14° F).

Crankcase oil capacity is approximately 0.75 L (0.8 qt.) for Model AT 330, 1.1 L (1.16 qt.) for Model VT 88 and 1.3 L (1.4 qt.) for Model VT 94.

**VALVE ADJUSTMENT.** Valve clearance should be checked and adjusted, if necessary, at 200 hour intervals. To check clearance, remove rocker arm cover and position piston at top dead center of compression stroke. Valve stem to rocker clearance for Model AT 330 should be 0.10 mm (0.004 inch) for intake valve and 0.15 mm (0.006 inch) for exhaust valve. Valve stem to rocker clearance for Models VT 88 and VT 94 should be 0.05 mm (0.002 inch) for in-

take valve and 0.10 mm (0.004 inch) for exhaust valve.

To adjust clearance, loosen jam nut (3—Fig. AC105) and turn adjusting screw

(4) in rocker arm (5) to obtain correct clearance. When correct clearance is obtained, hold adjustment screw in position while jam nut is tightened.

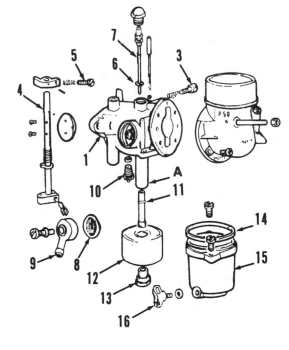

**Fig. AC103—Exploded view of float type carburetor used on Models VT 88 and VT 94.**

1. Carburetor body
3. Idle mixture screw
4. Throttle shaft
5. Throttle stop screw
6. Gasket
7. Idle jet
8. Screen (filter)
9. Filter housing
10. Fuel inlet needle valve
11. Nozzle
12. Float
13. Main jet
14. Gasket
15. Float bowl
16. Wing bolt

**Fig. AC104—To check float level, measure distance from fuel level (F) to top of bowl (E). Distance should be 32-34 mm (1.26-1.34 in.).**

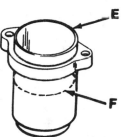

**Fig. AC105—Exploded view of Model AT 330 valve train assembly and governor components. Models VT 88 and VT 94 are similar except for the governor.**

1. Rocker arm shaft
2. Rocker arm shaft support
3. Jam nut
4. Adjustment bolt
5. Rocker arm
6. Retaining ring
7. Spring seat
8. Spring
9. Retainer
10. Keeper
11. Valve guide
12. Valve
13. Governor fork
14. Governor collar
15. Flyball
16. Camshaft & gear assy.
17. Push rod
18. Valve lifter (tappet)
19. Spring
20. Governor-to-carburetor link
21. Governor lever
22. Clamp nut
23. Spring
24. Pin
25. Governor shaft
26. Clamp bolt
27. Tension spring

**CRANKCASE BREATHER.** The crankcase breather should be removed and cleaned at 50 hour intervals. On Model AT 330, the breather is part of the valve tappet cover (Fig. AC109). On Models VT 88 and VT 94, the breather is located on the rocker arm cover (Fig. AC110). The flat face of diaphragm (3) must be installed toward rocker arm cover (4).

**GENERAL MAINTENANCE.** Check and tighten all loose bolts, nuts or clamps daily. Check for fuel or oil leakage and repair as necessary. Clean dust, dirt, grease or any foreign material from cylinder block and cylinder head cooling fins at 100 hour intervals, or more often when operating in extremely dirty conditions. Inspect fins for damage and repair as necessary.

## REPAIRS

**TIGHTENING TORQUES.** Recommended tightening torque specifications are as follows:

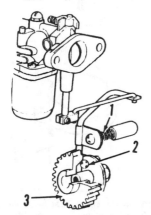

Fig. AC106—View of governor control linkage used on Models VT 88 and VT 94.

1. Spring
2. Flyball
3. Magneto drive gear

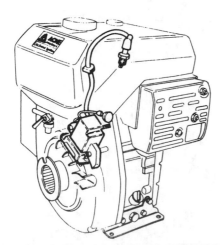

Fig. AC107—View showing location of Model AT 330 electronic ignition component parts.

Cylinder head:
AT 330 . . . . . . . . . . . . . . . .30 N·m
(22 ft.-lbs.)
VT 88, VT 94 . . . . . . . . . . .39 N·m
(29 ft.-lbs.)
Cylinder-to-crankcase:
VT 88, VT 94 . . . . . . . . . . .22 N·m
(16 ft.-lbs.)
Connecting rod:
AT 330 . . . . . . . . . . . . . . . .19 N·m
(14 ft.-lbs.)

Fig. AC108—Exploded view of magneto ignition unit used on Models VT 88 and VT 94.

1. Cap
2. Coil
3. Cover
4. Seal
5. Breaker points
6. Condenser
7. Housing

Fig. AC109—View of crankcase breather assembly (B) used on Model AT 330.

Fig. AC110—View of breather assembly used on Models VT 88 and VT 94.

1. Nut
2. Cap
3. Diaphragm
4. Rocker arm cover

Rocker arm cover . . . . . . . . . . .19 N·m
(14 ft.-lbs.)
Valve adjustment screw
jam nut . . . . . . . . . . . . . . . . .18 N·m
(13 ft.-lbs.)
Flywheel:
AT 330 . . . . . . . . . . . . . . .157 N·m
(116 ft.-lbs.)
VT 88, VT 94 . . . . . . . . . . .176 N·m
(130 ft.-lbs.)
Spark plug . . . . . . . . . . . . . . . .19 N·m
(14 ft.-lbs.)

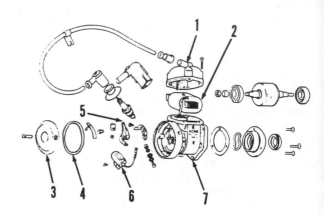

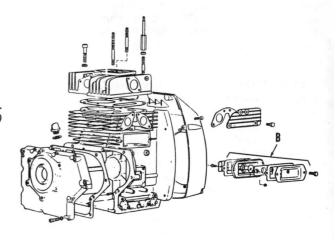

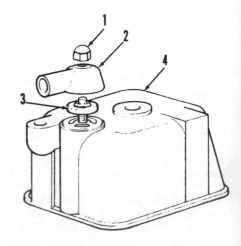

## CYLINDER HEAD AND VALVES.

The cylinder head on all models is an aluminum alloy casting with hardened steel valve seat inserts. To remove the cylinder head, first remove cooling shrouds and rocker arm cover. Loosen and remove cylinder head bolts in reverse order of sequence shown in Fig. AC112 for Model AT 330 or Fig. AC113 for Models VT 88 and VT 94. Do not loosen cylinder head bolts while engine is hot as cylinder head warpage may result.

Check cylinder head for distortion by placing head on a flat surface and using a feeler gage to determine warpage. If cylinder head is warped more than 0.5 mm (0.020 inch), resurface or renew head.

Install a new head gasket when installing cylinder head. Tighten cylinder head bolts evenly to specified torque following sequence shown in Fig. AC112 for Model AT 330 or Fig. AC113 for Models VT 88 and VT 94.

On all models, valve face and seat should be ground at 45 degree angle for intake and exhaust. Valve seat contact width should be 1.2-1.3 mm (0.047-0.051 inch).

Standard valve stem diameter on Model AT 330 is 6.955-6.970 mm (0.2738-0.2744 inch) for the intake and exhaust valves. Standard valve stem diameter on Models VT 88 and VT 94 is 8.965-8.987 mm (0.3530-0.3538 inch) for intake and exhaust valves.

The intake valve guide is cast iron and the exhaust valve guide is bronze. Valve guides may be renewed using suitable removal and installation tools. Standard valve guide inside diameter for Model AT 330 is 7.015-7.025 mm (0.2762-0.2766 inch). Standard valve guide inside diameter for Models VT 88 and VT 94 is 9.025-9.035 mm (0.3553-0.3557 inch).

Valve spring free length on Model AT 330 should be 37.5 mm (1.48 inch). If spring free length is 35 mm (1.38 inch) or less, renew spring.

Valve spring free length on Models VT 88 and VT 94 should be 54 mm (2.126 inches). It should require 24 Kg (53 lbs.) to compress valve spring to 40 mm (1.575 inches).

Maximum allowable clearance between rocker arm and rocker arm shaft is 0.15 mm (0.006 inch) on all models. If clearance exceeds specified dimension, renew rocker arm and/or rocker arm shaft.

### Model AT 330

**CONNECTING ROD.** On Model AT 330, the aluminum alloy connecting rod (6—Fig. AC114) rides directly on the crankpin journal. To remove the connecting rod and piston assembly, first remove the cylinder head and crankcase cover. Use care when crankcase cover is removed as governor flyweight balls will fall out of ramps in camshaft gear. Remove carbon and ring ridge, if present, from top of cylinder prior to removing piston. Remove connecting rod cap and push connecting rod and piston assembly out of cylinder head end of block. Remove piston pin retaining rings (1) and separate piston from connecting rod.

Clearance between the piston pin and connecting rod pin bore should be 0.016-0.039 mm (0.0006-0.0015 inch).

Connecting rod side play on crankpin journal should be 0.15-0.25 mm (0.006-0.010 inch). Connecting rod should be renewed if big end bore is worn or out-of-round more than 0.10 mm (0.004 inch). Specified clearance between connecting rod big end bore and crankshaft crankpin is 0.040-0.064 mm (0.0016-0.0025 inch).

Piston may be installed on connecting rod either way, however, when installing connecting rod and piston assembly in engine, triangle match marks on connecting rod and cap must be aligned and facing crankcase cover side of engine. Tighten connecting rod bolts evenly to specified torque. Use grease to retain governor flyweight balls in camshaft ramps or position crankcase in a suitable stand to aid crankcase cover installation. Tighten crankcase cover bolts to specified torque.

### Models VT 88, VT 94

On Models VT 88 and VT 94, the one-piece connecting rod (8—Fig. AC115) is installed on the crankpin journal (10) which is then pressed into the crankshaft halves (9 and 11). The connecting rod bearing is select fitted with the connecting rod and crankpin in three size classes, each having a tolerance of 0.004 mm (0.00016 inch). Connecting rod, bearing and crankshaft journal must be renewed as an assembly.

To remove connecting rod, first remove all cooling shrouds. Remove the cylinder head. Remove cylinder retaining nuts and pull cylinder up off the piston. Use a suitable puller to remove the flywheel. Remove crankcase cover. Withdraw crankshaft and connecting rod assembly from crankcase.

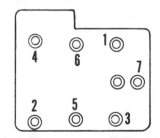

*Fig. AC112—Tighten cylinder head bolts on Model AT 330 following the sequence shown.*

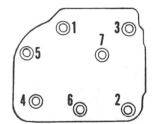

*Fig. AC113—Tighten cylinder head bolts on Models VT 88 and VT 94 following the sequence shown.*

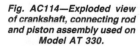
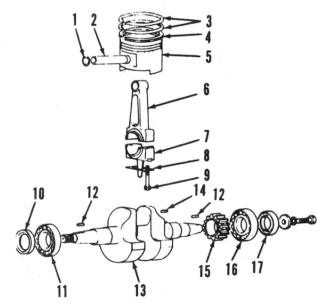

*Fig. AC114—Exploded view of crankshaft, connecting rod and piston assembly used on Model AT 330.*

1. Retaining ring
2. Piston pin
3. Compression rings
4. Oil control ring
5. Piston
6. Connecting rod
7. Connecting rod cap
8. Lock plate
9. Bolts
10. Oil seal
11. Main bearing
12. Key
13. Crankshaft
14. Crankshaft gear key
15. Crankshaft gear
16. Main bearing
17. Oil seal

The connecting rod should not be removed from the crankshaft unless suitable removal and installation tools are available. If connecting rod must be removed from crankshaft, first remove the piston from the connecting rod. The crankshaft halves must be separated using a suitable puller (such as Acme tool 365 126). To reassemble the crankshaft, use a suitable jig (Acme tool 365 130) and a press capable of 7 to 8 tons pressure. Press crankshaft together until there is 0.2-0.4 mm (0.008-0.016 inch) clearance between connecting rod and crankshaft shoulder. The two crankshaft halves must be concentric with the crankshaft connecting rod journal. Maximum off-center allowance is 0.05 mm (0.002 inch).

Clearance between piston pin and connecting rod pin bore should be 0.02-0.05 mm (0.0008-0.0019 inch).

Piston may be installed on connecting rod either way. When installing crankcase cover on crankcase, tighten bolts to specified torque in a crisscross pattern.

**PISTON, PIN AND RINGS.** On all models, refer to CONNECTING ROD paragraphs for piston removal and installation procedure.

On all models, compression ring end gap should be 0.25-0.40 mm (0.010-0.016 inch). Oil control ring end gap should be 0.30-0.50 mm (0.012-0.020 inch).

On Model AT 330, piston pin should be 0.005 mm (0.0002 inch) interference to 0.005 mm (0.0002 inch) loose fit in piston pin bore. On Models VT 88 and VT 94, piston pin should be 0.01-0.03 mm (0.0004-0.0012 inch) interference fit in piston pin bore. It may be necessary to heat piston slightly to aid in removal and installation of piston pin.

On Model AT 330, refer to Fig. AC116 for correct piston ring installation. On Models VT 88 and VT 94, refer to Fig. AC117 for correct piston ring installation. On all models, stagger ring end gaps equally around piston diameter. Lubricate piston, rings and cylinder with engine oil prior to reassembly.

**CYLINDER AND CRANKCASE.** On Model AT 330, cylinder and crankcase are an integral casting of aluminum alloy with a high density perlite cylinder sleeve cast as an integral part of the cylinder block. Standard cylinder bore diameter is 80.000-80.013 mm (3.1496-3.1501 inches).

On Models VT88 and VT 94, the cast iron cylinder is a separate casting and can be removed from the crankcase. Standard cylinder bore diameter is 88.000-88.013 mm (3.4646-3.4651 inches) for Model VT 88 and 94.000-94.013 mm (3.7008-3.7013 inches) for Model VT 94.

On all models, if cylinder bore is 0.06 mm (0.0024 inch) or more out-of-round or tapered, cylinder should be bored to

nearest oversize for which piston and rings are available.

On all models, crankshaft ball type main bearings should be a slight press fit in crankcase and crankcase cover bores. Renew bearings if they are loose, rough or damaged.

On Model AT 330, an oil slinger trough (2—Fig. AC118) is installed in the crankcase. If trough is removed, it must be securely repositioned in crankcase prior to engine reassembly.

**CRANKSHAFT.** On all models, the crankshaft is supported at each end by ball bearing type main bearings. Refer to Fig. AC114 for an exploded view of the crankshaft assembly used on Model AT 330 and to Fig. AC115 for an exploded view of the crankshaft assembly used on Models VT 88 and VT 94.

On Model AT 330, the crankshaft timing gear (15—Fig. AC114) is a press fit on crankshaft (13). Standard clearance between crankpin journal and connecting rod bore is 0.040-0.064 mm (0.0016-0.0025 inch). Standard crankpin journal diameter is 29.985-30.000 mm (1.1805-1.1811 inch). If crankshaft crankpin journal is worn or out-of-round 0.015 mm (0.0006 inch) or more, regrind crankpin or renew crankshaft. Standard main bearing journal diameter is 30.002-30.015 mm (1.1812-1.1817 inch). Main journals cannot be reground. Ball type main bearings should be a slight press fit on crankshaft journal. Renew bearings if they are rough, loose or damaged. Note that crankshaft should be supported on counterweights when pressing bearings or crankshaft timing gear onto or off of crankshaft to prevent bending crankshaft. When installing crankshaft, make certain crankshaft and camshaft gear timing marks are aligned.

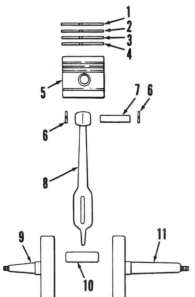

Fig. AC115—Exploded view of crankshaft, connecting rod and piston assembly used on Models VT 88 and VT 94.

1. Compression ring
2. Compression ring
3. Oil control ring
4. Oil control ring
5. Piston
6. Retaining ring
7. Piston pin
8. Connecting rod
9. Crankshaft half
10. Crankpin
11. Crankshaft half

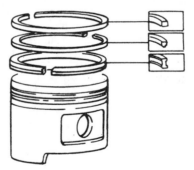

Fig. AC116—View of Model AT 330 piston and piston rings showing correct installation of rings. Be sure that notched side of second ring is toward bottom of piston.

Fig. AC117—On Models VT 88 and VT 94, install piston rings so notched side of second ring (2) faces down and the bevel on oil control rings (3 and 4) faces up.

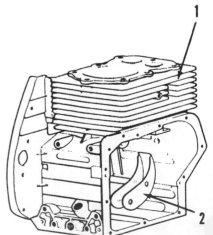

Fig. AC118—View of Model AT 330 cylinder block (1) and oil slinger trough (2). If trough is removed, it must be securely repositioned prior to assembly of engine.

Illustrations courtesy of Acme North America Corp.

On Models VT 88 and VT 94, the crankshaft is a three-piece assembly (Fig. AC115) which must be pressed together using special jig as described in CONNECTING ROD paragraphs. When installing crankshaft, make certain that crankshaft and camshaft gear timing marks are aligned.

## Model AT 330

**CAMSHAFT.** On Model AT 330, the camshaft and camshaft gear are an integral casting which rides in bearing bores in crankcase and crankcase cover. A compression release mechanism is mounted on the back side of camshaft gear and governor flyballs are located in ramps machined into face of camshaft gear. The spring loaded compression release mechanism should snap back against the camshaft when the weighted lever is pulled against spring tension and released.

Inspect camshaft journals and lobes and renew camshaft if worn, scored or damaged. Camshaft bearing journal diameters should be 15.973-15.984 mm (0.6289-0.6293 inch). Exhaust cam lobe height should be 26.290-26.320 mm (1.0350-1.0362 inch) and intake cam lobe height should be 19.590-19.620 mm (0.7713-0.7724 inch). Camshaft and tappets should be renewed as a set.

Governor flyballs should move smoothly across ramp faces without catching. When installing camshaft, retain governor flyballs in ramps with grease and make certain that camshaft

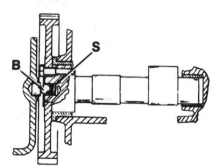

*Fig. AC119—On Models VT 88 and VT 94, camshaft axial thrust is controlled by a ball (B) and spring (S).*

and crankshaft gear timing marks are aligned.

## Models VT 88, VT 94

**CAMSHAFT.** On Models VT 88 and VT 94, the camshaft is supported at each end by renewable bushings. The camshaft gear has helical teeth to reduce noise and improve meshing. Camshaft axial thrust is controlled by a ball tensioned by a spring (Fig. AC119).

Inspect camshaft, tappets and bushings for excessive wear, scoring or other damage and renew as necessary. Maximum allowable clearance between camshaft journals and bushings is 0.20 mm (0.008 inch). New camshaft bushings must be reamed after installation to provide specified clearance of 0.025-0.086 mm (0.0010-0.0034 inch). Camshaft lobe standard height is 28.75 mm (1.132 inch) for intake and exhaust. Camshaft and tappets should be renewed as a set.

When installing camshaft, make certain that timing marks on camshaft gear and crankshaft gear are aligned.

# BRIGGS & STRATTON

**BRIGGS & STRATTON CORPORATION**
**Milwaukee, Wisconsin 53201**

## BRIGGS & STRATTON ENGINE IDENTIFICATION INFORMATION

In order to obtain correct service parts for Briggs & Stratton engines it is necessary to correctly identify engine model or series and provide engine serial number.

Briggs & Stratton model or series number also provides information concerning important mechanical features or optional equipment.

Refer to Fig. B1 for an explanation of each digit in relation to engine identification or description of mechanical features and options.

As an example, a 401417 series model number is broken down in the following manner.

40 — Designates 40 cubic inch displacement.
 1 — Designates design series 1.
 4 — Designates horizontal shaft, Flo-Jet carburetor and mechanical governor.
 1 — Designates flange mounting with plain bearings.
 7 — Designates electric starter, 12 volt gear drive with alternator.

| CUBIC INCH DISPLACEMENT | FIRST DIGIT AFTER DISPLACEMENT<br>BASIC DESIGN SERIES | SECOND DIGIT AFTER DISPLACEMENT<br>CRANKSHAFT, CARBURETOR GOVERNOR | THIRD DIGIT AFTER DISPLACEMENT<br>BEARINGS, REDUCTION GEARS & AUXILIARY DRIVES | FOURTH DIGIT AFTER DISPLACEMENT<br>TYPE OF STARTER |
|---|---|---|---|---|
| 6 | 0 | 0 - | 0 - Plain Bearing | 0 - Without Starter |
| 8 | 1 | 1 - Horizontal Vacu-Jet | 1 - Flange Mounting Plain Bearing | 1 - Rope Starter |
| 9 | 2 | 2 - Horizontal Pulsa-Jet | 2 - Ball Bearing | 2 - Rewind Starter |
| 10 | 3 | | | |
| 11 | 4 | 3 - Horizontal Flo-Jet (Pneumatic Governor) | 3 - Flange Mounting Ball Bearing | 3 - Electric - 110 Volt, Gear Drive |
| 13 | 5 | | | |
| 14 | 6 | 4 - Horizontal Flo-Jet (Mechanical Governor) | 4 - | 4 - Elec. Starter- Generator - 12 Volt, Belt Drive |
| 17 | 7 | | | |
| 19 | 8 | | | |
| 20 | 9 | 5 - Vertical Vacu-Jet | 5 - Gear Reduction (6 to 1) | 5 - Electric Starter Only - 12 Volt, Gear Drive |
| 23 | | | | |
| 24 | | | | |
| 25 | | 6 - | 6 - Gear Reduction (6 to 1) Reverse Rotation | 6 - Alternator Only * |
| 30 | | | | |
| 32 | | | | |
| | | 7 - Vertical Flo-Jet | 7 - | 7 - Electric Starter, 12 Volt Gear Drive, with Alternator |
| | | 8 - | 8 - Auxiliary Drive Perpendicular to Crankshaft | 8 - Vertical-pull Starter |
| | | 9 - Vertical Pulsa-Jet | 9 - Auxiliary Drive Parallel to Crankshaft | * Digit 6 formerly used for ''Wind-Up'' Starter on 60000, 80000 and 92000 Series |

*Fig. B1 — Explanation of model or series number of Briggs & Straton engines.*

A. First one or two digits indicate CUBIC INCH DISPLACEMENT.
B. First digit after displacement indicates BASIC DESIGN SERIES, relating to cylinder construction, ignition, general configuration, etc.
C. Second digit after displacement indicates POSITION OF CRANKSHAFT AND TYPE OF CARBURETOR.
D. Third digit after displacement indicates TYPE OF BEARINGS and whether or not engine is equipped with REDUCTION GEAR or AUXILIARY DRIVE.
E. Last digit indicates TYPE OF STARTER.

# BRIGGS & STRATTON

### BRIGGS & STRATTON CORPORATION
### Milwaukee, Wisconsin 53201

| Model | No. Cyls. | Bore | Stroke | Displacement | Power Rating |
|---|---|---|---|---|---|
| 170000, 171000 | 1 | 3 in. (76.2 mm) | 2.375 in. (60.3 mm) | 16.8 cu. in. (275.1 cc) | 7 hp. (5.2 kW) |
| 190000, 191000 | 1 | 3 in. (76.2 mm) | 2.75 in. (69.85 mm) | 19.44 cu. in. (318.5 cc) | 8 hp. (6 kW) |
| 220000 | 1 | 3.438 in. (87.3 mm) | 2.375 in. (60.3 mm) | 22.04 cu. in. (361.1 cc) | 10 hp. (7.5 kW) |
| 250000, 251000, 252000, 253000 | 1 | 3.438 in. (87.3 mm) | 2.625 in. (66.68 mm) | 24.36 cu. in. (399.1 cc) | 11 hp. (8.2 kW) |

Engines in this section are four-cycle, one cylinder engines and both vertical and horizontal crankshaft models are covered. Crankshaft is supported in main bearings which are an integral part of crankcase and cover or ball bearings which are a press fit on crankshaft. Cylinder block and bore are a single aluminum casting.

Connecting rod in all models rides directly on crankpin journal. Horizontal crankshaft models are splash lubricated by an oil dipper attached to connecting rod cap and vertical crankshaft models are lubricated by an oil slinger wheel located on governor gear.

Early models use a magneto type ignition system with points and condenser located underneath flywheel. Late models use a breakerless (Magnetron) ignition system.

A float type carburetor is used on all models and a fuel pump is available as optional equipment for some models.

Refer to **BRIGGS & STRATTON ENGINE IDENTIFICATION INFORMATION** section for engine identification and always give engine model and serial number when ordering parts or service material.

## MAINTENANCE

**SPARK PLUG.** Recommended spark plug is Champion J8 or equivalent. To decrease radio interference use Champion XJ8 or equivalent spark plug. Electrode gap for all models is 0.030 inch (0.762 mm).

**CARBURETOR.** One of two different Flo-Jet carburetors will be used. Carburetor may be identified as Flo-Jet I (Fig. B2) or Flo-Jet II (Fig. B3). Refer to appropriate figure for model being serviced.

FLO-JET I CARBURETOR. Initial adjustment procedure varies according to horsepower rating and crankshaft type.

For initial carburetor adjustment for 7, 8 and 10 horsepower horizontal crankshaft models, open idle mixture screw 1½ turns and main fuel mixture screw 2½ turns.

For initial carburetor adjustment for 11 horsepower horizontal crankshaft models, open idle mixture screw 1 turn and main fuel mixture screw 1½ turns.

For initial carburetor adjustment for all vertical crankshaft models, open idle mixture screw 1¼ turns and main fuel mixture screw 1½ turns.

Make final adjustment with engine at normal operating temperature and running. Place engine under load and adjust main fuel mixture screw for leanest mixture that will allow satisfactory acceleration and steady governor operation. Set engine at idle speed, no load and adjust idle mixture screw to obtain smoothest idle operation.

As each adjustment affects the other, adjustment procedure may have to be repeated.

FLO-JET II CARBURETOR. For initial adjustment open idle mixture screw 1 turn and main fuel mixture screw 1½ turns. Make final adjustment with engine at normal operating temperature and running. Place engine under load and adjust main fuel mixture screw for leanest mixture that will allow satisfactory acceleration and steady governor operation. Set engine at idle speed, no load and adjust idle mixture screw to obtain smoothest idle operation.

As each adjustment affects the other, adjustment procedure may have to be repeated.

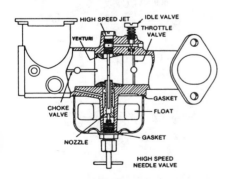

*Fig. B2—Cross-sectional view of Flo-Jet I carburetor.*

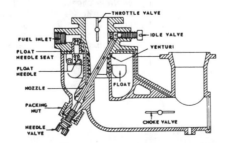

*Fig. B3—Cross-sectional view of Flo-Jet II carburetor. Before separating upper and lower body sections, loosen packing nut and power needle valve as a unit and use special screwdriver (tool number 19062) to remove nozzle.*

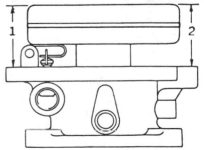

*Fig. B4—Dimension (2) must be the same as dimension (1) plus or minus 1/32-inch (0.79 mm).*

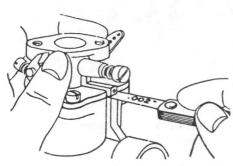

Fig. B5—Check upper body of Flo-Jet II carburetor for warpage as outlined in text.

To check float level for all models, invert carburetor body and float assembly. Refer to Fig. B4 for proper float level dimensions. Adjust by bending float lever tang that contacts inlet valve.

Check Flo-Jet II carburetor for upper body warpage using an 0.002 inch (0.0508 mm) feeler gage as shown in Fig. B5. If upper body is warped more than 0.002 inch (0.0508 mm) it must be renewed.

**CHOKE-A-MATIC CARBURETOR CONTROLS.** Engines may be equipped with a control unit with which carburetor choke, throttle and magneto grounding swtich are operated from a single lever (Choke-A-Matic carburetors).

To check operation of Choke-A-Matic controls, move control lever to "CHOKE" position; carburetor choke slide or plate must be completely closed. Move control lever to "STOP" position; magneto grounding switch should be

making contact. With control lever in "RUN", "FAST" or "SLOW" position, carburetor choke should be completely open. On units with remote controls, synchronize movement of remote lever to carburetor control lever by loosening screw (C–Fig. B6) and moving control wire housing (D) as required. Tighten screw to clamp housing securely. refer to Fig. B7 to check remote control wire movement.

**AUTOMATIC CHOKE (THERMO-STAT TYPE).** A thermostat operated choke is used on some models equipped with Flo-Jet II carburetor. To adjust choke linkage, hold choke shaft so thermostat lever is free. At room temperature, stop screw in thermostat collar should be located midway between thermostat stops. If not, loosen stop screw, adjust collar and tighten stop screw. Loosen set screw (S–Fig. B8) on thermostat lever. Slide lever on shaft to insure free movement of choke unit. Turn thermostat shaft clockwise until stop screw contacts thermostat stop. While holding shaft in this position, move shaft lever until choke is open exactly ⅛-inch (3 mm) and tighten lever set screw. Turn thermostat shaft counterclockwise until stop screw contacts thermostat stop as shown in Fig. B9. Manually open choke valve until it stops against top of choke link opening. At this time choke should be open at least 3/32-inch (2.38 mm), but not more than 5/32-inch (4 mm). Hold choke valve in wide open position and check position of counterweight lever. Lever should be in a horizontal position with free end towards right.

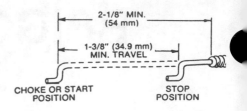

Fig. B7—For proper operation of Choke-A-Matic controls, remote control wire must extend to dimension shown in and have a minimum travel of 1⅜ inches (34.9 mm).

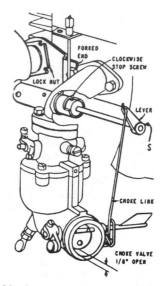

Fig. B8—Automatic choke used on some models equipped with Flo-Jet II carburetor showing unit in "HOT" position.

**FUEL TANK OUTLET.** Some models are equipped with a fuel tank outlet as shown in Fig. B10. On other engines, a fuel sediment bowl is incorporated with fuel tank outlet as shown in Fig. B11.

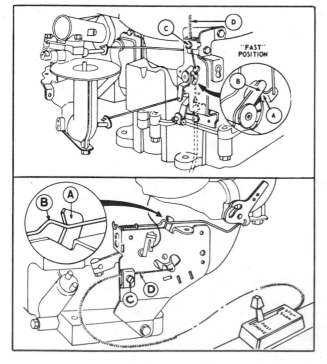

Fig. B6—On Choke-A-Matic controls shown, choke actuating lever (A) should just contact choke link or shaft (B) when control is at "FAST" position. If not, loosen screw (C) and move control wire housing (D) as required.

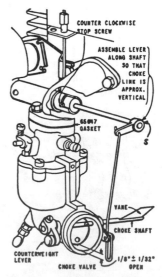

Fig. B9—Turn thermostat shaft counterclockwise until stop screw contacts thermostat stop as shown.

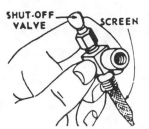

Fig. B10—Fuel tank outlet used on some B&S models.

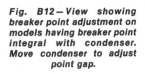

Fig. B12—View showing breaker point adjustment on models having breaker point integral with condenser. Move condenser to adjust point gap.

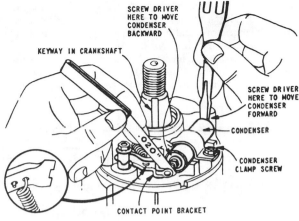

Clean any lint and dirt from tank outlet screens with a brush. Varnish or other gasoline deposits may be removed by use of a suitable solvent. Tighten packing nut or remove nut and shut-off valve, then renew packing if leakage occurs around shut-off valve stem.

**FUEL PUMP.** A fuel pump is available as optional equipment on some models. Refer to **SERVICING BRIGGS & STRATTON ACCESSORIES** section for service information.

**GOVERNOR.** Engines are equipped with a gear driven mechanical governor and governor unit is enclosed within engine crankcase and is driven from camshaft gear. Lubrication oil slinger is an integral part of governor unit on all vertical crankshaft engines. All binding or slack due to wear must be removed from governor linkage to prevent "hunting" or unsteady operation. To adjust carburetor to governor linkage, loosen clamp bolt on governor lever. Move link end of governor lever until carburetor throttle shaft is in wide open position. Using a screwdriver, rotate governor lever shaft clockwise as far as possible and tighten clamp bolt.

Governor gear and weight unit can be removed when engine is disassembled. Refer to exploded views of engines in Fig. B21, B22, B23 and B28. Remove governor lever, cotter pin and washer from outer end of governor lever shaft.

Slide governor lever out of bushing towards inside of engine. Governor gear and weight unit can now be removed. Renew governor lever shaft bushing in crankcase, if necessary and ream new bushing after installation to 0.2385-0.239 inch (6.05-6.07 mm).

**IGNITION SYSTEM.** Early models use a magneto system which incorporates a breaker point set and condenser. Late models use Magnetron breakerless ignition system. Refer to appropriate paragraph for model being serviced.

MAGNETO IGNITION. All models use breaker points and condenser located under flywheel.

One of two different types of ignition points as shown in Fig. B12 and B13 are used. Breaker point gap is 0.020 inch (0.508 mm) for all models with magneto ignition.

On each type, breaker contact arm is actuated by a plunger in a bore in engine crankcase which rides against a cam on engine crankshaft. Plunger can be removed after removing breaker points. Renew plunger if worn to a length of 0.870 inch (22.098 mm) or less. If breaker point plunger bore in crankcase is worn, oil will leak past plunger. Check bore with B&S gage, tool number 19055. If plug gage will enter bore ¼-inch (6.35

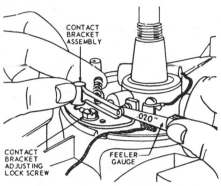

Fig. B13—Adjusting breaker point gap on models having breaker points separate from condenser.

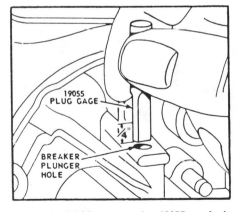

Fig. B14—If B&S gage number 19055 can be inserted in plunger bore ¼-inch (6.35 mm) or more, bore is worn and must be rebushed.

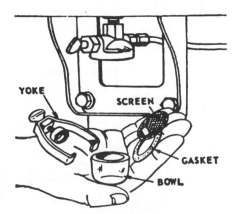

Fig. B11—Fuel sediment bowl and tank outlet used on some models.

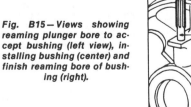

Fig. B15—Views showing reaming plunger bore to accept bushing (left view), installing bushing (center) and finish reaming bore of bushing (right).

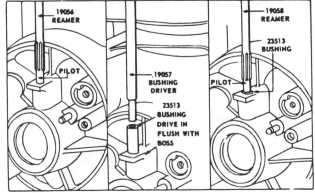

Illustrations courtesy of Briggs & Stratton Corp.

mm) or more, bore should be reamed and a bushing installed. Refer to Fig. B14. To ream bore and install bushing it will be necessary to remove breaker points, armature, ignition coil and crankshaft. Refer to Fig. B15 for steps in reaming bore and installation of bushing.

Plunger must be reinstalled with groove toward top (Fig. B15A) to prevent oil contamination in breaker point box.

For reassembly set armature to flywheel air gap at 0.010-0.014 inch (0.254-0.356 mm) for two-leg armature or 0.016-0.019 inch (0.406-0.483 mm) for

three-leg armature. Ignition timing is non-adjustable on these models.

MAGNETRON IGNITION. Magnetron ignition is a self-contained breakerless ignition system and flywheel does not need to be removed except to check or service keyways or crankshaft key.

To check spark, remove spark plug and connect spark plug cable to B&S tester, part number 19051 and ground remaining tester lead to engine cylinder head. Spin engine at 350 rpm or more. If spark jumps the 0.166 inch (4.2 mm) tester gap, system is functioning properly.

To remove armature and Magnetron module, remove flywheel shroud and armature retaining screws. Use a 3/16-inch (4.76 mm) diameter pin punch to release stop switch wire from module. To remove module, unsolder wires, push module retainer away from laminations and push module off. See Fig. B15B.

Resolder wires for reinstallation and use Permatex or equivalent to hold ground wires in position.

Adjust armature air gap to 0.010-0.014 inch (0.25-0.36 mm).

**LUBRICATION.** Horizontal crankshaft engines have a splash lubrication system provided by an oil dipper attached to connecting rod. Refer to Fig. B16 for view of various types of dippers used.

Vertical crankshaft engines are lubricated by an oil slinger wheel on governor gear which is driven by camshaft gear. See Fig. B16A.

Oils approved by manufacturer must meet requirements of API service classification SE or SF.

Use SAE 10W-40 oil for temperatures above 20°F (-7°C) and SAE 5W-30 oil for temperatures below 20°F (-7°C).

Check oil level at five hour intervals and maintain at bottom edge of filler plug. **DO NOT** overfill.

Recommended oil change interval for all models is every 25 hours of normal operation.

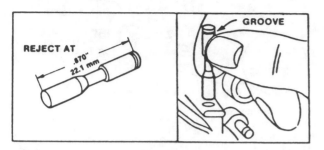

Fig. B15A—Insert plunger into bore with groove toward top.

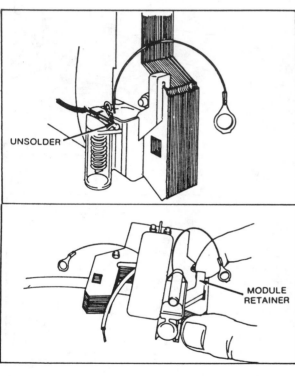

Fig. B15B—Wires must be unsoldered to remove Magnetron module.

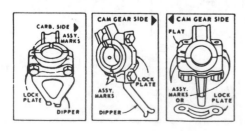

Fig. B16—Install connecting rod in engine as indicated according to type used. Note dipper installation on horizontal crankshaft engine connecting rod.

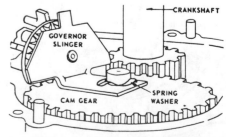

Fig. B16A—Vertical crankshaft engines are lubricated by an oil slinger mounted on governor gear assembly.

Illustrations courtesy of Briggs & Stratton Corp.

Crankcase oil capacity for 16.8 and 19.44 cubic inch (275.1 and 318.5 cc) engines is 2¼ pints (1.1 L) for vertical crankshaft models and 2¾ pints (1.3 L) for horizontal crankshaft models.

Crankcase oil capacity for 22.04 and 24.36 cubic inch (361.1 and 399.1 cc) engines is 3 pints (1.4 L) for both vertical and horizontal crankshaft models.

**CRANKCASE BREATHER.** A crankcase breather is built into engine valve cover. A partial vacuum must be maintained in crankcase to prevent oil from being forced out past oil seals and gaskets or past breaker point plunger or piston rings. Air can flow out of crankcase through breather, but a one-way valve blocks return flow, maintaining necessary vacuum. Breather mounting holes are offset one way. A vent tube connects breather to carburetor air horn for extra protection against dusty conditions.

## REPAIRS

**CYLINDER HEAD.** When removing cylinder head note location from which different length cap screws are removed as they must be reinstalled in their original positions.

Always use a new gasket when reinstalling cylinder head. Do not use sealer on gasket. Lubricate cylinder head bolt threads with graphite grease, install in correct locations and tighten in several even steps in sequence shown in Fig. B17 until 165 in.-lbs. (19 N·m) torque is obtained. Start and run engine until normal operating temperature is reached, stop engine and retighten head bolts as outlined.

It is recommended carbon and lead deposits be removed at 100 to 300 hour intervals, or whenever cylinder head is removed.

**CONNECTING ROD.** Connecting rod and piston are removed from cylinder head end of block as an assembly. Aluminum alloy connecting rod rides directly on induction hardened crankshaft crankpin journal. Rod should be rejected if crankpin bore is scored or out-of-round more than 0.0007 inch (0.0178 mm) or if piston pin bore is scored or out-of-round more than 0.0005 inch (0.0127 mm). Renew connecting rod if either crankpin journal or piston pin bore is worn to, or larger than, sizes given in chart.

### REJECT SIZES FOR CONNECTING ROD

| Model | Crankpin bore | Pin bore* |
|---|---|---|
| 170000, 171000 | 1.0949 in. (27.81 mm) | 0.674 in. (17.12 mm) |
| 190000, 191000 | 1.1265 in. (28.61 mm) | 0.674 in. (17.12 mm) |
| All other models | 1.252 in. (31.8 mm) | 0.802 in. (20.37 mm) |

*Piston pins of 0.005 inch (0.127 mm) oversize are available for service. Piston pin bore in rod can be reamed to this size if crankpin bore is within specifications.

Refer to Fig. B16, locate style rod used for model being serviced and install in engine as indicated.

Tighten connecting rod cap screws to 165 in.-lbs. (19 N·m) torque on 170000, 171000, 190000 and 191000 models and to 190 in.-lbs. (22 N·m) torque on all other models.

**PISTON, PIN AND RINGS.** Chrome plated aluminum piston used in aluminum bore (Kool-Bore) engines should be renewed if top ring side clearance in groove exceeds 0.007 inch (0.18 mm) or if piston is visibly scored or damaged. Piston should also be renewed, or pin bore reamed for 0.005 inch (0.127 mm) oversize pin, if pin bore is 0.673 inch (17.09 mm) or larger for 170000, 171000, 190000 and 191000 models or 0.801 inch (20.32 mm) or larger for all other models.

Renew piston pin if pin is 0.0005 inch (0.0127 mm) or more out-of-round or if pin is worn to a diameter of 0.671 inch (17.04 mm) or smaller for 170000, 171000, 190000 or 191000 models or 0.799 inch (20.29 mm) or smaller for all other models.

Ring end gap should be 0.010-0.025 inch (0.254-0.635 mm) and compression ring should be renewed if end gap is greater than 0.035 inch (0.89 mm) and oil control ring should be renewed if end gap is greater than 0.045 inch (1.14 mm).

Pistons used in 220000 and 250000 engines have a notch and a letter "F" stamped in piston (Fig. B17A) which must face flywheel side of engine after installation.

Pistons and rings are available in a variety of oversizes as well as standard.

A chrome ring set is available for slightly worn standard bore cylinders. Refer to note in **CYLINDER** section.

**CYLINDER.** Standard cylinder bore diameter is 2.999-3.000 inches (76.175-76.230 mm) for 170000, 171000, 190000 and 191000 models and 3.4365-3.4375 inches (87.287-87.313 mm) for all other models.

If cylinder is worn 0.003 inch (0.076 mm) or more, or out-of-round 0.0025 inch (0.0635 mm) or more, it should be

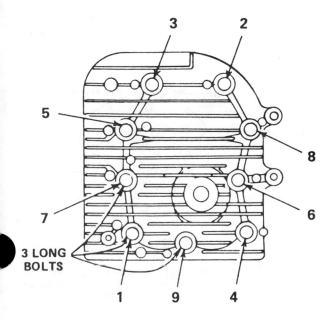

Fig. B17—Tighten cylinder head bolts in sequence shown. Note location of three long bolts.

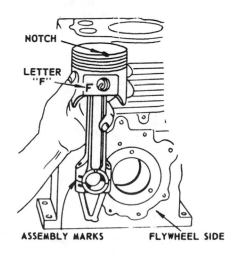

Fig. B17A—Notch and letter "F" stamped in piston of 220000 and 250000 engines must face flywheel side of engine after installation.

Illustrations courtesy of Briggs & Stratton Corp.

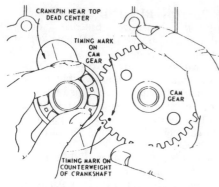

Fig. B18—Align timing mark on cam gear with mark on crankshaft counterweight on ball bearing equipped models.

resized to nearest oversize for which piston and rings are available.

**NOTE: A chrome piston ring set is available for slightly worn standard bore cylinders. No honing or cylinder deglazing is required for these rings. Cylinder bore can be a maximum of 0.005 inch (0.127 mm) oversize when using chrome rings.**

It is recommended a hone be used for resizing cylinders. Operate hone at 300-700 rpm with an up and down movement which will produce a 45° angle cross-hatch pattern. Clean cylinder after honing with oil or soap and water.

Approved hone for aluminum bore is Ammco No. 3956 for rough and finishing or Sunnen AN200 for rough and Sunnen AN500 for finishing.

**CRANKSHAFT AND MAIN BEARINGS.** Crankshaft may be supported at each end in main bearings which are an integral part of crankcase, cover or sump or ball bearing mains which are a press fit on crankshaft and fit into machined bores in crankcase, cover or sump.

Crankshaft for models with main bearings as an integral part of crankcase, cover or sump, should be renewed or reground if main bearing journals are worn to, or beyond crankshaft rejection specifications as listed in following chart.

## CRANKSHAFT REJECTION SIZES

| Model | Magneto end journal | Drive end journal |
|---|---|---|
| 170000, 190000 | 0.9975 in. (25.34 mm) | 1.1790 in. (29.95 mm) |
| 171000 Synchrobalanced 191000 Synchrobalanced | 1.1790 in. (29.95 mm) | 1.1790 in. (29.95 mm) |
| All other models | 1.376 in. (34.95 mm) | 1.376 in. (34.95 mm) |

Crankshaft for models with ball bearing main bearings should be renewed if new bearings are loose on journals. Bearings should be a press fit.

Crankshaft for all models should be renewed or reground if connecting rod crankpin journal diameter is worn to, or beyond crankshaft rejection specifications as listed in following chart.

## CRANKSHAFT REJECTION SIZES

| Model | Crankpin journal |
|---|---|
| 170000, 171000 Synchrobalanced | 1.090 in. (27.69 mm) |
| 190000, 191000 Synchrobalanced | 1.122 in. (28.50 mm) |
| All other models | 1.247 in. (31.67 mm) |

Connecting rod is available for crankshaft which has had connecting rod crankpin journal reground to 0.020 inch (0.508 mm) undersize.

Service main bearing bushings are available for 170000, 171000, 190000 and 191000 models with main bearings cast as an integral part of crankcase, cover or sump if main bearing bores are worn to, or beyond rejection sizes as listed in following chart. All other models with integral type main bearings must have crankcase, cover or sump renewed if they are worn to, or beyond rejection sizes listed in chart.

## MAIN BEARING REJECT SIZES

| Model | Magneto end bearing | Drive end bearing |
|---|---|---|
| 170000, 171000, 190000, 191000 | 1.0036 in. (25.491 mm) | 1.185 in. (30.099 mm) |
| Synchrobalanced models | 1.185 in. (30.099 mm) | 1.185 in. (30.099 mm) |
| All other models | 1.383 in. (35.128 mm) | 1.383 in. (35.128 mm) |

To install service main bushings it is necessary to ream main bearing bores to correct size. Main bearing tool kit, part number 19184, is available and contains all necessary reamers, guides, drivers and supports. Make certain all metal cuttings are removed before pressing bushings into place and align oil notches or holes as bushings are installed.

Ball bearing mains are a press fit on crankshaft and must be removed by pressing crankshaft out of bearing. Renew ball bearing if worn or rough. Expand new bearing by heating in oil and install on crankshaft with shield side towards crankpin journal.

Crankshaft end play should be 0.002-0.008 inch (0.0508-0.2032 mm). At least one 0.015 inch thick cover or sump gasket must be used. Additional cover gaskets in a variety of thicknesses are available if end play is greater than 0.008 inch (0.2032 mm) metal shims are available for use on crankshaft.

Place metal shims between crankshaft gear and cover or sump on models with integral type main bearings or between magneto end of crankshaft and crankcase on ball bearing equipped models.

When reinstalling crankshaft, make certain timing marks are aligned (Fig. B18 or B19) and if equipped with

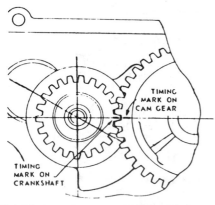

Fig. B19—Align timing marks on cam gear and crankshaft gear on plain bearing models.

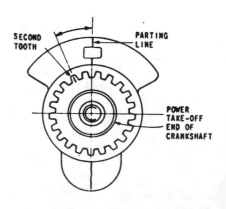

Fig. B20—Location of tooth to align with timing mark on cam gear if mark is not visible on crankshaft gear.

1. Cylinder block
2. Head gasket
3. Cylinder head
4. Connecting rod
5. Rod bolt lock
6. Rings
7. Piston
8. Rotocoil (exhaust valve)
9. Retainer clips
10. Piston pin
11. Intake valve
12. Exhaust valve
13. Retainers
14. Crankcase cover
15. Oil seal
16. Crankcase gasket
17. Main bearings
18. Key
19. Crankshaft
20. Camshaft
21. Tappet
22. Governor gear
23. Governor crank
24. Governor lever
25. Ground wire
26. Governor control plate
27. Spring
28. Governor rod
29. Spring
30. Nut

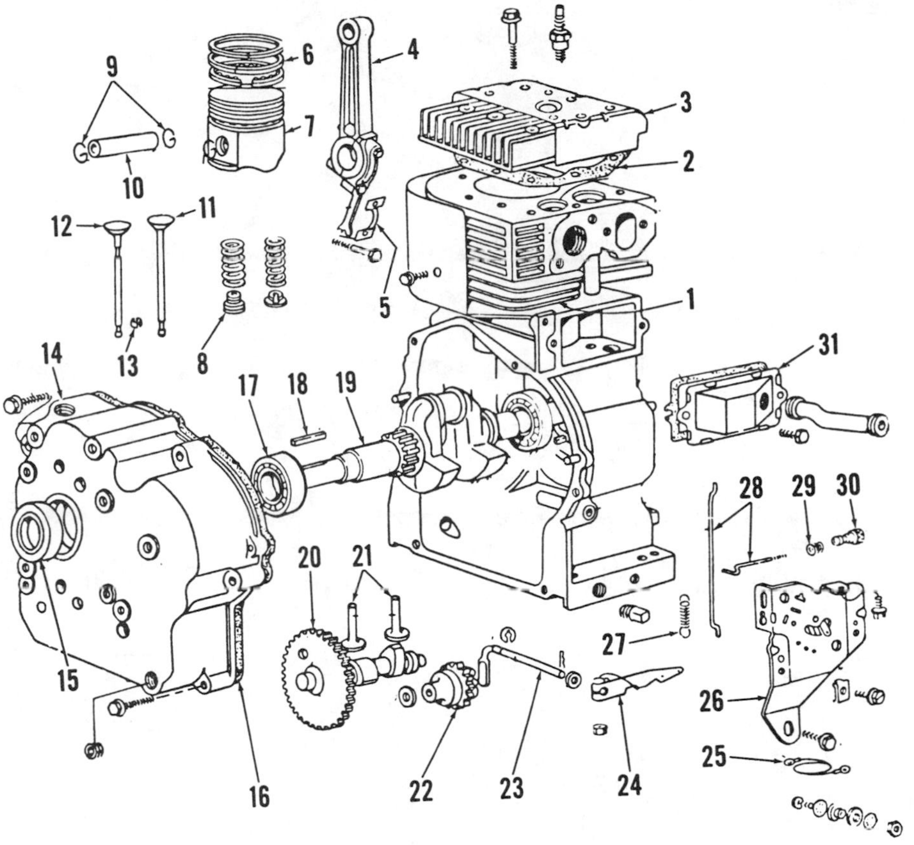

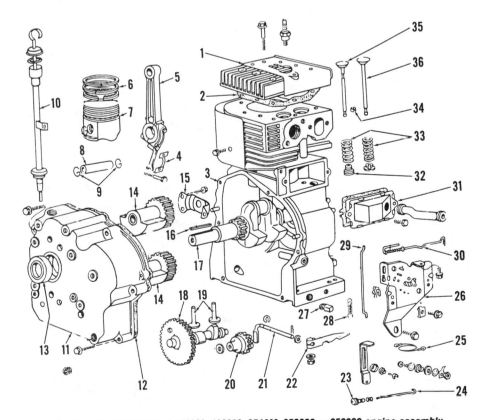

*Fig. B21 — Exploded view of 220000 engine assembly.*

1. Cylinder head
2. Head gasket
3. Cylinder block
4. Rod bolt lock
5. Connecting rod
6. Rings
7. Piston
8. Piston pin
9. Retainer clip
10. Dipstick assy.
11. Crankcase cover
12. Crankcase gasket
13. Oil seal
14. Counterweight & bearing assy.
15. Retainer
16. Key
17. Crankshaft
18. Camshaft assy.
19. Tappet
20. Governor gear
21. Governor crank
22. Governor lever
23. Governor nut & spring
24. Governor control rod
25. Ground wire
26. Governor control plate
27. Drain plug
28. Spring
29. Governor link
30. Choke link
31. Breather assy.
32. Rotocoil (exhaust valve)
33. Valve springs
34. Retainer
35. Exhaust valve
36. Intake valve

*Fig. B22 — Exploded view of 170000, 190000, 251000, 252000 or 253000 engine assembly.*

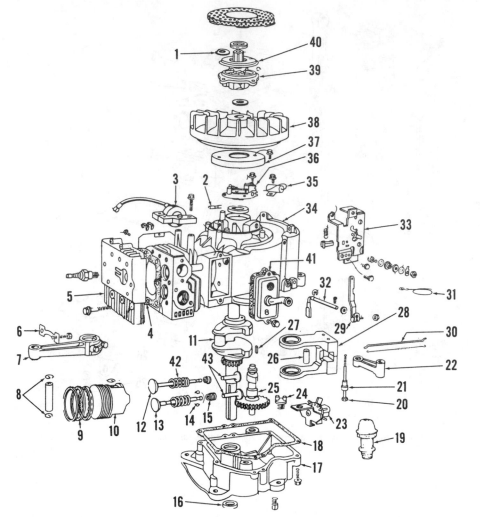

1. Thrust washer
2. Breaker point plunger
3. Armature assy.
4. Head gasket
5. Cylinder head
6. Rod screw lock
7. Connecting rod
8. Piston pin & retaining clips
9. Rings
10. Piston
11. Crankshaft
12. Intake valve
13. Exhaust valve
14. Retainer
15. Rotocoil (exhaust valve)
16. Oil seal
17. Oil sump (engine base)
18. Crankcase gasket
19. Oil minder
20. Cap screw (2 used)
21. Spacer (2 used)
22. Link
23. Governor & oil slinger
24. Plug
25. Cam gear assy.
26. Dowel pin (2 used)
27. Key
28. Counterweight assy.
29. Governor lever
30. Governor link
31. Ground wire
32. Governor crank
33. Choke-A-Matic control
34. Cylinder assy.
35. Condenser
36. Breaker points
37. Cover
38. Flywheel assy.
39. Clutch housing
40. Rewind starter clutch
41. Breather assy.
42. Valve spring
43. Tappets

**Fig. B23 — Exploded view of 171000, 191000, 251000 or 252000 Synchro-Balanced engine assembly.**

counter balance weights they must be positioned properly. See Fig. B27B.

**CAMSHAFT.** Camshaft gear is an integral part of camshaft which is supported at each end in bearing bores machined into crankcase, cover or sump.

Camshaft should be renewed if either journal is worn to a diameter of 0.498 inch (12.66 mm) or less or if cam lobes are worn or damaged.

Crankcase, cover or sump must be renewed if bearing bores are worn to 0.506 inch (12.852 mm) or larger or if tool number 19164 enters bearing bore ¼-inch (6.35 mm) or more.

Compression release mechanism on camshaft gear holds exhaust valve slightly open at very low engine rpm as a starting aid. Mechanism should work freely and spring holds actuator cam against pin.

When installing camshaft in engines with integral type main bearings, align timing mark on camshaft gear with timing mark on crankshaft gear (Fig. B19).

*Fig. B24 — View showing operating principle of Synchro-Balancer used on some vertical crankshaft engines. Counterweight oscillates in opposite direction of piston.*

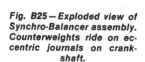

*Fig. B25 — Exploded view of Synchro-Balancer assembly. Counterweights ride on eccentric journals on crankshaft.*

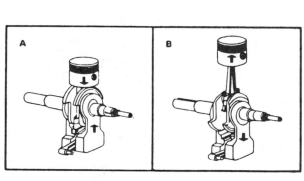

When installing camshaft in engines with ball bearing mains, align timing mark on camshaft gear with timing mark on crankshaft counterweight (Fig. B18).

If timing mark is not visible on crankshaft gear, align camshaft gear timing mark with second tooth to the left of crankshaft counterweight parting line (Fig. B20).

**VALVE SYSTEM.** Valve seats are machined directly into cylinder block and are ground at a 45° angle. Seat width should be 3/64 to 1/16-inch (1.19 to 1.58 mm).

Valve face is ground at 45° angle and valve should be renewed if margin is 1/64-inch (0.4 mm) or less after refacing.

Valve guides should be checked for wear using valve guide gage number 19151. If gage enters valve guide 5/16-inch (7.9 mm) or more, guides should be reamed using reamer number 19183 and bushing number 19192 should be installed.

Briggs & Stratton also has a tool kit, part number 19232, available for removing factory or field installed guide bushings so new bushing, part number 231218 may be installed.

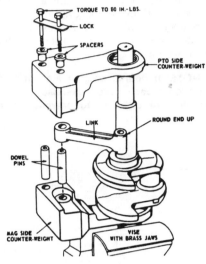

Fig. B26—Assemble balance units on crank shaft as shown. Install link with rounded edge on free end toward pto end of crankshaft.

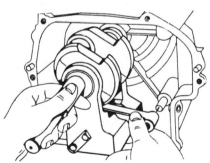

Fig. B27—When installing crankshaft and balancer assembly, place free end of link on anchor pin in crankcase.

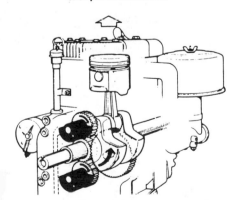

Fig. B27A—View of rotating counterbalance system used on some models. Counterweight gears are driven by crankshaft.

Fig. B27B—To properly align counterweights, remove two small screws from crankcase cover and insert ⅛-inch (3.18 mm) diameter locating pins.

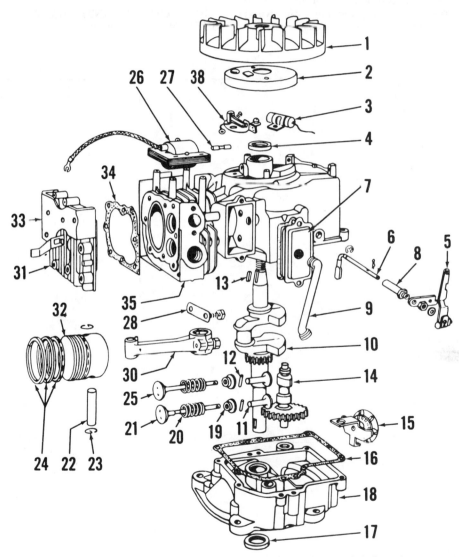

Fig. B28—Exploded view of 170000, 190000 and 220000 series vertical crankshaft engine assembly.

| | | |
|---|---|---|
| 1. Flywheel | 18. Oil sump (engine base) | 27. Breaker plunger |
| 2. Breaker point cover | 19. Valve spring retainer | 28. Rod bolt lock |
| 3. Condenser | 20. Valve spring | 30. Connecting rod |
| 4. Oil seal | 21. Exhaust valve | 31. Cylinder head |
| 5. Governor lever | 22. Piston pin | 32. Piston |
| 6. Governor crank | 23. Retaining rings | 33. Air baffle |
| 7. Breather & valve | 24. Piston rings | 34. Head gasket |
| 8. Bushing | 25. Intake valve | 35. Cylinder block |
| 9. Breather vent tube | 26. Armature & coil assy. | 38. Breaker points |
| 10. Crankshaft | | |
| 11. Tappets | | |
| 12. Valve retaining pins | | |
| 13. Flywheel key | | |
| 14. Camshaft and gear | | |
| 15. Governor & oil slinger assy. | | |
| 16. Gasket | | |
| 17. Oil seal | | |

Valve tappet gap (cold) for all models is 0.005-0.007 inch (0.13-0.18 mm) for intake valve and 0.009-0.011 inch (0.23-0.28 mm) for exhaust valve.

To adjust valves, piston must be at "top dead center" on compression stroke. Grind end of valve stem off squarely to obtain clearance.

**SYNCHRO—BALANCER.** All vertical crankshaft engines, except 220000 models, may be equipped with an oscillating Synchro-Balancer. Balance weight assembly rides on eccentric journals on crankshaft and moves in opposite direction of piston (Fig. B24).

To disassemble balancer unit, first remove flywheel, engine base, cam gear, cylinder head and connecting rod and piston assembly. Carefully pry off crankshaft gear and key. Remove the two cap screws holding halves of counterweight together. Separate weights and remove link, dowel pins and spacers. Slide weights from crankshaft (Fig. B25).

To reassemble, install magneto side weight on magneto end of crankshaft. Place crankshaft (pto end up) in a vise (Fig. B26). Install both dowel pins and place link on pin as shown. Note rounded edge on free end of link must be up. Install pto side weight, spacers, lock and cap screws. Tighten cap screws to 80 in.-lbs. (9 N·m) and secure with lock tabs. Install key and crankshaft gear with chamfer on inside of gear facing shoulder on crankshaft.

Install crankshaft and balancer assembly in crankcase, sliding free end of link on anchor pin as shown in Fig. B27. Reassemble engine.

**ROTATING COUNTERBALANCE SYSTEM.** All horizontal crankshaft engines, except 220000 models, may be equipped with two gear driven counterweights in constant mesh with crankshaft gear. Gears, mounted in crankcase cover, rotate in opposite direction of crankshaft (Fig. B27A).

To properly align counterweights when installing cover, remove two small screws from cover and insert 1/8-inch (3.18 mm) diameter locating pins through holes and into holes in counterweights as shown in Fig. B27B.

With piston at TDC, install cover assembly. Remove locating pins, coat threads of timing hole screws with non-hardening sealer and install screws with fiber sealing washers.

**NOTE: If counterweights are removed from crankcase cover, exercise care in handling or cleaning to prevent losing needle bearings.**

# BRIGGS & STRATTON

### BRIGGS & STRATTON CORPORATION
### Milwaukee, Wisconsin 53201

| Model | No. Cyls. | Bore | Stroke | Displacement | Power Rating |
|---|---|---|---|---|---|
| 19, 19D, 191000, 193000 | 1 | 3 in. (76.2 mm) | 2.625 in. (66.675 mm) | 18.56 cu. in. (304.1 cc) | 7.25 hp. (5.5 kW) |
| 200000 | 1 | 3 in. (76.2 mm) | 2.875 in. (73.025 mm) | 20.32 cu. in. (333 cc) | 8 hp. (6 kW) |
| 23, 23A, 23C, 23D, 23100, 233000 | 1 | 3 in. (76.2 mm) | 3.25 in. (82.55 mm) | 22.97 cu. in (376.5 cc) | 9 hp. (6.7 kW) |
| 243000 | 1 | 3.0625 in. (77.79 mm) | 3.25 in. (82.55 mm) | 23.94 cu. in. (392.3 cc) | 10 hp. (7.5 kW) |
| 300000, 301000 | 1 | 3.4375 in. (87.31 mm) | 3.25 in. (82.55 mm) | 30.16 cu. in. (494.2 cc) | 12 hp. (9 kW) |
| 302000 | 1 | 3.4375 in. (87.31 mm) | 3.25 in. (82.55 mm) | 30.16 cu. in. (494.2 cc) | 13 hp. (9.7 kW) |
| 320000 | 1 | 3.5625 in. (90.5 mm) | 3.25 in. (82.55 mm) | 32.4 cu. in. (531 cc) | 14 hp. (10.4 kW) |
| 325000 | 1 | 3.5625 in. (90.5 mm) | 3.25 in. (82.55 mm) | 32.4 cu. in. (531 cc) | 15 hp. (11.2 kW) |
| 326000 | 1 | 3.5625 in (90.5 mm) | 3.25 in. (82.55 mm) | 32.4 cu. in. (531 cc) | 16 hp. (11.9 kW) |

Engines in this section are four-cycle, one cylinder horizontal crankshaft models. Crankshaft is supported at each end in bushing type main bearings which are an integral part of main bearing support plate or by ball bearings which are pressed onto crankshaft and fit into machined bores of main bearing support plates. Cylinder block and crankcase are a single cast iron casting.

Connecting rod in all models rides directly on crankpin journal and is splash lubricated by an oil dipper attached to connecting rod cap.

Early models use a variety of magneto ignition systems with points and condenser either mounted externally or underneath flywheel. Late production models use Briggs & Stratton Magnetron breakerless ignition system.

A float type carburetor is used on all models and a fuel pump is available as optional equipment for some models.

Refer to **BRIGGS & STRATTON ENGINE IDENTIFICATION INFORMATION** section for engine identification and always give engine model and serial number when ordering parts or service material.

## MAINTENANCE

**SPARK PLUG.** Recommended spark plug is Champion J8 or equivalent. To decrease radio interference use Champion XJ8 or equivalent spark plug. Electrode gap for all models is 0.030 inch (0.762 mm).

**CARBURETOR.** All models are equipped with a two-piece float type carburetor. Refer to Fig. B30 for a cross-sectional view of carburetor and location of mixture adjustment screws.

For initial carburetor adjustment, open idle mixture screw 1 turn and main fuel mixture screw 1½ turns.

Make final adjustment with engine at normal operating temperature and running. Place engine under load and adjust main fuel mixture screw for leanest mixture that will allow satisfactory acceleration and steady governor operation. Set engine idle mixture screw to obtain smoothest idle operation.

As each adjustment affects the other, adjustment procedure may have to be repeated.

To check float level for all models, invert carburetor body and float assembly. Refer to Fig. B34 for proper float level dimensions. Adjust by bending float lever tang that contacts inlet valve.

Check carburetor for upper body warpage using a 0.002 inch (0.0508 mm)

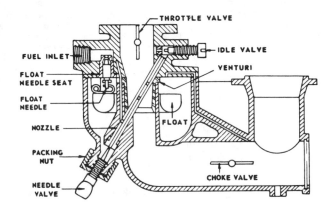

Fig. B30 — Cross-sectional view of typical two-piece float type carburetor. Before separating upper and lower carburetor bodies, loosen packing nut and unscrew needle valve and packing nut. Use special screwdriver, part number 19062 to remove nozzle.

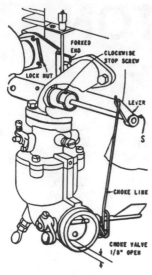

Fig. B31 — Typical B&S automatic choke unit in "HOT" position.

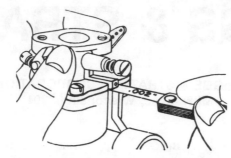

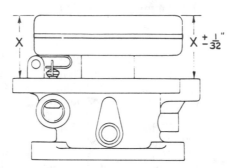

Fig. B33 — Check carburetor upper body for warpage. If a 0.002 inch (0.0508 mm) feeler gage can be inserted between upper and lower bodies as shown, renew upper body.

Fig. B34 — Check carburetor float setting as shown. Bend tang if necessary to adjust float level.

feeler gage as shown in Fig. B33. If upper body is warped more than 0.002 inch (0.0508 mm) it should be renewed.

**AUTOMATIC CHOKE.** A thermostat operated choke is used on some models. To adjust choke linkage, hold choke shaft so thermostat lever is free. At room temperature, stop screw in thermostat collar should be located midway between thermostat stops (Fig. B31 & B32). If not, loosen stop screw, adjust collar and tighten stop screw. Loosen set screw (S–Fig. B31) on thermostat lever. Slide lever on shaft to insure free movement of choke unit. Turn thermostat shaft clockwise until stop screw contacts thermostat stop. While holding shaft in this position, move shaft lever

until choke is open exactly ⅛-inch (3.18 mm) and tighten lever set screw. Turn thermostat shaft counter-clockwise until stop screw contacts thermostat stop as shown in Fig. B32. Manually open choke valve until it stops against top of choke link opening. At this time choke should be open at least 3/32-inch (2.25 mm), but not more than 5/32-inch (4 mm). Hold choke valve in wide open position and

check position of counterweight lever. Lever should be in a horizontal position with free end towards right.

**FUEL PUMP.** A diaphragm type fuel pump is available on some models as optional equipment. Refer to **SERVICING BRIGGS & STRATTON ACCESSORIES** section for service information.

**GOVERNOR.** All models are equipped with a gear driven mechanical governor. Governor weight unit is enclosed within engine and is driven by camshaft gear. Refer to Fig. B62, B63 or B64 for appropriate exploded view of model being serviced.

To adjust governor, loosen screw holding governor lever to governor crank (46–Fig. B62, B63 or B64). Push governor lever as far clockwise as it will go. Tighten screw holding governor lever to crank until it is snug, but not tight. Push governor lever as far counter-clockwise as it will go and tighten screw securely.

Governor gear and weight unit can be removed when engine is disassembled. Remove governor lever, cotter pin and washer from outer end of governor crank. Slide crank out of bushing towards inside of engine. Governor gear and weight unit will then slide off of shaft. Shaft can be removed from crankcase if necessary. Renew bushing in crankcase for governor crank if bushing is worn and ream new bushing, after installation, to 0.2385-0.239 inch (6.06-6.07 mm).

Refer to Figs. B35 through B38 for governor control and carburetor linkage hook-up and adjustments.

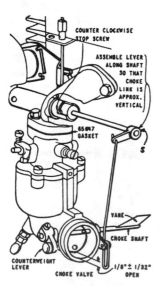

Fig. B32 — Typical B&S automatic choke in "COLD" position.

Fig. B35 — Views showing method of connecting different remote governor controls. Moving control lever will vary governor spring tension, thus varying engine governed speeds. For views showing remote throttle controls, refer to Fig. B36.

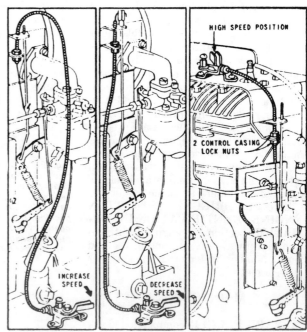

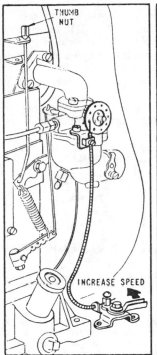

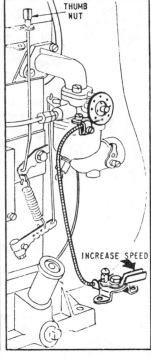

*Fig. B36—Views showing methods of connecting remote throttle controls. When control lever is in high speed position, governor controls speed of engine and governed speed is adjusted by turning thumb nut. Moving control lever to slow speed position moves carburetor throttle shaft stop to decrease engine speed.*

**IGNITION SYSTEM.** Early models use a variety of magneto type ignition systems and late production models use Briggs & Stratton Magnetron ignition system. Refer to appropriate paragraph for model being serviced.

MODEL 23C MAGNETO. Refer to Fig. B39 for exploded view of magneto used on 23C model engine. Condenser and breaker points are mounted on crankshaft bearing support plate (17) and are accessible after removing flywheel and magneto cover (7).

Breaker point gap is 0.020 inch (0.508 mm) and condenser capacity is 0.18-0.24 mfd.

Breaker point plunger can be removed from bore in crankshaft bearing support when breaker points are removed.

Plunger should be renewed if worn to a length of 0.870 inch (22.098 mm) or less. Plunger bore diameter may be checked using B&S plunger bore gage, part number 19055. If gage enters plunger bore ¼-inch (6.35 mm) or more, install service bushing, part number 23513. To install bushing, remove bearing support from engine, then ream bore using B&S reamer, part number 19056, drive bushing in flush with outer end of plunger bore with B&S driver, part number 19057 and use B&S reamer, part number 19058, to finish ream inside of bushing.

When reinstalling flywheel, inspect the soft metal key and renew if damaged in any way.

**NOTE: Renew key with correct B&S part. DO NOT substitute a steel key.**

After flywheel is installed and retaining nut tightened, check armature air gap and adjust as necessary to 0.022-0.026 inch (0.5588-0.6604 mm).

MODELS 19, 23 & 23A AND SERIES 191000 & 231000 MAGNETO. Refer to Fig. B40 for exploded view of magneto used on these models.

Condenser and breaker points are mounted externally in a breaker box located on carburetor side of engine and are accessible after removing breaker box cover (18–Fig. B40).

Breaker point gap is 0.020 inch (0.508 mm) and condenser capacity is 0.18-0.24 mfd.

When renewing breaker points, or if oil leak is noted, breaker shaft oil seal (16) should be renewed. To renew points and/or seal, turn engine so breaker point gap is at maximum. Remove terminal and breaker spring screws and loosen breaker arm retaining nut (19) so it is flush with end of shaft (29). Tap loosened nut lightly to free breaker arm (21) from taper on shaft, then remove breaker arm, breaker plate (22), pivot (23), insulating plate (24) and eccentric (17). Pry oil seal out and press new oil seal in with metal side out. Place breaker plate on insulating plate with dowel on breaker plate entering hole in insulator. Then, install unit with edges of plates parallel with breaker box as shown in Fig. B41. Turn breaker shaft clockwise as far as possible and install breaker arm while holding shaft in this position.

Breaker box can be removed without removing points or condenser. Refer to Fig. B42. Disassembly of unit is evident after removal from engine.

To renew ignition coil, engine flywheel must be removed. Disconnect coil ground and primary wires and pull spark plug wire from hole in back plate. Disengage clips (8–Fig. B40) retaining coil core (9) to armature (14) and push

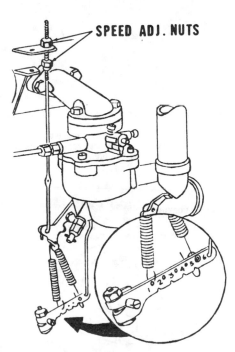

**SPEED ADJ. NUTS**

*Fig. B37—View showing governor linkage, springs and levers properly installed on 243000, 300000, 301000, 302000, 320000, 325000 and 326000 series engines. Refer to Fig. B38 for remote control hook-up.*

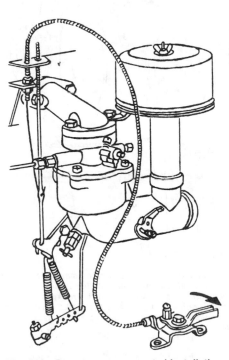

*Fig. B38—Remote governor control installation on 243000, 300000, 301000, 302000, 320000, 325000 and 326000 series engines.*

core from coil. Insert core in new coil with rounded side of core towards spark plug wire. Place coil and core on armature with retainer (7) between coil and armature. Reinstall core retaining clips, connect coil ground and primary wires and insert spark plug wire through hole in back plate.

Two types of magnetic rotors have been used. Refer to Fig. B43 for view of rotor retained by set screw and to Fig. B44 for rotor retained by clamp ring. If rotor is as shown in Fig. B44, refer to Fig. B45 when installing rotor on crankshaft.

If armature coil has been loosened or removed, rotor timing must be read-

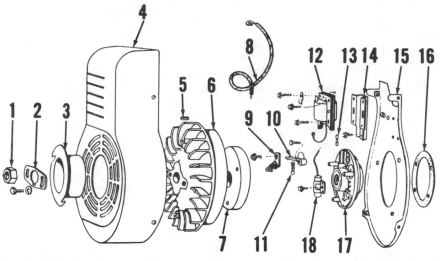

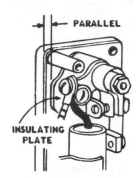

*Fig. B41—When installing breaker points on 19, 23 and 23A models and 191000 and 231000 series engines, be sure dowel on breaker point base enters hole in insulator. Place sides of base and insulating plate parallel with edge of breaker box.*

*Fig. B39 — Exploded view of magneto ignition system used on 23C model engines. Breaker points and condenser are mounted on bearing support (17) and are enclosed by cover (7) and flywheel. Points are actuated by plunger (13) which rides against breaker cam machined on engine crankshaft.*

1. Flywheel nut
2. Nut retainer
3. Starter pulley
4. Blower housing
5. Flywheel key
6. Flywheel
7. Breaker cover
8. Spark plug wire
9. Breaker point base
10. Breaker point arm
11. Breaker spring
12. Coil & armature
13. Plunger
14. Armature support
15. Back plate
16. Shim gasket
17. Bearing support
18. Condenser

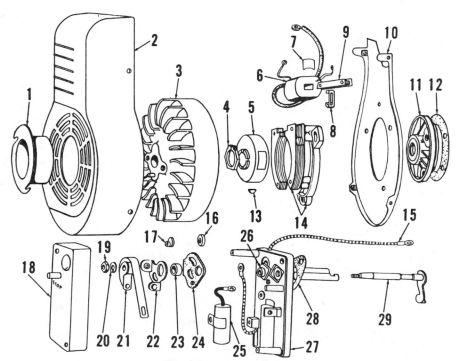

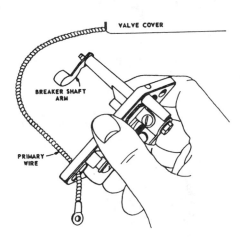

*Fig. B42—To remove breaker box on 19, 23 and 23A models and 191000 and 231000 series engines it is not necessary to remove points and condenser as unit may be removed as an assembly.*

*Fig. B40 — Exploded view of magneto ignition system used on 19, 23 and 23A models and 191000 and 231000 series engines. Flywheel is not keyed to crankshaft and may be installed in any position; however, on crank start models, flywheel should be installed as shown in Fig. B47. Breaker arm (21) is mounted on shaft (29) which is actuated by a cam on engine cam gear. See Fig. B48. Two different methods of attaching magneto rotor (5) to engine crankshaft have been used. Refer to Figs. B43 and B44.*

1. Starter pulley
2. Blower housing
3. Flywheel
4. Rotor clamp
5. Magneto rotor
6. Ignition coil
7. Coil clip
8. Coil core
9. Coil core
10. Back plate
11. Bearing support
12. Shim gasket
13. Rotor key
14. Armature
15. Primary coil lead
16. Shaft seal
17. Eccentric
18. Breaker box cover
19. Nut
20. Washer
21. Breaker point arm
22. Breaker point base
23. Pivot
24. Insulator
25. Condenser
26. Seal retainer
27. Breaker box
28. Gasket
29. Shaft

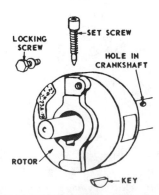

*Fig. B43—On some models, magneto rotor is fastened to crankshaft by a set screw which enters hole in shaft. Set screw is locked inplace by a second screw. Refer to Fig. B44 also.*

Illustrations courtesy of Briggs & Stratton Corp.

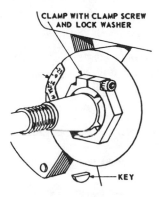

Fig. B44 — View showing magneto rotor fastened to crankshaft with clamp. Make certain split in clamp is between two slots in rotor as shown and check clearance between rotor and shoulder on crankshaft as shown in Fig. B45.

justed. With point gap adjusted to 0.020 inch (0.508 mm), connect a static timing light from breaker point terminal to ground (coil primary wire must be disconnected) and turn engine in normal direction of rotation until light goes on. Then, turn engine very slowly in same direction until light just goes out (breaker points start to open). At this time, engine model number on magneto rotor should be aligned with arrow on armature as shown in Fig. B46. If not, loosen armature core retaining cap screws and turn armature in slotted

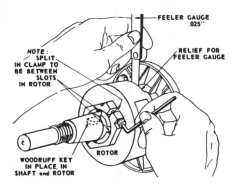

Fig. B45 — Models having magneto rotor clamped to crankshaft, position rotor 0.025 inch (0.635 mm) from shoulder on shaft, then tighten clamping screw.

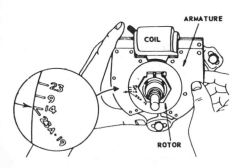

Fig. B46 — With engine turned so breaker points are just starting to open, align model number line on rotor with arrow on armature by rotating armature in slotted mounting holes.

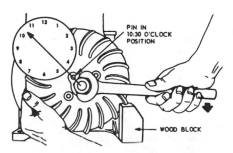

Fig. B47 — On 19, 23 and 23A models and 191000 and 231000 series engines equipped with crank starter, mount flywheel with pin in position shown with armature timing marks aligned.

mounting holes so arrow is aligned with appropriate engine model number on rotor. Tighten armature mounting screws.

To install flywheel on models with crank starter, place flywheel on crankshaft as shown in Fig. B47 with magneto timing marks aligned as in previous paragraph. On models not having a crank starter, flywheel may be installed in any position.

Magneto breaker points are actuated by a cam on a centrifugal weight mounted on engine camshaft (Fig. B48). When engine is being overhauled, or cam gear is removed, check action of advance spring and centrifugal weight unit by holding cam gear in position shown and pressing weight down. When weight is released, spring should return weight to its original position. If weight does not return to its original position, check weight for binding and renew spring.

MODELS 19D & 23D MAGNETO. Refer to Fig. B49 for exploded view of magneto system.

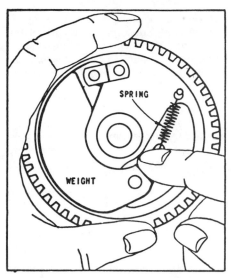

Fig. B48 — Check timing advance weight and spring on 19, 23 and 23A models and 191000 and 231000 series engines. Refer to text.

Condenser and breaker points are mounted externally in a breaker box located on carburetor side of engine and are accessible after removing cover (8 – Fig. B49).

Breaker point gap is 0.020 inch (0.508 mm) and condenser capacity is 0.18-0.24 mfd.

Installation of new breaker points is made easier by turning engine so points are open to their widest gap before removing old points. For method of adjusting breaker point gap, refer to Fig. B50.

**NOTE: When installing points, apply Permatex or equivalent sealer to retaining screw threads to prevent engine oil from leaking into breaker box.**

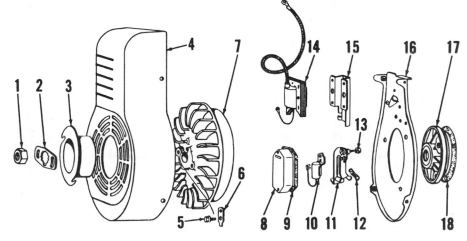

Fig. B49 — Exploded view of magneto ignition system used on 19D and 23D models. Magneto rotor (flywheel) to armature timing is adjustable as shown in Figs. B51 and B52. Adjust breaker points as in Fig. B50.

1. Flywheel nut
2. Nut retainer
3. Starter pulley
4. Blower housing
5. Key cap screw
6. Flywheel key
7. Flywheel
8. Breaker box cover
9. Gasket
10. Condenser
11. Breaker points
12. Breaker spring
13. Lock nut
14. Coil & armature assy.
15. Armature mounting bracket
16. Back plate
17. Bearing support
18. Shim gasket

Breaker points are actuated by a plunger that rides against breaker cam on engine cam gear. Plunger and plunger bushing in crankcase are renewable after removing cam gear and breaker points.

Magneto armature and ignition coil are mounted outside engine flywheel. Adjust armature air gap to 0.022-0.026 inch (0.5588-0.6604 mm).

If flywheel has been removed, magneto edge gap (armature timing) must be adjusted. With point gap adjusted to 0.020 inch (0.508 mm) and flywheel loosely installed on crankshaft, install flywheel key leaving retaining cap screw loose. Disconnect magneto primary wire and connect test light across breaker points. Turn flywheel in clockwise direction until breaker points just start to open (timing light goes out). Now, while making sure engine crankshaft does not turn, turn flywheel back slightly in counter-clockwise direction so edge of flywheel insert lines up with edge of armature as shown in Fig. B53. Tighten flywheel key screw, then tighten flywheel retaining nut to 110-118 ft.-lbs. (149-160 N·m) torque for 19D model and to 138-150 ft.-lbs. (187-203 N·m) torque on 23D model. Readjust armature air gap as necessary.

SERIES 193000, 200000, 233000, 243000, 300000, 301000, 302000, 320000, 325000 & 326000 MAGNETO. Refer to Fig. 54 for exploded view of magneto system.

Condenser and breaker points are mounted externally in a breaker box located on carburetor side of engine and are accessible after removing cover (8 – Fig. B54).

Breaker point gap is 0.020 inch (0.508 mm) and condenser capacity is 0.18-0.24 mfd.

Installation of new breaker points is made easier by turning engine so points

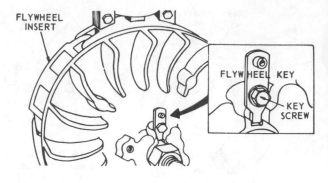

Fig. B51—On 19D and 23D models, magneto timing is adjusted by repositioning flywheel on crankshaft. Flywheel is then locked into proper position by tightening key screw and flywheel retaining nut. Refer also to Figs. B52 and B53.

Fig. B52—Turning engine in normal direction of rotation with flywheel and flywheel key loose. Turn engine slowly until breaker points are just starting to open. Refer to text and Figs. B51 and B53.

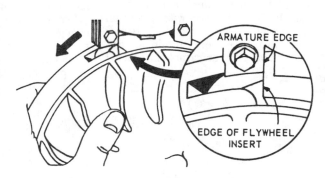

Fig. B53—Turn flywheel counter-clockwise on crankshaft until edge of flywheel insert is aligned with edge of armature as shown in insert. Then, tighten flywheel key screw and retaining nut.

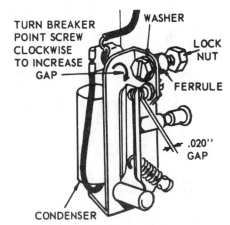

Fig. B50—On all models equipped with this type of breaker points, loosen lock nut and turn adjusting screw to obtain 0.020 inch (0.508 mm) gap.

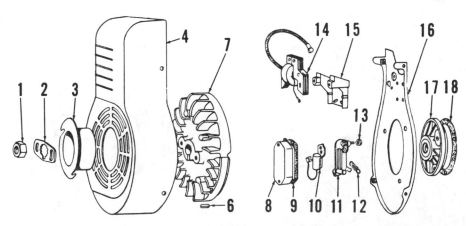

Fig. B54—Exploded view of 193000, 200000, 233000 and 243000 series magneto ignition system. Magneto used on 300000, 301000, 302000, 320000, 325000 and 326000 series is similar. Position of armature is adjustable to time armature with magneto rotor (flywheel) by moving armature mounting bracket (15) on slotted mounting holes. Refer to text and Figs. B55 and B56.

1. Flywheel nut
2. Nut retainer
3. Starter pulley
4. Blower housing
6. Flywheel key
7. Flywheel
8. Breaker box cover
9. Gasket
10. Condenser
11. Breaker points
12. Breaker spring
13. Lock nut
14. Coil & armature assy.
15. Armature mounting bracket
16. Back plate
17. Bearing support
18. Shim gasket

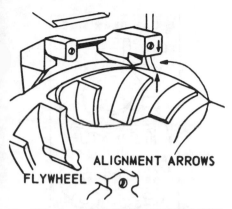

Fig. B55—On 193000, 200000, 233000, 243000, 300000, 301000, 302000, 320000, 325000 and 326000 series, time magneto by aligning armature core support so arrow on support is aligned with arrow on flywheel when breaker points are just starting to open. Refer to text for procedure.

are open to their widest gap before removing old points. For method of adjusting breaker point gap, refer to Fig. B50.

**NOTE: When installing points, apply Permatex or equivalent sealer to retaining screw threads to prevent engine oil from leaking into breaker box.**

Breaker points are actuated by a plunger that rides against breaker cam on engine cam gear. Plunger (43—Fig. B62) and plunger bushing (PB) are renewable after removing engine cam gear and breaker points. On 300000, 301000, 302000, 320000, 325000 and 326000 series engines breaker plunger and plunger bushing are similar to those shown in Fig. B62.

Magneto armature and ignition coil are mounted outside engine flywheel. Adjust armature air gap to 0.010-0.014 inch (0.254-0.3556 mm).

If flywheel has been removed, magneto edge gap (armature timing) must be adjusted. With point gap ad-

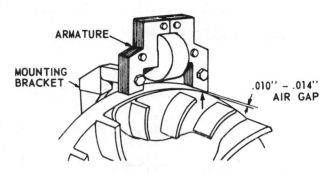

Fig. B57—After mounting bracket is properly installed (Fig. B56), install armature and coil assembly so there is a 0.010-0.014 inch (0.254-0.3556 mm) air gap between armature and flywheel.

justed to 0.020 inch (0.508 mm) and armature ignition coil assembly removed from bracket connect a test light across breaker points. Disconnect coil primary wire and slowly turn flywheel in a clockwise direction until light just goes out (breaker points start to open). At this time, arrow on flywheel should be exactly aligned with arrow on armature mounting bracket (Fig. B55). If not, mark position of bracket, remove flywheel and shift bracket on slotted mounting holes (Fig. B56) to bring arrows into alignment.

Reinstall flywheel, make certain arrows are aligned and tighten flywheel nut to 110-118 ft.-lbs. (149-160 N·m) torque on 193000 and 200000 series engines and 138-150 ft.-lbs. (187-203 N·m) torque on all remaining models. Readjust armature air gap as necessary.

MAGNETRON IGNITION. Magnetron ignition is a self-contained breakerless ignition system. Flywheel does not need to be removed except to change timing by moving armature bracket or to service crankshaft key or keyway.

To check spark, remove spark plug and connect spark plug cable to B&S tester, part number 19051 and ground remaining tester lead to engine cylinder head. Spin engine at 350 rpm or more. If spark jumps the 0.166 inch (4.2 mm) tester gap, system is functioning properly.

To remove armature and Magnetron module, remove flywheel shroud and armature retaining screws. Use a 3/16-inch diameter pin punch to release stop switch wire from module. To remove module, unsolder wires, push module retainer away from laminations and push module off. See Fig. B57A.

Resolder wires for reinstallation and use Permatex or equivalent gasket sealer to hold ground wires in position.

To set timing for gasoline fuel operation, install armature bracket so mounting screws are centered in slots (Fig. B57B).

To set timing for kerosene fuel operation, install armature bracket as far to left as possible (Fig. B57B).

Adjust armature air gap to 0.010-0.014 inch (0.25-0.36 mm).

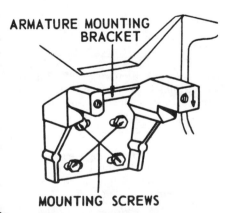

Fig. B56—View with flywheel removed on 193000, 200000, 233000, 243000, 300000, 301000, 302000, 320000, 325000 and 326000 series engines. Magneto is timed by shifting armature mounting bracket on slotted mounting holes. Refer to Fig. B55 also.

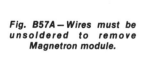

Fig. B57A—Wires must be unsoldered to remove Magnetron module.

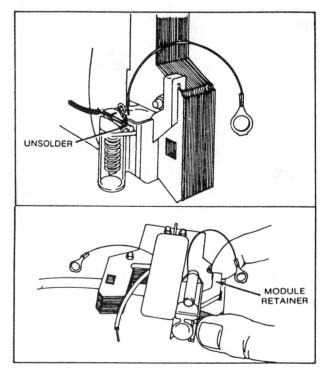

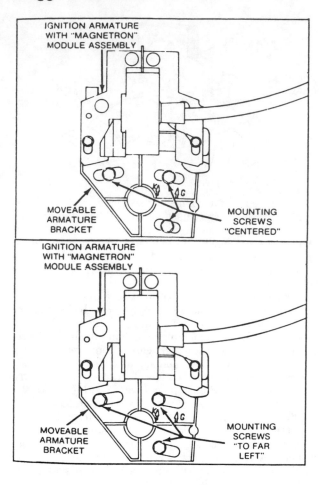

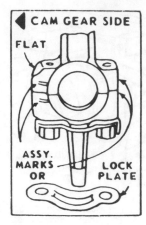

Fig. B57B—Upper view shows position of armature bracket for gasoline fuels and lower view shows position of armature bracket for kerosene fuels for correct ignition timing.

Fig. B59—Drawing of lower end of connecting rod showing clearance flat and assembly marks.

It is recommended carbon and lead deposits be removed at 100 to 300 hour intervals, or whenever cylinder head is removed.

**CONNECTING ROD.** Connecting rod and piston are removed from cylinder head end of block as an assembly. Aluminum alloy connecting rod rides directly on induction hardened crankshaft crankpin journal. Rod should be renewed if crankpin bore is scored or out-of-round more than 0.0007 inch (0.0178 mm) or if piston pin bore is scored or out-of-round more than 0.0005 inch (0.0127 mm).

**LUBRICATION.** All models are splash lubricated by an oil dipper attached to connecting rod.

Oils approved by manufacturer must meet requirements of API service classification SE or SF.

Use SAE 10W-40 oil for temperatures above 20° F (-7° C) and SAE 5W-30 oil for temperatures below 20° F (-7° C).

Check oil level at five hour intervals and maintain at bottom edge of filler plug. **DO NOT** overfill.

Recommended oil change interval for all models is every 25 hours of normal operation.

Crankcase oil capacity for 18.56 and 20.32 cubic inch models is 3 pints (1.4 L) and crankcase oil capacity for all remaining models is 4 pints (1.9 L).

**REPAIRS**

**CYLINDER HEAD.** When removing cylinder head note location from which different cap screws are removed as they must be reinstalled in their original positions.

Always use a new gasket when reinstalling cylinder head. Do not use sealer on gasket. Lubricate cylinder head bolt threads with graphite grease, install in correct locations and tighten in several even steps in sequence shown in Fig. B58 according to type head used. Tighten cap screws to 190 in.-lbs. (22 N·m) torque. Start and run engine until normal operating temperature is reached, stop engine and retighten head bolts as outlined.

### REJECT SIZES FOR CONNECTING ROD

| Model | Crankpin bore | Pin bore* |
|---|---|---|
| 19, 19D | | |
| 190000 | 1.001 in. (25.43 mm) | 0.674 in. (17.12 mm) |
| 200000 | 1.127 in. (28.63 mm) | 0.674 in. (17.12 mm) |
| 23, 23A, 23C, 23D, | | |
| 230000 | 1.189 in. (30.20 mm) | 0.736 in. (18.69 mm) |
| 240000 | 1.314 in. (33.38 mm) | 0.674 in. (17.12 mm) |
| 300000, 320000 | 1.314 in. (33.38 mm) | 0.802 in. (20.37 mm) |

*Piston pins of 0.005 inch (0.127 mm) oversize are available for service. Piston pin bore in rod can be reamed to this size if crankpin bore is within specification.

Connecting rod running clearance to crankpin is 0.0015-0.0045 inch (0.0381-0.1143 mm) and piston pin to rod pin bore clearance is 0.0005-0.0015 inch (0.0127-0.0381 mm).

On 300000, 301000, 302000, 320000, 325000 or 326000 series engines, notch on top of piston and letter "F" on side of piston must be on the same side as assembly marks on connecting rod

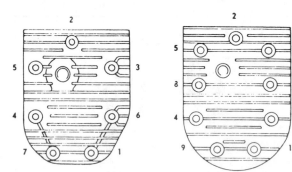

Fig. B58—View of two different cylinder heads used. Tighten cylinder head cap screws in sequence shown to 190 in.-lbs. (22 N·m) torque. Refer to text for tightening procedure.

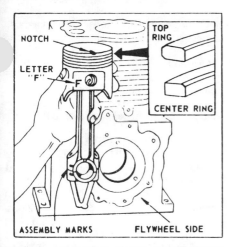

Fig. B60 — On 300000, 301000, 302000, 320000, 325000 and 326000 series, assemble piston to connecting rod with notch and stamped letter "F" on piston to same side as assembly marks on rod. Install assembly in cylinder with assembly marks to flywheel side of crankcase.

(Fig. B60). Install assembly in cylinder with assembly marks to flywheel side of crankcase.

On all remaining models connecting rod position in crankcase is determined by clearance flat on rod as shown in Fig. B59. Assemble cap to rod by matching assembly marks.

Tighten connecting rod cap screws on all models to 190 in.-lbs. (22 N·m) torque.

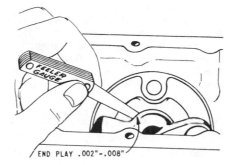

Fig. B61 — Crankshaft end play can be checked as shown on models with main bearings which are an integral part of bearing support plates.

**PISTON, PIN AND RINGS.** Aluminum alloy piston has two compression and one oil control ring and piston should be renewed if top ring side clearance in groove exceeds 0.007 inch (0.178 mm) or if piston is visibly scored or damaged. Piston should also be renewed or pin bore reamed for 0.005 inch (0.127 mm) oversize pin if pin bore is worn to, or greater than rejection diameter shown in chart.

## PISTON PIN BORE REJECTION SIZES

| Model | Pin bore |
|---|---|
| 19, 19D, 191000, 193000, 200000, 243000 | 0.673 in. (17.09 mm) |
| 23, 23A, 23C, 23D, 231000, 233000 | 0.736 in. (18.69 mm) |
| 300000, 301000, 302000, 320000, 325000, 326000 | 0.801 in. (20.35 mm) |

Renew piston pin if pin is 0.0005 inch (0.0127 mm) out-of-round or if pin is worn to, or below, rejection diameter shown in chart.

## PISTON PIN REJECTION SIZES

| Model | Pin diameter |
|---|---|
| 19, 19D, 191000, 193000, 200000, 243000 | 0.671 in. (17.04 mm) |
| 23, 23A, 23C, 23D, 231000, 233000 | 0.734 in. (18.64 mm) |
| 300000, 301,000, 302,000, 320000, 325000, 326000 | 0.799 in. (20.30 mm) |

Ring end gap should be 0.010-0.025 inch (0.245-0.635 mm) and compression ring should be renewed if end gap is greater than 0.030 inch (0.75 mm). Oil control ring should be renewed if end gap is greater than 0.035 inch (0.90 mm).

Pistons used in 300000, 301000, 302000, 320000, 325000 and 326000 series engines have a notch and a letter "F" stamped in piston (Fig. B60) which

19. Spark plug
20. Air baffle
21. Cylinder head
22. Gasket
23. Breather tube
24. Breather
25. Main bearing plate
26. Gasket
27. Engine crankcase & cylinder block
28. Governor shaft
29. Governor gear & weight unit
30. Camshaft
31. Cam gear
32. Tappets
33. Camshaft plug
34. Engine base
35. Gasket
36. Valves
37. Valve spring washers
38. Valve springs
39. Spring retainers or "Roto-Caps"
40. Keepers
41. Gasket
42. Tappet chamber cover
43. Breaker point plunger
44. Governor control lever
45. Governor spring
46. Governor crank
47. Governor control rod
48. Governor link
49. Governor lever
50. Output drive key
51. Crankshaft
52. Ball bearings (on models so equipped)
53. Oil dipper
54. Rod bolt lock
55. Connecting rod
56. Piston pin retaining rings
57. Piston pin
58. Piston
59. Piston rings

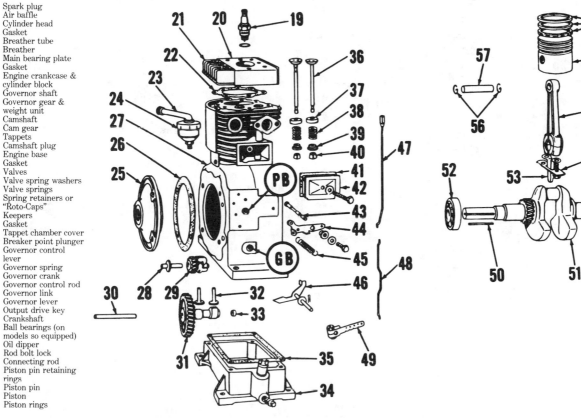

Fig. B62 — Exploded view of typical 19D, 23D, 193000, 200000, 233000 and 243000 engines. Breaker plunger bushing (PB) and governor crank bushing (GB) in engine crankcase (27) are renewable. Breaker plunger (43) rides against a lobe on cam gear (31).

must face flywheel side of engine after installation.

Pistons and rings are available in a variety of oversizes as well as standard.

**CYLINDER.** Standard cylinder bore diameters are listed in following chart.

## STANDARD CYLINDER BORE SIZES

| Model | Standard bore |
|---|---|
| 19, 19D, 191000, 193000, 23, 23A, 23C, 23D, 200000 231000, 233000 | 2.999-3.000 in. (76.18-76.23 mm) |
| 243000 | 3.0615-3.0625 in. (76.79-76.82 mm) |
| 300000, 301000, 302000 | 3.4365-3.4375 in. (87.287-87.313 mm) |
| 320000, 325000, 326000 | 3.5615-3.5625 in. (90.462-90.488 mm) |

If cylinder is worn 0.003 inch (0.076 mm) or more, or out-of-round 0.0025 inch (0.0635 mm) or more, it should be resized to nearest oversize for which piston and rings are available.

**NOTE: A chrome piston ring set is available for slightly worn standard bore cylinders. No honing or cylinder deglazing is required for these rings. Cylinder bore can be a maximum of 0.005 inch (0.127 mm) oversize when using chrome rings.**

**CRANKSHAFT AND MAIN BEARINGS.** Crankshaft may be supported at each end in main bearings which are an integral part of crankcase and sump or ball bearing mains which are a press fit on crankshaft and fit into machined bores in main bearing support plates.

Crankshaft for models with main bearings as an integral part of main bearing support plates should be renewed or reground if journals are out-of-round 0.0007 inch (0.0178 mm) or more or if main bearing journals are worn to a diameter of 1.179 inch (29.95 mm) or less for 19, 19D models and 191000, 193000 and 200000 series engines or 1.3759 inch (34.95 mm) or less for 23, 23A, 23C, 23D models and 231000 and 233000 series engines.

Crankshaft for models with ball bearing main bearings should be renewed if

new bearings are loose on journals. Bearings should be a press fit.

Crankshaft for all models should be renewed or reground if connecting rod crankpin journal diameter is worn to, or beyond crankshaft rejection specifications as listed in following chart.

## CRANKSHAFT REJECTION SIZES

| Model | Crankpin journal |
|---|---|
| 19, 19D, 190000, 193000 | 0.996 in. (25.30 mm) |
| 200000 | 1.122 in. (28.5 mm) |
| 23, 23A, 23C, 23D, 230000 | 1.184 in. (30.07 mm) |
| 243000, 300000, 301000, 302000, 320000, 325000, 326000 | 1.309 in. (33.25 mm) |

Connecting rod is available for crankshaft which has had connecting rod crankpin journal reground to 0.020 inch (0.508 mm) undersize.

Crankshaft main bearing support plates should be renewed if integral type

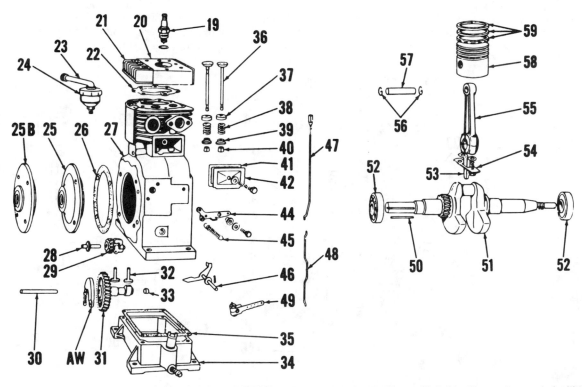

*Fig. B63–Exploded view of typical model 19, 23, 23A, 191000 and 231000 engines equipped with "Magna-Matic" ignition system; note ignition advance weight (AW). Refer to Fig. B40 for ignition system.*

| | | | |
|---|---|---|---|
| AW. Advance weight | 26. Gasket | 34. Engine base | 42. Tappet chamber cover |
| 19. Spark plug | 27. Engine crankcase & cylinder block | 35. Gasket | 44. Governor control lever |
| 20. Air baffle | 28. Governor shaft | 36. Valves | 45. Governor spring |
| 21. Cylinder head | 29. Governor gear & weight unit | 37. Valve spring washers | 46. Governor crank |
| 22. Head gasket | 30. Camshaft | 38. Valve springs | 47. Governor control rod |
| 23. Breather tube | 31. Cam gear | 39. Spring retainers or "Roto-Caps" | 48. Governor link |
| 24. Breather | 32. Tappets | 40. Keepers | 49. Governor lever |
| 25. Bearing plate (plain bushing) | 33. Camshaft plug | 41. Gasket | 50. Output drive key |
| 25B. Bearing plate (ball bearing) | | | 51. Crankshaft |
| | | | 52. Ball bearings (on models so equippe) |
| | | | 53. Oil dipper |
| | | | 54. Rod bolt lock |
| | | | 55. Connecting rod |
| | | | 56. Retaining rings |
| | | | 57. Piston pin |
| | | | 58. Piston |
| | | | 59. Piston rings |

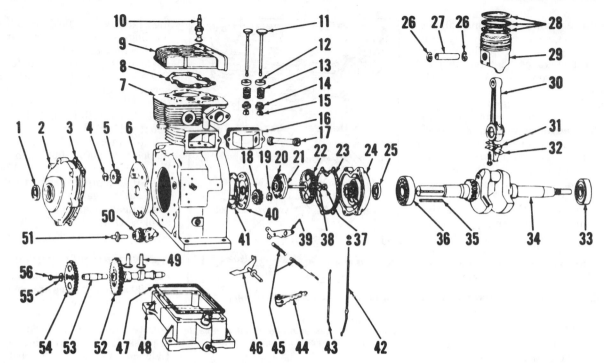

**Fig. B64 — Exploded view of typical 300000, 301000, 302000, 320000, 325000 and 326000 engines. Balance weights, ball bearings and cover assemblies (2 and 24) are serviced only as assemblies.**

1. Oil seal
2. Cover & balance assy. (pto end)
3. Gasket
4. "E" ring
5. Idler gear
6. Bearing support plate
7. Cylinder block
8. Head gasket
9. Cylinder head
10. Spark plug
11. Valves
12. Spring caps
13. Valve springs
14. Spring retainer or rotator
15. Keepers
16. Breather assy.
17. Breather tube
18. Idler gear
19. "E" ring
20. Shim
21. Cam bearing
22. Balancer drive gear
23. Gasket
24. Cover & balance assy. (magneto end)
25. Oil seal
26. Retaining rings
27. Piston pin
28. Piston rings
29. Piston
30. Connecting rod
31. Oil dipper
32. Rod bolt lock
33. Ball bearing
34. Crankshaft
35. Key
36. Ball bearing
37. Drive gear bolt
38. Belleville washer
39. Governor control lever
40. Bearing support plate
41. Shim
42. Control rod
43. Link
44. Governor lever
45. Governor springs
46. Governor crank assy.
47. Gasket
48. Engine base
49. Valve lifters
50. Governor gear & weight unit
51. Governor shaft
52. Cam gear
53. Camshaft
54. Balancer drive gear
55. Belleville washer
56. Drive gear bolt

main bearing bores are 0.0007 inch (0.0178 mm) or more out-of-round or if bearing bore diameter is 1.185 inches (30.1 mm) or more for 19, 19D models or 191000, 193000 and 200000 series engines, 1.382 inches (34.1 mm) or more for 23, 23A, 23C, 23D models or 243000, 300000, 301000, 302000, 320000, 325000 and 326000 series engines. No service bushings for main bearings are available.

Ball bearing mains are a press fit on crankshaft and must be removed by pressing crankshaft out of bearing. Renew ball bearing if worn or rough. Expand new bearing by heating in oil and install on crankshaft with shield side towards crankpin journal.

Crankshaft end play should be 0.002-0.008 inch (0.0508-0.2032 mm) and is adjusted by varying thickness of gaskets between flywheel main bearing support plate and crankcase. Gaskets are available in a variety of thicknesses.

When reinstalling crankshaft, make certain timing marks are aligned (Fig. B66 or B67).

### CAM GEAR AND CAMSHAFT. On 19, 19D, 23, 23A, 23C, 23D models or

191000, 193000, 200000, 233000 and 243000 series engines timing gear and camshaft lobes are cast as an integral part and are referred to as a "cam gear". Cam gear turns on a stationary shaft referred to as a "camshaft". Reject camshaft if it is worn to, or less than, a diameter of 0.4968 inch (12.62 mm).

On all remaining models cam gear and lobes are an integral part referred to as a "cam gear". Camshaft (53 – Fig. B64) rotates with cam gear (52). Camshaft rides in a bearing in cylinder block on pto end of engine and journal on cam-gear rides in renewable cam bearing (21)

on magneto end of engine.

Renew camshaft if worn to 0.6145 inch (15.6 mm) or less and renew cam gear (52) if journal diameter is 0.8105 inch (20.6 mm) or less or if lobes are worn to a diameter of 1.184 inches (30.1 mm) or less for 300000, 301000 and 302000 series engines or 1.215 inches (30.9 mm) or less for 320000, 325000 and 326000 series engines.

When installing cam gear in 300000, 301000, 302000, 320000, 325000 and 326000 series engines, cam gear end play should be 0.002-0.008 inch (0.0508-0.2032 mm) and is controlled by

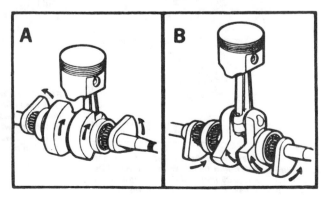

**Fig. B65 — Synchro-Balance weights rotate in opposite direction of crankshaft counterweights on 300000, 301000, 302000, 320000, 325000 and 326000 series engines.**

using different thickness shims (20) between cam gear bearing (21) and crankcase. Shims are available in a variety of thicknesses. Tighten cam gear bearing cap screws to 85 in.-lbs. (10 N·m) torque.

On models with "Magna-Matic" ignition system, cam gear is equipped with an ignition advance weight (AW–Fig. B63). A tang on advance weight contacts breaker shaft lever (29–Fig. B40) each camshaft revolution. On all other models, breaker plunger rides against a lobe on cam gear.

On models with "Easy-Spin" starting intake cam lobe is machined to hold intake valve slightly open during part of compression stroke, thereby relieving compression and making engine easier to start due to increased cranking speed.

**NOTE: To check compression on models with "Easy-Spin" starting, engine must be turned backwards.**

"Easy-Spin" cam gears can be identified by two holes drilled in web of gear. If part number of older cam gear and "Easy-Spin" cam gear are the same except for an "E" following new part number, gears are interchangeable.

On all models, align timing marks on crankshaft gear and cam gear when reassembling engine.

**GOVERNOR WEIGHT UNIT.** Tangs on governor weights should be square and smooth and weights should operate freely. If not, renew gear and weight assembly (29–Figs. B62 and B63) or (50–Fig. B64). Renew governor shaft if worn or scored.

**GOVERNOR CRANK.** With governor weight and gear unit removed,

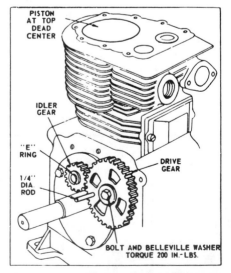

Fig. B66—View showing balancer drive gear (magneto end) being timed. With piston at TDC, insert 1/4-inch (6.35 mm) rod through timing hole in crankshaft bearing support plate.

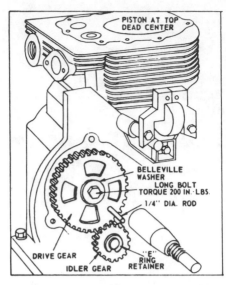

Fig. B67—View showing balancer drive gear on pto end of crankshaft being timed. With piston at TDC, insert 1/4-inch (6.35 mm) rod through timing hole in gear and into locating hole in crankshaft bearing support plate.

remove and inspect governor crank; renew if worn. Governor crank should be free fit in bushing in engine crankcase with minimum bushing to crank clearance. Renew bushing if new crank fits loosely. Bushing should be finish reamed to 0.2385-0.239 inch (6.0579-6.0706 mm) after installation.

**VALVE SYSTEM.** All models are equipped with renewable exhaust valve seat inserts and intake valve seats are machined directly into cylinder block. Seats are ground at a 45° angle and seat width should be 3/64 to 1/16-inch (1.2 to 1.6 mm). Service exhaust and intake valve seat inserts are available and are installed using special tools available from Briggs & Stratton.

Valve face is ground at a 45° angle and valve should be renewed if margin is 1/64-inch (0.4 mm) or less after refacing.

Valve guides should be checked for wear using valve guide gage number 19151. If gage enters valve guide 5/16-inch (7.9 mm) or more, guides should be reamed using reamer number 19183 and bushing number 19192 should be installed.

Briggs & Stratton also had a tool kit, part number 19232, available for remov-

ing factory or field installed guide bushings so new bushing, part number 231218 may be installed.

Valve tappet gap (cold) for 19, 19D models or 191000, 193000 and 200000 series engines is 0.007-0.009 inch (0.18-0.23 mm) for intake valves and 0.014-0.016 inch (0.36-0.41 mm) for exhaust valves.

Valve tappet gap (cold) for all other models is 0.007-0.009 inch (0.18-0.23 mm) for intake valves and 0.17-0.19 inch (0.43-0.48 mm) for exhaust valves.

To adjust valves, piston must be at "top dead center" on compression stroke. Grind end of valve stem off squarely to obtain clearance.

**SYNCHRO-BALANCER.** 300000, 301000, 302000, 320000, 325000 and 326000 series engines are equipped with rotating balance weights at each end of crankshaft. Balancers are geared to rotate in opposite direction of crankshaft counterweights (Fig. B65). Balance weights, ball bearings and covers (2 and 24–Fig. B64) are serviced as assemblies only.

Balancers are driven from idler gears (5 and 18) which are driven by gears (22 and 54). Drive gears (22 and 54) are bolted to camshaft. To time balancers, first remove cover and balancer assemblies (2 and 24). Position piston at TDC. Loosen bolts (37 and 56) until drive gears will rotate on cam gear and camshaft. Insert a 1/4-inch (6.35 mm) rod through timing hole in each drive gear and into locating holes in main bearing support plates as shown in Figs. B66 and B67. With piston at TDC and 1/4-inch (6.35 mm) rods in place, tighten drive gear bolts to a torque of 200 in.-lbs. (23 N·m). Remove 1/4-inch (6.35 mm) rods. Remove timing hole screws (Fig. B68) and insert 1/8-inch (3.18 mm) rods through timing holes and into hole in balance weights. Then, with piston at TDC, carefully slide cover assemblies into position. Tighten cap screws in pto end cover to 200 in.-lbs. (23 N·m) torque and tighten magneto end cover cap screws to 120 in.-lbs. (14 N·m) torque. Remove 1/8-inch (3.18 mm) rods. Coat threads of timing hole screws with Permatex or equivalent and install screws with fiber sealing washers.

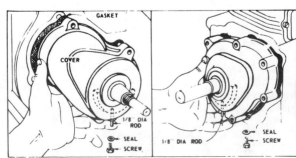

Fig. B68—Insert 1/8-inch (3.18 mm) rod through timing hole in covers and into hole in balance weights when installing cover assemblies. Piston must be at TDC.

# BRIGGS & STRATTON

### BRIGGS & STRATTON CORPORATION
#### Milwaukee, Wisconsin 53201

| Model | No. Cyls. | Bore | Stroke | Displacement | Power Rating |
|-------|-----------|------|--------|--------------|--------------|
| 400400 | 2 | 3.44 in. (87.3 mm) | 2.16 in. (54.8 mm) | 40 cu. in. (656 cc) | 14 hp. (10.4 kW) |
| 400700 | 2 | 3.44 in. (87.3 mm) | 2.16 in. (54.8 mm) | 40 cu. in. (656 cc) | 14 hp. (10.4 kW) |
| 401400 | 2 | 3.44 in. (87.3 mm) | 2.16 in. (54.8 mm) | 40 cu. in. (656 cc) | 16 hp. (11.9 kW) |
| 401700 | 2 | 3.44 in. (87.3 mm) | 2.16 in. (54.8 mm) | 40 cu. in. (656 cc) | 16 hp. (11.9 kW) |
| 402400 | 2 | 3.44 in. (87.3 mm) | 2.16 in. (54.8 mm) | 40 cu. in. (656 cc) | 16 hp. (11.9 kW) |
| 402700 | 2 | 3.44 in. (87.3 mm) | 2.16 in. (54.8 mm) | 40 cu. in. (656 cc) | 16 hp. (11.9 kW) |
| 421400 | 2 | 3.44 in. (87.3 mm) | 2.28 in. (57.9 mm) | 42.33 cu. in. (694 cc) | 18 hp. (13.4 kW) |
| 421700 | 2 | 3.44 in. (87.3 mm) | 2.28 in. (57.9 mm) | 42.33 cu. in. (694 cc) | 18 hp. (13.4 kW) |
| 422400 | 2 | 3.44 in. (87.3 mm) | 2.28 in. (57.9 mm) | 42.33 cu. in. (694 cc) | 18 hp. (13.4 kW) |
| 422700 | 2 | 3.44 in. (87.3 mm) | 2.28 in. (57.9 mm) | 42.33 cu. in. (694 cc) | 18 hp. (13.4 kW) |

Engines in this section are four cycle, two cylinder opposed, horizontal or vertical crankshaft engines. Crankshaft is supported at each end in bearings which are an integral part of crankcase and sump or cover, DU type bearings or ball bearings which fit into machined bores of crankcase and sump or cover. Cylinder block and crankcase are a single aluminum casting, however, some models are equipped with cast iron cylinder liners which are an integral part of aluminum casting.

Connecting rods for all models ride directly on crankpin journals. Vertical crankshaft models are lubricated by a gear driven oil slinger and horizontal crankshaft models are splash lubricated by an oil dipper attached to number one cylinder connecting rod cap.

Early models are equipped with a flywheel magneto ignition with points, condenser and coil mounted externally on engine. Late models are equipped with "Magnetron" breakerless ignition.

All models use a float type carburetor with an integral fuel pump.

Always give engine model and serial number when ordering parts or service material.

## MAINTENANCE

**SPARK PLUG.** Recommended spark plug for all models is Champion RJ12. Electrode gap is 0.030 inch (0.762 mm).

**CARBURETOR.** A downdraft float type carburetor with integral fuel pump is used. Refer to Fig. B70.

For initial carburetor adjustment, open idle mixture screw (11 – Fig. B70) and main fuel mixture screw (10) 1½-turns each.

Make final adjustments with engine at operating temperature and running. Place governor speed control lever in idle position and hold throttle lever against idle stop. Turn idle speed adjusting screw (15) to obtain 1400 rpm. Adjust idle mixture screw to obtain smoothest idle operation. Hold throttle shaft in closed position and readjust idle speed adjusting screw (15) to obtain 900 rpm. Release throttle. Move remote control to a position where a ⅛-inch (3.18

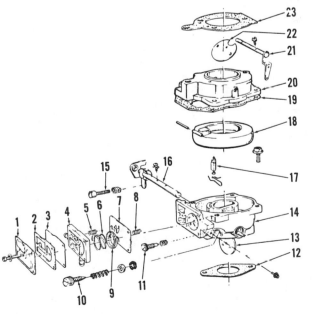

**Fig. B70 — Exploded view of downdraft Flo-Jet with integral fuel pump.**

1. Diaphragm cover
2. Gasket
3. Damping diaphragm
4. Pump body
5. Spring
6. Pump spring
7. Diaphragm
8. Spring
9. Spring cap
10. Needle valve assy.
11. Idle valve assy.
12. Mounting gasket
13. Throttle valve
14. Lower carburetor body
15. Throttle adjustment screw
16. Throttle shaft
17. Fuel inlet valve
18. Float
19. Carburetor body gasket
20. Upper carburetor body
21. Choke shaft
22. Choke valve
23. Air cleaner gasket

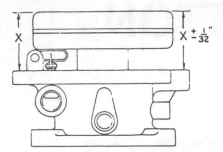

*Fig. B70A—Check carburetor float setting as shown. Bend tang if necessary to adjust float level.*

mm) diameter pin can be inserted through two holes in governor control plate (Fig. B71). With remote control in governed idle position, bend tab "A", Fig. B72, to obtain 1400 rpm. Remove pin. Place governor speed control lever in fast position and adjust main fuel mixture screw for leanest mixture that will allow satisfactory acceleration and steady governor operation.

To disassemble carburetor, remove idle and main fuel mixture screws. Remove fuel pump body and upper carburetor body. Remove float assembly and fuel inlet valve. Inlet valve seat is a press fit in upper carburetor body. Use a self threading screw to remove seat. New seat should be pressed into upper carburetor body until flush with body. Remainder of carburetor disassembly is evident after inspection and reference to Fig. B70.

If necessary to renew throttle shaft bushings, use a ¼ x 20 tap to remove old bushings. Press new bushings in using a vise and ream with a 7/32-inch drill if throttle shaft binds.

To check float level, invert carburetor body and float assembly. Refer to Fig. B70A for proper float level dimensions. Adjust by bending float lever tang that contacts inlet valve. Reassemble carburetor by reversing disassembly procedure.

Correct choke and speed control operation is dependent upon proper ad-

justment of remote controls. To adjust choke, place control lever in "CHOKE" position. Loosen casing clamp screw. Move casing and wire until choke is completely closed and tighten screw.

**FUEL PUMP.** All parts of the vacuum-diagram type pump are serviced separately. When disassembling pump, care must be taken to prevent damage to pump body (plastic housing) and diaphragm. Inspect diaphragm for punctures, wrinkles or wear. All mounting surfaces must be free of nicks, burrs and debris.

To assemble pump, position diaphragm on carburetor. Place spring and cup on top of diaphragm. Install flapper valve springs. Carefully place pump body, remaining diaphragm, gasket and cover plate over carburetor casting and install mounting screws. Tighten screws in a staggered sequence to avoid distortion.

**GOVERNOR.** All models are equipped with a gear driven mechanical governor. Governor gear and weight assembly is enclosed within the engine and is driven by the camshaft gear. Refer to Fig. B74.

To adjust governor, loosen nut holding governor lever to governor shaft. Push governor lever counter-clockwise until throttle is wide open. Hold lever in this position while rotating governor shaft counter-clockwise as far as it will go. Tighten governor lever nut to a torque of 100 in.-lbs. (11 N·m).

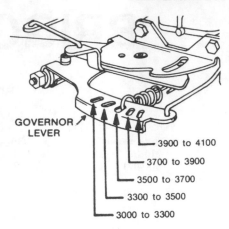

*Fig. B73—Governor spring should be installed with end loops as shown. Install loop in appropriate governor lever hole for engine speed (rpm) desired.*

To adjust top governed speed, first adjust carburetor and governed idle as previously outlined. Install governor spring end loop in appropriate hole in governor lever for desired engine rpm as shown in Fig. B73. Check engine top governed speed using an accurate tachometer.

Governor gear and weight assembly can be removed when engine is disassembled. Loosen nut and remove governor lever. To obtain maximum clearance, rotate crankshaft until timing mark on crankshaft gear is at approximately 10 o'clock position. Remove "E" ring and thick washer on outer end of governor shaft. Carefully slide shaft down past crankshaft.

**CAUTION: Be careful not to bind shaft against crankshaft as governor lower bearing could be damaged.**

To reassemble governor shaft, refer to Fig. B75 and reverse disassembly procedure.

Governor gear assembly located on cover (Fig. B74) is serviced as an

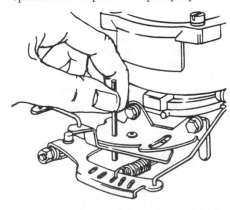

*Fig. B71—Insert a 1/8-inch (3.18 mm) diameter pin through the two holes in governor control plate to correctly set governed idle position.*

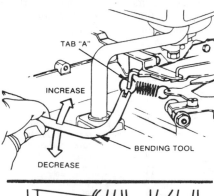
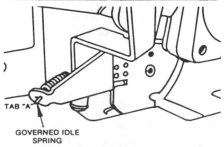

*Fig. B72—With governor plate locked with a 1/8-inch (3.18 mm) pin (Fig. B71), bend tab "A" to obtain 1400 rpm. Upper view is for a horizontal crankshaft engine and lower view is for vertical crankshaft engines.*

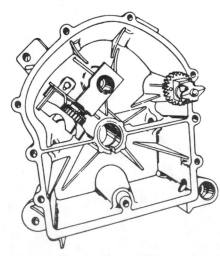

*Fig. B74—Typical governor gear assembly position in crankcase.*

Illustrations courtesy of Briggs & Stratton Corp.

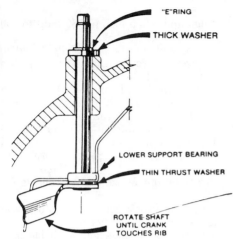

Fig. B75 — Cross-sectional view of governor shaft assembly. To disassemble, remove "E" ring and washer. Carefully guide shaft down past crankshaft.

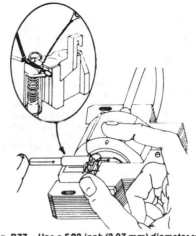

Fig. B77 — Use a 5/32-inch (3.97 mm) diameter rod to release wires from "Magnetron" module. Refer to text.

assembly only. Note that there is a thrust washer installed between cover and governor gear assembly.

**IGNITION SYSTEM.** Early production engines were equipped with a flywheel type magneto ignition with points, condenser and coil located externally on engine and late production engines are equipped with "Magnetron" breakerless ignition system.

Refer to appropriate paragraph for model being serviced.

FLYWHEEL MAGNETO IGNITION. Flywheel magneto system consists of a permanent magnet cast into flywheel, armature and coil assembly, breaker points and condenser.

Breaker points and condenser are located under or behind intake manifold and are protected by a metal cover which must be sealed around edges and at wire entry location to prevent entry of dirt or moisture.

Breaker point gap should be 0.020 inch (0.508 mm) for all models.

Breaker points are actuated by a plunger (Fig. B76) which is installed with the smaller diameter end toward breaker points. Renew plunger if length

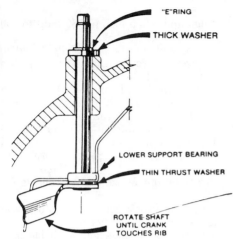

Fig. B76 — Plunger must be renewed if plunger length (A) is worn to 1.115 inch (28.32 mm) or less.

is 1.115 inch (28.32 mm) or less. Renew plunger seal by installing seal on plunger (make certain it is securely attached) and installing seal and plunger assembly into plunger bore. Slide seal over plunger boss until seated against casting at base of boss.

Armature air gap should be 0.010-0.014 inch (0.25-0.36 mm) and is adjusted by loosening armature retaining bolts and moving armature as necessary on slotted holes. Tighten armature retaining bolts.

MAGNETRON IGNITION. "Magnetron" ignition consists of permanent magnets cast in flywheel and a self-contained transistor module mounted on ignition armature.

To check ignition, attach B&S tester number 19051 to each spark plug lead and ground tester to engine. Spin flywheel rapidly. If spark jumps the 0.166 inch (4.2 mm) tester gap, ignition system is operating satisfactorily.

Armature air gap should be 0.008-0.012 inch (0.20-0.30 mm) and is adjusted by loosening armature retaining bolts and moving armature as necessary on slotted holes. Tighten armature bolts.

Flywheel does not need to be removed to service "Magnetron" ignition except to check condition of flywheel key or keyway.

To remove Magnetron module from armature, remove stop switch wire, module primary wire and armature primary wire from module by using a 5/32 inch (3.97 mm) rod to release spring and retainer (Fig. B77). Remove spring and retainer clip. Unsolder wires. Remove module by pulling out on module retainer while pushing down on module until free of armature laminations.

During reinstallation, use 60/40 rosin

core solder and make certain all wires are held firmly against coil body with tape or "Permatex" number 2 or equivalent gasket sealer.

**LUBRICATION.** Vertical crankshaft models are splash lubricated by a gear driven oil slinger (5 – Fig. B79) and horizontal crankshaft models are splash lubricated by an oil dipper attached to number one connecting rod.

Oils approved by manufacturer must meet requirements of API service classification SE or SF.

Use SAE 10W-40 oil for temperatures above 20°F (−7°C) and SAE 5W-20 oil for temperatures below 20°F (-7°C).

Check oil at regular intervals and maintain at "FULL" mark on dipstick. Dipstick should be pushed or screwed in completely for accurate measurement. **DO NOT** overfill.

Recommended oil change interval for all models is every 25 hours of normal operation.

Crankcase oil capacity for early production engines is 3.5 pints (1.65 L) and for late production engines oil capacity is 3 pints (1.42 L). Check oil level with dipstick. **DO NOT** overfill.

**CRANKCASE BREATHER.** Crankcase breathers are built into engine valve covers. Horizontal crankshaft models have a breather valve in each cover assembly and vertical crankshaft models have only one breather in cover of number one cylinder.

Breathers maintain a partial vacuum in crankcase to prevent oil from being forced out past oil seals and gaskets or past breaker point plunger or piston rings.

Fiber disc of breather assembly must not be stuck or binding. A 0.045 inch (1.14 mm) wire gage **SHOULD NOT** enter space between fiber disc valve and body. Check with gage at 90° intervals around fiber disc.

When installing breathers make certain side of gasket with notches is toward crankshaft.

## REPAIRS

**CYLINDER HEADS.** When removing cylinder heads, note locations from which different length bolts are removed as they must be reinstalled in their original positions.

Always use a new gasket when reinstalling cylinder head. Do not use sealer on gasket. Lubricate cylinder head bolt threads with graphite grease, install in correct locations and tighten in several even steps in sequence shown in Fig. B78 to 160 in.-lbs. (18 N·m) torque.

It is recommended carbon and lead deposits be removed at 100 to 300 hour

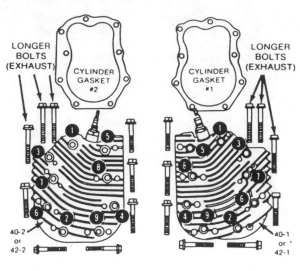

CYLINDER GASKET #2

CYLINDER GASKET #1

LONGER BOLTS (EXHAUST)

LONGER BOLTS (EXHAUST)

40-2 or 42-2

40-1 or 42-1

*Fig. B78 — Note locations of various length cylinder head bolts. Long bolts are used around exhaust valve area. Bolts should be tightened in sequence shown to 160 in.-lbs. (18 N·M) torque.*

intervals, or whenever cylinder head is removed.

**CONNECTING RODS.** Connecting rods and pistons are removed from cylinder head end of block as an assembly. Aluminum alloy connecting rods ride directly on crankpin journals.

Connecting rod should be renewed if crankpin bearing bore measures 1.627 inches (41.33 mm) or more, if bearing surfaces are scored or damaged or if pin bore measures 0.802 inch (20.37 mm) or more.

**NOTE: A 0.005 inch (0.127 mm) oversize piston pin is available.**

Connecting rod should be installed on piston so oil hole in connecting rod is toward cam gear side of engine when notch on top of piston is toward flywheel with piston and rod assembly installed. Make certain match marks on connecting rod and cap are aligned and if equipped with an oil dipper (horizontal crankshaft models), it should be installed on number one rod. Install special washers and nuts and tighten to 190 in.-lbs. (22 N·m) torque for all models.

**PISTONS, PINS AND RINGS.** Pistons used in engines with cast iron cylinder liners (Series 400400, 400700, 402400, 402700, 422400 and 422700) have an "L" on top of piston. A chrome plated aluminum piston is used in aluminum bore (Kool-Bore) cylinders. Due to the different cylinder bore materials, pistons **WILL NOT** interchange.

Pistons for all models should be renewed if they are scored or damaged, or if a 0.007 inch (0.178 mm) feeler gage can be inserted between a new top ring and ring groove. If piston pin bore measures 0.801 inch (20.35 mm) or more, piston should be renewed or pin bore reamed for 0.005 inch (0.127 mm) oversize pin.

Piston ring end gaps for 401400, 401700, 421400 and 421700 models should be 0.035 inch (0.889 mm) for compression rings and 0.045 inch (1.14 mm) for oil control ring. Piston ring end gaps for 400400, 400700, 402400, 402700, 422400 and 422700 models should be 0.030 inch (0.762 mm) for compression

rings and 0.035 inch (0.889 mm) for oil control ring.

Ring end gaps on all models should be staggered around diameter of piston during installation.

On all models pistons must be installed with notch on top of piston toward flywheel side of engine.

Standard piston pin diameter is 0.799 inch (20.30 mm) and should be renewed if worn or out-of-round more than 0.0005 inch (0.0127 mm). A 0.005 inch (0.127 mm) oversize pin is available for all models.

Piston pin is a slip fit in both piston and connecting rod bores.

Pistons and rings are available in a variety of oversizes as well as standard and as with pistons, ring sets for aluminum bore (Kool-Bore) engines and ring sets for engines with cast iron cylinder liners should not be interchanged.

**CYLINDERS.** Cylinder bores may be either aluminum, or a cast iron liner which is an integral part of the cylinder block casting. Pistons and rings for aluminum cylinders and cast iron

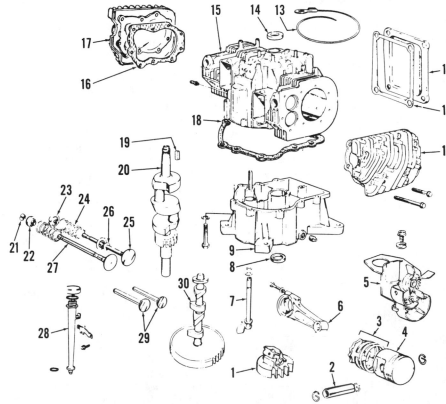

*Fig. B79 — Exploded view of vertical crankshaft engine assembly.*

| | | |
|---|---|---|
| 1. Governor gear | 9. Crankcase assy. | 17. Cylinder head |
| 2. Piston pin & clips | 10. Cylinder head | 18. Crankcase gasket |
| 3. Piston rings | 11. Crankcase cover plate | 19. Key |
| 4. Piston | 12. Crankcase gasket | 20. Crankshaft |
| 5. Oil slinger assy. | 13. Ground wire | 21. Retainer |
| 6. Connecting rod | 14. Oil seal | 22. Rotocoil (exhaust valve) |
| 7. Governor shaft assy. | 15. Cylinder assy. | 23. Retainer (intake valve) |
| 8. Oil seal | 16. Head gasket | 24. Valve spring (2) |
| | | 25. Intake valve |
| | | 26. Seal and retainer assy. |
| | | 27. Exhaust valve |
| | | 28. Oil dipstick assy. |
| | | 29. Valve tappet (2) |
| | | 30. Camshaft assy. |

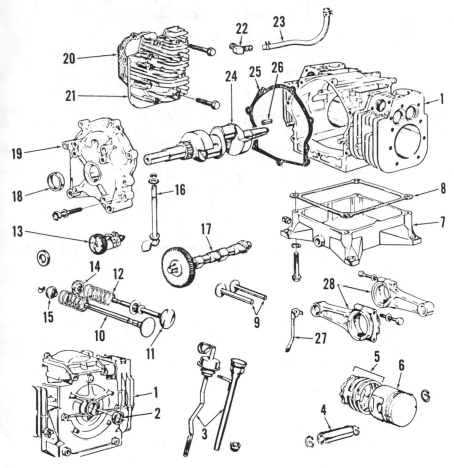

**Fig. B80 — Exploded view of horizontal crankshaft engine assembly.**

| | | |
|---|---|---|
| 1. Cylinder assy. | 8. Crankcase gasket | 15. Rotocoil (exhaust |
| 2. Oil seal | 9. Valve tappets | valve) |
| 3. Dipstick assy. | 10. Exhaust valve | 16. Governor shaft |
| 4. Piston pin & | 11. Intake valve | 17. Camshaft |
| retainer clips | 12. Valve spring | 18. Oil seal |
| 5. Piston rings | 13. Governor gear assy. | 19. Crankcase cover |
| 6. Piston | 14. Intake valve | 20. Head gasket |
| 7. Crankcase | retainer | 21. Cylinder head |

| | |
|---|---|
| 22. Elbow connector | |
| 23. Fuel line | |
| 24. Crankshaft | |
| 25. Crankcase cover | |
| gasket | |
| 26. Key | |
| 27. Oil dipper | |
| 28. Connecting rods | |

cylinders **SHOULD NOT** be interchanged. Series 400400, 400700, 402400, 402700, 422400 and 422700 have cast iron cylinder liners as an integral part of cylinder block casting and Series 401400, 401700, 421400 and 421700 have aluminum cylinder bores

(Kool-Bore). Standard cylinder bore for all models is 3.4365-3.4375 inches (87.29-87.31 mm) and should be resized using a suitable hone (B&S part number 19205 for aluminum bore or 19211 for cast iron bore) if more than 0.003 inch (0.0762 mm) oversize or 0.0015 inch (0.0381 mm) out-of-round for cast iron liner engines or 0.003 inch (0.0762 mm)

oversize or 0.0025 inch (0.0635 mm) out-of-round for aluminum bore (Kool-Bore) engines. Resize to nearest oversize for which piston and rings are available.

**CRANKSHAFT AND MAIN BEARINGS.** Crankshaft may be supported at each end in main bearings which are an integral part of crankcase, cover or sump, DU type bearings or ball bearing mains which are a press fit on crankshaft and fit into machined bores in crankcase, cover or sump.

To remove crankshaft from engines with integral type or DU type main bearings, remove necessary air shrouds. Remove flywheel and front gear cover or sump. Remove cam gear making certain valve tappets clear camshaft lobes. Remove crankshaft.

To reinstall crankshaft, reverse removal procedure making certain timing marks are aligned as shown in Fig. B81.

To remove crankshaft from engines with ball bearing main bearing, remove any necessary air shrouds and remove flywheel. Remove front gear cover or sump. Compress exhaust and intake valve springs on number two cylinder to provide clearance for camshaft lobes. Remove crankshaft and camshaft together.

To reinstall crankshaft, reverse removal procedure making certain timing marks are aligned as shown in Fig. B82.

Crankshaft for models with integral type or DU type main bearings should be renewed if main bearing journals measure 1.376 inches (34.95 mm) or less. Crankshaft should be renewed or reground if crankpin journal measures 1.622 inches (41.15 mm) or less. Connecting rods for 0.020 inch (0.508 mm) undersize journal is available.

Ball bearing main bearing is a press fit on crankshaft and must be removed by pressing crankshaft out of bearing. Renew ball bearing if worn or rough. Expand new bearing by heating in oil and install with shield side towards crankpin.

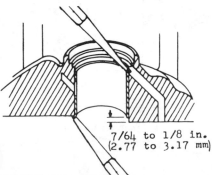

7/64 to 1/8 in. (2.77 to 3.17 mm)

**Fig. B80A — Press DU type bearings in until 7/64 to ⅛-inch (2.77 to 3.17 mm) from thrust surface. Make certain oil holes are aligned and stake bearing as shown. Refer to text.**

**Fig. B81 — Align timing marks as shown on models having main bearings as an integral part of crankcase, cover or sump. Refer to text.**

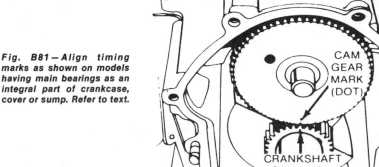

CAM GEAR MARK (DOT)

CRANKSHAFT TIMING MARK

Integral type main bearings should be reamed out and service bushings installed if 0.0007 inch (0.0178 mm) or more out-of-round or if they measure 1.383 inches (35.13 mm) or more in diameter. Special reamers are available from Briggs & Stratton.

DU type main bearings should be renewed if 0.0007 inch (0.0178 mm) or more out-of-round or if they measure 1.383 inches (35.13 mm) or more in diameter.

Worn DU type bearings are pressed out of bores using B&S cylinder support number 19227 and driver number 19226. Make certain oil holes in bearings will align with holes in block and cover or sump and press bearings in until they are 7/64 to 1/8-inch (2.77 to 3.17 mm) below thrust face. Stake bearings into place. See Fig. B80A.

If ball bearing is loose in crankcase, cover or sump bores, crankcase, cover or sump must be renewed.

Crankshaft end play for all models should be 0.002-0.008 inch (0.050-0.200 mm). At least one 0.015 inch cover or sump gasket must be used. Additional cover gaskets of 0.005 and 0.009 inch thickness are available if end play is less than 0.002 inch (0.050 mm). If end play is over 0.008 inch (0.200 mm), metal shims are available for use on crankshaft between crankshaft gear and cover or sump.

**NOTE: Thrust washer can not be used on double ball bearing engines.**

**CAMSHAFT.** Camshaft and camshaft gear are an integral part which ride in journals at each end of camshaft.

To remove camshaft, refer to appropriate **CRANKSHAFT AND MAIN BEARING** section for model being serviced.

Camshaft should be renewed if gear teeth or lobes are worn or damaged or if bearing journals measure 0.623 inches (15.82 mm) or less.

**VALVE SYSTEM.** Valves seat in renewable inserts pressed into cylinder head surfaces of block. Valve seat width should be 3/64 to 1/16-inch (1.17 to 1.57 mm) and are ground at a 45° angle for exhaust seats and a 30° angle for intake valves. If seats are loose or damaged, renew seat as shown in Fig. B83 and grind to correct angle.

Valves should be refaced at 45° angle for exhaust valves and 30° angle for intake valves. Valves should be renewed if margin is less than 1/64-inch (0.10 mm). See Fig. B84.

When reinstalling valves, note exhaust valve spring is shorter, has heavier diameter coils and is usually painted red. Intake valve has a stem seal

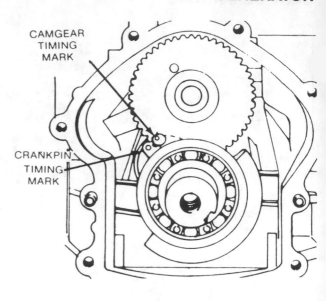

Fig. B82 – Align timing marks as shown on models having ball bearing type main bearings. Refer to text.

which should be renewed if valve has been removed.

Valve guides should be checked for wear using valve guide gage (B&S tool number 19151). If gage enters guide 5/16-inch (7.9 mm) or more, guide should be reconditioned or renewed.

To recondition aluminum valve guides use B&S tool kit number 19232. Place reamer (B&S tool number 19231) and guide (B&S tool number 19234) in worn guide and center with valve seat. Mark reamer 1/16-inch (1.57 mm) above top edge of service bushing (Fig. B85). Ream worn guide until mark on reamer is flush with top of guide bushing. **DO**

**NOT** ream completely through guide. Place service bushing, part number 231218, on driver (B&S tool number 19204) so grooved end of bushing will enter guide bore first. Press bushing into guide until it bottoms. Finish ream completely through guide using B&S reamer number 19233 and guide number 19234. Lubricate reamer with kerosene during reaming procedure.

To renew brass or sintered iron guides, use tap (B&S tool number 19264) to thread worn guide bushing approximately 1/2-inch (12.7 mm) deep. **DO NOT** thread more than 1 inch (25.4 mm) deep. Install puller washer (B&S tool

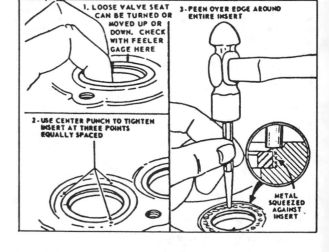

Fig. B83 – Loose inserts can be tightened or renewed as shown. If a 0.005 inch (0.127 mm) feeler gage can be inserted between seat and seat bore, renew cylinder.

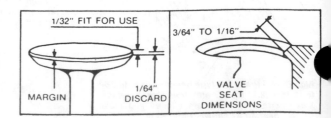

Fig. B84 – View showing correct valve face and seat dimensions. Refer to text.

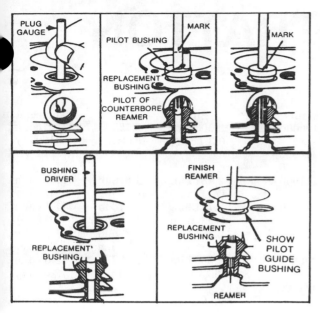

*Fig. B85 — View showing correct procedure for reconditioning valve guides. Refer to text.*

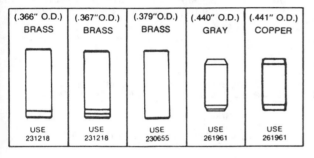

*Fig. B86 — Use correct bushing as indicated when renewing valve guide bushings.*

number 19240) on puller screw (B&S tool number 19238) and thread screw and washer assembly into worn guide. Center washer on valve seat and tighten puller nut (B&S tool number 19239) against washer, continue to tighten while keeping threaded screw from turning (Fig. B85) until guide has been removed. Identify guide using Fig. B86 and find appropriate replacement. Place correct service bushing on driver (B&S tool number 19024) so the two grooves on service bushings number 231218 are down. Remaining bushing types can be installed either way. Press bushing in until it bottoms. Finish ream with B&S reamer number 19233 and reamer guide number 19234. Lubricate reamer with kerosene and ream completely through new service bushing.

To adjust valves for all models, cylinder to be adjusted must be at "top dead center" on compression stroke. If springs are installed, stem end clearance should be 0.007-0.009 inch (0.18-0.23 mm) for exhaust valves and 0.004-0.006 inch (0.10-0.15 mm) for intake valves. Without springs, stem end clearance should be 0.009-0.0011 inch (0.23-0.28 mm) for exhaust valves and 0.006-0.008 inch (0.15-0.23 mm) for intake valves. If clearance is less than specified, grind end of stem as necessary. If clearance is excessive, grind valve seat deeper as necessary.

# BRIGGS & STRATTON

| Model | Bore | Stroke | Displacement |
|---|---|---|---|
| 6B, | | | |
| Early 60000 | 2.3125 in. | 1.500 in. | 6.3 cu. in. |
| | (58.7 mm) | (38.1 mm) | (103 cc) |
| Late 60000 | 2.3750 in. | 1.500 in. | 6.7 cu. in. |
| | (60.3 mm) | (38.1 mm) | (109 cc) |
| 8B, 80000 | 2.3750 in. | 1.750 in. | 7.8 cu. in. |
| | (60.3 mm) | (44.5 mm) | (127 cc) |
| 81000, 82000 | 2.3125 in. | 1.750 in. | 7.8 cu. in. |
| | (60.3 mm) | (44.5 mm) | (127 cc) |
| 92000, 93500 | 2.5625 in. | 1.750 in. | 9.0 cu. in. |
| | (65.1 mm) | (44.5 mm) | (148 cc) |
| 94000, 94500, | | | |
| 94900 | 2.5625 in. | 1.750 in. | 9.0 cu. in. |
| | (65.1 mm) | (44.5 mm) | (148 cc) |
| 95500 | 2.5625 in. | 1.750 in. | 9.0 cu. in. |
| | (65.1 mm) | (44.5 mm) | (148 cc) |

| Model | Bore | Stroke | Displacement |
|---|---|---|---|
| 100000 | 2.5000 in. | 2.125 in. | 10.4 cu. in. |
| | (63.1 mm) | (54.0 mm) | (169 cc) |
| 110900, | | | |
| 111200, | | | |
| 111900 | 2.7813 in. | 1.875 in. | 11.4 cu. in. |
| | (70.6 mm) | (47.6 mm) | (186 cc) |
| 112200 | 2.7813 in. | 1.875 in. | 11.4 cu. in. |
| | (70.6 mm) | (47.6 mm) | (186 cc) |
| 113900 | 2.7813 in. | 1.875 in. | 11.4 cu. in. |
| | (70.6 mm) | (47.6 mm) | (186 cc) |
| 130000 | 2.5625 in. | 2.438 in. | 12.6 cu .in. |
| | (65.1 mm) | (60.9 mm) | (206 cc) |
| 131400, | | | |
| 131900 | 2.5625 in. | 2.438 in. | 12.6 cu. in. |
| | (65.1 mm) | (60.9 mm) | (206 cc) |
| 140000 | 2.7500 in. | 2.375 in. | 14.1 cu. in. |
| | (69.9 mm) | (60.3 mm) | (231 cc) |

| CUBIC INCH DISPLACEMENT | FIRST DIGIT AFTER DISPLACEMENT — BASIC DESIGN SERIES | SECOND DIGIT AFTER DISPLACEMENT — CRANKSHAFT, CARBURETOR GOVERNOR | THIRD DIGIT AFTER DISPLACEMENT — BEARINGS, REDUCTION GEARS & AUXILIARY DRIVES | FOURTH DIGIT AFTER DISPLACEMENT — TYPE OF STARTER |
|---|---|---|---|---|
| 6 | 0 | 0 - | 0 - Plain Bearing | 0 - Without Starter |
| 8 | 1 | 1 - Horizontal Vacu-Jet | 1 - Flange Mounting Plain Bearing | 1 - Rope Starter |
| 9 | 2 | 2 - Horizontal Pulsa-Jet | 2 - Ball Bearing | 2 - Rewind Starter |
| 10 | 3 | 3 - Horizontal Flo-Jet (Pneumatic Governor) | 3 - Flange Mounting Ball Bearing | 3 - Electric - 110 Volt, Gear Drive |
| 11 | 4 | 4 - Horizontal Flo-Jet (Mechanical Governor) | 4 - | 4 - Elec. Starter-Generator - 12 Volt, Belt Drive |
| 13 | 5 | 5 - Vertical Vacu-Jet | 5 - Gear Reduction (6 to 1) | 5 - Electric Starter Only - 12 Volt, Gear Drive |
| 14 | 6 | 6 - | 6 - Gear Reduction (6 to 1) Reverse Rotation | 6 - Alternator Only * |
| 17 | 7 | 7 - Vertical Flo-Jet | 7 - | 7 - Electric Starter, 12 Volt Gear Drive, with Alternator |
| 19 | 8 | 8 - | 8 - Auxiliary Drive Perpendicular to Crankshaft | 8 - Vertical-pull Starter |
| 20 | 9 | 9 - Vertical Pulsa-Jet | 9 - Auxiliary Drive Parallel to Crankshaft | |
| 23 | | | | |
| 24 | | | | |
| 25 | | | | |
| 30 | | | | |
| 32 | | | | |

\* Digit 6 formerly used for "Wind-Up" Starter on 60000, 80000 and 92000 Series

## EXAMPLES

To identify Model 100202:

| 10 | 0 | 2 | 0 | 2 |
|---|---|---|---|---|
| 10 Cubic Inch | Design Series 0 | Horizontal Shaft - Pulsa-Jet Carburetor | Plain Bearing | Rewind Starter |

Similarly, a Model 92998 is described as follows:

| 9 | 2 | 9 | 9 | 8 |
|---|---|---|---|---|
| 9 Cubic Inch | Design Series 2 | Vertical Shaft - Pulsa-Jet Carburetor | Auxiliary Drive Parallel to Crankshaft | Vertical Pull Starter |

*Fig. BS59—Explanation of numerical code used by Briggs & Stratton to identify engine and optional equipment from model number.*

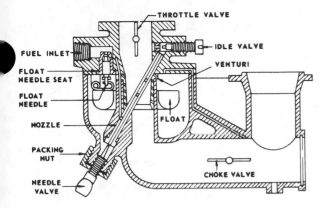

THROTTLE VALVE

IDLE VALVE

FUEL INLET

VENTURI

FLOAT
NEEDLE SEAT

FLOAT
NEEDLE

FLOAT

NOZZLE

PACKING
NUT

NEEDLE
VALVE

CHOKE VALVE

*Fig. BS60—Cross-sectional view of typical B&S "two-piece" carburetor. Before separating upper and lower body sections, loosen packing nut and unscrew nut and needle valve as a unit. Then, using special screwdriver, remove nozzle. Refer to text.*

## MAINTENANCE

**SPARK PLUG.** Recommended spark plug for all models is a Champion 8, or equivalent. If resistor type plugs are necessary to decrease radio interference, use Champion RJ8 or equivalent. Electrode gap for all models is 0.030 inch (0.76 mm).

**CAUTION: Briggs & Stratton does not recommend using abrasive blasting method to clean spark plugs as this may introduce some abrasive material into the engine which could cause extensive damage.**

## ENGINE IDENTIFICATION

Engines covered in this section have aluminum cylinder blocks with either plain aluminum cylinder bore or with a

cast iron sleeve integrally cast into the block.

Early production of the 60000 model engine were of the same bore and stroke as the 6B model engine. The bore on 60000 engine was changed from 2.3125 inches (58.7 mm) to 2.3750 inches (60.3 mm) at serial number 5810060 on engines with plain aluminum bore, and at serial number 5810030 on engines with a cast iron sleeve.

Refer to Fig. BS59 for chart explaining engine numerical code to identify engine model. Always furnish correct engine model and model number when ordering parts or service information.

**FLOAT TYPE (FLO-JET) CARBURETORS.** Three different float type carburetors are used. They are called a "two-piece" (Fig. BS60), a small "one-piece" (Fig. BS64) or a large "one-piece" (Fig. BS66) carburetor depending upon the type of construction.

Float type carburetors are equipped with adjusting needles for both idle and power fuel mixtures. Counterclockwise rotation of the adjusting needles richens the mixture. For initial starting adjustment, open the main needle valve (power fuel mixture) 1½ turns on the two-piece carburetor and 2½ turns on the small one-piece carburetor. Open the idle needle ½ to ¾ turn on the two-piece carburetor and 1½ turns on the small one-piece carburetor. On the large one-piece carburetor, open both needle valves 1-1/8 turns.

Make final adjustments with engine at operating temperature and running. Set the speed control for desired operating speed, turn main needle clockwise until engine misses, and then turn it counterclockwise just past the smooth operating point until the engine begins to run unevenly. Return the speed control to idle position and adjust the idle speed

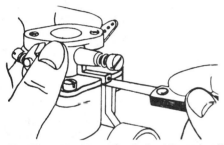

*Fig. BS61—Checking upper body of "two-piece" carburetor for warpage. Refer to text.*

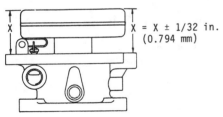

X = X ± 1/32 in. (0.794 mm)

*Fig. BS62—Carburetor float setting should be within specifications shown. To adjust float setting, bend tang with needlenose pliers as shown in Fig. BS63.*

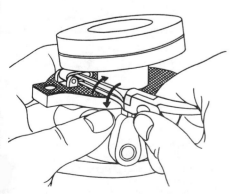

*Fig. BS63—Bending tang with needlenose pliers to adjust float setting. Refer to Fig. BS62 for method of checking float setting.*

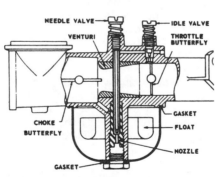

NEEDLE VALVE

IDLE VALVE

VENTURI

THROTTLE BUTTERFLY

CHOKE BUTTERFLY

GASKET

FLOAT

NOZZLE

GASKET

*Fig. BS64—Cross-sectional view of typical B&S small "one-piece" float type carburetor. Refer to Fig. BS65 for disassembly views.*

*Fig. BS65—Disassembling the small "one-piece" float type carburetor. Pry out welch plug, remove choke butterfly (disc), remove choke shaft and needle valve; venturi can then be removed as shown in left view.*

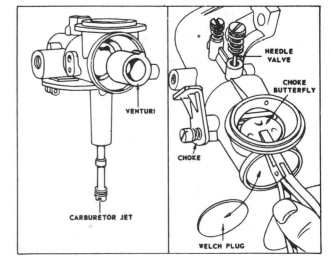

VENTURI

CARBURETOR JET

NEEDLE VALVE

CHOKE BUTTERFLY

CHOKE

WELCH PLUG

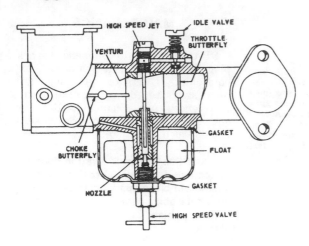

Fig. BS66—Cross-sectional view of B&S large "one-piece" float type carburetor.

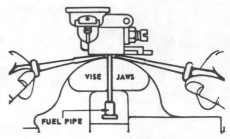

Fig. BS68—Removing brass fuel feed pipe from suction type carburetor. Press new brass pipe into carburetor until it projects 2-9/32 to 2-5/16 inches (57.9-58.7 mm) from carburetor face. Nylon fuel feed pipe is threaded into carburetor.

stop screw until the engine idles at 1750 rpm. Adjust the idle needle valve until the engine runs smoothly. Reset the idle speed stop screw if necessary. The engine should then accelerate without hesitation. If engine does not accelerate properly, turn the main needle valve counterclockwise slightly to provide a richer fuel mixture.

The float setting on all float type carburetors should be within dimensions shown in Fig. BS62. If not, bend the tang on float as shown in Fig. BS63 to adjust float setting. If any wear is visible on the inlet valve or the inlet valve seat, install a new valve and seat assembly. On large one-piece carburetors, the renewable inlet valve seat is pressed into the carburetor body until flush with the body.

NOTE: The upper and lower bodies of the two-piece float type carburetor are locked together by the main nozzle. Refer to cross-sectional view of carburetor in Fig. BS60. Before attempting to separate the upper body from the lower body, loosen packing nut and unscrew nut and needle valve. Then, using special screwdriver (B&S tool 19061 or 19062), remove nozzle.

If a 0.002 inch (0.05 mm) feeler gage can be inserted between upper and lower bodies of the two-piece carburetor as shown in Fig. BS61, the upper body is warped and should be renewed.

Check the throttle shaft for wear on all float type carburetors. If 0.010 inch (0.25 mm) or more free play (shaft-to-bushing clearance) is noted, install new throttle shaft and/or throttle shaft bushings. To remove worn bushings, turn a ¼ inch × 20 tap into bushing and pull bushing from body casting with the tap. Press new bushings into casting by using a vise and if necessary, ream bushings with a 7/32 inch drill bit.

**SUCTION TYPE (VACU-JET) CARBURETORS.** A typical suction type (Vacu-Jet) carburetor is shown in Fig. BS67. This type carburetor has only one fuel mixture adjusting needle. Turning the needle clockwise leans the air:fuel mixture. Adjust suction type carburetors with fuel tank approximately one-half full and with the engine at operating temperature and running at approximately 3000 rpm, no-load. Turn needle valve clockwise until engine begins to run unevenly from a too rich air:fuel mixture. This should result in a correct adjustment for full load operation. Adjust idle speed to 1750 rpm.

To remove the suction type carburetor, first remove carburetor and fuel tank as an assembly, then remove carburetor from fuel tank. When reinstalling carburetor on fuel tank, use a new gasket and tighten retaining screws evenly.

The suction type carburetor has a fuel feed pipe extending into fuel tank. The pipe has a check valve to allow fuel to feed up into the carburetor but prevents fuel from flowing back into the tank. If check valve is inoperative and cleaning in alcohol or acetone will not free the check valve, renew the fuel feed pipe. If feed pipe is made of brass, remove as shown in Fig. BS68. Using a vise, press new pipe into carburetor so it extends from 2-9/32 to 2-5/16 inches (57.9-58.7 mm) from carburetor body. If pipe is made of nylon (plastic), screw pipe out of carburetor body with wrench. When installing new nylon feed pipe, be careful not to overtighten.

NOTE: If soaking carburetor in cleaner for more than one-half hour, be sure to remove all nylon parts and "O" ring, if used, before placing the carburetor in cleaning solvent.

**PUMP TYPE (PULSA-JET) CARBURETORS.** The pump type (Pulsa-Jet) carburetor is basically a suction type carburetor incorporating a fuel

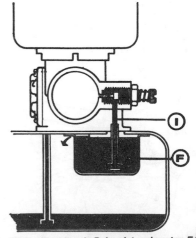

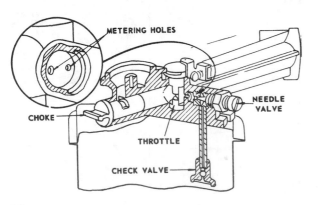

Fig. BS67—Cutaway view of typical suction type (Vacu-Jet) carburetor. Inset shows fuel metering holes which are accessible for cleaning after removing needle valve. Be careful not to enlarge the holes when cleaning them.

Fig. BS69—Fuel flow in Pulsa-Jet carburetor. Fuel pump incorporated in carburetor fills constant level sump (F) below carburetor and excess fuel flows back into tank. Fuel is drawn from sump through inlet (I) past fuel mixture adjusting needle by vacuum in carburetor.

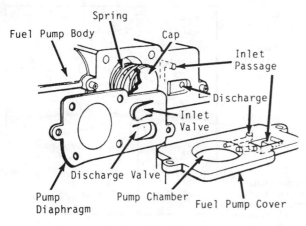

Fig. BS70—Exploded view of fuel pump that is incorporated in Pulsa-Jet carburetor except those used on 82900, 92900, 94900, 110900. 111900, 112200 and 113900 models; refer to Fig. BS71.

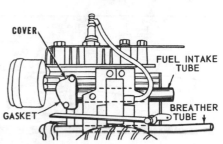

Fig. BS74—On 82000 models, intake tube is threaded into intake port of engine; a gasket is placed between intake port cover and intake port.

pump to fill a constant level fuel sump in top of fuel tank. Refer to schematic view in Fig. BS69. This makes a constant air:fuel mixture available to engine regardless of fuel level in tank. Adjustment of the pump type carburetor fuel mixture needle valve is the same as outlined for suction type carburetors in previous paragraph, except that fuel level in tank is not important.

To remove the pump type carburetor, first remove the carburetor and fuel tank as an assembly; then, remove carburetor from fuel tank. When reinstalling carburetor on fuel tank, use a new gasket or pump diaphragm as required and tighten retaining screws evenly.

Fig. BS70 shows an exploded view of the pump unit used on all carburetors except those for Models 82900, 92900, 94900, 110900, 111900, 112200 and 113900 the pump diaphragm is placed between the carburetor and fuel tank as shown in Fig. BS71.

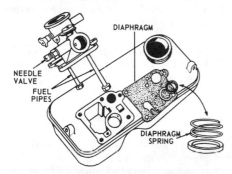

Fig. BS71—On Models 82900, 92900, 94900, 110900, 111900, 112200 and 113900, pump type (Pulsa-Jet) carburetor diaphragm is installed between carburetor and fuel tank.

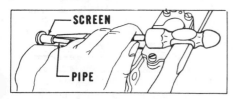

Fig. BS72—To renew screen housing on pump type carburetor with brass feed pipes, drive old screen housing from pipe as shown. To hold pipe, clamp lightly in a vise.

The pump type carburetor has two fuel feed pipes. The long pipe feeds fuel into the pump portion of the carburetor from which fuel then flows to the constant level fuel sump. The short pipe extends into the constant level sump and feeds fuel into the carburetor venturi via fuel mixture needle valve.

As check valves are incorporated in the pump diaphragm, fuel feed pipes on pump type carburetors do not have a check valve. However, if the fuel screen in lower end of pipe is broken or clogged and cannot be cleaned, the pipe or screen housing can be renewed. If pipe is made of nylon, pipe snaps into place and considerable force is required to remove or install pipe. Be careful to not damage new pipe. If pipe is made of brass, clamp pipe lightly in a vise and drive old screen housing from pipe with a screwdriver or small chisel as shown in Fig. BS72. Drive a new screen housing onto pipe with a soft faced hammer.

**NOTE: If soaking carburetor in cleaner for more than one-half hour, be sure to remove all nylon parts and "O" ring, if used, before placing carburetor in cleaning solvent.**

**NOTE: On engine Models 82900, 92900, 94900, 95500, 110900, 111900 and 113900, be sure air cleaner retaining screw is in place if engine is being operated (during tests) without air cleaner installed. If screw is not in place, fuel will lift up**

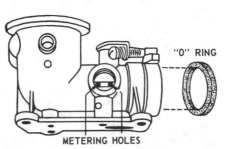

Fig. BS73—Metering holes in pump type carburetors are accessible for cleaning after removing fuel mixture needle valve. On models with intake pipe, carburetor is sealed to pipe with "O" ring.

through the screw hole and enter carburetor throat as the screw hole leads directly into the constant level fuel sump.

**INTAKE TUBE.** Models 82000, 92000, 93500, 94500, 94900, 95500, 100900, 110900, 111900, 113900, 130900 and 131900 have an intake tube between carburetor and engine intake port. Carburetor is sealed to intake tube with an "O" ring as shown in Fig. BS73.

On Model 82000 engines, the intake tube is threaded into the engine intake port. A gasket is used between the engine intake port cover and engine casting. Refer to Fig. BS74.

On Models 92000, 93500, 94500, 94900, 95500, 110900, 111900 and 113900 engines, the intake tube is bolted to the engine intake port and gasket is used between the intake tube and engine casting. Refer to Fig. BS75. On Models 100900, 130900 and 131900 intake tubes are attached to engine in similar manner.

**CHOKE-A-MATIC CARBURETOR CONTROLS.** Engines equipped with float, suction or pump type carburetors may be equipped with a control unit with which the carburetor choke, throttle and magneto grounding switch are operated from a single lever (Choke-A-Matic carburetors). Refer to Figs. BS76 through BS82 for views showing the different

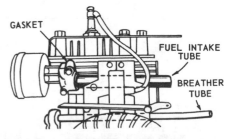

Fig. BS75—On 92000, 93500, 94500, 94900, 95500, 110900, 111900 and 113900 models, fuel intake tube is bolted to engine intake port and a gasket is placed between tube and engine. On 100900, 130900 and 131900 vertical crankshaft models, intake tube and gasket are similar.

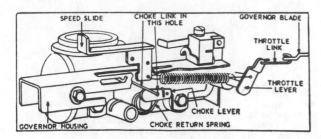

Fig. BS76—Choke-A-Matic control on float type carburetor. Remote control can be attached to speed slide.

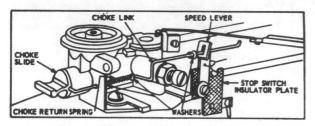

Fig. BS77—Typical Choke-A-Matic control on suction type carburetor. Remote control can be attached to speed lever.

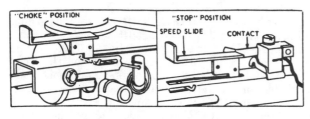

Fig. BS78—Choke-A-Matic control in choke and stop positions on float carburetor.

Fig. BS79—Choke-A-Matic controls in choke and stop positions on suction carburetor. Bend choke link if necessary to adjust control.

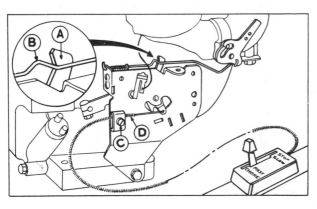

Fig. BS80—On Choke-A-Matic control shown, choke actuating lever (A) should just contact choke link or shaft (B) when control is at "FAST" position. If not, loosen screw (C) and move control wire housing (D) as required.

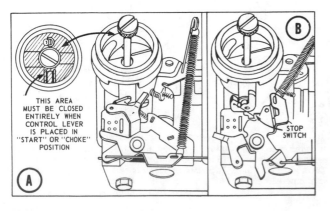

Fig. BS81—When Choke-A-Matic control is in "START" or "CHOKE" position, choke must be completely closed as shown in view A. When control is in "STOP" position, arm should contact stop switch (view B).

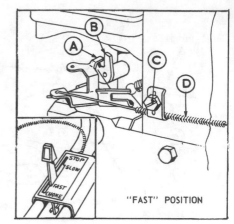

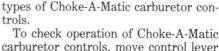

"FAST" POSITION

Fig. BS82—On Choke-A-Matic controls shown, lever (A) should just contact choke shaft arm (B) when control is in "FAST" position. If not, loosen screw (C) and move control wire housing (D) as required, then tighten screw.

types of Choke-A-Matic carburetor controls.

To check operation of Choke-A-Matic carburetor controls, move control lever to "CHOKE" position. Carburetor choke slide or plate must be completely closed. Then, move control lever to "STOP" position. Magneto grounding switch should be making contact. With the control lever in "RUN", "FAST" or "SLOW" position, carburetor choke should be completely open. On units with remote controls, synchronize movement of remote lever to carburetor control lever by loosening screw (C—Fig. BS80 or Fig. BS82) and moving control wire housing (D) as required; then, tighten screw to clamp the housing securely. Refer to Fig. BS83 to check remote control wire movement.

**AUTOMATIC CHOKE (THERMOSTAT TYPE).** A thermostat operated choke is used on some models equipped with the two-piece carburetor. To adjust choke linkage, hold choke shaft so thermostat lever is free. At room temperature, stop screw in thermostat collar should be located midway between thermostat stops. If not, loosen stop screw, adjust the collar and tighten stop screw. Loosen set screw (S—Fig.

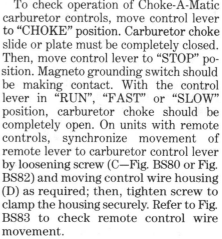

Fig. BS83—For proper operation of Choke-A-Matic controls, remote control wire must extend to dimension shown and have a minimum travel of 1-3/8 inches (34.9 mm).

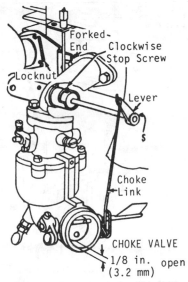

Fig. BS84—Automatic choke used on some models equipped with "two-piece" Flo-Jet carburetor showing unit in "HOT" position.

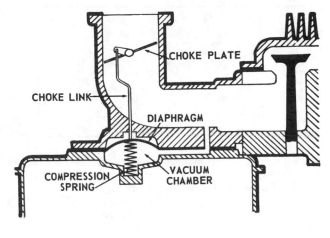

*Fig. BS86—Diagram showing vacuum operated automatic choke used on some 92000, 93500, 94500, 94900, 95500, 110900, 111900 and 113900 model vertical crankshaft engines in closed (engine not running) position.*

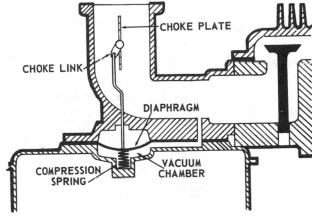

*Fig. BS87—Diagram showing vacuum operated automatic choke in open (engine running) position.*

BS84) on thermostat lever. Then, slide lever on shaft to ensure free movement of choke unit. Turn thermostat shaft clockwise until stop screw contacts thermostat stop. While holding shaft in this position, move shaft lever until choke is open exactly 1/8 inch (3.17 mm) and tighten lever set screw. Turn thermostat shaft counterclockwise until stop screw contacts thermostat stop as shown in Fig. BS85. Manually open choke valve until it stops against top of choke link opening. At this time, choke valve should be open at least 3/32 inch (2.38 mm), but not more than 5/32 inch (3.97 mm). Hold choke valve in wide open position and check position of counter-

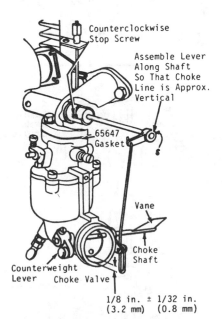

Fig. BS85—Automatic choke on "two-piece" Flo-Jet carburetor in "COLD" position.

weight lever. Lever should be in a horizontal position with free end towards right.

**AUTOMATIC CHOKE (VACUUM TYPE).** A spring and vacuum operated automatic choke is used on some 92000, 93500, 94500, 94900, 95500, 110900, 111900 and 113900 vertical crankshaft engines. A diaphragm under carburetor is connected to the choke shaft by a link. The compression spring works against the diaphragm, holding choke in closed position when engine is not running. See Fig. BS86. As engine starts, increased vacuum works against the spring and pulls the diaphragm and choke link down, holding choke in open (running) position shown in Fig. BS87.

During operation, if a sudden load is applied to engine or a lugging condition develops, a drop in intake vacuum occurs, permitting choke to close partially. This provides a richer fuel mixture to meet the condition and keeps the engine running smoothly. When the load condition has been met, increased vacuum returns choke valve to normal running (fully open) position.

**FUEL TANK OUTLET.** Small models with float type carburetors are equipped with a fuel tank outlet as

Fig. BS88—Fuel tank outlet used on smaller engines with float type carburetor.

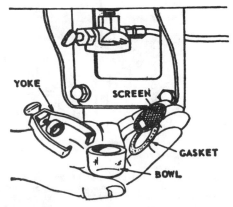

Fig. BS89—Fuel tank outlet used on larger B&S engines.

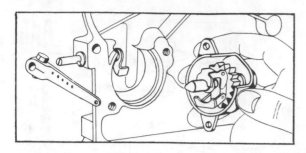

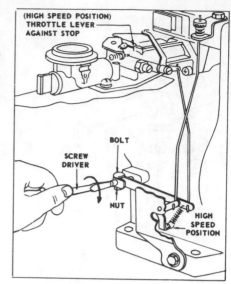

*Fig. BS90—Removing governor unit (except on 100000, 112200, 130000, 131000 and late 140000 models) from inside crankcase cover on horizontal crankshaft models. Refer to Fig. BS91 for exploded view of governor.*

*Fig. BS95—Linkage adjustment on 100000, 112200, 130000, 131000 and late 140000 horizontal crankshaft mechanical governor models; refer to text for procedure.*

shown in Fig. BS88. On larger engines, a fuel sediment bowl is incorporated with the fuel tank outlet as shown in Fig. BS89. Clean any lint and dirt from tank outlet screens with a brush. Varnish or other gasoline deposits may be removed by using a suitable solvent. Tighten packing nut or remove nut and shut-off valve, then renew packing if leakage occurs around shut-off valve stem.

**FUEL PUMP.** A fuel pump is available as optional equipment on some models. Refer to SERVICING BRIGGS & STRATTON ACCESSORIES section

in this manual for fuel pump service information.

**GOVERNOR.** All models are equipped with either a mechanical (flyweight type) or an air vane (pneumatic) governor. Refer to the appropriate paragraph for model being serviced.

**Mechanical Governor.** Three different designs of mechanical governors are used.

On all engines except 100000, 112200, 130000, 131000 and all late 140000 models, a governor unit as shown in Fig. BS90 is used. An exploded view of this governor unit is shown in Fig. BS91. The governor housing is attached to inner

side of crankcase cover and the governor gear is driven from the engine camshaft gear. Use Figs. BS90 and BS91 as a

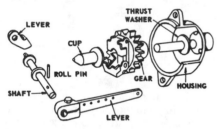

*Fig. BS91—Exploded view of governor unit used on all horizontal crankshaft models (except 100000, 112200, 130000, 131000 and late 140000 models) with mechanical governor. Refer also to Figs. BS90 and BS92.*

*Fig. BS93—Installing crankcase cover on 100000, 112200, 130000, 131000 and late 140000 models with mechanical governor. Governor crank (C) must be in position shown. A thrust washer (W) is placed between governor (G) and crankcase cover.*

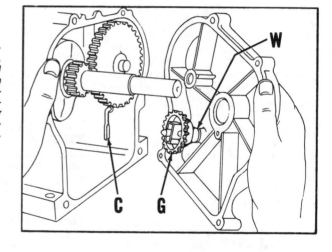

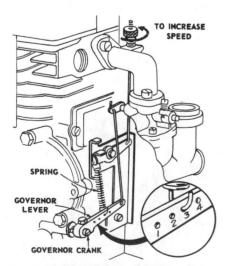

*Fig. BS92—View of governor linkage used on horizontal crankshaft mechanical governor models except 100000, 112200, 130000, 131000 and late 140000 models. Governor spring should be hooked in governor lever as shown in inset.*

*Fig. BS94—Cutaway drawing of governor and linkage used on 100000, 112200, 130000, 131000 and late 140000 horizontal crankshaft models.*

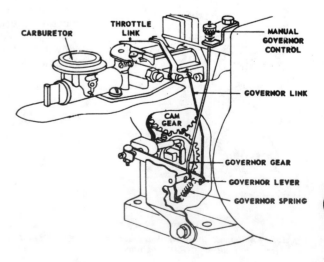

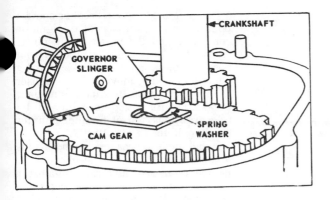

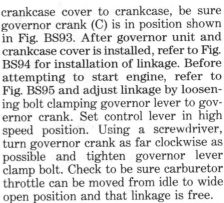

Fig. BS96—View showing 95500, 100000, 113900, 130000, 131000 and 140000 vertical crankshaft model mechanical governor unit. Drawing is of lower side of engine with oil sump (engine base) removed; note spring washer location used on 100000, 130000 and 131000 models only.

disassembly and assembly guide. Renew any parts that bind or show excessive wear. After governor is assembled, refer to Fig. BS92 and adjust linkage by loosening screw clamping governor lever to governor crank. Turn governor lever counterclockwise so carburetor throttle is in wide open position. Hold lever and turn governor crank as far counterclockwise as possible, then tighten screw clamping lever to crank. Governor crank can be turned with screwdriver. Check linkage to be sure it is free and that the carburetor throttle will move from idle to wide open position.

On 100000, 112200, 130000, 131000 and late 140000 horizontal crankshaft models, the governor gear and weight unit (G—Fig. BS93) is supported on a pin in engine crankcase cover and the governor crank is installed in a bore in the engine crankcase. A thrust washer (W) is placed between governor gear and crankcase cover. When assembling crankcase cover to crankcase, be sure governor crank (C) is in position shown in Fig. BS93. After governor unit and crankcase cover is installed, refer to Fig. BS94 for installation of linkage. Before attempting to start engine, refer to Fig. BS95 and adjust linkage by loosening bolt clamping governor lever to governor crank. Set control lever in high speed position. Using a screwdriver, turn governor crank as far clockwise as possible and tighten governor lever clamp bolt. Check to be sure carburetor throttle can be moved from idle to wide open position and that linkage is free.

On vertical crankshaft 95500, 100000, 113900, 130000, 131000 and 140000 with mechanical governor, the governor weight unit is integral with the lubricating oil slinger and is mounted on lower end of camshaft gear as shown in Fig. BS96. With engine upside down, place governor and slinger unit on camshaft gear as shown, place spring washer (100000, 130900 and 131900 models only) on camshaft gear and install engine base on crankcase. Assemble linkage as shown in Fig. BS97; then, refer to Fig. BS98 and adjust linkage by loosening governor lever to governor crank clamp bolt, place control lever in high speed position, turn governor crank with screwdriver as far as possible and tighten governor lever clamping bolt.

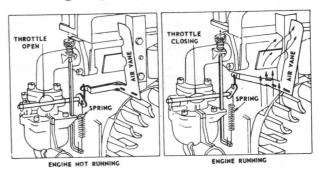

Fig. BS99—Views showing operating principle of air vane (pneumatic) governor. Air from flywheel fan acts against air vane to overcome tension of governor spring; speed is adjusted by changing spring tension.

Fig. BS97—Schematic drawing of 95500, 100000, 113900, 130000, 131000 and 140000 mechanical governor and linkage used on vertical crankshaft mechanical governed models.

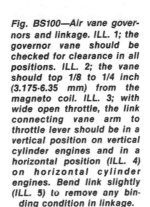

Fig. BS98—View showing adjustment of 95500, 100000, 113900, 130000, 131000 and 140000 vertical crankshaft mechanical governor; refer to text for procedure.

Fig. BS100—Air vane governors and linkage. ILL. 1; the governor vane should be checked for clearance in all positions. ILL. 2; the vane should top 1/8 to 1/4 inch (3.175-6.35 mm) from the magneto coil. ILL. 3; with wide open throttle, the link connecting vane arm to throttle lever should be in a vertical position on vertical cylinder engines and in a horizontal position (ILL. 4) on horizontal cylinder engines. Bend link slightly (ILL. 5) to remove any binding condition in linkage.

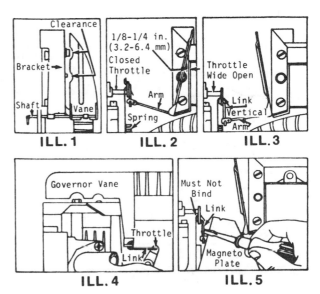

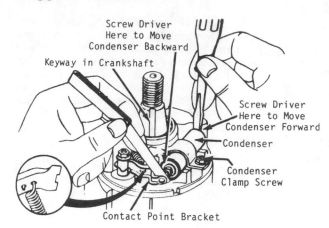

Fig. BS101—View showing breaker point adjustment on models having breaker point integral with condenser. Move condenser to adjust point gap.

**Air Vane (Pneumatic) Governor.** Refer to Fig. BS99 for views showing typical air vane governor operation.

The vane should stop 1/8 to 1/4 inch (3.17-6.35 mm) from the magneto coil (Fig. BS100, illustration 2) when the linkage is assembled and attached to carburetor throttle shaft. If necessary to adjust, spring the vane while holding the shaft. With wide open throttle, the link from the air vane arm to the carburetor throttle should be in a vertical position on horizontal crankshaft models and in a horizontal position on vertical crankshaft models. (Fig. BS100, illustration 3 and 4.) Check linkage for binding. If binding condition exists, bend links slightly to correct. Refer to Fig. BS100, illustration 5.

NOTE: Some engines are equipped with a nylon governor vane which does not require adjustment.

**MAGNETO.** Breaker contact gap on all models with magneto type ignition is 0.020 inch (0.51 mm). On all models except "Sonoduct" (vertical crankshaft with flywheel below engine), breaker points and condenser are accessible after removing engine flywheel and breaker cover. On "Sonoduct" models, breaker points are located below breaker cover on top side of engine.

Fig. BS102—Adjustment of breaker point gap on models having breaker point separate from condenser.

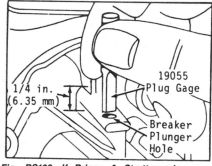

Fig. BS103—If Briggs & Stratton plug gage #19055 can be inserted in breaker plunger bore a distance of 1/4 inch (6.35 mm) or more, bore is worn and must be rebushed.

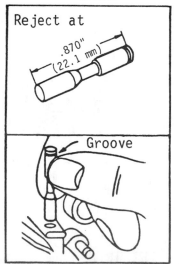

Fig. BS102A—Insert plunger into bore with groove toward top. Refer to text.

On some models, one breaker contact point is an integral part of the ignition condenser and the breaker arm pivots on a knife edge retained by a slot in pivot post. On these models, breaker contact gap is adjusted by moving the condenser as shown in Fig. BS101. On other models, breaker contact gap is adjusted by relocating position of breaker contact bracket. Refer to Fig. BS102.

On all models, breaker contact arm is actuated by a plunger held in a bore in engine crankcase and rides against a cam on engine crankshaft. Plunger can be removed after removing breaker points. Renew plunger if worn to a length of 0.870 inch (22.1 mm) or less. Plunger must be installed with grooved end to the top to prevent oil seepage (Fig. BS102A). Check breaker point plunger bore wear in crankcase with B&S plug gage 19055. If plug gage will enter bore 1/4 inch (6.35 mm) or more, bore should be reamed and a bushing installed. Refer to Fig. BS103 for method of checking bore and to Fig. BS104 for steps in reaming bore and installing bushing if bore is worn. To ream bore and install bushing, it is necessary that the breaker points, armature and ignition coil and the crankshaft be removed.

On "Sonoduct" models, armature and ignition coil are inside flywheel on bottom side of engine. Armature air gap is correct if armature is installed flush with mounting boss as shown in Fig. BS105. On all other models, armature and ignition coil are located outside flywheel. Armature-to-flywheel air gap should be as follows:

Models 6B, 8B, 6000 —
Two Leg Armature . . . 0.006-0.010 in. (0.15-0.25 mm)
Three Leg Armature . . 0.012-0.016 in. (0.30-0.41 mm)

Models 80000, 82000, 92000, 93000, 93500, 94000, 94500, 94900, 95000, 95500, 110000, 110900 —
Two Leg Armature . . . 0.006-0.010 in. (0.15-0.25 mm)
Three Leg Armature . . 0.012-0.016 in. (0.30-0.41 mm)

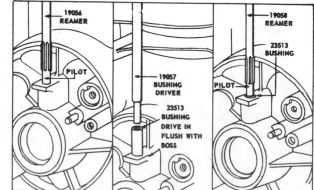

Fig. BS104—Views showing reaming plunger bore to accept bushing (left view), installing bushing (center) and finish reaming bore (right) of bushing.

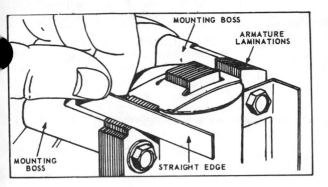

Fig. BS105—On "Sonoduct" models, align armature core with mounting boss for proper magneto air gap.

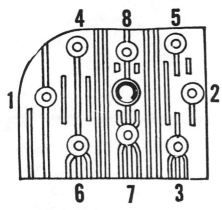

Fig. BS106—Cylinder head screw tightening sequence. Long screws are used in positions 2, 3 and 7.

Models 111200, 111900,
112200, 113900 . . . . . . . . 0.010-0.016 in.
(0.25-0.41 mm)
Models 100000,
130000, 131000 –
  Two Leg Armature . . . 0.010-0.014 in.
  (0.25-0.36 mm)
  Three Leg Armature . . 0.012-0.016 in.
  (0.30-0.41 mm)
Model 140000 –
  Two Leg Armature . . . 0.010-0.014 in.
  (0.25-0.36 mm)
  Three Leg Armature . . 0.016-0.019 in.
  (0.41-0.48 mm)

**MAGNETRON IGNITION.** Magnetron ignition is a self-contained breakerless ignition system. Flywheel does not need to be removed except to check or service keyways or crankshaft key.

To check spark, remove spark plug. Connect spark plug cable to B&S tester, part 19051, and ground remaining tester lead to cylinder head. Spin engine at 350 rpm or more. If spark jumps the 0.166 inch (4.2 mm) tester gap, system is functioning properly.

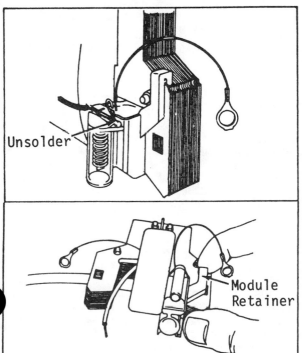

Fig. BS105A—Wires must be unsoldered to remove Magnetron ignition module. Refer to text.

To remove armature and Magnetron module, remove flywheel shroud and armature retaining screws. Use a 3/16 inch (4.76 mm) diameter pin punch to release stop switch wire from module. To remove module, unsolder wires, push module retainer away from laminations and remove module. See Fig. BS105A. Resolder wires for reinstallation and use Permatex or equivalent to hold ground wires in position.

Adjust armature air gap to 0.010-0.014 inch (0.25-0.36 mm).

**LUBRICATION.** Vertical crankshaft engines are lubricated by an oil slinger wheel driven by the cam gear. On early 6B and 8B models, the oil slinger wheel was mounted on a bracket attached to the crankcase. Renew the bracket if pin on which gear rotates is worn to 0.490 inch (12.45 mm). Renew steel bushing in hub or gear if worn. On later model vertical crankshaft engines, the oil slinger wheel, pin and bracket are an integral unit with the bracket being retained by the lower end of the engine camshaft.

On 100000, 130000 and 131000 models, a spring washer is placed on lower end of camshaft between bracket and oil sump boss. Renew the oil slinger assembly if teeth are worn on slinger gear or gear is loose on bracket. On horizontal crankshaft engines, a splash system (oil dipper on connecting rod) is used for engine lubrication.

Check oil level at five hour intervals and maintain at bottom edge of filler plug or to FULL mark on dipstick.

Recommended oil change interval for all models is every 25 hours of normal operation.

Manufacturer recommends using oil with an API service classification of SE or SF. Use SAE 10W-40 oil for temperatures above 20°F (−7°C) and SAE 5W-30 oil for temperatures below 20°F (−7°C).

Crankcase capacity for all aluminum cylinder engines with displacement of 9 cubic inches (147 cc) or below, is 1¼ pint (0.6 L).

Crankcase capacity of vertical crankshaft aluminum cylinder engines with displacement of 11 cubic inches (186 cc) is 1¼ pint (0.6 L).

Crankcase capacity for all 10 and 13 cubic inch (169 and 203 cc) vertical crankshaft, aluminum cylinder engines is 1¾ pint (0.8 L).

Crankcase capacity for all 10 and 13 cubic inch (169 and 203 cc) horizontal crankshaft, aluminum cylinder engines is 1¼ pint (0.6 L).

Crankcase capacity for 14 cubic inch (231 cc) vertical crankshaft, aluminum cylinder engines is 2¼ pint (1.1 L).

Crankcase capacity for 14 cubic inch (231 cc) horizontal crankshaft, aluminum cylinder engines if 2¾ pint (1.3 L).

Crankcase capacity for engines with cast iron cylinder liner is 3 pints (1.4 L).

**CRANKCASE BREATHER.** The crankcase breather is built into the

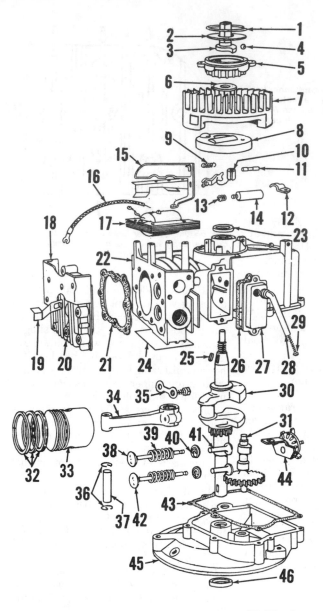

**Fig. BS107—Exploded view of typical vertical crankshaft model with air vane (pneumatic) governor. To remove flywheel, remove blower housing and starter unit; then, unscrew starter clutch housing (5). Flywheel can then be pulled from crankshaft.**

1. Snap ring
2. Washer
3. Ratchet
4. Steel balls
5. Starter clutch
6. Washer
7. Flywheel
8. Breaker cover
9. Breaker point spring
10. Breaker arm & pivot
11. Breaker plunger
12. Condenser clamp
13. Coil spring (primary wire retainer)
14. Condenser
15. Governor air vane & bracket assy.
16. Spark plug wire
17. Armature & coil assy.
18. Air baffle
19. Spark plug grounding switch
20. Cylinder head
21. Cylinder head gasket
22. Cylinder block
23. Crankshaft oil seal
24. Cylinder shield
25. Flywheel key
26. Gasket
27. Breather & tappet chamber cover
28. Breather tube assy.
29. Coil spring
30. Crankshaft
31. Cam gear & shaft
32. Piston rings
33. Piston
34. Connecting rod
35. Rod bolt lock
36. Piston pin retaining rings
37. Piston pin
38. Intake valve
39. Valve springs
40. Valve spring keepers
41. Tappets
42. Exhaust valve
43. Gasket
44. Oil slinger assy.
45. Oil sump (engine base)
46. Crankshaft oil seal

engine valve cover. The mounting holes are offset so the breather can only be installed one way. Rinse breather in solvent and allow to drain. A vent tube connects the breather to the carburetor air horn on certain model engines for extra protection against dusty conditions.

### REPAIRS

**CYLINDER HEAD.** When removing cylinder head, be sure to note the position from which each of the different length screws was removed. If screws are not reinstalled in the same holes when installing the head, it will result in screws bottoming in some holes and not enough thread contact in others. Lubricate the cylinder head screws with graphite grease before installation. Do not use sealer on head gasket. When installing cylinder head, tighten all screws lightly and then retighten them in sequence shown in Fig. BS106 to 165 in.-lbs.

(19 N·m) on 140000 models and to 140 in.-lbs. (16 N·m) on all other models. Run the engine for 2 to 5 minutes to allow it to reach operating temperature and retighten the head screws again following the sequence and torque value specified.

**NOTE: When checking compression on models with "Easy-Spin" starting, turn engine opposite the direction of normal rotation. See CAMSHAFT paragraph.**

**OIL SUMP REMOVAL, AUXILARY PTO MODELS.** On 92000, 93500, 94900, 95500, 110900 or 113900 models with auxiliary pto, one of the oil sump (engine base) to cylinder retaining screws is installed in the recess in sump for the pto auxiliary drive gear. To remove the oil sump, refer to Fig. BS108 and remove the cover plate (upper view) and then remove the shaft stop (lower left view). The gear and shaft can then be moved as shown in lower right view

to allow removal of the retaining screws. Reverse procedure to reassemble.

**OIL BAFFLE PLATE.** 140000 MODEL ENGINES. Model 140000 engines with mechanical governor have a baffle located in the cylinder block (crankcase). When servicing these engines, it is important that the baffle be correctly installed. The baffle must fit tightly against the valve tappet boss in the crankcase. Check for this before installing oil sump (vertical crankshaft models) or crankcase cover (horizontal crankshaft models).

**CONNECTING ROD.** The connecting rod and piston are removed from cylinder head end of block as an assembly. The aluminum alloy connecting rod rides directly on the induction hardened crankpin. The rod should be rejected if the crankpin hole is scored or out-of-round more than 0.0007 inch (0.018 mm) or if the piston pin hole is scored or out-of-round more than 0.0005 inch (0.013 mm). Wear limit sizes are given in the following chart. Reject the connecting rod if either the crankpin or piston pin hole is worn to, or larger than the sizes given in the following table.

### REJECT SIZES FOR CONNECTING ROD

| Model | Bearing Bore | Pin Bore |
|---|---|---|
| 6B, 60000 . . . | 0.876 in. (22.25 mm) | 0.492 in. (12.50 mm) |
| 8B, 80000 . . . | 1.001 in. (25.43 mm) | 0.492 in. (12.50 mm) |
| 82000, 92000, 11000 . . . | 1.001 in. (25.43 mm) | 0.492 in. (12.50 mm) |
| 92000, 93500 . . . | 1.001 in. (25.43 mm) | 0.492 in. (12.50 mm) |
| 94500, 94900, 95500 . . . | 1.001 in. (25.43 mm) | 0.492 in. (12.50 mm) |
| 100000 . . | 1.001 in. (25.43 mm) | 0.555 in. (14.10 mm) |
| 110900, 111200, 111900, 112200, 113900, 130000, 131000 . . | 1.001 in. (25.43 mm) | 0.492 in. (12.50 mm) |
| 140000 . . | 1.095 in. (27.81 mm) | 0.674 in. (17.12 mm) |

**NOTE: Piston pins of 0.005 inch (0.13 mm) oversize are available for service. Piston pin hole in rod can be reamed to this size if crankpin hole in rod is within specifications.**

Connecting rod must be reassembled so match marks on rod and cap are aligned. Tighten connecting rod bolts to 165 in.-lbs. (19 N·m) for 140000 models and to 100 in.-lbs. (11 N·m) for all other models.

**PISTON, PIN AND RINGS.** Pistons for use in engines having aluminum bore ("Kool Bore") are not interchangeable with those for use in cylinders having cast-iron sleeve. Pistons may be identified as follows: Those for use in cast-iron sleeve cylinders have a plain, dull aluminum finish, have an "L" stamped on top and use an oil ring expander. Those for use in aluminum bore cylinders are chrome plated (shiny finish), do not have an identifying letter and do not use an oil ring expander.

Reject pistons showing visible signs of wear, scoring and scuffing. If, after cleaning carbon from top ring groove, a new top ring has a side clearance of 0.007 inch (0.18 mm) or more, reject the piston. Reject piston or hone piston pin hole to 0.005 inch (0.13 mm) oversize if pin hole is 0.0005 inch (0.013 mm) or more out-of-round, or is worn to a diameter of 0.554 inch (14.07 mm) or more on 100000 engines, 0.673 inch (17.09 mm) or more on 140000 engines, or 0.491 inch (12.47 mm) or more on all other models.

If the piston pin is 0.0005 inch (0.013 mm) or more out-of-round, or is worn to a diameter of 0.552 inch (14.02 mm) or smaller on 100000 engines, 0.671 inch (17.04 mm) or smaller on 140000 engines, or 0.489 inch (12.42 mm) or smaller on all other models, reject pin.

The piston ring gap for new rings should be 0.010-0.025 inch (0.25-0.64 mm) for models with aluminum cylinder bore and 0.010-0.018 inch (0.25-0.46 mm) for models with cast-iron cylinder bore. On aluminum bore engines, reject compression rings having an end gap of 0.035 inch (0.80 mm) or more and reject oil rings having an end gap of 0.045 inch (1.14 mm) or more. On cast-iron bore engines, reject compression rings having an end gap of 0.030 inch (0.75 mm) or more and reject oil rings having an end gap of 0.035 inch (0.90 mm) or more.

Pistons and rings are available in several oversizes as well as standard.

A chrome ring set is available for slightly worn standard bore cylinders. Refer to note in CYLINDER section.

**CYLINDER.** If cylinder bore wear is 0.003 inch (0.76 mm) or more or is 0.0025 inch (0.06 mm) or more out-of-round, cylinder must be rebored to next larger oversize.

The standard cylinder bore sizes for each model are given in the following table.

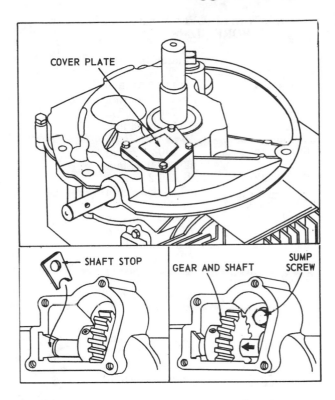

*Fig. BS108—To remove oil sump (engine base) on 92000, 93500, 94900, 95500, 110900 or 113900 models with auxiliary pto, remove cover plate, shaft stop and retaining screw as shown.*

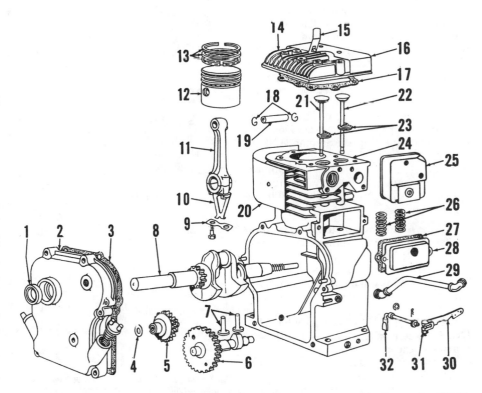

*Fig. BS109—Exploded view of 100000 horizontal crankshaft engine assembly. Except for 112200, 130000, and late 140000 models, other horizontal crankshaft models with mechanical governor will have governor unit as shown in Fig. BS90; otherwise, construction of all other horizontal crankshaft models is similar.*

| | |
|---|---|
| 1. Crankshaft oil seal | 12. Piston |
| 2. Crankcase cover | 13. Piston rings |
| 3. Gasket | 14. Cylinder head |
| 4. Thrust washer | 15. Spark plug ground switch |
| 5. Governor assy. | 16. Air baffle |
| 6. Cam gear & shaft | 17. Cylinder head gasket |
| 7. Tappets | 18. Piston pin retaining rings |
| 8. Crankshaft | 19. Piston pin |
| 9. Rod bolt lock | 20. Air baffle |
| 10. Oil dipper | 21. Exhaust valve |
| 11. Connecting rod | 22. Intake valve |

| | |
|---|---|
| 23. Valve spring retainers | |
| 24. Cylinder block | |
| 25. Muffler | |
| 26. Valve springs | |
| 27. Gaskets | |
| 28. Breather & tappet chamber cover | |
| 29. Breather pipe | |
| 30. Governor lever | |
| 31. Clamping bolt | |
| 32. Governor crank | |

## STANDARD CYLINDER BORE SIZES

| Model | Cylinder diameter |
|---|---|
| 6B, | |
| Early 60000 | 2.3115-2.3125 in. |
| | (58.71-58.74 mm) |
| Late 60000 | 2.3740-2.3750 in. |
| | (60.30-60.33 mm) |
| 8B, 80000, | |
| 81000, 82000 | 2.3740-2.3750 in. |
| | (60.30-60.33 mm) |
| 92000, 93500 | 2.5615-2.5625 in. |
| | (65.06-65.09 mm) |
| 94500, 94900, | |
| 95500 | 2.5615-2.5625 in. |
| | (65.06-65.09 mm) |
| 100000 | 2.4990-2.5000 in. |
| | (63.47-63.50 mm) |
| 110900, 111200, | |
| 111900, 112200, | |
| 113900 | 2.7802-2.7812 in. |
| | (70.62-70.64 mm) |
| 130000, 131400, | |
| 131900 | 2.5615-2.5625 in. |
| | (65.06-65.09 mm) |
| 140000 | 2.7490-2.7500 in. |
| | (69.82-69.85 mm) |

A hone is recommended for resizing cylinders. Operate hone at 300-700 rpm and with an up and down movement that will produce a 45° crosshatch pattern. Clean cylinder after honing with oil or soap suds. Always check availability of oversize piston and ring sets before honing cylinder.

Approved hones are as follows: For aluminum bore, use Ammco 3956 for rough and finishing or Sunnen AN200 for rough and Sunnen AN500 for finishing. For sleeved bores, use Ammco 4324 for rough and finishing, or Sunnen AN100 for rough and Sunnen AN300 for finishing.

**NOTE: A chrome piston ring set is available for slightly worn standard bore cylinders. No honing or cylinder deglazing is required for these rings. The cylinder bore can be a maximum of 0.005 inch (0.01 mm) oversize when using chrome rings.**

**CRANKSHAFT AND MAIN BEARINGS.** Except where equipped with ball bearings, the main bearings are an integral part of the crankcase and cover or sump. The bearings are renewable by reaming out the crankcase and cover or sump bearing bores and installing service bushings. The tools for reaming the crankcase and cover or sump, and for installing the service bushings are available from Briggs & Stratton. If the bearings are scored, out-of-round 0.0007 inch (0.018 mm) or more, or are worn to or larger than the reject sizes in the following table, ream the bearings and install service bushings.

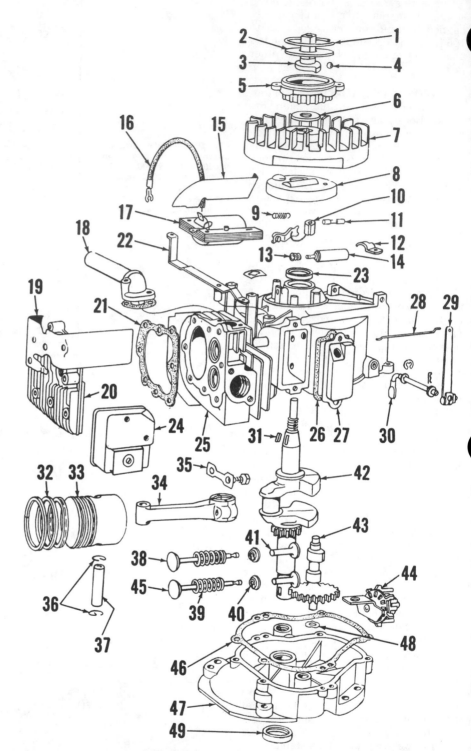

**Fig. BS110—Exploded view of 100000 vertical crankshaft engine with mechanical governor. Models 130000, 131000 and 140000 vertical crankshaft models with mechanical governor are similar. Refer to Fig. BS107 for typical vertical crankshaft model with air vane governor.**

1. Snap ring
2. Washer
3. Starter ratchet
4. Steel balls
5. Starter clutch
6. Washer
7. Flywheel
8. Breaker cover
9. Breaker arm spring
10. Breaker arm & pivot
11. Breaker plunger
12. Condenser clamp
13. Primary wire retainer spring
14. Condenser
15. Air baffle
16. Spark plug wire
17. Armature & coil assy.
18. Intake pipe
19. Air baffle
20. Cylinder head
21. Cylinder head gasket
22. Linkage lever
23. Crankshaft oil seal
24. Muffler
25. Cylinder block
26. Gasket
27. Breather & tappet chamber cover
28. Governor link
29. Governor crank
30. Governor crank
31. Flywheel key
32. Piston rings
33. Piston
34. Connecting rod
35. Rod bolt lock
36. Piston pin retaining rings
37. Piston pin
38. Intake valve
39. Valve springs
40. Valve spring retainers
41. Tappets
42. Crankshaft
43. Cam gear
44. Governor & oil slinger assy.
45. Exhaust valve
46. Gasket
47. Oil sump (engine base)
48. Thrust washer
49. Crankshaft oil seal

### MAIN BEARING REJECT SIZES

| Model | Magneto Bearing | PTO Bearing |
|---|---|---|
| 6B, 60000 . . . | 0.878 in. (22.30 mm) | 0.878 in.* (22.30 mm)* |
| 8B, 80000 . . . | 0.878 in. (22.30 mm) | 0.878 in.* (22.30 mm)* |
| 81000, 82000 . . . | 0.878 in. (22.30 mm) | 0.878 in.* (22.30 mm)* |
| 92000, 93500, 94000, 94500, 94900, 95500 . . . | 0.878 in. (22.30 mm) | 0.878 in.* (22.30 mm)* |
| 100000 . . | 0.878 in. (22.30 mm) | 1.003 in. (25.48 mm) |
| 110900, 111200, 111900, 112200, 113900 . . | 0.878 in. (22.30 mm) | 0.878 in.* (22.30 mm)* |
| 130000, 131000 . . | 0.878 in. (22.30 mm) | 1.003 in. (25.48 mm) |
| 140000 . . | 1.004 in. (25.50 mm) | 1.185 in. (30.10 mm) |

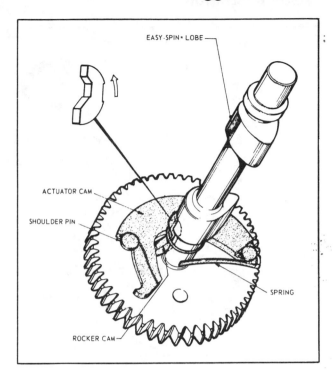

*Fig. BS111—Compression release camshaft used on 111200, 111900 and 112200 engines. At cranking speed, spring holds actuator cam inward against the rocker cam and rocker cam is forced above exhaust cam surface.*

*All models equipped with auxiliary drive unit have a main bearing rejection size for main bearing at pto side of 1.003 inch (25.48 mm)

Bushings are not available for all models; therefore, on some models the sump must be renewed if necessary to renew the bearing.

Main bearing journal diameter for flywheel and pto side main bearing journals on all standard models with plain bearings, except 130000, 131000 and 140000 models, is 0.873 inch (22.17 mm). On models equipped with auxiliary drive unit, pto side main bearing journal diameter is 0.998 inch (25.35 mm). Main bearing journal diameter at flywheel side of 130000 and 131000 models is 0.873 inch (22.17 mm) and journal diameter at pto side is 0.998 inch (25.35 mm). Main bearing journal diameter at flywheel side of 140000 model is 0.997 inch (25.32 mm) and journal diameter at pto side is 1.179 inch (29.95 mm).

Crankpin journal rejection size for 6B and 60000 models is 0.870 inch (22.10 mm), crankpin journal rejection size for 140000 model is 1.090 inch (27.69 mm) and crankpin journal rejection size for all other models is 0.996 inch (25.30 mm).

Ball bearing mains are a press fit on the crankshaft and must be removed by pressing the crankshaft out of the bearing. Reject ball bearing if worn or rough.

Expand new bearing by heating it in oil and install it on crankshaft with seal side towards crankpin journal.

Crankshaft end play on all models is 0.002-0.008 inch (0.05-0.20 mm). At least one 0.015 inch cover or sump gasket must be in place. Additional gaskets in several sizes are available to aid in end play adjustment. If end play is over 0.008 inch (0.20 mm), metal shims are available for use on crankshaft between crankshaft gear and cylinder.

Refer to VALVE TIMING section for proper timing procedure when installing crankshaft.

**CAMSHAFT.** The camshaft and camshaft gear are an integral casting on all models. The camshaft and gear should be inspected for wear on the journals, cam lobes and gear teeth. On all standard models except 110900, 111200, 111900, 112200 and 113900 models, if camshaft journal diameter is 0.498 inch (12.65 mm) or less, reject camshaft.

On 110900, 111200, 111900, 112200 and 113900 models, camshaft journal rejection size for journal on flywheel side is 0.436 inch (11.07 mm) and for pto side journal rejection size is 0.498 inch (12.65 mm).

On models with "Easy-Spin" starting, the intake cam lobe is designed to hold the intake valve slightly open on part of the compression stroke. Therefore, to check compression, the engine must be turned backwards.

"Easy-Spin" camshafts (cam gears) can be identified by two holes drilled in the web of the gear. Where part number of an older cam gear and an "Easy-Spin" cam gear are the same (except for an "E" following the "Easy-Spin" part number), the gears are interchangeable.

The camshaft used on 11200, 11900 and 112200 models is equipped with the "Easy-Spin" intake lobe and a mechanically operated compression release on the exhaust lobe. With engine stopped or at cranking speed, the spring holds the actuator cam weight inward against the rocker cam. See Fig. BS111. The rocker cam is held slightly above the exhaust cam surface which in turn holds the exhaust valve open slightly during compression stroke. This compression release greatly reduces the power needed for cranking.

When engine starts and rpm is increased, the actuator cam weight moves outward overcoming the spring pressure. See Fig. BS111A. The rocker cam is rotated below the cam surface to provide normal exhaust valve operation.

Refer to VALVE TIMING section for proper timing procedure when installing camshaft.

**VALVE SYSTEM.** Valve tappet clearance (cold) for 94500, 94900 and 95500 models is 0.004-0.006 inch (0.10-0.15 mm) for intake valve and 0.007-0.009 inch (0.18-0.23 mm) for exhaust valve.

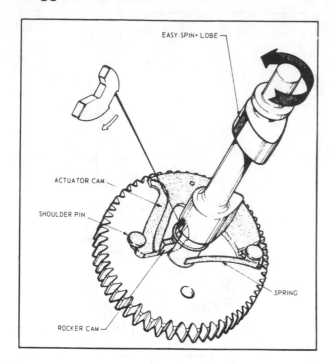

Fig. BS111A—When engine starts and rpm is increased, actuator cam weight moves outward allowing rocker cam to rotate below exhaust cam surface.

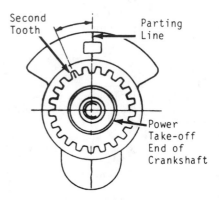

Fig. BS113—Align timing marks on cam gear and crankshaft gear on plain bearing models.

Valve tappet clearance (cold) for all other models is 0.005-0.007 inch (0.13-0.18 mm) for intake valve and 0.009-0.011 inch (0.23-0.28 mm) for exhaust valve.

Valve tappet clearance is adjusted on all models by carefully grinding off end of valve stem to increase clearance or by grinding valve seats deeper and/or renewing valve or lifter to decrease clearance.

Valve face and seat angle should be ground at 45°. Renew valve if margin is 1/16 inch (1.588 mm) or less. Seat width should be 3/64 to 1/16 inch (1.119-1.588 mm).

The valve guides on all engines with aluminum blocks are an integral part of the cylinder block. To renew the valve guides, they must be reamed out and a bushing installed.

On all models except 140000 model, ream guide out first with reamer (B&S part 19064) only abut 1/16 inch (1.588 mm) deeper than length of bushing (B&S part 63709). Press in bushing with driver (B&S part 19065) until it is flush with top of guide bore. Finish ream the

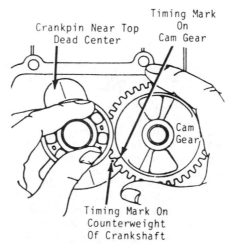

Fig. BS112—Align timing marks on cam gear with mark on crankshaft counterweight on ball bearing equipped models.

bushing with a finish reamer (B&S part 19066).

On 140000 model, ream guide out with reamer (B&S part 19183) only about 1/16 inch (1.588 mm) deeper than top end of flutes on reamer. Press in

Fig. BS114—Location of tooth to align with timing mark on cam gear if mark is not visible on crankshaft gear.

bushings with soft driver as bushing is finish reamed at the factory.

**VALVE TIMING.** On engines equipped with ball bearing mains, align the timing mark on the cam gear with the timing mark on the crankshaft counterweight as shown in Fig. BS112. On engines with plain bearings, align the timing marks on the camshaft gear with the timing mark on the crankshaft gear as shown in Fig. BS113. If the timing mark is not visible on the crankshaft gear, align the timing mark on the camshaft gear with the second tooth to the left of the crankshaft counterweight parting line as shown in Fig. BS114.

# BRIGGS & STRATTON

## SERVICING BRIGGS & STRATTON ACCESSORIES

### REWIND STARTER

**OVERHAUL.** To renew broken rewind spring, proceed as follows: Grasp free outer end of spring (S – Fig. B90) and pull broken end from starter housing. With blower housing removed, remove tangs (T) and remove starter pulley from housing. Untie knot in rope (R) and remove rope and inner end of broken spring from pulley. Apply a small amount of grease on inner face of pulley, thread inner end of spring through notch in starter housing, engage inner end of spring in pulley hub and place bar in pulley hub and turn pulley approximately 13½ turns in a counter-clockwise direction as shown in Fig. B91. Tie wrench to blower housing with wire to hold pulley so hole (H) in pulley is aligned with rope guide (G) in housing as shown in Fig. B92. Hook a wire in inner end of rope and thread rope through guide and hole in pulley; then, tie a knot in rope and release the pulley allowing spring to wind rope into pulley groove.

To renew starter rope only, it is not generally necessary to remove starter pulley and spring. Wind up spring and install new rope as outlined in preceding paragraph.

Two different types of starter clutches have been used; refer to exploded view of early production unit in Fig. B93 and exploded view of late production unit in Fig. B94. Outer end of late production ratchet (refer to cutaway view in Fig. B95) is sealed with a felt and a retaining plug and a rubber ring is used to seal ratchet to ratchet cover.

To disassemble early type starter clutch unit, refer to Fig. B93 and proceed as follows: Remove snap ring (3) and lift ratchet (5) and cover (4) from starter housing (7) and crankshaft. Be careful not to lose the steel balls (6).

Starter housing (7) is also flywheel retaining nut; to remove housing, first remove screen (2) and using Briggs & Stratton flywheel wrench No. 19114, unscrew housing from crankshaft in counter-clockwise direction. When reinstalling housing, be sure spring washer (8) is placed on crankshaft with cup (hollow) side towards flywheel, then install starter housing and tighten securely. Reinstall rotating screen. Place ratchet on crankshaft and into housing and insert the steel balls. Reinstall cover and retaining snap ring.

To disassemble late starter clutch unit, refer to Fig. B94 and proceed as follows: Remove rotating screen (2) and starter ratchet cover (4). Lift ratchet (5) from housing and crankshaft and extract the steel balls (6). If necessary to

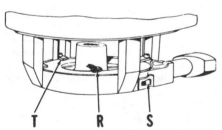

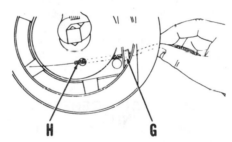

**Fig. B92 – Threading starter rope through guide (G) in blower housing and hole (H) in starter pulley with wire hooked in end of rope.**

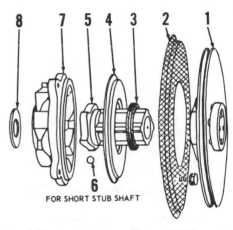

FOR SHORT STUB SHAFT

**Fig. B94 – Exploded view of late production sealed starter clutch unit. Late unit can be used with "short stub shaft" only; refer to Fig. B96.**

1. Starter rope pulley
2. Rotating screen
3. Snap ring
4. Ratchet cover
5. Starter ratchet

6. Steel balls
7. Clutch housing (flywheel nut)
8. Spring washer

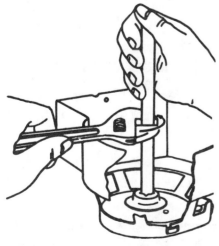

**Fig. B90 – View of rewind starter showing rope (R), spring end (S) and retaining tangs (T).**

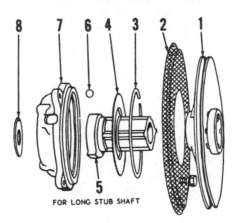

FOR LONG STUB SHAFT

**Fig. B93 – Exploded view of early production starter clutch unit; refer to Fig. B96 for "long stub shaft". A late type unit (Fig. B94) should be installed when renewing "long" crankshaft with "short" (late production) shaft.**

1. Starter rope pulley
2. Rotating screen
3. Snap ring
4. Ratchet cover
5. Starter ratchet

6. Steel balls
7. Clutch housing (flywheel nut)
8. Spring washer

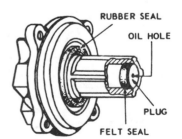

RUBBER SEAL

OIL HOLE

PLUG

FELT SEAL

**Fig. B95 – Cut-away showing felt seal and plug in end of late production starter ratchet (5 – Fig. B94).**

**Fig. B91 – Using square shaft and wrench to wind up the rewind starter spring. Refer to text.**

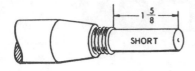

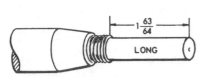

*Fig. B96—Crankshaft with short, 1⅝-inch (31.275 mm) stub (top view) must be used with late production starter clutch assembly. Early crankshaft (bottom view) can be modified by cutting off the 1-63/64 inch (50.4 mm) stub end to the dimension shown in top view and beveling end of shaft to allow installation of late type clutch unit.*

remove housing (7), hold flywheel and unscrew housing in counter-clockwise direction using Briggs & Stratton flywheel wrench No. 19114. When installing housing, be sure spring washer (8) is in place on crankshaft with cup (hollow) side towards flywheel; then, tighten housing securely. Inspect felt seal and plug in outer end of ratchet; renew ratchet if seal or plug are damaged as these parts are not serviced separately. Lubricate the felt with oil and place ratchet on crankshaft. Insert the steel balls and install ratchet cover, rubber seal and rotating screen.

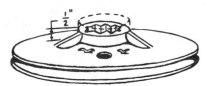

*Fig. B97—When installing late type starter clutch unit as a replacement for early type, either install new starter rope pulley or cut hub of old pulley to ½-inch (12.7 mm) as shown.*

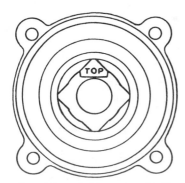

*Fig. B98—When installing blower housing and starter assembly, turn starter ratchet so word "TOP" stamped on outer end of ratchet is towards engine cylinder head.*

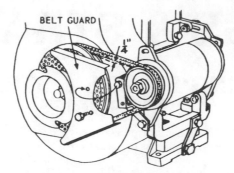

*Fig. B99—View showing starter generator belt adjustment on models so equipped. Refer to text.*

**NOTE: Crankshafts used with early and late starter clutches differ; refer to Fig. B96.**

If renewing early (long) crankshaft with late (short) shaft, also install late type starter clutch unit. If renewing early starter clutch with late type unit, crankshaft must be shortened to dimension shown for short shaft in Fig. B96; also, hub or starter rope pulley must be shortened to ½-inch (12.7 mm) dimension shown in fig. B97. Bevel end of crankshaft after removing the approximate ⅜-inch (9.52 mm) from shaft.

When installing blower housing and starter assembly, turn starter ratchet so word "TOP" on ratchet is toward engine cylinder head.

### 12-VOLT STARTER-GENERATOR UNITS

Combination starter-generator functions as a cranking motor when starting switch is closed. When engine is operating and with starting switch open, unit operates as a generator. Generator output and circuit voltage for battery and various operating requirements are controlled by a current-voltage regulator. On units where voltage regu-

lator is mounted separately from generator unit, do not mount regulator with cover down as regulator will not function in this position. To adjust belt tension, apply approximately 30 pounds (13.6 kg) pull on generator adjusting flange and tighten mounting bolts. Belt tension is correct when a pressure of 10 pounds (44.5 N) applied midway between pulleys will deflect belt ¼-inch (6.35 mm). See Fig. B99. On units equipped with two drive belts, always renew belts in pairs. A 50-ampere capacity battery is recommended. Starter-generator units are intended for use in temperatures above 0° F (-18° C). Refer to Fig. B100 for exploded view of starter-generator. Parts and service on starter-generator are available at authorized Delco-Remy service stations.

### GEAR-DRIVE STARTERS

Two types of gear drive starters may be used, a 110 volt AC starter or a 12 volt DC starter. Refer to Figs. B101 and B102 for exploded views of starter motors. A properly grounded receptacle should be used with power cord connected to 110 volt AC starter motor. A 32 ampere hour capacity battery is recommended for use with 12 volt DC starter motor.

To renew a worn or damaged flywheel ring gear, drill out retaining rivets using a 3/16-inch drill. Attach new ring gear using screws provided with new ring gear.

To check for correct operation of starter motor, remove starter motor from engine and place motor in a vise or other holding fixture. Install a 0-5 amp ammeter in power cord to 110 volt AC starter motor. On 12 volt DC motor, connect a 12 volt battery to motor with a 0-50 amp ammeter in series with positive line from battery to starter motor. Connect a tachometer to drive end of starter. With starter activated on

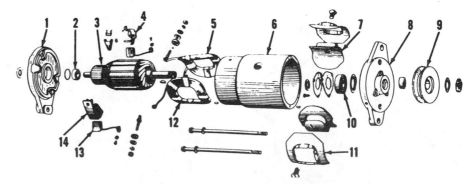

*Fig. B100—Exploded view of Delco-Remy starter-generator unit used on some B&S engines.*

1. Commutator end frame
2. Bearing
3. Armature
4. Ground brush holder
5. Field coil L.H.
6. Frame
7. Pole shoe
8. Drive end frame
9. Pulley
10. Bearing
11. Field coil insulator
12. Field coil R.H.
13. Brush
14. Insulated brush holder

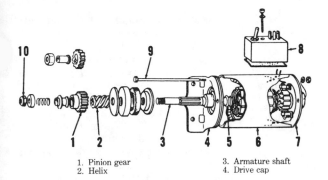

*Fig. B101 — Exploded view of 110 volt AC starter motor. Twelve volt DC starter motor is similar. Rectifier and switch unit (8) is used on 110 volt motor only.*

1. Pinion gear
2. Helix
3. Armature shaft
4. Drive cap
5. Thrust washer
6. Housing
7. End cap
8. Rectifier and switch unit
9. Bolt
10. Nut

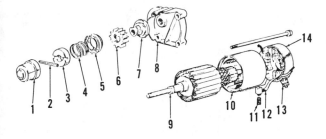

*Fig. B102 — Exploded view of 12 volt 4-brush starter motor.*

1. Cap
2. Roll pin
3. Retainer
4. Pinion spring
5. Spring cup
6. Starter gear
7. Clutch assy.
8. Drive end cap assy.
9. Armature
10. Housing
11. Spring
12. Brush assy.
13. Battery wire terminal
14. Commutator end cap assy.

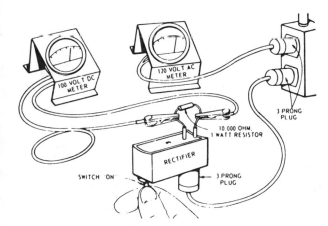

*Fig. B103 — View of test connections for 110 volt rectifier. Refer to text for procedure.*

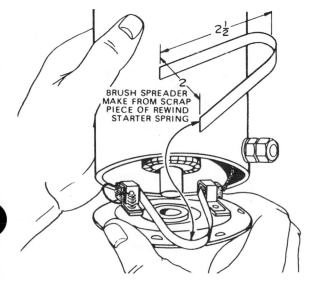

*Fig. B104 — Tool shown may be fabricated to hold brushes when installing motor end cap.*

110 volt motor, starter motor should turn at 5200 rpm minimum with a maximum current draw of 3½ amps. The 12 volt motor should turn at 6200 rpm minimum with a current draw of 16 amps maximum. If starter motor does not operate satisfactorily, check operation of rectifier or starter switch. If rectifier and starter switch are good, disassemble and inspect starter motor.

To check rectifier used on 110 volt AC starter motor, remove rectifier unit from starter motor. Solder a 10,000 ohm 1 watt resistor to DC internal terminals of rectifier as shown in Fig. B103. Connect a 0-100 range DC voltmeter to resistor leads. Measure voltage of AC outlet to be used. With starter swtich in "OFF" position, a zero reading should be shown on DC voltmeter. With starter switch in "ON" position, DC voltmeter should show a reading that is 0-14 volts lower than AC line voltage measured previously. If voltage drop exceeds 14 volts, renew rectifier unit.

Disassembly of starter motor is self-evident after inspection of unit and referral to Fig. B101 and B102. Note position of bolts during disassembly so they can be installed in their original positions during reassembly. When reassembling motor lubricate end cap bearings with SAE 20 oil. Be sure to match drive cap keyway to stamped key in housing when sliding armature into motor housing. Brushes may be held in their holders during installation by making a brush spreader tool from a piece of metal as shown in Fig. B104. Splined end of helix (2 – Fig. B101) must be towards end of armature shaft as shown in Fig. B105. Tighten armature shaft nut to 170 in.-lbs. (19 N·m).

## FLYWHEEL ALTERNATORS

### 4 Amp. Non-Regulated Alternator

Some engines are equipped with the 4 ampere non-regulated flywheel alter-

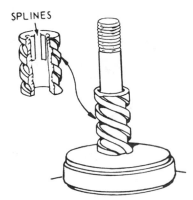

*Fig. 105 — Install helix on armature so splines of helix are toward outer end of shaft as shown.*

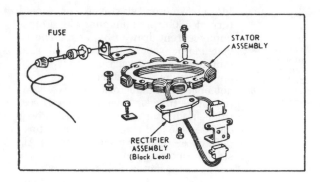

*Fig. B106 — Stator and rectifier assemblies used on the 4 ampere non-regulated flywheel alternator. Fuse is 7½ amp AGC or 3AG.*

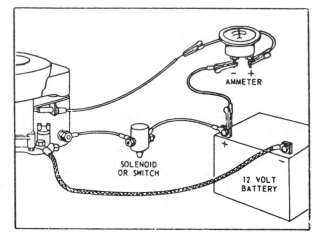

*Fig. B107 — Install ammeter as shown for output test.*

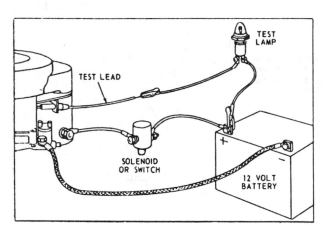

*Fig. B108 — Connect a test lamp as shown to test for shorted stator or defective rectifier. Refer to text.*

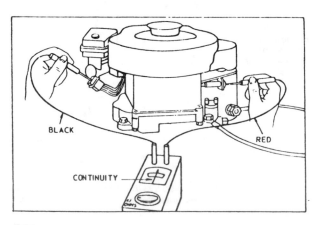

*Fig. B109 — Use an ohmmeter to check condition of stator. Refer to text.*

nator shown in Fig. B106. A solid state rectifier and 7½ amp fuse is used with this alternator.

If battery is run down and no output from alternator is suspected, first check the 7½ amp fuse. If fuse is good, clean and tighten all connections. Disconnect charging lead and connect an ammeter as shown in Fig. B107. Start engine and check for alternator output. If ammeter shows no charge, stop engine, remove ammeter and install a test lamp as shown in Fig. B108. Test lamp should not light. If it does light, stator or rectifier is defective. Unplug rectifier plug under blower housing. If test lamp does not go out, stator is shorted.

If shorted stator is indicated, use an ohmmeter and check continuity as follows: Touch one test lead to lead inside of fuse holder as shown in Fig. B109. Touch remaining test lead to each of the four pins in rectifier connector. Unless ohmmeter shows continuity at each of the four pins, stator winding is open and stator must be renewed.

If defective rectifier is indicated, unbolt and remove flywheel blower housing with rectifier. Connect one ohmmeter test lead to blower housing and remaining test lead to the single pin connector in rectifier connector. See Fig. B110. Check for continuity, then reverse leads and again test for continuity. If tests show no continuity in either direction or continuity in both directins, rectifier is faulty and must be renewed.

## 7 Amp. Regulated Alternator

A 7 amp regulated flywheel alternator is used witn 12 volt gear drive starter motor on some models. Alternator is equipped with a solid state rectifier and regulator. An isolation diode is also used on most models.

If engine will not start, using electric start system and trouble is not in starting motor, install an ammeter in circuit as shown in Fig. B112. Start engine manually. Ammeter should indicate charge. If ammeter does not show battery charging taking place, check for defective wiring and if necessary proceed with troubleshooting.

If battery charging occurs with engine running, but battery does not retain charge, then isolation diode may be defective. The isolation diode is used to prevent battery drain if alternator circuit malfunctions. After troubleshooting diode, remainder of circuit should be inspected to find reason for excessive battery drain. To check operation of diode, disconnect white lead of diode from fuse holder and connect a test lamp from the diode white lead to negative terminal of battery. Test lamp should not light. If test lamp lights, diode is defective. Disconnect test lamp and disconnect red

Illustrations courtesy of Briggs & Stratton Corp.

lead of diode. Test continuity of diode with ohmmeter by connecting leads of ohmmeter to leads of diode then reverse lead connections. Ohmmeter should show continuity in one direction and an open circuit in the other direction. If readings are incorrect, then diode is defective and must be renewed.

To troubleshoot alternator assembly, proceed as follows: Disconnect white lead of isolation diode from fuse holder and connect a test lamp between positive terminal of battery and fuse holder on engine. Engine must not be started. With connections made, test lamp should not light. If test lamp does light, stator, regulator or rectifier is defective. Unplug rectifier-regulator plug under blower housing. If lamp remains lighted, stator is grounded. If lamp goes out, regulator or rectifier is shorted.

If previous test indicated stator is grounded, check stator leads for defects and repair if necessary. If shorted leads are not found, renew stator. Check stator for an open circuit as follows: Using a ohmmeter, connect positive lead to fuse holder as shown in Fig. B113 and negative lead to one of the pins in rectifier and regulator connector. Check each of the four pins in connector. Ohmmeter should show continuity at each pin, if not, then there is an open in stator and stator must be renewed.

To test rectifier, unplug rectifier and regulator connector plug and remove blower housing from engine. Using an ohmmeter check for continuity between connector pins connected to black wires and blower housing as shown in Fig. B114. Be sure good contact is made with metal of blower housing. Reverse ohmmeter leads and check continuity again. Ohmmeter should show a continuity reading for one direction only on each plug. If either pin shows a continuity reading for both directions, or if either pin shows no continuity for either direction, then rectifier must be renewed.

To test regulator unit, repeat procedure used to test rectifier unit except connect ohmmeter lead to pins connected to red wire and white wire. If ohmmeter shows continuity in either direction for red lead pin, regulator is defective an must be renewed. White lead pin should read as an open on ohmmeter in one direction and a weak reading in the other direction. Otherwise, regulator is defective and must be renewed.

## 10 Amp Regulated Alternator

Engines may be equipped with a 10 ampere flywheel alternator and a solid state rectifier-regulator. To check charging system, disconnect charging lead from battery. Connect a DC

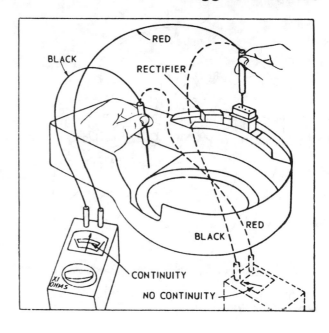

*Fig. B110—If ohmmeter shows continuity in both directions or in neither direction, rectifier is defective.*

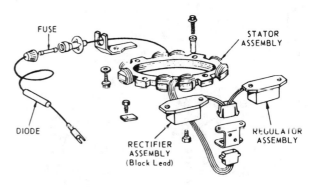

*Fig. B111—Stator, rectifier and regulator assemblies used on the 7 ampere regulated flywheel alternator.*

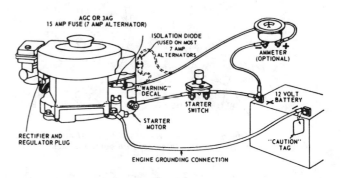

*Fig. B112—Typical wiring used on engines equipped with 7 ampere flywheel alternator.*

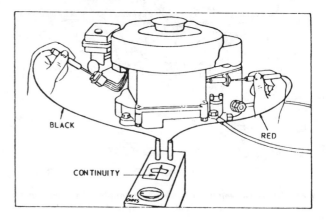

*Fig. B113—Use an ohmmeter to check condition of stator. Refer to text.*

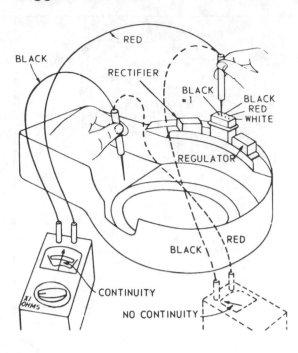

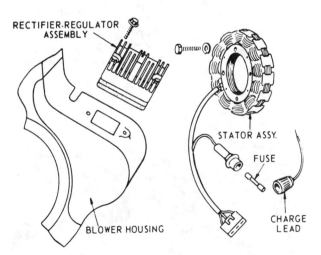

voltmeter between charging lead and ground as shown in Fig. B116. Start engine and operate at 3600 rpm. A voltmeter reading of 14 volts or above indicates alternator is functioning. If reading is less than 14 volts, stator or rectifier-regulator is defective.

To test stator, disconnect stator plug from rectifier-regulator. Operate engine at 3600 rpm and connect AC voltmeter leads to AC terminals in stator plug as shown in Fig. B117. Voltmeter reading above 20 volts indicates stator is good. A reading less than 20 volts indicates stator is defective.

To test rectifier-regulator, make certain charging lead is connected to battery and stator plug is connected to rectifier-regulator. Check voltage across battery terminals with DC voltmeter (Fig. B118). If voltmeter reading is 13.8 volts or higher, reduce battery voltage by connecting a 12 volt load lamp across battery terminals. When battery voltage is below 13.5 volts, start engine and operate at 3600 rpm. Voltmeter reading should rise. If battery is fully charged,

*Fig. B114 — Be sure good contact is made between ohmmeter test lead and metal cover when checking rectifier and regulator.*

*Fig. B115 — View of 10 ampere regulated flywheel alternator stator and rectifier-regulator used on some engines.*

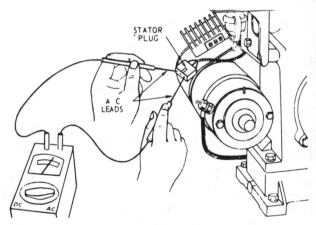

*Fig. B117 — AC voltmeter is used to test stator.*

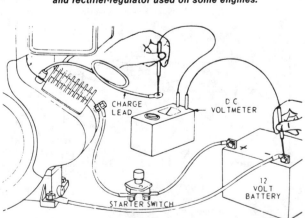

*Fig. B116 — DC voltmeter is used to determine if alternator is functioning. Refer to text.*

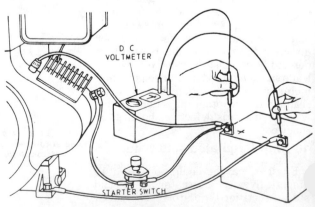

*Fig. B118 — Check battery voltage with DC voltmeter. Refer to text for rectifier-regulator test.*

reading should rise above 13.8 volts. If voltage does not increase or if voltage reading rises above 14.7 volts, rectifier-regulator is defective and must be renewed.

## FUEL PUMP

A diaphragm type fuel pump is available on many models as optional equipment. Refer to Fig. B119 for exploded view of pump.

To disassemble pump, refer to Figs. B119 and B120; then, proceed as follows: Remove clamp (1), fuel bowl (2), gasket (3) and screen (4). Remove screws retaining upper body (9) to lower body (16). Pump valves (5) and gaskets (6) can now be removed. Drive pin (14) out to either side of body (16), then press diaphragm (10) against spring (11) as shown in view A, Fig. B120, and remove lever (13). Diaphragm and spring (11—Fig. B119) can now be removed.

To reassemble, place diaphragm spring in lower body and place diaphragm on spring, being sure spring enters cup on bottom side of diaphragm and slot in shaft is at right angle to pump lever. Then, compressing diaphragm against spring as in view A, Fig. B120, insert hooked end of lever with hole in lower body and drive pin into place. Then, insert lever spring (15) into body and push outer end of spring into place over hook on arm of lever as shown in view B. Hold lever downward as shown in view C while tightening screws holding upper body to lower body. When installing pump on engine, apply a liberal amount of grease on lever (13) at point where it contacts groove in crankshaft.

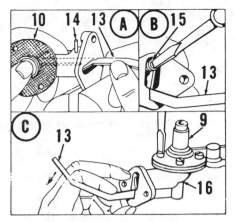

Fig. B120—Views showing disassembly and reassembly of diaphragm type fuel pump. Refer to text for procedure and to Fig. B119 for exploded view of pump and for legend.

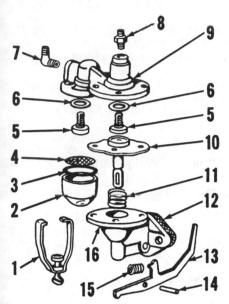

Fig. B119—Exploded view of diaphragm type fuel pump used on some B&S engines.

| | |
|---|---|
| 1. Yoke assy. | 9. Fuel pump head |
| 2. Filter bowl | 10. Pump diaphragm |
| 3. Gasket | 11. Diaphragm spring |
| 4. Filter screen | 12. Gasket |
| 5. Pump valves | 13. Pump lever |
| 6. Gaskets | 14. Lever pin |
| 7. Elbow fitting | 15. Spring |
| 8. Connector | 16. Fuel pump body |

## BRIGGS & STRATTON SPECIAL TOOLS

The following special tools are available from Briggs & Stratton Central Service Distributors.

### TOOL KITS

19184 – Main bearing tool kit for Series 170000, 171000, 190000 and 191000

19228 – Main bearing tool kit for all twin-cylinder engines with integral and DU type main bearings

19232 – Valve guide puller/reamer kit for Series 170000, 190000, 220000, 233000, 243000, 250000, 300000, 326000 and twin-cylinder engines

19138 – Valve seat insert puller for all models and series so equipped

19205 – Cylinder hone kit for all aluminum bore (Kool-Bore) engines

19211 – Cylinder hone kit for all cast iron cylinder engines

19237 – Valve seat cutter kit for all models and series

### PLUG GAGES

19055 – Check breaker plunger bore on Model 23C and Series 170000, 190000, 220000 and 251000.

19117 – Check main bearing bore on Models 19, 19D, 23, 23A, 23C, 23D and Series 190000, 200000 and 230000

19151 – Check valve guide bores on Models 19, 19D, 23, 23A, 23C, 23D and Series 170000, 190000, 200000, 230000, 240000, 250000, 300000, 320000 and twin-cylinder engines

19164 – Check camshaft bearings on Series 170000, 190000, 220000 and 250000

19178 – Check main bearing bore on Series 170000, 171000, 190000 and 191000

19219 – Check integral or DU type main bearing bore on twin-cylinder engines

### REAMERS

19056 – Breaker plunger bushing reamer for 170000, 190000, 220000 and 251000

19058 – Finish reamer for breaker plunger bushing for Series 170000, 190000, 220000 and 251000

19172 – Counter bore reamer for main bearings for Series 170000 and 190000

19173 – Finish reamer for main bearings for Series 170000 and 190000

19174 – Counter bore reamer for main bearings for Series 170000, 171000, 190000 and 191000

19175 – Finish reamer for main bearings for Series 170000, 171000, 190000 and 191000

19183 – Valve guide reamer for bushing installation for Models 19, 19D, 23, 23A, 23C, 23D and Series 170000, 190000, 200000, 220000, 230000, 240000, 251000, 300000 and 320000

19231 – Valve guide bushing reamer for Models 19 and 23 and Series 170000, 190000, 200000, 220000, 230000, 240000, 250000, 300000, 320000 and twin-cylinder engines

19233 – Valve guide bushing finish reamer for Models 19 and 23 and Series 170000, 190000, 200000, 220000, 230000, 240000, 250000, 300000, 320000 and twin-cylinder engines

## PILOTS

19096 – Main bearing reamer pilot for Series 170000, 171000, 190000 and 191000

19127 – Expansion pilot for valve seat counterbore cutter for Models 19 and 23 and Series 170000, 190000, 200000, 220000, 230000, 240000, 300000, 320000 and twin-cylinder engines

19130 – "T" handle for 19127 pilot

## REAMER GUIDE BUSHING

19168 – Guide bushing for main bearing reaming for Series 170000 and 190000

19169 – Guide bushing for main bearing reaming for Series 170000 and 190000

19170 – Guide bushing for main bearing reaming for Series 170000 and 190000

19171 – Guide bushing for main bearing reaming for Series 170000 and 190000

19192 – Guide bushing for valve guide reaming for Models 19 and 23 and Series 170000, 190000, 200000, 220000, 230000, 240000, 251000, 300000 and 320000

19201 – Guide bushing for main bearing reaming for Series 171000 and 191000

19222 – Guide bushing for tool kit 19228 for twin-cylinder engines

19234 – Guide bushing for tool kit 19232 for twin-cylinder engines

## PILOT GUIDE BUSHINGS

19168 – Pilot guide bushing for main bearing reaming for Series 170000 and 190000

19169 – Pilot guide bushing for main bearing reaming for Series 170000 and 190000

19220 – Pilot guide bushing for main bearing reaming for tool kit 19228 for twin-cylinder engines

## COUNTERBORE CUTTERS

19131 – Counterbore valve seat cutter for Models 19, 19D, 23, 23A, 23D and Series 190000, 200000, 230000, 240000 and 243000

## CRANKCASE SUPPORT

19123 – To support crankcase when removing or installing main bearings on Series 170000 and 190000

19227 – To support crankcase when removing or installing DU type main bearings on twin-cylinder engines

## DRIVERS

19057 – To install breaker plunger bushing on Series 170000, 190000, 220000 and 251000

19136 – To install valve seat inserts on all models and series

19179 – To install main bearings on Series 170000 and 190000

19226 – To install main bearing on twin cylinder engines

## FLYWHEEL PULLERS

19068 – Flywheel removal on Models 19, 19D, 23, 23A, 23C, and 23D and Series 190000, 200000, 230000, 240000, 300000 and 320000

19165 – Flywheel removal on Series 170000, 190000 and 250000

19203 – Flywheel removal on Models 19 and 23 and Series 190000, 200000, 220000, 230000, 240000, 250000, 251000, 300000, 320000 and twin-cylinder engines

## VALVE SPRING COMPRESSOR

19063 – Valve spirng compressor for all models and series

## PISTON RING COMPRESSOR

19230 – Ring compressor for all models and series

## STARTER WRENCH

19114 – All models and series with rewind starter

19161 – All models and series with rewind starter (to be used with ½-inch drive torque wrench)

## CLUTCH WRENCH

19244 – To remove, install and torque rewind starter clutches

## BENDING TOOL

19229 – Governor tang bending tool

## SPARK TESTER

19051 – Test ignition spark on all models and series

---

# BRIGGS & STRATTON CENTRAL SERVICE DISTRIBUTORS

(Arranged Alphabetically by States)

**These franchised firms carry extensive stocks of repair parts. Contact them for name of the nearest service distributor who may have the parts you need.**

BEBCO, Incorporated
2221 Second Avenue, South
**Birmingham, Alabama 35233**

Power Equipment Company
3141 North 35th Avenue, Unit 101
**Phoenix, Arizona 85107**

Pacific Power Equipment Company
1565 Adrain Road
**Burlingame, California 94010**

Power Equipment Company
1045 Cindy Lane
**Carpinteria, California 93013**

Pacific Power Equipment Company
5000 Oakland Street
**Denver, Colorado 80239**

Spencer Engine Incorporated
1114 West Cass Street
**Tampa, Florida 33606**

Sedco Incorporated
1414 Red Plum Road
**Norcross, Georgia 30093**

Small Engine Clinic, Incorporated
98019 Kam Highway
**Aiea, Hawaii 96701**

Midwest Engine Warehouse
515 Romans Road
**Elmhurst, Illinois 60126**

Medart Engines & Parts of Kansas
15500 West 109th Street
**Lenexa, Kansas 66215**

Commonwealth Engine, Incorporated
1361 South 15th Street
**Louisville, Kentucky 40210**

Suhren Engine Company, Inc.
8330 Earhart Boulevard
**New Orleans, Louisiana 70118**

Grayson Company of Louisiana, Inc.
100 Fannin Street
**Shreveport, Louisiana 71101**

W. J. Connell Company
65 Green Street
**Foxboro, Massachusetts 02035**

Carl A. Anderson Inc. of Minnesota
2737 West Service Road
**Eagan, Minnesota 55121**

Medart Engines & Parts
100 Larkin Williams Ind. Ct.
**Fenton, Missouri 63026**

Original Equipment, Incorporated
905 Second Avenue, North
**Billings, Montana 59101**

Carl A. Anderson Inc. of Nebraska
7410 "L" Street
**Omaha, Nebraska 68127**

W. J. Connel Co., Bound Brook Div.
Chimney Rock Road, Route 22
**Bound Brook, New Jersey 08805**

Power Equipment Company
7209 Washington Street, North East
**Alburquerque, New Mexico 87109**

AEA, Incorporated
700 West 28th Street
**Charlotte, North Carolina 28206**

Gardner, Incorporated
1150 Chesapeake Avenue
**Columbus, Ohio 43212**

American Electric Ignition Company
124 North West Eighth
**Oklahoma City, Oklahoma 73101**

Brown & Wiser, Incorporated
9991 South West Avery Street
**Tualatin, Oregon 97062**

Pitt Auto Electric Company
2900 Stayton Street
**Pittsburgh, Pennsylvania 15212**

Automotive Electric Corporation
3250 Millbranch Road
**Memphis, Tennessee 38116**

American Electric Ignition Co., Inc.
618 Jackson Street
**Amarillo, Texas 79101**

Grayson Company, Incorporated
1234 Motor Street
**Dallas, Texas 75207**

Engine Warehouse, Incorporated
7415 Empire Central Drive
**Houston, Texas 77040**

Frank Edwards Company
110 South 300 West
**Salt Lake City, Utah 84101**

RBI Corporation
101 Cedar Ridge Drive
**Ashland, Virginia 23005**

Bitco Western, Incorporated
4030 1st Avenue, South
**Seattle, Washington 98134**

Air Cooled Engine Supply
South 127 Walnut Street
**Spokane, Washington 99204**

Wisconsin Magnetic Incorporated
4727 North Teutonia Avenue
**Milwaukee, Wisconsin 53209**

## CANADIAN DISTRIBUTORS

Canadian Curtiss-Wright, Ltd.
2-3571 Viking Way
**Richmond, British Columbia
V6V 1W1**

Canadian Curtiss-Wright, Ltd.
89 Paramount Road
**Winnipeg, Manitoba R2X 2W6**

Canadian Curtiss-Wright, Ltd.
1815 Sismet Road
**Mississauga, Ontario L4W 1P9**

Canadian Curtiss-Wright, Ltd.
8100-N Trans Canada Hwy.
**St. Laurent, Quebec H4S 1M5**

# KAWASAKI

**KAWASAKI MOTOR CORP.**
**P.O. Box 504**
**Shakopee, MN 55379**

| Model | No. Cyls. | Bore | Stroke | Displacement |
|-------|-----------|------|--------|--------------|
| FA76 | 1 | 52 mm | 36 mm | 76 cc |
| | | (2.05 in.) | (1.42 in.) | (4.64 cu. in.) |
| FA130 | 1 | 62 mm | 43 mm | 129 cc |
| | | (2.44 in.) | (1.69 in.) | (7.9 cu. in.) |
| FA210 | 1 | 72 mm | 51 mm | 207 cc |
| | | (2.83 in.) | (2.01 in.) | (12.7 cu. in.) |

Kawasaki FA series engines are four-stroke, single-cylinder, air-cooled, horizontal crankshaft engines. Engine model number is located on the cooling shroud just above the rewind starter. Engine model number and serial number are both stamped on crankcase cover.

## MAINTENANCE

**SPARK PLUG.** Recommended spark plug is NGK BM6A or BMR6A or equivalent. Spark plug should be removed, checked and cleaned after every 50 hours of operation. Renew spark plug if electrode is corroded or damaged. Spark plug electrode gap should be set at 0.7-0.8 mm (0.028-0.031 in.).

**AIR FILTER.** The engine is equipped with a foam type air filter element. The air filter element should be removed and cleaned after every 50 hours of operation, or more often if operating in extremely dusty conditions. To clean element, wash in a nonflammable cleaning solvent and gently squeeze element dry. Oil element with clean engine oil and gently squeeze out the excess oil.

**FUEL VALVE.** A filter screen and sediment bowl are mounted on bottom of the fuel shut-off valve. Sediment bowl and screen should be removed and cleaned after every 50 hours of operation, or more often if water or dirt is visible in the sediment bowl.

**CARBURETOR.** Engine is equipped with a float-type carburetor (Fig. KA101). Main fuel mixture is controlled by fixed main jet (8).

Initial adjustment of idle mixture needle (5) is 1⅛ turns out from a lightly seated position. Recommended engine idle speed is 1500-1700 rpm. To adjust idle speed, run engine until it reaches normal operating temperature, then turn idle speed screw (7) to obtain specified engine idle speed. Adjust idle mixture screw (7) so engine runs at maximum idle speed, then turn mixture screw out (counterclockwise) an additional ¼ turn.

To adjust engine maximum operating speed, refer to GOVERNOR paragraphs.

To check carburetor float level, remove the carburetor from the engine. Remove the float bowl and invert carburetor body. Surface of float should be parallel to carburetor body (Fig. KA102). The plastic float used on Model FA76D engine is nonadjustable and must be renewed if float level is incorrect. On other models, bend float tab to adjust.

**Overhaul.** To disassemble carburetor, remove fuel bowl retaining screw (15--Fig. KA101) and fuel bowl (14). Re-

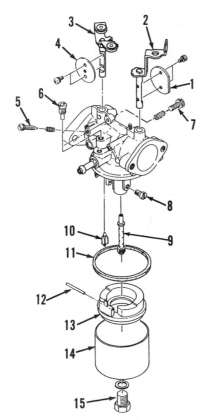

Fig. KA101—Exploded view of float type carburetor.

1. Choke valve
2. Choke shaft
3. Throttle shaft
4. Throttle valve
5. Idle mixture adjustment screw
6. Pilot jet
7. Idle speed adjustment screw
8. Main jet
9. Main nozzle
10. Fuel inlet valve
11. Gasket
12. Pin
13. Float
14. Float bowl
15. Bolt

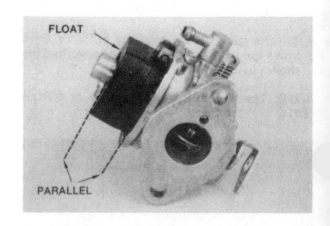

Fig. KA102—To check float level, hold carburetor so tab of float touches fuel inlet needle valve. Float should be parallel with surface of carburetor body.

move float pin (12), float (13) and fuel inlet needle (10). Remove throttle and choke shaft assemblies after unscrewing throttle and choke plate retaining screws. Remove idle mixture screw (5), idle mixture jet (6), main jet (8) and main fuel nozzle (9). Note that main jet must be removed before the nozzle as the jet blocks the nozzle.

Clean carburetor using suitable carburetor cleaner. When assembling the carburetor note the following. Place a small drop of thread locking compound such as Loctite 242 or equivalent on throttle and choke plate retaining screws.

**GOVERNOR.** A gear driven flyweight governor assembly is located inside the engine crankcase. To adjust external linkage, place engine throttle control in idle position. Move governor lever (1--Fig. KA103) to fully open carburetor throttle valve. Loosen governor lever clamp bolt and rotate governor shaft (2) clockwise as far as possible. While holding governor shaft and lever in this position, tighten clamp bolt.

Engine speed with generator loaded from 2/3 to full capacity should be 3600 rpm. To adjust, turn the throttle cable adjusting nuts (3) as necessary.

**IGNITION SYSTEM.** Early production engines are equipped with a magneto ignition with breaker points. The ignition coil is located just above the flywheel and the breaker point set and condenser are located behind the flywheel. On these engines, the ignition timing is adjusted by changing the breaker point gap.

To adjust points, first remove air cleaner, control panel, muffler cover, recoil starter and flywheel blower housing. Remove flywheel retaining nut and tap flywheel with a plastic or rubber mallet to remove flywheel from crankshaft. Connect ohmmeter or battery powered test light leads to black wire from the condenser and to ground. Loosely reinstall flywheel on crankshaft and rotate until edges of flywheel and ignition coil core are aligned as shown in Fig. KA104. At this point ohmmeter needle should move or test light should go out, indicating that points are just opening. If necessary, loosen breaker point retaining screw and move stationary point until correct ignition timing is obtained. Turn crankshaft until points are fully open and measure point gap using a feeler gauge. Point gap should be 0.3-0.4 mm (0.012-0.016 inch).

Later production engines are equipped with a transistorized ignition system which has no moving parts except the flywheel. No adjustment is re-

quired on this ignition system. Correct ignition timing will be obtained when ignition coil unit (Fig. KA105) is installed so that air gap between coil laminations and outer surface of flywheel is 0.5 mm (0.020 inch).

**VALVE ADJUSTMENT.** Clearance between tappet and end of valve stem should be checked after every 200 hours

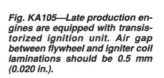

Fig. KA103--View of external governor linkage. Refer to text for adjustment procedure.

Fig. KA104—Ignition breaker points should just start to open when edge of flywheel is aligned with ignition coil core as shown.

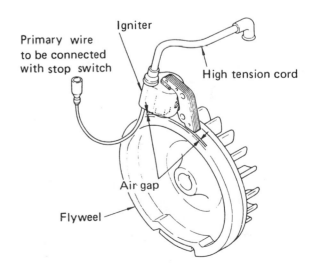

Fig. KA105—Late production engines are equipped with transistorized ignition unit. Air gap between flywheel and igniter coil laminations should be 0.5 mm (0.020 in.).

of operation. To check clearance, remove muffler cover and valve spring chamber cover. Turn crankshaft to position piston at top dead center on compression stroke. Measure valve clearance with engine cold, using a feeler gauge. Specified clearance for FA76D engine is 0.10-0.20 mm (0.004-0.008 in.) for intake and 0.10-0.30 mm (0.004-0.012 in.) for exhaust. Specified clearance for FA130D and FA210D engines is 0.12-0.18 mm

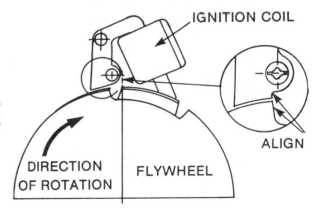

(0.005-0.007 in.) for intake and 0.10-0.24 mm (0.004-0.009 in.) for exhaust.

To adjust valve clearance, valves must be removed from engine. If clearance is too large, grind valve seat or renew valve to reduce clearance. If clearance is too small, grind end of valve stem to increase clearance. Refer to VALVE SYSTEM paragraphs in REPAIR section for additional information.

**CYLINDER HEAD AND COMBUSTION CHAMBER.** Cylinder head, combustion chamber and top of piston should be cleaned and carbon and other deposits removed after every 200 hours of operation. Refer to REPAIRS section for service procedure.

**LUBRICATION.** The engine is splash lubricated by an oil slinger mounted on bottom of connecting rod cap. Engine oil level should be checked prior to each operating interval. To check oil level, insert dipstick into oil filler hole until plug touches the first threads. Do not screw dipstick plug in to check oil level. Oil level should be maintained between the reference marks on oil dipstick.

Engine oil should be changed after the first 20 hours of operation and every 50 hours thereafter. Crankcase capacity is 0.32 L (0.34 qt.) for Model FA76D, 0.50 L (0.53 qt.) for Model FA130D and 0.60 L (0.63 qt.) for Model FA210D.

Engine oil should meet or exceed latest API service classification. Normally, SAE 30 oil is recommended for use when temperatures are above 32° F (0° C). SAE 10W-30 oil is recommended when temperatures are below 32° F (0° C).

**AIR FILTER.** The engine is equipped with a foam type air filter element. The air filter element should be removed and cleaned after every 50 hours of operation, or more often if operating in extremely dusty conditions. To clean element, wash in a nonflammable cleaning solvent and gently squeeze

element dry. Oil element with clean engine oil and gently squeeze out the excess oil.

**GENERAL MAINTENANCE.** Check and tighten all loose bolts, nuts or clamps prior to each period of operation. Check for fuel or oil leakage and repair if necessary.

Clean dirt, dust, grease or any foreign material from cylinder head and cylinder block cooling fins after every 100 hours of operation, or more often if operating in extremely dirty conditions.

## REPAIRS

**TIGHTENING TORQUES.** Recommended tightening torques are as follows:

Connecting rod:
FA76D . . . . . . . . . . . . . . . . . . 7 N•m
(60 in.-lbs.)
FA130D, FA210D . . . . . . . . 12 N•m
(105 in.-lbs.)
Crankcase cover . . . . . . . . . . . . 6 N•m
(50 in.-lbs.)
Cylinder head:
FA76D . . . . . . . . . . . . . . . . . . 7 N•m
(60 in.-lbs.)
FA130D, FA210D . . . . . . . . 20 N•m
(175 in.-lbs.)
Flywheel nut:
FA76D . . . . . . . . . . . . . . . . . . 34 N•m
(25 ft.-lbs.)
FA130D, FA210D . . . . . . . . 62 N•m
(46 ft.-lbs.)

**CYLINDER HEAD.** To remove cylinder head, first remove the air cleaner, muffler cover, control panel and fuel tank. Remove rewind starter, flywheel blower fan housing and cylinder head shroud. Loosen cylinder head screws gradually in ¼-turn increments until all screws are loose enough to be removed by hand. Remove cylinder head and gasket from the engine.

Remove spark plug and clean carbon and other deposits from cylinder head. Place cylinder head on a flat surface and check gasket sealing surface for distortion using a feeler gauge. Renew cylinder head if head is warped more than 0.15 mm (0.006 in.) on Model FA76D, 0.25 mm (0.010 in.) on Model FA130D, or 0.30 mm (0.012 in.) on Model FA210D.

Reinstall cylinder head using a new head gasket. Tighten cylinder head screws evenly in three steps to specified torque following tightening sequence shown in Fig. KA106 or KA107.

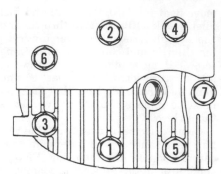

Fig. KA107—When installing cylinder head on Model FA130D, tighten head bolts gradually following sequence shown.

**VALVE SYSTEM.** The valves are accessible after removing cylinder head and valve spring chamber cover (23--Fig. KA108). Compress valve springs (26) and remove valve spring retainers (27).

Minimum allowable valve margin is 0.55 mm (0.022 in.). Valve face and seat angles are 45 degrees. Grinding the valve face is not recommended. Recommended valve seating width is 0.5-1.1 mm (0.020-0.043 in.) on Model FA76D or 1.0-1.6 mm (0.039-0.063 in.) for Models FA130D and FA210D.

Minimum valve stem diameter for Model FA76D is 5.462 mm (0.2150 in.) for intake valve and 5.435 mm (0.2140 in.) for exhaust valve. Minimum valve stem diameter for Models FA130D and FA210D is 5.960 mm (0.2346 in.) for intake valve and 5.940 mm (0.2339 in.) for exhaust valve.

Valve guides are integral with cylinder block. Specified valve guide maximum inside diameter for Model FA76D is 5.617 mm (0.2211 in.) for intake valve guide and 5.595 mm (0.2203 in.) for exhaust valve guide. Valve guide maximum inside diameter for Models FA130D and FA210D is 6.117 mm (0.2408 in.) for intake valve guide and 6.095 mm (0.2400 in.) for exhaust valve guide. Cylinder block should be renewed if valve guides are worn beyond specified limits.

Specified valve spring minimum free length for Model FA76D is 20.5 mm (0.81 in.) for both valve springs. Valve spring minimum free length for Models FA130D and FA210D is 23.5 mm (0.93 in.) for both valve springs.

**CAMSHAFT.** Engine must be separated from generator unit to remove camshaft. Camshaft (29--Fig. KA108) is supported directly in bores of the aluminum crankcase and crankcase cover. To remove camshaft, first remove crankcase cover. Turn cylinder block upside down so tappets fall away from cam lobes, then remove camshaft and gear

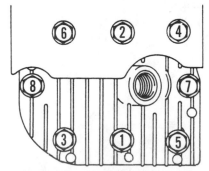

Fig. KA106—When installing cylinder head on Models FA76D and FA210D, tighten head bolts gradually following sequence shown.

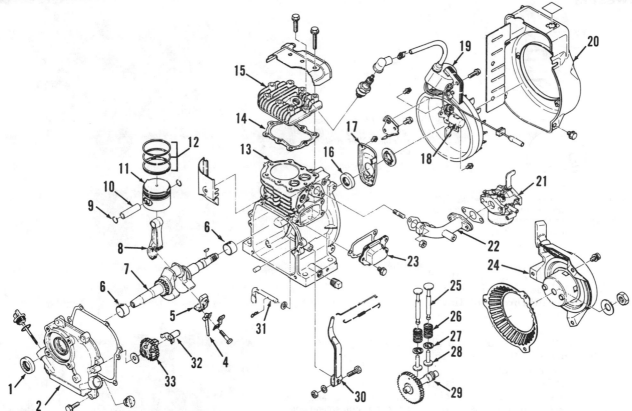

*Fig. KA108—Exploded view of typical early production engine assembly. Late production engines are similar except that transistorized ignition unit is located outside the flywheel and breaker points are not used. Note that main bearing (6) in crankcase cover (2) is a ball bearing on some engines.*

| | | |
|---|---|---|
| 1. Oil seal | 9. Retaining ring | 15. Cylinder head | 21. Carburetor | 27. Retainer |
| 2. Crankcase cover | 10. Piston pin | 16. Oil seal | 22. Intake manifold | 28. Tappet |
| 4. Oil slinger | 11. Piston | 17. Breaker point cover | 23. Tappet chamber cover | 29. Camshaft |
| 5. Rod cap | 12. Piston rings | 18. Condenser & breaker points | 24. Recoil starter assy. | 30. Governor lever |
| 6. Main bearings | 13. Crankcase & cylinder assy. | 19. Ignition coil assy. | 25. Valve | 31. Governor shaft |
| 7. Crankshaft | 14. Head gasket | 20. Blower housing | 26. Valve spring | 32. Sleeve |
| 8. Connecting rod | | | | 33. Governor assy. |

from crankcase. Identify tappets (28) so they can be reinstalled in their original positions, then remove tappets from crankcase.

Minimum camshaft journal diameter is 9.945 mm (0.3915 in.) for Model FA76D, 11.937 mm (0.4670 in.) for Model FA130D and 12.935 mm (0.5093 in.) for Model FA210D. Maximum allowable clearance between camshaft journals and bearing bores is 0.10 mm (0.004 in.). If clearance is excessive, renew camshaft and/or crankcase and cover.

Make certain that camshaft lobes are smooth and free of scoring, pitting and other damage. Minimum lobe height for intake and exhaust lobes is 17.35 mm (0.683 in.) for Model FA76D, 23.25 mm (0.915 in.) for Model FA130D and 26.35 mm (1.037 in.) for Model FA210D. If camshaft is renewed, the tappets should also be renewed.

When installing camshaft, make certain that crankshaft gear and camshaft gear timing marks are aligned (Fig. KA109).

**PISTON, PIN AND RINGS.** Engine must be separated from generator unit to remove piston (11--Fig. KA108). Pis-

ton and connecting rod are removed as an assembly after cylinder head (15) and crankcase cover (2) have been removed. Carbon deposits and ring ridge, if present, should be removed from top of cylinder prior to removing piston. Remove connecting rod cap (5) and oil slinger (4), then push piston and connecting rod assembly out top of cylinder. Remove retaining rings (9) and push pin (10) out of piston to separate piston from connecting rod.

Renew piston if top or second piston ring side clearance in ring groove exceeds 0.15 mm (0.006 in.). Specified ring end gap is 0.15-0.35 mm (0.006-0.014 in.). If ring end gap exceeds 1.0 mm (0.040 in.), rebore cylinder oversize or renew cylinder.

Measure piston diameter at bottom of piston skirt and 90 degrees from pin bore. Specified standard diameter is 51.90 mm (2.0433 in.) for Model FA76D, 61.94 mm (2.4385 in.) for Model FA130D

*Fig. KA109—Timing marks on crankshaft gear and camshaft gear must be aligned during installation.*

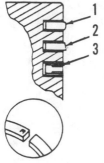

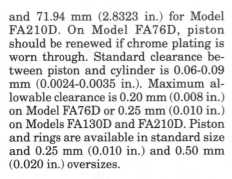

Fig. KA110—Piston rings must be installed with manufacturer's mark on end of ring facing upward. Note location and shape of top compression ring (1), second compression ring (2) and oil control ring (3).

Fig. KA112—Install piston pin retaining rings with "tail" pointing either straight up or down.

Fig. KA113—Stagger piston rings on piston so end gaps are positioned as shown.

and 71.94 mm (2.8323 in.) for Model FA210D. On Model FA76D, piston should be renewed if chrome plating is worn through. Standard clearance between piston and cylinder is 0.06-0.09 mm (0.0024-0.0035 in.). Maximum allowable clearance is 0.20 mm (0.008 in.) on Model FA76D or 0.25 mm (0.010 in.) on Models FA130D and FA210D. Piston and rings are available in standard size and 0.25 mm (0.010 in.) and 0.50 mm (0.020 in.) oversizes.

When installing rings on piston, make sure that manufacturer's mark on face of ring is toward piston crown (Fig. KA110). Note that top piston ring has a barrel face and second piston ring has a taper face.

Install piston on connecting rod so the side of connecting rod which is marked "MADE IN JAPAN" is toward the "M" stamped on piston pin boss (Fig. KA111). Secure piston pin with new retaining rings. Position the retaining rings in piston so the "tail" points straight up or down (Fig. KA112), otherwise inertia produced at top and bottom of piston stroke will cause retaining rings to compress and possibly fall out of piston pin bore. Stagger piston ring end gaps around piston so that top ring

and oil control ring end gaps are positioned away from valves (Fig. KA113) and end gap of second ring is positioned toward the valves.

**CONNECTING ROD.** Engine must be separated from generator unit to remove connecting rod. Piston and connecting rod are removed as an assembly. Refer to PISTON, PIN AND RINGS paragraphs for removal and installation procedure. Remove retaining rings and

push pin out of piston to separate piston from connecting rod.

The connecting rod rides directly on crankshaft crankpin journal. Inspect connecting rod and renew if piston pin bore or crankpin bearing bore are scored or damaged. Standard clearance between piston pin and connecting rod small end bore is 0.006-0.025 mm (0.0002-0.0010 in.); maximum allowable clearance is 0.050 mm (0.002 in.). Standard clearance between connecting rod big end bore and crankshaft crankpin is 0.030-0.056 mm (0.0012-0.0022 in.). Renew connecting rod and/or crankshaft if clearance exceeds 0.070 mm (0.0028 in.).

Refer to following table for connecting rod small and big end maximum allowable diameters:

Small end diameter (max.):
FA76D . . . . . . . . . . . . . . 11.035 mm
(0.4344 in.)
FA130D . . . . . . . . . . . . . 13.035 mm
(0.5134 in.)
FA210D . . . . . . . . . . . . . 15.040 mm
(0.5912 in.)
Big end diameter (max.):
FA76D . . . . . . . . . . . . . 20.057 mm
(0.7897 in.)

Fig. KA111—Piston and connecting rod should be assembled so that lettering on rod and "M" mark on underside of piston are on the same side.

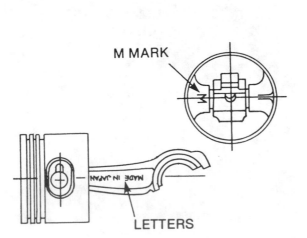

M MARK

LETTERS

FA130D . . . . . . . . . . . . . 24.072 mm
(0.9478 in.)
FA210D . . . . . . . . . . . . 27.072 mm
(1.0658 in.)

Connecting rod side play on crankpin should not exceed 0.5 mm (0.012 in.).

When installing connecting rod cap, make sure identification marks on side of cap are aligned with marks on connecting rod. Install oil slinger (4—Fig. KA108) and new lock plate making sure that slinger arm and bendable tabs of lock plate are toward open side of crankcase. Tighten connecting rod screws to 7 N•m (60 in.-lbs.) on Model FA76D and to 12 N•m (105 in.-lbs.) on Models FA130D and FA210D. Bend tabs of lock plate against flat of connecting rd screws.

**GOVERNOR.** The internal centrifugal flyweight governor (33--Fig. KA108) is located in the crankcase cover and is driven by the camshaft gear. Governor assembly should not be removed from crankcase cover unless renewal is necessary. The governor gear will be damaged as it is pulled off the cover stub shaft, and must be replaced with a new governor flyweight assembly.

Adjust governor external linkage as outlined in MAINTENANCE section.

**CRANKSHAFT, MAIN BEARINGS AND SEALS.** On some engines, the crankshaft (7—Fig. KA108) is supported at each end in bushing type main bearings (6) which are pressed into crankcase and crankcase cover. Some engines use a pressed-in bushing in the crankcase and a ball bearing in the crankcase cover to support crankshaft.

To remove crankshaft, remove flywheel, cylinder head, crankcase cover, camshaft, tappets, connecting rod and piston assembly. Withdraw crankshaft from crankcase. Drive oil seals (1 and

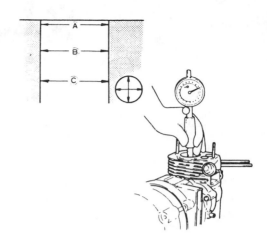

*Fig. KA114—To check cylinder bore for wear or out-of-round, measure bore in six different locations as shown.*

16) out of crankcase and crankcase cover.

Minimum crankpin diameter is 19.95 mm (0.7854 in.) for Model FA76D, 23.92 mm (0.9417 in.) for Model FA130D and 26.92 mm (1.0598 in.) for Model FA210D. Maximum allowable clearance between crankpin and connecting rod big end bore is 0.070 mm (0.003 in.) on Model FA76D or 0.10 mm (0.004 in.) for all other models.

Maximum allowable clearance between crankshaft main journals and main bearing bushings is 0.13 mm (0.005 in.). When renewing main bearing bushings, make sure that oil hole in bushing is aligned with oil hole in crankcase and crankcase cover.

Crankshaft oil seals should be pressed into crankcase and crankcase cover until face of seal is flush with outer surface of crankcase and cover bores.

Lubricate oil seals, bushings and crankshaft journals with engine oil prior to reassembly. Be sure that crankshaft gear and camshaft gear timing marks are aligned during installation as shown in Fig. KA109. Tighten crank-

case cover mounting screws evenly to 6 N•m (50 in.-lbs.) torque.

**CYLINDER AND CRANKCASE.** Cylinder and crankcase are an integral casting. On Model FA76D, standard cylinder bore diameter is 52.00 mm (2.047 in.) and wear limit is 52.07 mm (2.050 in.). On Model FA130D, standard cylinder bore diameter is 62.00 mm (2.441 in.) and wear limit is 62.07 mm (2.444 in.). On Model FA210D, standard cylinder bore diameter is 72.00 mm (2.835 in.) and wear limit is 72.07 mm (2.838 in.). Cylinder bore should be measured at six different locations as shown in Fig. KA114. If cylinder is out-of-round more than 0.05 mm (0.0002 in.), rebore cylinder to appropriate oversize or renew cylinder.

**NOTE: The cylinder block is made of an aluminum-silicon alloy and resizing requires special honing process. Special honing stones are available from Sunnen Products Company. Follow manufacturer's instruction for proper operation.**

# KAWASAKI

| Model | No. Cyls. | Bore | Stroke | Displacement |
|-------|-----------|------|--------|--------------|
| KF24 | 1 | 56 mm | 40 mm | 98 cc |
| | | (2.20 in.) | (1.58 in.) | (6 cu. in.) |
| KF34 | 1 | 60 mm | 47 mm | 132 cc |
| | | (2.36 in.) | (1.85 in.) | (8.1 cu. in.) |
| KF53 | 1 | 66 mm | 53 mm | 181 cc |
| | | (2.60 in.) | (2.09 in.) | (11 cu. in.) |
| KF64 | 1 | 72.0 mm | 60 mm | 244 cc |
| | | (2.83 in.) | (2.36 in.) | (14.9 cu. in.) |
| KF68 | 1 | 74 mm | 60 mm | 258 cc |
| | | (2.91 in.) | (2.36 in.) | (15.7 cu. in.) |

Kawasaki KF series engines are four-stroke, single-cylinder, air-cooled, horizontal crankshaft engines. Engine model number is located on the cooling shroud just above the rewind starter. Model number and serial number are both stamped on crankcase cover.

## MAINTENANCE

**SPARK PLUG.** Recommended spark plug for Models KF24 and KF53 is NGK BR6HS or equivalent. Recommended spark plug for Models KF34, KF64 and KF68 is NGK BR6S or equivalent.

Spark plug should be removed, cleaned and checked for damage after every 100 hours of operation. Specified electrode gap is 0.6-0.7 mm (0.024-0.27 inch).

**AIR FILTER.** The engine is equipped with a foam type air filter element. The air filter element should be removed and cleaned after every 50 hours of operation, or more often if operating in extremely dusty conditions. To clean element, wash in a nonflammable cleaning solvent and gently squeeze element dry. Oil element with clean engine oil and gently squeeze out the excess oil.

**FUEL VALVE.** A filter screen and sediment bowl is mounted on bottom of the fuel shut-off valve. Sediment bowl and screen should be removed and cleaned after every 50 hours of operation, or more often if water or dirt is visible in the sediment bowl.

**CARBURETOR.** All engines are equipped with a float type carburetor (Fig. KA201). Main fuel mixture is controlled by a fixed main jet (8).

Initial adjustment of idle mixture needle (6) is 1-1/8 turns out from a lightly seated position for Model KF24, 7/8 turn out for Model KF34, 3/4 turn out for Model KF53 and 1-1/8 turns out for Models KF64 and KF68. Recommended engine idle speed is 1500-1800 rpm. To adjust idle speed, run engine until it reaches normal operating temperature and turn idle adjusting screw (7) as necessary. Adjust idle mixture needle (6) so engine runs smoothly and accelerates without "stumbling".

To check carburetor float level, remove the carburetor from the engine. Remove the float bowl and invert carburetor body so that tab on float just touches fuel inlet needle; surface of float should be parallel to carburetor body (Fig. KA202). To adjust float setting, carefully bend the float tab as necessary.

**GOVERNOR.** A gear driven flyweight governor assembly is located inside the engine crankcase. To adjust governor external linkage, place engine throttle control in idle position. Move governor lever (1—Fig. KA203) to fully open carburetor throttle valve. Loosen governor lever clamp bolt (2) and rotate governor shaft counterclockwise as far as possible. While holding governor shaft and lever in this position, tighten clamp bolt.

Governor spring (Fig. KA204) must be hooked in lower hole in governor lever for 3600 rpm (60 Hz) engine operation. Engine speed with generator loaded at 2/3 to full capacity should be 3600 rpm. To adjust, turn the throttle cable adjusting nuts as necessary.

**IGNITION SYSTEM.** Early production engines are equipped with a magneto ignition with breaker points. Late production Model KF53 is equipped with transistor (TSI) ignition system and Models KF34 and KF68 are equipped with capacitor discharge (CDI) electronic type ignition system. Refer to appropriate paragraph for model being serviced.

**Breaker-Point Ignition.** The breaker point set and condenser are located behind the flywheel and the ignition coil may be located just above the flywheel or behind the flywheel. The ignition timing is adjusted by changing the breaker point gap. Flywheel must be removed to service or adjust breaker points.

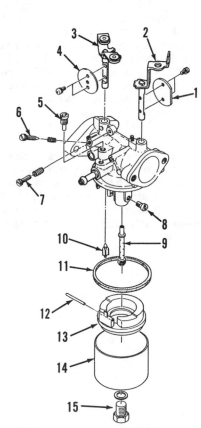

*Fig. KA201—Exploded view of float type carburetor used on all models.*

1. Choke valve
2. Choke shaft
3. Throttle shaft
4. Throttle valve
5. Pilot jet
6. Idle mixture adjustment screw
7. Idle speed screw
8. Main jet
9. Main nozzle
10. Fuel inlet valve
11. Gasket
12. Pin
13. Float
14. Float bowl
15. Bolt

To adjust points, first remove air cleaner, control panel, muffler cover, recoil starter and flywheel blower housing. Remove flywheel retaining nut and use a suitable puller to remove flywheel from crankshaft. Remove breaker point cover, then reinstall flywheel loosely on crankshaft and rotate until "P" mark (Fig. KA205) on flywheel is aligned with timing mark on crankcase or blower housing mounting case. Carefully remove flywheel without rotating crankshaft. Breaker points should just begin to open at this point if timing is correct. If necessary, loosen breaker point retaining screw and move stationary point until correct ignition timing is obtained. Turn crankshaft until points are fully open and measure point gap using a feeler gage. Point gap should be 0.3-0.4 mm (0.012-0.016 inch).

**Transistorized Ignition System.** The transistorized ignition system has no moving parts except the flywheel. No adjustment is required on this ignition system. Correct engine timing will be obtained when ignition coil unit (Fig. KA206) is installed so that air gap between coil laminations and outer surface of flywheel is 0.5 mm (0.020 inch). Ignition system should produce spark at spark plug when cranking speed reaches 300 rpm.

Ignition unit can be checked by measuring resistance of coils using an ohmmeter. To check primary coil resistance, connect one ohmmeter lead to primary wire connector and other test lead to core laminations (Fig. KA207). Resistance should be approximately 0.6 ohm. To check secondary coil resistance, connect one ohmmeter lead to high tension (spark plug) wire and other test lead to core laminations. Resistance should be approximately 11.2K ohms. If test results are significantly higher or lower than specified resistance, renew ignition unit.

**Capacitor Discharge System.** The capacitor discharge ignition system consists of an exciter coil located behind the flywheel, a pulser coil located just above the flywheel and a CDI igniter unit (Fig. KA208). The only moving part of this system is the flywheel, and no adjustments are required. CDI system should generate a spark at the spark plug when engine cranking speed reaches 200 rpm.

Resistance of exciter coil and pulser coil may be checked using an ohmmeter. Excitor coil resistance should be approximately 480 ohms and pulser coil resistance should be approximately 70 ohms. If test results are significantly higher or lower than specified resistance, renew components as necessary.

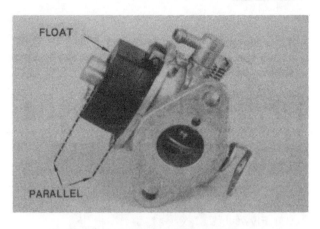

*Fig. KA202—To check float level, hold carburetor so tab of float touches fuel inlet needle valve. Float should be parallel with surface of carburetor body.*

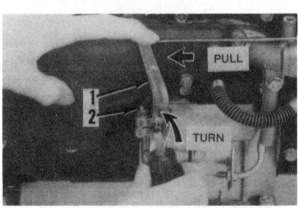

*Fig. KA203—View of external governor linkage. Refer to text for adjustment procedure.*

*Fig. KA204—Governor spring must be installed in lower hole in governor lever for 3600 rpm (60 Hz) operation. Refer to text for speed adjustment procedure.*

*Fig. KA205—On models equipped with a breaker-point ignition system, "P" mark on flywheel should align with mark on crankcase or blower housing mounting plate when breaker points just to start to open for correct ignition timing.*

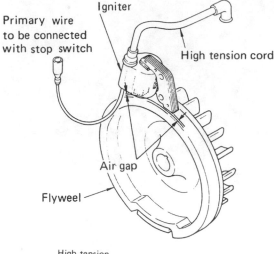

Primary wire to be connected with stop switch

Igniter

High tension cord

Air gap

Flyweel

Fig. KA206—Some engines are equipped with transistorized ignition unit mounted outside the flywheel. Air gap between flywheel and igniter coil laminations should be 0.5 mm (0.020 in.).

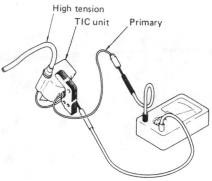

High tension

TIC unit

Primary

Fig. KA207—On transistorized ignition unit, resistance of primary and secondary coils may be checked with an ohmmeter. Refer to text.

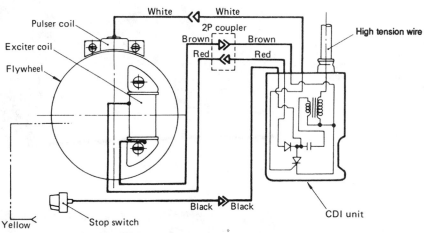

White   White

2P coupler

Pulser coil

Exciter coil

Flywheel

Brown   Brown

Red   Red

High tension wire

Black   Black

Yellow

Stop switch

CDI unit

Fig. KA208—Wiring schematic for CDI ignition unit used on some models.

| | To be connected with positive ⊕ terminal of the tester | | | | |
|---|---|---|---|---|---|
| Lead wire | Red | Black | White | Brown | Black (H.T.) |
| To be connected with negative ⊖ terminal of the tester — Red | | Continuity | Infinite | Infinite | Infinite |
| Black | Infinite | | Infinite | Infinite | Infinite |
| White | Continuity | Continuity | | Continuity | Continuity |
| Brown | Continuity | Continuity | Continuity | | Continuity |
| Black (H.T.) | Continuity | Continuity | Continuity | Continuity | |

Fig. KA209—Condition of CDI ignition unit may be checked using an ohmmeter to check for correct continuity between wires of the system as indicated in above chart.

CDI system may also be checked using an ohmmeter to test for continuity or infinite resistance following chart shown in Fig. KA209. Renew components as necessary if indicated continuity or infinite resistance readings are not obtained.

**VALVE ADJUSTMENT.** Clearance between tappet and end of valve stem should be checked after every 200 hours of operation. To check clearance, remove muffler cover and valve spring chamber cover. Turn crankshaft to position piston at top dead center on compression stroke. Measure valve clearance using a feeler gage. On Models KF24 and KF34, specified clearance (engine cold) is 0.10-0.34 mm (0.004-0.013 inch) for intake and exhaust. On Models KF53, KF64 and KF68, specified clearance (engine cold) is 0.12-0.18 mm (0.005-0.007 inch) for intake and 0.10-0.34 mm (0.004-0.013 inch) for exhaust.

To adjust valve clearance, valves must be removed from engine. If clearance is too large, grind valve face and/or seat to reduce clearance. If clearance is too small, grind end of valve stem to increase clearance. Refer to VALVE SYSTEM paragraphs in REPAIRS section.

**CYLINDER HEAD AND COMBUSTION CHAMBER.** Cylinder head combustion chamber and top of piston should be cleaned and carbon and other deposits removed after every 200 hours of operation. Refer to REPAIRS section for service procedure.

**LUBRICATION.** All models are splash lubricated by an oil slinger mounted on bottom of the connecting rod cap. Engine oil level should be checked prior to each operating interval. To check oil level, insert dipstick into oil filler hole until plug touches the first threads. Do not screw dipstick plug in to check oil level. Oil level should be maintained between the reference marks on oil dipstick.

Engine oil should be changed after the first 20 hours of operation and every 50 hours thereafter. Crankcase capacity is 0.45 L (0.48 qt.) for Model KF24, 0.50 L (0.53 qt.) for Model KF34, 0.70 L (0.74 qt.) for Model KF53 and 0.90 L (0.95 qt.) for Models KF64 and KF68.

Manufacturer recommends using oil with API service classification of SE or SF. SAE 10W-40, 10W-50, 20W-40 or 20W-50 oil is recommended for use in all temperatures.

**GENERAL MAINTENANCE.** Check and tighten all loose bolts, nuts or clamps prior to each period of operation. Check for fuel or oil leakage and repair if necessary.

Clean dirt, dust, grease or any foreign material from cylinder head and cylinder block cooling fins after every 100 hours of operation, or more often if operating in extremely dirty conditions.

### REPAIRS

**TIGHTENING TORQUES.** Recommended tightening torques are as follows:

Spark plug . . . . . . . . . . . . . . . . .27 N·m
(20 ft.-lbs.)

Cylinder head bolts
KF24, KF34 . . . . . . . . . . . . .20 N·m
(15 ft.-lbs.)

KF53, KF64, and
KF68 . . . . . . . . . . . . . . . . . .25 N·m
(19 ft.-lbs.)

Connecting rod bolts
KF24 . . . . . . . . . . . . . . . . . .10 N·m
(7 ft.-lbs.)

KF34 . . . . . . . . . . . . . . . . . .17 N·m
(12.5 ft.-lbs.)

KF53 . . . . . . . . . . . . . . . . . .18 N·m
(13 ft.-lbs.)

KF64, KF68 . . . . . . . . . . . . .20 N·m
(14.5 ft.-lbs.)

Crankcase cover bolts
KF64, KF68 . . . . . . . . . . . . .15 N·m
(11 ft.-lbs.)
All other models . . . . . . . . . .6 N·m
(4 ft.-lbs.)

Flywheel nut
KF24 . . . . . . . . . . . . . . . . . .42 N·m
(31 ft.-lbs.)
All other models . . . . . . . . .62 N·m
(46 ft.-lbs.)

**CYLINDER HEAD.** To remove cylinder head, first remove the air cleaner, muffler cover, control panel and fuel tank. Remove recoil starter, flywheel blower fan housing and cylinder head shroud. Loosen cylinder head bolts gradually in 1/4 turn increments until all bolts are loose enough to be removed by hand. Remove cylinder head and gasket from the engine.

Remove spark plug and clean carbon and other deposits from cylinder head. Place cylinder head on a flat surface and check gasket sealing surface for warpage using a feeler gage. If head is warped more than 0.25 mm (0.010 inch), cylinder head must be renewed.

Reinstall cylinder head using a new head gasket. Tighten cylinder head bolts evenly to specified torque following tightening sequence shown in Fig. KA210.

**CONNECTING ROD.** The engine must be separated from generator unit to remove connecting rod. Piston and connecting rod are removed as an assembly after cylinder head and crankcase cover have been removed. Carbon deposits

and ring ridge, if present, should be removed from top of cylinder prior to removing piston. Remove connecting rod cap and oil slinger, then push piston and connecting rod assembly out top of cylinder. Remove retaining rings and push pin out of piston to separate piston from connecting rod.

The connecting rod rides directly on crankshaft crankpin journal. Inspect connecting rod and renew if piston pin bore or crankpin bearing bore are scored or damaged. Standard clearance between piston pin and connecting rod small end bore is 0.006-0.025 mm (0.00024-0.00098 inch). Renew connecting rod and pin if clearance exceeds

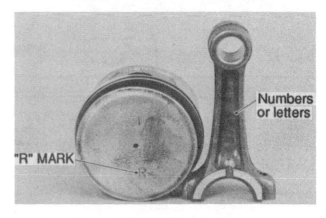

*Fig. KA210—Tighten and loosen cylinder head bolts following sequence shown.*

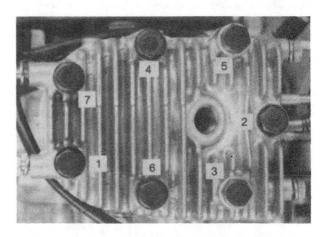

*Fig. KA211—When assembling piston and connecting rod, make sure that numbered or lettered side of rod is on same side as "R" mark on piston crown.*

*Fig. KA212—Piston pin retaining ring should be positioned with the "tail" pointing straight up or down.*

0.050 mm (0.002 inch). Standard clearance between connecting rod big end bore and crankshaft crankpin is 0.030-0.056 mm (0.0012-0.0022 inch). Renew connecting rod and/or crankshaft if clearance exceeds 0.070 mm (0.0028 inch). Connecting rod side play on crankpin should not exceed 0.5 mm (0.012 inch).

Install piston on connecting rod so the "R" mark on piston crown is on same side as numbered or lettered side of connecting rod (Fig. KA211). Secure piston pin with new retaining rings. Position the retaining rings in piston so the "tail" points straight up or down (Fig. KA212), otherwise inertia produced at the top and bottom of piston stroke will cause retaining rings to compress and possibly fall out of piston pin bore. Lubricate piston and cylinder with engine oil, then install piston and connecting rod assembly so lettered side of connecting rod and "R" mark on piston is toward open side of crankcase.

Install connecting rod cap making sure that identification marks on side of cap are aligned with marks on connecting rod. Be sure that oil slinger arm and bendable tabs of lockplate are toward open side of crankcase. Tighten connecting rod bolts to specified torque and lock by bending tabs of lockplate. When installing crankcase cover, tighten cap screws to specified torque following tightening sequence shown in Fig. KA215.

**PISTON, PIN AND RINGS.** Piston and connecting rod are removed as an

Fig. KA213—Piston rings must be installed with manufacturer's mark on end of ring facing upward.

assembly. Refer to CONNECTING ROD paragraphs for removal and installation procedure.

After removing piston rings and separating piston from connecting rod, carefully clean carbon and other deposits from piston surface and ring lands.

**CAUTION: Extreme care should be exercised when cleaning ring lands. Do not damage squared edges or widen ring grooves. If ring lands are damaged, piston must be renewed.**

To check piston ring groove wear, install new rings in top and second grooves and measure side clearance between ring and piston groove using a feeler gage. If side clearance exceeds 0.15 mm (0.006 inch), renew piston.

Install new piston rings squarely in cylinder and measure ring end gap using a feeler gage. Standard ring end gap is 0.15-0.35 mm (0.006-0.014 inch). If ring end gap exceeds 1.0 mm (0.040 inch), rebore cylinder oversize or renew cylinder.

Standard piston diameter, measured at bottom of piston skirt and 90 degrees from pin bore, for Model KF24 is 55.93 mm (2.202 inches) and wear limit is 55.75 mm (2.195 inches). Standard piston diameter for Model KF34 is 59.95

mm (2.360 inches) and wear limit is 59.75 mm (2.352 inches). Standard piston diameter for Model KF53 is 65.94 mm (2.596 inches) and wear limit is 65.75 mm (2.588 inches). Standard piston diameter for Models KF64 and KF68 is 71.94 mm (2.932 inches) and wear limit is 71.75 mm (2.825 inches).

Standard clearance between piston and cylinder is 0.04-0.05 mm (0.0016-0.0020 inch) for all models. If clearance exceeds 0.25 mm (0.010 inch), rebore cylinder for oversize piston or renew cylinder.

When installing rings on piston, make sure that manufacturer's mark (Fig. KA213) on end of ring faces upward. Refer to CONNECTING ROD paragraphs to properly assemble piston to connecting rod. Lubricate piston and cylinder with engine oil prior to installation. Stagger piston ring end gaps around piston so that top ring and oil control ring end gaps are positioned away from valves (Fig. KA214) and end gap of second ring is positioned toward the valves.

**CYLINDER AND CRANKCASE.** Cylinder and crankcase are an integral casting. To check cylinder bore for wear and out-of-round, cylinder should be measured at six different locations as shown in Fig. 216. If any two measurements vary by more than 0.15 mm (0.006 inch) or if cylinder diameter exceeds specified wear limit, cylinder should be rebored to appropriate oversize or renewed. Refer to following cylinder bore specifications:

| Model | Standard Dia. | Wear Limit |
|---|---|---|
| KF24 . . . . . . | 56.00 mm (2.205 in.) | 56.15 mm (2.210 in.) |
| KF34 . . . . . . | 60.00 mm (2.362 in.) | 60.15 mm (2.360 in.) |
| KF53 . . . . . . | 66.00 mm (2.598 in.) | 66.15 mm (2.604 in.) |
| KF64, KF68 | 72.00 mm (2.835 in.) | 72.15 mm (2.840 in.) |

**CRANKSHAFT, MAIN BEARINGS AND SEALS.** The crankshaft is supported at each end in ball type main bearings (3—Fig. KA220 or KA221). To remove crankshaft, remove flywheel, cylinder head, crankcase cover, connecting rod and piston assembly, camshaft and tappets. Use a soft mallet to tap crankshaft from crankcase. Drive oil seals (2) out of crankcase and crankcase cover.

Inspect crankshaft and bearings for wear and damage and renew as necessary. To facilitate removal and installation of main bearings, heat crankcase and end cover in the area around the bearings. Crankshaft oil seals should be pressed into crankcase and crankcase

Fig. KA214—Stagger piston rings on piston so end gaps are positioned as shown.

Fig. KA215—Tighten crankcase cover retaining bolts to specified torque following sequence shown.

cover until face of seal is flush with outer surface of crankcase and cover bores.

Lubricate oil seals, bearings and crankshaft journals with engine oil prior to reassembly. Be sure that crankshaft gear and camshaft gear timing marks are aligned during installation as shown in Fig. KA217. When installing crankcase cover, tighten cap screws to specified torque following tightening sequence shown in Fig. KA215.

**CAMSHAFT AND BEARINGS.** On some models, camshaft (23—Fig. KA220 or KA221) is supported directly in bores of the aluminum crankcase and crankcase cover, while on other models, camshaft is supported at each end in ball type bearings which are pressed into crankcase and crankcase cover. To remove camshaft, first remove crankcase cover. Turn cylinder block upside down so tappets fall away from cam lobes, then remove camshaft and gear from crankcase.

On engines not equipped with ball bearings, maximum allowable clearance between camshaft journals and housing bores is 0.10 mm (0.004 inch). On engines equipped with ball bearings, the bearings should turn freely without rough spots. Make certain that camshaft lobes and tappets are smooth and free of scoring, pitting and other damage. If camshaft is renewed, the tappets should also be renewed. Refer to the following cam lobe height specifications:

Fig. KA216—To check cylinder bore for wear or out-of-round, measure bore in six different locations as shown.

Fig. KA217—Camshaft gear and crankshaft gear timing marks must be aligned as shown during crankshaft and camshaft installation.

| Model | Standard Height | Wear Limit |
|---|---|---|
| KF24 | 25 mm (0.984 in.) | 24.50 mm (0.964 in.) |
| KF34 | 27.50 mm (1.083 in.) | 27.00 mm (1.063 in.) |
| KF53 | 32.65 mm (1.286 in.) | 32.15 mm (1.266 in.) |
| KF64, KF68 | 33.72 mm (1.328 in.) | 33.22 mm (1.308 in.) |

When installing camshaft, make certain that crankshaft gear and camshaft gear timing marks are aligned (Fig. KA217).

**GOVERNOR.** The internal centrifugal flyweight governor is located on the camshaft gear (Fig. KA218). Governor assembly should be checked for free movement of governor weights. Governor should be renewed if weights, pivot pins or mounting bracket is worn or damaged. When installing governor, be sure that notch in sleeve fits on pin in cam gear and that flyweights are hooked over edge of sleeve as shown in Fig. KA218. Adjust governor external linkage as outlined in MAINTENANCE section.

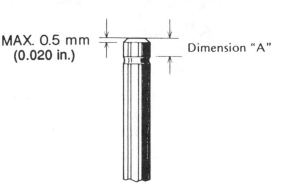

Fig. KA218—Governor flyweight assembly is mounted on camshaft gear.

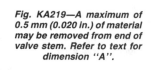

MAX. 0.5 mm (0.020 in.)    Dimension "A"

Fig. KA219—A maximum of 0.5 mm (0.020 in.) of material may be removed from end of valve stem. Refer to text for dimension "A".

**VALVE SYSTEM.** Clearance between valve stem end and valve tappet (engine cold) should be 0.10-0.34 mm (0.004-0.013 inch) for intake and exhaust on Models KF24 and KF34. On Models KF53, KF64 and KF68, specified clearance (engine cold) is 0.12-0.18 mm (0.005-0.007 inch) for intake and 0.10-0.34 mm (0.004-0.013 inch) for exhaust. If clearance is not within specified range, valves must be removed and end of stems ground off to increase clearance or seats ground deeper to reduce clearance.

CAUTION: A maximum of 0.5 mm (0.020 inch) material may be removed from end of valve stem; otherwise, valve failure may occur.

Valve must be renewed if stem length (A—Fig. KA219), measured from shoulder of keeper groove, is less than specified limits listed in following table.

| Stem Length "A" | Standard | Service Limit |
|---|---|---|
| Model KF24 | 3.0 mm (0.118 in.) | 2.5 mm (0.098 in.) |
| Model KF34 | 3.3 mm (0.130 in.) | 2.8 mm (0.110 in.) |
| Model KF53 | 5.6 mm (0.220 in.) | 5.1 mm (0.200 in.) |
| Model KF64 | 3.8 mm (0.150 in.) | 3.3 mm (0.130 in.) |
| Model KF68 | 3.9 mm (0.154 in.) | 3.4 mm (0.134 in.) |

To remove valves, first remove cylinder head and the valve spring chamber cover and breather hose assembly. Use a screwdriver or other suitable tool to compress the valve spring, then remove valve spring retainer.

Clean carbon deposits off valve head and stem, then inspect valve for damage or distortion. Valve should be renewed if valve head edge thickness is less than 0.70 mm (0.028 inch).

Valve face and seat angles are 45 degrees. Recommended valve seating width is 1.0-1.5 mm (0.040-0.060 inch).

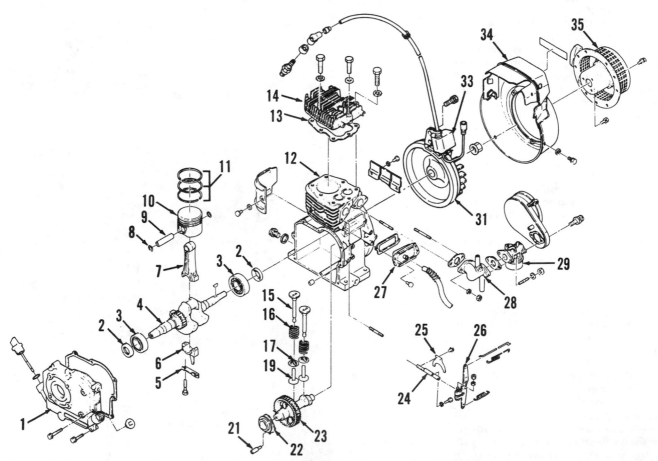

Fig. KA220—Exploded view of late production Model KF53 engine equipped with transistorized ignition. Early production engines were similar except that magneto ignition was used.

1. Crankcase cover
2. Oil seals
3. Main bearings
4. Crankshaft
5. Lockplate
6. Rod cap & oil slinger
7. Connecting rod
8. Retaining ring
9. Piston pin
10. Piston
11. Piston rings
12. Cylinder & crankcase assy.
13. Head gasket
14. Cylinder head
15. Valve
16. Valve spring
17. Retainer
19. Tappet
21. Pin
22. Governor sleeve
23. Camshaft & gear assy.
24. Governor shaft
25. Fork
26. Governor lever
27. Tappet chamber cover
28. Intake manifold
29. Carburetor
31. Flywheel
33. Ignition unit
34. Blower housing
35. Recoil starter

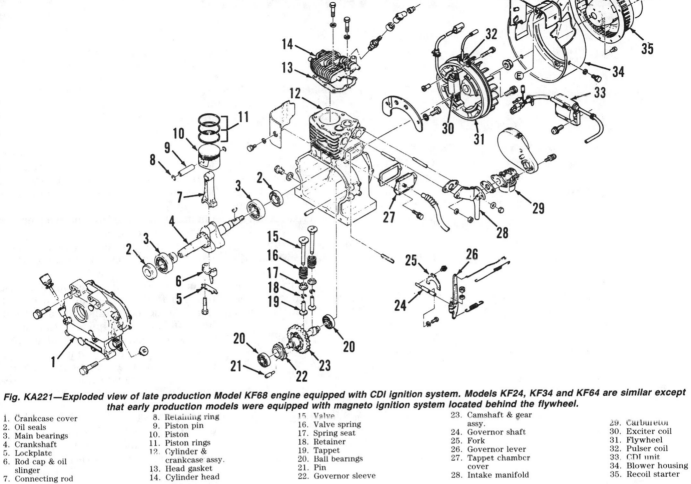

**Fig. KA221—Exploded view of late production Model KF68 engine equipped with CDI ignition system. Models KF24, KF34 and KF64 are similar except that early production models were equipped with magneto ignition system located behind the flywheel.**

1. Crankcase cover
2. Oil seals
3. Main bearings
4. Crankshaft
5. Lockplate
6. Rod cap & oil slinger
7. Connecting rod
8. Retaining ring
9. Piston pin
10. Piston
11. Piston rings
12. Cylinder & crankcase assy.
13. Head gasket
14. Cylinder head
15. Valve
16. Valve spring
17. Spring seat
18. Retainer
19. Tappet
20. Ball bearings
21. Pin
22. Governor sleeve
23. Camshaft & gear assy.
24. Governor shaft
25. Fork
26. Governor lever
27. Tappet chamber cover
28. Intake manifold
29. Carburetor
30. Exciter coil
31. Flywheel
32. Pulser coil
33. CDI unit
34. Blower housing
35. Recoil starter

# KAWASAKI

| Model | No. Cyls. | Bore | Stroke | Displacement |
|-------|-----------|------|--------|--------------|
| KF82 | 1 | 80 mm (3.15 in.) | 68 mm (2.68 in.) | 341 cc (20.8 cu. in.) |
| KF100 | 1 | 85 mm (3.35 in.) | 70 mm (2.76 in.) | 397 cc (24.2 cu. in.) |
| KF150 | 1 | 95 mm (3.74 in.) | 82 mm (3.23 in.) | 581 cc (35.4 cu. in.) |

Kawasaki KF series engines are four-stroke, single-cylinder, air-cooled, horizontal crankshaft engines. Engine model number is located on cooling shroud just above the rewind starter. Model number and serial number are both stamped into crankcase cover (Fig. KA300). Always furnish engine model and serial numbers when ordering parts.

## MAINTENANCE

**SPARK PLUG.** Recommended spark plug for Model KF82 is a Champion L88C or equivalent. Recommended spark plug for all other models is a Champion L90 or equivalent.

On all models, spark plug should be removed and cleaned and electrode gap set at 0.6-0.7 mm (0.024-0.027 inch) after every 100 hours of operation. Renew spark plug if electrode is burned or damaged.

**AIR FILTER.** Models KF82 and KF150 are equipped with washable pleated felt type air filter element. Model KF100 is equipped with a dry type filter element with an outer foam precleaner.

On all models, air filter should be removed and cleaned at 50 hour intervals. Clean dry type element using low pressure compressed air blown from the inside toward the outside. Foam or felt elements can be washed in a solution of mild detergent and water. Allow elements to completely air dry before reinstalling.

**CARBURETOR.** All models are equipped with a float type carburetor (Fig. KA301) which has an idle fuel mixture adjustment screw (6). Main fuel mixture is controlled by a fixed main jet (8).

Initial adjustment of idle mixture needle (6) is 1-1/8 turns out from a lightly seated position for Model KF82, 1 turn out for Model KF100, 1-1/4 turns out for Model KF150. Recommended engine idle speed is 1500-1800 rpm. To adjust idle speed, run engine until it reaches normal operating temperature and turn idle speed adjusting screw (7) as necessary. Adjust idle mixture needle (6) so engine runs smoothly and accelerates without "stumbling."

To check carburetor float level, remove the carburetor from the engine. Remove the float bowl and invert carburetor body so that tab on float just touches fuel inlet needle; surface of float should be parallel to carburetor body. If float is equipped with a metal tang which contacts fuel inlet needle, carefully bend tang to adjust float. If float is a plastic assembly, float height is nonadjustable and float must be renewed if float height is incorrect.

**GOVERNOR.** The mechanical flyweight type governor is located internally in the engine with the flyweight assembly attached to camshaft gear. To adjust external governor linkage, first make certain linkage is in good condition and tension spring (4—Fig. KA302) is not stretched. With engine stopped, loosen clamp bolt (5) and push governor lever (3) to fully open throttle. Rotate

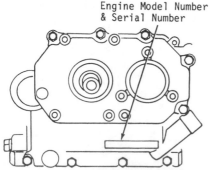

Fig. KA300—Engine model and serial numbers are located on crankcase cover as shown.

Engine Model Number & Serial Number

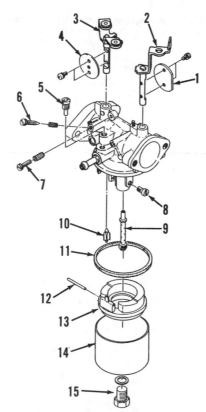

Fig. KA301—Exploded view of float type carburetor typical of all models.

1. Choke valve
2. Choke shaft
3. Throttle shaft
4. Throttle valve
5. Pilot jet
6. Idle mixture adjusting screw
7. Idle speed adjusting screw
8. Main jet
9. Main nozzle
10. Fuel inlet valve
11. Gasket
12. Pin
13. Float
14. Float bowl
15. Bolt

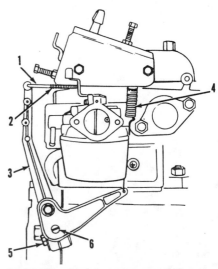

Fig. KA302—View of governor linkage similar to linkage used on all models. Refer to text for adjustment procedure.

1. Governor-to-carburetor rod
2. Spring
3. Governor lever
4. Governor spring
5. Clamp bolt
6. Governor shaft

governor shaft (6) clockwise for Model KF82 or counterclockwise for Models KF100 and KF150 as far as possible, then retighten clamp bolt while holding lever and shaft in this position.

Engine speed with generator operating under load should be 3600 rpm.

**IGNITION SYSTEM.** Early models are equipped with a breaker-point type ignition system. Late Models KF82 and KF100 are equipped with a capacitor discharge type ignition system, and late Model KF150 is equipped with a transistorized ignition system. Refer to the appropriate paragraph for model being serviced.

**Breaker-Point Ignition.** On breaker-point type ignition system, the ignition coil is located behind the flywheel and the breaker-point set and condenser is located on the crankcase on the carburetor side of engine. To obtain correct ignition timing, adjust breaker-points so that they just begin to open when the "P" mark on the flywheel is aligned with alignment mark on the crankcase (Fig. KA303). Turn crankshaft until points are fully open and measure point gap using a feeler gage. Point gap should be 0.3-0.4 mm (0.012-0.016 inch).

Models KF100 and KF150 are equipped with a mechanical spark advance mechanism. This system functions when centrifugal force overcomes spring tension and breaker-point mounting plate rotates to advance timing. Initial timing advance should be 10 degrees BTDC at 1000 rpm for Model KF100 and 8 degrees BTDC at 1000 rpm for Model KF150. Total timing advance should be 25 degrees BTDC at 2000 rpm for Model KF100 and 25 degrees BTDC at 2800 rpm for Model KF150.

**Capacitor Discharge System.** Some KF82 and KF100 models are equipped with a capacitor discharge (CDI) ignition system. The only moving part of this system is the flywheel and no adjustments are required. CDI system should fire the spark plug when engine cranking speed reaches 200 rpm.

CDI ignition unit can be checked using an ohmmeter. Connect one lead of ohmmeter to terminal end of the brown lead and remaining test lead to terminal end of red lead. Ohmmeter should register 370 ohms for Model KF82 and 260 ohms for Model KF100. Connect one ohmmeter test lead to terminal end of brown wire and remaining test lead to terminal end of white lead. Ohmmeter reading should register 65 ohms for Models KF82 and KF100.

**Transistorized Ignition System.** Some Model KF150 engines are equipped with a transistorized ignition system. The only moving part of this system is the flywheel and no adjustments are required. Transistorized ignition system should fire the spark plug when engine cranking speed reaches 200 rpm.

Transistorized ignition unit can be checked using an ohmmeter. Connect one ohmmeter lead to the terminal end of read lead and the remaining test lead to the core laminations. Ohmmeter should register 1.4 ohms. Connect one ohmmeter lead to terminal end of the white lead and the remaining ohmmeter lead to core laminations. Ohmmeter should register 65 ohms.

**VALVE ADJUSTMENT.** Clearance between tappets and valve stem ends should be checked after every 200 hours of operation. To check clearance, remove tappet chamber cover and rotate crankshaft to position piston at top dead center on compression stroke. Use a feeler gage to measure clearance (engine cold), which should be 0.22 mm (0.009 inch) for intake and exhaust valves. Valve clearance is adjusted by installing valve tappet caps of different thicknesses. Refer to VALVE SYSTEM paragraphs in REPAIRS section for service procedure.

**LUBRICATION.** Models KF82 and KF100 are splash lubricated by an oil slinger mounted on connecting rod cap. Model KF150 is pressure lubricated by a trochoid type oil pump located at the front of the crankcase behind the flywheel.

On all models, oil level should be checked prior to each operating interval. Oil level should be maintained at top edge of reference marks on gage with oil fill plug just touching first threads. Do not screw oil fill plug and gage in to check oil level.

Manufacturer recommends oil with an API service classification SE or SF. SAE 10W-40, 10W-50, 20W-40 or 20W-50 oil is recommended for use in all temperatures.

Oil should be changed after the first 20 hours of operation and every 50 hours of operation thereafter. Crankcase capacity is approximately 1.2 L (1.3 qts.) for Model KF82, 1.8 L (1.9 qts.) for Model KF100 and 2.2 L (2.3 qts.) for Model KF150.

**CYLINDER HEAD AND COMBUSTION CHAMBER.** Carbon and other combustion deposits should be removed from cylinder head combustion chamber and piston after every 300 hours of operation. Refer to CYLINDER HEAD paragraphs under REPAIRS section for service procedure.

**GENERAL MAINTENANCE.** Check and tighten all loose bolts, nuts and clamps prior to each period of operation. Check for fuel or oil leakage and repair as necessary. Clean dust, dirt, grease or any foreign material from cylinder head and cylinder block cooling fins after every 100 hours or operation or more frequently if operating in extremely dirty conditions.

## REPAIRS

**TIGHTENING TORQUES.** Recommended tightening torques are as follows:
Connecting rod
  KF82, KF100 . . . . . . . . . . .39-41 N·m
               (29-30 ft.-lbs.)
  KF150 . . . . . . . . . . . . . . . .64-74 N·m
               (47-54 ft.-lbs.)
Crankcase . . . . . . . . . . . . . .20-22 N·m
               (15-16 ft.-lbs.)
Cylinder head
  KF82 . . . . . . . . . . . . . . . .22-24 N·m
               (16-18 ft.-lbs.)
  KF100 . . . . . . . . . . . . . . .34-39 N·m
               (25-29 ft.-lbs.)
  KF150:
    M10 bolts . . . . . . . . . . . . .34-39 N·m
               (25-29 ft.-lbs.)
    M12 bolts . . . . . . . . . . . . .59-69 N·m
               (44-50 ft.-lbs.)
Flywheel . . . . . . . . . . . . . .84-88 N·m
               (62-65 ft.-lbs.)

**CYLINDER HEAD.** To remove cylinder head, first remove cylinder head shroud. Clean engine to prevent entrance of foreign material. Loosen cylinder head bolts in 1/4-turn increments following sequence shown in Fig. KA304 until all bolts are loose enough to remove by hand.

Remove spark plug and clean carbon and other combustion deposits from cylinder head. Place cylinder head on a flat surface and check entire sealing surface for warpage. Renew cylinder head if it is warped more than 0.25 mm (0.010 inch).

Reinstall cylinder head using a new head gasket. Tighten head bolts evenly

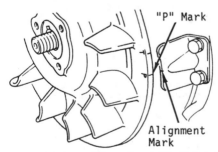

*Fig. KA303—On models equipped with a breaker-point ignition system, "P" mark on flywheel should align with alignment mark on bracket bolted on crankcase when points just start to open for correct ignition timing.*

to specified torque following sequence shown in Fig. KA304.

**CONNECTING ROD.** Connecting rod used on Models KF82 and KF150 rides directly on the crankshaft crankpin. Connecting rod used on Model KF100 is equipped with renewable insert type connecting rod bearings.

To remove connecting rod, engine must first be separated from generator unit. Remove all cooling shrouds, cylin-

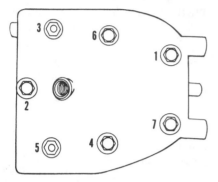

Fig. KA304—Cylinder head bolts should be loosened and tightened following the sequence shown.

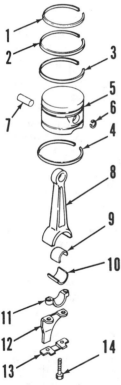

Fig. KA305—Exploded view of piston and connecting rod assembly. Note that Models KF82 and KF150 do not have bearing inserts (9 and 10). Model KF82 is equipped with only one oil control ring (3).

| | |
|---|---|
| 1. Compression ring | 9. Bearing insert (KF100) |
| 2. Compression ring | 10. Bearing insert (KF100) |
| 3. Oil control ring | 11. Connecting rod cap |
| 4. Oil control ring (KF100 & KF150) | 12. Oil dipper (KF82 & KF100) |
| 5. Piston | 13. Lockplate |
| 6. Retaining ring | 14. Bolts |
| 7. Piston pin | |
| 8. Connecting rod | |

der head and crankcase cover. Remove cylinder retaining nuts and carefully pull cylinder assembly up off piston. Remove connecting rod cap, then withdraw connecting rod and piston assembly. Remove piston pin retaining rings, push out piston pin and separate piston from connecting rod as required.

Clearance between piston pin and connecting rod pin bore should be 0.05 mm (0.002 inch) or less. If clearance is greater than specified, renew pin and/or connecting rod.

Clearance between crankshaft crankpin and connecting rod big end bore (KF82 and KF150) or bearing insert (KF100) should be 0.11 mm (0.004 inch) or less. If clearance is excessive, renew connecting rod or bearing insert and/or crankshaft.

Side clearance between connecting rod and crankshaft should be 0.5 mm (0.020 inch) or less. If clearance is greater than specified, renew connecting rod and/or crankshaft.

When assembling connecting rod and piston, be sure that ''R'' mark stamped on top of piston is toward side of connecting rod with the Japanese characters.

On Model KF100, the connecting rod bearing insert with the hole should be installed in the connecting rod and the plain bearing insert should be installed in the connecting rod cap.

On all models, connecting rod (8—Fig. KA305) and connecting rod cap (11) match marks must be aligned. Install connecting rod on the crankshaft with the side having the Japanese characters facing toward flywheel side of engine. Install oil slinger on Model KF82 and KF100 as shown in Fig. KA306.

Reverse disassembly procedure for reassembly. Tighten crankcase bolts to specified torque following sequence shown in Fig. KA309. Tighten cylinder retaining nuts following sequence shown in Fig. KA308. Tighten cylinder head bolts following sequence shown in Fig. KA304.

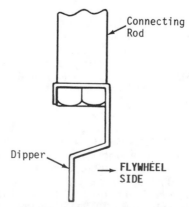

Fig. KA306—On Models KF82 and KF100, install oil dipper on connecting rod as shown.

**PISTON, PIN AND RINGS.** Piston can be removed from connecting rod after cylinder head and cylinder have been removed. Piston used in Model KF82 is equipped with two compression rings and one oil control ring. Piston used on Models KF100 and KF150 is equipped with two compression rings and two oil control rings.

After separating piston and connecting rod, carefully remove rings and clean carbon and other combustion deposits from piston crown and ring lands.

CAUTION: Extreme care should be exercised when cleaning piston ring lands. Do not damage squared edges or widen piston ring grooves. If piston ring lands are damaged, piston must be renewed.

Clearance between piston skirt and cylinder should be 0.25 mm (0.010 inch) or less. If clearance is greater than specified, renew piston and/or recondition cylinder bore.

Clearance between piston pin and piston pin bore should be 0.05 mm (0.002 inch) or less. If clearance is greater than specified, renew piston pin and piston.

Side clearance between new piston rings and piston ring grooves should be 0.15 mm (0.006 inch) or less. If clearance is greater than specified, renew piston.

Position each ring squarely in cylinder bore and measure ring end gap using a feeler gage. End gap should be 1.0 mm (0.039 inch) or less. If clearance is excessive, recondition cylinder bore for installation of oversize piston and rings.

Install piston rings on piston as shown in Fig. KA307. Stagger ring end gaps around diameter of piston; however, do not position ring end gaps above or below piston pin center line.

Piston should be installed on connecting rod as outlined in CONNECTING ROD paragraphs. Lubricate piston and cylinder bore with engine oil, then install cylinder and tighten retaining nuts following sequence shown in Fig. KA308. Tighten cylinder head bolts following sequence shown in Fig. KA304.

**CYLINDER AND CRANKCASE.** Cylinder is a separate casting and can be removed from the crankcase assembly as outlined under CONNECTING ROD paragraphs.

Standard cylinder inside diameter is 79.85-80.00 mm (3.145-3.150 inches) for Model KF82, 84.85-85.00 mm (3.341-3.346 inches) for Model KF100 and 94.85-95.00 mm (3.735-3.740 inches) for Model KF150.

**CRANKSHAFT AND MAIN BEARINGS.** Crankshaft is supported by ball bearings at each end. To remove crankshaft, remove all metal shrouds, flywheel, fan housing and cylinder head. Remove crankcase cover and remove piston and connecting rod assembly. Remove dynamic balancer shaft. Turn crankcase upside down so tappets fall away from camshaft, then remove camshaft and crankshaft. Remove ball type main bearings as necessary.

Crankshaft crankpin standard diameter is 32 mm (1.260 inches). If diameter is less than 31.95 mm (1.258 inches), renew crankshaft.

Main bearings should be a light press fit on crankshaft journals and in bearing bores of crankcase and crankcase cover.

Oil seal in crankcase cover should be pressed in until seal is 5 mm (0.2 inch) below flush with cover seal bore. Oil seal in crankcase should be pressed in until seal is flush to 1 mm (0.040 inch) below flush.

Make certain crankshaft gear, camshaft gear and dynamic balancer gear timing marks are correctly aligned when reassembling. Refer to Fig. KA310 and Fig. KA311.

**CAMSHAFT AND BEARINGS.** Camshaft is supported at each end by ball bearings. Refer to CRANKSHAFT AND MAIN BEARINGS paragraphs for camshaft removal and installation.

Camshaft ball bearing should be a light press fit on camshaft and in camshaft bearing bores machined into crankcase and crankcase cover. Inspect camshaft lobes and tappets for excessive wear, scoring or pitting and renew as necessary. Camshaft and tappets should be renewed as a set.

On Model KF82, make certain that governor sleeve and governor assembly are correctly positioned on camshaft during installation.

**GOVERNOR.** All models are equipped with a mechanical flyweight type governor. Governor flyweight assembly on Model KF82 is located on the camshaft gear. Models KF100 and KF150 are equipped with a governor gear and flyweight assembly located on a stub shaft pressed into crankcase cover.

Refer to GOVERNOR paragraphs in MAINTENANCE section for adjustment of governor external linkage.

**OIL PUMP.** Model KF150 is equipped with a trochoid type oil pump which is located behind the flywheel. Oil pump can be removed after removing flywheel and the four screws retaining oil pump and cover assembly to crankcase.

**VALVE SYSTEM.** Clearance between valve stem and valve tappet (engine cold) should be 0.22 mm (0.009 inch) for intake and exhaust valves. Clearance is adjusted by installing tappet caps of different thicknesses. Tappet caps are identified by the number stamped on the top of the cap. Refer to Fig. KA312 and to the chart in Fig. KA313 for cap identification and part number of available caps.

Valve face and seat angle is 45 degrees for intake and exhaust. Standard valve seat width is 1.0 mm (0.040 inch). If seat width exceeds 1.6 mm (0.063 inch), seat must be narrowed.

Clearance between valve stem and valve guide should be 0.12 mm (0.005 inch) or less. If clearance is greater than specified, renew valve and/or valve guide.

**DYNAMIC BALANCER.** All models are equipped with a dynamic balancer assembly (Fig. KA314) located in the crankcase. Balancer shaft is supported at each end in ball bearings. Ball bearings should be a light press fit on balancer shaft and in bearing bores machined into crankcase and crankcase cover. Make certain balancer shaft gear and crankshaft drive gear timing marks are aligned as shown in Fig. KA310 during balancer installation.

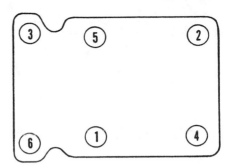

Fig. KA308—Tighten cylinder retaining nuts following sequence shown.

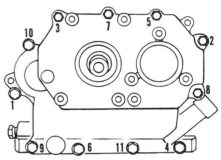

Fig. KA309—Tighten crankcase retaining bolts following sequence shown. Model KF150 is equipped with an additional retaining bolt located between bolts 1 and 10. Additional retaining bolt is number 12 in tightening sequence.

Fig. KA310—Dynamic balancer must be timed to crankshaft. Align timing marks as shown.

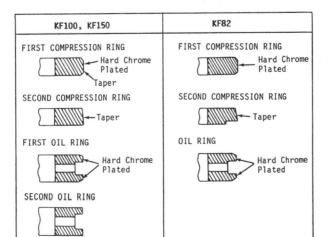

Fig. KA307—Rings should be installed on piston as shown.

| KF100, KF150 | KF82 |
|---|---|
| FIRST COMPRESSION RING — Hard Chrome Plated, Taper | FIRST COMPRESSION RING — Hard Chrome Plated |
| SECOND COMPRESSION RING — Taper | SECOND COMPRESSION RING — Taper |
| FIRST OIL RING — Hard Chrome Plated | OIL RING — Hard Chrome Plated |
| SECOND OIL RING | |

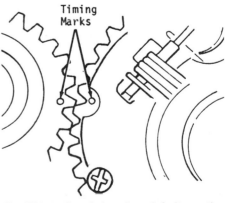

Fig. KA311—Camshaft and crankshaft gear timing marks must be aligned as shown during crankshaft and camshaft installation.

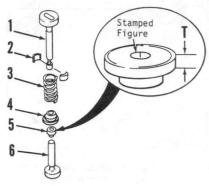

Fig. KA312—Exploded view of valves and related parts. Tappet cap (5) is available in a variety of thicknesses (T) to adjust valve clearance. Refer to Fig. KA313.

Fig. KA313—Chart showing Kawasaki part numbers, tappet cap identification numbers and tappet cap thicknesses required to adjust valve clearance on all models.

| Parts Number | Thickness (m/m) | Figure Stamped |
|---|---|---|
| 316670-2260A | 2.70 | 1 |
| 316670-2261A | 2.75 | 2 |
| 316670-2262A | 2.80 | 3 |
| 316670-2263A | 2.85 | 4 |
| 316670-2264A | 2.90 | 5 |
| 316670-2265A | 2.95 | 6 |
| 316670-2266A | 3.00 | 7 |
| 316670-2267A | 3.05 | 8 |
| 316670-2268A | 3.10 | 9 |
| 316670-2269A | 3.15 | 10 |
| 316670-2270A | 3.20 | 11 |
| 316670-2271A | 3.25 | 12 |
| 316670-2272A | 3.30 | 13 |
| 316670-2273A | 3.35 | 14 |
| 316670-2274A | 3.40 | 15 |
| 316670-2275A | 3.45 | 16 |
| 316670-2276A | 3.50 | 17 |
| 316670-2277A | 3.55 | 18 |
| 316670-2278A | 3.60 | 19 |
| 316670-2279A | 3.65 | 20 |

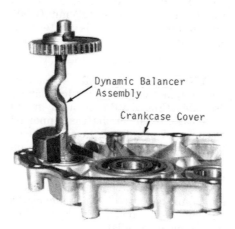

Fig. KA314—View showing dynamic balancer assembly.

# KAWASAKI

| Model | No. Cyls. | Bore | Stroke | Displacement |
|-------|-----------|------|--------|--------------|
| FG150 | 1 | 64 mm (2.51 in.) | 47 mm (1.85 in.) | 151 cc (9.2 cu. in.) |
| FG200 | 1 | 71 mm (2.79 in.) | 51 mm (2.01 in.) | 201 cc (12.3 cu. in.) |
| FG300 | 1 | 78 mm (3.07 in.) | 62 mm (2.44 in.) | 296 cc (18.07 cu. in.) |

Kawasaki FG series engines are four-stroke, single-cylinder, air-cooled gasoline engines. Engine model number is located on the shroud adjacent to rewind starter, and engine model number and serial number are located on crankcase cover (Fig. KA400). Always furnish engine model and serial numbers when ordering parts or service information.

## MAINTENANCE

**SPARK PLUG.** Recommended spark plug for Models FG150 and FG200 is NGK BP4HS or equivalent. Recommended spark plug for Model FG300 is Champion L92YC or equivalent.

Spark plug should be removed, cleaned and inspected after every 100 hours of operation. Renew spark plug if burned or damaged. Electrode gap should be set at 0.6-0.7 mm (0.024-0.027 inch) on all models.

**AIR FILTER.** The air filter element should be removed and cleaned after every 50 hours of operation, or more frequently if operating in extremely dirty conditions.

To remove filter elements (2 and 3—Fig. KA401), unsnap air filter cover (1) and pull out elements. Clean elements in nonflammable solvent, squeeze out excess solvent and allow to air dry. Soak elements in SAE 30 engine oil and squeeze out excess oil. Reinstall by reversing removal procedure.

**FUEL FILTER.** A fuel filter screen is located in the sediment bowl below fuel shut-off valve (Fig. KA402). Sediment bowl and screen should be cleaned after every 50 hours of operation. To remove sediment bowl, shut off fuel valve and unscrew sediment bowl from valve body. Clean bowl and filter screen using suitable solvent. When reinstalling, make certain that gasket is in position before installing sediment bowl.

**CARBURETOR.** All models are equipped with a float type carburetor (Fig. KA403 or Fig. KA404). Low speed fuel:air mixture is controlled by the pilot air jet and high speed fuel:air mixture is controlled by a fixed main jet. Carburetor adjustment should be checked whenever poor or erratic performance is noted.

Engine idle speed is adjusted by turning throttle stop screw (1) clockwise to increase idle speed or counterclockwise to decrease idle speed. Initial adjustment of pilot screw (2) from a lightly seated position is 7/8 turn open for Model FG150, 1 turn open for Model FG200 and 1-1/4 turns open for Model FG300. On all models, final adjustment should be made with engine at operating temperature and running. Adjust pilot screw to obtain the smoothest idle and acceleration.

Standard pilot jet size is #37.5 for Model FG150, #45 for Model FG200 and #45 for Model FG300. Standard main jet size is #72.5 for Model FG150, #87.5 for Model FG200 and #91.3 for Model FG300.

To check float level, carburetor must be removed from engine. Float should be parallel to carburetor body when carburetor is held so float tab just touches fuel inlet valve needle. To adjust, carefully bend float tab.

**GOVERNOR.** A gear driven flyweight governor assembly is located inside engine crankcase. To adjust external gover-

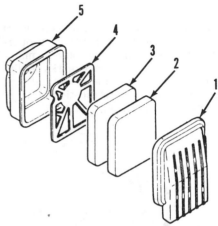

*Fig. KA401—Exploded view of air cleaner assembly.*

1. Cover
2. Element
3. Element
4. Plate
5. Housing (case)

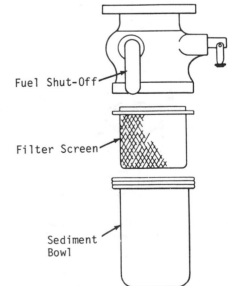

Fuel Shut-Off

Filter Screen

Sediment Bowl

*Fig. KA402—Exploded view of fuel filter and sediment bowl assembly used on most models.*

Engine Model And Serial Number Location

*Fig. KA400—View showing location of engine model number and serial number.*

nor linkage, place engine throttle control in idle position. Make certain that all linkage is in good condition and that tension spring (4—Fig. KA405) is not stretched. Loosen clamp bolt (5) and rotate governor shaft (6) clockwise as far as possible and move governor lever (3) to fully open carburetor throttle. Then tighten clamp bolt while holding governor lever and shaft in these positions.

Start engine and apply load to generator. Engine speed should be 3600 rpm with generator loaded to approximately full capacity.

**IGNITION SYSTEM.** All models are equipped with a transistor ignition system and no regular maintenance is required. Ignition timing is not adjustable. Ignition coil is located outside the flywheel. Air gap between ignition coil and flywheel magnets should be 0.3 mm (0.010 inch).

**LUBRICATION.** Engine oil level should be checked prior to each operating interval. Oil level should be maintained between reference marks on dipstick with dipstick just touching first threads. Do not screw dipstick in when checking oil level (Fig. KA406).

Manufacturer recommends oil with an API service classification of SE or SF. SAE 10W-40, 10W-50, 20W-40 or 20W-50 oil is recommended for use in all temperatures.

Oil should be changed after the first 20 hours of operation and every 50 hours of operation thereafter. Crankcase capacity is 0.5 L (1.0 pt.) for Model FG150, 0.7 L (1.5 pt.) for Model FG200 and 0.75 L (1.6 pt.) for Model FG300.

**VALVE ADJUSTMENT.** Clearance between valve stem ends and tappets should be checked after every 200 hours of operation. To check clearance, remove tappet chamber cover and rotate crankshaft to position piston at top dead center on compression stoke. Measure clearance using a feeler gage. Specified clearance (engine cold) is 0.15 mm (0.006 inch) for intake valve and 0.22 mm (0.009 inch) for exhaust valve.

If clearance is not as specified, valves must be removed and end of stems ground off to increase clearance or seats ground deeper to reduce clearance. Refer to VALVE SYSTEM paragraphs in REPAIRS SECTION for service procedure.

**CYLINDER HEAD AND COMBUSTION CHAMBER.** Cylinder head, combustion chamber and piston should be cleaned and carbon and other deposits removed after every 200 hours of operation. Refer to REPAIRS section for service procedure.

**GENERAL MAINTENANCE.** Check and tighten all loose bolts, nuts or clamps prior to each interval of operation. Check for fuel or oil leakage and repair as necessary.

Clean dust, dirt, grease or any foreign material from cylinder head and cylinder block cooling fins after every 100 hours of operation, or more frequently if operating in extremely dirty conditions.

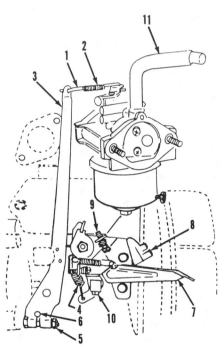

*Fig. KA405—View of external governor linkage. Refer to text for adjustment procedure.*

1. Governor-to-carburetor rod
2. Spring
3. Governor lever
4. Governor spring
5. Clamp bolt
6. Governor shaft
7. Throttle lever
8. Throttle plate
9. Maximum speed adjusting screw
10. Idle speed adjusting screw
11. Throttle pivot

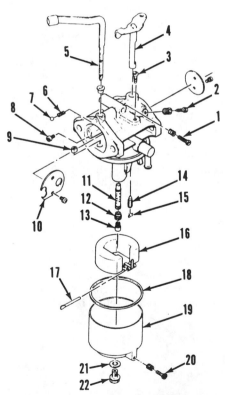

*Fig. KA403—Exploded view of float type carburetor used on Model FG150 engine.*

1. Idle speed screw
2. Pilot screw
3. Pilot jet
4. Carburetor body
5. Nozzle
6. Bleed tube
7. Drain screw
8. Gasket
9. Float bowl
10. Plug
11. Gasket
12. Float pin
13. Float
14. Fuel inlet needle
15. Throttle shaft
16. Main jet
17. Choke shaft

*Fig. KA404—Exploded view of float type carburetor used on Models FG200 and FG300.*

1. Idle speed screw
2. Pilot screw
3. Pilot air jet
4. Throttle shaft
5. Choke shaft
6. Spring
7. Ball
8. Air jet
9. Collar
10. Choke valve
11. Nozzle
12. Main jet holder
13. Main jet
14. Fuel inlet needle
15. Clip
16. Float
17. Float pin
18. Gasket
19. Float bowl
20. Drain screw
21. Gasket
22. Bolt

*Fig. KA406—Oil plug and gage should not be screwed into crankcase when checking oil level.*

### REPAIRS

**TIGHTENING TORQUES.** Recommended tightening torques are as follows:

Spark plug . . . . . . . . . . . . . . . . . 27 N·m
(20 ft.-lbs.)
Cylinder head bolts . . . . . . 23-24 N·m
(17-18 ft.-lbs.)
Connecting rod bolts . . . . . 20-21 N·m
(15-16 ft.-lbs.)
Crankcase cover bolts . . . . . 11-13 N·m
(8-10 ft.-lbs.)
Flywheel nut . . . . . . . . . . . . 59-64 N·m
(44-47 ft.-lbs.)

**CYLINDER HEAD.** To remove cylinder head, first remove cylinder head shroud. Clean engine to prevent entrance of foreign material. Loosen cylinder head bolts in 1/4-turn increments following sequence shown in Fig. KA407 until all bolts are loose enough to remove by hand.

Remove spark plug and clean carbon and other deposits from cylinder head. Place cylinder head on a flat surface and check entire sealing surface for warpage. Renew cylinder head if warpage exceeds 0.25 mm (0.010 inch).

Reinstall cylinder head using a new head gasket. Tighten head bolts evenly to specified torque in sequence shown in Fig. KA407.

**CONNECTING ROD.** Engine must be separated from generator unit to remove connecting rod. Piston and connecting rod are removed as an assembly after removing cylinder head and splitting crankcase. Remove connecting rod cap bolts and remove rod cap (8—Fig. KA408). Remove carbon and ring ridge, if present, from top of cylinder before removing piston. Push piston and connecting rod assembly out through top of cylinder block. Remove snap rings (4), push piston pin (5) out and separate piston from connecting rod if necessary.

The connecting rod rides directly on crankshaft crankpin journal. Inspect connecting rod and renew if piston pin bore or crankpin bearing bore are scored or damaged. Refer to the following specifications for maximum allowable connecting rod wear limits. Renew connecting rod as necessary.

Small end pin bore:
FG150 . . . . . . . . . . . . . . . . . . 13.043 mm
(0.5135 in.)
FG200 . . . . . . . . . . . . . . . . . 15.027 mm
(0.5916 in.)
FG300 . . . . . . . . . . . . . . . . . 16.046 mm
(0.6317 in.)
Big end crankpin bore:
FG150 . . . . . . . . . . . . . . . . . 24.553 mm
(0.9667 in.)
FG200 . . . . . . . . . . . . . . . . . 27.050 mm
(1.0650 in.)

FG300 . . . . . . . . . . . . . . . 30.050 mm
(1.1831 in.)
Big end width (min.)
FG150 . . . . . . . . . . . . . . . . 23.30 mm
(0.917 in.)
FG200 . . . . . . . . . . . . . . . . 24.30 mm
(0.957 in.)
FG300 . . . . . . . . . . . . . . . . 26.10 mm
(1.028 in.)

On all models, clearance between connecting rod small end bore and piston pin should not exceed 0.05 mm (0.002 inch). Clearance between connecting rod big end bore and crankshaft crankpin should not exceed 0.10 mm (0.004 inch). Connecting rod side play on crankshaft should not exceed 0.7 mm (0.028 inch).

Piston should be installed on connecting rod so the "R" mark on top of piston is on the unmarked side of the connecting rod. Install rod and piston assembly in cylinder block with numbered side (EC—Fig. KA408) of connecting rod facing away from flywheel end of crankshaft. Match marks (AM) on rod and cap must be aligned. Tighten connecting rod bolts evenly to specified torque.

**PISTON, PIN AND RINGS.** Piston and connecting rod are removed as an assembly as outlined in CONNECTING ROD paragraphs.

After separating piston and connecting rod, carefully remove rings and clean carbon and other deposits from piston surface and ring lands.

**CAUTION: Extreme care should be exercised when cleaning ring lands. Do not damage squared edges or widen ring grooves. If ring lands are damaged, piston must be renewed.**

Measure piston diameter at point 6 mm (0.24 inch) above bottom of skirt and 90 degrees from piston pin bore center line. Minimum allowable piston diameter is 63.810 mm (2.2526 inches) for Model FG150, 70.810 mm (2.7878 inches) for Model FG200 or 77.81 mm (3.0138 inches) for Model FG300. Clearance between piston and cylinder bore should not exceed 0.25 mm (0.010 inch). Renew piston if excessively worn, scored or damaged.

Maximum allowable piston pin bore inside diameter is 13.043 mm (0.5135 inch) for Model FG150, 15.027 mm (0.5916 inch) for Model FG200 and 16.038 mm (0.6314 inch) for Model FG300. Minimum allowable piston pin outside diameter is 12.986 mm (0.5113 inch) for Model FG150, 14.977 mm (0.5896 inch) for Model FG200 and 15.987 mm (0.6294 inch) for Model FG300.

To check piston ring groove wear, insert new ring in ring groove and use a feeler gage to measure side clearance between ring and ring land. Renew piston if side clearance exceeds 0.17 mm (0.007 inch) for all models.

Install rings squarely in cylinder bore and measure ring end gap using a feeler gage. Maximum allowable end gap is 0.8 mm (0.031 inch) for Model FG150 or 1.0 mm (0.040 inch) for Models FG200

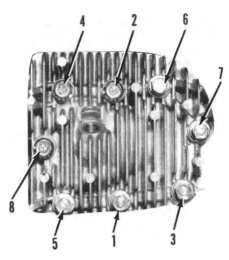

*Fig. KA407—Tighten and loosen cylinder head bolts following sequence shown.*

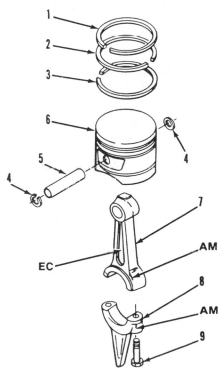

*Fig. KA408—Exploded view of piston and connecting rod assembly. When assembling, be sure match marks (AM) on connecting rod and cap are aligned.*

| | |
|---|---|
| 1. Compression ring | 6. Piston |
| 2. Compression ring | 7. Connecting rod |
| 3. Oil control ring | 8. Connecting rod cap |
| 4. Retaining ring | 9. Connecting rod |
| 5. Piston pin | bolts |

and FG300. If end gap is excessive, recondition cylinder bore for installation of oversize piston and rings.

When installing rings on piston, make sure that manufacturer's mark on end of ring faces upward. Refer to CONNECTING ROD paragraphs to properly assemble piston to connecting rod. Lubricate piston and cylinder with engine oil prior to installation. Stagger piston ring end gaps equally around piston.

**CYLINDER AND CRANKCASE.** Cylinder and crankcase are an integral casting. Standard cylinder bore diameter is 63.98-64.00 mm (2.519-2.520 inches) for Model FG150, 70.98-71.00 mm (2.794-2.795 inches) for Model FG200 or 77.98-78.00 mm (3.070-3.071 inches) for Model FG300. If cylinder diameter exceeds specified limit or if bore is out-of-round more than 0.05 mm (0.002 inch), recondition or renew cylinder. Piston and rings are available in several oversizes as well as standard size.

When installing crankshaft cover, tighten bolts evenly to specified torque following sequence shown in Fig. KA409 or Fig. KA410.

**CRANKSHAFT AND MAIN BEARINGS.** Crankshaft is support by ball bearings at each end. To remove crankshaft, remove fan housing, flywheel, cylinder head and crankcase cover. Remove connecting rod cap and remove piston and connecting rod assembly. Turn crankcase upside down so tappets fall away from camshaft, then withdraw camshaft and tappets. Remove crankshaft from crankcase. It may be necessary to heat crankcase cover and crankcase slightly to remove crankshaft bearings. Remove crankshaft seals using suitable puller.

Inspect crankshaft for wear, scoring or other damage. Crankshaft crankpin journal minimum diameter is 24.447 mm (0.9625 inch) for Model FG150, 26.95 mm (1.061 inches) for Model FG200 and

29.95 mm (1.1791 inches) for Model FG300. On all models, crankshaft runout should not exceed 0.05 mm (0.002 inch).

Ball bearing main bearings should be a light press fit on crankshaft and in bearing bores in crankcase and crankcase cover. Bearings should spin smoothly and have no rough spots. It may be necessary to heat crankcase or crankcase cover slightly to install bearings.

When installing crankshaft, make certain crankshaft gear and balancer drive gear timing marks are aligned on Model FG300. On all models, make sure crankshaft gear and camshaft gear timing marks are aligned as shown in Fig. KA411.

**CAMSHAFT AND BEARINGS.** Camshaft is supported at crankcase end in a bearing which is an integral part of crankcase casting and is supported at crankcase cover end in a ball bearing. Refer to CRANKCASE AND BEARINGS paragraphs for camshaft removal.

Inspect camshaft and tappets for excessive wear, scoring, pitting or other damage and renew as necessary. Camshaft and tappets should be renewed as a set. Refer to the following table for wear limit specifications:

**Model FG150**
Cam lobe height
  Intake . . . . . . . . . . . . . . . .27.3 mm
                                (1.075 in.)
  Exhaust . . . . . . . . . . . . . . .27.1 mm
                                (1.067 in.)
Bearing journal diameter
  Pto side . . . . . . . . . . . . .14.940 mm
                                (0.5882 in.)
  Flywheel side . . . . . . . .14.942 mm
                                (0.5883 in.)

**Model FG200**
Cam lobe height
  Intake & exhaust . . . . . . . .31.7 mm
                                (1.248 in.)
Bearing journal diameter
  Pto side . . . . . . . . . . . . .14.966 mm
                                (0.5892 in.)
  Flywheel side . . . . . . . .16.942 mm
                                (0.6670 in.)

**Model FG300**
Cam lobe height
  Intake & exhaust . . . . . . .32.67 mm
                                (1.286 in.)
Bearing journal diameter
  Pto side . . . . . . . . . . . . .19.955 mm
                                (0.7856 in.)
  Flywheel side . . . . . . . .19.960 mm

Camshaft bearing bore in crankcase wear limit is 15.043 mm (0.5922 inch) for Model FG150, 17.043 mm (0.6710 inch) for Model FG200 or 20.056 mm (0.7869 inch) for Model FG300.

Camshaft ball bearing should spin smoothly with no rough spots or looseness. Bearing should be a light press fit on camshaft and in crankcase cover bore. It may be necessary to heat crankcase cover slightly to install ball bearing.

Lubricate tappets and camshaft with engine oil prior to installation. When installing camshaft, make certain camshaft gear and crankshaft gear timing marks are aligned as shown in Fig. KA411.

**GOVERNOR.** The internal centrifugal flyweight governor is driven by the camshaft gear. Refer to GOVERNOR paragraph in MAINTENANCE section for external governor adjustments.

To remove governor assembly, remove external linkage, shrouds and crankcase cover. Remove governor gear cover mounting screws and remove cover. Lift gear assembly from governor shaft.

To reinstall governor assembly, reverse removal procedure.

**VALVE SYSTEM.** Clearance between valve stem and valve tappet (engine cold) should be 0.15 mm (0.006 inch) for intake valve and 0.22 mm (0.009 inch) for exhaust valve. If clearance is not as specified, valves must be removed and end of stems ground off to increase clearance or seats ground deeper to reduce clearance.

Valve face and seat angle is 45 degrees for intake and exhaust valves. Valve seat

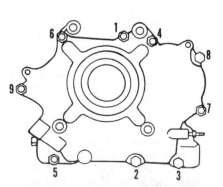

Fig. KA409—On Models FG150 and FG200, tighten crankcase cover bolts to specified torque in sequence shown.

Fig. KA410—On Model FG300, tighten crankcase cover bolts to specified torque following sequence shown.

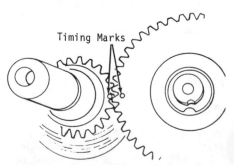

Fig. KA411—Crankshaft gear and camshaft gear timing marks must be aligned during crankshaft or camshaft installation.

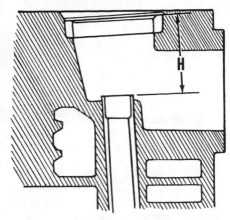

Fig. KA412—On Model FG300, install new valve guide so there is 26 mm (1.02 in.) distance (H) between top of valve seat and top of guide. Refer to text.

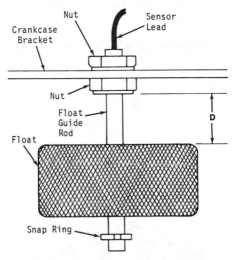

Fig. KA413—On models equipped with low oil sensor, sensor float and switch assembly is mounted on crankcase cover. Float gap (D) must be 9.5-15.5 mm (0.37-0.61 in.). Refer to text.

width should be 1.0-1.6 mm (0.039-0.063 inch). Valve should be renewed if valve head margin is less than 0.6 mm (0.024 inch).

Refer to the following table for valve stem and valve guide wear limit specifications:

### Model FG150
Valve stem OD (min.)
Intake . . . . . . . . . . . . . . . .5.948 mm
(0.2342 in.)
Exhaust . . . . . . . . . . . . . .5.935 mm
(0.2336 in.)
Valve guide ID (max.)
Intake . . . . . . . . . . . . . . . .6.078 mm
(0.2393 in.)
Exhaust . . . . . . . . . . . . . .6.085 mm
(0.2395 in.)

### Model FG200
Valve stem OD (min.)
Intake . . . . . . . . . . . . . . . .6.942 mm
(0.2733 in.)
Exhaust . . . . . . . . . . . . . .6.952 mm
(0.2737 in.)
Valve guide ID (max.)
Intake . . . . . . . . . . . . . . . .7.072 mm
(0.2784 in.)
Exhaust . . . . . . . . . . . . . .7.075 mm
(0.2785 in.)

### Model FG300
Valve stem OD (min.)
Intake . . . . . . . . . . . . . . . .7.442 mm
(0.2930 in.)
Exhaust . . . . . . . . . . . . . .7.425 mm
(0.2923 in.)
Valve guide ID (max.)
Intake . . . . . . . . . . . . . . . .7.572 mm
(0.2981 in.)
Exhaust . . . . . . . . . . . . . .7.575 mm
(0.2982 in.)

Use suitable tools to remove old guide and press in new guide. On Model FG300, press new guide into valve guide bore until top of guide is 26 mm (1.02 inch) below top of valve seat (Fig. KA412).

### Models So Equipped

**LOW OIL SENSOR.** Float type low oil sensor is located inside the crankcase (Fig. KA413). To remove or check float gap (D), it is necessary to remove crankcase cover.

To check oil level sensor switch, remove bearing plate and float cover. Disconnect electrical leads and connect ohmmeter test leads to switch leads. Slide float to top of shaft. Ohmmeter should indicate infinite resistance. Slide float slowly down the shaft until ohmmeter just deflects. Gap between top of float and lower nut (Fig. KA413) should be 9.5-15.5 mm (0.37-0.61 inch). If not, renew switch.

### Model FG300

**ENGINE BALANCER.** The engine balancer shaft is supported at crankcase side in a bearing which is an integral part of crankcase casting and is supported at crankcase cover side in a ball bearing.

Engine balancer bearing journal diameters should not be less than 17.951 mm (0.7067 inch) at each end. Ball bearing should be a light press fit on balancer shaft journal and in crankcase cover bore. It may be necessary to heat the crankcase cover slightly to remove bearing.

When installing balancer, make certain balancer drive gear and crankshaft gear timing marks are aligned.

# KOHLER

**KOHLER COMPANY**
**Kohler, Wisconsin 53044**

| Model | Bore | Stroke | Displacement |
|---|---|---|---|
| K90, K91 | 2³⁄₈ in. | 2 in. | 8.9 cu. in. |
| | (60.3 mm) | (50.8 mm) | (145 cc) |

## ENGINE IDENTIFICATION

Kohler engine identification and serial number decals are located on engine shrouding (Fig. K1). The model number designates displacement (in cubic inches), digit after displacement on late models indicates number of cylinders and the letter suffix indicates the specific version. No suffix indicates a rope start model. Therefore, K91 would indicate a Kohler (K), 9 cubic inch (9), single-cylinder (1) engine with rope start (no suffix). Suffix interpretation is as follows:

| | |
|---|---|
| C | Clutch model |
| A | Oil pan type |
| EP | Generator set |
| P | Pump model |
| R | Gear reduction |
| S | Electric start |
| T | Retractable start |
| G | Housed with fuel tank |
| H | Housed without fuel tank |

Always furnish engine model, specification and serial numbers when ordering parts or service material.

## MAINTENANCE

**SPARK PLUG.** Recommended spark plug for all models is Champion J8, or equivalent.

**Fig. K1—View showing nameplate and serial number plate on Kohler engine. Refer to text.**

After every 100 hours of operation, the spark plug should be removed and cleaned and the electrode gap should be checked. Electrode gap should be 0.025 inch (0.7 mm).

**NOTE: Manufacturer does not recommend the use of abrasive grit type spark plug cleaners. Improper cleaning allows entrance of grit into cylinder causing premature cylinder wear and damage.**

Renew spark plug if electrode is severely burnt or if other damage is apparent. Tighten spark plug to 18-22 ft.-lbs. (24-30 N·m).

**CARBURETOR.** Refer to Fig. K2 for an exploded view of Carter Model N carburetor used on early production models. Late production K91 engines are equipped with the Kohler carburetor shown in Fig. K3. Carburetor adjust-

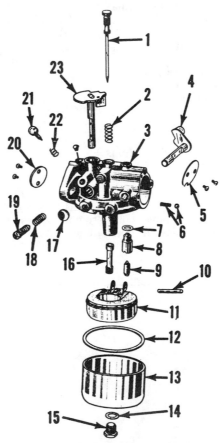

**Fig. K2—Exploded view of typical Carter Model N carburetor used on early production K90 and K91 engines.**

| | | | |
|---|---|---|---|
| 1. | Main fuel needle | 13. | Float bowl |
| 2. | Spring | 14. | Sealing washer |
| 3. | Carburetor body | 15. | Retainer |
| 4. | Choke shaft | 16. | Main jet |
| 5. | Choke plate | 17. | Plug |
| 6. | Choke detent | 18. | Spring |
| 7. | Sealing washer | 19. | Idle stop screw |
| 8. | Inlet valve seat | 20. | Throttle plate |
| 9. | Inlet valve | 21. | Idle fuel needle |
| 10. | Float pin | 22. | Spring |
| 11. | Float | 23. | Throttle shaft |
| 12. | Gasket | | |

**Fig. K3—Exploded view of Kohler carburetor used on late production engines.**

| | | | |
|---|---|---|---|
| 1. | Main fuel needle | 9. | Inlet valve seat |
| 2. | Spring | 10. | Inlet valve |
| 3. | Carburetor body assy. | 11. | Float pin |
| 4. | Spring | 12. | Float |
| 5. | Idle speed stop screw | 13. | Gasket |
| 6. | Spring | 14. | Float bowl |
| 7. | Idle fuel needle | 15. | Sealing washer |
| 8. | Sealing washer | 16. | Bowl retainer |

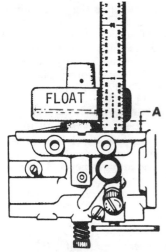

Fig. K3A—Float height should be 11/64 inch (4.36 mm) measured at (A). Bend tang on float arm to adjust.

ment should be checked whenever poor or erratic engine operation is apparent.

Engine idle speed is 1200 rpm and is adjusted by turning idle speed screw (19—Fig. K2 or 5—Fig. K3) clockwise to increase idle speed or counterclockwise to decrease idle speed.

Initial adjustment of idle mixture screw is 1½ turns out from a lightly seated position. Initial setting of main fuel mixture screw is 2 turns out from a lightly seated position. Make final adjustments with engine at normal operating temperature and running. Place engine under load and adjust main fuel mixture screw for leanest setting that will allow satisfactory acceleration and steady governor operation. Set engine at idle speed, no-load, and adjust idle mixture screw to obtain smoothest idle operation.

As each adjustment affects the other, adjustment procedure may have to be repeated.

To check float level, invert carburetor throttle body and float assembly. There

should be 11/64 inch (4.36 mm) clearance between free side of float and machined surface of body casting (Fig. K3A). Carefully bend float lever tang that contacts inlet valve as necessary to provide correct clearance.

**FUEL FILTER.** A fuel filter screen is located in sediment bowl below fuel shut off valve. Fuel filter screen and sediment bowl should be removed and cleaned after every 100 hours of operation. To remove, loosen bail nut, swing bail out of the way and remove sediment bowl, filter screen and gasket. Make certain sediment bowl gasket is in place before reinstalling sediment bowl.

**AIR FILTER.** Engines may be equipped with either an oil bath air filter (Fig. K4) or a dry element filter (Fig. K5). Refer to appropriate paragraph for model being serviced.

**Oil Bath Type.** Oil bath air filter should be serviced at 25 hour intervals of normal operation. To service, remove cover, lift element from bowl and drain oil. Thoroughly clean bowl and cover and rinse element in clean solvent. Allow element to dry, then lightly coat element and fill bowl to correct level with the same grade and type oil used in crankcase (see LUBRICATION section).

**NOTE: Do not use air pressure to force dry element as element material may be damaged.**

Make certain gasket is in place on air horn and reinstall air filter assembly.

**Dry Type Filter.** Element should be removed and cleaned after every 50 hours of operation under normal operating conditions. Tap element lightly on a flat surface to dislodge loose dirt and foreign material.

FILL TO LEVEL MARK
WITH SAME OIL AS ENGINE

Fig. K4—View of oil bath type air cleaner. Refer to text for service intervals and procedure.

Fig. K5—Exploded view of dry type air cleaner (filter). Foam precleaner (3) is optional. Refer to text for service interval and procedure.

1. Wing nut
2. Cover
3. Precleaner
4. Filter
5. Adapter

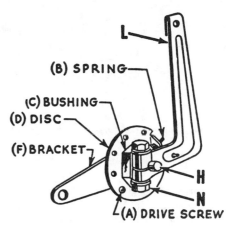

Fig. K6—Points of adjustment of governor on Series K90 and K91 engines.

**NOTE: Do not wash element or use air pressure to blow dirt out as element material may be damaged.**

Renew element after every 150 hours of operation or whenever element is damaged or excessive dirt cannot be correctly removed.

Some dry type filters may be equipped with a foam type precleaner which fits over dry element and extends dry element service intervals. Precleaner should be cleaned and re-oiled after every 25 hours of operation. To clean, remove precleaner and wash in a mild detergent and water solution. Rinse thoroughly and squeeze away excess water. Allow to air dry, then soak precleaner in the same type and grade oil as used in engine crankcase (see LUBRICATION section). Squeeze out excess oil and reinstall precleaner.

**CAUTION: Do not wring element or use air pressure to dry as element material may be stretched or damaged.**

**CRANKCASE BREATHER.** The crankcase breather is attached to cylinder block on carburetor side of engine on early production model and to valve cover plate on late production model. Crankcase breather is designed to maintain a slight vacuum in crankcase to eliminate oil leakage at seals.

On late production model, crankcase breather should be disassembled, reed clearance checked and filter element cleaned after every 500 hours of operation. Reed clearance should be 1/64 to 1/32 inch (0.40-0.80 mm) between reed valve and its seat.

**GOVERNOR.** A centrifugal flyball type governor mounted within the crankcase and driven by the camshaft gear is used on all models. To adjust governed speed, first sychronize linkage by loosening clamp bolt nut (N—Fig. K6), turn shaft (H) counterclockwise

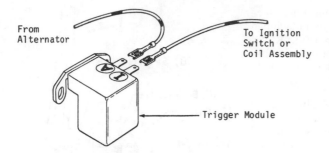

**Fig. K7—View of trigger module used with CD ignition system. Terminal "A" must be connected to alternator and terminal "I" must be connected to ignition switch.**

until internal resistance is felt. Pull arm (L) completely to the left (away from carburetor) and tighten the clamp bolt nut. To increase or decrease maximum engine speed, vary the tension of governor spring (B). On engine with remote throttle control, spring (B) tension is varied by moving bracket (F) up or down. On engines without remote throttle control, rotate disc (D) after loosening bushing (C) to vary spring (B) tension.

**IGNITION SYSTEM.** Engines may be equipped with a magneto ignition system, capacitor-discharge (CDI) or battery ignition system. Refer to appropriate paragraph for model being serviced.

**Magneto and Battery Ignition System.** Breaker point cover should be removed and condenser and breaker points checked and renewed or adjusted after every 500 hours of operation. Breaker point gap is adjusted to 0.020 inch (0.51 mm) on all models.

Late production engine has a timing port in left side of bearing plate and a timing light is used to precisely set timing. There are two marks on flywheel, a "T" for top dead center and either "S", "SP" or "20" mark for the spark point. With engine running at 1200 to 1800 rpm, light should flash when spark point mark ("S", "SP" or "20") is centered in timing port. Vary point gap to align marks.

**CD Ignition System.** Capacitor-discharge (CDI) ignition system does not have breaker points and timing is not adjustable. A trigger module (Fig. K7) is used instead of breaker points and must

be installed correctly to prevent damage to internal components. Terminal marked "A" must be connected to the alternator and terminal "I" must be connected to the ignition switch. DO NOT reverse these leads.

If a faulty trigger module is suspected, remove module from engine. To test trigger module diodes, connect one lead

of an ohmmeter to the "I" terminal and connect remaining lead to the "A" terminal. Observe ohmmeter reading. Reverse leads. Observe ohmmeter reading. Ohmmeter should indicate continuity with leads in one position only. If ohmmeter indicates continuity when connected both ways, or an open circuit for both connections, renew module.

To test trigger module SCR switch, connect one ohmmeter lead to "I" terminal and remaining lead to trigger module mounting bracket. If ohmmeter indicates continuity, reverse the leads as ohmmeter must indicate an open circuit initially for this test. Lightly tap module magnet with a metal object. This should activate the SCR switch and ohmmeter should indicate continuity. If ohmmeter indicates continuity when initially attached to module in both directions or if SCR switch will not activate as previous-

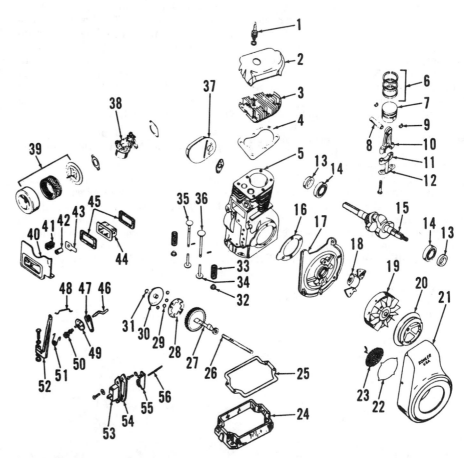

**Fig. K9—Exploded view of K90 and K91 engine. Breaker points (55) are actuated by cam on right end of camshaft (27) through push rod (56).**

| | | |
|---|---|---|
| 1. Spark plug | 15. Crankshaft | 29. Steel balls | 43. Breather reed |
| 2. Air baffle | 16. Gasket | 30. Thrust cone | 44. Breather plate |
| 3. Cylinder head | 17. Bearing plate | 31. Snap ring | 45. Gaskets |
| 4. Head gasket | 18. Magneto | 32. Spring retainer | 46. Governor shaft |
| 5. Cylinder block | 19. Flywheel | 33. Valve spring | 47. Bracket |
| 6. Piston rings | 20. Pulley | 34. Valve tappets | 48. Link |
| 7. Piston | 21. Shroud | 35. Exhaust valve | 49. Speed disc |
| 8. Piston pin | 22. Screen retainer | 36. Intake valve | 50. Bushing |
| 9. Retaining rings | 23. Screen | 37. Muffler | 51. Governor spring |
| 10. Connecting rod | 24. Oil pan | 38. Carburetor | 52. Governor lever |
| 11. Rod cap | 25. Gasket | 39. Air cleaner assy. | 53. Breaker cover |
| 12. Rod bolt lock | 26. Camshaft pin | 40. Valve cover | 54. Gasket |
| 13. Oil seal | 27. Camshaft | 41. Filter | 55. Breaker points |
| 14. Ball bearing | 28. Flyball retainer | 42. Breather seal | 56. Push rod |

**Fig. K8—Tighten cylinder head cap screws to 200 in.-lbs. (22.6 N·m) in sequence shown.**

Illustrations courtesy of Kohler Co.

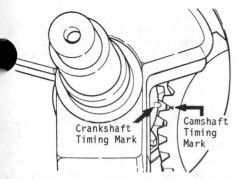

*Fig. K10—Make certain timing marks are aligned during crankshaft and camshaft installation.*

ly described, the trigger module may be considered defective and should be renewed.

When installing the trigger module, set air gap between trigger assembly and flywheel projection at 0.005-0.010 inch (0.13-0.25 mm).

**VALVE ADJUSTMENT.** Valve stem clearance should be checked after every 500 hours of operation. Clearance for intake valve stem should be 0.005-0.009 inch (0.13-0.23 mm) and clearance for exhaust valve stem should be 0.011-0.015 inch (0.28-0.38 mm). If clearance is not as specified, refer to REPAIR section for valve service procedure.

**CYLINDER HEAD AND COMBUSTION CHAMBER.** Cylinder head should be removed and carbon and lead deposits cleaned after every 500 hours of operation. Refer to REPAIR section for cylinder head removal procedure.

**LUBRICATION.** Engine oil should be checked daily and oil level maintained between the "F" and "L" mark on dipstick. Insert threaded plug type dipstick only until threads touch opening; do not screw in to check oil level. On models with extended oil fill tube, push dipstick all the way down in tube to obtain reading.

Manufacturer recommends oil having API service classification SE or SF. Use SAE 5W-30 oil for temperatures below 32°F (0°C) and use SAE 30W oil for temperatures above 32°F (0°C).

Oil should be changed after the first 5 hours of operation and at 25 hour intervals thereafter. Crankcase oil capacity is 0.5 quart (0.47 L).

**GENERAL MAINTENANCE.** Check and tighten all loose bolts, nuts or clamps daily. Check for fuel and oil leakage and repair if necessary. Clean cooling fins and external surfaces at 50 hour intervals.

## REPAIR

**TIGHTENING TORQUES.** Recommended tightening torque specifications are as follows:

Spark plug . . . . . . . . . . . . . . 18-22 ft.-lbs.
(24-30 N·m)
Flywheel nut . . . . . . . . . . . . 40-50 ft.-lbs.
(54-68 N·m)
Cylinder head . . . . . . . . . . . . 200 in.-lbs.*
(22.6 N·m)*
Connecting rod . . . . . . . . . . . 140 in.-lbs.*
(15.8 N·m)*

*With threads lightly lubricated.

**CYLINDER HEAD.** To remove cylinder head, first remove all necessary metal shrouds. Clean engine to prevent entrance of foreign material and remove cylinder head retaining bolts.

Always use a new head gasket when installing cylinder head. Tighten cylinder head bolts evenly and in graduated steps using the sequence shown in Fig. K8 until specified torque is obtained.

**CONNECTING ROD.** The aluminum alloy connecting rod rides directly on the crankpin journal. Connecting rod and piston are removed as an assembly after cylinder head and oil pan (engine base) have been removed. Remove the two connecting rod bolts and connecting rod cap and push piston and rod assembly out the top of block. Remove snap rings (9—Fig. K9) and push pin (8) out of piston (7). Separate connecting rod and piston.

Before installing connecting rod, check the following dimensions to ensure rod is suitable for service.

**Small End Inside Diameter.** Connecting rod piston pin bore is 0.5630-0.5633 inch (14.300-14.308 mm) and piston pin-to-connecting rod pin bore clearance should be 0.0007-0.0008 inch (0.018-0.020 mm). Renew connecting rod if dimensions are not as specified.

**Big End Inside Diameter.** Connecting rod big end (bearing) diameter for standard connecting rod is 0.9384-0.9387 inch (23.835-23.843 mm) and connecting rod-to-crankpin journal clearance should be 0.0010-0.0025 inch (0.025-0.063 mm). If dimensions are not as specified or if bearing clearance exceeds 0.0035 inch (0.090 mm), renew connecting rod and/or recondition crankpin journal.

**Connecting Rod Side Clearance.** With crankshaft removed from crankcase, mount connecting rod on crankpin journal.

Use a suitable feeler gage to measure clearance between crankshaft and flat thrust surface on connecting rod big end. Side play should be 0.005-0.016 inch (0.13-0.41 mm). If clearance is not as specified, renew connecting rod.

Piston may be installed on connecting rod either way (always use new retaining rings), however, connecting rod cap and connection rod match marks must align and be towards flywheel side of engine after installation. Tighten connecting rod bolts to specified torque.

**PISTON, PIN AND RINGS.** The aluminum alloy piston is fitted with two compression rings and one oil control ring. Refer to CONNECTING ROD section for removal and installation procedure.

After separating piston and connecting rod, carefully remove rings and clean carbon and lead deposits from piston surface and ring lands.

**CAUTION: Extreme care should be exercised when cleaning ring lands. Do not damage squared edges or widen ring grooves. If ring lands or grooves are damaged, piston must be renewed.**

Measure piston diameter just below oil control ring groove 90° from piston pin center. Standard piston diameter is 2.369-2.371 inches (60.173-60.223 mm) and minimum piston diameter is 2.366 inches (60.096 mm). Renew piston if dimensions are not as specified.

Install piston in cylinder bore before installing rings and use a suitable feeler gage to measure clearance between piston thrust surface (90° from piston pin) and cylinder bore. Clearance should be 0.0035-0.0060 inch (0.089-0.152 mm). If dimension is not as specified, renew piston and/or recondition cylinder bore.

Piston pin fit in piston pin bore should be 0.0002 inch (0.005 mm) interference to 0.0002 inch (0.005 mm) loose. Piston pin is available in oversize if dimension is not as specified.

If piston ring-to-piston groove side clearance exceeds 0.006 inch (0.152 mm) for compression rings or 0.0015-0.0035 inch (3.81-8.89 mm) for oil control ring, renew piston.

Piston ring end gap for compression rings, measured with ring squarely installed in cylinder bore, is 0.007-0.017 inch (0.18-0.43 mm) for new bore and 0.007-0.027 inch (0.18-0.68 mm) for used bore. If dimensions are not as specified, renew piston and/or recondition cylinder bore.

Install piston rings, which are marked, with marked side toward top of piston. If compression ring has a groove or bevel on outside surface, install ring with groove or bevel down. If groove or bevel is on inside surface of compression ring, install ring with groove or bevel up. Stagger ring end gaps equally around circumference of piston before installation.

**CYLINDER AND CRANKCASE.** Cylinder and crankcase are integral castings. Standard cylinder bore diameter is 2.3745-2.3755 inches (60.31-60.33 mm). If cylinder bore exceeds 2.378 inches (60.40 mm), cylinder taper exceeds 0.003 inch (0.076 mm) or if cylinder is out-of-round more than 0.005 inch (0.13 mm), recondition cylinder bore to nearest oversize for which piston and rings are available.

**CRANKSHAFT, MAIN BEARINGS AND SEALS.** The crankshaft is supported at each end by a ball bearing type main bearing. Renew bearings (14 – Fig. K9) if excessively rough or loose. Crankshaft end play should be 0.0038-0.0228 inch (0.096-0.579 mm) and is adjusted by varying number and thickness of shim gaskets (16).

Standard crankpin journal diameter is 0.9355-0.9360 inch (23.762-23.774 mm) and minimum diameter is 0.9350 inch (23.75 mm). Maximum crankpin journal out-of-round is 0.0005 inch (0.013 mm) and maximum journal taper is 0.001 inch (0.025 mm). If crankpin dimensions are not as specified, renew or recondition crankshaft.

Main bearings should be a light press fit on crankshaft journals and in crankcase and bearing plate bores. If not, renew bearings and/or crankshaft or crankcase and bearing plate.

Front and rear crankshaft oil seals should be pressed into seal bores so outside edge of seal is 1/32 inch (0.80 mm) below seal bore surface.

Make certain crankshaft gear and camshaft gear timing marks are aligned (Fig. K10) as crankshaft and camshaft are reinstalled.

**CAMSHAFT AND BEARINGS.** The hollow camshaft and integral cam gear (27—Fig. K9) rotate on pin (26). Camshaft can be removed after removing bearing plate (17) and crankshaft, then drive pin (26) out towards bearing plate side of crankcase. When reinstalling camshaft, make certain camshaft gear and crankshaft gear timing marks are aligned (Fig. K10).

Camshaft pin (26) is a press fit in closed (crankcase) side and a slip fit with 0.0005-0.0012 inch (0.013-0.030 mm) clearance in bearing plate side.

Camshaft-to-camshaft pin clearance should be 0.001-0.0025 inch (0.025-0.063 mm). Camshaft end play should be 0.005-0.020 inch (0.13-0.50 mm) and is controlled by varying number and thickness of shim washers between camshaft and bearing plate side of

crankcase.

Governor flyball retainer (28), flyballs (29), thrust cone (30) and snap ring (31) are attached to, and rotate with, the camshaft assembly.

**VALVE SYSTEM.** Clearance between valve stem and valve tappet (cold) should be 0.005-0.009 inch (0.13-0.23 mm) for the intake valve and 0.011-0.015 inch (0.28-0.38 mm) for the exhaust valve. If clearance is not as specified, remove valves. To increase clearance, shorten valve length by grinding end of stem. To reduce clearance, grind valve seat.

The exhaust valve seats on a renewable seat insert and the intake valve seat is machined directly into cylinder block surface. Valve face and seat angle is 45° and seat width should be 0.037-0.045 inch (0.94-1.14 mm) for intake and exhaust valve.

Minimum valve stem diameter is 0.2478 inch (6.29 mm) for intake valve and 0.2458 inch (6.24 mm) for exhaust valve.

Valve stem clearance in guide should be 0.0005-0.0020 inch (0.013-0.050 mm) for intake valve and 0.0020-0.0035 inch (0.050-0.089 mm) for exhaust valve. Excessive valve stem-to-guide clearance is corrected by reaming guides and installing valves with 0.005 inch (0.13 mm) oversize stems.

# KOHLER

**KOHLER COMPANY**
**Kohler, Wisconsin 53044**

| Model | No. Cyls. | Bore | Stroke | Displacement | Power Rating |
|---|---|---|---|---|---|
| K-141* | 1 | 2.875 in. (73.025 mm) | 2.5 in. (63.5 mm) | 16.22 cu. in. (266 cc) | 6.25 hp. (4.7 kW) |
| K-141** | 1 | 2.9375 in. (74.613 mm) | 2.5 in. (63.5 mm) | 16.9 cu. in. (277.7 cc) | 6.25 hp. (4.7 kW) |
| K-160 | 1 | 2.875 in. (73.025 mm) | 2.5 in. (63.5 mm) | 16.22 cu. in. (266 cc) | 7 hp. (5.2 kW) |
| K-161* | 1 | 2.875 in. (73.025 mm) | 2.5 in. (63.5 mm) | 16.22 cu. in. (266 cc) | 7 hp. (5.2 kW) |
| K-161** | 1 | 2.9375 in. (74.613 mm) | 2.5 in. (63.5 mm) | 16.9 cu. in. (277.7 cc) | 7 hp. (5.2 kW) |
| K-181 | 1 | 2.9375 in. (74.613 mm) | 2.75 in. (69.85 mm) | 18.6 cu. in. (305.4 cc) | 8 hp. (6 kW) |
| KV-181 | 1 | 2.9375 in. (74.613 mm) | 2.75 in. (69.85 mm) | 18.6 cu. in. (305.4 cc) | 8 hp. (6 kW) |
| K-241 | 1 | 3.25 in. (82.55 mm) | 2.875 in. (73.025 mm) | 23.9 cu. in. (390.8 cc) | 10 hp. (7.5 kW) |
| K-301 | 1 | 3.375 in. (85.725 mm) | 3.25 in. (82.55 mm) | 29.07 cu. in. (476.5 cc) | 12 hp. (8.9 kW) |
| K-321 | 1 | 3.5 in. (88.9 mm) | 3.25 in. (82.55 mm) | 31.27 cu. in. (512.4 cc) | 14 hp. (10.4 kW) |
| K-341 | 1 | 3.75 in. (95.25 mm) | 3.25 in. (82.55 mm) | 35.89 cu. in. (588.2 mm) | 16 hp. (11.9 kW) |

\* Designates early production (before 1970) engines.

\*\* Designates late production (after 1970) engines.

All engines in this section are one cylinder, four-cycle engines. All models except KV-181 have horizontal crankshafts. Model KV-181 has a vertical crankshaft.

Model K-141 has a ball bearing main at pto end of crankshaft and a bushing type main bearing at flywheel end. Model KV-181 has a ball bearing main at pto end of crankshaft and a needle roller bearing main at flywheel end. All remaining models have ball bearing mains at each end of crankshaft.

Connecting rod on all models rides directly on crankpin journal. All models except KV-181 are splash lubricated. Model KV-181 has an oil circulating system which lubricates upper main bearing and crankpin.

Various models may be equipped with either a battery type ignition system, a magneto type system, each with externally mounted breaker points or a solid state breakerless ignition system.

Either a side draft or an updraft carburetor is used depending on model and application.

Special engine accessories or equipment is noted by a suffix letter on the engine model number. Key to suffix letters is as follows:
C – Over-center lever operated clutch.
P – Crankcase and crankshaft machined for direct drive applications.
R – Reduction drive unit.
S – Starter-generator unit or gear drive.
T – Rewind starter.

## MAINTENANCE

**SPARK PLUG.** Recommended spark plug for K-241, K-301, K-321 and K-341 engines is Champion H10 or equivalent. All other models use Champion J8 or equivalent. Spark plug gap should be 0.025 inch (0.635 mm) for all models.

If radio noise reduction is needed use Champion EH10 or equivalent for Model K-241, K-301, K-321 and K-341 engines and Champion EJ8 or equivalent in remaining models. Spark plug gap should be 0.020 inch (0.508 mm) for all models.

**CARBURETOR.** Model K-141 engines are equipped with a Tillotson "E" series updraft carburetor. Early production K-161, K-181, K-241 and K-301 engines are equipped with Carter "N" model side draft carburetors. Late production K-161, K-181, K-241 and K-301 and all KV-181, K-321 and K-341 engines are equipped with Kohler side draft carburetors. Refer to appropriate paragraph for carburetor service information.

CARTER CARBURETOR. Refer to Fig. KO1 for exploded view of typical "N" model Carter carburetor.

For initial adjustment, open idle fuel needle 1½ turns and open main fuel needle 2 turns. Make final adjustment with engine at normal operating temperature and running. Place engine under load and adjust main fuel needle for leanest setting that will allow satisfactory acceleration and steady governor operation. Set engine at idle speed, no load and adjust idle mixture screw to obtain smoothest idle operation.

As each adjustment affects the other, adjustment procedure may have to be repeated.

To check float level, invert carburetor throttle body and float assembly. There should be 13/64-inch (5.159 mm) clearance between free side of float and machined surface of body casting. Carefully bend float lever tang that contacts inlet valve as necessary to provide correct measurement.

KOHLER CARBURETOR. Refer to Fig. KO2 for exploded view of Kohler carburetor. For initial adjustment, open main fuel needle 2 turns and open idle fuel needle 1¼ turns. Make final adjustment with engine at normal operating temperature and running. Place engine under load and adjust main fuel needle to leanest mixture that will allow satisfactory acceleration and steady governor operation.

Adjust idle speed stop screw to maintain an idle speed of 1000 rpm, then adjust idle fuel needle for smoothest idle operation. As each adjustment affects the other, adjustment procedure may have to be repeated.

To check float level, invert carburetor body and float assembly. There should be 11/64-inch (4.366 mm) clearance between machined surface of body casting and free end of float. Carefully bend float lever tang that contacts inlet valve as necessary to provide correct measurement.

TILLOTSON CARBURETOR. Refer to Fig. KO3 for exploded view of Tillotson "E" series carburetor similar to that used on Model K-141 engine. Design of choke shaft lever will differ from that shown in exploded view.

For initial adjustment, open idle fuel mixture needle ¾-turn and open main fuel mixture needle 1 turn. Make final adjustment with engine at normal operating temperature and running. Place engine under load and adjust main fuel needle to leanest mixture that will allow satisfactory acceleration and steady governor operation. Slow engine to idle speed and adjust idle mixture screw to obtain smoothest idle operation. As each adjustment affects the other, adjustment procedure may have to be repeated.

To check float level, invert carburetor cover and float assembly. There should be 1-5/64 inches (27.384 mm) clearance between free side of float and machined surface of cover. See Fig. KO4. Carefully bend float lever tang that contacts inlet valve as necessary to provide correct measurement.

AUTOMATIC CHOKE. Some models equipped with Kohler or Carter carburetors are also equipped with an automatic choke. "Thermostatic" type shown in Fig. KO5 may be used on either electric start or manual start models. "Electric-Thermostatic" type shown in Fig. KO6 can be used only on electric start models. Automatic choke adjustment should be made with engine cold.

THERMOSTATIC TYPE. To adjust "Thermostatic" type, loosen lock screw (Fig. KO5) and rotate adjustment bracket as necessary to obtain correct amount of choking. Tighten lock screw. In cold temperature, choke should be closed. At 70° F. (21° C) choke should be partially open and choke lever should be in vertical position. Start engine and allow it to run until normal operating temperature is reached. Choke should be fully open when engine is at normal operating temperature.

ELECTRIC-THERMOSTATIC TYPE. To adjust "Electric-Thermostatic" type, refer to Fig. KO6 and move choke lever arm until hole in brass cross shaft is aligned with slot in bearings. Insert a No. 43 (0.089 inch) drill through

shaft until drill is engaged in notch in base of choke. Loosen clamp bolt on lever arm and move link end of lever arm upward until choke disc is closed to desired position. Tighten clamp bolt while holding lever arm in this position. Remove drill. Choke should be fully closed when starting a cold engine and should be fully open when engine is running at normal operating temperature.

GOVERNOR. All models are equipped with a gear driven flyweight governor that is located inside engine crankcase. Maximum recommended governed engine speed is 3800 rpm on KV-181 engines and 3600 rpm for all other engines. Recommended idle speed of 1000 rpm is controlled by adjustment of throttle stop screw on carburetor.

Before attempting to adjust engine governed speed, synchronize governor linkage as follows: On Models K-141,

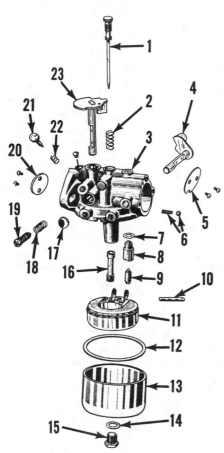

Fig. KO1—Exploded view of typical Carter "N" model carburetor used on early production K-160, K-161, K-181, K-241 and K-301 engines.

| 1. Main fuel needle | 13. Float bowl |
|---|---|
| 2. Spring | 14. Sealing washer |
| 3. Carburetor body | 15. Retainer |
| 4. Choke shaft | 16. Main jet |
| 5. Choke disc | 17. Plug |
| 6. Choke detent | 18. Spring |
| 7. Sealing washer | 19. Idle stop screw |
| 8. Inlet valve seat | 20. Throttle disc |
| 9. Inlet valve | 21. Idle fuel needle |
| 10. Float pin | 22. Spring |
| 11. Float | 23. Throttle shaft |
| 12. Gasket | |

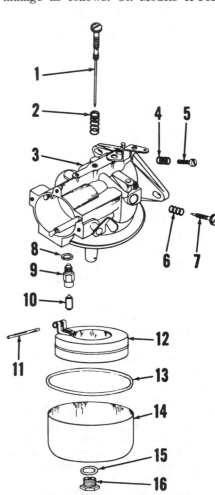

Fig. KO2—Exploded view of Kohler carburetor used on late production K-161, K-181, K-241 and K-301 and all KV-181, K-321 and K-341 engines.

| 1. Main fuel needle | 9. Inlet valve seat |
|---|---|
| 2. Spring | 10. Inlet valve |
| 3. Carburetor body assy. | 11. Float pin |
| 4. Spring | 12. Float |
| 5. Idle speed stop screw | 13. Gasket |
| 6. Spring | 14. Float bowl |
| 7. Idle fuel needle | 15. Gasket |
| 8. Sealing washer | 16. Bowl retainer |

Illustrations courtesy of Kohler Co.

K-160, K-161 and K-181, loosen bolt clamping governor arm (G–Fig. KO7) to governor cross shaft (F) and turn governor cross shaft counter-clockwise as far as possible. While holding cross shaft in this position, move governor arm away from carburetor to limit of linkage travel and tighten clamping bolt.

To synchronize governor linkage on KV-181, follow procedure for K-141, K-160, K-161 and K-181 and refer to Fig. KO9.

On Models K-241, K-301, K-321 and K-341, refer to Fig. KO10 and loosen governor arm hex nut. Rotate governor cross shaft counter-clockwise as far as possible, move governor arm away from carburetor to limit of linkage travel, then tighten hex nut on arm.

To adjust maximum governed speed on Models K-241, K-301, K-321 and K-341, adjust position of stop so speed control lever contacts stop at desired engine speed. On Models K-141, K-160, K-161 and K-181, loosen governor shaft bushing nut (C-Fig. KO7) and move throttle bracket (E) to increase or decrease governed speed. See Fig. KO8 for direction of movement. Model KV-181 governed speed is adjusted by loosening set screw in speed control bracket and moving high speed stop shown in Fig. KO9. Adjust throttle cable so control handle is in full throttle position when drive pin in throttle is contacting bracket.

On Models K-241, K-301, K-321 and K-341, governor sensitivity is adjusted by moving governor spring to alternate holes in governor arm and speed lever.

Governor unit is accessible after removing engine crankshaft and camshaft. Governor gear and flyweight assembly turns on a stub shaft pressed into engine crankcase. Renew gear and weight assembly if gear teeth or any part of assembly is excessively worn. Desired clearance of governor gear to stud shaft is 0.0025-0.0055 inch (0.0635-0.1397 mm) for Models K-141, K-160, K-161 and K-181 and

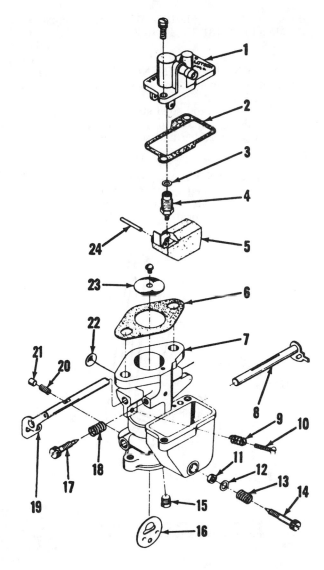

Fig. KO3 – Exploded view of Tillotson "E" series carburetor used on K-141 engine. Refer to Fig. KO4 for method of checking float level.

1. Float bowl cover
2. Gasket
3. Gasket
4. Inlet valve assy.
5. Float
6. Gasket
7. Carburetor body
8. Choke shaft
9. Spring
10. Idle speed stop screw
11. Packing
12. Washer
13. Spring
14. Main fuel needle
15. Screw plug
16. Choke disc
17. Idle fuel needle
18. Spring
19. Throttle shaft
20. Spring
21. Choke shaft detent
22. Plug
23. Throttle disc
24. Float pin

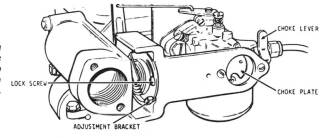

Fig. KO5 – View showing "Thermostatic" automatic choke used on some engines equipped with Kohler or Carter carburetors.

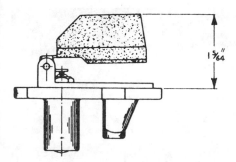

Fig. KO4 – Check float setting on Tillotson carburetor by inverting throttle body and float assembly and measuring distance from free end of float to edge of cover as shown. Float setting is correct if distance is 1-5/64 inch (27.384 mm).

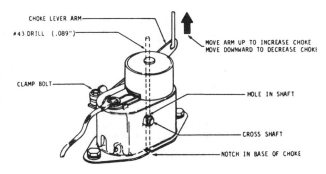

Fig. KO6 – View of "Electric-Thermostatic" automatic choke used on some engines equipped with Kohler or Carter carburetor.

Illustrations courtesy of Kohler Co.

0.0005-0.002 inch (0.0127-0.0508 mm) for Models KV-181, K-241, K-301, K-321 and K-341.

**IGNITION AND TIMING.** Three types of ignition systems used are as follows: Battery ignition, flywheel magneto ignition and solid state breakerless ignition. Refer to appropriate paragraphs for timing and service procedures.

BATTERY AND MAGNETO IGNITION. On engines equipped with either battery or magneto ignition systems, breaker points are located externally on engine crankcase as shown in Fig. KO11. Breaker points are actuated by a cam through a push rod.

On early models, breaker cam is driven by camshaft gear through an automatic advance mechanism as shown in Fig. KO12. At cranking speed, ignition occurs at 3° BTDC on Models K-141, K-160, K-161 and K-181 or 3° ATDC on Models K-241 and K-301. At operating speeds, centrifugal force of the advance weights overcomes spring

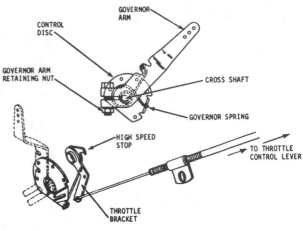

Fig. KO9 — View of governor linkage used on KV-181 model.

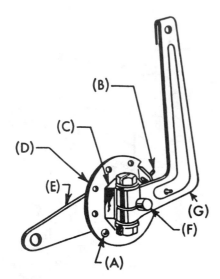

Fig. KO7 — Drawing showing governor lever and related parts on K-141, K-160, K-161 and K-181 models.

A. Drive pin
B. Governor spring
C. Bushing nut
D. Speed control disc
E. Throttle bracket
F. Governor Shaft
G. Governor lever

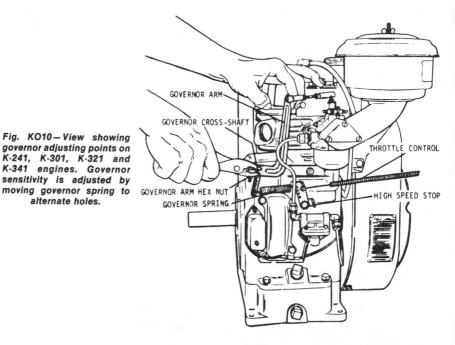

Fig. KO10 — View showing governor adjusting points on K-241, K-301, K-321 and K-341 engines. Governor sensitivity is adjusted by moving governor spring to alternate holes.

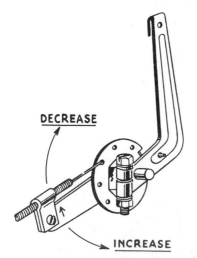

Fig. KO8 — On K-141, K-160, K-161 and K-181, loosen bushing nut (C — Fig. KO7) and move throttle bracket as shown to change governed speed of engine.

Fig. KO11 — Ignition breaker points are located externally on engine crankcase on all models equipped with magneto or battery ignition systems.

Fig. KO12—View showing alignment marks on ignition breaker cam and camshaft gear on early models equipped with automatic timing advance.

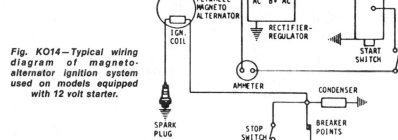

Fig. KO14—Typical wiring diagram of magneto-alternator ignition system used on models equipped with 12 volt starter.

tension and advances timing of breaker cam so ignition occurs at 20° BTDC on all models.

Late Model K-141, K-160, K-161, K-181, K-241 and K-301 and all Models K 321 and K-341 engines are equipped with an automatic compression release (see **CAMSHAFT** paragraph) and do not have an automatic timing advance. Ignition occurs at 20° BTDC at all engine speeds.

Initial breaker point gap on Model KV-181 is 0.015 inch (0.3810 mm) and is 0.020 inch (0.508 mm) for all remaining models. Breaker point gap should be varied to obtain exact ignition timing as follows: With a static timing light, disconnect coil and condenser leads from breaker point terminal and attach one timing light lead to terminal and ground remaining lead. Remove button plug from timing sight hole and turn engine so piston has just completed its compression stroke and "DC" mark (models with automatic compression release) on flywheel appears in sight hole. Loosen breaker point adjustment screw and adjust points to closed position (timing light will be on); slowly move breaker point base towards engine until timing light goes out. Tighten adjustment screw and check point setting by turning engine in normal direction of rotation until timing light goes on. Continue to turn engine very slowly until timing light goes out. "DC" mark (models with automatic timing advance) or "SP" mark (models with automatic compression release) should now be in register with

sight hole. Disconnect timing light leads, connect leads from coil and condenser to breaker point terminal and install breaker cover.

If a power timing light is available, more accurate timing can be obtained by adjusting points with engine running. Breaker point gap should be adjusted so

light flash causes "SP" timing mark (20° BTDC) to appear centered in sight hole when engine is running above 1500 rpm, when checking breaker point setting with a power timing light on models with automatic timing advance.

Typical wiring diagrams of magneto ignition systems are shown in Figs.

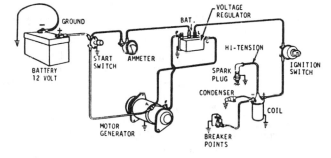

Fig. KO15—Typical wiring diagram of battery ignition system used on models equipped with 12 volt starter-generator.

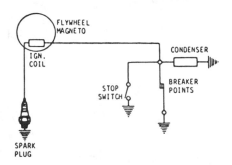

Fig. KO13—Typical wiring diagram of magneto ignition system used on manual start engines.

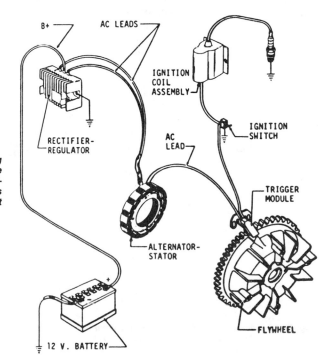

Fig. KO16—Typical wiring diagram of solid state breakerless-alternator ignition system used on models equipped with 12 volt starter.

# Kohler

KO13 and KO14. Fig. KO15 is typical wiring diagram of battery ignition system.

SOLID STATE BREAKERLESS IGNITION. Breakerless-alternator ignition system uses solid state devices which eliminate need for mechanically operated breaker points. Ignition timing is non-adjustable. The only adjustment is trigger module to flywheel trigger projection air gap. To adjust air gap, rotate flywheel until projection is adjacent to trigger module. Loosen trigger retaining screws and move trigger until an air gap of 0.005-0.010 inch (0.127-0.254 mm) is obtained. Make certain flat surfaces on trigger and projection are parallel, then tighten retaining screws. Refer to Fig. KO16 for typical wiring diagram of breakerless ignition system.

The four main components of a breakerless ignition system are ignition winding (on alternator stator), trigger module, ignition coil assembly and special flywheel with trigger projection. System also includes a conventional spark plug, high tension lead and ignition switch. Ignition winding is separate from battery charging AC windings on alternator stator.

Trigger module includes three diodes, a resistor, a sensing coil and magnet plus the SCR (electronic switch). Trigger module has two clip-on type terminals. See Fig. KO17. Terminal marked "A" must be connected to ignition winding on alternator stator. "I" terminal must be connected to ignition switch. Improper connection will cause damage to electronic devices.

Ignition coil assembly includes capacitor and a pulse transformer arrangement similar to a conventional high tension coil. Flywheel has a special projection for triggering ignition.

If ignition trouble exists and spark plug is known to be good, the following tests should be made to determine which component is at fault. A flashlight type continuity tester can be used to test ignition coil assembly and trigger module.

To test coil assembly, remove high tension lead (spark plug wire) from coil. Insert one tester lead in coil terminal and remaining tester lead to coil mounting bracket. Tester light should be on. Connect one tester lead to coil mounting bracket and remaining tester lead to ignition switch wire of coil. Tester light should be out. Renew ignition coil if either test shows wrong results.

To test trigger module, connect one tester lead to AC inlet (A – Fig. KO17) and remaining lead to ignition lead terminal (I). Check for continuity. Reverse tester leads and again check for continuity. Test light must be on in one test only. The second test is made by connecting one tester lead to trigger module mounting bracket and other lead to AC inlet (A). Check for continuity. Reverse tester leads and again check for continuity. Again, test light must be on in one test only. The third test is made by connecting POSITIVE lead of tester to (I) terminal of trigger and connect remaining tester lead to trigger module mounting bracket. Test light should be off. Rotate flywheel in normal direction of rotation. When trigger projection on flywheel passes trigger module, test light should turn on. Renew trigger module if any of the three tests show wrong results.

If ignition trouble still exists after ignition coil assembly and trigger module tests show they are good, ignition winding on alternator stator is faulty. In this event, renew stator.

**LUBRICATION.** All models are splash lubricated except Model KV-181 which has an oil circulating system. See Fig. KO18. This vertical crankshaft engine has two oil sumps. Outer sump (7) is oil reservoir and inner sump (13) is a low level oil drain back sump. Oil is drawn from outer sump (7) through main gallery (6), oil passage (5) in crankcase and passage (4) in bearing plate. After lubricating upper main bearing (3), oil flows through thrust bearing (2) which acts as a centrifugal oil pump and is forced through tapered hole in crankpin (9) to lubricate connecting rod. Throw-off oil lubricates piston, camshaft, gears and lower main bearing (14). This drain back oil is then taken from inner sump (13) through oil pick-up assembly (10) and reed valve (11) and returned via gallery (12) to outer sump (7).

Maintain crankcase oil level at full mark on dipstick, but do not overfill. High quality detergent oil having API classification "SF" is recommended.

Use SAE 30 oil in temperatures above 32° F. (0° C), SAE 10W-30 oil in temperatures between 32° F (0° C) and 0° F (-18° C) and SAE 5W-20 oil in temperatures below 0° F (-18° C).

**CRANKCASE BREATHER.** Refer to Figs. KO19 and KO20. A reed valve assembly is located in valve spring compartment to maintain a partial vacuum in crankcase and thus reduce leakage of oil at bearing seals. If a slight amount of crankcase vacuum (5-10 inches water or ½-1 inch mercury) is not present, reed valve is faulty or engine has excessive blow-by past rings and/or valves or worn oil seals.

**AIR CLEARNER.** Engines may be

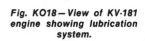

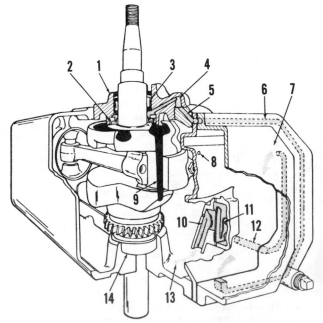

**Fig. KO18 – View of KV-181 engine showing lubrication system.**

1. Bearing plate
2. Thrust bearing
3. Upper main bearing (needle)
4. Oil passage in bearing plate
5. Oil passage in crankcase
6. Main oil gallery
7. Outer sump
8. Vent hole (0.030 inch)
9. Tapered hole in crankpin
10. Oil pick-up assy.
11. Reed valve
12. Oil return gallery
13. Inner sump
14. Lower main bearing (ball)

*Fig. KO17 – Wires must be connected to trigger module as shown. Reversing connections will damage electronic devices.*

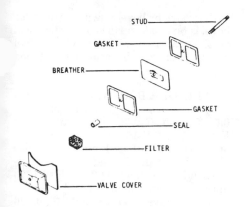

*Fig. KO19—Exploded view of crankcase breather assembly used on K-141, K-160, K-161, K-181 and KV-181 models.*

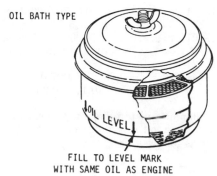

FILL TO LEVEL MARK
WITH SAME OIL AS ENGINE

*Fig. KO21-View of oil bath type air cleaner.*

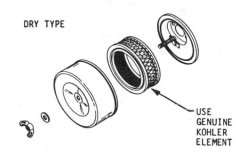

USE
GENUINE
KOHLER
ELEMENT

*Fig. KO22—Exploded view of dry element type air cleaner.*

equipped with either an oil bath type air cleaner shown in Fig. KO21 or a dry element type shown in Fig. KO22.

Oil bath air cleaner should be serviced every 25 hours of normal operation or more often if operating in extremely dusty conditions. To service oil bath type, remove complete cleaner assembly from engine. Remove cover and element from bowl. Empty used oil from bowl, then clean cover and bowl in solvent. Clean element in solvent and allow to drip dry. Lightly re-oil element. Fill bowl to oil level marked on bowl using same grade and weight oil as used in engine crankcase. Renew gaskets as necessary when reassembling and reinstalling unit.

Dry element type air cleaner should be cleaned every 100 hours of normal operation or more frequently if operating in dusty conditions. Remove dry element and tap element lightly on a flat surface to remove surface dirt. Do not wash element or attempt to clean element with compressed air. Renew element if extremely dirty or if it is bent, crushed or otherwise damaged. Make certain sealing surfaces of element seal effectively against back plate and cover.

## REPAIRS

**TIGHTENING TORQUES.** Recom-

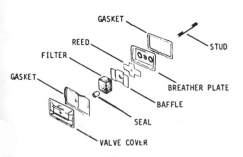

*Fig. KO20—Exploded view of crankcase breather assembly used on K-241, K-301, K-321 and K-341 models.*

mended tightening torques are as follows:

Spark plug . . . . . . . . . . . . . . .22 ft.-lbs.
(30 N·m)
Connecting rod
cap screws* . . . . .See CONNECTING
ROD section
Cylinder head
cap screws*
K-141,
K-160,
K-161,
K-181,
KV-181 . . . . . . . . . . . . . . . .20 ft.-lbs.
(27 N·m)
K-241,
K-301,
K-321,
K-341 . . . . . . . . . . . . . . . . .30 ft.-lbs.
(41 N·m)
Flywheel retaining nut
Models with ⅝-inch nut . . . .60 ft.-lbs.
(81 N·m)
Models with ¾-inch nut . . .100 ft.-lbs.
(136 N·m)
*With threads lubricated.

**CONNECTING ROD.** Connecting rod and piston unit is removed after removing oil pan and cylinder head. Aluminum alloy connecting rod rides directly on crankpin. Connecting rod with 0.010 inch (0.254 mm) undersize crankpin bore is available for reground crankshaft. Oversize piston pins are available on some models. Desired running clearances are as follows:

Connecting rod
to crankpin . . . . . . . . . .0.001-0.002 in.
(0.0254-0.0508 mm)
Connecting rod to piston pin,
K-141,
K-160,
K-161,
K-181,
KV-181 . . . . . . . . . .0.0006-0.0011 in.
(0.0152-0.0279 mm)
K-241,
K-301,
K-321,
K-341 . . . . . . . . . .0.0003-0.0008 in.
(0.0076-0.0203 mm)

Rod side play
on crankpin
K-141,
K-160,
K-161,
K-181,
KV-181 . . . . . . . . . . . . .0.005-0.016 in.
(0.1270-0.4064 mm)
K-241,
K-301,
K-321,
K-341 . . . . . . . . . . . . . .0.007-0.016 in.
(0.1778-0.4064 mm)

Standard crankpin diameter is 1.1855-1.1860 inches (30.1117-30.1244 mm) on Models K-141, K-160, K-161, K-181 and KV 181 and 1.4995-1.5000 inches (38.0873-38.1000 mm) on Models K-241, K-301, K-321 and K-341.

Assemble piston to rod so arrow on piston (when so marked) faces away from valves and match marks (Fig. KO23) on rod and cap are aligned and towards camshaft side of engine. Always use new retainer rings to secure piston pin in piston.

Kohler recommends connecting rod cap screws be overtorqued to one specification, loosened, then tightened

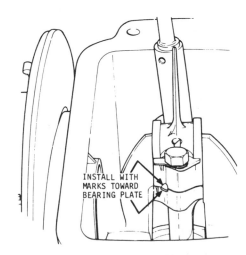

INSTALL WITH
MARKS TOWARD
BEARING PLATE

*Fig. KO23—When installing connecting rod and piston unit, be sure marks on rod and cap are aligned and are facing toward flywheel side of engine.*

to correct torque specification. Tightening specifications are as follows:

K-141,
K-160,
K-161,
K-181,
KV-181 (Overtorque) . . . . . . . .20 ft.-lbs.
(27 N·m)
(Final torque) . . . . . . . . . . . . .17 ft.-lbs.
(23 N·m)
K-241,
K-301,
K-321,
K-341 (Overtorque) . . . . . . . .28 ft.-lbs.
(38 N·m)
(Final torque) . . . . . . . . . . . .24 ft.-lbs.
(32 N·m)

**NOTE: Torque specifications for connecting rod cap screws are with lubricated threads.**

**PISTON, PIN AND RINGS.** Aluminum alloy piston is fitted with two compression rings and one oil control ring. Renew piston if scored or if side clearance of new ring in piston top groove exceeds 0.005 inch (0.1270 mm). Pistons and rings are available in oversizes of 0.010, 0.020 and 0.030 inch (0.254, 0.508 and 0.762 mm) as well as standard sizes.

Piston pin fit in piston bore should be from 0.0001 inch (0.0025 mm) interference to 0.0003 inch (0.0076 mm) loose on Models K-141, K-160, K-161, K-181 and KV-181 and 0.0002-0.0003 inch (0.0051-0.0076 mm) loose on Models K-241, K-301, K-321 and K-341. Standard piston pin diameter is 0.6248 inch (15.8699 mm) on Models K-141, K-160, K-161, K-181 and KV-181, 0.8592 inch (21.8237 mm) on Model K-241 and 0.8753 inch (22.2326 mm) on Models K-301, K-321 and K-341. Piston pins are available in oversize of 0.005 inch (0.127 mm) on some models. Always renew piston pin retaining rings.

Recommended piston to cylinder bore clearance measured at thrust side at bottom of skirt is as follows:

K-141, K-160,
K-161, K-181,
KV-181 . . . . . . . . . . . . .0.0045-0.0065 in.
(0.1143-0.1651 mm)
K-241, K-301 . . . . . . . . . .0.003-0.004 in.
(0.0762-0.1016 mm)
K-321, K-341 . . . . . . .0.0035-0.0045 in.
(0.0889-0.1143 mm)

Recommended piston to cylinder bore clearance measured at thrust side just below oil ring is as follows:

K-141, K-160,
K-161, K-181,
KV-181 . . . . . . . . . . . . . .0.006-0.0075 in.
(0.1524-0.1905 mm)

K-241 . . . . . . . . . . . . . . .0.0075-0.0085 in.
(0.1905-0.2159 mm)
K-301 . . . . . . . . . . . . . .0.0065-0.0095 in.
(0.1651-0.2413 mm)
K-321,
K-341 . . . . . . . . . . . . . . . . .0.007-0.010 in.
(0.1778-0.254 mm)

Minimum piston diameters measured just below oil ring 90° from piston pin are as follows:

K-141*
K-160
K-161* . . . . . . . . . . . . . . . . . . . .2.866 in.
(72.7964 mm)
K-181
K-141**
K-161**
KV-181 . . . . . . . . . . . . . . . . . .2.9275 in.
(74.3585 mm)
K-241 . . . . . . . . . . . . . . . . . .3.2400 in.
(82.296 mm)
K-301 . . . . . . . . . . . . . . . . . .3.3640 in.
(85.4456 mm)
K-321 . . . . . . . . . . . . . . . . . .3.4885 in.
(88.6079 mm)
K-341 . . . . . . . . . . . . . . . . . .3.7385 in.
(94.9579 mm)

*Designates early production (before 1970) engines.

**Designates late production (after 1970) engines.

Renew any piston which does not meet specifications.

Kohler recommends piston rings always be renewed whenever they are removed. Piston ring specifications are as follows:

Ring end gap,
K-141, K-160,
K-161, K-181,
KV-181 . . . . . . . . . . . . .0.007-0.017 in.
(0.1778-0.4318 mm)
K-241, K-301,
K-321, K-341 . . . . . . .0.010-0.020 in.
(0.254-0.508 mm)
Ring side clearance (compression rings),
K-141, K-160,
K-161, K-181,
KV-181 . . . . . . . . . . . .0.0025-0.004 in.
(0.0635-0.1016 mm)
Ring side clearance (oil control ring),
K-141, K-160,
K-161, K-181,
KV-181 . . . . . . . . . . . .0.001-0.0025 in.
(0.0254-0.0635 mm)
K-241, K-301,
K-321, K-341 . . . . . . .0.001-0.003 in.
(0.0254-0.0762 mm)

If compression ring has a groove or bevel on outside surface, install ring

with groove or bevel down. If groove or bevel is on inside surface of compression ring, install ring with groove or bevel up. Oil control ring can be installed either side up.

**CYLINDER BLOCK.** If cylinder wall is scored or bore is tapered more than 0.0025 inch (0.0635 mm) on Models K-141, K-160, K-161, K-181 and KV-181 or 0.0015 inch (0.0381 mm) on Models K-241, K-301, K-321 and K-341, or out-of-round more than 0.005 inch (0.1270 mm), cylinder should be honed to nearest suitable oversize of 0.010, 0.020 or 0.030 inch (0.254, 0.508 or 0.762 mm) for which piston and rings are available.

Standard cylinder bore diameters are as follows:

K-141*
K-160
K-161* . . . . . . . . . . . . . . . . . . .2.875 in.
(73.025 mm)
K-141**
K-161**
K-181
KV-181 . . . . . . . . . . . . . . . . . .2.9375 in.
(74.613 mm)
K-241 . . . . . . . . . . . . . . . . . . .3.251 in.
(82.58 mm)
K-301 . . . . . . . . . . . . . . . . . . .3.375 in.
(85.725 mm)
K-321 . . . . . . . . . . . . . . . . . . .3.500 in.
(88.90 mm)
K-341 . . . . . . . . . . . . . . . . . . .3.750 in.
(95.725 mm)

*Designates early production (before 1970) engines.
**Designates late production (after 1970) engines.

**CYLINDER HEAD.** Always use a new gasket when installing cylinder head. Cylinder head should be checked for warpage by placing cleaned head on flat surface. If clearance between head and plate exceeds 0.003 inch (0.0762 mm) when checked with feeler gage, renew head. Tighten cylinder head cap screws evenly and in steps using correct sequence shown in Fig. KO24, KO25 or KO25A.

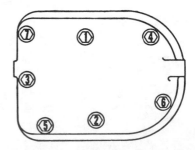

*Fig. KO24—On K-141, K-160, K-161, K-181 and KV-181 models, tighten cylinder head cap screws evenly, in sequence shown, to 15-20 ft.-lbs. (20-27 N·m) torque.*

Illustrations courtesy of Kohler Co.

**CRANKSHAFT.** Crankshaft on standard K-141 engine is supported by ball bearing in crankcase at pto end of shaft and a bushing type bearing in bearing plate at flywheel end. Model KV-181 has a roller bearing at flywheel end of crankshaft and a ball bearing at opposite end. Crankshaft on all other models is supported in two ball bearings.

On Model K-141 engine with bushing type main bearing, renew crankshaft and/or bearing plate if crankshaft journal and bearing are excessively worn or scored. Recommended crankshaft journal to bushing running clearance is 0.001-0.0025 inch (0.0254-0.0635 mm).

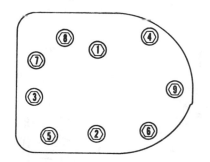

*Fig. KO25—On K-241, K-301 and K-321 models, tighten cylinder head cap screws evenly, in sequence shown, to 25-30 ft.-lbs. (34-41 N·m) torque.*

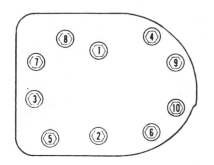

*Fig. KO25A—On K-341 models, tighten cylinder head cap screws evenly, in sequence shown, to 25-30 ft.-lbs. (34-41 N·m) torque.*

On all models, renew ball bearing type mains if excessively loose or rough. Crankshaft end play should be 0.003-0.028 inch (0.0762-0.7112 mm) on KV-181 models, 0.003-0.020 inch (0.0762-0.508 mm) on K-241, K-301, K-321 and K-341 models, and 0.002-0.023 inch (0.0508-0.5842 mm) on all other models. End play is controlled by thickness of bearing plate gaskets. Gaskets are available in thicknesses of 0.010 and 0.020 inch. Install ball bearing mains with sealed side towards crankpin.

Crankpin journal may be reground to 0.010 inch (0.254 mm) undersize for use of undersize connecting rod if journal is scored or out-of-round. Standard crankpin diameter is 1.1855-1.1860 inch (30.1117-30.1244 mm) on Models K-141, K-160, K-161, K-181 and KV-181 and 1.4995-1.500 inch (38.0873-38.10 mm) on Models K-241, K-301, K-321 and K-341.

When installing crankshaft, align timing marks on crankshaft and camshaft gear as shown in Fig. KO26.

**NOTE: On Models K-241, K-301, K-321 and K-341 equipped with dynamic balancer, refer to DYNAMIC BALANCER paragraph for installation and timing of balancer gears.**

On all models, Kohler recommends crankshaft seals be installed in crankcase and bearing plate after crankshaft and bearing plate are installed. Carefully work oil seals over crankshaft and drive seals into place with hollow driver that contacts outer edge of seals.

**CAMSHAFT.** The hollow camshaft and integral camshaft gear turn on a pin that is a slip fit in flywheel side of crankcase and a drive fit in closed side of crankcase. Remove and install pin from open side (bearing plate side) of crankcase. Desired camshaft to pin running clearance is 0.0005-0.003 inch (0.0127-0.0762 mm) on Models K-141, K-160, K-161, K-181 and KV-181 and 0.001-0.0035 inch (0.0254-0.0889 mm) on

all other models. Desired camshaft end play of 0.005-0.010 inch (0.1270-0.254 mm) is controlled by use of 0.005 and 0.010 inch thick spacer washers between camshaft and cylinder block at bearing plate side of crankcase.

On early models, ignition spark advance weights, springs and weight pivot pins are renewable separately from camshaft gear. When reinstalling camshaft in engine, be sure breaker cam is cor-

*Fig. KO27—Views showing operation of camshaft with automatic compression release. In view 1, spring (C) has moved control lever (D) which moves cam lever (B) upward so tang (T) is above exhaust cam lobe. This tang holds exhaust valve open slightly on a portion of compression stroke to relieve compression while cranking engine. At engine speeds of 650 rpm or more, centrifugal force moves control lever (D) outward allowing tang (T) to move below lobe surface as shown in view 2.*

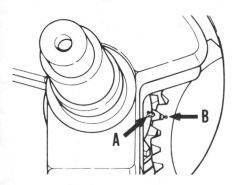

*Fig. KO26—When installing crankshaft, make certain timing mark (A) on crankshaft is aligned with timing mark (B) on camshaft gear.*

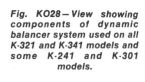

*Fig. KO28—View showing components of dynamic balancer system used on all K-321 and K-341 models and some K-241 and K-301 models.*

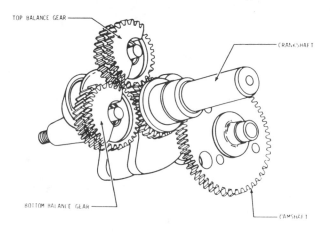

rectly installed. Spread springs as shown in Fig. KO12 and install breaker cam on tangs of flyweights with timing mark on cam and camshaft gear aligned.

Late production camshafts do not have automatic timing advance mechanism shown in Fig. KO12, but are equipped with automatic compression release mechanism shown in Fig. KO27. The automatic compression release mechanism holds exhaust valve slightly open during first part of compression stroke, reducing compression pressure and allowing easier cranking of engine. Refer to Fig. KO27 for operational details. At speeds above 650 engine rpm, compression release mechanism is inactive. Service procedures remain the same as for early production camshaft units except for the difference in timing advance and compression release mechanisms. Service kits are available for adding automatic compression release to early production engines.

To check compression on engine equipped with automatic compression release, engine must be cranked at 650 rpm or higher to overcome compression

release action. A reading can also be obtained by rotating flywheel in reverse direction with throttle wide open. Compression reading should be 110-120 psi (758-827kPa) on an engine in top mechanical condition. When compression reading falls below 100 psi (689 kPa), it indicates leaking rings or valves.

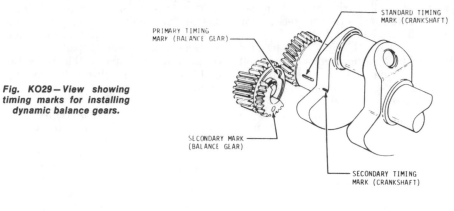

Fig. KO29—View showing timing marks for installing dynamic balance gears.

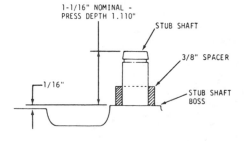

Fig. KO28A—If stub shaft boss is approximately 7/16-inch (10.113 mm) above main bearing boss, press new shaft in until it is 0.735 inch (18.669 mm) above stub shaft boss.

Fig. KO28B—If stub shaft boss is approximately 1/16-inch (1.588 mm) above main bearing boss, press new shaft in until it is 1.110 inch (28.194 mm) above stub shaft boss and use a 3/8-inch spacer between block and gear.

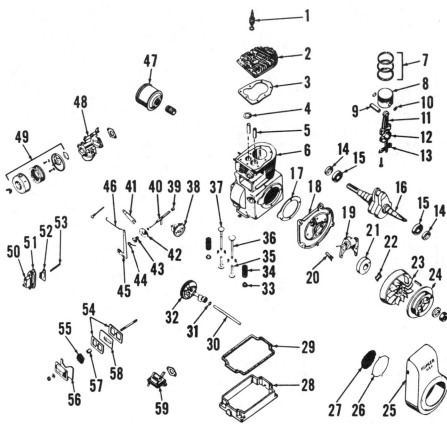

Fig. KO30—Exploded view of K-181 model basic engine assembly. Models K-141, K-160 and K-161 are similar. Standard K-141 engine is equipped with a bushing type main bearing in bearing plate (18).

1. Spark plug
2. Cylinder head
3. Head gasket
4. Exhaust valve seat
5. Valve guide
6. Cylinder block
7. Piston rings
8. Piston
9. Piston pin
10. Retaining rings
11. Connecting rod
12. Rod cap
13. Rod bolt lock
14. Oil seal
15. Ball bearing
16. Crankshaft
17. Gasket
18. Bearing plate
19. Magneto
20. Condenser
21. Magneto rotor
22. Wave washer
23. Flywheel
24. Pulley
25. Shroud
26. Screen retainer
27. Screen
28. Oil pan
29. Gasket
30. Camshaft pin
31. Shim washer
32. Camshaft
33. Spring retainer
34. Valve spring
35. Valve tappet
36. Intake valve
37. Exhaust valve
38. Governor gear & weight assy.
39. Needle bearing
40. Governor shaft
41. Bracket
42. Speed disc
43. Bushing
44. Governor spring
45. Governor lever
46. Link
47. Muffler
48. Carburetor
49. Air cleaner assy.
50. Breaker cover
51. Gasket
52. Breaker points
53. Push rod
54. Gaskets
55. Filter
56. Valve cover
57. Breather seal
58. Reed plate
59. Fuel pump

Illustrations courtesy of Kohler Co.

**VALVE SYSTEM.** Valve tappet gap (cold) is as follows:
Models K-241, K-301, K-321 and K-341
  Intake . . . . . . . . . . . . . .0.008-0.010 in.
                  (0.2032-0.254 mm)
  Exhaust . . . . . . . . . . .0.017-0.020 in.
                  (0.4318-0.508 mm)
Models K-141, K-160, K-161,
  K-181 and KV-181
  Intake . . . . . . . . . . . . .0.006-0.008 in.
                  (0.1524-0.2032 mm)
  Exhaust . . . . . . . . . . .0.015-0.017 in.
                  (0.3810-0.4318 mm)

Correct valve tappet gap is obtained by grinding ends of valve stems on Models K-141, K-160, K-161, K-181 and KV-181. Be sure to grind end square and remove all burrs from end of stem after grinding. Models K-241, K-301, K-321 and K-341 have adjustable tappets.

The exhaust valve seats on a renewable seat insert on all models. Intake valve seats directly on a machined seat in cylinder block on some models. On all Models K-321 and K-341 engines and some other models, a renewable intake valve seat insert is used. Valve face and seat angle is 45°. Desired seat width is 1/32 to 1/16-inch (0.794 to 1.588 mm).

Renewable valve guides are used on all models. Intake valve stem to guide clearance should be 0.001-0.0025 inch (0.0254-0.0635 mm) and exhaust valve stem to guide clearance should be 0.0025-0.004 inch (0.0635-0.1016 mm) on all models. Ream valve guides after installation to obtain correct inside diameter of 0.312-0.313 inch (7.9248-7.9502 mm).

**DYNAMIC BALANCER.** A dynamic balance system (Fig. KO28) is used on some K-241 and K-301 engines and on all K-321 and K-341 engines. The two balance gears, equipped with needle bearings, rotate on two stub shafts which are pressed into bosses on pto side of crankcase. Snap rings secure gears on stub shafts and shim spacers are used to control gear end play. Balance gears are driven by crankshaft in opposite direction of crankshaft rotation. Use following procedure to install and time dynamic balancer components.

To renew stub shafts, press old shafts out and discard. If stub shaft boss is approximately 7/16-inch (10.113 mm) above main bearing boss (see Fig. KO28A), press new shaft in until it is 0.735 inch (18.669 mm) above stub shaft boss. On blocks where stub shaft boss is approximately 1/16-inch (1.588 mm) above main bearing boss (Fig. KO28B), press new shaft in until it is 1.110 inch (27.19 mm) above stub shaft boss and use a 3/8-inch (9.525 mm) spacer between block and gear.

To install top balance gear-bearing assembly, first place one 0.010 inch shim on stub shaft, install top gear assembly on shaft making certain timing marks are facing flywheel side of crankcase. Install one 0.005 inch, one 0.010 inch and one 0.020 inch shim spacers in this order and install snap ring. Using a feeler gage, check gear end play. Proper end play of balance gear is 0.005-0.010 inch.

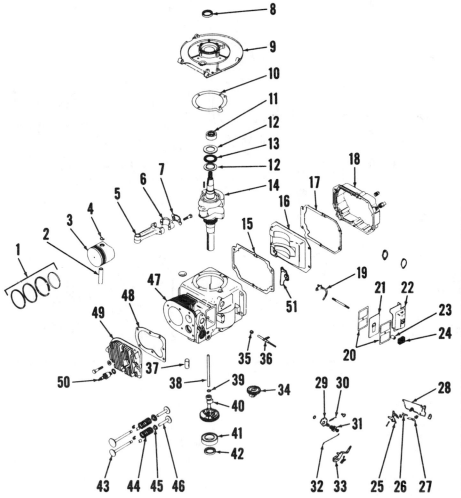

1. Piston rings
2. Piston pin
3. Piston
4. Retaining rings
5. Connecting rod
6. Rod cap
7. Rod bolt lock
8. Oil seal
9. Bearing plate
10. Gasket
11. Needle bearing
12. Thrust washers
13. Thrust bearing
14. Crankshaft
15. Gasket
16. Inner sump cover
17. Gasket
18. Outer sump
19. Breather drain tube
20. Gaskets
21. Reed plate
22. Valve cover
23. Breather seal
24. Filter
25. Throttle lever
26. Speed control
27. Stop switch
28. Bracket
29. Speed disc
30. Throttle link
31. Bushing
32. Governor link
33. Governor lever
34. Governor gear & weight assy.
35. Needle bearing
36. Governor cross-shaft
37. Governor shaft
38. Camshaft pin
39. Shim washer
40. Camshaft assembly
41. Ball bearing
42. Oil seal
43. Valve
44. Valve spring
45. Spring retainer
46. Valve tappet
47. Cylinder block
48. Head gasket
49. Cylinder head
50. Spark plug
51. Oil pick-up assy.

*Fig. KO31 — Exploded view of KV-181 model basic engine assembly.*

Add or remove 0.005 inch thick spacers as necessary to obtain correct end play.

**NOTE: Always install the 0.020 inch thick spacer next to snap ring.**

Install crankshaft in crankcase and align primary timing mark on top balance gear with standard timing mark on crankshaft. See Fig. KO29. With primary timing marks aligned, engage crankshaft gear 1/16-inch into narrow section of top balance gear and rotate

crankshaft to align timing marks on camshaft gear and crankshaft as shown in Fig. KO26. Press crankshaft into crankcase until it is seated firmly into ball bearing main.

Rotate crankshaft until crankpin is approximately 15° past bottom dead center. Install one 0.010 inch shim on stub shaft, align secondary timing mark on bottom balance gear with secondary timing mark on crankshaft counterweight. See Fig. KO29. Install gear

assembly onto stub shaft. If properly timed, secondary timing mark on bottom balance gear will be aligned with standard timing mark on crankshaft after gear is fully on stub shaft. Install one 0.005 inch and one 0.020 inch shim, then install snap ring. Check bottom balance gear end play and add or remove 0.005 inch thick spacers as required to obtain proper end play of 0.005-0.010 inch. Make certain the 0.020 inch shim is used against the snap ring.

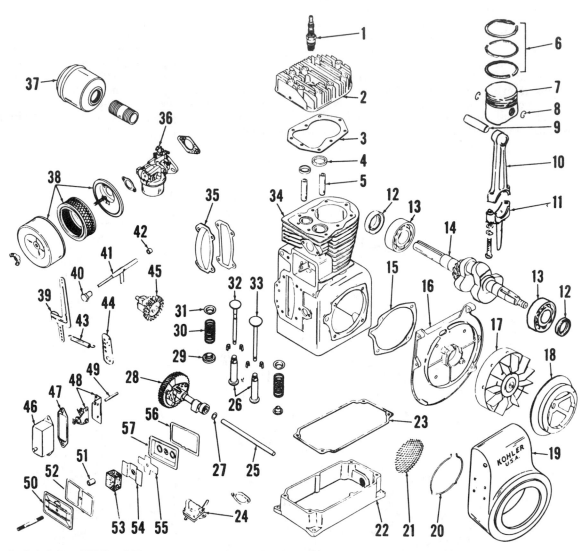

**Fig. KO32 – Exploded view of K-241 or K-301 basic engine assembly. Models K-321 and K-341 are similar. Refer to Fig. KO28 for dynamic balancer used on all K-321 and K-301 models.**

| | | | |
|---|---|---|---|
| 1. Spark plug | 16. Bearing plate | 30. Valve spring | 44. Speed lever |
| 2. Cylinder head | 17. Flywheel | 31. Spring retainer | 45. Governor gear & weight unit |
| 3. Head gasket | 18. Pulley | 32. Exhaust valve | 46. Breaker cover |
| 4. Valve seat insert | 19. Shroud | 33. Intake valve | 47. Gasket |
| 5. Valve guide | 20. Screen retainer | 34. Cylinder block | 48. Breaker point assy. |
| 6. Piston rings | 21. Screen | 35. Camshaft cover | 49. Push rod |
| 7. Piston | 22. Oil pan | 36. Carburetor | 50. Valve cover |
| 8. Retaining rings | 23. Gasket | 37. Muffler | 51. Breather seal |
| 9. Piston pin | 24. Fuel pump | 38. Air cleaner assy. | 52. Gasket |
| 10. Connecting rod | 25. Camshaft pin | 39. Governor lever | 53. Filter |
| 11. Rod cap | 26. Valve tappets | 40. Bushing | 54. Baffle |
| 12. Oil seal | 27. Shim washer | 41. Governor shaft | 55. Reed |
| 13. Ball bearing | 28. Camshaft | 42. Needle bearing | 56. Gasket |
| 14. Crankshaft | 29. Valve rotator | 43. Governor spring | 57. Breather plate |
| 15. Gasket | | | |

Illustrations courtesy of Kohler Co.

# KOHLER

## SERVICING KOHLER ACCESSORIES

### RETRACTABLE STARTERS

Fairbanks-Morse or Eaton retractable starters are used on some Kohler engines. When servicing starters, refer to appropriate following paragraph.

### Fairbanks-Morse

**OVERHAUL.** To disassemble starter, remove retainer ring, retainer washer, brake spring, friction washer, friction shoe assembly and second friction washer as shown in Fig. KO100. Hold rope handle in one hand and cover in the other and allow rotor to rotate to unwind recoil spring preload. Lift rotor from cover, shaft and recoil spring. Note winding direction of recoil spring and rope for aid in reassembly. Remove recoil spring from cover and unwind rope from rotor.

When reassembling unit, lubricate recoil spring, cover shaft and its bore in rotor with "Lubriplate" or equivalent. Install rope on rotor and rotor to shaft, then engage recoil spring inner end hook. Preload recoil spring four turns, then install middle flange and mounting flange. Check friction shoe sharp ends and renew if necessary. Install friction washers, friction shoe assembly, brake spring, retainer washer and retainer ring. Make certain friction shoe assembly is installed properly for correct starter rotation. If properly installed, sharp ends of friction shoe plates will extend when rope is pulled.

Starter operation can be reversed by winding rope and recoil spring in opposite direction and turning friction shoe assembly upside down. See Fig. KO101 for counter-clockwise assembly.

### Eaton

**OVERHAUL.** To disassemble starter, first release tension of rewind spring as follows: Hold starter assembly with pulley facing up. Pull starter rope until notch in pulley is aligned with rope hole in cover. Use thumb pressure to prevent pulley from rotating. Engage rope in notch of pulley and slowly release thumb pressure to allow spring to unwind until all tension is released.

When removing rope pulley, use extreme care to keep starter spring confined in housing. Check starter spring for breaks, cracks or distortion. If starter spring is to be renewed, carefully

remove it from housing, noting direction of rotation of spring before removing. Exploded view of clockwise starter is shown in Fig. KO102.

Check pawl, brake, spring, retainer and hub for wear and renew as necessary. If starter rope is worn or frayed, remove from pulley, noting

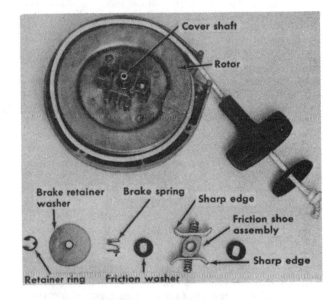

**Fig. KO100 – Fairbanks-Morse retractable starter with friction shoe assembly removed.**

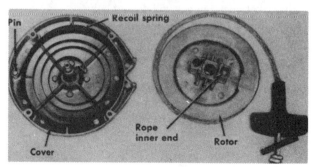

**Fig. KO101 – View showing recoil spring and rope installed for counter-clockwise starter operation.**

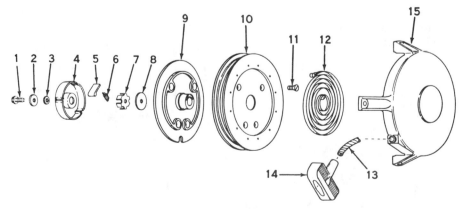

**Fig. KO102 – Exploded view of Eaton retractable starter assembly.**

| | | |
|---|---|---|
| 1. Retainer screw | 6. Spring | 11. Screw |
| 2. Brake washer | 7. Brake | 12. Recoil spring |
| 3. Spacer | 8. Thrust washer | 13. Rope |
| 4. Retainer | 9. Pulley hub | 14. Handle |
| 5. Pawl | 10. Pulley | 15. Starter housing |

direction it is wrapped on pulley. Renew rope and install pulley in housing, aligning notch in pulley assembly in housing, align notch in pulley hub with hook in end of spring. Use a wire bent to form a hook to aid in positioning spring in hub.

After securing pulley assembly in housing, engage rope in notch and rotate pulley at least two full turns in same direction it is pulled to properly preload starter spring. Pull rope to fully extended position. Release handle and if spring is properly preloaded, rope will fully rewind.

Before installing starter on engine, check teeth in starter driven hub (165-Fig. KO103) for wear and renew hub if necessary.

## 12-VOLT STARTER-GENERATOR

A combination 12-volt starter-generator manufactured by Delco-Remy is used on some Kohler engines. Starter-generator functions as a cranking motor when starting switch is closed. When engine is operating and with starting switch open, unit operates as a generator. Generator output and circuit voltage for battery and various operating requirements are controlled by a current-voltage regulator.

Kohler recommends starter-generator belt tension be adjusted until about 10 pounds pulling pressure (4.5 kg) applied midway between pulleys will deflect belt ½-inch (12.7 mm).

To determine cause of abnormal operation starter-generator should be given a "no-load" test or a "generator output" test. Generator output test can be performed with starter-generator on or off engine. No-load test must be made with starter-generator removed from engine. Refer to Fig. KO104 for exploded view of starter-generator assembly. Parts are available from Kohler as well as authorized Delco-Remy service stations.

Starter-generator brush spring tension for all models should be 24-32 oz. (0.68-0.91 kg).

Starter-generator and regulator service test specifications are as follows:

### Starter-Generators 1101940, 1101970, 1101973 & 1101980.

Field draw –
 Amperes . . . . . . . . . . . . . . . . 1.52-1.62
 Volts . . . . . . . . . . . . . . . . . . . . . . . 12
Cold output –
 Amperes . . . . . . . . . . . . . . . . . . . . . 12
 Volts . . . . . . . . . . . . . . . . . . . . . . . . 12
 Rpm . . . . . . . . . . . . . . . . . . . . . . 4950
No-load test –
 Volts . . . . . . . . . . . . . . . . . . . . . . . 11
 Amperes (max.) . . . . . . . . . . . . . . . . 18
 Rpm (min.) . . . . . . . . . . . . . . . . . 2500
 Rpm (max.) . . . . . . . . . . . . . . . . . 2900

### Starter-Generators 1101932, 1101948, 1101968, 1101972 & 1101974.

Field draw –
 Amperes . . . . . . . . . . . . . . . . 1.45-1.57
 Volts . . . . . . . . . . . . . . . . . . . . . . . 12
Cold output –
 Amperes . . . . . . . . . . . . . . . . . . . . . 10
 Volts . . . . . . . . . . . . . . . . . . . . . . . . 14
 Rpm . . . . . . . . . . . . . . . . . . . . . . 5450
No-load test –
 Volts . . . . . . . . . . . . . . . . . . . . . . . 11
 Amperes (max.) . . . . . . . . . . . . . . . . 17
 Rpm (min.) . . . . . . . . . . . . . . . . . 2500
 Rpm (max.) . . . . . . . . . . . . . . . . . 2900

### Starter-Generator 1101951 & 1101967.

Field draw –
 Amperes . . . . . . . . . . . . . . . . 1.52-1.62
 Volts . . . . . . . . . . . . . . . . . . . . . . . 12
Cold output –
 Amperes . . . . . . . . . . . . . . . . . . . . . 15
 Volts . . . . . . . . . . . . . . . . . . . . . . . . 14
 Rpm . . . . . . . . . . . . . . . . . . . . . . 3400
No-load test –
 Volts . . . . . . . . . . . . . . . . . . . . . . . 11
 Amperes (max.) . . . . . . . . . . . . . . . . 14
 Rpm (min.) . . . . . . . . . . . . . . . . . 1650
 Rpm (max.) . . . . . . . . . . . . . . . . . 1950

### Starter-Generator 1101996.

Field draw –
 Amperes . . . . . . . . . . . . . . . . 1.52-1.62
 Volts . . . . . . . . . . . . . . . . . . . . . . . 12
Cold output –
 Amperes . . . . . . . . . . . . . . . . . . . . . 12
 Volts . . . . . . . . . . . . . . . . . . . . . . . . 14
 Rpm . . . . . . . . . . . . . . . . . . . . . . 4950
No-load test –
 Volts . . . . . . . . . . . . . . . . . . . . . . . 11
 Amperes (max.) . . . . . . . . . . . . . . . . 18
 Rpm (min.) . . . . . . . . . . . . . . . . . 2500
 Rpm (max.) . . . . . . . . . . . . . . . . . 2900

### Regulators 1118984, 1118988 &1118999.

Ground polarity . . . . . . . . . . . . Negative
Cut-out relay –
 Air gap . . . . . . . . . . . . . . . . . 0.020 in.
 (0.508 mm)
 Point gap . . . . . . . . . . . . . . . . 0.020 in.
 (0.508 mm)
Closing voltage,
 range . . . . . . . . . . . . . . . . . 11.8-14.0
 Adjust to . . . . . . . . . . . . . . . . . . 12.8
Voltage regulator –
 Air gap . . . . . . . . . . . . . . . . . 0.075 in.
 (1.905 mm)
 Setting voltage,
 range . . . . . . . . . . . . . . . . . 13.6-14.5
 Adjust to . . . . . . . . . . . . . . . . . 14.0

### Regulator 1118985.

Ground polarity . . . . . . . . . . . . Positive
Cut-out relay –
 Air gap . . . . . . . . . . . . . . . . . 0.020 in.
 (0.508 mm)
 Point gap . . . . . . . . . . . . . . . . 0.020 in.
 (0.508 mm)
Closing voltage,
 range . . . . . . . . . . . . . . . . . 11.8-14.0
 Adjust to . . . . . . . . . . . . . . . . . . 12.8
Voltage regulator –
 Air gap . . . . . . . . . . . . . . . . . 0.075 in.
 (1.905 mm)
 Setting voltage,
 range . . . . . . . . . . . . . . . . . 13.6-14.5
 Adjust to . . . . . . . . . . . . . . . . . 14.0

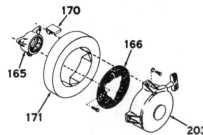

*Fig. KO103—View showing retractable starter and starter hub.*

165. Starter hub
166. Screen
170. Bracket
171. Air director
203. Retractable starter assy.

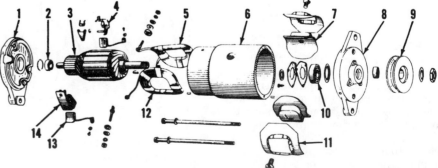

*Fig. KO104— Exploded view of typical Delco-Remy starter generator assembly.*

1. Commutator end frame
2. Bearing
3. Armature
4. Ground brush holder
5. Field coil (L.H.)
6. Frame
7. Pole shoe
8. Drive end frame
9. Pulley
10. Bearing
11. Field coil insulator
12. Field coil (L.H.)
13. Brush
14. Insulated brush holder

## 12-VOLT GEAR DRIVE STARTERS

Four types of gear drive starters are used on Kohler engines. Refer to Figs. KO105, KO106, KO107 and KO108 for exploded view of starter motors and drives.

**TWO BRUSH COMPACT TYPE.** To disassemble starting motor, clamp mounting bracket in a vise. Remove through-bolts (H – Fig. KO105) and slide commutator end plate (J) and frame assemble (A) off armature. Clamp steel armature core in a vise and remove Bendix drive (E), drive end plate (F), thrust washer (D) and spacer (C) from armature (B).

Renew brushes if unevenly worn or worn to a length of 5/16-inch (7.938 mm) or less. To renew ground brush (K), drill out rivet, then rivet new brush lead to end plate. Field brush (P) is soldered to field coil lead.

Reassemble by reversing disassembly procedure. Lubricate bushings with a light coat of SAE 10 oil. Inspect Bendix drive pinion and splined sleeve for damage. If Bendix is in good condition, wipe clean and install completely dry. Tighten Bendix drive retaining nut to a torque of 130-150 in.-lbs. (15-18 N·m). Tighten through-bolts (H) to a torque of 40-55 in.-lbs. (4-7 N·m).

**PERMANENT MAGNET TYPE.** To disassemble starting motor, clamp mounting bracket in a vise and remove through-bolts (19 – Fig. KO106). Carefully slide end cap (10) and frame (11) off armature. Clamp steel armature core in a vise and remove nut (18), spacer (17), anti-drift spring (16), drive assembly (15), end plate (14) and thrust washer (13) from armature (12).

The two input brushes are part of terminal stud (6). Remaining two brushes (9) are secured with cap screws. When reassembling, lubricate bushings with American Bosch lubricant #LU3001 or equivalent. Do not lubricate starter drive. Use rubber band to hold brushes in position until started in commutator, then cut and remove rubber band. Tighten through-bolts to a torque of 80-95 in.-lbs. (8-10 N·m) and nut (18) to a torque of 90-110 in.-lbs. (11-12 N·m).

**FOUR BRUSH BENDIX DRIVE TYPE.** To disassemble starting motor, remove screws securing drive end plate (K – Fig. KO107) to frame (I). Carefully withdraw armature and drive assembly from frame assembly. Clamp steel armature core in a vise and remove Bendix drive retaining nut, then remove drive unit (A), end plate (K) and thrust washer from armature (J). Remove cover (H) and screws securing end plate (E) to

frame. Pull field brushes (C) from brush holders and remove end plate assembly.

The two ground brush leads are secured to end plate (E) and the two field brush leads are soldered to field coils. Renew brush set if excessively worn.

Inspect bushing (L) in end plate (K) and renew bushing if necessary. When reassembling, lubricate bushings with light coat of SAE 10 oil. Do not lubricate Bendix drive assembly.

Note starter may be reinstalled with Bendix in engaged or disengaged posi-

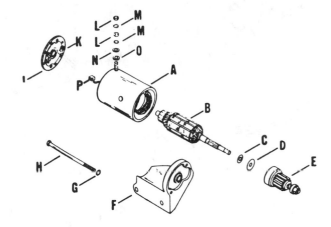

**Fig. KO105 — Exploded view of two brush compact gear drive starting motor.**

A. Frame & field coil assy.
B. Armature
C. Spacer
D. Thrust washer
E. Bendix drive assy.
F. Drive end plate & mounting bracket
G. Lockwasher
H. Through-bolt
J. Commutator end plate
K. Ground brush
L. Terminal nuts
M. Lockwashers
N. Flat washer
O. Insulating washer
P. Field brush

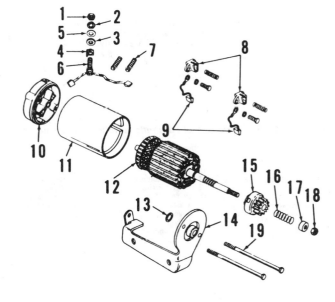

**Fig. KO106 — Exploded view of permanent magnet type starting motor.**

1. Terminal nut
2. Lockwasher
3. Insulating washer
4. Terminal insulator
5. Flat washer
6. Terminal stud & input brushes
7. Brush springs (4 used)
8. Brush holders
9. Brushes
10. Commutator end cap
11. Frame & permanent magnets
12. Armature
13. Thrust washer
14. Drive end plate & mounting bracket
15. Drive assy.
16. Anti-drift spring
17. Spacer
18. Nut
19. Through-bolts

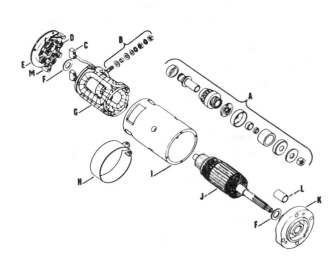

**Fig. KO107 — Exploded view of conventional four brush starting motor with Bendix drive.**

A. Bendix drive assy.
B. Terminal stud set
C. Field brushes
D. Brush springs
E. Commutator end plate
F. Thrust washers
G. Field coils
H. Cover
I. Frame
J. Armature
K. Drive end plate
L. Bushing
M. Ground brushes

tion. Do not attempt to disengage Bendix if it is in the engaged position.

## FOUR BRUSH SOLENOID SHIFT TYPE.

To disassemble starting motor, refer to Fig. KO108; then unbolt and remove solenoid switch assembly (items 1 through 6). Remove through-bolts (23), end plate (24) and frame (30) with brushes (26), brush holders (27 and 29) and field coil assembly (33). Remove screws retaining center bearing (21) to drive housing (12), remove shift lever pivot bolt, raise shift lever (9) and carefully withdraw armature and drive assembly. Drive unit (16) and center bearing (21) can be removed from armature (22) after snap ring (14) and retainer (15) are removed. Drive out shift lever pin and separate plunger (7), seal (8) and shift lever (9) from drive housing. Any further disassembly is obvious after examination of unit. Refer to Fig. KO108. Renew brushes (26), center bearing (21) and bushings in end plate (24) and drive housing (12) as necessary.

## FLYWHEEL ALTERNATORS

**3 AMP ALTERNATOR.** The 3 amp alternator consists of a permanent magnet ring with five or six magnets on flywheel rim, a stator assembly attached to crankcase and a diode in charging output lead. See Fig. KO109.

To avoid possible damage to charging system, the following precautions must be observed:

1. Negative post of battery must be connected to engine ground and correct battery polarity must be observed at all times.

2. Prevent alternator leads (AC) from touching or shorting.

3. Remove battery or disconnect battery cables when recharging battery with battery charger.

4. Do not operate engine for any length of time without a battery in system.

5. Disconnect plug before electric welding is done on equipment powered by and in common ground with engine.

TROUBLESHOOTING. Defective conditions and possible causes are as follows:

1. No output. Could be caused by:
   A. Faulty windings in stator.
   B. Defective diode.
   C. Broken lead wire.
2. No lighting. Could be caused by:
   A. Shorted stator wiring.
   B. Broken lead.

If "no output" condition is the trouble, run following tests:

1. Connect ammeter in series with charging lead. Start engine and run at 2400 rpm. Ammeter should register 2 amp charge. Run engine at 3600 rpm. Ammeter should register 3 amp charge.

2. Disconnect battery charge lead from battery, measure resistance of lead to ground with an ohmmeter. Reverse ohmmeter leads and take another reading. One reading should be about mid-scale with meter set at R x 1. If both readings are high, diode or stator is open.

3. Expose diode connections on battery charge lead. Check resistance on stator side to ground. Reading should be

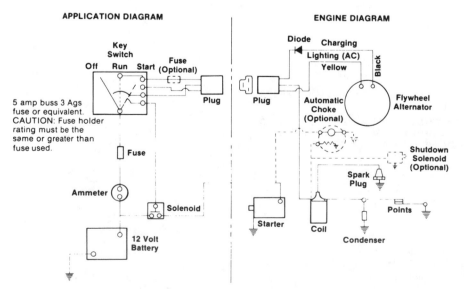

**Fig. KO108 — Exploded view of conventional four brush starting motor with solenoid shift engagement.**

1. Switch cover
2. Spring
3. Contact disc
4. Gasket
5. Coil assy.
6. Return spring
7. Plunger
8. Seal
9. Shift lever
10. Bushing
11. Lubrication wick
12. Drive housing
13. Drive end thrust washer
14. Snap ring
15. Retainer
16. Drive unit
17. Spring
18. Shift collar
19. Snap ring
20. Brake washer
21. Center bearing
22. Armature
23. Through-bolt
24. End plate
25. Thrust washer
26. Brush (4 used)
27. Insulated brush holder
28. Brush spring
29. Ground brush
30. Frame
31. Field coil insulator
32. Pole shoe
33. Field coil assy.

**Fig. KO109 — Typical electrical wiring diagram for engines equipped with 3 amp alternator.**

1 ohm. If 0 ohms, winding is shorted. If infinity ohms, stator winding is open or lead wire is broken.

If "no lighting" condition is the trouble, use an AC voltmeter and measure open circuit voltage from lighting lead to ground with engine running at 3000 rpm. If 15 volts, wiring may be shorted.

Check resistance of lighting lead to ground. If 0.5 ohms, stator is good, 0 ohms indicates shorted stator and a reading of infinity indicates stator is open or lead is broken.

**3/6 AMP ALTERNATOR.** The 3/6 amp alternator consists of a permanent magnet ring with six magnets on flywheel rim, a stator assembly attached to crankcase and two diodes located in battery charging lead and auxiliary load lead. See Fig. KO110.

To avoid possible damage to charging system, the following precautions must be observed.

1. Negative post of battery must be connected to engine ground and correct battery polarity must be observed at all times.

2. Prevent alternator leads (AC) from touching or shorting.

3. Do not operate for any length of time without a battery in system.

4. Remove battery or disconnect battery cables when recharging battery with battery charger.

5. Disconnect plug before electric welding is done on equipment powered by and in common ground with engine.

TROUBLESHOOTING. Defective conditions and possible causes are as follows:

1. No output. Could be caused by:
   A. Faulty windings in stator.
   B. Defective diode.
   C. Broken lead.
2. No lighting. Could be caused by:
   A. Shorted stator wiring.
   B. Broken lead.

If "no output" condition is the trouble, run the following tests:

1. Disconnect auxiliary load lead and measure voltage from lead to ground with engine running 3000 rpm. If 17 volts or more, stator is good.

2. Disconnect battery charging lead from battery. Measure voltage from charging lead to ground with engine running at 3000 rpm. If 17 volts or more, stator is good.

3. Disconnect battery charge lead from battery and auxiliary load lead from switch. Measure resistance of both leads to ground. Reverse ohmmeter leads and take readings again. One reading should be infinity and the other reading should be about mid-scale with meter set at R x 1. If both readings are low, diode is shorted. If both readings are high, diode or stator is open.

4. Expose diode connections on battery charging lead and auxiliary load lead. Check resistance on stator side of diodes to ground. Readings should be 0.5 ohms. If reading is 0 ohms, winding is shorted. If infinity ohms, stator winding is open or lead wire is broken.

If "no lighting" condition is the trouble, disconnect lighting lead and measure open circuit voltage with AC voltmeter from lighting lead to ground with engine running at 3000 rpm. If 22 volts or more, stator is good. If less than 22 volts, wiring may be shorted.

Check resistance of lighting lead to ground. If 0.5 ohms, stator is good, 0 ohms reading indicates shorted stator

and an infinity reading indicates an open stator winding or broken lead wire.

**10 AND 15 AMP ALTERNATOR.** Either a 10 or 15 amp alternator is used on some engines. Alternator output is controlled by a solid state rectifier-regulator.

To avoid possible damage to charging system, the following precautions must be observed:

1. Negative post of battery must be connected to engine ground and correct battery polarity must be observed at all times.

2. Rectifier-regulator must be con-

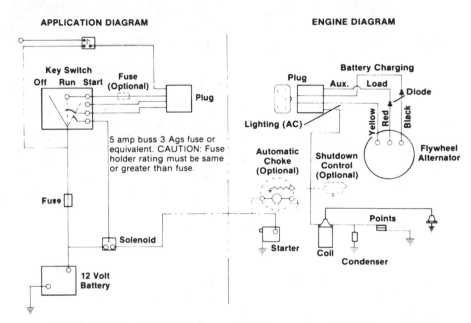

*Fig. KO110 — Typical electrical wiring diagram for engines equipped with 3/6 amp alternator.*

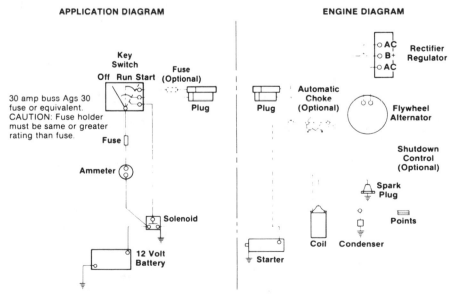

*Fig. Ko111 — Typical electrical wiring diagram for engines equipped with 15 amp alternator and breaker point ignition. The 10 amp alternator is similar.*

nected in common ground with engine and battery.

3. Disconnect leads at rectifier-regulator if electric welding is to be done on equipment in common ground with engine.

4. Remove battery or disconnect battery cables when recharging battery with battery charger.

5. Do not operate engine with battery disconnected.

6. Make certain AC leads are prevented from being grounded at all times.

OPERATION. Alternating current (AC) produced by alternator is changed to direct current (DC) in rectifier-regulator. See Fig. KO112. Current regulation is provided by electronic devices which "sense" counter-voltage created by battery to control or limit charging rate. No adjustments are possible on alternator charging system. Faulty components must be renewed. Refer to the following troubleshooting paragraph to help locate possible defective parts.

TROUBLESHOOTING. Defective conditions and possible causes are as follows:

1. No output. Could be caused by:
   A. Faulty windings in stator.
   B. Defective diode(s) in rectifier.
   C. Rectifier-regulator not properly grounded.
2. Full charge-no regulation. Could be caused by:
   A. Defective rectifier-regulator.
   B. Defective battery.

If "no output" condition is the trouble, disconnect B+ cable from rectifier-regulator. Connect a DC voltmeter between B+ terminal on rectifier-regulator and engine ground. Start

engine and operate at 3600 rpm. DC voltage should be above 14 volts. If reading is above 0 volts but less than 14 volts, check for defective rectifier-regulator. If reading is 0 volts, check for defective rectifier-regulator or defective stator by disconnecting AC leads from rectifier-regulator and connecting an AC voltmeter to the two AC leads. Check AC voltage with engine running at 3600 rpm. If reading is less than 20 volts (10 amp alternator) or 28 volts (15 amp alternator), stator is defective. If reading is more than 20 volts (10 amp alternator) or 28 volts (15 amp alternator), rectifier-regulator is defective.

If "full charge-no regulation" is the condition, use a DC voltmeter and check B+ to ground with engine operating at 3600 rpm. If reading is over 14.7 volts, rectifier-regulator is defective. If reading is under 14.7 volts but over 14.0 volts, alternator and rectifier-regulator are satisfactory and battery is probably defective (unable to hold a charge).

**30 AMP ALTERNATOR.** A 30 amp flywheel alternator consisting of a permanent field magnet ring (on flywheel) and an alternator stator (on bearing plate on single cylinder engines or gear cover on two cylinder engines) is used on some models. Alternator output is controlled by a solid state rectifier-regulator.

To avoid possible damage to charging system, the following precautions must be observed:

1. Negative post of battery must be connected to engine ground and correct battery polarity must be observed at all times.

2. Rectifier-regulator must be connected in common ground with engine and battery.

3. Disconnect wire from rectifier-

regulator terminal marked "BATT. NEG." if electric welding is to be done on equipment in common ground with engine.

4. Remove battery or disconnect battery cables when recharging battery with battery charger.

5. Do not operate engine with battery disconnected.

6. Make certain AC leads are prevented from being grounded at all times.

OPERATION. Alternating current (AC) produced by alternator is carried by two black wires to full wave bridge rectifier where it is changed to direct current (DC). Two red stator wires serve to complete a circuit from regulator to secondary winding in stator. A zener diode is used to sense battery voltage and it controls a Silicon Controlled Rectifier (SCR). SCR functions as a switch to allow current to flow in secondary winding in stator when battery voltage gets above a specific level.

An increase in battery voltage increases current flow in secondary winding in stator. This increased current flow in secondary winding brings about a corresponding decrease in AC current in primary winding, thus controlling output.

When battery voltage decreases, zener diode shuts off SCR and no current flows to secondary winding. At this time, maximum AC current is produced by primary winding.

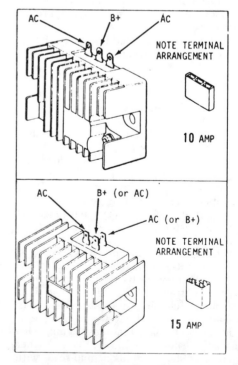

Fig. KO113—Rectifier-regulators used with 10 amp and 15 amp alternators. Although similar in appearance, units must not be interchanged.

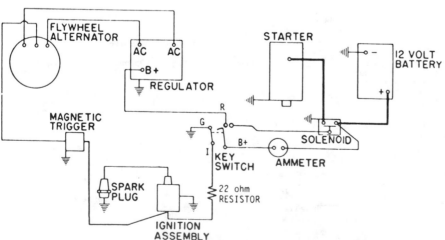

Fig. KO112—Typical electrical wiring diagram for engine equipped with 15 amp flywheel alternator and breakerless ignition system. The 10 amp alternator is similar.

Illustrations courtesy of Kohler Co.

TROUBLESHOOTING. Defective conditions and possible causes are as follows:

1. No output. Could be caused by:
   A. Faulty windings in stator.
   B. Defective diode(s) in rectifier.
2. No charge (when normal load is applied to battery). Could be caused by:
   A. Faulty secondary winding in stator.
3. Full charge-no regulation. Could be caused by:
   A. Faulty secondary winding in stator.
   B. Defective regulator.

If "no output" condition is the trouble, check stator windings by disconnecting all four stator wires from rectifier-regulator. Check resistance on R x 1 scale of ohmmeter. Connect ohmmeter leads to the two red stator wires. About 2.0 ohms should be noted. Connect ohmmeter leads to the two black stator wires. Approximately 0.1 ohm should be noted. If readings are not at test values, renew stator. If ohmmeter readings are correct, stator is good and trouble is in rectifier-regulator. Renew rectifier-regulator.

If "no charge when normal load is applied to battery" is the trouble, check stator secondary winding by disconnecting red wire from "REG" terminal on rectifier-regulator. Operate engine at 3600 rpm. Alternator should now charge at full output. If full output of at least 30 amps is not attained, renew stator.

If "full charge-no regulation" is the trouble, check stator secondary winding by removing both red wires from rectifier-regulator and connecting ends of these two wires together. Operate engine at 3600 rpm. A maximum 4 amp charge should be noted. If not, stator secondary winding is faulty. Renew stator. If maximum 4 amp charge is noted, stator is good and trouble is in rectifier-regulator. Renew rectifier-regulator.

Refer to Fig. KO114 and KO115 for correct rectifier-regulator wiring connections.

**CLUTCH.** Some models are equipped with either a dry disc clutch (Fig. KO116) or a wet type clutch (Fig. KO118). Both type clutches are lever operated. Refer to the following paragraphs for adjustment procedure.

DRY DISC TYPE. A firm pressure should be required to engage over-center linkage. If clutch is slipping, remove nameplate (Fig. KO116) and locate adjustment lock by turning flywheel. Release clutch, back out adjusting lock screw, then turn adjusting spider clockwise until approximately 20 pounds (9 kg) pull is required to snap clutch over-center. Tighten adjusting

lock screw. Every 50 hours, lubricate clutch bearing collar through inspection cover opening.

WET TYPE CLUTCH. To adjust wet type clutch, remove nameplate and use a screwdriver to turn adjusting ring (Fig.

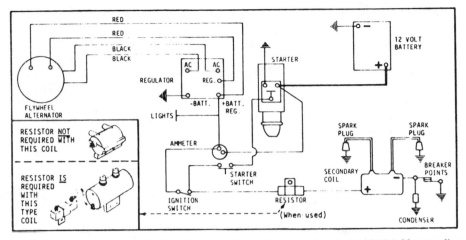

Fig. KO114—Typical electrical wiring diagram of two cylinder engine equipped with 30 amp alternator charging system. The 30 amp alternator on single cylinder engines is similar.

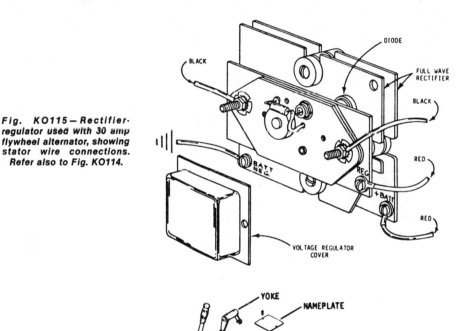

Fig. KO115—Rectifier-regulator used with 30 amp flywheel alternator, showing stator wire connections. Refer also to Fig. KO114.

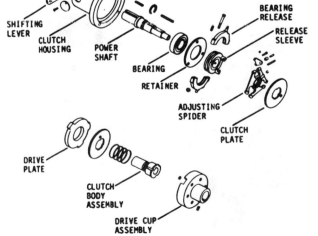

Fig. KO116—Exploded view of dry disc type clutch used on some models.

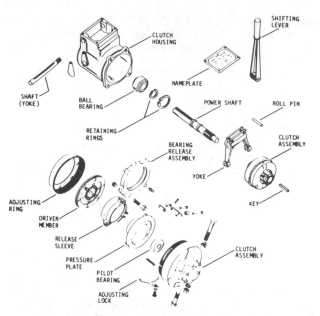

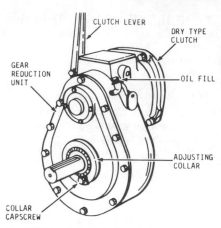

Fig. KO118—Exploded view of wet type clutch used on some models.

Fig. KO121—Output shaft end play on combination clutch and reduction drive must be adjusted to 0.0015-0.003 inch (0.0381-0.0762 mm). To adjust end play, loosen cap screw and rotate adjusting collar.

KO118) in clockwise direction until a pull of 40-50 pounds (18-23 kg) at hand grip lever is required to snap clutch over-center.

**NOTE: Do not pry adjusting lock away from adjusting ring as spring type lock prevents adjusting ring from backing off during operation.**

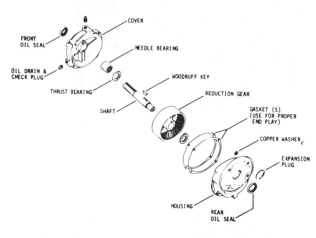

Fig. KO119—Exploded view of gear reduction drive used on some models.

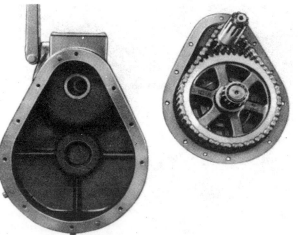

Fig. KO120—Combination clutch and chain type reduction drive used on some models.

Change oil after each 100 hours of normal operation. Fill housing to level plug opening with non-detergent oil. Use SAE 30 oil in temperatures above 50° F (10° C), SAE 20 oil in temperatures 50° F (10° C) to freezing and SAE 10 oil in temperatures below freezing.

**REDUCTION DRIVE (GEAR TYPE).** The 6:1 ratio reduction gear unit (Fig. KO119) is used on some models. To remove unit, first drain lubricating oil, then unbolt cover from housing. Remove cover and reduction gear. Unbolt and remove gear housing from engine. Separate reduction gear, shaft and thrust washer from cover. Renew oil seals and needle bearings (bronze bushings on early units) as necessary.

When reassembling, wrap tape around gear on crankshaft to protect oil seal and install gear housing. Use new copper washers on two cap screws on inside of housing. Wrap tape on shaft to prevent keyway from damaging cover oil seal and install thrust washer, shaft and reduction gear in cover. Install cover and gear assembly using new gaskets as required to provide a shaft end play of 0.001-0.006 inch (0.0254-0.1524 mm). Gaskets are available in a variety of thicknesses. Fill unit to oil check plug opening with same grade oil as used in engine.

**CLUTCH AND REDUCTION DRIVE (CHAIN TYPE).** Some models are equipped with a combination clutch and reduction drive unit. Clutch is a dry type and method of adjustment is the same as for clutch shown in Fig. KO116. Clutch release collar should be lubricated each 50 hours of normal operation. Remove clutch cover for access to lubrication fitting. Reduction drive unit is a chain and sprocket type. See Fig. KO120. Fill reduction housing to level hole with same grade oil as used in engine. Capacity is 3 pints (1.4 L) and should be changed each 50 hours of normal operation. The tapered roller bearings on output shaft should be adjusted to provide 0.0015-0.003 inch (0.0381-0.0762 mm) shaft end play. Adjustment is by means of a collar which is locked in position by a 5/16-inch cap screw. See Fig. KO121.

Illustrations courtesy of Kohler Co.

## KOHLER CENTRAL ENGINE DISTRIBUTORS
### (Arranged Alphabetically by States)
These franchised firms carry extensive stocks of repair parts. Contact them
for name of dealer in their area who will have replacement parts.

Auto Electric & Carburetor Company
Phone: (205) 323-7113
2625 4th Avenue, South
**Birmingham, Alabama 35233**

Industry Services, Incorporated
Phone: (907) 562-2621
4113 Ingra Street
**Anchorage, Alaska 99503**

Charlie C. Jones, Incorporated
Phone: (602) 272-5621
2440 West McDowell Road
**Phoenix, Arizona 85009**

Generator Equipment Company
Phone: (213) 731-2401
3409 West Jefferson Boulevard
**Los Angeles, California 90018**

H. G. Makelim Company
Phone: (415) 873-4753
219 Shaw Road
**South San Francisco, California 94080**

Spitzer Industrial Products Company
Phone: (303) 287-3414
6601 North Washington Street
**Denver, Colorado 80229**

Spencer Engine, Incorporated
Phone: (813) 253-6035
1114 West Cass Street
**Tampa, Florida 33606**

Sedco, Incorporated
Phone: (404) 925-4706
1414 Red Plum Road
**Norcross, Georgia 30093**

Small Engine Clinic
Phone: (808) 488-0711
98019 Kam Highway, Box 98
**Aiea, Hawaii 96701**

Midwest Engine Warehouse
Phone: (312) 833-1200
515 Romans Road
**Elmhurst, Illinois 60126**

Medart Engines & Parts of Kansas
Phone: (913) 888-8828
15500 West 109th Street
**Lenexa, Kansas 66215**

The Grayson Company of Louisiana
Phone: (318) 222-3211
100 Fannin Street
**Shreveport, Louisiana 71101**

C. V. Foster Equipment Company
Phone: (301) 235-3351
2502 Harford Road
**Baltimore, Maryland 21218**

W. J. Connell Company
Phone: (617) 543-3600
65 Green Street
**Foxboro, Massachusetts 02035**

Carl A. Anderson, Inc. of Minnesota
Phone: (612) 542-2010
2737 South Lexington Avenue
**Eagan, Minnesota 55121**

Medart Engines & Parts
Phone: (314) 343-0505
100 Larkin Williams Industrial Crt.
**Fenton, Missouri 63026**

Original Equipment, Incorporated
Phone: (406) 245-3081
905 Second Avenue, North
**Billings, Montana 59101**

Carl A Anderson, Inc. of Nebraska
Phone: (402) 339-4944
7410 "L" Street
**Omaha, Nebraska 68127**

Power Distributors, Incorporated
Phone: (201) 225-5922
102 Mayfield Avenue
**Edison, New Jersey 08817**

Spitzer Engines & Parts
Phone: (505) 842-6472
1016 Third Street, North West
**Albuquerque, New Mexico 87103**

AEA, Incorporated
Phone: (704) 377-6991
700 West 28 Street
**Charlotte, North Carolina 28206**

Gardner, Incorporated
Phone: (614) 488-7951
1150 Chesapeake Avenue
**Columbus, Ohio 43212**

MICO, Incorporated
Phone: (918) 627-1448
7450 East 46th Place
**Tulsa, Oklahoma 74145**

Truck & Industrial Equipment Company
Phone: (503) 234-8401
7 Northeast Oregon Street
**Portland, Oregon 97232**

Pitt Auto Electric Company
Phone: (412) 766-9112
2900 Stayton Street
**Pittsburgh, Pennsylvania 15212**

Automotive Electric Corporation
Phone: (901) 345-0300
3250 Millbranch Road
**Memphis, Tennessee 38116**

Tri-State Equipment Company
Phone: (915) 532-6931
410 South Cotton Street
**El Paso, Texas 79901**

Waukesha-Pearch Industries, Inc.
Phone: (713) 723-1050
12320 South Main Street
**Houston, Texas 77035**

Diesel Electric Service & Supply Co.
Phone: (801) 972-1836
652 West 1700 South
**Salt Lake City, Utah 84104**

RBI Corporation
Phone: (804) 798-1541
P.O. Box 9378
**Richmond, Virginia 23227**

Northwest Motor Parts & Mfg. Co.
Phone: (206) 624-4448
2930 Sixth Avenue South
**Seattle, Washington 98134**

Air-Cooled Engine Supply
Phone: (509) 624-8926
127 South Walnut
**Spokane, Washington 99204**

Wisconsin Magneto, Incorporated
Phone: (414) 445-2800
4727 North Teutonia Avenue
**Milwaukee, Wisconsin 53209**

## CANADIAN DISTRIBUTORS
Power Electric & Equipment Co., Ltd.
Phone: (403) 236-1515
4250 80th Avenue, South East
**Calgary, Alberta T2C 3A2**

Coast Dieselec, Ltd.
Phone (604) 872-5201
1920 Main Street
**Vancouver, British Columbia V5T 3B9**

Yetman's Ltd.
Phone: (204) 586-8046
949 Jarvis Avenue
**Winnipeg, Manitoba R2X 0A1**

W. N. White Company, Ltd.
Phone: (902) 443-5000
2-213-215 Bedford Highway
**Halifax, Nova Scotia B3M 2J9**

Suntester Equipment, Ltd.
Phone: (416) 624-6200
5466 Timberlea Boulevard
**Mississauga, Ontario L4W 2T7**

Suntester Equipment, Ltd.
Phone: (514) 636-1921
2081 Chartier Avenue
**Dorval, P.Q. H9P 1H3**

# KUBOTA

**550 W. Artesia Blvd.
Compton, CA 90220**

| Model | No. Cyls. | Bore | Stroke | Displacement | Power Rating |
|---|---|---|---|---|---|
| GS90 | 1 | 52 mm (2.05 in.) | 40 mm (1.57 in.) | 84 cc (5.13 cu. in.) | 1.64 kW (2.2 hp) |
| GS130 | 1 | 60 mm (2.36 in.) | 46 mm (1.81 in.) | 130 cc (7.93 cu. in.) | 1.69 kW (2.3 hp) |
| GS280 | 1 | 73 mm (2.87 in.) | 66 mm (2.60 in.) | 276 cc (16.8 cu. in.) | 5.14 kW (7.0 hp) |

## MAINTENANCE

**SPARK PLUG.** Recommended spark plug is a NGK BR4HS for all models. Spark plug should be removed and cleaned and electrode gap set at 0.6-0.7 mm (0.024-0.028 inch) after every 100 hours of operation. Renew spark plug if electrode is worn or damaged.

**CAUTION: Caution should be exercised if abrasive type spark plug cleaner is used. Inadequate cleaning procedure may allow the abrasive cleaner to be deposited in engine cylinder, accelerating wear and part failure.**

**CARBURETOR.** All models are equipped with a float type carburetor. Refer to Figs. KU100 and KU101. Main fuel mixture is controlled by a fixed main jet (7).

Initial adjustment of fuel mixture needle (6—Fig. KU100 or KU101) from a lightly seated position is 1-1/2 turns open for Model GS90, 1-3/8 turns open for Model GS130 or 1-1/4 turns open for Model GS230. Engine idle speed should be set at 1000 rpm with engine at normal operating temperature using idle speed screw (5).

With carburetor body held in inverted position, float (14) should be parallel with carburetor body. Carefully bend float lever tang to adjust.

**AIR FILTER.** All models are equipped with a foam type air filter element. The air filter element should be removed and cleaned at 50 hour intervals. To clean element, wash in a nonflammable cleaning solvent and gently squeeze element dry. Oil element with clean engine oil and gently squeeze out the excess oil.

**GOVERNOR.** All models are equipped with a centrifugal governor which is located internally in engine. On Model GS90, governor assembly is mounted directly on the camshaft gear. On Models GS130 and GS280, governor assembly is mounted on a separate governor gear which is driven by the camshaft gear.

To adjust governor linkage, loosen clamp bolt (3—Fig. KU102). Completely open throttle valve by pushing or pulling governor lever (2). Use a screwdriver to turn governor shaft (1) fully clockwise. Tighten clamp bolt (3) while holding lever and shaft in position.

Start engine and run until normal operating temperature is reached. Move speed control lever (8) to low speed position and adjust low speed stop screw (6) as required to obtain idle speed of 800-1200 rpm. Move speed control lever to operating position and adjust high speed stop screw (7) to obtain no-load speed of about 3850 rpm.

**IGNITION SYSTEM.** All models are equipped with a transistorized ignition system. The only moving part of this system is the flywheel. No adjustment is required when unit is correctly installed.

**VALVE ADJUSTMENT.** Valve stem to tappet clearance (engine cold) for all models is 0.08-0.13 mm (0.003-0.005 inch) for intake and exhaust valves. To check clearance, remove tappet chamber cover and breather assembly. Set piston at top dead center on compression stroke. Use a feeler gage to measure clearance (C—Fig. KU103) between

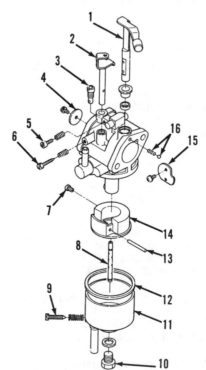

**Fig. KU100—Exploded view of carburetor used on Model GS90.**

1. Choke shaft
2. Throttle shaft
3. Pilot jet
4. Throttle plate
5. Throttle stop screw
6. Mixture screw
7. Main jet
8. Main nozzle
9. Drain screw
10. Bolt
11. Float bowl
12. Gasket
13. Pin
14. Float
15. Choke plate
16. Choke detent spring & ball

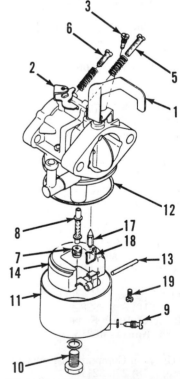

**Fig. KU101—Exploded view of carburetor used on Models GS130 and GS280.**

1. Choke shaft
2. Throttle shaft
3. Pilot jet
4. Throttle stop screw
5. Throttle stop screw
6. Mixture screw
7. Main jet
8. Main nozzle
9. Drain screw
10. Bolt
11. Float bowl
12. Gasket
13. Pin
14. Float
17. Inlet needle valve
18. Clip
19. Set screw

Illustrations courtesy of Kubota Tractor Corp.

end of valve stem (1) and tappet (2). On all models, grind end of valve stem to increase clearance or grind valve seat deeper or renew valve or tappet to decrease clearance. Refer to the REPAIRS section for service procedure.

**CYLINDER HEAD AND COMBUSTION CHAMBER.** Cylinder head, combustion chamber and piston should be cleaned and combustion deposits removed after every 500 hours of operation. Refer to REPAIRS section for service procedure.

**LUBRICATION.** All models are splash lubricated by a dipper attached to connecting rod cap.

Oil level should be checked prior to each operating interval. Oil level should be maintained at top edge of reference marks on gage with oil fill plug just touching first threads. Do not screw oil fill plug and gage in to check oil level.

Manufacturer recommends using oil with API service classification of SE or SF. Use SAE 30 oil when ambient temperature is above 15° C (59° F), SAE 20 oil when ambient temperature is between 15° C (59° F) and -10° C (14° F) and SAE 10W-30 oil when ambient temperature is below -10° C (14° F).

Oil should be changed after the first 20 hours of operation and every 50 hours of operation thereafter. Crankcase capacity is approximately 0.35 L (0.75 pt.) for Model GS90, 0.55 L (1.1 pt.) for Model GS130 and 0.90 L (1.9 pt.) for Model GS280.

**GENERAL MAINTENANCE.** Check and tighten all loose bolts, nuts or clamps prior to each day of operation. Check for fuel or oil leakage and repair if necessary.

Clean dust, dirt, grease or any foreign material from cylinder head and cylinder block cooling fins after every 100 hours of operation. Inspect fins for damage and repair if necessary.

## REPAIRS

**TIGHTENING TORQUES.** Recommended tightening torques are as follows:

Connecting rod:
GS90 . . . . . . . . . . . . . . . .3.9-5.6 N·m
(3-4 ft.-lbs.)
GS130 . . . . . . . . . .9.8-13.7 N·m
(7-10 ft.-lbs.)
GS280 . . . . . . . . . .16.7-22.6 N·m
(12-17 ft.-lbs.)

Crankcase:
GS90 . . . . . . . . . . . . . .3.4-4.9 N·m
(2.5-3.5 ft.-lbs.)
GS130 . . . . . . . . . . .7.8-12.7 N·m
(6-9 ft.-lbs.)
GS280 . . . . . . . . . . .13.7-19.6 N·m
(10-14 ft.-lbs.)

Cylinder head:
GS90 . . . . . . . . . . . . .10.8-13.7 N·m
(8-10 ft.-lbs.)
GS130 . . . . . . . . . . .19.6-29.4 N·m
(15-21 ft.-lbs.)
GS280 . . . . . . . . . . .34.3-46.1 N·m
(25-34 ft.-lbs.)

Spark plug . . . . . . . . . . .9.8-24.5 N·m
(7-18 ft.-lbs.)

**CYLINDER HEAD.** Cylinder head may be removed after removing fuel tank and air cowling. Clean head mating surface and check flatness using straightedge and feeler gage. Renew cylinder head if warpage exceeds 0.4 mm (0.016 inch).

Renew head gasket when installing cylinder head. Install cylinder head gasket with gasket liner facing up (smoother side down). Tighten cylinder head bolts in two steps following tightening sequence shown in Fig. KU104. Specified final torque is 10.8-13.7 N·m (8-10 ft.-lbs.) for Model GS90, 19.6-29.4 N·m (15-21 ft.-lbs.) for Model GS130 and 34.3-46.1 N·m (25-34 ft.-lbs.) for Model GS280.

**CONNECTING ROD.** Connecting rod and piston assembly may be removed after removing cylinder head and crankcase cover. Prior to separating piston and connecting rod assembly, mark piston so it can be installed on connecting rod correctly.

The connecting rod rides directly on the crankshaft crankpin journal. Standard diameter of connecting rod crankshaft bearing bore is 19.500-19.521 mm (0.7678-0.7685 inch) for Model GS90, 24.000-24.021 mm (0.9449-0.9457 inch) for Model GS130 and 30.000-30.025 mm (1.1811-1.1821 inch) for Model GS280.

Specified clearance between connecting rod bearing bore and crankpin journal is 0.020-0.054 mm (0.0008-0.0021 inch) for Model GS90 and 0.018-0.054 mm (0.0007-0.0021 inch) for all other models. On all models, if clearance is 0.1 mm (0.004 inch) or more, renew connecting rod and/or crankshaft.

Standard piston pin bore diameter in connecting rod is 12.015-12.025 mm (0.4731-0.4737 inch) for Model GS90, 13.015-13.025 mm (0.5124-0.5128 inch) for Model GS130 and 18.015-18.025 mm (0.7093-0.7096 inch) for Model GS280. On all models if clearance between piston pin and connecting rod pin bore exceeds 0.1 mm (0.004 inch), renew pin and/or connecting rod.

Side clearance between connecting rod and crankshaft should be 0.2-0.9 mm (0.008-0.035 inch) for all models. If clearance is 1.5 mm (0.059 inch) or more, renew connecting rod and/or crankshaft.

On Model GS90, connecting rod is installed on crankshaft with "90" casting mark (Fig. KU105) facing toward generator end of crankshaft. Connecting rod cap is installed with oil dipper facing direction shown in Fig. KU105.

On Models GS130 and GS280, connecting rod is installed on crankshaft with the Japanese casting mark side toward flywheel end of crankshaft (Fig. KU106). Install connecting rod cap with

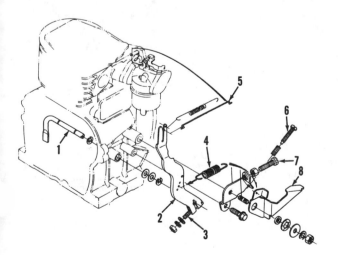

**Fig. KU102—View of typical governor linkage.**
1. Governor shaft
2. Governor lever
3. Clamp bolt
4. Governor spring
5. Connecting rod
6. Low speed stop screw
7. High speed stop screw
8. Speed control lever

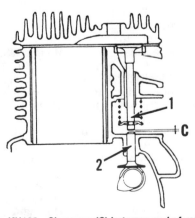

**Fig. KU103—Clearance (C) between end of valve stem (1) and tappet (2) should be 0.08-0.13 mm (0.003-0.005 in.) for intake and exhaust. Clearance can be measured with a feeler gage after removing tappet chamber and breather assembly.**

Illustrations courtesy of Kubota Tractor Corp.

smooth machined surfaces on rod and rod cap aligned.

**PISTON, PIN AND RINGS.** Piston on all models is equipped with two compression rings and one oil control ring.

Mark piston before removing from connecting rod to insure piston reassembly in same direction as removed. Note that removal and installation of piston pin will be easier if piston is first heated in hot water or oil.

After separating piston and connecting rod, carefully remove rings and clean carbon and other combustion deposits from piston surface and ring lands.

**CAUTION: Extreme care should be exercised when cleaning ring lands. Do not damage squared edges or widen ring grooves. If ring lands are damaged, piston must be renewed.**

Standard piston diameter, measured 90 degrees from piston pin bore and 15 mm (0.59 inch) up from the bottom of the skirt, is 51.95-51.98 mm (2.0453-2.3606 inches) for Model GS90, 59.94-59.96 mm (2.3598-2.3606 inches) for Model GS130 and 72.94-72.96 mm (2.8716-2.8724 inches) for Model GS280.

Clearance between piston skirt and cylinder should be 0.04-0.08 mm (0.0016-0.0031 inch) for Models GS90 and GS130 and 0.06-0.10 mm (0.0024-0.0039 inch) for Model GS280. If clearance is greater than specified, renew piston and/or recondition cylinder bore.

Clearance between piston pin and piston pin bore in piston on all models should be 0.04 mm (0.0016 inch) or less. If clearance is greater than specified, renew piston and pin.

Measure side clearance between new piston rings and piston ring grooves using a feeler gage to determine if ring grooves are excessively worn. Standard ring side clearance for Model GS90 is 0.015-0.050 mm (0.0006-0.0020 inch) for top ring and 0.010-0.045 mm (0.0004-0.0018 inch) for second ring. Standard ring side clearance for Models GS130 and GS280 is 0.02-0.06 mm (0.0008-0.0024 inch) for top and second ring. On all models, if clearance between ring and ring grooves in piston is 0.1 mm (0.004 inch) or more, renew piston.

Position each ring squarely in cylinder bore and measure ring end gap using a feeler gage. On Model GS90, ring end gap should be 0.1-0.3 mm (0.004-0.012 inch) for top and second ring and 0.3-0.9 mm (0.012-0.035 inch) for oil control ring. On Model GS130, ring end gap should be 0.25-0.45 mm (0.010-0.018 inch) for top and second ring and 0.2-0.4 mm (0.008-0.016 inch) for oil control ring. On Model GS280, ring end gap

Fig. KU104—Cylinder head bolt tightening sequence for Model GS90. Models GS130 and GS280 are similar.

should be 0.2-0.4 mm (0.008-0.016 inch) for all rings.

Install rings on piston with the "N" marked side of ring toward top of piston and stagger ring end gaps at 120 degree intervals around piston. Gap of top compression ring should be positioned so that it will be on side of piston facing away from intake and exhaust valves. Refer to Fig. KU107 for correct installation of piston rings.

Lubricate piston and cylinder with engine oil prior to installation of piston.

**CYLINDER/CRANKCASE ASSEMBLY.** The cylinder is an integral part of the crankcase casting. Models GS130 and GS280 are standard upright cylinders (Fig. KU109) while Model GS90 cylinder is at a slant (Fig. KU108).

Standard cylinder bore inside diameter is 52.00-52.02 mm (2.0472-2.0480 inches) for Model GS90, 60.00-60.02 mm (2.3622-2.3701 inches) for Model GS130 and 73.00-73.02 mm (2.8740-2.8748 inches) for Model GS280. If Model GS90 cylinder bore diameter is 52.12 mm (2.052 inches) or more, renew cylinder. On all other models, if difference between maximum and minimum wear areas is 0.1 mm (0.004 inch) or more, recondition or renew cylinder.

**CRANKSHAFT AND MAIN BEARINGS.** The crankshaft (6—Fig. KU108 or KU109) is supported at each end by ball bearing type main bearings. Ball bearings should be a light press fit in bearing bores and on crankshaft.

Crankshaft end play should be 0.2 mm (0.008 inch) or less on all models. End play is controlled by shims (4) on crankcase cover end of crankshaft.

Standard crankshaft crankpin journal diameter is 19.467-19.480 mm (0.7664-0.7669 inch) for Model GS90, 23.967-23.982 mm (0.9436-0.9442 inch) for Model GS130 and 29.967-29.982 mm (1.1798-1.1804 inch) for Model GS280.

When installing crankshaft in engine block, make certain crankshaft and camshaft timing marks are aligned as shown in Fig. KU110.

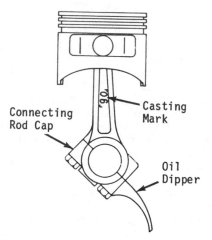

Fig. KU105—View of GS90 piston and connecting rod assembly with connecting rod cap correctly installed. Install connecting rod in cylinder so "90" casting mark faces crankcase cover side of engine.

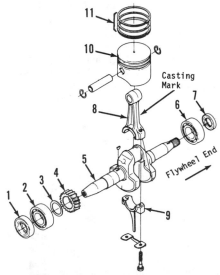

Fig. KU106—Exploded view of GS130 and GS280 connecting rod and crankshaft assembly. Install connecting rod in cylinder so casting mark faces flywheel end of crankshaft.

1. Oil seal
2. Main bearing
3. Shim
4. Crankshaft gear
5. Crankshaft
6. Main bearing
7. Oil seal
8. Connecting rod
9. Rod cap
10. Piston
11. Piston rings

**CAMSHAFT AND BEARINGS.** On all models, camshaft can be removed after removing crankcase cover. Mark tappets for installation in same location as removed.

Intake and exhaust lobe height is 16.82 mm (0.662 inch) for Model GS90, and minimum allowable height is 16.72 mm (0.658 inch). On Model GS130, specified intake and exhaust lobe height is 24.65 mm (0.9705 inch), and minimum allowable lobe height is 24.55 mm

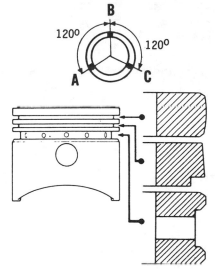

*Fig. KU107—View showing correct installation of piston rings. Stagger ring end gaps around outside diameter of piston at 120 degree intervals (A, B and C).*

(0.9665 inch). On Model GS280, specified intake and exhaust lobe height is 30.2 mm (1.189 inch), and minimum allowable lobe height is 30.1 mm (1.185 inch). If camshaft is renewed, it is recommended that valve tappets be renewed also.

Camshaft end play is controlled by varying thickness of shims (15—Fig. KU108 or Fig. KU109) on crankcase cover side of camshaft. Maximum allowable end play is 0.2 mm (0.008 inch). When installing camshaft, make certain camshaft gear and crankshaft gear timing marks are aligned as shown in Fig. KU110.

**VALVE SYSTEM.** Valve stem to tappet clearance for all models is 0.08-0.13 mm (0.003-0.005 inch) for intake and exhaust valves. Valve clearance should be checked at 100 hour intervals. On all models, grind end of valve stem to increase clearance, or grind valve seat deeper or renew valve or tappet to decrease clearance.

Valve face and seat angle is 45 degrees and standard valve face width is 0.70-0.80 mm (0.028-0.031 inch).

Refer to the following chart for valve specifications:

### Model GS90

Intake Valve
Stem OD . . . . . . . . . 4.968-4.980 mm
(0.1956-0.1961 in.)

Exhaust Valve
Stem OD . . . . . . . . . 4.950-4.965 mm
(0.1949-0.1956 in.)

Valve Guide ID—
Intake and Exhaust 5.020-5.035 mm
(0.1977-0.1982 in.)

Valve Stem-to-Guide Clearance—
Intake, Desired . . . . 0.040-0.067 mm
(0.0016-0.0026 in.)

Wear Limit . . . . . . . . . . . . . 0.10 mm
(0.004 in.)

Exhaust, Desired . . . 0.055-0.085 mm
(0.0022-0.0033 in.)

Wear Limit . . . . . . . . . . . . . 0.10 mm
(0.004 in.)

### Model GS130

Intake Valve
Stem OD . . . . . . . . . 5.968-5.980 mm
(0.2350-0.2354 in.)

Exhaust Valve
Stem OD . . . . . . . . . 5.940-5.960 mm
(0.2339-0.2346 in.)

Valve Guide ID—
Intake and Exhaust . 6.010-6.035 mm
(0.2366-0.2376 in.)

Valve Stem-to-Guide Clearance—
Intake, Desired . . . . 0.030-0.067 mm
(0.0012-0.0026 in.)

Wear Limit . . . . . . . . . . . . . 0.10 mm
(0.004 in.)

Exhaust, Desired . . . 0.050-0.095 mm
(0.0020-0.0037 in.)

Wear Limit . . . . . . . . . . . . . 0.10 mm
(0.004 in.)

### Model GS280

Intake Valve
Stem OD . . . . . . . . . 6.960-6.975 mm
(0.2740-0.2746 in.)

Exhaust Valve
Stem OD . . . . . . . . . 6.950-6.965 mm
(0.2736-0.2742 in.)

Valve Guide ID—
Intake and Exhaust . 7.010-7.035 mm
(0.2760-0.2770 in.)

Valve Stem-to-Guide Clearance—
Intake, Desired . . . . 0.035-0.075 mm
(0.0014-0.0030 in.)

Wear Limit . . . . . . . . . . . . . 0.10 mm
(0.004 in.)

Exhaust, Desired . . . 0.045-0.085 mm
(0.0018-0.0033 in.)

Wear Limit . . . . . . . . . . . . . 0.10 mm
(0.004 in.)

Standard intake and exhaust valve spring free length is 21.3-21.7 mm (0.839-0.854 inch) for Model GS90, 30.8-31.3 mm (1.213-1.232 inches) for Model GS130 and 32.8-33.3 mm (1.291-1.311 inch) for Model GS280. If valve spring free length is 21.0 mm (0.827 inch) or less for Model GS90, 30.5 mm (1.201 inch) or less for Model GS130 or 32.5 mm (1.280 inch) or less for Model GS280, renew valve spring.

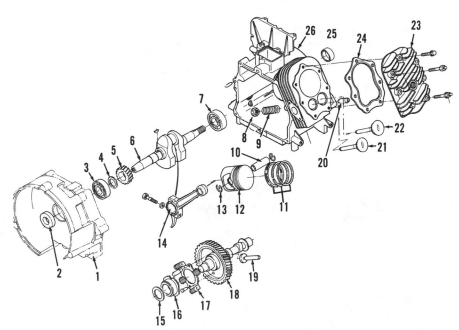

*Fig. KU108—Exploded view of Model GS90 engine components.*

| | | |
|---|---|---|
| 1. Crankcase cover | 8. Spring retainer | 15. Shim |
| 2. Oil seal | 9. Valve spring | 16. Governor sleeve |
| 3. Bearing | 10. Piston pin | 17. Governor flyweight |
| 4. Shim | 11. Piston rings | assy. |
| 5. Gear | 12. Piston | 18. Camshaft & gear |
| 6. Crankshaft | 13. Retaining ring | assy. |
| 7. Bearing | 14. Connecting rod | 19. Tappet |

| | |
|---|---|
| 20. Valve guide | 24. Gasket |
| 21. Exhaust valve | 25. Oil seal |
| 22. Intake valve | 26. Cylinder block |
| 23. Cylinder head | |

Illustrations courtesy of Kubota Tractor Corp.

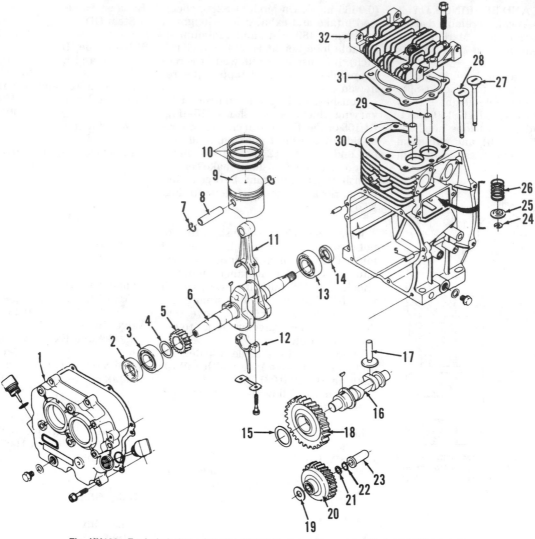

**Fig. KU109—Exploded view of Model GS130 engine components. Model GS280 is similar.**

| | | |
|---|---|---|
| 1. Crankcase cover | 12. Rod cap & oil dipper | 22. Snap ring |
| 2. Oil seal | 13. Bearing | 23. Governor sleeve |
| 3. Bearing | 14. Oil seal | 24. Retainer clip |
| 4. Shim | 15. Shim | 25. Spring retainer |
| 5. Gear | 16. Camshaft | 26. Valve spring |
| 6. Crankshaft | 17. Tappet | 27. Intake valve |
| 7. Retaining ring | 18. Camshaft gear | 28. Exhaust valve |
| 8. Piston pin | 19. Thrust washer | 29. Valve guides |
| 9. Piston | 20. Governor gear & | 30. Cylinder block |
| 10. Piston ring | flyweight assy. | 31. Gasket |
| 11. Connecting rod | 21. Washer | 32. Cylinder head |

**Fig. KU110—When installing crankshaft (6) and camshaft (18), make
certain timing marks (TM) are aligned as shown.**

# KUBOTA

| Model | No. Cyls. | Bore | Stroke | Displacement | Power Rating |
|-------|-----------|------|--------|--------------|--------------|
| GN550 | 1 | 42 mm (1.65 in.) | 40 mm (1.57 in.) | 55 cc (3.36 cu. in.) | 0.88 kW (1.2 hp) |
| GN1850 | 1 | 67 mm (2.64 in.) | 52 mm (2.05 in.) | 183 cc (11.17 cu. in.) | 2.57 kW (3.5 hp) |
| GN2500 | 1 | 72 mm (2.83 in.) | 62 mm (2.44 in.) | 252 cc (15.38 cu. in.) | 3.68 kW (4.9 hp) |

## MAINTENANCE

**SPARK PLUG.** Recommended spark plug is a NGK BM4A for Model GN550 or a BP4HS-10 for Models GN1850 and GN2500. Spark plug should be removed and cleaned and electrode gap set after every 100 hours of operation. Electrode gap should be 0.6-0.7 mm (0.024-0.028 inch) for Model GN550 and 0.9-1.0 mm (0.036-0.039 inch) for Models GN1850 and GN2500. Renew spark plug if electrode is burned or damaged.

**CAUTION: Caution should be exercised if abrasive type spark plug cleaner is used. In-** adequate cleaning procedure may allow the abrasive cleaner to be deposited in engine cylinder accelerating wear and part failure.

**AIR FILTER.** All models are equipped with a foam type air filter element. The air filter element should be removed and cleaned at 50 hour intervals. To clean element, wash in a nonflammable cleaning solvent and gently squeeze element dry. Oil element with clean engine oil and gently squeeze out the excess oil.

**CARBURETOR.** All models are equipped with a float type carburetor.

Refer to Figs. KU200 and KU201. Main fuel mixture is controlled by a fixed main jet (15).

Initial adjustment of fuel mixture needle (4—Fig. KU200 or KU201) from a lightly seated position is 1 turn open for Models GN550 and GN1850 and 1-1/4 turns open for Model GN2500. Engine idle speed should be set at 1000 rpm with engine at operating temperature using idle speed screw (5).

With carburetor body held in inverted position, float (11) should be parallel with carburetor body. Carefully bend float lever tang to adjust.

**GOVERNOR.** All models are equipped with a centrifugal governor which is located internally in engine. Governor gear is driven by the camshaft gear. To adjust governor linkage, loosen governor lever clamp bolt (3—Fig. KU202). Completely open throttle valve by pushing or pulling governor lever (4). Use a screwdriver to turn governor shaft (2) fully clockwise, then tighten clamp bolt (3) while holding lever and shaft in position.

Start engine and run until normal operating temperature is reached. Move speed control lever (7) to low speed position and adjust low speed stop screw (9) as required to obtain idle speed of 800-1200 rpm. Move speed control lever to operating position and adjust high speed stop screw (8) to obtain no-load speed of about 3850 rpm.

**IGNITION SYSTEM.** All models are equipped with a transistorized ignition system. The only moving part of this system is the flywheel. No adjustment is necessary other than setting the air gap between ignition unit core and flywheel magnet to 0.5 mm (0.020 inch) when installing the ignition unit.

**VALVE ADJUSTMENT.** Valve stem to tappet clearance (C—Fig. KU203) for intake and exhaust valves is 0.08-0.15 mm (0.003-0.006 inch) for Models GN550 and GN1850 and 0.07-0.13 mm (0.003-0.005 inch) for Model GN2500. Valve stem clearance should be checked at 1000 hour intervals. To check clearance, re-

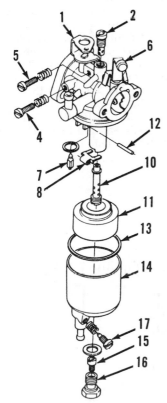

**Fig. KU200-Exploded view of carburetor used on Model GN550.**

1. Throttle shaft
2. Pilot jet
4. Mixture screw
5. Throttle stop screw
6. Choke shaft
7. Fuel inlet needle
8. Float arm
10. Main nozzle
11. Float
12. Pin
13. Gasket
14. Float bowl
15. Main jet
16. Bolt
17. Drain screw

**Fig. KU201—Exploded view of carburetor used on Models GN1850 and GN2500.**

1. Throttle shaft
2. Pilot jet
3. Throttle plate
4. Mixture screw
5. Throttle stop screw
6. Choke shaft
7. Fuel inlet needle
9. Choke plate
10. Main nozzle
11. Float
12. Pin
13. Gasket
14. Float bowl
15. Main jet
16. Bolt
17. Drain screw

Illustrations courtesy of Kubota Tractor Corp.

move tappet chamber cover and breather assembly. Set piston at top dead center on compression stroke. Use a feeler gage to measure clearance between end of valve stem (1) and tappet (2). On all models, grind end of valve stem to increase clearance, or grind valve seat deeper or renew valve or tappet to decrease clearance. Refer to the REPAIRS section for service procedure.

**CYLINDER HEAD AND COMBUSTION CHAMBER.** Cylinder head, combustion chamber and piston should be cleaned and combustion deposits removed every 500 hours of operation. Refer to REPAIRS section for service procedure.

**LUBRICATION.** All models are splash lubricated by a dipper attached to connecting rod cap.

Oil level should be checked prior to each operating interval. Oil level should be maintained at top edge of reference marks on gage with oil fill plug just touching first threads. Do not screw oil fill plug and gage in to check oil level.

Manufacturer recommends using oil with API service classification of SE or SF. Use SAE 30 oil when ambient temperature is above 15° C (59° F), SAE 20 oil when ambient temperature is between 15° C (59° F) and -10° C (14° F) and SAE 10W-30 oil when ambient temperature is below -10° C (14° F).

Oil should be changed after the first 20 hours of operation and every 50 hours of operation thereafter. Crankcase capacity is 0.32 L (0.34 qt.) for Model GN550, 0.60 L (0.63 qt.) for Model GN1850 and 0.85 L (0.9 qt.) for Model GN2500.

**GENERAL MAINTENANCE.** Check and tighten all loose bolts, nuts or clamps prior to each day of operation. Check for fuel or oil leakage and repair as necessary. Clean dust, dirt, grease or any foreign material from cylinder head and cylinder block cooling fins after every 100 hours of operation.

## REPAIRS

**TIGHTENING TORQUES.** Recommended tightening torques are as follows:

Connecting rod:
GN550 . . . . . . . . . . . . . . .3.9-5.6 N·m
(3-4 ft.-lbs.)
GN1850 . . . . . . . . . . . .13.7-19.6 N·m
(10-14.5 ft.-lbs.)
GN2500 . . . . . . . . . . .16.7-22.6 N·m
(12.3-17 ft.-lbs.)
Crankcase:
GN550 . . . . . . . . . . . . . . .2.8-3.0 N·m
(2.1-2.3 ft.-lbs.)
GN1850, GN2500 . . . . .13.7-19.6 N·m
(10-14.5 ft.-lbs.)
Cylinder head:
GN550 . . . . . . . . . . . . . .8.7-8.9 N·m
(6.4-6.6 ft.-lbs.)
GN1850 . . . . . . . . . . .19.6-29.4 N·m
(14.5-21.7 ft.-lbs.)
GN2500 . . . . . . . . . . .34.3-39.2 N·m
(25-29 ft.-lbs.)
Spark plug . . . . . . . . . . . .9.8-24.5 N·m
(7-18 ft.-lbs.)

**CYLINDER HEAD.** Cylinder head may be removed after removing fuel tank and air cowling. Clean head mating surface and check for distortion using a straightedge and feeler gage. Renew cylinder head if warpage exceeds 0.4 mm (0.016 inch).

Renew head gasket when installing cylinder head. Install cylinder head gasket with gasket liner facing up (smoother side down). Tighten cylinder head bolts in two steps following tightening sequence shown in Fig. KU204. Specified final torque is 8.7-8.9 N·m (6.4-6.6 ft.-lbs.) for Model GN550, 19.6-29.4 N·m (14.5-21.7 ft.-lbs.) for Model GN1850 and 34.3-39.2 N·m (25.3-29.0 ft.-lbs.) for Model GN2500.

**CONNECTING ROD.** Connecting rod and piston assembly is removed from cylinder head end of engine after first removing cylinder head and crankcase cover. Prior to separating piston and connecting rod assembly, mark piston so it can be installed on connecting rod correctly.

The connecting rod rides directly on the crankshaft crankpin journal. Standard diameter of connecting rod crankpin bearing bore is 18.000-18.018 mm (0.7087-0.7094 inch) for Model GN550, 25.500-25.521 mm (1.0039-1.0047 inch) for Model GN1850 and 27.000-27.021 mm (1.0629-1.0638 inch) for Model GN2500.

Specified clearance between connecting rod crankpin bearing bore and crankshaft crankpin journal is 0.016-0.045 mm (0.0006-0.0018 inch) for Model GN550 and 0.018-0.054 mm (0.0007-0.0021 inch) for Models GN1850 and GN2500. On all models, if clearance is 0.1 mm (0.004 inch) or more, renew connecting rod and/or crankshaft.

Standard piston pin bore diameter in connecting rod is 13.025-13.035 mm (0.5128-0.5132 inch) for Model GN550, 15.015-15.025 mm (0.5911-0.5915 inch) for Model GN1850 and 16.510-16.528 mm (0.6500-0.6507 inch) for Model GN2500.

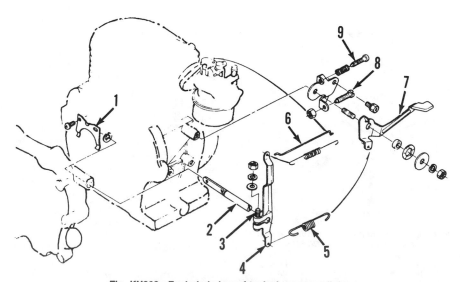

*Fig. KU202—Exploded view of typical governor linkage.*

| | | |
|---|---|---|
| 1. Governor rocker | 4. Governor lever | 7. Speed control lever |
| 2. Governor shaft | 5. Governor spring | 8. High speed stop screw |
| 3. Clamp bolt | 6. Link | 9. Low speed stop screw |

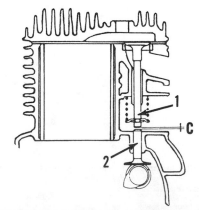

*Fig. KU203—Clearance (C) between valve stem end (1) and tappet (2) can be checked with a feeler gage after removing tappet chamber cover and breather assembly.*

If clearance between piston pin and connecting rod pin bore is 0.08 mm (0.003 inch) or more for Model GN550 or 0.1 mm (0.004 inch) or more for Models GN1850 and GN2500, renew pin and/or connecting rod.

On all models, connecting rod is installed on crankshaft with the Japanese casting mark side toward flywheel end of crankshaft (Fig. KU206). Install connecting rod cap with smooth machined surfaces on rod and rod cap aligned.

**PISTON, PIN AND RINGS.** Piston on all models is equipped with two compression rings and one oil control ring.

Mark piston before removing from connecting rod to insure piston reassembly in same direction as removed. Note that removal and installation of piston pin will be easier if piston is first heated in hot oil or water to expand pin bore in piston.

After separating piston and connecting rod, carefully remove rings and clean carbon and other combustion deposits from piston surface and ring lands.

**CAUTION: Extreme care should be exercised when cleaning ring lands. Do not damage squared edges or widen ring grooves. If ring lands are damaged, piston must be renewed.**

Standard piston diameter, measured 90 degrees from piston pin bore and 15 mm (0.59 in.) up from the bottom of the skirt is 41.975-41.995 mm (1.6526-1.6533 inch) for Model GN550, 66.93-66.95 mm (2.6350-2.6358 inch) for Model GN1850 and 71.90-71.92 mm (2.8307-2.8315 inch) for Model GN2500.

Clearance between piston thrust surface and cylinder should be 0.05-0.09 mm (0.0020-0.0035 inch) for Models GN550 and GN1850 and 0.06-0.010 mm (0.0024-0.0039 inch) for Model GN2500. If clearance is greater than specified, renew piston and/or recondition cylinder bore.

Maximum allowable clearance between piston pin and piston pin bore in piston on all models is 0.05 mm (0.002 inch). If clearance is greater than specified, renew piston pin and/or piston.

Measure side clearance between new piston rings and piston ring grooves using a feeler gage to determine if ring grooves are excessively worn. Standard clearance between rings and ring grooves in piston for Model GN550 is 0.015-0.050 mm (0.0006-0.0020 inch) for top and second rings and 0.010-0.045 mm (0.0004-0.0018 inch) for bottom (oil) ring.

Standard side clearance between rings and ring grooves in piston for Model GN1850 is 0.02-0.06 mm (0.0008-0.0024 inch) for all rings.

Standard side clearance between rings and ring grooves in piston for Model GN2500 is 0.05-0.07 mm (0.0019-0.0028 inch) for top and second rings and 0.02-0.06 mm (0.0008-0.0024 inch) for bottom (oil) ring.

On all models, if clearance between ring and ring grooves in piston exceeds 0.1 mm (0.004 inch), renew piston.

Position each piston ring squarely in cylinder bore and measure ring end gap using a feeler gage. Specified ring end gaps should be 0.15-0.35 mm (0.006-0.014 inch) for Model GN550 and 0.2-0.4 mm (0.008-0.016 inch) for Models GN1850 and GN2500.

Install rings on piston with the "N" marked side of ring toward top of piston and stagger ring end gaps at 120 degree intervals around diameter of piston as shown in Fig. KU205.

Lubricate piston and cylinder with engine oil prior to installation of piston.

**CYLINDER/CRANKCASE ASSEMBLY.** The cylinder is an integral part of the crankcase casting.

Standard cylinder bore inside diameter is 42.000-42.025 mm (1.6535-1.6545 inch) for Model GN550, 67.000-67.025 mm (2.6378-2.6388 inch) for Model GN1850 and 71.98-72.00 mm (2.8339-

2.8346 inch) for Model GN2500. On all models, if difference between maximum and minimum wear areas is 0.2 mm (0.008 inch) or more, recondition or renew cylinder.

**CRANKSHAFT AND MAIN BEARINGS.** On all models, crankshaft is supported at each end by ball bearing type main bearings. Ball bearings should be a light press fit in bearing bores and on crankshaft.

Crankshaft end play should be 0.02-0.1 mm (0.001-0.004 inch) for all models. If end play is 0.2 mm (0.008 inch) or more, vary thickness of shim (3—Fig. KU206) to obtain correct end play. Shims are available in a variety of thicknesses.

Standard crankshaft crankpin journal diameter is 17.973-17.984 mm (0.7076-0.7080 inch) for Model GN550, 25.467-25.482 mm (1.0026-1.0032 inch) for Model GN1850 and 26.967-26.982 mm (1.0617-1.0623 inch) for Model GN2500.

When installing crankshaft in engine block, make certain that timing marks on crankshaft gear and camshaft gear are aligned.

**CAMSHAFT AND BEARINGS.** On all models, camshaft (5—Fig. KU207) can be removed after removing crankcase cover. Mark tappets (6) for installation in same location as removed.

Intake and exhaust lobe height is 16.25 mm (0.6398 inch) for Model GN550, 27.2 mm (1.0709 inch) for Model GN1850 and 30.0 mm (1.1811 inch) for Model GN2500. If camshaft lobe height is 16.13 mm (0.635 inch) or less for Model GN550, 27.1 mm (1.067 inch) or less for

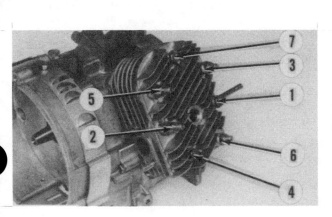

Fig. KU204—Tighten cylinder head bolts in two steps following tightening sequence shown.

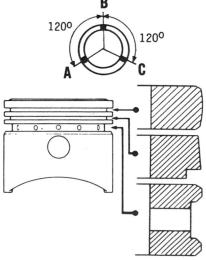

*Fig. KU205—View showing correct installation of piston rings. Stagger ring end gaps around outside diameter of piston at 120 degree intervals (A, B and C).*

Model GN1850 or 29.9 mm (1.177 inch) or less for Model GN2500, renew camshaft. If camshaft is renewed, it is recommended that valve tappets be renewed also.

Camshaft end play is controlled by varying thickness of shims (3—Fig. KU207) located between camshaft gear and bearing. When installing camshaft, make certain camshaft gear and crankshaft gear timing marks are aligned.

**VALVE SYSTEM.** Valve stem to tappet clearance for intake and exhaust valves is 0.08-0.15 mm (0.003-0.006 inch) for Models GN550 and GN1850 and 0.07-0.13 mm (0.003-0.005 inch) for Model GN2500. On all models, grind end of valve stem to increase clearance or grind valve seat deeper to decrease clearance.

Valve face and seat angle is 45 degrees and standard valve face width is 0.700-0.800 mm (0.0276-0.0315 inch).

## Model GN550

Valve Stem OD—
Intake and Exhaust 3.968-3.980 mm
(0.1562-0.1567 in.)
Valve Guide ID—
Intake and Exhaust 4.000-4.012 mm
(0.1575-0.1580 in.)
Valve Stem-to-Guide Clearance—
Desired . . . . . . . . . . 0.020-0.044 mm
(0.0008-0.0017 in.)

Wear Limit . . . . . . . . . . . . . 0.10 mm
(0.004 in.)

## Model GN1850

Intake Valve
Stem OD . . . . . . . . . 6.960-6.975 mm
(0.2740-0.2746 in.)
Exhaust Valve
Stem OD . . . . . . . . . 6.950-6.965 mm
(0.2740-0.2746 in.)
Valve Guide ID—
Intake and Exhaust . 7.010-7.035 mm
(0.2760-0.2770 in.)
Valve Stem-to-Guide Clearance—
Intake, Desired . . . . 0.035-0.075 mm
(0.0014-0.0030 in.)
Wear Limit . . . . . . . . . . . . . 0.10 mm
(0.004 in.)
Exhaust, Desired . . . 0.045-0.085 mm
(0.0018-0.0033 in.)

Wear Limit . . . . . . . . . . . . . 0.10 mm
(0.004 in.)

## Model GN2500

Intake Valve
Stem OD . . . . . . . . . 6.960-6.975 mm
(0.2740-0.2746 in.)
Exhaust Valve
Stem OD . . . . . . . . . 6.950-6.965 mm
(0.2736-0.2742 in.)
Valve Guide ID—
Intake and Exhaust . 7.000-7.015 mm
(0.2756-0.2762 in.)
Valve Stem-to-Guide Clearance—
Intake, Desired . . . . 0.025-0.055 mm
(0.0010-0.0022 in.)
Wear Limit . . . . . . . . . . . . . 0.10 mm
(0.004 in.)
Exhaust, Desired . . . 0.035-0.065 mm
(0.0014-0.0026 in.)
Wear Limit . . . . . . . . . . . . . 0.10 mm
(0.004 in.)

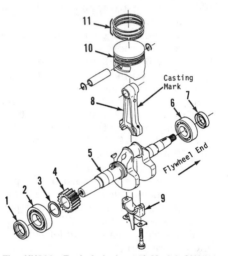

*Fig. KU206—Exploded view of Model GN2500 crankshaft/piston assembly. Models GN550 and GN1850 are similar.*

1. Seal
2. Bearing
3. Shim
4. Gear
5. Crankshaft
6. Bearing
7. Seal
8. Connecting rod
9. Connecting rod cap
10. Piston
11. Rings

Standard intake and exhaust valve spring free length is 21.2-21.8 mm (0.835-0.858 inch) for Model GN550, 29.5-30.0 mm (1.161-1.181 inch) for Model GN1850 and 32.8-33.3 mm (1.291-1.311 inch) for Model GN2500. Renew valve spring if free length is 19.9 mm (0.783 inch) or less for Model GN550, 29.2 mm (1.150 inch) or less for Model GN1850 or 32.5 mm (1.280 inch) or less for Model GN2500.

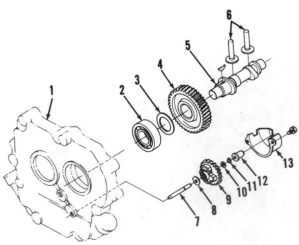

*Fig. KU207—Exploded view of Model GN2500 camshaft assembly and governor assembly. On Models GN550 and GN1850, governor flyweights are mounted on the camshaft gear.*

1. Crankcase cover
2. Bearing
3. Shim
4. Camshaft gear
5. Camshaft
6. Tappets
7. Governor shaft
8. Thrust washer
9. Governor assy.
10. Washer
11. Snap ring
12. Governor sleeve
13. Cover

# LOMBARDINI

**LOMBARDINI ENGINE, INC.**
**3402 Oakcliff Road, B-2**
**Doraville, Georgia 30340**

| Model | Cyls. | Bore | Stroke | Displ. |
|-------|-------|------|--------|--------|
| 530 | 1 | 82 mm | 68 mm | 359 cc |
| | | (3.228 in.) | (2.677 in.) | (22 cu. in.) |

The Lombardini Model 530 engine is a four-stroke, air-cooled diesel engine. Cleanliness during operation and servicing is required to prevent contamination of fuel system and possible damage. Metric fasteners are used throughout engine.

## MAINTENANCE

**GOVERNOR.** Model 530 is equipped with a flyweight centrifugal type governor which is attached to back of oil pump drive gear as shown in Fig. L1-1. Oil pump drive gear (1) is driven by crankshaft which rotates governor flyweight assembly (G). The flyweights are interlocked with sleeve (5) to move fork (7) and rotate attached shaft. As

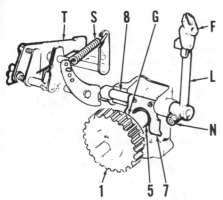

*Fig. L1-1— View of governor mechanism. Refer to text for operation.*

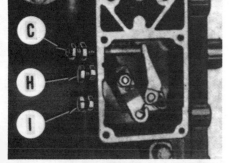

*Fig. L1-2— Turn screw (I) to adjust idle speed and screw (H) to adjust high idle speed. Torque control screw (C) must be adjusted as outlined in text.*

fork shaft rotates, governor lever (L) forces arm (F) against a pin in the fuel injection pump control sleeve thereby changing fuel flow to cylinder. Throttle lever (T) operates through governor spring (S) to control engine speed.

Idle speed is adjusted by idle speed screw (I—Fig. L1-2). Idle speed should be 1000-1050 rpm on engines required to run at idle. Maximum governed speed is adjusted by turning high speed screw (H). Maximum governed speed under load should be 3600 rpm.

**INJECTION PUMP TIMING.** Injection pump timing is adjusted using shim gaskets (G—Fig. L1-3) between pump body and mounting surface on crankcase. To check injection pump timing, unscrew delivery line (D) fitting from delivery union (1—Fig. L1-4). Unscrew delivery union and remove valve (3), spacer (4) and spring then screw delivery union (1) into pump body. Move throttle control lever to full speed position. Rotate engine in normal direction (clockwise at flywheel end) so piston is on compression stroke. Note fuel in delivery union will spill out of union. Stop engine rotation at moment fuel ceases to spill out of union. Timing dot (R—Fig. L1-5) on fan plate should align with injection timing dot (I) on fan shroud. Correct injection timing is 29°20'-31°20'BTDC. To advance injec-

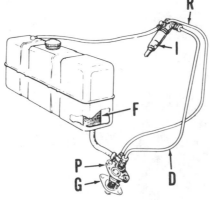

*Fig. L1-3—Diagram of fuel system.*

tion timing, remove shim gaskets (G—Fig L1-3); installing shim gaskets will retard injection timing. Shim gaskets are available in thicknesses of 0.1, 0.3 and 0.5 mm (0.004, 0.012 and 0.020 in.). Reinstall removed pump components after checking injection timing. Tighten injection pump retaining screws to 29 N·m (22 ft.-lbs.).

**VALVE GAP ADJUSTMENT.** An overhead type valve system is used. Valve gap may be adjusted after removing rocker arm cover. Valve gap should be 0.15 mm (0.006 in.) for both valves with engine cold. Note that there are two adjusting screws on exhaust valve rocker arm. Adjusting screw (V—Fig. L1-6) nearer rocker arm shaft is used to adjust valve clearance while outer screw (C) adjusts compression release gap.

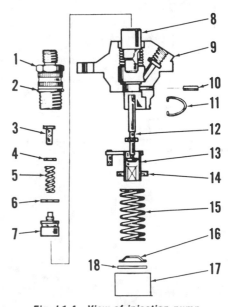

*Fig. L1-4—View of injection pump.*

| | |
|---|---|
| 1. Delivery union | 10. Pin |
| 2. "O" ring | 11. Clip |
| 3. Spring guide | 12. Plunger |
| 4. Shim | 13. Control sleeve |
| 5. Spring | 14. Spring seat |
| 6. Gasket | 15. Spring |
| 7. Delivery valve | 16. Spring retainer |
| 8. Barrel | 17. Tappet |
| 9. Pump body | 18. Spacer |

**COMPRESSION RELEASE ADJUSTMENT.** The exhaust valve is held open to ease engine starting by turning compression release lever (L–Fig. L1-6). The compression release is adjusted by turning outer adjusting screw (C) in exhaust valve rocker arm. Adjust gap between adjusting screw and compression release shaft AFTER adjusting exhaust valve gap. With compression release lever in off position, clearance between adjusting screw and shaft should be 0.9-1.1 mm (0.035-0.043 in.).

**FUEL SYSTEM.** Refer to Fig. L1-3 for a view of fuel system. Fuel system cleanliness is necessary to prevent damage to tight fitting components in injection pump and fuel injector. Care must be used when servicing system not to allow entry of foreign objects. Fuel tank filter (F) should be renewed after every 300 hours of operation.

Fuel must be free of water. Water contaminated fuel will damage injection pump and injector due to acid formation and lack of lubricity.

**LUBRICATION.** Recommended

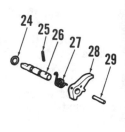

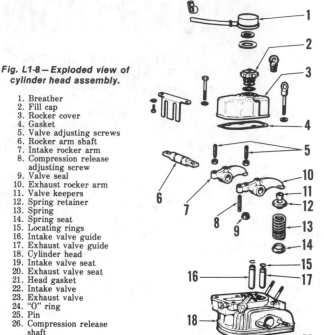

Fig. L1-8 — Exploded view of cylinder head assembly.

1. Breather
2. Fill cap
3. Rocker cover
4. Gasket
5. Valve adjusting screws
6. Rocker arm shaft
7. Intake rocker arm
8. Compression release adjusting screw
9. Valve seal
10. Exhaust rocker arm
11. Valve keepers
12. Spring retainer
13. Spring
14. Spring seat
15. Locating rings
16. Intake valve guide
17. Exhaust valve guide
18. Cylinder head
19. Intake valve seat
20. Exhaust valve seat
21. Head gasket
22. Intake valve
23. Exhaust valve
24. "O" ring
25. Pin
26. Compression release shaft
27. Spring
28. Compression release lever
29. Pin

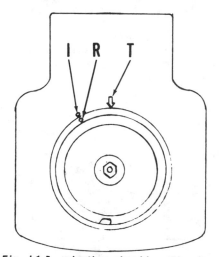

Fig. L1-5—injection should occur when timing dot (R) of fan plate is aligned with injection timing dot (I) on fan shroud. Piston is at TDC when timing dot (R) and arrow (T) are aligned.

engine oil is graded CD with SAE 10W used in temperatures below 32°F., SAE 20W used from 32° to 70°F. and SAE 40 used above 70°F. Oil sump capacity is one quart.

A renewable oil filter is located in side of engine block. Manufacturer recommends removing filter (22–Fig. L1-11) and installing a new filter after every 300 hours of operation.

## REPAIRS

**CYLINDER HEAD AND VALVE SYSTEM.** Manufacturer does not recommend removing a hot cylinder head as deformation may result. Note that different thicknesses of cylinder

head gasket are used and same thickness as original must be installed except as noted in CYLINDER HEAD GASKET section.

Valve face angle is 45 degrees and minimum valve head margin is 0.5 mm (0.020 in.). Valve seat angle should be 45 degrees with a seat width of 1.4-1.6 mm (0.055-0.063 in.). Valve seats are renewable and must be installed with head heated to 160°-180°C. (320°-356°F.). Valve seals are used on intake valves. Valve stem diameter is 6.98-7.00 mm (0.2748-0.2756 in.) while valve guide diameter should be 7.03-7.05

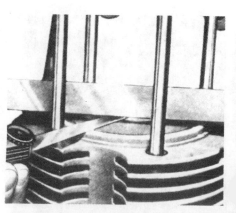

Fig. L1-6—With compression lever (L) in off position, turn adjusting screw (C) so clearance is 0.9-1.1 mm (0.035-0.043 in.) between screw and shaft. Adjusting screw (V) determines valve clearance.

Fig. L1-9—Measure piston height in cylinder and refer to text for cylinder head gasket thickness.

mm (0.2768-0.2776 in.). Desired valve stem clearance is 0.03-0.07 mm (0.0011-0.0027 in.). Valve guides are renewable and oversize guides are available. Note locating ring around top of each guide. The cylinder head should be heated to 160°-180°C. (320°-356°F.) when installing valve guides.

Valve spring free length should be 42 mm (1.653 in.). Valve spring pressure should be 23 kg. at 32 mm (50.6 lbs. at 1.260 in.).

The rocker arms are supported by rocker arm shaft (6 – Fig. L1-8). Desired clearance between shaft and rocker arms is 0.03-0.06 mm (0.001-0.002 in.). Renew shaft and rocker arms if clearance exceeds 0.1 mm (0.004 in.).

The compression relief valve mechanism is mounted in the cylinder head. Diameter of shaft (26) should be 9.37-10.00 mm (0.369-0.394 in.) while the lobe height should be 8.45-8.50 mm (0.333-0.335 in.).

Tighten cylinder head nuts in a crosswise pattern to 39.2 N·m (29 ft.-lbs.).

**CYLINDER HEAD GASKET.** If the cylinder block, piston, connecting rod or crankshaft have been renewed, then piston height at top dead center (TDC) must be measured. Measure from piston crown to gasket seating surface of cylinder as shown in Fig. L1-9. Subtract measurement from 0.6-0.7 mm (0.024-0.028 in.) to determine required gasket thickness. Cylinder head gaskets are available in thicknesses of 0.5 mm (0.020 in.), 0.6 mm (0.024 in.), 0.7 mm (0.028 in.) and 0.8 mm (0.032 in.). Tighten cylinder head nuts to 39.2 N·m (29 ft.-lbs.).

**FLYWHEEL.** The flywheel is retained by a left-hand nut and a suitable puller should be used to break free the taper fit between flywheel and crankshaft end. Tighten flywheel nut to 147 N·m (108 ft.-lbs.).

The electric starter ring gear is mounted on flywheel. To remove or install ring gear, heat inner circumference of ring gear and drive gear on or off. Be sure ring gear is properly seated on flywheel.

**CONNECTING ROD.** Connecting rod and piston must be removed as a unit after cylinder head and oil pan are removed.

The connecting rod small end is fitted with a renewable bushing. Clearance between piston pin and bushing should be 0.015-0.030 mm (0.0006-0.0012 in.). An insert type bearing is used in connecting rod big end. Desired rod bearing clearance is 0.03-0.06 mm (0.0018-0.0024 in.) while maximum allowable clearance is 0.20 mm (0.008 in.). Big end bearings are available in undersizes of 0.25 mm (0.010 in.) and 0.50 mm (0.020 in.).

The connecting rod is fractured to form the rod cap and serrations must match perfectly when installing cap on rod. Tighten connecting rod screws to 33.3 N·m (25 ft.-lbs.),

Refer to CYLINDER HEAD GASKET section for selection of correct head gasket thickness.

**PISTON, PIN AND RINGS.** Piston and connecting rod must be removed as a unit after cylinder head and oil pan are removed.

The piston may be equipped with two or three compression rings and an oil control ring. Piston ring end gap is 0.25-0.40 mm (0.010-0.016 in.) for all compression rings and 0.20-0.35 mm (0.008-0.014 in.) for the oil ring. Maximum side clearance is 0.22 mm (0.009 in.) for top compression ring, 0.17 mm (0.007 in.) for second, and if used, third compression ring and 0.12 mm (0.005 in.) for oil control ring.

Clearance between piston pin and bushing should be 0.015-0.030 mm (0.0006-0.0012 in.). Renew pin if excessively worn or damaged.

Piston to cylinder wall clearance should be 0.11-0.14 mm (0.0043-0.0055 in.) with a maximum allowable clearance of 0.28 mm (0.011 in.). When determining clearance, measure piston diameter 2 mm (0.079 in.) from bottom of piston skirt perpendicular to piston pin. Piston

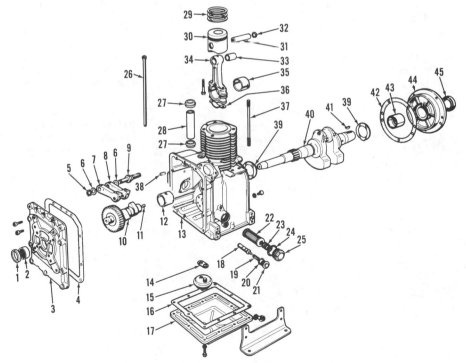

*Fig. L1-11—Exploded view of engine.*

| | | | |
|---|---|---|---|
| 1. Seal | 11. Plug | 22. Oil filter | 34. Connecting rod |
| 2. Roller bearing | 12. Bushing | 23. Spring | 35. Rod bearing |
| 3. Crankcase cover | 13. Engine block | 24. "O" ring | 36. Lock plate |
| 4. Gasket | 14. Gasket | 25. Plug | 37. Studs |
| 5. Snap ring | 15. Oil pickup | 26. Push rods | 38. Dowel pins |
| 6. Washer | 16. Gasket | 27. Seal | 39. Thrust washers |
| 7. Exhaust cam | 17. Oil pan | 28. Push rod tube | 40. Crankshaft |
| follower | 18. Oil pressure relief | 29. Piston rings | 41. Key |
| 8. Intake cam | valve | 30. Piston | 42. Gasket |
| follower | 19. Spring | 31. Piston pin | 43. Bushing |
| 9. Stud | 20. Gasket | 32. Snap ring | 44. Support |
| 10. Camshaft | 21. Plug | 33. Bushing | 45. Seal |

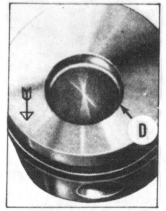

*Fig. L1-10—Install piston so depression (D) is nearer flywheel side of engine. Some pistons may have an arrow on crown and arrow must point towards flywheel.*

Illustrations courtesy of Lombardini Engine, Inc.

and rings are available in oversizes of 0.50 mm (0.020 in.) and 1.0 mm (0.040 in.).

When installing piston, note that depression (D – Fig. L1-10) is closer to one side of piston. Install piston so depression side of piston is nearer flywheel. Some pistons also have an arrow embossed in piston crown as shown in Fig. L1-10. Properly installed, arrow on piston crown will point towards flywheel.

Refer to CYLINDER HEAD GASKET section for selection of correct head gasket thickness.

**CAMSHAFT, CAM FOLLOWERS AND PUSH RODS.** The camshaft rides directly in crankcase cover and crankcase bulkhead. Cam followers (7 and 8 – Fig. L1-11) pivot on stud (9) and transfer motion to push rods (26) which pass through tube (28) to rocker arms. In addition to lobes for the valves, a lobe is ground on the camshaft to actuate the fuel injection pump.

Oil passages in camshaft may be cleaned after removing plug (11). Be sure plug is securely reinstalled. Lobe height for intake and exhaust valves should be 33.14-33.15 mm (1.3047-1.3051 in.) while lobe height for injection pump should be 33.99-34.00 mm (1.3381-1.3386 in.). Camshaft bearing journal diameters are 19.937-19.950 mm (0.7849-0.7854 in.) and 25.937-25.950 mm (1.0211-1.0216 in.). Desired clearance between cam followers and pivot stud (9 – Fig. L1-11) is 0.03-0.06 mm (0.001-0.002 in.) with a maximum clearance of 0.1 mm (0.004 in.).

Install camshaft so timing marks (M – Fig. L1-12) are aligned. If timing marks are absent from gears, proceed as follows: Position piston at top dead center (TDC) then install camshaft so intake cam follower is on opening side of cam lobe and exhaust cam follower is on closing side of cam lobe. If necessary, remesh gears so cam followers are at same height. Mark gears for future reference.

Depth of camshaft in crankcase must not be greater than 0.10 mm (0.004 in.) as measured from thrust face (TF – Fig. L1-12) to crankcase gasket surface (G). Camshaft end play should be 0.10-0.30 mm (0.004-0.012 in.) and is adjusted by varying thickness of crankcase cover gasket (4 – Fig. L1-11). Apply "Loctite" to crankcase cover (3) screws and tighten to 29 N·m (22 ft.-lbs.).

The push rods are contained in tube (28) and must cross between cam followers and rocker arms. Push rod nearer cylinder connects intake cam follower and rocker arm while outer push rod connects exhaust cam follower and rocker arm.

**CRANKSHAFT AND CRANKCASE.** The crankshaft is supported by bushing (12 – Fig. L1-11) in the crankcase bulkhead, bushing (43) in support (44) and by roller bearing (2) in the crankcase cover.

Desired bearing clearance for center and flywheel end main bearings is 0.03-0.06 mm (0.0011-0.0024 in.). Crankshaft journal diameter for center and flywheel end main bearings is 39.99-40.00 mm (1.5744-1.5748 in.). Center and flywheel end main bearings are available in undersizes of 0.5 mm (0.020 in.) and 1.0 mm (0.040 in.). Crankshaft journal diameter at pto end is 27.94-28.00 mm (1.100-1.102 in.). Crankshaft must be renewed if pto end journal is worn more than 0.10 mm (0.004 in.).

The crankshaft has drilled oil passages to circulate oil. Expansion plugs located adjacent to crankpin may be removed to clean oil passages, however, new plugs must be installed securely.

Thrust washer (39) thickness should be 2.31-2.36 mm (0.0909-0.0929 in.). Crankshaft end play should be 0.10-0.20 mm (0.004-0.008 in.). End play is adjusted by changing thickness of gasket (4) which is available in thicknesses of 0.10 mm (0.004 in.), 0.20 mm (0.008 in.), 0.30 mm (0.012 in.) and 0.40 mm (0.016 in.). Tighten nuts retaining support (44) to 29 N·m (22 ft.-lbs.). Apply "Loctite" to crankcase cover (3) screws and tighten to 29 N·m (22 ft.-lbs.).

**GOVERNOR.** Governor components must move freely for proper governor operation. Governor spring (S – Fig. L1-1) free length should be 56.9-57.0 mm (2.240-2.244 in.). At a spring length of 71.9-72.0 mm (2.831-2.835 in.), spring tension should be 1.4-1.6 kg (3.0-3.5 lbs.).

Spindle (8 – Fig. L1-13) diameter is 7.95 mm (0.313 in.). Desired clearance between spindle and bores in oil pump body (13) is 0.06-0.10 mm (0.002-0.004 in.) with a maximum allowable clearance of 0.15 mm (0.006 in.).

Hook governor spring end in second hole from governor lever end.

**OIL PUMP.** Clearance between gears and walls of pump body must not exceed 0.15 mm (0.006 in.). Renew oil pump if components are excessively worn or damaged. Tighten pump mounting screws evenly to 11.8 N·m (104 in.-lbs.).

**INJECTION PUMP.** Refer to Fig. L1-4 for view of injection pump. Disassembly and reassembly is evident after inspection of pump and referral to Fig. L1-4. Note that slot in barrel (8) must align with pin (10) and helix in plunger (12) must face pin (10).

The following tests may be used to check injection pump if necessary test equipment is available. With a suitable pressure gage connected to delivery union (1), operate pump. With control sleeve (13) at mid-point, pump pressure should be at least 29400 kPa (4260 psi) while pump pressure should be at least 39200 kPa (5680 psi) with control sleeve in maximum fuel position. To check delivery valve, move control sleeve (13) to mid-point position and operate pump. After maximum pressure is reached, pressure should drop sharply to a pressure of 2940-4900 kPa (427-710 psi) less than maximum pressure if delivery valve is operating properly. Maximum delivery rate of pump is 44-46 cc at 1800 rpm for 1000 pump strokes.

Outside diameter of tappet (17) is 27.96-27.98 mm (1.1007-1.1015 in.) while maximum allowable clearance in tappet guide bore is 0.10 mm (0.004 in.). Thickness of spacer (18) should be 3.45-3.55 mm (0.1358-0.13898 in.).

When installing injection pump, place shim gaskets (G – Fig. L1-3) under pump then engage control sleeves (13 – Fig.

**Fig. L1-12—View of camshaft and crankshaft timing marks (M). Measure depth of camshaft thrust face (TF) from crankcase gasket surface (G) as outlined in text.**

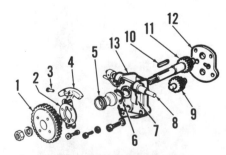

**Fig. L1-13—Exploded view of governor and oil pump assembly.**

| | |
|---|---|
| 1. Drive gear | |
| 2. Governor frame | 8. Spindle |
| 3. Pins | 9. Gear |
| 4. Weights | 10. Key |
| 5. Sleeve | 11. Gear & shaft |
| 6. Stop | 12. Cover |
| 7. Fork | 13. Oil pump body |

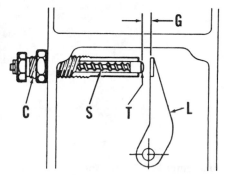

*Fig. L1-14—View of torque control screw. Refer to text for adjustment.*

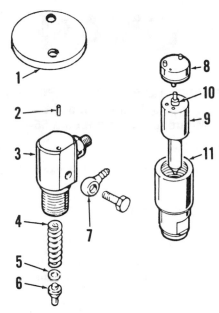

*Fig. L1-15—Exploded view of injector.*

1. Clamp plate
2. Dowel pin
3. Nozzle body
4. Spring
5. Shim
6. Spring seat
7. Return line fitting
8. Spacer
9. Nozzle tip
10. Valve
11. Nozzle holder nut

L1-4) pin with governor arm (F–Fig. L1-3). Tighten pump retaining screws to 29 N·m (22 ft.-lbs.). Loosen clamp nut (N) then move throttle lever (T) to full throttle. Push governor lever (L) in until it stops thus moving injection pump control sleeve to maximum delivery. Tighten clamp nut (N).

Torque control screw (C–Figs. L1-2 and L1-14) is used to allow additional fuel usage under high torque load. The tip (T–Fig. L1-14) is backed by spring (S). Tip (T) must travel 0.2-0.3 mm (0.008-0.012 in.) when 400-430 grams (14-15 oz.) is forced against tip. To adjust torque control screw, run engine at high idle with no load and turn screw so there is 2.1-2.3 mm (0.083-0.091 in.) gap between tip (T) and lever (L). Tighten locknut.

Refer to INJECTION PUMP TIMING section to time injection pump.

**INJECTION NOZZLE.** To remove injection nozzle, first remove dirt from nozzle, injection line, return line and cylinder head. Disconnect return line and high pressure line and immediately cap or plug all openings. Unscrew clamp nuts and lift off clamp plate (1–Fig. L1-15) being careful not to lose dowel pin (2). Injector may now be carefully removed from cylinder head. Do not lose shims between injector and cylinder head.

To disassemble injector, clamp nozzle body (3) in a vise with nozzle tip pointing upward. Remove nozzle holder nut (11). Remove nozzle tip (9) with valve (10) and spacer (8). Invert nozzle body (3) and remove spring seat (6), shim (5) and spring (4). Thoroughly clean all parts in a suitable solvent. Clean inside orifice end of nozzle tip with a wooden cleaning stick. The 0.20 mm (0.008 in.) diameter orifice spray holes may be cleaned by inserting a cleaning wire of proper size. Cleaning wire should be slightly smaller than spray holes. When reassembling injector make certain all components are clean and wet with clean diesel fuel oil. Tighten nozzle holder nut (11) to 60-90 N·m (44-66 ft.-lbs.). Tighten injector clamp plate nuts to 12 N·m (104 in.-lbs.). If accessible, measure protrusion of nozzle into combustion chamber. Nozzle tip should protrude 2.5-3.0 mm (0.094-0.118 in.) above adjacent combustion chamber surface. Adjust position of nozzle using shims between injector and cylinder head which are available in thicknesses of 0.5 mm (0.020 in.) and 1.0 mm (0.040 in.).

If a suitable test stand is available, injector operation may be checked.

**WARNING. Fuel leaves the injection nozzle with sufficient force to penetrate the skin. When testing, keep yourself clear of nozzle spray.**

Opening pressure with a new spring (4–Fig. L1-15) should be 20580-22540 kPa (2986-3271 psi) while opening pressure with a used spring should be 19600-21560 kPa (2844-3129 psi). Opening pressure is adjusted by varying number and thickness of shims (5). Valve should not show leakage at orifice spray holes for 10 seconds at 17640 kPa (2560 psi).

**ELECTRIC STARTER.** Early models are equipped with a Prestolite MGL-4002A electric starter while later models are equipped with Bosch starter B.001.214.002.

The Prestolite starter is secured by clamps to the cylinder block and a rubber spacer ring between spacer pinion housing and steel stamped backplate is used to properly locate starter. Rubber spacer ring thickness should be 14.5-15.5 mm (0.571-0.610 in.). The Bosch starter is bolted to a cast aluminum backplate.

**ALTERNATOR.** An alternator is mounted on flywheel end of engine to recharge battery. The stator is secured to the engine crankcase while a ring of magnets is carried by the flywheel. Note wiring schematic in Fig. L1-16. The magnet ring may be removed from flywheel if faulty. Stator and rotor are available only as an assembly.

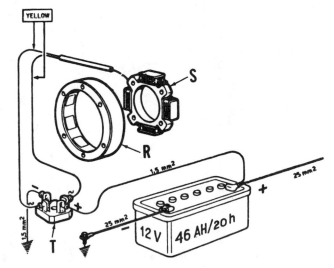

*Fig. L1-16—Alternator wiring schematic.*

R. Rotor
S. Stator
T. Rectifier

# ONAN

**A DIVISION OF ONAN CORPORATION**
1400 73rd Avenue N.E.
Minneapolis, Minnesota 55432

| Model | No. Cyls. | Bore | Stroke | Displacement | Power Rating |
|-------|-----------|------|--------|--------------|--------------|
| LK | 1 | 3.25 in. (82.55 mm) | 3.0 in. (76.2 mm) | 24.9 cu. in. (407.8 cc) | 5 hp. (3.7 kW) |
| LKB | 1 | 3.25 in. (82.55 mm) | 3.0 in. (76.2 mm) | 24.9 cu. in. (407.8 cc) | 8.5 hp. (6.3 kW) |

Models LK and LKB engines are one cylinder, four-cycle, horizontal shaft engines. Crankshaft is supported at each end in precision type sleeve bearings.

Connecting rod in LK model is aluminum alloy, rides directly on crankpin journal and may be splash lubricated if in early model or pressure lubricated in later models.

Connecting rod in LKB model has renewable precision inserts which are pressure lubricated by a gear type oil pump driven off of crankshaft gear.

Various ignition systems utilizing externally mounted points and condenser with generating coil located under flywheel and high tension coil externally located or a combination generating and ignition coil under flywheel are used.

A float type side draft carburetor which may be equipped with an automatic choke is used on all models.

LK engines maximum speed should be 1500 rpm when used on 50 cycle electric generating units and 1800 rpm when used on 60 cycle electric generating units.

LKB engines maximum speed should be 3000 rpm when used on 50 cycle electric generating units and 3600 rpm when used on 60 cycle electric generating units.

## MAINTENANCE

**SPARK PLUG.** Recommended spark plug is Champion H8 or equivalent. Electrode gap is 0.025 inch (0.635 mm) for gasoline and 0.018 inch (0.4572 mm) for LP-Gas or natural gas fuel.

**CARBURETOR.** The same carburetor with a variety of modifications for use with gasoline, LP-Gas, natural gas or a combination of fuels is used on all models. Fig. O1 shows an exploded view of carburetor equipped for gasoline and Fig. O2 shows a carburetor for combination gas or gasoline fuel. Unnecessary parts are eliminated when unit is equipped for single fuel operation.

For initial adjustment, open both the main fuel mixture and the idle fuel mixture screws to 1 to 1¼ turns. Make final adjustment with engine at operating temperature and running. Place engine under load and adjust main fuel needle to leanest mixture that will allow satisfactory acceleration and steady governor operation while maintaining correct generating frequency required.

Run engine at idle speed, no load and adjust idle mixture screw for smoothest idle operation. As each adjustment affects the other, adjustment procedure may have to be repeated.

Throttle idle stop screw should be adjusted to clear throttle shaft stop by 1/32-inch (0.794 mm) when operating at desired speed and no-load condition. This helps prevent erratic governor operation under varying load conditions.

Carburetors used on LK models equipped for LP-Gas and natural gas operation beginning with specification letter "E" have no idle adjustment.

To check float level of gasoline fuel carburetor, invert carburetor body and float assembly. There should be 11/64-inch (4.4 mm) clearance between gasket surface of body and nearest edge of float. Adjust float by bending float lever tang that contacts inlet valve.

**AUTOMATIC CHOKE.** A variety of automatic choke styles have been used which vary in construction.

Some manual start models use a counterweighted choke which is closed

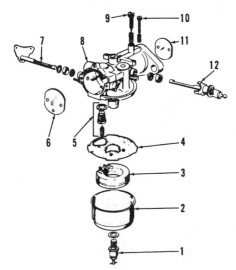

*Fig. O1 — Exploded view of gasoline carburetor showing component parts.*

1. Main adjusting screw
2. Fuel bowl
3. Float
4. Gasket
5. Needle valve & seat
6. Choke valve
7. Choke shaft
8. Carburetor body
9. Idle mixture needle
10. Idle speed screw
11. Throttle valve
12. Throttle shaft

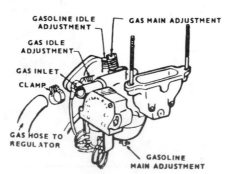

*Fig. O2 — Installed view of carburetor equipped for gasoline and gaseous fuel.*

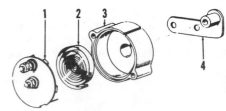

*Fig. O3 — Exploded view of electric automatic choke used on some models.*

Illustrations courtesy of Onan Corp.

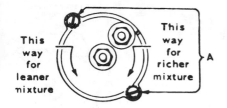

*Fig. O4 — Schematic view showing choke adjustment.*

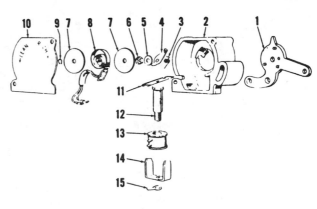

*Fig. O5 — Exploded view of solenoid-operated electric automatic choke used on some models.*

1. Mounting plate
2. Body
3. Spring
4. Lever
5. Washer
6. Pal nut
7. Insulator
8. Heater assembly
9. Snap ring
10. Cover
11. Armature
12. Core
13. Coil
14. Frame
15. Terminal

when engine is stopped, but is opened by air stream when engine is running.

Remote control units are equipped with an electric thermal action choke (Fig. O3). When unit starts, electric current begins heating element in choke housing which causes choke to open. At temperature of 70° F (21°C) choke should be approximately 1/8-inch (3.175 mm) open before engine is started and fully open after engine reaches operating temperature and is running.

Extreme temperature conditions may require choke adjustment. Loosen screws (A – Fig. O4) and rotate housing for leaner or richer setting as shown and retighten screws.

Some models are equipped with thermomagnetic choke (Fig. O5). An electric solenoid closes choke for starting and a bi-metal heating element opens choke as engine reaches operating temperature. A continuous flow of electric current exists in heating element while engine is running. Heating element resistance should be 30.6-37.4 ohms. Choke body mounting screw (Fig. O6) can be loosened and body rotated to adjust choke for best starting performance.

**GOVERNOR.** Governor ball and cup assembly is located on end of camshaft. Refer to **CAMSHAFT** section for assembly information.

For linkage adjustment with engine not running, governor spring should hold carburetor throttle in full open position. Control link (2 – Fig. O7) should just be able to be connected with throttle plate in full open position.

Generating unit engines are governed at a constant speed at all loads, with a minimum of change during variation of load.

Governed speed for LK model should be 1500 rpm for 50 cycle electric units and 1800 rpm for 60 cycle electric units.

Governed speed for LKB model should be 3000 rpm for 50 cycle electric units and 3600 rpm for 60 cycle electric units.

A voltmeter can be used to check governed speed. Maximum permissible no load voltage is 126 volts for 120 volt units and 252 volts for 240 volt units. Minimum recommended full load voltage is 110 volts and 220 volts respectively. Preferred voltage spread is 5 volts for 120 volt unit and 9 volts for 240

volt unit. To adjust governed speed, make certain carburetor and governor control link are correctly adjusted and engine and generator are at normal operating temperature. With all electrical load removed, turn speed adjusting nut (N) to obtain exact rated voltage of 120 or 240 volts. Apply a full electrical load and recheck voltage. Adjust sensitivity screw (S) to obtain least variation in voltage without fluctuation during changes in load. Changing sensitivity adjustment may necessitate a change in speed adjustment.

**IGNITION SYSTEM.** Several different types of ignition systems have been used. Early LK models with

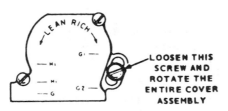

*Fig. O6 — Choke adjustment procedure.*

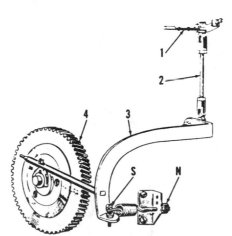

*Fig. O7 — View of governor unit showing component parts.*

1. Throttle shaft
2. Link
3. Governor arm
4. Cam gear

N. Speed adjusting nut
S. Sensitivity adjusting screw

manual start were equipped with an energy transfer magneto system with a generating coil mounted underneath flywheel and a separate high tension ignition coil mounted externally on engine (Fig. O10).

LK remote start units use a battery ignition system with high tension coil mounted externally on engine.

Late LK and all LKB models use a combined generating and high tension coil mounted underneath flywheel (Fig. O11).

All models have externally mounted condenser and breaker points. Breaker point gap is 0.020 inch (0.508 mm) on all models (Fig. O8).

Timing gear cover is marked at "TC",

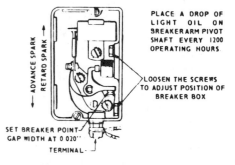

*Fig. O8 — View of ignition breaker box to show timing and point adjustment procedure.*

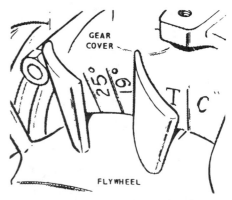

*Fig. O9 — Most models are furnished with flywheel timing marks as shown.*

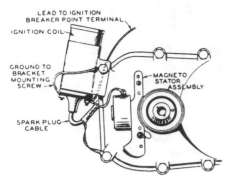

*Fig. O10 — View of magneto and coil installation used on early magneto units.*

19° and 25° as shown in Fig. O9 and flywheel has a timing mark. Timing is set at 19° BTDC on all LK models and is adjusted by moving breaker box up to retard timing or down to advance timing. Timing on LKB engines is set at 24° BTDC with engine running at rated speed.

Manual start LKB engines are equipped with a centrifugal advance mechanism which provides retarded timing for starting only. Static spark should occur at 5° BTDC on these models and advance should occur at 800 rpm. Fig. O14 shows an exploded view of centrifugal advance mechanism located on rear (generator) end of camshaft and is accessible for service by removing crankcase housing end plug or camshaft. If end plug is removed do not dent when reinstalling as interference with weight mechanism will occur.

**LUBRICATION.** Early production engines are splash lubricated by oil dipper on rod cap. When assembling, make certain oil dipper is installed to splash oil toward camshaft side of engine.

Late production engines are pressure lubricated by a gear type pump driven by crankshaft gear. Pump parts are not serviced separately so entire pump must

be renewed if worn or damaged. Clearance between pump gear and crankshaft gear should be 0.005 inch (0.127 mm) and normal operating pressure should be 25 psi (173 kPa).

Oils approved by manufacturer must meet requirements of API service classification "SE" or "SE/CC".

Use SAE 30 oil in temperatures of 32° F (0° C) to 90° F (32° C), 10W-40 or 5W-30 oil in temperatures of 0° F (-18° C) to 32° F (0° C) and 5W-30 oil in temperatures below 0° F (-18° C).

Check oil level with engine stopped and maintain level at top of fill plug.

Change oil every 25 hours of normal operation. Crankcase capacity is 2 quarts (1.9 L) for all models.

## REPAIRS

**TIGHTENING TORQUES.** Recommended tightening torques are as follows:

| | |
|---|---|
| Connecting rod | 24-26 ft.-lbs. |
| | (33-35 N·m) |
| Cylinder head | 27-29 ft.-lbs. |
| | (37-39 N·m) |
| Flywheel | 40-45 ft.-lbs. |
| | (54-61 N·m) |
| Gear cover | 15-20 ft.-lbs. |
| | (20-27 N·m) |
| Oil base | 43-48 ft.-lbs. |
| | (57-65 N·m) |
| Rear bearing plate | 20-25 ft.-lbs. |
| | (27-34 N·m) |

**CONNECTING ROD.** Connecting rod and piston unit is removed from above after removal of cylinder head and oil base.

LK model aluminum alloy rod rides directly on crankpin journal and should have a rod to journal clearance of 0.002-0.003 inch (0.0508-0.0762 mm) and should have side clearance of 0.013-0.038 inch (0.3302-0.9652 mm).

Models with forged rod and renewable bearing insert should have rod bearing to journal clearance of 0.0005-0.002 inch (0.0127-0.0508 mm) and should have

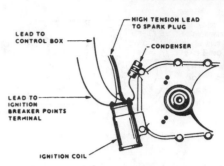

*Fig. O13 — Coil mounting and wiring used on late battery ignition models.*

side clearance of 0.002-0.016 inch (0.0508-0.4064 mm).

Aluminum alloy rods or bearing inserts for forged rods are available in a variety of sizes for undersize crankshaft journals as well as standard.

When assembling engines with splash lubrication system make certain oil dipper on rod cap is installed to splash oil toward camshaft side of engine.

**PISTON, PIN AND RINGS.** An aluminum, cam ground piston is fitted with two compression rings and one oil control ring. Piston ring end gap should be 0.010-0.023 inch (0.254-0.5842 mm).

The floating type piston pin should be a hand push fit in piston and thumb push fit in connecting rod at 72° F (22° C) and is retained in piston by two snap rings. Pin is available in 0.002 inch (0.0508 mm) oversize as well as standard.

Standard cylinder bore is 3.249-3.250 inches (82.53-82.55 mm) and recommended piston skirt clearance is 0.0015-0.0035 inch (0.0381-0.0889 mm) when measured at bottom of skirt. Pistons and rings are available in a variety of oversizes as well as standard.

Some engines were factory equipped with 0.005 inch (0.127 mm) oversize pistons during manufacture and are identified by an "E" suffix on serial number. Standard rings are used as 0.005 inch (0.127 mm) oversize rings are not available.

**CRANKSHAFT, BEARINGS AND SEALS.** Crankshaft is supported at each end by precision type sleeve bearings located in cylinder block housing

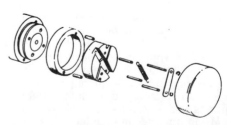

*Fig. O14 — Exploded view of centrifugal advance mechanism used on Series LKB with manual start.*

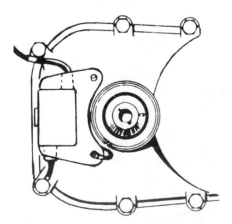

*Fig. O11 — On late magneto units, high tension coil is combined with generating coil as shown.*

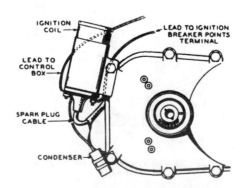

*Fig. O12 — View of coil mounting used on early battery ignition units.*

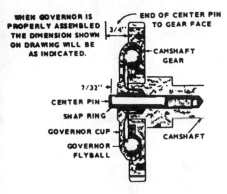

WHEN GOVERNOR IS PROPERLY ASSEMBLED THE DIMENSION SHOWN ON DRAWING WILL BE AS INDICATED.

END OF CENTER PIN TO GEAR FACE

3/4"

CAMSHAFT GEAR

7/32"

CENTER PIN

SNAP RING

GOVERNOR CUP

GOVERNOR FLYBALL

CAMSHAFT

*Fig. O15—Cross-sectional view of camshaft gear and governor.*

*Fig. O17—Exploded view of crankshaft and associated parts.*

*Fig. O18—Exploded view of camshaft, cam gear and governor weight unit.*

*Fig. O16—Exploded view of piston and connecting rod assembly used on Series LKB. Models without pressure lubrication do not have precision bearing inserts and rod cap is equipped with oil dipper.*

and rear bearing plate. Main bearing journal standard diameter should be 1.9995-2.000 inches (50.7873-50.800 mm) and bearing operating clearance should be 0.0025-0.0038 inch (0.0635-0.0965 mm). Main bearings are available in 0.002, 0.010, 0.020 and 0.030 inch (0.0508, 0.254, 0.508 and 0.762 mm) oversize as well as standard.

Oil holes in bearing and bore **MUST** be aligned when bearings are installed and

bearings should be pressed into bores so inside edge of bearing is 1/16 to 3/32-inch (1.589-2.381 mm) back from inside end of bore to allow clearance for radius of crankshaft. Oil grooves in thrust washers must face crankshaft and washers two alignment notches must fit over lock pins. Thrust washers must be in good condition or excessive crankshaft end play will result.

Recommended crankshaft end play should be 0.006-0.012 inch (0.1524-0.3048 mm) and is adjusted by varying thickness of gaskets between bearing plate and cylinder block. Gaskets are available in a variety of thicknesses.

It is recommended rear bearing plate and front timing gear cover be removed for seal installation. Rear seal is pressed in until flush with seal bore and old style front seal is driven inward 31/32-inch (24.606 mm) and new thin, open face seal is driven 1-7/64 inches (28.179 mm) from mounting face of cover.

## CAMSHAFT, BEARINGS AND GOVERNOR. 
Camshaft is supported at each end by precision sleeve type bearings pressed into bearing bores. Make certain oil holes in bearings and bores are aligned during installation. Recommended bearing to camshaft clearance is 0.0015-0.003 inch (0.0381-0.0762 mm).

Governor weight unit is mounted on camshaft gear and governor cup rides on center pin pressed into camshaft (Fig. O15) so it extends ¾-inch (19 mm) from front face of camshaft gear. Governor cup is retained on shaft by snap ring and cup should have 7/32-inch (5.556 mm) movement on shaft. Governor cup is prevented from rotating by a pin in

timing gear cover which should enter metal lined hole in cup during cover installation.

Camshaft gear is a press fit on camshaft and a camshaft gear thrust washer between camshaft gear and cylinder block is used to adjust end play of camshaft to 0.003 inch (0.0762 mm). Make certain timing marks on camshaft gear and crankshaft gear are aligned after installation.

**VALVE SYSTEM.** Valve seats are renewable insert type and are available in a variety of oversizes as well as standard. Seats are ground to 45° angle and seat width should be 1/32 to 3/64-inch (0.794 to 1.191 mm).

Valves should be ground to 44° angle to provide an interference angle of 1°. Stellite valves and seats are available and roto-caps are standard on both intake and exhaust valves.

Recommended valve stem to guide clearance is 0.001-0.0025 inch (0.254-0.0635 mm) for intake valves and 0.0025-0.004 inch (0.0635-0.1016 mm) for exhaust valves. Renewable valve guides are shouldered and are pushed out from above. Early models use a gasket between guide and cylinder block.

Recommended valve tappet gap (cold) for intake and exhaust valves on LK engines is 0.015-0.017 inch (0.3810-0.4318 mm) and on LKB engines gap should be 0.010-0.012 inch (0.254-0.305 mm). Adjustment is made by turning tappet adjusting screw as required with engine at TDC on compression stroke.

Valve tappet clearance to bores should be 0.015-0.017 inch (0.3810-0.4318 mm).

# ONAN

### A DIVISION OF ONAN CORPORATION
#### 1400 73rd Avenue, N.E.
#### Minneapolis, Minnesota 55432

| Model | No. Cyls. | Bore | Stroke | Displacement | Power Rating |
|-------|-----------|------|--------|--------------|--------------|
| BF | 2 | 3.125 in. (79.38 mm) | 2.625 in. (66.68 mm) | 40.3 cu. in. (660 cc) | 16 hp. (11.9 kW) |
| BG | 2 | 3.250 in. (82.55 mm) | 3 in. (76.2 mm) | 49.8 cu. in. (815.7 cc) | 18 hp. (13.4 kW) |
| B43M, B43E | 2 | 3.250 in. (82.55 mm) | 2.620 in. (66.55 mm) | 43.3 cu. in. (712.4 cc) | 16 hp. (11.9 kW) |
| B43G | 2 | 3.250 in. (82.55 mm) | 2.620 in. (66.55 mm) | 43.3 cu. in. (712.4 cc) | 18 hp. (13.4 kW) |
| B48G | 2 | 3.250 in. (82.55 mm) | 2.875 in. (73.03 mm) | 47.7 cu. in. (781.7 cc) | 20 hp. (14.9 kW) |
| B48M | 2 | 3.250 in. (82.55 mm) | 2.875 in. (73.03 mm) | 47.7 cu. in. (781.7 cc) | 18 hp. (13.4 kW) |

Engines in this section are four-cycle, twin-cylinder opposed, horizontal crankshaft type in which crankshaft is supported at each end in precision sleeve type bearings.

Connecting rods ride directly on crankshaft journals and all except early BF engines are pressure lubricated by a gear type oil pump driven by crankshaft gear. Early BF engines were splash lubricated.

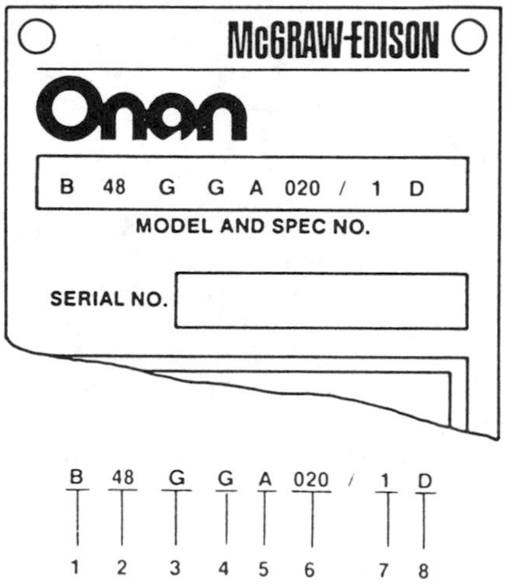

All models use a battery ignition system consisting of points, condenser, coil, battery and spark plug. Timing is adjustable only by varying point gap.

All models use a single down draft float type carburetor and may have a vacuum operated fuel pump attached to carburetor or mounted separately according to model and application.

Refer to model and specification number (Fig. O19) for engine model identification and interpretation.

Always give model, specification and serial numbers when ordering parts or service information.

## MAINTENANCE

**SPARK PLUG.** Recommended spark plug is Champion H8 or equivalent. Electrode gap for all models using gasoline fuel is 0.025 inch (0.635 mm). Electrode gap for all models using vapor (butane, LP or natural gas) fuel is 0.018 inch (0.457 mm).

**CARBURETOR.** According to model and application all engines use a single down draft float type carburetor manufactured by Marvel Schebler, Walbro or Nikki. Exploded view of each

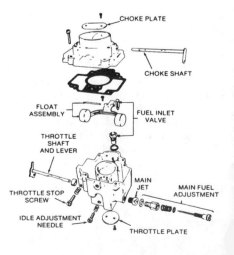

**McGRAW-EDISON**

**Onan**

| B 48 G G A 020 / 1 D |
|---|

**MODEL AND SPEC NO.**

**SERIAL NO.**

B 48 G G A 020 / 1 D
1 2 3 4 5 6 7 8

Fig. O19 — Typical model, specification and serial number plate on Onan engine showing digit interpretation.

1. Identification of basic engine series.
2. Displacement (in cubic inches).
3. Engine duty cycle.
4. Fuel required.
5. Cooling system designation.
6. Power rating (in BHP)
7. Designated optional equipment.
8. Production modifications.

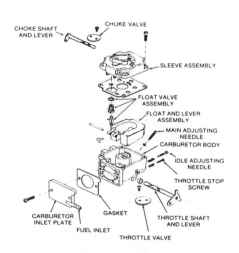

Fig. O20 — Exploded view of Marvel Schebler carburetor showing component parts and their relative positions.

Fig. O20A — Exploded view of Walbro carburetor showing component parts and their relative positions.

Illustrations courtesy of Onan Corp.

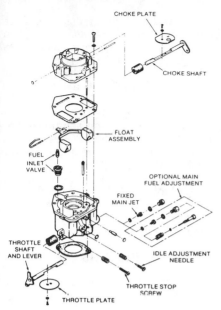

Fig. O20B—Exploded view of Nikki carburetor showing component parts and their relative positions.

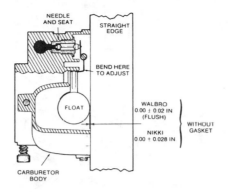

Fig. O22—To adjust float level on Walbro and Nikki carburetors, position carburetor as shown. Clearance should be measured from machined surface without gasket to float edge.

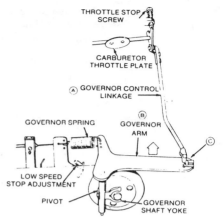

Fig. O25—View of front pull variable speed governor linkage.

carburetor is shown in either Fig. O20, Fig. O20A or Fig. O20B. Refer to appropriate Figure for model being serviced.

For initial adjustment of Marvel Schebler carburetor refer to Fig. O20. Open idle mixture screw 1 turn and main fuel mixture screw 1¼ turns.

For initial adjustment of Walbro carburetor refer to Fig. O20A. Open idle mixture screw 1-1/8 turns. If equipped with optional main fuel adjustment open main fuel mixture screw 1½ turns.

For initial adjustment of Nikki carburetor refer to Fig. O20B. Open idle mixture screw ¾-turn. If equipped with optional main fuel adjustment open main fuel mixture screw 1½ turns.

Make final adjustment to all models with engine at operating temperature and running. Place engine under load and adjust main fuel mixture screw to leanest mixture that will allow satisfactory acceleration and steady governor operation.

Run engine at idle speed, no load and adjust idle mixture screw for smoothest idle operation. As each adjustment affects the other, adjustment procedure may have to be repeated.

To check float level of Marvel Schebler Carburetor refer to Fig. O21. Invert carburetor throttle body and float assembly. There should be 1/8-inch (3.175 mm) clearance between gasket and float as shown. Adjust float by bending float lever tang that contacts inlet valve.

To check float level of Walbro or Nikki carburetor refer to Fig. O22. Position carburetor as shown. Walbro carburetor should have 0.00 inch, plus or minus 0.02 inch (0.00 mm, plus or minus 0.508 mm) clearance and Nikki carburetor should have 0.00 inch, plus or minus 0.028 inch (0.00 mm, plus or minus 0.7112 mm) clearance between machined surface without gasket, to float edge. Adjust float by bending float lever tang that contacts inlet valve.

All models use a pulsating diaphragm fuel pump. Refer to **SERVICING ONAN ACCESSORIES** section for service information.

**GOVERNOR.** All models use a flyball weight governor located under timing gear on camshaft gear. Various linkage arrangements allow for either fixed speed operation or variable speed operation.

**FIXED SPEED.** Fixed speed linkage is used on electric generators and welders or any application which requires a single constant speed under varying load conditions. Fig. O24 shows typical linkage arrangement.

Governor link length should be adjusted so stop on throttle shaft almost touches stop on side of carburetor with engine stopped and governor spring under tension.

Sensitivity is adjusted by varying tension of governor spring and changing location of governor spring on governor arm.

Engine speed should be set to manufacturers specifications and output of generators should be checked with a frequency meter.

**VARIABLE SPEED.** Variable speed linkage is used where it is necessary to maintain a wide range of engine speeds under varying load conditions. Moving a control lever determines engine speed and governor maintains this speed under varying loads.

Front pull (Fig. O25) and side pull

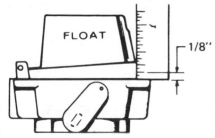

Fig. O21—Float valve setting for Marvel Schebler carburetor. Measure between inverted float and surface of gasket.

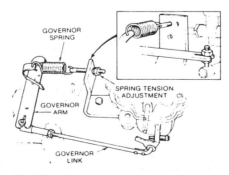

Fig. O24—View of fixed speed governor linkage.

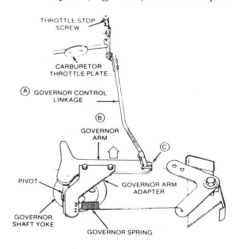

Fig. O26—View of side pull variable speed governor linkage.

(Fig. O26) linkage arrangements are used and service procedures are similar.

For correct governor link installation on either front or side pull system, disconnect linkage (A) from hole (C) and push linkage and governor arm (B) toward carburetor as far as they will go. While held in this position insert end of linkage into nearest hole in governor arm. If between two holes insert in next hole out.

Normal factory setting is in third hole from pivot for side pull linkage and second hole from pivot for front pull linkage.

Sensitivity is increased by connecting governor spring closer to pivot and decreased by connecting governor spring further from pivot.

Engine speed should be set to manufacturers specifications and speed should be checked after linkage adjustments.

**IGNITION.** A battery type ignition system is used on all models. Breaker point box is non-movable and only means of changing timing is to vary point gap slightly. Static timing should be checked if point gap is changed or new points are installed. Use a continuity test light across breaker points to check timing and remove air intake hose from blower housing for access to timing marks. Refer to ignition specifications as follows:

| Model | Initial Point Gap | Static Timing BTDC |
|---|---|---|
| BF | 0.025 in. (0.635 mm) | 21° |
| BF* | 0.025 in. (0.635 mm) | 21° |
| BF** | 0.025 in. (0.635 mm) | 26° |
| BG, B43M, B48M | 0.021 in. (0.533 mm) | 21° |
| B48G | 0.020 in. (0.508 mm) | 20° |
| B48G* | 0.016 in. (0.406 mm) | 16° |

*Specification letter "C" and after.
**"Power Drawer" models.

**LUBRICATION.** Early Model BF is splash lubricated and all other models are pressure lubricated by a gear type pump driven by crankshaft gear. Internal pump parts are not serviced separately so entire pump must be renewed if worn or damaged. Clearance between pump gear and crankshaft gear should be 0.002-0.005 inch (0.0508-0.127 mm) and normal operating pressure should be 30 psi (207 kPa). Oil pressure is regulated by an oil pressure

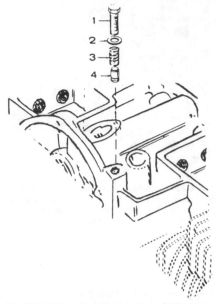

**Fig. O27 — View showing location of oil pressure relief valve and spring.**

1. Cap screw
2. Sealing washer
3. Spring
4. Valve

relief valve (Fig. O27) located in engine block near timing gear cover. Spring (3) free length should be 1.00 inch (25.4 mm) and valve (4) diameter should be 0.3365-0.3380 inch (8.55-8.59 mm).

Oils approved by manufacturer must meet requirements of API service classification "SE" or "SE/CC".

Use SAE 5W-30 oil in below freezing temperatures and SAE 20W-40 oil in above freezing temperatures.

Check oil level with engine stopped and maintain at "FULL" mark on dipstick. **DO NOT** overfill.

Recommended oil change intervals for

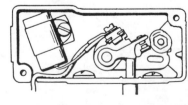

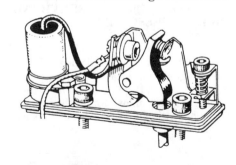

**Fig. O28 — Two different type point boxes are used according to model and application. Upper illustration shows top adjust point models while lower illustration shows side adjust point model.**

models without oil filter is every 25 hours of normal operation. If equipped with a filter, change oil at 50 hour intervals and filter every 100 hours.

Crankcase capacity without filter is 3.5 pints (1.66 L) for BG model, 4 pints (1.86 L) for BF model, 1.5 quarts (1.4 L) for B43E, B43G and B48G models. If filter is changed add an additional 0.5 pint (0.24 L) for BF and BG models and an additional 0.5 quart (0.47 L) for B43E, B43G and B48G models. Oil filter is screw on type and gasket should be lightly lubricated before installation. Screw filter on until gasket contacts, then turn an additional ½-turn. Do not overtighten.

## REPAIRS

**TIGHTENING TORQUES.** Recommended tightening torques are as follows:

Cylinder heads—
BF,
BG . . . . . . . . . . . . . . . . . .14-16 ft.-lbs.
(19-22 N·m)

B43E,*
B43G,*
B48G,*
B43M,*
B48M* . . . . . . . . . . . . . . .16-18 ft.-lbs.
(22-24 N·m)

*If graphoil gasket is used, tighten to 14-16 ft.-lbs. (19-22 N·m) torque.

Rear bearing plate . . . . . . .25-27 ft.-lbs.
(34-37 N·m)

Connecting rod—
BG,
BF . . . . . . . . . . . . . . . . . .14-16 ft.-lbs.
(19-22 N·m)

All others . . . . . . . . . . . . .12-14 ft.-lbs.
(16-19 N·m)

Flywheel . . . . . . . . . . . . . .35-40 ft.-lbs.
(48-54 N·m)

Oil base . . . . . . . . . . . . . .18-23 ft.-lbs.
(24-31 N·m)

Timing cover . . . . . . . . . . .8-10 ft.-lbs.
(11-13 N·m)

Oil pump . . . . . . . . . . . . . . .7-9 ft.-lbs.
(10-12 N·m)

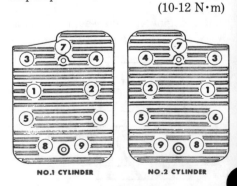

**Fig. O29 — Torque sequence shown is for all models even though spark plug may be in different location in head.**

Illustrations courtesy of Onan Corp.

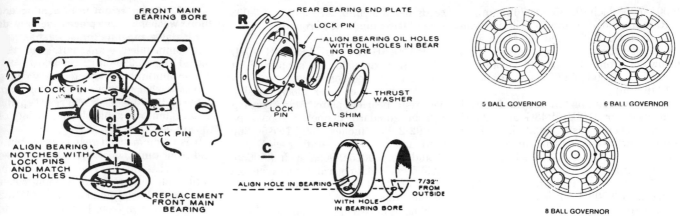

Fig. O31 — Alignment of precision main camshaft bearing at rear (View R). Note shim use for end play adjustment. View C shows placement of camshaft bearings.

Fig. O32 — Flyballs must be arranged in pockets as shown, according to total number used to obtain governor sensitivity desired. Refer to text.

**CYLINDER HEADS.** It is recommended cylinder heads be removed and carbon cleaned at 200 hour intervals.

**CAUTION: Cylinder heads should not be unbolted from block when hot due to danger of head warpage.**

When installing cylinder heads torque bolts in sequence shown in Fig. O29 in gradual, even steps until correct torque for model being serviced is obtained. Note spark plug location on some models varies but torque sequence is the same.

**CONNECTING RODS.** Connecting rod and piston are removed from cylinder head surface end of block after removing cylinder head and oil base. Connecting rods ride directly on crankshaft journal on all models and standard journal diameter should be 1.6252-1.6260 inches (41.28-41.30 mm). Connecting rods are available for 0.005, 0.010, 0.020, 0.030 and 0.040 inch (0.127, 0.254, 0.508, 0.762 and 1.016 mm) undersize crankshafts as well as standard.

Connecting rod to crankshaft journal running clearance should be 0.002-0.0033 inch (0.0508-0.0838 mm) and side play should be 0.002-0.016 inch (0.0508-0.406 mm).

When reinstalling connecting rods make certain rods are installed with rod bolts off-set toward outside of block and tighten to specified torque.

**PISTON, PIN AND RINGS.** Aluminum pistons are fitted with two compression rings and one oil control ring. Pistons in BF models should be renewed if top compression ring has 0.008 inch (0.2032 mm) or more side clearance and pistons in all other models should be renewed if 0.004 inch (0.1016 mm) or more side clearance is present.

Pistons and rings are available in 0.005, 0.010, 0.020, 0.030 and 0.040 inch

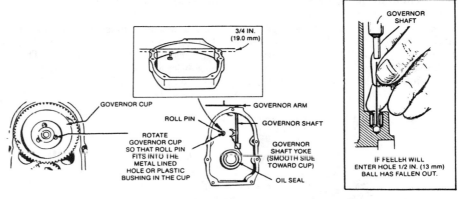

Fig. O33 — View of governor shaft, timing cover and governor cup showing assembly details. Refer to text.

(0.127, 0.254, 0.508, 0.762 and 1.016 mm) oversize as well as standard.

Recommended piston to cylinder wall clearance for BF engines when measured 0.10 inch (2.54 mm) below oil control ring 90° from pin should be

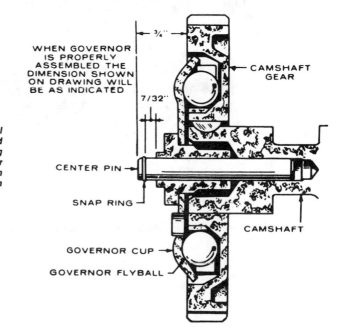

Fig. O34 — Cross-sectional view of camshaft gear and governor assembly showing correct dimensions for center pin extension from camshaft. Further detail in text.

0.001-0.003 inch (0.0254-0.0762 mm).

Recommended piston to cylinder wall clearance for B43M and B48M engines when measured 0.35 inch (8.89 mm) below oil control ring 90° from pin should be 0.0033-0.0053 inch (0.0838-0.1346 mm).

Recommended piston to cylinder wall clearance for B43E, B43G and B48G engines when measured 0.35 inch (8.89 mm) below oil control ring 90° from pin should be 0.0044-0.0064 inch (0.1118-0.1626 mm).

Ring end gap should be 0.010-0.020 inch (0.254-0.508 mm) clearance in standard bore.

Piston pin to piston bore clearance should be 0.0002-0.0004 inch (0.0051-0.0102 mm) and pin to rod clearance should be 0.0002-0.0007 inch (0.0051-0.0178 mm) for all models.

Standard cylinder bore of BF engine is 3.1245-3.1255 inches (79.36-79.39 mm) and standard cylinder bore for all other models is 3.249-3.250 inches (82.53-82.55 mm).

Engines should be rebored if taper exceeds 0.005 inch (0.127 mm) or if 0.003 inch (0.0762 mm) out-of-round.

**CRANKSHAFT, BEARINGS AND SEALS.** Crankshaft is supported at each end by precision type sleeve bearings located in cylinder block housing and rear bearing plate. Main bearing journal standard diameter should be 1.9992-2.000 inches (50.7797-50.800 mm) and bearing operating clearance should be 0.0025-0.0038 inch (0.0635-0.0965 mm). Main bearings are available for a variety of undersize crankshafts as well as standard.

Oil holes in bearing and bore **MUST** be aligned when bearings are installed and bearings should be pressed into bores so inside edge of bearing is 1/16 to 3/32-inch (1.588 to 2.381 mm) back from inside end of bore to allow clearance for radius of crankshaft. Oil grooves in thrust washers must face crankshaft. The two alignment notches on thrust washers must fit over lock pins. Thrust washers must be in good condition or excessive crankshaft end play will result.

NOTE: Replacement front bearing and thrust washer is a one piece assembly, do not install a separate thrust washer.

Recommended crankshaft end play should be 0.006-0.012 inch (0.1524-0.3048 mm) and is adjusted by varying thrust washer thickness or placing shim between thrust washer and rear bearing plate. Shims are available in a variety of thicknesses. See Fig. O31.

It is recommended rear bearing plate and front timing gear cover be removed for seal renewal. Rear seal is pressed in until flush with seal bore. Timing cover seal should be driven in until it is 1-1/32 inch (26.19 mm) from mounting face to cover.

**CAMSHAFT, BEARINGS AND GOVERNOR.** Camshaft is supported at each end in precision type sleeve bearings. Make certain oil holes in bearings are aligned with oil holes in block, press front bearing in flush with outer surface of block and press rear bearing in until flush with bottom of counterbore. Camshaft to bearing clearance should be 0.0015-0.0030 inch (0.0381-0.0762 mm).

Camshaft end play should be 0.003 inch (0.0762 mm) and is adjusted by varying thickness of shim located between camshaft timing gear and engine block.

Camshaft timing gear is a press fit on end of camshaft and is designed to accommodate 5, 8 or 10 flyballs arranged as shown in Fig. O32. Number of flyballs is varied to alter governor sensitivity. Fewer flyballs are used on engines with variable speed applications and greater number of flyballs are used for continuous speed application. Flyballs must be arranged in "pockets" as shown according to total number used. Make certain timing mark on cam gear aligns with timing mark on crankshaft gear when reinstalling.

Center pin (Fig. O34) should extend ¾-inch (19 mm) from end of camshaft to allow 7/32-inch (5.6 mm) in-and-out movement of governor cup. If distance is incorrect remove pin and press new pin in to correct depth.

Always make certain governor shaft pivot ball is in timing cover by measuring as shown in Fig. O33 inset and check length of roll pin which engages bushed hole in governor cup. This pin must extend 25/32-inch (19.844 mm) from timing cover mating surface.

**VALVE SYSTEM.** Valve seats are renewable insert type and are available in a variety of oversizes as well as standard. Seats are ground at a 45° angle and seat width should be 1/32 3/64-inch (0.794 to 1.191 mm).

Valves should be ground at a 44° angle to provide an interference angle of 1°. Stellite valves and seats are

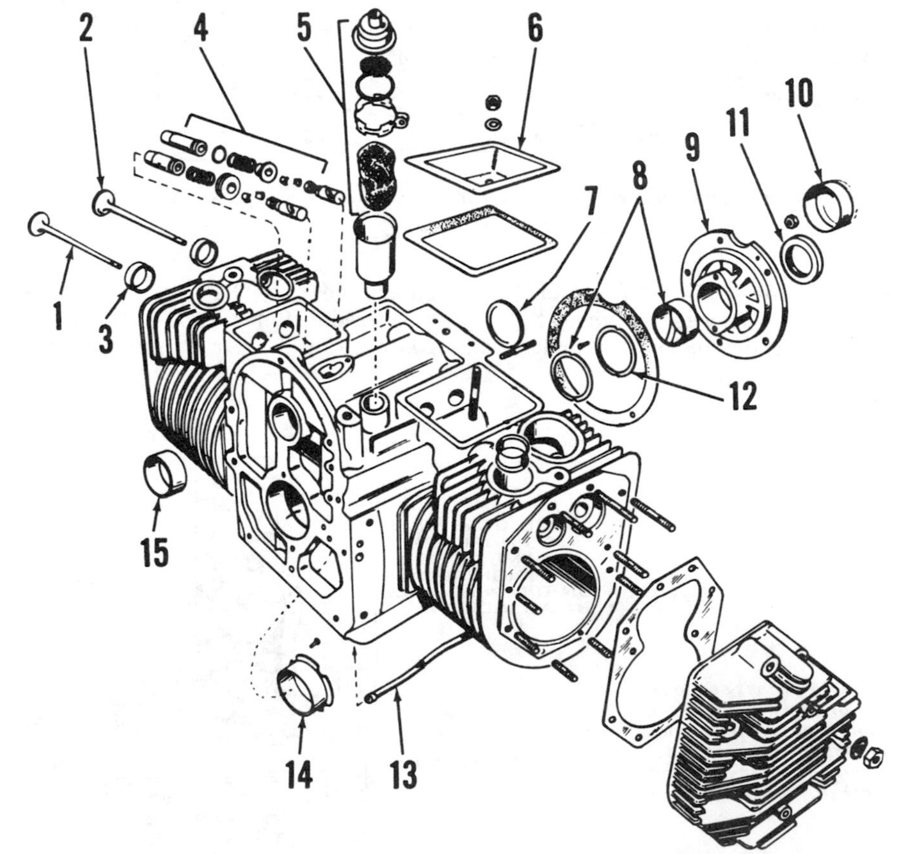

*Fig. O35—Exploded view of cylinder and crankcase assembly typical of all models.*

1. Exhaust valve
2. Intake valve
3. Seat insert (2)
4. Guide, spring, tappet group
5. Crankcase breather group
6. Valve compartment cover
*7. Camshaft expansion plug
8. Two-piece main bearing
9. Rear main bearing plate
*10. Timing control cover
11. Crankshaft seal
12. Crankshaft thrust shim
13. Oil tube
14. One-piece main bearing
15. Camshaft bearing (2)
*Plug is installed on engine without timing control.

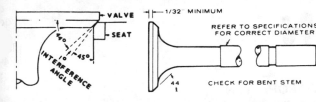

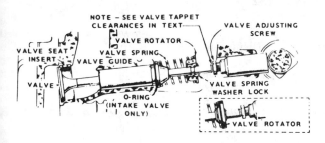

*Fig. O36 — Principal valve service specifications illustrated. Refer to text for further detail.*

*Fig. O37 — Typical valve train on all models. Refer to text.*

available and rotocaps are available for exhaust valves.

Recommended valve stem to guide clearance is 0.001-0.0025 inch (0.0254-0.0635 mm) for intake valves and 0.0025-0.004 inch (0.0635-0.1016 mm) for exhaust valve. Renewable valve guides are shouldered and are pushed out from above. Early models use a gasket between guide and cylinder block and some models may be equipped with valve stem seal on intake valve which must be renewed if valve is removed.

Recommended valve tappet gap (cold) for B43E model is 0.003 inch (0.0762 mm) for intake valves and 0.010 inch (0.254 mm) for exhaust valves. Recommended valve tappet gap (cold) for all remaining models is 0.008 inch (0.2032 mm) for intake valves and 0.013 inch (0.3302 mm) for exhaust valves.

Adjustment is made by turning tappet adjusting screw as required and valves of each cylinder must be adjusted with cylinder at "top dead center" on compression stroke. At this position both valves will be fully closed.

# ONAN

## A DIVISION OF ONAN CORPORATION
### 1400 73rd Avenue N.E.
### Minneapolis, Minnesota 55432

| Model | No. Cyls. | Bore | Stroke | Displacement | Power Rating |
|---|---|---|---|---|---|
| NH | 2 | 3.563 in. (90.50 mm) | 3 in. (76.2 mm) | 60 cu. in. (980.3 cc) | 25 hp. (18.6 kW) |
| NHA | 2 | 3.563 in. (90.50 mm) | 3 in. (76.2 mm) | 60 cu. in. (980.3 cc) | 18 hp. (13.4 kW) |
| NHB | 2 | 3.563 in. (90.50 mm) | 3 in. (76.2 mm) | 60 cu. in. (980.3 cc) | 20 hp. (14.9 kW) |
| NHC | 2 | 3.563 in. (90.50 mm) | 3 in. (76.2 mm) | 60 cu. in. (980.3 cc) | 25 hp. (18.6 kW) |
| T-260G | 2 | 3.563 in. (90.50 mm) | 3 in. (76.2 mm) | 60 cu. in. (980.3 cc) | 24 hp. (17.9 kW) |

Engines in this section are four-cycle, twin cylinder opposed, horizontal crankshaft engines. Crankshaft is supported at each end in precision sleeve type bearings.

Connecting rods in T-260G engines ride directly on crankshaft journal and all remaining models use renewable precision insert rod bearings. All models are pressure lubricated by a gear type oil pump and engines are equipped with "spin-on" oil filter with a by-pass.

Most models are equipped with a battery ignition system, however a magneto system is available for manual start models.

Side draft or down draft carburetor may be used according to model or application and fuel is supplied by either a mechanically operated or vacuum operated fuel pump.

Refer to Fig. O38 or O38A for interpretation of engine specification and model numbers. Always give specification, model and serial numbers when ordering parts or service material.

## MAINTENANCE

**SPARK PLUG.** Recommended spark plug for NHC and NHCV models with specification letter A through C is Onan part number 167-0240 for non-resistor plug and 167-0247 for a resistor type plug or their equivalent. Recommended spark plug for all remaining models is Onan part number 167-0291 or equivalent. Electrode gap for all models using vapor (butane, LP or natural gas) fuel is 0.018 inch (0.4572 mm) and gap

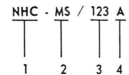

**Fig. O38 – Interpretation of engine model and specification number as an aid to identification of various engines.**

1. General engine model identification
2. Specific type:
   S – Manual starting
   MS – Electric starting
3. Optional equipment identification
4. Specification letter which advances with factory production modifications

**Fig. O38A – Interpretation of engine model and specification number as an aid to identification of various engines.**

1. General engine model identification
2. Number of cylinders
3. Cubic inch displacement
4. Engine duty cycle (M = medium duty)
5. Fuel required (G = gasoline)
6. Cooling system description (A = air cooling pressure)
7. BHP rating
8. Optional equipment identification
9. Specification letter which advances with factory production modifications

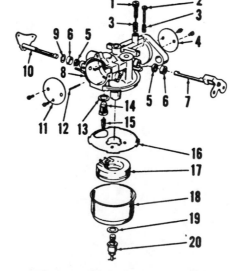

**Fig. O39 – Exploded view of Onan side draft carburetor used on NH and NHC model engines.**

1. Idle mixture needle
2. Throttle stop screw
3. Springs (2)
4. Throttle plate
5. Shaft seal (2)
6. Seal retainer (2)
7. Throttle shaft
8. Body
9. Washer
10. Choke shaft
11. Choke plate
12. Float pin
13. Washer
14. Inlet valve seat
15. Inlet valve needle
16. Body gasket
17. Float
18. Float bowl
19. Washer
20. Main fuel valve

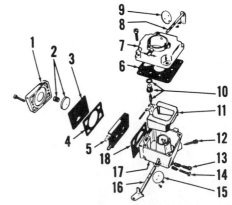

**Fig. O40 – Exploded view of downdraft carburetor used on NHA, NHB and NHC models.**

1. Pump cover
2. Diaphragm plunger & spring
3. Diaphragm
4. Gasket
5. Pump body
6. Gasket
7. Air horn
8. Choke shaft
9. Choke plate
10. Inlet valve
11. Float
12. Main fuel needle
13. Idle mixture needle
14. Throttle stop screw
15. Throttle plate
16. Throttle shaft
17. Carburetor body
18. Gasket

Illustrations courtesy of Onan Corp.

for models using gasoline for fuel is 0.025 inch (0.635 mm).

**CARBURETOR.** According to model and application, engine may be equipped with either a side draft or a downdraft carburetor. Refer to appropriate paragraph for model being serviced.

SIDE DRAFT CARBURETOR. Side draft carburetor is used on some NH and NHC model engines. Refer to Fig. O39 for exploded view of carburetor and location of mixture adjustment screws.

For initial carburetor adjustment, open idle mixture screw and main fuel mixture screw to 1 to 1½ turns. Make final adjustments with engine at normal operating temperature and running. Place engine under load and adjust main fuel mixture screw for leanest setting that will allow satisfactory acceleration and steady governor operation. Set engine at idle speed, no load and adjust idle mixture screw for smoothest idle operation.

As each adjustment affects the other, adjustment procedure may have to be repeated.

To check float level refer to Fig. O41. Invert carburetor throttle body and float assembly. There should be 1/8-inch (3.2 mm), plus or minus 1/16-inch (1.6 mm), clearance between gasket and free end of float. See Fig. O41.

Adjust float by bending float lever tang that contacts inlet valve.

DOWNDRAFT CARBURETOR. Downdraft carburetor is used on T-260G model and some NH and NHC models. Refer to Fig. O40 for exploded view of carburetor and location of mixture adjustment screws.

For initial carburetor adjustment, open idle mixture screw 1-3/8 to 1-5/8 turns and open main fuel mixture screw 1¼ to 1½ turns. Make final adjustments with engine at normal operating temperature and running. Place engine under load and adjust main fuel mixture screw for leanest setting that will allow satisfactory acceleration and steady

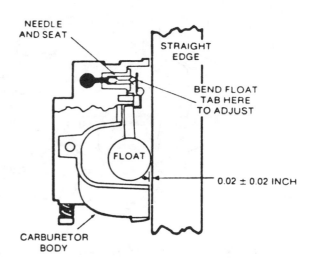

*Fig. O42 — View showing use of straightedge to check float level on downdraft type carburetor. Float level clearance should be 0.00-0.04 inch (0.00-1.02 mm).*

governor operation. Set engine at idle speed, no load and adjust idle mixture screw for smoothest idle operation.

As each adjustment affects the other, adjustment procedure may have to be repeated.

To check float level refer to Fig. O42. Position carburetor as shown. There should be 0.00-0.04 inch (0.00-1.02 mm) clearance between float and straight-edge.

Adjust float by bending float lever tang that contacts inlet valve.

**CAUTION: Remove float assembly to adjust float level. Failure to do so could result in deformation of inlet needle and seat.**

To adjust float drop, refer to Fig. O43 and adjust float drop to 0.20 inch (5.08 mm).

Various mechanical, electrical and pulse type fuel pumps are used according to model and application. Refer to

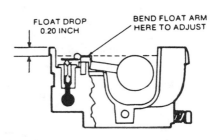

*Fig. O43 — View showing proper procedure for checking and adjusting float drop on downdraft style carburetor.*

SERVICING ONAN ACCESSORIES section for service information.

**GOVERNOR.** All models use a flyball weight governor located under timing gear cover on camshaft gear. Various linkage arrangements allow for either fixed speed operation or variable speed operation.

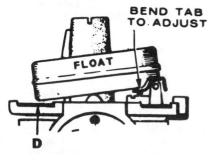

*Fig. O41 — Side draft carburetor inlet valve setting. Dimension (D), measured between inverted float and gasket surface is ⅛-inch (3.2 mm), plus or minus 1/16-inch (1.6 mm). Refer to text.*

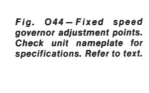

*Fig. O44 — Fixed speed governor adjustment points. Check unit nameplate for specifications. Refer to text.*

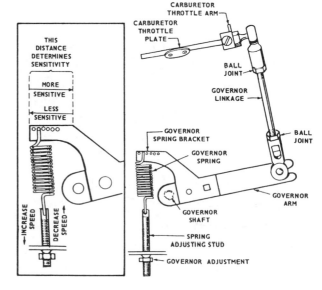

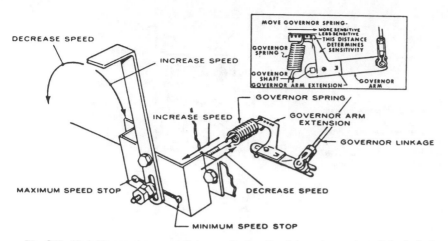

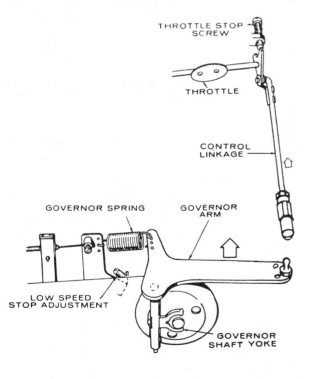

Fig. O45 — *Variable speed governor linkage adjustment points and procedure. Refer to text.*

**FIXED SPEED.** Fixed speed linkage is used on electric generators and welders or any application which requires a single constant speed under varying load conditions. Fig. O44 shows typical linkage arrangement.

Engines with fixed speed governors start at wide open throttle and engine speed is pre-set at factory to 2400 rpm unless special ordered with different specified speed.

With engine stopped, adjust length of linkage connecting throttle arm to governor arm so stop on carburetor throttle lever is 1/32-inch (0.794 mm) from stop boss. This allows immediate governor control at engine start and synchronizes travel of governor arm and throttle shaft.

Engine speed is determined by governor spring tension. Increasing spring tension results in higher engine speeds and decreasing spring tension results in lower engine speeds. Adjust spring tension as necessary by turning nut on spring adjusting stud. See Fig. O44.

Governor sensitivity is determined by location of governor spring at governor spring bracket. Refer to Fig. O44 for adjustment.

VARIABLE SPEED. Variable speed linkage is used where it is necessary to maintain a wide range of different engine speeds under varying load conditions. Moving a control lever determines engine speed and governor maintains this speed under varying loads.

To adjust variable speed governor linkage adjust throttle stop screw on carburetor so engine idles at 1100 rpm, then adjust governor spring tension so engine idles at 1500 rpm when manual control lever is at minimum speed position (Fig. O45) or "Bowden" cable control knob (Fig. O46) is at first notch (low speed) position.

Adjust sensitivity with engine running at minimum speed to obtain smoothest engine operation by moving governor spring outward on extension arm to decrease sensitivity or inward to increase sensitivity. Refer to Fig. O45 or O46.

Maximum full load speed should not exceed 3000 rpm for continuous operation. To adjust, apply full load to engine and move control lever or knob to maximum speed position and adjust set screw in bracket slot to stop lever travel or turn knob until desired speed is attained.

**IGNITION SYSTEM.** Most models are equipped with battery ignition system, however a magneto system is available for manual start models. Refer to appropriate paragraph for model being serviced.

BATTERY IGNITION. Breaker points are located in a non-movable breaker box (Fig. O47) and timing is adjusted by varying point gap slightly. Static timing should be checked if point gap is changed or new points are installed.

Remove air intake hose from blower housing on pressure cooled engines or remove sheet metal plug in air shroud over right cylinder of "Vacu-Flo" engines to gain access to timing marks.

Fig. O46 — *View of variable speed governor linkage when "Bowden" cable is used. Refer to text for adjustment procedure.*

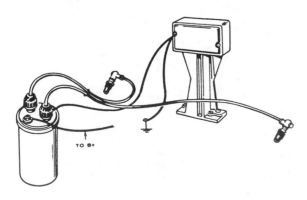

Fig. O47 — *View of external wiring of engines with battery ignition. Breaker box is non-adjustable and timing is set by varying point gap slightly.*

Illustrations courtesy of Onan Corp.

Initial point gap is 0.016 inch (0.41 mm) and static timing should be checked by connecting a test light across ignition points and rotating engine in direction of normal rotation (clockwise) until light comes on and then just goes out. This should be 20° BTDC and is obtained by varying point gap.

MAGNETO IGNITION. Magneto ignition system is used on manual start models and a stop button wired across breaker points is used to ground primary circuit to stop engine (Fig. O48).

To prevent engine recoil starter damage during starting, spark is retarded to 3° ATDC and advance mechanism shown in Fig. O49 automatically advances timing to 22° BTDC for normal operation.

Initial point gap is 0.020 inch (0.508 mm) and running timing should be checked using a timing light. Timing should be 22° BTDC at 1500 rpm and is adjusted by varying point gap.

If spark advance does not respond when engine is over 1500 rpm, or if advance is sluggish, remove cup-shaped cover (9 – Fig. O49) at rear of camshaft on engine block and check advance mechanism condition. Clean assembly thoroughly and renew worn parts as necessary.

LUBRICATION. All models are pressure lubricated by a gear type pump driven by crankshaft gear. Internal pump parts are not serviced separately so entire pump must be renewed if worn or damaged. Clearance between pump gear and crankshaft gear should be 0.002-0.005 inch (0.0508-0.127 mm) and

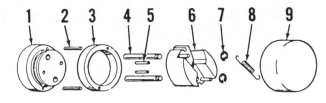

Fig. O49 – Exploded view of spark advance (timing control) used on NH model engine.

1. Camshaft
2. Cam roll pin
3. Control cam
4. Groove pin
5. Roll pin
6. Weights (2)
7. Retainers (2)
8. Control spring
9. Cover

normal operating pressure should be 30 psi (207 kPa). Oil pressure is regulated by an oil pressure relief valve located on top of engine block near timing gear cover. Relief valve spring free length should be 1.00 inch (25.4 mm) and valve diameter should be 0.3365-0.3380 inch (8.55-9.59 mm).

Oils approved by manufacturer must meet requirements of API service classification "SE" or "SE/CC".

Use SAE 5W-30 oil in below freezing temperatures and SAE 20W-40 oil in above freezing temperatures.

Check oil level with engine stopped and maintain at "FULL" mark on dipstick. **DO NOT** overfill.

Recommended oil change interval for T-260G models is every 50 hours of normal operation and change filter at 100 hour intervals. T-260G model oil capacity is 2.5 quarts (2.4 L) without filter change and 3 quarts (2.8 L) with filter change.

Recommended oil change interval for all other models is every 100 hours of normal operation and change filter at 200 hour intervals. Oil capacity is 3.5 quarts (3.3 L) without filter change and 4 quarts (3.8 L) with filter change.

Oil filter is screw on type and gasket should be lightly lubricated before installation. Screw filter on until gasket contacts, then turn an additional ½-turn. Do not overtighten.

## REPAIRS

TIGHTENING TORQUES. Recommended tightening torques are as follows:

Cylinder head . . . . . . . . See **CYLINDER HEAD** section
Rear bearing plate . . . . . . . 25-28 ft.-lbs. (34-38 N·m)
Connecting rod –
    Nodular iron rod . . . . . . . 27-29 ft.-lbs. (34-39 N·m)
    Aluminum rod . . . . . . . . . 14-16 ft.-lbs. (19-22 N·m)
Flywheel cap screws . . . . . . 35-40 ft.-lbs. (48-54 N·m)
Gear case cover . . . . . . . . . . 8-10 ft.-lbs. (11-13 N·m)
Oil pump . . . . . . . . . . . . . . . . 7-9 ft.-lbs. (10-12 N·m)
Intake manifold . . . . . . . . . 20-23 ft.-lbs. (27-31 N·m)
Exhaust manifold . . . . . . . . 20-23 ft.-lbs. (27-31 N·m)
Other 3/8-inch
    cylinder block nuts . . . . . 18-23 ft.-lbs. (24-31 N·m)

CYLINDER HEADS. It is recommended cylinder heads be removed and carbon cleaned at 200 hour intervals (400 hours if using unleaded fuel).

**CAUTION: Cylinder heads should not be unbolted from block when hot due to danger of head warpage.**

Cylinder heads are retained to block by studs and nuts. Late model engines use compression washers between nuts and hardened flat washers on top six studs and these should be installed on early models during service.

Onan recommends testing the top six original equipment studs on each cylinder to make certain they are not pulling out of threads in cylinder block.

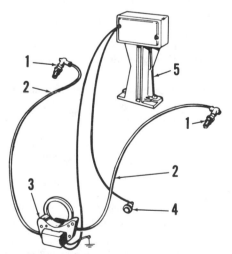

Fig. O48 – Magneto ignition used on NH model. Stop button (4) grounds breaker points to halt engine.

1. Spark plugs
2. Plug leads
3. Magneto coil
4. Stop button
5. Breaker box & stand

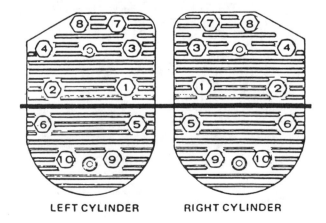

Fig. O50 – View of cylinder heads showing correct torque sequence and location of the six studs to be checked, above solid line, as outlined in text.

LEFT CYLINDER          RIGHT CYLINDER

To test studs, remove nuts and compression washers from top six studs (Fig. O50) leaving hardened flat washers in place. Reinstall nuts and tighten to 30 ft.-lbs. (40 N·m) torque. Studs with weak threads will pull out of head before 30 ft.-lbs. (40 N·m) torque is reached.

A special stepped renewal stud, part number 520-0912 and a drilling fixture, part number 420-0398, are available from Onan to service damaged stud hole threads.

To install stepped renewal stud, remove cylinder head and gasket. Examine head and block surface. If head or block is warped or has a depression of more than 0.005 inch (0.127 mm) it may be resurfaced with a maximum total of 0.010 inch (0.254 mm) of material removed. Remove studs from holes with damaged threads and install drilling fixture (Fig. O52) securing it in position with nuts and flat washers on studs indicated.

**NOTE: Some engines may require addition of flat washers between block and fixture plate to clear sheet metal scroll backing plate. If so, be certain drilling fixture remains parallel with head surface of block.**

Place bushing with small hole in it which is furnished with drilling plate, in plate hole over stud hole with damaged threads. Use a 27/64-inch drill bit to drill damaged threads out of block. Holes at side of block should be drilled through to the fourth cooling fin and holes at top of block should penetrate into corresponding intake or exhaust valve port. Manifolds should be removed during drilling procedure. Remove small bushing and install bushing with larger hole and lock in place. Use a ½-13 tap to thread hole for new stepped stud. Repeat process for any other hole with damaged threads. Remove any ridge around holes using a flat file or a 45° chamfer tool. When using chamfer tool chamfer depth should be 1/32-1/16 inch (0.794-1.588 mm). Apply "Loctite 242" to stepped stud threads and install in holes making certain entire stepped portion is below gasket surface of cylinder block.

**NOTE: Stepped studs installed in holes corresponding with exhaust ports must have 3/16-1/4 inch (4.763-6.350 mm) cut off large end of stud so it does not extend into exhaust port.**

Remove any metal particles from engine ports and place new graphoil type gasket on cylinder head and install gasket and head on studs at the same time to avoid damaging gasket.

**NOTE: Graph-oil type gaskets become soft and gummy at temperatures above 100° F (38° C). Avoid installation or removal if engine temperature exceeds this.**

Install hardened flat washers on all studs, two compression washers on each of the top six long studs (Fig. O51) so outside edges of compression washers are in contact with each other. Install nuts and tighten in several even steps until top six studs reach 12 ft.-lbs. (16 N·m) torque and the bottom studs reach 15 ft.-lbs. (20 N·m) torque.

Tighten in sequence shown in Fig. O50 and recheck all nuts after initial torque has been reached.

**CAUTION: Too much torque will flatten compression washers and could result in engine damage.**

Recheck torque before engine has a total of 50 hours operation.

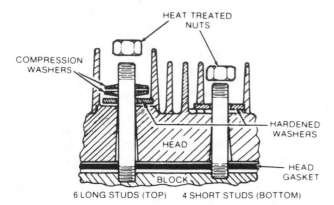

Fig. O51 — View showing correctly installed compression washers. Compression washers are installed on top six (long) studs only. Hardened flat washers are installed on all studs.

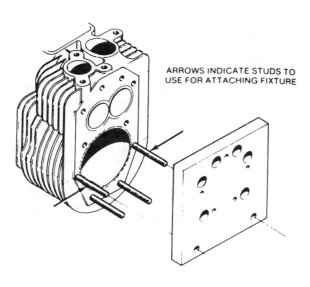

Fig. O52 — Special drilling fixture is required to drill out damaged threads in stud holes. Fixture part number is 420-398.

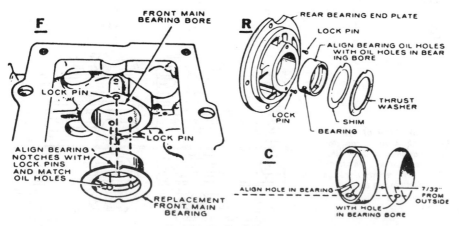

Fig. O53 — Alignment of precision main and camshaft bearings. One-piece bearing is used at front (view F), two-piece bearing at rear (view R). Note shim use for end play adjustment. View C shows placement of camshaft bearings.

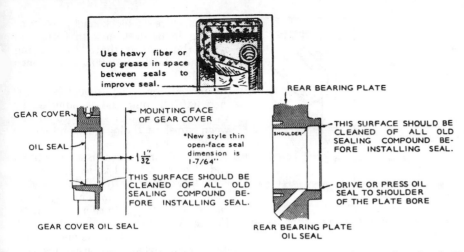

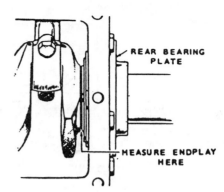

Fig. O54 – View of crankshaft end play measurement procedure. Refer to text.

Fig. O53A – View showing correct seal installation procedure. Fill cavity between lips of seal with heavy grease to improve sealing efficiency.

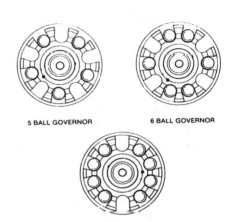

5 BALL GOVERNOR       6 BALL GOVERNOR

8 BALL GOVERNOR

Fig. O55 – View showing correct spacing of governor flyballs according to total number of balls used.

**CONNECTING ROD.** Connecting rod and piston are removed from cylinder head surface end of block after removing cylinder head and oil base.

T-260G model connecting rod rides directly on crankpin journal and all remaining models have renewable insert type rod bearings and a renewable piston pin bushing.

Standard crankpin journal diameter is 1.6252 1.6260 inches (41.28-41.30 mm) for all models and connecting rod or bearing inserts are available in a variety of sizes for undersize crankshafts as well as standard.

Connecting rod to crankshaft journal running clearance for T-260G engine or engines with aluminum rod is 0.002-0.003 inch (0.051-0.076 mm) and running clearance for bearings in nodular iron rod should be 0.0005-0.0028 inch (0.013-0.071 mm).

Piston and rod assembly should be installed so rod bolts are offset toward outside of cylinder block. If rod has an oil hole, hole must be toward camshaft. Tighten connecting rod cap screws to specified torque.

**PISTON, PIN AND RINGS.** NH model engines with specification letters "A" through "C" use a three ring strut type piston which should have 0.0015-0.0035 inch (0.0381-0.0889 mm) clearance between cylinder and piston when measured 0.10 inch (2.54 mm) below oil control ring 90° from piston pin. If side clearance of top ring in piston groove exceeds 0.006 inch (0.1524 mm) piston should be renewed. NH models with specification letter D or after and all T-260G engines use a three ring piston. Piston to cylinder clearance should be 0.007-0.009 inch (0.178-0.229 mm) when measured 0.10 inch (2.54 mm) below oil control ring 90° from piston pin.

Top ring side clearance in piston groove should not exceed 0.004 inch (0.1016 mm) for T-260G models and should be 0.002-0.008 inch (0.0508-0.2032 mm) clearance for NH models with specification letter D or after.

Ring end gap for all models is 0.010-0.020 inch (0.254-0.508 mm).

Piston pin to piston pin bore clearance is 0.0001-0.0005 inch (0.0025-0.0127 mm) for all models and pin to rod clearance is 0.0002-0.0008 inch (0.005-0.020 mm) for aluminum rod and 0.00005-0.00055 inch (0.0013-0.014 mm) for bushing in nodular iron rod.

Standard piston pin diameter is 0.7500-0.7502 inch (19.05-19.06 mm) for all models and pins are available in 0.002 inch (0.0508 mm) oversize for aluminum rod.

Standard cylinder bore diameter for all models is 3.5625-3.5635 inches (90.49-90.51 mm) and cylinders should be bored to nearest oversize for which

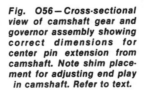

Fig. O56 – Cross-sectional view of camshaft gear and governor assembly showing correct dimensions for center pin extension from camshaft. Note shim placement for adjusting end play in camshaft. Refer to text.

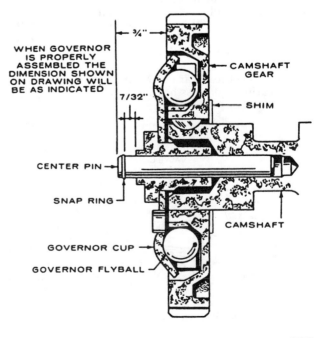

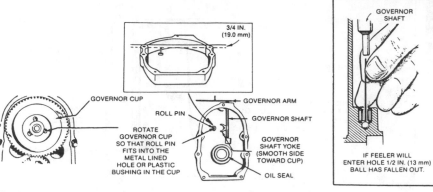

Fig. O57—View of governor shaft, timing gear cover and governor cup showing assembly details. Refer to text.

piston and rings are available if cylinder is scored or out-of-round more than 0.003 inch (0.0762 mm) or if taper exceeds 0.005 inch (0.127 mm). Pistons and rings are available in a variety of oversizes as well as standard.

**CRANKSHAFT, BEARINGS AND SEALS.** Crankshaft is supported at each end by precision type sleeve bearings located in cylinder block housing and rear bearing plate. Main bearing journal standard diameter should be 1.9992-2.000 inches (50.7797-50.800 mm) and bearing operating clearance should be 0.0025-0.0038 inch (0.0635-0.0965 mm) for T-260G model and 0.0015-0.0043 inch (0.0381-0.1092 mm) for all other models. Main bearings are available for a variety of undersize crankshafts as well as standard.

Oil holes in bearing and bore **MUST** be aligned when bearings are installed and rear bearing should be pressed into bearing plate until flush, or recessed into bearing plate 1/64-inch (0.406 mm). Make certain bearing notches are aligned with lock pins (Fig. O53) during installation.

Apply "Loctite Bearing Mount" to front bearing and press bearing in until flush with block. Make certain bearing notches are aligned with lock pins (Fig. O53) during installation.

NOTE: Replacement front bearing and thrust washer is a one piece assembly, do not install a separate thrust washer.

Rear thrust washer is installed with oil grooves toward crankshaft. Measure crankshaft end play as shown in Fig. O54 and add or remove shims or renew thrust washer (Fig. O53) as necessary to obtain 0.005-0.009 inch (0.13-0.23 mm) end play.

It is recommended rear bearing plate and front timing gear cover be removed for seal installation. Rear seal is pressed in until flush with seal bore. Timing cover seal should be driven in until it is 1-1/32 inch (26.19 mm) from mounting face of cover for old style seal and 1-7/64 inch (28.18 mm) from mounting face of cover for new style, thin, open faced seal. See Fig. O53A.

**CAMSHAFT, BEARINGS AND GOVERNOR.** Camshaft is supported at each end in precision type sleeve bearings. Make certain oil holes in bearings are aligned with oil holes in block (Fig. O53), press front bearing in flush with outer surface of block and press rear bearing in until flush with bottom of counterbore. Camshaft to bearing clearance should be 0.0015-0.0030 inch (0.0381-0.0762 mm).

Camshaft end play should be 0.003 inch (0.0762 mm) and is adjusted by varying thickness of shim located between camshaft timing gear and engine block. See Fig. O56.

Camshaft timing gear is a press fit on end of camshaft and is designed to accomodate 5, 8 or 10 flyballs arranged as shown in Fig. O55. Number of flyballs is varied to alter governor sensitivity. Fewer flyballs are used on engines with variable speed applications and greater number of flyballs are used for continuous (fixed) speed application. Flyballs must be arranged in "pockets" as shown according to total number used.

Center pin (Fig. O56) should extend 3/4-inch (19 mm) from end of camshaft to allow 7/32-inch (5.6 mm) in-and-out movement of governor cup. If distance is incorrect, remove pin and press new pin in to correct depth.

Always make certain governor shaft pivot ball is in timing cover by measuring as shown in Fig. O57 inset and check length of roll pin which engages bushed hole in governor cup. This pin must extend to within 3/4-inch (19 mm) of timing gear cover mating surface.

Make certain timing mark on camshaft gear is aligned with timing mark on crankshaft gear during installation.

**VALVE SYSTEM.** Valve seats are renewable insert type and are available in a variety of oversizes as well as standard. Seats are ground at a 45° angle and seat width should be 1/32 to 3/64-inch (0.794 to 1.191 mm).

Valves should be ground at a 44° angle to provide an interference angle of 1°.

Recommended valve stem to guide clearance is 0.001-0.0025 inch (0.0254-0.0635 mm) for intake valves and 0.0025-0.004 inch (0.0635-0.1016 mm) for exhaust valves. Renewable shouldered valve guides are pushed out from head surface end of block. An "O" ring is installed on intake valve guide of some models and a valve stem seal is also available for intake valve. See Fig. O58.

Recommended valve tappet gap (cold) is 0.003 inch (0.0762 mm) for intake valves and 0.010 inch (0.254 mm) for exhaust valves.

Adjustment is made by turning tappet adjusting screw as required and valves of each cylinder must be adjusted with cylinder at "top dead center" on compression stroke. At this position both valves will be fully closed.

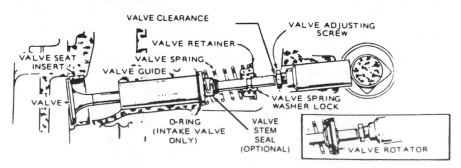

Fig. O58—Valve train typical of NH and T-260G engines. Refer to text.

Illustrations courtesy of Onan Corp.

# ONAN

### A DIVISION OF ONAN CORPORATION
#### 1400 73rd Avenue N.E.
#### Minneapolis, Minnesota 55432

| Model | No. Cyls. | Bore | Stroke | Displacement | Power Rating |
|-------|-----------|------|--------|--------------|--------------|
| CCK | 2 | 3.25 in. (82.55 mm) | 3 in. (76.2 mm) | 49.8 cu. in. (815.7 cc) | 12.9 hp. (9.6 kW) |
| CCKA | 2 | 3.25 in. (82.55 mm) | 3 in. (76.2 mm) | 49.8 cu. in. (815.7 cc) | 16.5 hp. (12.3 kW) |
| CCKB | 2 | 3.25 in. (82.55 mm) | 3 in. (76.2 mm) | 49.8 cu. in. (815.7 cc) | 20 hp. (14.9 kW) |

Engines in this section are four-cycle, twin cylinder opposed, horizontal crankshaft engines. Crankshaft is supported at each end in precision sleeve type bearings.

CCK and CCKA engines prior to specification letter "D" have aluminum connecting rods which ride directly on crankshaft journal.

CCK and CCKA engines with specification letter "D" or after and all other models have forged steel rods equipped with renewable bearing inserts.

All models are pressure lubricated by a gear type oil pump. A "spin-on" oil filter is available.

Most models are equipped with a battery ignition system, however a magneto system is available for manual start models.

Engines may be equipped with a side draft or downdraft carburetor which may be used with gasoline or vapor (butane, LP or natural gas) fuels according to model and application. Mechanical, pulsating diaphragm or electric fuel pumps are available.

Refer to Fig. O59 for interpretation of engine specification and model numbers.

CCKA engines may be visually distin-guished by the cup-shaped advance mechanism cover, rather than the flat-shaped expansion plug, in rear camshaft opening just below ignition point breaker box.

High compression heads are identified by a 3/32-inch radius boss located on corner of head nearest spark plug.

Always give specification, model and serial numbers when ordering parts or service material.

## MAINTENANCE

**SPARK PLUG.** Recommended spark plug is Onan part number 167-0241 for non-resistor plug and 167-0237 for a resistor type plug or their equivalent. Electrode gap for all models using vapor (butane, LP or natural gas) fuel is 0.018 inch (0.4572 mm) and gap for models using gasoline for fuel is 0.025 inch (0.635 mm).

**CARBURETOR.** According to model and application, engine may be equipped with either a side draft or a downdraft carburetor which uses either gasoline or vapor (butane, LP or natural gas) as fuel. Refer to appropriate paragraph for model being serviced.

SIDE DRAFT CARBURETOR (GASOLINE). Side draft carburetor is used on some CCKB models. Refer to Fig. O60 for exploded view of carburetor and location of mixture adjustment screws.

For initial carburetor adjustment, open idle mixture screw and main fuel

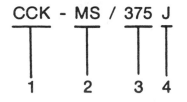

**Fig. O59—Interpretation of engine model and specification number as an aid to identification number of various engines.**

CCK - MS / 375 J

1. General engine model identification
2. Specific type
   S—Manual starting
   MS—Electric starting
3. Optional equipment identification
4. Specification letter which advances with factory production modifications.

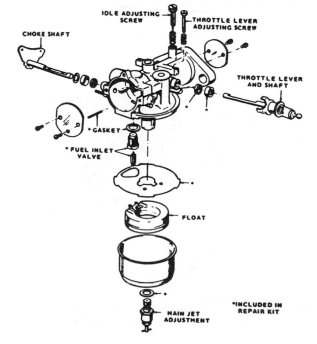

**Fig. O60—Exploded view of side draft (gasoline fuel) carburetor used on some models.**

CHOKE SHAFT

IDLE ADJUSTING SCREW

THROTTLE LEVER ADJUSTING SCREW

THROTTLE LEVER AND SHAFT

*GASKET

*FUEL INLET VALVE

FLOAT

MAIN JET ADJUSTMENT

*INCLUDED IN REPAIR KIT

mixture screw 1 to 1½ turns. Make final adjustments with engine at normal operating temperature and running. Place engine under load and adjust main fuel mixture screw for leanest setting that will allow satisfactory acceleration and steady governor operation. Set engine at idle speed, no load and adjust idle mixture screw for smoothest idle operation.

As each adjustment affects the other, adjustment procedure may have to be repeated.

To check float level, refer to Fig. O61. Invert carburetor throttle body and float assembly. Float clearance should be 11/64-inch (4.366 mm) for models prior to specification letter "H" and 1/8 to 3/16-inch (3.2 to 4.8 mm) for models with specification letter "H" or after.

Adjust float by bending float lever tang that contacts inlet valve.

**SIDE DRAFT CARBURETOR (VAPOR FUEL/GASOLINE).** A side draft carburetor designed to use either gasoline or vapor (butane, LP or natural gas) fuel is available for some models.

For initial adjustment, refer to Fig. O62 for location of appropriate adjustment screws and make initial and final adjustments as outlined in **SIDE DRAFT CARBURETOR (GASOLINE)** section.

To change system from gasoline fuel operation to vapor fuel operation, follow the procedure below.

Shut off gasoline supply valve.

Install lock wire on choke.

Set spark plug gap at 0.018 inch (0.4572 mm).

Lock float in position.*

Open supply valve for vapor fuel.

*It may be necessary to remove float assembly and inlet valve on models which are not equipped with a float locking mechanism for extended vapor fuel use.

To change system from vapor fuel operation to gasoline fuel operation, follow the procedure below.

Close vapor fuel supply valve.

Remove lock wire from choke.

Set spark plug gap at 0.025 inch (0.635 mm).

Unlock carburetor float.*

Open gasoline fuel supply valve.

*Replace float assembly and inlet valve if they were removed because carburetor was not equipped with float lock.

To check float level, refer to procedure outlined in **SIDE DRAFT CARBURETOR (GASOLINE)** section.

**DOWNDRAFT CARBURETOR (GASOLINE).** A downdraft carburetor is used on most CCK and CCKA models. Refer to Fig. O63 for exploded view of carburetor and location of mixture adjustment screws.

For initial carburetor adjustment, open idle mixture screw 1 turn. If equipped with main fuel adjustment open main fuel mixture screw about 2 turns.

Make final adjustments with engine at normal operating temperature and running. Place engine under load and adjust main fuel mixture screw (if equipped) for leanest setting that will allow satisfactory acceleration and steady governor operation. Onan special tool number 420-0169 (Fig. O64) aids main fuel mixture screw adjustment. Set engine at idle speed, no load and adjust idle mixture screw for smoothest idle operation.

As each adjustment affects the other, adjustment procedure may have to be repeated.

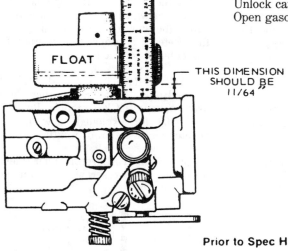

Prior to Spec H

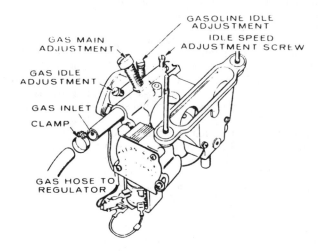

GASOLINE IDLE ADJUSTMENT
IDLE SPEED ADJUSTMENT SCREW
GAS MAIN ADJUSTMENT
GAS IDLE ADJUSTMENT
GAS INLET CLAMP
GAS HOSE TO REGULATOR

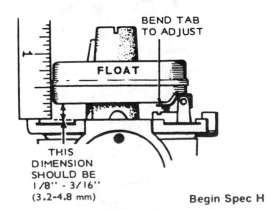

BEND TAB TO ADJUST
FLOAT
THIS DIMENSION SHOULD BE 1/8" - 3/16" (3.2-4.8 mm)

Begin Spec H

Fig. O61 — Measure and set float level as shown noting upper view is for engines with specification letter prior to "H" and lower view is for engines with specification letter "H" and after.

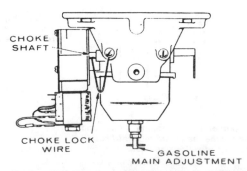

CHOKE SHAFT
CHOKE LOCK WIRE
GASOLINE MAIN ADJUSTMENT

Fig. O62 — View showing location of mixture adjustment screws on carburetor designed for either vapor (butane, LP or natural gas) fuel or gasoline fuel. Refer to text.

Illustrations courtesy of Onan Corp.

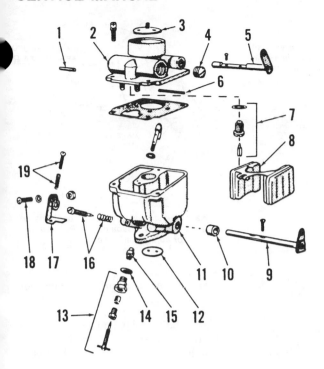

*Fig. O63 — Exploded view of downdraft carburetor used on CCK and CCKA models.*

1. Choke stop pin
2. Cover assy.
3. Choke valve
4. Gas inlet plug
5. Choke shaft
6. Float pin
7. Inlet valve assy.
8. Float
9. Throttle shaft
10. Shaft bushing
11. Body assy.
12. Throttle plate
13. Main fuel valve
14. Gasket
*15. Main jet
16. Idle mixture adjustment needle
17. Idle stop lever
18. Clamp screw
19. Stop screw & spring
*Either 13 or 15 is used.

*Fig. O64 — View showing use of Onan main fuel adjusting tool, part number 420-0169, used to adjust main fuel mixture needle. Refer to text.*

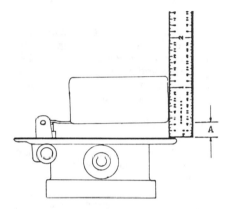

*Fig. O65 — View of typical float assembly of down draft type carburetor. Clearance at "A" should be ¼-inch (6.35 mm) if float is made of metal and 5/16-inch (7.94 mm) if float is made of styrofoam. Clearance is measured with gasket in place.*

To check float level, refer to Fig. O65. Invert carburetor throttle body and float assembly. Float clearance should be ¼-inch (6.35 mm) for metal float or 5/16-inch (7.94 mm) for styrofoam float when measured from gasket to closest edge of float as shown.

Adjust float by bending float lever tang that contacts inlet valve.

**DOWNDRAFT CARBURETOR (VAPOR FUEL/GASOLINE).** A downdraft carburetor designed to use either gasoline or vapor (butane, LP or natural gas) fuel is available for some models.

For initial and final carburetor adjustment, refer to Fig. O66 for appropriate mixture screws and adjust as outlined in **DOWNDRAFT CARBURETOR (GASOLINE)** section.

To change system from gasoline fuel operation to vapor fuel operation, follow the procedure below.

Shut off gasoline supply valve.

Install lock wire on choke.

Set spark plug gap at 0.018 inch (0.4572 mm).

Lock float in position.*

Open supply valve for vapor fuel.

*It may be necessary to remove float assembly and inlet valve on models which are not equipped with a float locking mechanism for extended vapor fuel use.

To change system from vapor fuel operation to gasoline fuel operation, follow the procedure below.

Close vapor fuel supply valve.

Remove lock wire from choke.

Set spark plug gap at 0.025 inch (0.635 mm).

Unlock carburetor float.*

Open gasoline fuel supply valve.

*Replace float assembly and inlet valve if they were removed because carburetor was not equipped with float lock.

To check float level, refer to procedure outlined in **DOWNDRAFT CARBURETOR (GASOLINE)** section.

**CHOKE.** One of three types of automatic choke may be used according to model and application. Refer to appropriate paragraph for model being serviced.

THERMAL MAGNETIC CHOKE. This choke (Fig. O67) uses a strip heating element and a heat-reactive bi-metallic spring to control choke valve position. When engine is cranked, solenoid shown at lower portion of choke body is activated and choke valve is closed, fully or partially, according to temperature conditions. When engine

starts and runs, solenoid is released and bi-metallic spring controls choke opening. Refer to Fig. O70 for table of choke settings based on ambient temperature and adjustment procedure. Adjustment is made by rotating entire cover as shown.

BI-METALLIC CHOKE. This choke (Fig. O68) uses an electric heating element located inside its cover to activate

*Fig. O66 — View showing location of mixture screw adjustments on combination vapor (butane, LP or natural gas) fuel and gasoline fuel downdraft carburetor.*

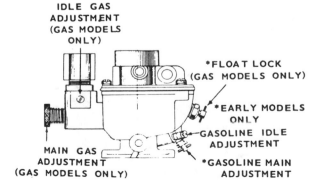

IDLE GAS ADJUSTMENT (GAS MODELS ONLY)

*FLOAT LOCK (GAS MODELS ONLY)

*EARLY MODELS ONLY

GASOLINE IDLE ADJUSTMENT

*GASOLINE MAIN ADJUSTMENT

MAIN GAS ADJUSTMENT (GAS MODELS ONLY)

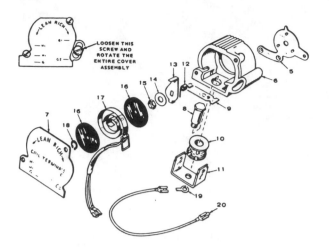

Fig. O67—Exploded view of thermal magnetic choke used on some engines.

5. Mounting plate
6. Body housing
7. Cover
8. Solenoid core
9. Armature
10. Solenoid coil
11. Frame
12. Lever spring
13. Lever
14. Washer
15. Pal nut
16. Insulator disc (2)
17. Heater & spring
18. Retainer (to shaft)
19. Ground terminal
20. Choke lead wire

Fig. O69—View showing various locations of component parts for Sisson choke. Refer to text for adjustment procedure.

a bi-metallic spring which opens or closes choke valve according to temperature. Electric current is supplied to heating element by leads from generator exciter. Refer to Fig. O70 for choke settings according to ambient temperature and adjustment procedure. Adjustment is made by rotating cover to open or close choke valve.

SISSON CHOKE. This choke (Fig. O69) uses a bi-metallic strip which reacts to manifold temperatures and a magnetic solenoid which is wired in series with starter switch circuit.

To adjust, pull choke lever up and insert a 1/16-inch diameter rod through shaft hole to engage notch in mounting flange and lock shaft against rotation. Loosen choke lever clamp screw and adjust choke lever so choke plate is completely closed, or not more than 5/16-inch (7.94 mm) open. Tighten clamp screw and remove 1/16-inch rod.

FUEL PUMP. Various mechanical, electric and pulse type fuel pumps are used according to model and application. Refer to **SERVICING ONAN ACCESSORIES.**

GOVERNOR. All models use a flyball weight governor located under timing gear cover on camshaft gear. Various linkage arrangements allow for either fixed speed operation or variable speed operation.

Some engines may be equipped with vacuum operated speed booster, auto-matic idle control or two-speed electric solenoid controls. Refer to **SERVICING ONAN ACCESSORIES** for service.

Before any governor linkage adjustment is made, make certain all worn or binding linkage is repaired and carburetor is properly adjusted. Clean plastic ball joints, but do not lubricate. Beginning with specification letter "J", metal ball joints are used and these should be cleaned and lubricated with graphite.

FIXED SPEED. Fixed speed linkage is used on electric generators and welders or any application which requires a single constant speed under varying load conditions. Fig. O71 shows typical linkage arrangement.

Engines with fixed speed governors start at wide open throttle and engine speed is pre-set at factory to 2400 rpm unless special ordered with different specified speed.

With engine stopped, adjust length of linkage connecting throttle arm to governor arm so stop on carburetor throttle lever is 1/32-inch (0.794 mm) from stop boss. This allows immediate governor control at engine start and synchronizes travel of governor arm and throttle shaft.

Engine speed is determined by governor spring tension. Increasing spring tension results in higher engine speeds and decreasing spring tension results in lower engine speeds. Adjust spring tension as necessary by turning nut on spring adjusting stud. See Fig. O71.

Governor sensitivity is determined by location of governor spring at governor spring bracket. Refer to Fig. O71. Governor sensitivity is increased by shifting sliding clip toward governor shaft (prior to specification letter "D", turn adjusting stud clockwise). Decrease

| AMBIENT TEMP. | CHOKE SETTING |
|---|---|
| 60° F (16° C) | 1/8-in. (3.2 mm) |
| 65° F (18° C) | 9/64-in. (3.6 mm) |
| 70° F (21° C) | 5/32-in. (4 mm) |
| 75° F (24° C) | 11/64-in. (4.4 mm) |
| 80° F (27° C) | 3/16-in. (4.8 mm) |
| 85° F (29° C) | 13/64-in. (5.2 mm) |
| 90° F (32° C) | 7/32-in. (5.6 mm) |
| 95° F (35° C) | 15/64-in. (6 mm) |
| 100° F (38° C) | 1/4-in. (6.4 mm) |

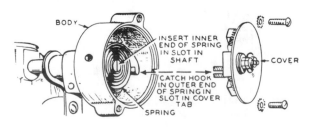

Fig. O68—Bi-metallic choke used on some models. Refer to text.

Fig. O70—Let engine cool at least one hour before adjusting choke setting. Drill bits may be used as gages to set choke opening according to ambient temperature.

sensitivity by shifting sliding clip toward linkage end of governor arm (prior to specification letter "D", turn adjusting stud counter-clockwise).

VARIABLE SPEED. Variable speed linkage is used where it is necessary to maintain a wide range of different engine speeds under varying load conditions. Moving a control lever determines engine speed and governor maintains this speed under varying loads. See Fig. O71A.

To adjust variable speed governor linkage adjust throttle stop screw on carburetor so engine idles at 1450 rpm, then adjust governor spring tension so engine idles at 1500 rpm when manual control lever is at minimum speed position.

Adjust sensitivity with engine running at minimum speed to obtain smoothest engine operation by moving governor spring outward on extension arm to decrease sensitivity or inward to increase sensitivity. Refer to Fig. O71A.

Maximum full load speed should not exceed 3000 rpm for continuous operation. To adjust, apply full load to engine and move control lever to maximum speed position. Adjust set screw in bracket slot to stop lever travel at desired speed.

**IGNITION SYSTEM.** Most models are equipped with battery ignition

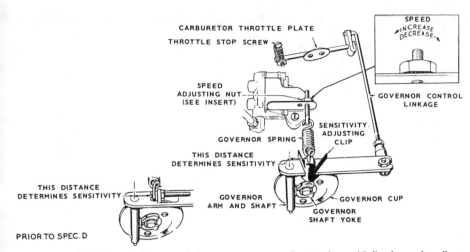

Fig. O71 – View of typical governor linkage arrangement used on engines with fixed speed application.

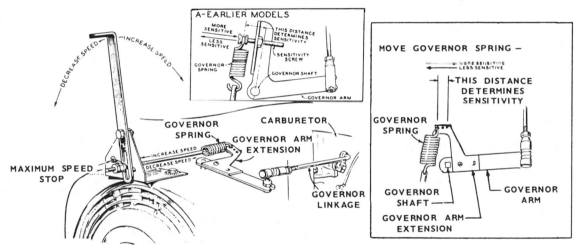

Fig. O71A – View of typical governor linkage arragement used on engines with variable speed application.

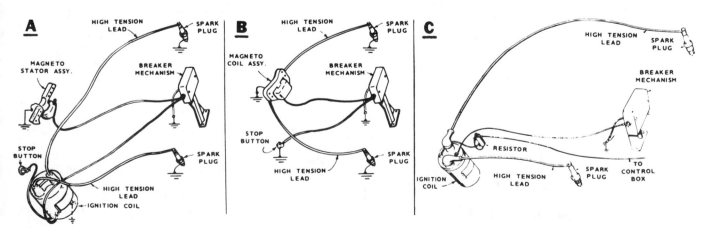

Fig. O72 – Layout of ignition wiring arrangements for CCK series engines. View A shows magneto ignition without spark advance mechanism. View B shows flywheel magneto system with spark advance mechanism. View C shows battery ignition wiring. Refer to text for timing and service procedures.

Illustrations courtesy of Onan Corp.

system, however a magneto system is available for manual start models. Ignition timing specifications are stamped on crankcase adjacent to breaker box.

Refer to appropriate paragraph for model being serviced.

BATTERY IGNITION. Ignition points and condenser are located in a breaker box at center, rear of engine. Refer to Fig. O72 for view of ignition wiring layout and Fig. O72A for view of points box and timing information.

To check timing, remove air intake hose from blower housing on pressure cooled engines or remove sheet metal plug in air shroud over right cylinder of "Vacu-Flo" engines to gain access to timing marks. Make certain points are set at 0.020 inch (0.51 mm) for all models.

Static timing is checked by connecting a test light across ignition points and rotating engine in direction of normal rotation (clockwise) until light comes on, then just goes out. This should be at 19° BTDC for all CCK models, 20° BTDC for CCKA and CCKB models without automatic spark advance and 1° ATDC for CCKA model with automatic spark advance (Fig. O73). Adjust by moving breaker box as required (Fig. O72A).

To check running timing, connect timing light to either spark plug and start and run engine. Operate engine at 1500 rpm or over. Light should flash when flywheel mark is aligned with 19° BTDC mark for all CCK models, 20° BTDC mark for CCKA model without automatic spark advance, 24° BTDC for CCKA model with automatic spark advance and 24° BTDC mark for CCKB models using vapor (butane, LP or natural gas) fuel. Adjust by moving breaker box as required (Fig. O72A).

MAGNETO IGNITION. If engine is equipped with automatic spark advance mechanism (Fig. O73), magneto coil, mounted on gear cover (B–Fig. O72), contains both primary and secondary windings.

If engine is not equipped with automatic spark advance, stator contains primary (low voltage) windings only and a separate coil is used to develop secondary voltage (A–Fig. O72).

Ignition points and condenser are located in a breaker box at center, rear of engine. Refer to Fig. O72 for view of ignition wiring layout and Fig. O72A for view of points box and timing information.

To check timing, remove air intake hose from blower housing on pressure cooled engines or remove sheet metal plug in air shroud over right cylinder of "Vacu-Flo" engines to gain access to timing marks. Make certain points are set at 0.020 inch (0.51 mm) for all models.

Static timing is checked by connecting a test light across ignition points and rotating engine in direction of normal rotation (clockwise) until light comes on, then just goes out. This should be at 19° BTDC for CCK model, 1° ATDC for CCKA model with automatic spark advance, 20° BTDC for CCKA model without automatic spark advance, 5° BTDC for CCKB model with electric start and 5° BTDC to 1° ATDC for CCKB model with manual start. Adjust by moving breaker box as required (Fig. O72A).

To check running timing, connect timing light to either spark plug and start and run engine. Operate engine at 1500 rpm or over. Light should flash when flywheel mark is aligned with 19° BTDC mark for all CCK models, 20° BTDC for CCKA model without automatic spark advance, 24° BTDC for CCKA model with automatic spark advance and 24° BTDC mark for CCKB models. Adjust by moving breaker box as required (Fig. O72A).

LUBRICATION. All models are pressure lubricated by a gear type pump driven by crankshaft gear. Internal pump parts are not serviced separately so entire pump must be renewed if worn or damaged. Clearance between pump gear and crankshaft gear should be 0.002-0.005 inch (0.0508-0.127 mm) and normal operating pressure should be 30 psi (207 kPa). Oil pressure is regulated by an oil pressure relief valve located on top of engine block near timing gear cover. Relief valve spring free length should be 2-5/16 inch (58.74 mm) and valve diameter should be 0.3365-0.3380 inch (8.55-8.59 mm).

Oils approved by manufacturer must meet requirements of API service classification "SE" or "SE/CC".

Use SAE 5W-30 oil in below freezing temperatures and SAE 20W-40 oil in above freezing temperatures.

Check oil level with engine stopped and maintain at "FULL" mark on dipstick. DO NOT overfill.

Recommended oil change interval for all models is every 100 hours of normal operation and change filter at 200 hour intervals. Oil capacity for all models with electric start is 3.5 quarts (3.3 L) without filter change and 4 quarts (4 L) with filter change. Oil capacity for all manual start models is 3 quarts (2.8 L) without filter change and 3.5 quarts (3.3 L) with filter change.

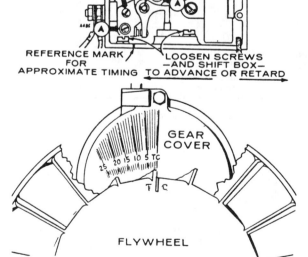

PLACE DROP OF OIL ON PIVOT POINT WHENEVER NEW POINTS ARE INSTALLED.

REFERENCE MARK FOR APPROXIMATE TIMING

LOOSEN SCREWS —AND SHIFT BOX— TO ADVANCE OR RETARD

GEAR COVER

25 20 15 10 5 TC

T C

FLYWHEEL

Fig. O72A – View showing location of points in breaker box of all models and location of timing marks on pressure cooled engines.

1. Camshaft
2. Cam roll pin
3. Control cam
4. Groove pin
5. Roll pin
6. Weights (2)
7. Retainers (2)
8. Control spring
9. Cover

Fig. O73 – Exploded view of spark advance (timing control) used on some series engines.

Oil filter is a screw on type and gasket should be lightly lubricated before installation. Screw filter on until gasket contacts, then turn an additional ½-turn. Do not overtighten.

**CRANKCASE BREATHER.** Engines are equipped with a crankcase breather (Fig. O73A) which helps maintain a partial vacuum in crankcase during engine operation to help control oil loss and to ventilate crankcase.

Clean at 200 hour intervals of normal engine operation. Wash valve in suitable solvent and pull baffle out of breather tube to clean. Reinstall valve with perforated disk toward engine.

## REPAIRS

**TIGHTENING TORQUES.** Recommended tightening torques are as follows:

Cylinder head . . . . . . . . . .29-31 ft.-lbs.
(39-42 N·m)
Connecting rod –
Aluminum . . . . . . . . . . .24-26 ft.-lbs.
(33-35 N·m)
Forged steel . . . . . . . . . .27-29 ft.-lbs.
(37-39 N·m)
Rear bearing plate . . . . . . .20-25 ft.-lbs.
(27-34 N·m)
Flywheel . . . . . . . . . . . . . .35-40 ft.-lbs.
(48-54 N·m)
Oil base . . . . . . . . . . . . . .43-48 ft.-lbs.
(58-65 N·m)
Blower housing screws . . . . .8-10 ft.-lbs.
(11-14 N·m)
Exhaust manifold . . . . . . . .15-20 ft.-lbs.
(20-27 N·m)
Intake manifold . . . . . . . . .15-20 ft.-lbs.
(20-27 N·m)
Timing gear cover . . . . . . .10-13 ft.-lbs.
(14-18 N·m)
Valve cover nut . . . . . . . . . . .4-8 ft.-lbs.
(6-11 N·m)
Starter bolts . . . . . . . . . . . .25-28 ft.-lbs.
(34-38 N·m)
Magneto stator screws . . . .15-20 ft.-lbs.
(20-27 N·m)
Spark plug . . . . . . . . . . . . .25-30 ft.-lbs.
(34-41 N·m)

**CYLINDER HEAD.** Cylinder heads should be removed and carbon and lead deposits cleaned at 200 hour intervals (400 hours if using unleaded fuel) or as a reduction in engine power or excessive pre-ignition occurs.

**CAUTION: Cylinder heads should not be unbolted from block when hot due to danger of head warpage.**

Always install new head gaskets and torque retaining cap screws in 5 ft.-lbs. (7 N·m) steps in sequence shown (Fig. O74) until 29-31 ft.-lbs. (39-42 N·m) torque is obtained. Operate engine at normal operating temperature and light

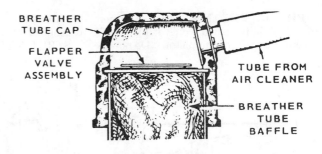

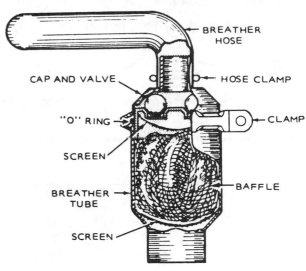

*Fig. O73A — Crankcase breather assembly should be cleaned at 200 hour intervals of normal engine operation. Upper breather is used on engines before specification letter "H" and lower breather is used on engines after specification letter "H".*

load for a short period, allow to cool and re-torque head bolts.

**CONNECTING ROD.** Connecting rod and piston are removed from cylinder head surface end of block after removing cylinder head and oil base.

Connecting rods for CCK and CCKA engines with specification letter "C" or before are aluminum rods which ride directly on crankpin journal and piston pin operates in unbushed bore of connecting rod.

Connecting rods for CCK and CCKA engines with specification letter "D" or after and all other models are equipped with precision type insert rod bearings and piston pin operates in renewable bushing pressed into forged steel connecting rod.

Standard crankpin journal diameter is 1.6252-1.6260 inches (41.28-41.30 mm) for all models and connecting rod or bearing inserts are available in a variety of sizes for undersize crankshafts as well as standard.

Connecting rod to crankshaft journal running clearance for aluminum rod should be 0.002-0.0033 inch (0.0508-0.0838 mm) and running clearance for bearings in forged steel rod should be 0.0005-0.0023 inch (0.0127-0.0584 mm).

Connecting rod caps should be reinstalled on original rod with raised lines (witness marks) aligned and con-

necting rod caps must be facing oil base after installation. Tighten to specified torque.

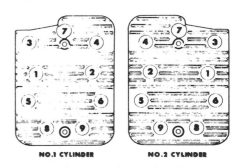

*Fig. O74 — Tightening sequence for CCK series cylinder head cap screws. Procedure is outlined in text.*

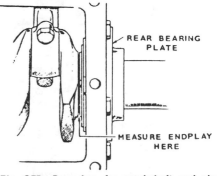

*Fig. O75 — Procedure for crankshaft end play measurement. Refer to text.*

**PISTON, PIN AND RINGS.** Engines may be equipped with one of three different type three ring pistons. Do not intermix different type pistons in engines during service.

**STRUT TYPE PISTON.** Strut type piston may be visually identified by struts cast in underside of piston which run parallel to pin bosses. Piston to cylinder clearance when measured just below oil ring, 90° from pin, should be 0.0025-0.0045 inch (0.0635-0.1143 mm).

**CONFORMATIC PISTON.** Conformatic piston may be visually identified by smooth contours in underside of piston and no struts are visible. Top ring groove is 5/32-inch (3.969 mm) from top of piston and slots on opposite sides of piston, behind oil control ring allow oil return and expansion. Piston to cylinder clearance when measured just below oil ring, 90° from pin, should be 0.0015-0.0035 inch (0.0381-0.0889 mm).

**VANASIL PISTON.** Vanasil type piston may be visually identified by smooth contours in underside of piston and no struts are visible. Top ring groove is 9/32-inch (7.144 mm) from top of piston and round holes behind oil control ring allow oil return and expansion. Piston to cylinder clearance when measured just below oil ring, 90° from pin, should be 0.006-0.008 inch (0.1524-0.2032 mm).

Side clearance of top ring in piston groove for all models should be 0.002-0.008 inch (0.051-0.203 mm).

Ring end gap for all models should be 0.010-0.023 inch (0.254-0.584 mm) and end gaps should be staggered around piston circumference.

Piston pin to piston pin bore clearance is 0.0001-0.0005 inch (0.0025-0.0127 mm) for all models and pin to rod clearance is 0.0002-0.0008 inch (0.005-0.020 mm) for aluminum rod and 0.00005-0.00055 inch (0.001-0.014 mm) for bushing in forged steel rod.

Standard piston pin diameter is 0.7500-0.7502 inch (19.05-19.06 mm) for all models and pins are available in 0.002 inch (0.0508 mm) oversize for aluminum rod.

Standard cylinder bore diameter for all models is 3.249-3.250 inches (82.53-82.55 mm) and cylinders should be bored to nearest oversize for which piston and rings are available if cylinder is scored or out-of-round more than 0.003 inch (0.0762 mm) or taper exceeds 0.005 inch (0.127 mm). Pistons and rings are available in a variety of oversizes as well as standard.

**CRANKSHAFT, BEARINGS AND SEALS.** Crankshaft is supported at each end by precision type sleeve bearings located in cylinder block housing and rear bearing plate. Main bearing journal standard diameter should be 1.9992-2.000 inches (50.7797-50.800 mm) and bearing operating (running) clearance for early model flanged type bearing (before specification letter "F")

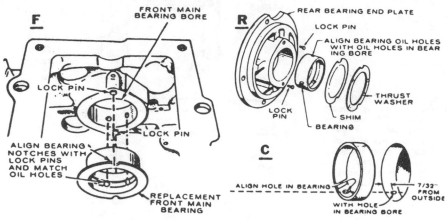

Fig. O76—Alignment of precision main and camshaft bearings. One-piece bearing is used at front (view F), two-piece bearing at rear (view R). Note shim use for end play adjustment. View C shows placement of camshaft bearing.

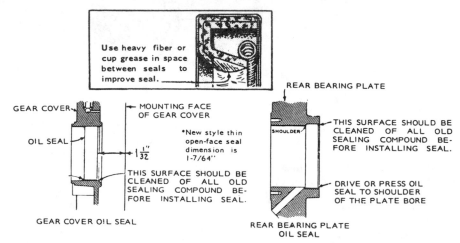

Fig. O77—View showing correct seal installation procedure. Fill cavity between lips of seal with heavy grease to improve sealing efficiency.

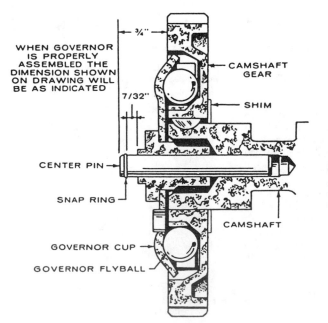

Fig. O78—Cross-sectional view of camshaft gear and governor assembly showing correct dimensions for center pin extension from camshaft. Note shim placement for adjusting end play of camshaft. Refer to text.

should be 0.002-0.003 inch (0.0508-0.0762 mm) and for all remaining models clearance should be 0.0024-0.0042 inch (0.061-0.107 mm). Main bearings are available for a variety of sizes for undersize crankshafts as well as standard.

Oil holes in bearing and bore **MUST** be aligned when bearings are installed and rear bearing should be pressed into bearing plate until flush, or recessed into bearing plate 1/64-inch (0.40 mm). Make certain bearing notches are aligned with lock pins (Fig. O76) during installation.

Apply "Loctite Bearing Mount" to front bearing and press bearing in until flush with block. Make certain bearing notches are aligned with lock pins (Fig. O76) during installation.

**NOTE: Replacement front bearing and thrust washer is a one piece assembly, do not install a separate thrust washer.**

Rear thrust washer is installed with oil grooves toward crankshaft. Measure crankshaft end play as shown in Fig. O75 and add or remove shims or renew thrust washer (Fig. O76) as necessary to obtain 0.006-0.012 inch (0.15-0.30 mm) end play.

It is recommended rear bearing plate and front timing gear cover be removed for seal installation. Rear seal is pressed in until flush with seal bore. Timing cover seal should be driven in until it is 1-1/32 inch (26.19 mm) from mounting face of cover for old style seal and 1-7/64 inch (28.18 mm) from mounting face of cover for new style, thin, open faced seal. See Fig. O77.

## CAMSHAFT, BEARINGS AND GOVERNOR.

Camshaft is supported at each end in precision type sleeve bearings. Make certain oil holes in bearings are aligned with oil holes in block (Fig. O76), press front bearing in flush with

outer surface of block and press rear bearing in until flush with bottom of counterbore. Camshaft to bearing clearance should be 0.0015-0.0030 inch (0.0381-0.0762 mm).

Camshaft end play should be 0.003-0.012 inch (0.0762-0.305 mm) and is adjusted by varying thickness of shim located between camshaft timing gear and engine block. See Fig. O78.

Camshaft timing gear is a press fit on end of camshaft and is designed to accomodate 5, 8 or 10 flyballs arranged as shown in Fig. O79. Number of flyballs is varied to alter governor sensitivity. Fewer flyballs are used on engines with variable speed applications and greater number of flyballs are used for continuous (fixed) speed application. Flyballs must be arranged in "pockets" as shown according to total number used.

Center pin (Fig. O78) should extend 3/4-inch (19 mm) from end of camshaft to allow 7/32-inch (5.6 mm) in-and-out movement of governor cup. If distance is incorrect, remove pin and press new pin in to correct depth.

Always make certain governor shaft pivot ball is in timing cover by measuring as shown in Fig. O80 inset and check length of roll pin which engages bushed hole in governor cup. This pin must extend to within 3/4-inch (19 mm) of timing gear cover mating surface. See Fig. O80.

Make certain timing mark on cam-

shaft gear is aligned with timing mark on crankshaft gear during installation.

**VALVE SYSTEM.** Valve seats are renewable insert type and are available in a variety of oversizes as well as standard. Seats are ground at a 45° angle and seat width should be 1/32 to 3/64-inch (0.794 to 1.191 mm).

Valves should be ground at a 44° angle to provide an interference angle of 1°.

Recommended valve stem to guide clearance is 0.001-0.0025 inch (0.0254-0.0635 mm) for intake valves and 0.0025-0.004 inch (0.0635-0.1016 mm) for exhaust valves. Renewable shouldered valve guides are pushed out from head surface end of block. An "O" ring is installed on intake valve guide of some models and a valve stem seal is also available for intake valve. See Fig. O81.

Recommended valve tappet gap (cold) is 0.010-0.012 inch (0.254-0.305 mm) for both intake and exhaust valves on early models. Current models should have valve tappet gap (cold) of 0.006-0.008 inch (0.152-0.203 mm) for intake valves and 0.015-0.017 inch (0.381-0.432 mm) for exhaust valves.

Adjustment is made by turning tappet adjusting screw as required and valves of each cylinder must be adjusted with cylinder at "top dead center" on compression stroke. At this position both valves will be fully closed.

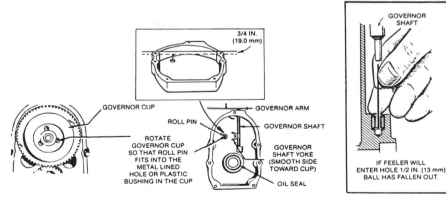

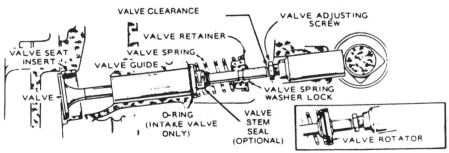

**Fig. O80 — View of governor shaft, timing gear cover and governor cup showing assembly details. Refer to text.**

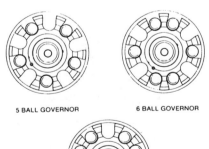

5 BALL GOVERNOR

6 BALL GOVERNOR

8 BALL GOVERNOR

**Fig. O79 — View showing correct spacing of governor flyballs according to total number of balls used.**

**Fig. O81 — View of typical valve train. Refer to text.**

# ONAN
### Division of Onan Corporation
### 1400 73rd Ave. N.E.
### Minneapolis, Minnesota 55432

| Model | Cyls. | Bore | Stroke | Displ. |
|-------|-------|------|--------|--------|
| AJ | 1 | 2¾ in. | 2½ in. | 14.9 cu. in. |
|   |   | (70 mm) | (63.5 mm) | (244 cc) |
| AK | 1 | 2½ in. | 2½ in. | 12.2 cu. in. |
|   |   | (63.5 mm) | (63.5 mm) | (250 cc) |

## MAINTENANCE

**SPARK PLUG.** Recommended plug is 14 mm Champion H-8 or equivalent. Electrode gap is 0.025 inch for gasoline; 0.018 inch for LP-Gas or natural gas fuel.

**CARBURETOR (GASOLINE).** Refer to Fig. 090 for exploded view of typical Carter Model N carburetor used on early AJ and AK engines or to Fig. O91 for exploded view of Walbro carburetor used on later engines. Clockwise rotation of main fuel needle or idle fuel needle leans the fuel mixture.

For initial adjustment of Carter carburetor, open main fuel needle approximately 2½ turns and open idle fuel needle about one turn. On Walbro carburetor, initial setting of idle mixture screw is one turn open and main fuel needle is 1½ turns open. Make final adjustments with engine running at operating temperature.

With engine operating at full rated load, turn main fuel needle in slowly until engine begins to lose speed (or light plant voltage starts to drop), then turn needle out until engine will carry full load, or at lowest charging rate for battery charging plant, turn idle fuel needle in slowly until engine loses speed, then turn needle out to point of smoothest engine operation.

To help prevent governor "hunting" under changes in load, adjust throttle stop screw (2) as follows: With engine operating at rated speed at no load, turn throttle stop screw in until it just touches throttle lever, then turn screw out one turn.

Float level (L–Fig. 092) is measured from free end of float to gasket surface of carburetor body. Float level should be 11/64 inch on Carter carburetors and 1/8 inch on Walbro carburetors.

**AUTOMATIC CHOKE.** Some gasoline fuel models are equipped with an automatic electrically operated choke. Refer to exploded view of the choke assembly in Fig. 093. When cold, the bimetal element (2) turns the carburetor choke to closed position. After engine is started, heat from electric

**Fig. O90—Exploded view of typical Carter Model N carburetor as used on gasoline fuel engines. Refer to Fig. O92 for checking float level.**

1. Choke shaft
2. Throttle stop screw
3. Spring
4. Welch plug
5. Throttle plate
6. Spring
7. Idle fuel needle
8. Idle passage plug
9. Throttle shaft
10. Main fuel needle
11. Spring
12. Carburetor body
13. Choke plate
14. Spring
15. Friction ball, choke shaft
16. Gasket
17. Inlet valve seat
18. Inlet valve
19. Float pin
20. Float
21. Gasket
22. Float bowl
23. Gasket
24. Screw

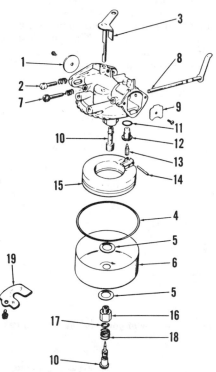

**Fig. O91—Exploded view of Walbro used on later engines. Fuel baffle (19) suppresses fuel slosh on rough surface engines. Components (7 through 19) are not used on gas carburetor model.**

1. Throttle valve
2. Idle speed screw
3. Throttle shaft
4. Gasket
5. Gasket
6. Float bowl
7. Idle mixture screw
8. Choke shaft
9. Choke valve
10. Nozzle
11. Gasket
12. Valve seat
13. Fuel inlet valve
14. Float pin
15. Float
16. Nut
17. "O" ring
18. Spring
19. Main fuel needle

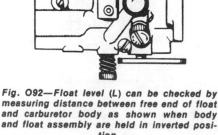

**Fig. O92—Float level (L) can be checked by measuring distance between free end of float and carburetor body as shown when body and float assembly are held in inverted position.**

heating coil in cover (1) causes bimetal element to move choke to open position.

At temperature of 70°F., the choke should be approximately ⅛-inch from fully closed position. Extreme temperature variation may require adjustment of the choke; refer to Fig. 094 for choke adjustment.

FUEL PUMP. Refer to Fig. 095 for exploded view of the gasoline fuel lift pump assembly. All parts except upper

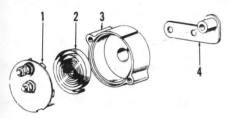

Fig. O93—Exploded view of automatic choke unit available on gasoline models. Refer to Fig. O94 for adjusting unit.

1. Cover & heating element assy.
2. Bimetal element
3. Housing
4. Mounting bracket

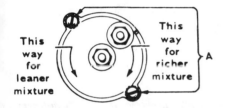

Fig. O94—To adjust gasoline carburetor automatic choke unit, loosen cover retaining screws and turn cover as required. Refer to text.

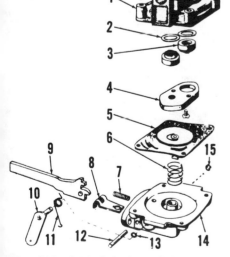

Fig. O95—Exploded view of fuel pump assembly used on some models.

Upper body
Gaskets
Valves
4. Valve retainer
5. Diaphragm
6. Spring
7. Spring
8. Link
9. Arm
10. Primer lever
11. Spring
12. Pin
13. "O" ring
14. Lower body
15. Snap ring

and lower bodies (1 and 14) are serviced separately. Actuating primer lever (10) will pump fuel into carburetor.

GASEOUS FUEL CONVERSIONS. All engines in this group are convertible to operation on gaseous (natural gas, butane or L-P) fuels. Adjustments to idle circuit and to main fuel system are essentially the same as for gasoline carburetors. Except as specially modified for gaseous fuel service, carburetors are identical to those used for operation on regular gasoline. In a dual-fuel (gaseous-gasoline) system, changeover steps are as follows:

**To gasoline:**

Close gaseous fuel supply valve, open gasoline valve.

Remove lock wire from choke.

Set spark plug gap at 0.025 in.

Reinstall carburetor float and inlet needle valve.

**To gaseous fuel:**

Shut off gasoline supply valve.

Install lock wire on choke.

Set spark plug gap at 0.018 in.

Remove carburetor float and inlet needle valve (extended operation only).

For additional details regarding types of gaseous fuel regulators and L-P gas

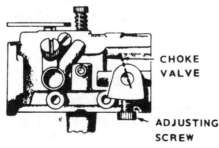

Fig. O98—View showing choke adjusting screw location on some LP-Gas or natural gas carburetors.

operation; refer to SERVICING ONAN ACCESSORIES.

Choke position is determined by counterweight on choke shaft. On some models, a choke adjusting screw is provided; refer to Fig. 098. The weighted choke should just close, but be free to open with air stream through carburetor when engine is running. Turn adjusting screw in to reduce choking.

**GOVERNOR.** Engines which drive electric plants are governed at 1500, 1600, 1800, 2400, 2600, 3000 or 3600 rpm as indicated on generating plant nameplate.

To adjust governor linkage, proceed as follows: On models with automatic idle control (Fig. O100), move control toggle switch to "off" position. With engine stopped, the tension of the governor spring should hold the throttle arm in the wide open position and the throttle lever on the carburetor throttle shaft should just clear the carburetor body by not more than 1/32-inch. This setting can be obtained by adjusting the ball joint on the governor control linkage shown at upper left in Fig. O99.

On industrial engines, start engine and adjust no load speed 50-100 rpm higher than desired full load speed by turning the speed adjusting nut. Apply full load and if speed drop is too great correct by adjusting sensitivity screw to move end of governor spring closer to governor shaft. If engine tends to hunt, adjustment is too close. Any change in sensitivity adjustment will require a speed readjustment. For governor repairs refer to CAMSHAFT and GOVERNOR paragraph.

On AC generating plants, connect a voltmeter across generator output terminals. With plant operating at no load, adjust engine speed so that voltmeter reading is 126 volts on 120 volt plant, or 252 volts on a 240 volt plant. Then, when a full rated load is connected to

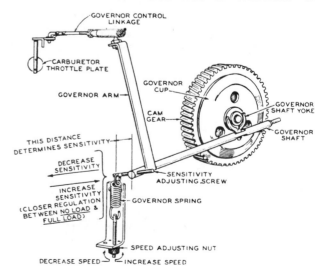

Fig. O99—Schematic view of governor and linkage showing speed, sensitivity and surge adjustments.

generator, voltmeter reading should not fall below 108 volts on a 120 volt plant, or 216 volts on a 240 volt plant. If voltage drop is excessive, turn sensitivity adjustment screw (See Fig. 099) in (clockwise) to increase governor sensitivity. If voltage remains above limits at full load, but voltmeter reading is unsteady (governor "hunts"), turn sensitivity adjustment screw counterclockwise to decrease governor sensitivity. Any change in the sensitivity adjustment will require a compensating change in the speed adjustment nut. On 115-volt DC direct service plants, governor can be adjusted following procedure outlined for AC generating plants.

On models with automatic idle control, refer to Fig. O100 and proceed as follows: Set idle control to "OFF" position and loosen screw (F) so there is no tension on spring (E). Slip socket of flexible joint (A) from ball on lever (B). With locknuts (H) loosened, adjust governor for normal 3600 rpm operation as outlined in previous paragraph for AC generating plants. Then, tighten locknuts (H) with spring (E) as close as possible to end of sensitivity lever.

**MAGNETO AND TIMING.** Refer to Fig. O101 for view of flywheel type

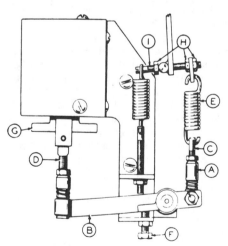

**Fig. O100—View of automatic idle control used on some generating plants. Adjustment procedure is outlined in text.**

magneto used on some engines. Breaker points are located under flywheel. Recommended breaker point gap of 0.020-0.022 inch can be obtained by loosening the points assembly retaining screw and shifting the point set. Timing on engines operated at 1800 rpm or less should be 19 degrees BTC; engines operated above 1800 rpm should be timed at 25 degrees BTC. Magneto back plate has elongated mounting screw holes to permit timing adjustment. Breaker points should just start to open when index mark on flywheel is aligned with the correct degree mark on gear cover. Air gap between coil pole shoes and flywheel should not be less than 0.010 inch and not more than 0.015 inch.

**BATTERY IGNITION AND TIMING.** Some engines are equipped with a battery ignition system. The ignition breaker points and condenser are located on side of engine. Timing is adjusted by varying the breaker point gap. The point gap can be varied from 0.016 to 0.024 inch in order to attain a spark advance of 19 to 25 crankshaft degrees BTC. Recommended timing is 19 degrees BTC on engines that run at speeds of 1800 rpm and slower; 25 degrees BTC on engines that run above 1800 rpm. Decreasing the point gap retards timing and increasing gap advances timing. A reference mark and 19 and 25 degree marks on flywheel can be seen through an opening in the blower housing.

**LUBRICATION.** Recommended oil is API classification SE. During break-in or for operation in temperatures below 32°F., CC rated oils may be used. With air temperatures below 0°F., use SAE 5W-30 weight oil. From 0°F. to 32°F., use 5W-30 or 10W-30, and from 32°F. to 90°F. use SAE 30.

Crankcase capacity is 3.5 pints for stationary units and 2.5 pints for portable units. Pressure lubrication is optional. Pressure lubricated engines utilize a gear type oil pump, an oil intake cup and a non-adjustable oil pressure relief valve. If pump is to be removed it must

be turned off the intake pipe. If the oil pump fails install a complete new pump.

## REPAIRS

**TIGHTENING TORQUES.** Recommended tightening torque values are as follows. All values are in ft.-lbs.

Connecting Rod . . . . . . . . . . . . . . . .10-12
Cylinder Head . . . . . . . . . . . . . . . . . .24-26
Gear Cover . . . . . . . . . . . . . . . . . .15-20
Oil Base . . . . . . . . . . . . . . . . . . . . . .25-30
Oil pump mounting screws . . . . . . . . .7-9

**PISTON, PIN AND RINGS.** The aluminum piston is fitted with two compression rings and one oil control ring.

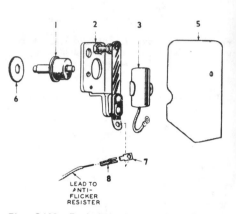

**Fig. O102—Exploded view of "anti-flicker" points used on some generating plant engines. Lead is attached to anti-flicker resister in generator and has no connection with engine ignition system. Adjust anti-flicker breaker point (2) gap to 0.020 in.**

1. Plunger assembly
2. Point set
3. Condenser (0.5 mfd.)
5. Cover
6. Gasket
7. Terminal
8. Terminal

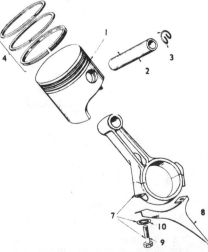

**Fig. O103—Exploded view of piston and connecting rod assembly. Note position of oil dipper (8).**

1. Piston & pin assy.
2. Piston pin
3. Retaining rings
4. Piston rings
7. Connecting rod assy.
8. Oil dipper
9. Rod cap screws
10. Lock washers

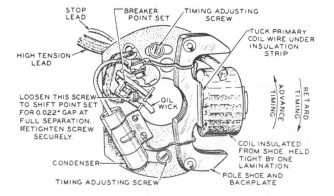

STOP LEAD
BREAKER POINT SET
TIMING ADJUSTING SCREW
TUCK PRIMARY COIL WIRE UNDER INSULATION STRIP
HIGH TENSION LEAD
LOOSEN THIS SCREW TO SHIFT POINT SET FOR 0.022" GAP AT FULL SEPARATION. RETIGHTEN SCREW SECURELY.
OIL WICK
CONDENSER
TIMING ADJUSTING SCREW
COIL INSULATED FROM SHOE. HELD TIGHT BY ONE LAMINATION
POLE SHOE AND BACKPLATE
ADVANCE TIMING
RETARD TIMING

**Fig. O101—View of magneto ignition system. Timing is adjustable by rotating magneto assembly on slotted mounting holes.**

Illustrations courtesy of Onan Corp.

Tapered compression rings should be installed with word "TOP" or other identifying mark up. Recommended piston ring end gap for all rings is 0.006-0.024 inch for Model AJ and 0.006-0.018 inch for Model AK.

Desired piston skirt to cylinder bore clearance is 0.003-0.005 inch for Model AK engines and 0.006-0.008 inch for Model AJ engines. Standard cylinder bore is 2.502-2.503 inches for Model AK and 2.754-2.755 inches for Model AJ engines. Pistons and piston rings are available in oversizes of 0.010, 0.020, 0.030 and 0.040 inch and in standard size for both models.

The floating type piston pin is retained by snap rings. Pin should be a hand push fit in piston and a thumb push fit in connecting rod at 72°F. Piston pin is available in standard size and in 0.002 inch oversize.

**CONNECTING ROD.** Rod and piston unit is removed from above. The aluminum rod rides directly on the crankshaft crankpin. Crankpin diameter is 1.3742-1.3750 inch. Recommended bearing clearance is 0.0015-0.0025 inch. Rod assembly is available in undersizes of 0.010, 0.020 and 0.030 inch as well as standard size. Side play on the crankpin

should be 0.012-0.035 inch. Note that dipper is installed so as to splash oil towards the camshaft side of engine on splash lubricated models.

**CRANKSHAFT, BEARINGS AND SEALS.** The crankshaft rides in two renewable sleeve type bearings. Some models require flange type bearings (bushings) which must be pressed into bore from inside of block or bearing plate. Bearings used in early production engines required line boring or reaming after installation; however, current service parts are precision type bearings which require no reaming and are used to renew the earlier type bearings.

Crankshaft main bearing journal diameter is 1.6860-1.6865 inches; desired journal to bearing running clearance is 0.003-0.004 inch. Renew bearings if clearance is excessive. If crankshaft main journals are worn, journals may be ground undersize as bearings are available in undersizes of 0.002, 0.010, 0.020 and 0.030 inch as well as standard size. Desired crankshaft end play is 0.008-0.012 inch. Obtain desired end play by varying thickness of gaskets (4—Fig. O104) used between bearing plate and crankcase.

When installing new crankshaft main bearings, be sure oil hole in bearing sleeve is aligned with oil supply hole in bearing bore. On splash lubricated engines, oil hole will be upward. On pressure lubricated engines, oil hole will be opposite from camshaft. Bearing plate and crankcase should be heated 200°F. in oven or in hot water before pressing bearings into place.

Renewal of front oil seal requires removal of the timing gear cover. Rear oil seal removal requires removal of bearing plate. Open side (lip) of oil seals must be installed to inside of engine. Rear seal should be flush with face of

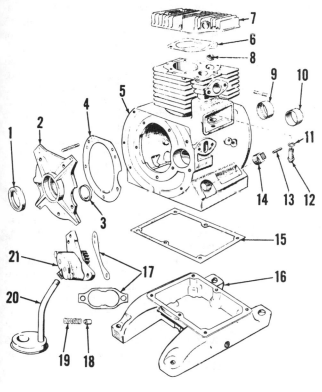

**Fig. O104—Exploded view of crankcase assembly.**

1. Oil seal
2. Bearing plate
3. Plug
4. Gaskets
5. Cylinder block
6. Gasket
7. Cylinder head
8. Exhaust valve seat insert
9. Crankshaft bearings
10. Camshaft bearings
11. Gasket
12. Valve guides
13. Dowel pin (timing gear cover)
14. Oil filler plug
15. Gasket
16. Oil pan
17. Gaskets
18. Oil pressure relief valve
19. Relief valve spring
20. Oil intake tube & screen
21. Oil pump assembly

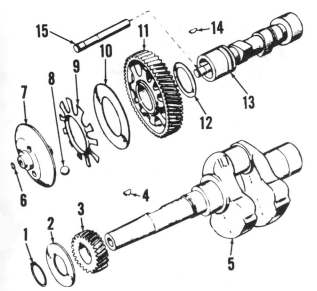

**Fig. O105—Exploded view of crankshaft, camshaft and governor units. Pin (15) is pressed into camshaft (13); refer to Fig. O106. Items (9) and (10) not used on all models.**

1. Snap ring
2. Washer
3. Crankshaft gear
4. Woodruff key
5. Crankshaft
6. Snap ring
7. Governor cup
8. Steel balls
9. Spacer
10. Plate
11. Camshaft gear
12. Thrust washer
13. Camshaft
14. Woodruff key
15. Pin

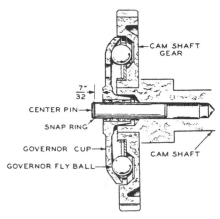

**Fig. O106—When governor cup is pushed in tight against camshaft gear, there should be 7/32-inch clearance between snap ring and cup.**

Illustrations courtesy of Onan Corp.

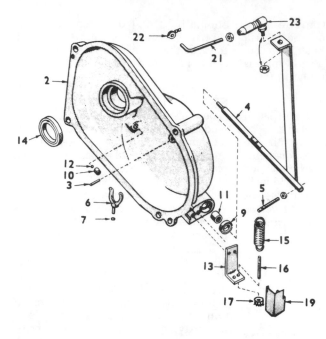

Fig. O107—Exploded view of timing gear cover assembly. Pin (3) engages a hole in governor cup.

2. Cover
3. Roll pin
4. Governor arm & shaft
5. Sensitivity adjustment stud
6. Yoke
7. Snap ring (not all models)
9. Oil seal
10. Bearing
11. Bearing
12. Thrust ball
13. Spring bracket
14. Crankshaft oil seal
15. Governor spring
16. Spring adjusting stud
17. Adjusting nut
19. Spring cover
21. Governor link
22. Clip
23. Balljoint

boss. Use shim stock or pilot sleeve to avoid damage to seals when installing timing gear cover or bearing plate. Seal the mating surface of seal and seal bore with Permatex.

**CAMSHAFT AND GOVERNOR.** Cam gear is a tight press fit on shaft and should be removed from engine as a single unit with shaft. Unit can be removed after first removing cylinder head, gear cover, valves, tappets (fuel pump if used) and the crankshaft gear lock ring and washer. Early camshaft bearings were babbitt-lead lined bushings which can be renewed using latest precision type bearings that do not need to be align bored or reamed after installation. Recommended running clearance is 0.0015-0.0030 inch. Install bushings with oil groove at top. Front bushing should be installed flush with cylinder block; rear bushing flush with bottom of counter bore for expansion plug. Shaft should have minimum end play of 0.003 inch (measured at front bearing) which is controlled by a thrust washer behind the cam gear.

Governor weight unit is mounted on front face of cam gear as shown in Fig. O106. Make sure the distance from outer face of cup sleeve (bushing), or cup itself if bushing is flush, to inner face of snap ring is 7/32-inch as shown when cup is held against fly balls. If less than specified amount, grind end of sleeve as required being sure to remove all burrs from sleeve bore after grinding. If dimension is more than specified amount, press pin further into camshaft. Engines designed for 3600 rpm have 5 fly balls; others have 10.

When installing the gear cover, make sure pin (3 – Fig. O107) in cover engages any one of the 3 holes in governor cup (7 – Fig. O105).

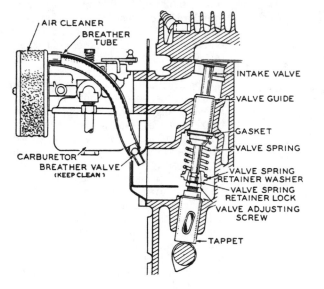

AIR CLEANER
BREATHER TUBE
INTAKE VALVE
VALVE GUIDE
GASKET
VALVE SPRING
CARBURETOR
BREATHER VALVE (KEEP CLEAN)
VALVE SPRING RETAINER WASHER
VALVE SPRING RETAINER LOCK
VALVE ADJUSTING SCREW
TAPPET

Fig. O108—Cross-sectional view of engine valve system and breather.

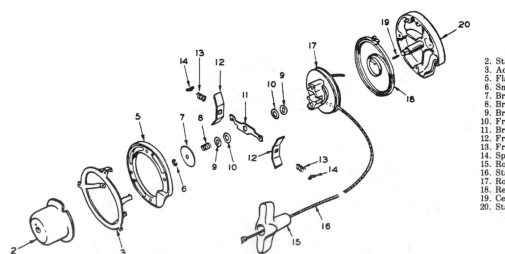

2. Starter cup
3. Adapter ring
5. Flange
6. Snap ring
7. Brake washer
8. Brake spring
9. Brake washer
10. Friction washer
11. Brake lever
12. Friciton shoes
13. Friction shoe springs
14. Spring retainers
15. Rope handle
16. Starter rope
17. Rope pulley
18. Rewind spring
19. Centering pin
20. Starter cover

Fig. O109—Exploded view of rewind starter assembly used on some models. Refer to Fig. O20 for installation of friction shoes (12).

**TIMING GEARS.** Timing gears should always be renewed in pairs, never separately. To remove cam gear first remove camshaft and gear as a unit as per preceding paragraph then press gear off shaft. Crankshaft gear can be removed by using two No. 10-32 screws threaded into holes in gear to push gear from shaft. The "O" marks on gears must be in register for correct valve timing.

**VALVE SYSTEM.** Valve tappet clearance for both intake and exhaust valves is 0.010-0.012 inch cold. Obtain recommended clearance by turning the self locking adjusting screws as needed. Valve face angle is 44°, valve seat angle is 45° and the seat width 1/32-3/64 inch. Renewal of valve seat inserts requires the use of special equipment and should not be attempted unless same is available. Valve stem clearance in guides is 0.0010-0.0025 inch for intake valves; 0.0025-0.0040 inch for exhaust valves. Install valve guides so that shoulder on guide is flush against gasket at valve guide openings in cylinder block casting. Valve tappets are also replaceable from the valve chamber after removing the valve assemblies. Valves are properly timed when timing mark on crankshaft gear registers with timing mark on camshaft gear.

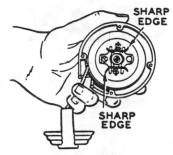

Fig. O110—Install friction shoe and lever assembly with sharp edges of friction shoes pointing in direction shown.

## RECOIL STARTER

Refer to Fig. O109 for exploded view of friction shoe type recoil starter. To disassemble starter, hold rope pulley (17) securely with thumb and remove four screws securing ring (3) and flange (5) to cover (20). Remove ring and flange and release thumb pressure enough to allow spring to rotate pulley until spring (18) is unwound. Remove snap ring (6), washer (7), spring (8), slotted washer (9), and fiber washer (10). Lift out friction shoe assembly (11, 12, 13 and 14), then

remove remaining washers. Withdraw rope pulley (17) from cover. Remove rewind spring from cover if necessary and note direction of windings.

When reassembling, lubricate rewind spring, cover shaft and center bore in rope pulley with a light coat of "Lubriplate" or equivalent. Install rewind spring so that windings are in same direction as removed spring. Install rope on pulley and place pulley on cover shaft. Make certain that inner and outer ends of spring are correctly hooked on cover and rotor. Pre-load the rewind spring by rotating the rope pulley two full turns. Hold pulley in pre-load position and install flange (5) and ring (3). Check sharp end of friction shoes (12) and sharpen or renew as necessary. Install washers (9 and 10), friction shoe assembly, spring (8), washer (7) and snap ring (6). Make certain that friction shoe assembly is installed properly for correct starter rotation. Refer to Fig. O110. If properly installed, sharp ends of friction shoes will extend when rope is pulled.

Remove brass centering pin (19) from cover shaft, straighten pin if necessary, then reinsert pin ⅓ of its length into cover shaft. When installing starter on engine, centering pin will align starter with hole in starter cup retaining cap screw.

# ONAN

## SERVICING ONAN ACCESSORIES

### STARTERS

**MANUAL STARTER.** Engines of LK series and larger, when so equipped, are fitted with manual "Readi-Pull" starter shown in exploded view in Fig. O140. For convenience, direction of starter rope pull may be adjusted to suit special cases by loosening clamps which hold starter cover (5) in position on its mounting ring (20), then turning cover (5) so rope (2) exits in desired direction.

Mounting ring (20) must be firmly attached to engine blower housing which must be as rigid as possible. If blower housing is damaged or if mounting holes for starter are misshaped or worn it may be necessary to renew entire blower housing. See Fig. O141 for cross-section detail of starter mounting ring attachment to blower housing.

To attach starter to earlier production engines, refer to Fig. O142, and use a pair of 10-penny (3 inch) common nails passed through holes in cover to insert into recesses in heads of special screws which retain ratchet wheel to engine flywheel. In later production (after specification D), spirol pin (12A – Fig. O140) is centered upon and engages drilled head screws (17) which secure ratchet wheel (22) and rope sheave hub bearing (16) for alignment during assembly.

Common repairs to manual starter can be made with minimum disassembly. Mounting ring (20) is left in place and only cover assembly (5) need be removed after its four clamps (19) are released. To renew starter rope (2), remove cover (5) from mounting ring (20), release clamp (15) to remove old rope from sheave (10). Then rotate sheave (10) in normal direction of crankshaft rotation so as to tighten spring (8) all the way. Align rope hole in sheave with rope slot in cover, secure new rope by its clamp (15), and when sheave (10) is released, spring (8) will recoil and wind rope on sheave. If renewal of recoil spring (8) is required, sheave (10) must be lifted out of cover (5). Remove starter rope (2) from sheave. Starting at outer end, wrap new spring (8) into a coil small enough to fit into recess in cover with loop at inner end of spring engaging roll pin (7) in cover. It may be necessary to secure wound-up spring with a temporary restraint such as a wire while do-

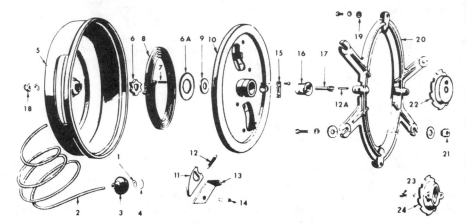

**Fig. O140 – Exploded view of "Ready-Pull" manual starter which is installed on LK and larger engines.**

| | | | |
|---|---|---|---|
| 1. Rope retainer | 7. Roll pin | 13. Ratchet arm (2) | 20. Mounting ring |
| 2. Starter rope | 8. Recoil spring | 14. Pivot roll pin (2) | 21. Speed grip nut |
| 3. Starter grip | 9. Thrust washer | 15. Rope clamp | 22. Ratchet wheel |
| 4. Grip plug | 10. Rope sheave | 16. Hub bearing | (late) |
| 5. Cover | 11. Pawl (2) | 17. Recessed screw | 23. Special cap screw |
| 6. Anti-backlash cog | 12. Pawl spring | 18. Flexlock nut | 24. Ratchet wheel |
| 6A. Spring washer | 12A. Spiral pin | 19. Washer | (early) |

ing so. Then reinstall rope sheave (10) so tab on sheave fits into loop at outer end of recoil spring. Install thrust washer (9) and spring washer (6A) if starter is so equipped.

Whenever starter is disassembled for any service, it is advisable to add a small amount of grease to factory-packed sheave hub bearing (16) and to clean and lubricate pawls (11) and ratchet arms (13) at pivot and contact points. If ratchet arms (13) require renewal due to wear, pawls (11) must first be removed. If kept clean, securely mounted and lightly lubricated, this starter will give long term reliable service.

**AUTOMOTIVE TYPE ELECTRIC STARTERS.** Two styles of battery-driven electric starter motors are used on ONAN engines. Bendix-drive starter shown at A – Fig. O143 is designed to engage teeth of flywheel ring gear when starter switch is depressed to close cir-

**Fig. O141 – Cross-section of mounting ring bolt to show housing attachment detail.**

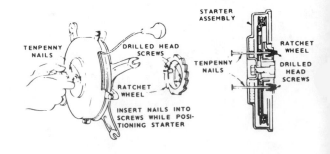

**Fig. O142 – Technique for mounting older style starter to ratchet wheel on engine flywheel. Refer to text.**

Illustrations courtesy of Onan Corp.

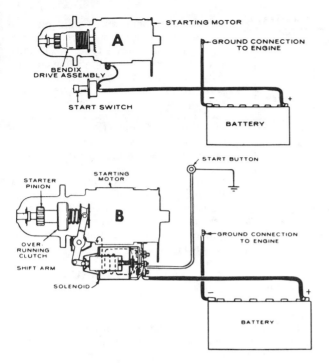

Fig. O143 — Views of Bendix-drive starter and solenoid-shift starter in basic electric circuits. Refer to text.

per selection of spacer washers fitted to armature shaft as installed in housing. To adjust pinion clearance against pinion stop (Fig. O145), remove mounting screws which attach solenoid magnetic coil to front bracket and pinion housing assembly and select a proper thickness of fiber packing gaskets to set required clearance. Be sure plunger is pressed inward as shown when measuring.

Starting motor brushes require renewal when worn away by 0.3 inch (7.62 mm). Original brush length is 0.55 inches (13.97 mm).

Commutator must be clean and free from oil. Use No. 00 sandpaper to clean and lightly polish segments of commutator; never use emery cloth or any abrasive which may have a metallic content. Starter motor commutators do not need to have mica separators between segments undercut. Mica may be flush with surface.

## STARTER MOTOR TESTS

**Armature Short Circuit.** Place armature in growler as shown in Fig. O146 and hold a hack saw blade or similar piece of thin steel stock above and parallel to core. Turn growler "ON". A short circuit is indicated by vibration of blade and attraction to core. If this condition appears, renew armature.

**Grounded Armature.** Check each segment of commutator for grounding to shaft (or core) using ohmmeter setup as shown in Fig. O147. A low (R×1 scale) continuity reading indicates armature is grounded and renewal is necessary.

It is good procedure to mount armature on a test bench or between lathe centers to check for runout of commutator shaft. If shaft is worn badly,

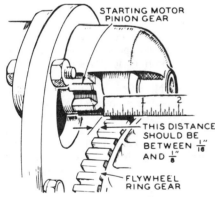

Fig. O144 — Check starter pinion to ring gear clearance as shown. Adjustment details in text.

cuit causing starter motor to turn. Engagement of starter pinion with ring gear by means of spiral shaft screw within Bendix pinion is cushioned by action of its coiled drive spring so starting

motor can absorb sudden loading shock of engagement. Engine manufacturer recommends complete Bendix drive unit be renewed in case of failure, however, if a decision is made to overhaul starter drive by obtaining parts from manufacturer of starter (Prestolite), be sure correct drive spring is used. Length of spring is critical to mesh and engagement of starter pinion to flywheel ring gear. There are no procedures for adjustment of this starter.

Service is generally limited to cleaning and careful lubrication, renewal of starting motor brushes (4 used), brush tension springs and starter motor and drive housing bearings. See STARTER MOTOR TESTS for electrical check-out procedures for starter armature and field windings.

Solenoid-shift style starter (B – Fig. O143) uses a coil solenoid to shift starter pinion into mesh with flywheel ring gear and an over-running (one-way) clutch to ease disengagement of pinion from flywheel as engine starts and runs.

**NOTE: Starter clutch will burn out if held in contact with flywheel for long periods and starter switch must be released quickly as engine starts.**

All parts of starter are available for service. Solenoid unit and starter clutch are renewed as complete assemblies.

Refer to Figs. O144 and O145 for adjustment check points to measure flywheel ring gear to starter pinion clearance and gap between pinion and pinion stop on starter shaft. Starter pinion to ring gear clearance (Fig. O144) is adjusted during starter assembly by pro-

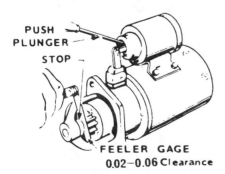

Fig. O145 — Measure clearance between starter pinion and pinion stop with feeler gage as shown.

Fig. O146 — Use of growler to check armature for short circuit. Follow procedure in text.

*Fig. O147 — Test for grounded armature commutator by placing ohmmeter test probes as shown. Probe shown touching shaft may also be held against core. See text.*

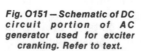

*Fig. O150 — Schematic for bench test of starting motor. Note ammeter is connected in series with load and voltmeter is connected "across" the load in parallel. Refer to text for test values.*

failure or other cause, unit may be started by use of a manual rope starter. In some cases, recoil type "Readi-Pull" starter, as covered in preceding section, may be furnished for standby use.

In case exciter cranking system will not operate, isolate starter solenoid switch and battery from DC field windings and perform a routine continuity check of all components by use of a volt-ohmmeter. See Fig. O151 for possible test points. If problem does not become apparent as caused by battery (low voltage), defective starter solenoid, short or open circuit in lead wires or DC brushes, it will be necessary to check out generator in detail and may involve factory service.

renewal is recommended. If commutator runout exceeds 0.004 inch (0.1016 mm), reface by turning.

**Grounded Field Coils.** Refer to Fig. O148 and touch one ohmmeter probe to a clean, unpainted spot on frame and the other to connector as shown, after unsoldering field coil shunt wire. A low range reading (R×1 scale) indicates grounded coil winding. Be sure to check for possible grounding at connector lead which can be corrected, while grounded field coil cannot be repaired and calls for renewal.

**Open in Field Coils.** Use procedure shown in Fig. O149 and check all four

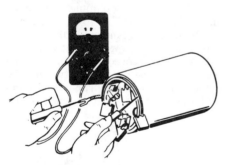

*Fig. O148 — Use ohmmeter as shown to check field coil for suspected internal grounding.*

brush holders for continuity. If there is no continuity or if a high resistance reading appears, renewal is necessary.

**No-Load Test.** When starter is considered ready to return to service, connect motor on bench top as shown in Fig. O150. Acceptable test readings are:

Minimum speed . . . . . . . . . . . . 3700 rpm
Voltmeter reading . . . . . . . . . 11.5 Volts
Maximum current draw . . . 60 Amperes

If starter motor does not check out as satisfactory on this test, make further checks for:

Weak brush springs
Brushes not squarely seated
Dirty commutator
Poor electrical connections. May be caused by "cold" or corroded solder joints.
Tight armature. Not sufficient end play. End play should be 0.004-0.020 inch (0.1016-0.508 mm).
Open or ground in field coil
Short circuit, open or ground in armature.

**EXCITER CRANKING.** Exciter cranking, with cranking torque furnished by switching battery current through a separate series winding of generator field coils and DC brushes, using DC portion of generator armature is wired as shown in Fig. O151.

This starting procedure may be standard for LK or CCK series, or for NH or JB models. In cases where exciter cranking is inoperative, due to battery

## BATTERY CHARGING

**FLYWHEEL ALTERNATOR.** This battery charging system is simple and basically trouble-free. Flywheel-mounted permanent magnet rotor provides a rotating magnetic field to induce AC voltage in fixed stator coils. Current is then routed through a two-step mechanical regulator to a full-wave rectifier which converts this regulated alternating current to direct current for battery charging. Later models are equipped with a fuse between negative (−) side of rectifier and ground to protect rectifier from accidental reversal of battery polarity. See schematic Fig. O153. Maintenance services are limited to keeping components clean and ensuring wire connections are secure.

TESTING. Check alternator output by connecting an ammeter in series between positive (+), red terminal of rectifier and ignition switch. Refer to Fig. O153. At 1800 engine rpm, a discharged battery should cause about 8 amps to register on a meter so connected. As battery charge builds up, current should decrease. Regulator will switch from high charge to low charge at about 14½ volts with low charge current of about 2 amps. Switch from low charge to high charge occurs at about 13 volts. If output is inadequate, test as follows:

Check rotor magnetism with a piece of steel. Attraction should be strong.

Check stator for grounds after dis-

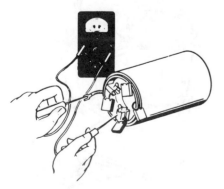

*Fig. O149 — Ohmmeter used to check for breaks or opens in field coil windings. Be sure to check lead wires.*

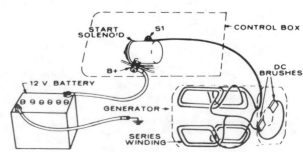

*Fig. O151 — Schematic of DC circuit portion of AC generator used for exciter cranking. Refer to text.*

connecting by grounding each of the three leads through a 12-V test lamp. If grounding is indicated by lighted test lamp, renew stator assembly.

To check stator for shorts or open circuits, use an ohmmeter of proper scale connected across open leads to check for correct resistance values. Identify leads by reference to schematic.

From lead 7 to lead 8 . . . . . . . 0.25 ohms
From lead 8 to lead 9 . . . . . . . 0.95 ohms
From lead 9 to lead 7 . . . . . . . 1.10 ohms

Variance by over 25% from these values calls for renewal of stator.

RECTIFIER TESTS. Use an ohmmeter connected across a pair of terminals as shown in Fig. O154. All rectifier leads should be disconnected when testing. Check directional resistance through each of the four diodes by comparing resistance reading when test leads are reversed. One reading should be much higher than the other.

**NOTE: Forward-backward ratio of a diode is on the order of several hundred ohms.**

If a 12-V test lamp is used instead of an ohmmeter, bulb should light, but dimly. Full bright or no light at all indicates diode being tested is defective.

Voltage regulator may be checked for high charge rate by installing a jumper lead across regulator terminals (B and C – Fig. O153). With engine running, battery charge rate should be about 8 amperes. If charge rate is low, alternator or its wiring is defective.

If charge rate is correct (near 8 amps), defective regulator or its power circuit is indicated. To check, use a 12-V test lamp

to check input at regulator terminal (A). If lamp lights, showing adequate input, regulator is defective and should be renewed.

**NOTE: Regulator, being mechanical, is sensitive to vibration. Be sure to mount it on bulkhead or firewall separate from engine for protection from shock and pulsating motion.**

Engine should not be run with battery disconnected, however, this alternator system will not be damaged if battery terminal should be accidentally separated from binding post.

## FUEL SYSTEMS

**ELECTRIC FUEL PUMP.** Some engines may be furnished with Bendix Electric Fuel Pump, code R-8 or after.

Maintenance service to these electric pumps is generally limited to simple disassembly and cleaning of removable components, not electrical overhaul. Stored gasoline is prone to deterioration and gum residues formed can foul internal precision units of a fuel system so as to cause sticking and sluggish operation of functional parts.

Refer to Fig. O155 for sequence of disassembly, beginning with cover (1) by turning 5/8-inch hex to release cover from bayonet lugs after fuel lines have been disconnected. Wash parts in solvent and blow dry using air pressure. Renew damaged or deteriorated parts, especially gasket (2) or filter element (4). Use needle-nose pliers to remove retainer (5) and withdraw remainder of parts (6 through 10) from plunger tube (11). Clean and dry each item and inspect carefully for wear or damage. Plunger (10) calls for special attention. Clean rough spots very gently using crocus cloth if necessary. Clean bore of plunger tube (11) thoroughly and blow dry. For best results, use a swab on a stick to remove stubborn deposits from inside tube.

During reassembly, check fit of plunger (10) in tube (11). Full, free, in-and-out motion with no binding or sticking is required. If movement of plunger does not produce an audible click, interrupter assembly within housing (12) is defective. Renew entire pump.

If all parts appear serviceable, reassemble pump parts in order shown.

Pump output pressure can be raised or lowered by selection of a different plunger return spring (9). Consult authorized parts counter for special purpose spring or other renewable electric fuel pump parts.

**NOTE: Seal at center of pump case mounting bracket retains a dry gas in pump electrical system. Be sure this seal is not damaged during disassembly for servicing.**

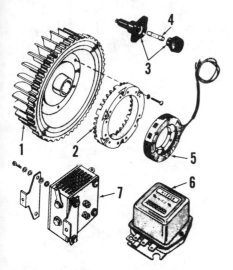

**Fig. O152 — Typical flywheel alternator shown in exploded view. In some models, regulator (6) and rectifier (7) are combined in a single unit.**

1. Flywheel
2. Rotor
3. Fuse holder
4. Fuse
5. Stator & leads
6. Regulator
7. Rectifier assy.

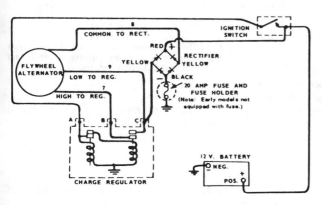

**Fig. O153 — Schematic of flywheel alternator circuits for location of test and check points. Refer to text for procedures.**

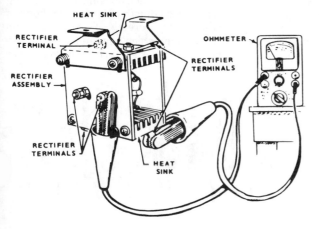

**Fig. O154 — Test each of four diodes in rectifier using Volt-Ohmmeter hook-up as shown. See text for procedure.**

Illustrations courtesy of Onan Corp.

**MECHANICAL FUEL PUMP.** A mechanical fuel pump (Fig. O155A) is used on some models. Pump operation may be checked by disconnecting fuel line at carburetor and slowly cranking engine by hand. Fuel should discharge from line.

**CAUTION: Make certain engine is cool and there is nothing present to ignite discharged fuel.**

To recondition pump, scribe a locating mark across upper and lower pump bodies and remove retaining screws. Noting location for reassembly, remove upper pump body, valve plate screw and washer, valve retainer, valves, valve springs and valve gasket. To remove lower diaphragm, hold mounting bracket and press down on diaphragm to compress spring. Turn bracket 90° to unhook diaphram. Clean all parts thoroughly.

To reassemble, hold pump cover with diaphragm surface up. Place valve gasket and assembled valve springs and valves in cavity. Assemble valve retainer and lock in position by inserting and tightening valve retainer screw. To reassemble lower diaphragm section hold mounting bracket and press down on diaphragm to compress spring. Turn bracket 90° to hook diaphragm. Assemble pump upper and lower bodies, but do not tighten screws. Push pump lever to its limit of travel, hold in this position and tighten the four screws. This prevents stretching diaphragm. Reinstall pump.

**PULSATING DIAPHRAGM FUEL PUMP.** A pulsating diaphragm type fuel pump (Fig. O155B) is used on some models. Pump may be mounted directly to side of carburetor or at a remote mounting location.

Pump relies on a combination of crankcase pressure and spring pressure for correct operation.

Refer to Fig. O155B for disassembly and reassembly noting air bleed hole (10) must be open for correct pump operation.

**HOOF GOVERNOR.** This governor is flyweight-operated, camgear driven to control engine rpm at all points of engine speed range.

Pressure generated by centrifugal force of revolving flyweights is exerted against thrust sleeve and bearing shown in Fig. O156 so as to impart rotary motion to governor rocker shaft through rocker shaft lever. Governor mechanism is liberally lubricated from engine crankcase and severe wear to parts is unusual. If erratic governor performance is traced to its speed sensor (flyweight mechanism), refer to Fig. O156 and proceed as follows:

**Fig. O155 — Exploded view of electric fuel pump used on some engine models.**

1. Cover
2. Cover gasket
3. Magnet
4. Filter element
5. Retainer spring
6. Washer
7. "O" ring
8. Cup valve
9. Plunger spring
10. Plunger
11. Plunger tube
12. Pump housing

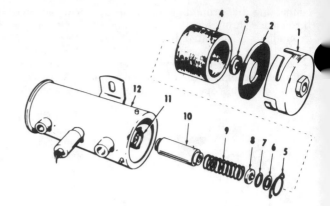

Disconnect linkages.

Unbolt governor body flange and remove governor from engine.

Entire shaft assembly with gear and front cover can be removed after backing out the one cover to body screw.

At this point, any abnormal condition should be apparent when all sludge and contaminated oil is flushed away. Further disassembly is not likely to be needed. Flyweights, pivots and limiting stops should be checked. If necessary, remove retainers and flyweight pins to disassemble governor completely. Thrust bearing and washers can be removed from thrust sleeve after removal of lock ring. After flyweight assembly and thrust sleeve are removed from shaft and shaft is withdrawn, shaft ball bearing can be pressed from cover. Necessary renewal parts should be ordered from manufacturer of governor. Check nameplate for details.

ADJUSTMENTS. All preliminary adjustments should be made to fuel system. Be sure 1/32-inch (0.794 mm) gap is set between stop pin and throttle stop screw as low speed-no load setting. See Fig. O157. Be sure governor bumper screw does not restrict governor action.

Set low speed limits first, then high speed to specifications using most accurate means available to measure engine rpm. Refer to nameplate of electric generator sets for correct speed settings. Increase or decrease spring tension at adjustment points shown in Fig. O157 to set low and high speeds correctly.

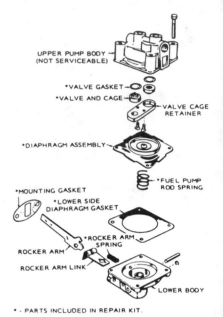

UPPER PUMP BODY (NOT SERVICEABLE)

\*VALVE GASKET

\*VALVE AND CAGE

VALVE CAGE RETAINER

\*DIAPHRAGM ASSEMBLY

\*FUEL PUMP ROD SPRING

\*MOUNTING GASKET

\*LOWER SIDE DIAPHRAGM GASKET

\*ROCKER ARM SPRING

ROCKER ARM

ROCKER ARM LINK

LOWER BODY

\* - PARTS INCLUDED IN REPAIR KIT.

**Fig. O155A — Exploded view of mechanical type fuel pump used on some models. Refer to text for assembly information.**

Operate engine through entire speed range with and without load to determine need for sensitivity adjustment. Proper sensitivity adjustment should result in constant stable engine speed over a complete range of load condition with no hunting or stumbling as load varies. Deviation from rated speed should not exceed 50 rpm. When making sensitivity adjustment, note condition of adjustment screw. If serrations on screw body which engage matching serrations in lever slot are badly worn so as

**Fig. O155B — Exploded view of pulsating diaphragm fuel pump used on some models. Hole (10) must be open before reassembly.**

1. Pump cover
2. Gasket
3. Reed valve
4. Valve body
5. Gasket
6. Diaphragm
7. Pump plate
8. Spring
9. Pump base
10. Air bleed hole

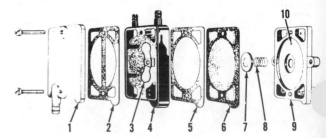

Illustrations courtesy of Onan Corp.

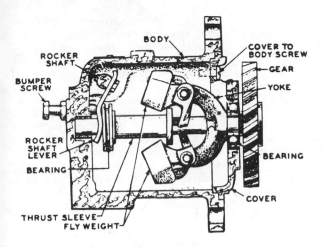

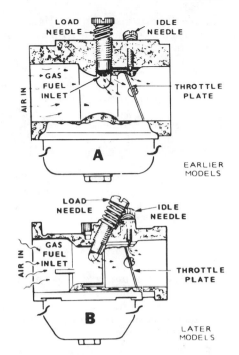

Fig. O156 — Cross-section view of Hoof governor used on older four-cylinder gasoline models (early JC). Refer to text for description of parts functions.

Fig. O158 — Cross-section view of carburetor designed for gaseous fuel only. A-Early style. B-Late style.

not to hold adjustment, renewal of parts is in order. As in all governor adjustments, movement of sensitivity adjuster away from its pivot control point will decrease sensitivity.

**GASEOUS FUEL OPERATION.** This section is concerned only with exclusive operation on gaseous (vapor-type) fuels such as natural gas, methane, butane, propane or mixtures such as LPG. See CCK engine section for dual-fuel (gas-gasoline) operations.

Only one style carburetor is used for constant gaseous fuel operation. Fig. O158 shows a functional cross-section of this carburetor in both early and late models. Note adjustment points. Adjustment of vapor fuel carburetors is essentially the same as for gasoline-fueled models. It will be noted idle adjustment has only a limited effect on performance of these engines as they normally operate at rated rpm. If carburetor is entirely out of adjustment, to such extent engine will not start or run, proceed as follows:

Turn idle adjustment and main adjustment needles inward until lightly seated. Then, open idle needle from one to two turns and crank engine while opening main adjustment needle a little at a time until engine starts. As with gasoline carburetors, final adjustment is made to provide a smooth running engine at both idle speed and rated rpm.

GASEOUS FUELS. It should be noted operation on gaseous fuels sometimes involves changes in types of fuel available for use and not all gaseous fuels have the same power potential. Rating of these fuels is based on their heat output measured in BTU's per cubic foot of volume.

**Butane.** Butane develops about 3200 BTU/cu. ft. and is rated on a par with gasoline as a fuel. It is generally compounded with other gases for regular use.

**Propane.** This gas, rated at about 2500 BTU/cu. ft., will also perform near the level of gasoline. It is also frequently mixed with other gases to suit special

circumstances of its use.

**LPG.** Liquified Petroleum Gas, LPG, sometimes called "bottled gas" is specially compounded to meet requirements for energy (heat) output and for reliability of performance in different climates. LPG is a variable proportion of butane, propane and other hydrocarbons generally "tailor-made" for a particular requirement. As with butane and propane gases, no de-rating of engine is required for its use.

**Natural Gas.** The principal component of natural gas is methane (sometimes called marsh gas) which has a heat output of about 500 BTU/cu. ft.

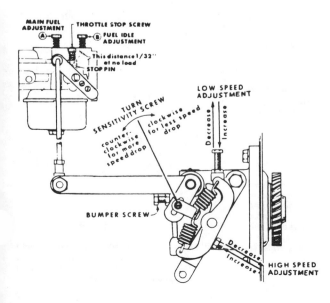

Fig. O157 — View of external linkage and adjustments for Hoof governor. Refer to text.

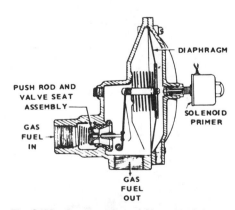

Fig. O159 — Cross-section of Algas regulator used on early vapor (butane, LP or natural gas) fuel systems. Note installation of optional solenoid controlled primer. Refer to text for primer adjustment.

As delivered to users, natural gas will contain 80-95% methane with propane and ethane making up the difference. Heat values based on such compounding will range from 900-1200 BTU/cu. ft.

**Manufactured Gas.** This gas, as delivered in city mains by utility companies, is usually a coal or coke distillation product, with additives, capable of 475-550 BTU output per cubic foot. Internal combustion engines using this gas must be severely de-rated (about 50%) because of this low heat level. Easy availability, with no storage problems and no sensitivity to temperature extremes may cause manufactured gas to be a desirable engine fuel in some circumstances, even in spite of its somewhat greater cost.

**NOTE: All coal and coke gases are toxic. They contain enough carbon monoxide to be poisonous when inhaled. Leakage of this gas is hazardous above and beyond the danger from fire.**

For comparison purposes, available gaseous fuels are rated by a percentage figure against pump grade regular gasoline:

## FUEL COMPARISON TABLE

| GASEOUS FUEL | GASOLINE |
|---|---|
| LPG (Butane, Propane) | 98-100% |
| 1100 BTU gas* | 80-95% |
| 850 BTU gas* | 80-85% |
| 600 BTU gas* | 70-75% |
| 450 BTU gas* | 50-60% |

*Check supplier for BTU rating of fuel gas used.

**GAS REGULATOR.** Engines which operate on gaseous fuels require a pressure regulator be installed in supply line to carburetor. Regulators used are highly sensitive demand types, opening only when fuel is called for by vacuum at carburetor inlet. Ease of starting is dependent upon rapid cranking for high vacuum at intake manifold. Reduced or disconnected engine load during cranking helps. If manual instead of electric cranking is employed, gas used should have a rating of 800 BTU/cu. ft. or higher. Supply line to regulator should be shut off when engine is being serviced to prevent accidental starting if engine

is turned over as part of test procedures. Factory furnished flexible tubing (hose) between carburetor and regulator should never be replaced by a rigid fuel line. Engine vibration must not be transmitted to regulator. Typical regulators used are shown in Fig. O159 and Fig. O160.

REGULATOR TESTING. For a quick operational check of regulator response, blow into vent hole in regulator cover. Hissing sound will indicate release of gas and that diaphragm is reacting to open regulator valve.

For proper operation of a gaseous fuel system, required pressure tests should be made by use of a "U-tube" water manometer having a minimum working range of 14 inches. A commercial model manometer as shown in Fig. O161 is recommended, however, a length of clear plastic tubing of about ⅜-inch inside diameter, formed into a "U" shape comparable to that shown in the figure may be attached to a board and will serve to make pressure measurements. Pour water (with coloring added if hard to see) in open end of tube so it rises into each leg of "U" for 7 or 8 inches or to ZERO level if commercial (scaled) manometer is used. It should be kept in mind pressure is read as the **difference** in fluid level in tube legs. Space between levels, high and low, measured in inches, is pressure of gas in inches of water. For conversion purposes, 1.73 inches of water is equal to one ounce of gas pressure. One inch of water lift or displacement is equal to 0.58 ounces of pressure. When testing or adjusting systems, be sure to identify unit of measure used on gas bottle gages or in utility gas mains and convert where necessary.

To test regulator for leaking, proceed as follows:

Refer to Fig. O160. Turn off gas supply valve.

Remove ⅛-inch pipe plug and connect manometer as shown in Fig. O161. Detach carburetor gas hose from regulator outlet.

Open supply valve (plug valve shown in Fig. O160) and quickly cover open regulator outlet with your hand. Observe manometer while alternately

covering and uncovering regulator outlet. Pressure reading will be constant if regulator valve is closing properly. If manometer reading drops slightly or fluctuates for each time hand is removed from outlet, turn adjustment screw (G–Fig. O161) located just above inlet line inward, a little at a time, until reading is constant when outlet is repeatedly covered and uncovered.

If regulator does not respond to this lock-off screw adjustment and leaking persists, remove and carefully disassemble regulator body. A repair kit furnished with parts for renewal of both valve and diaphragm of regulator is available.

When regulator tests satisfactorily, close supply valve, remove manometer, bleed air from supply line and replace test-hole pipe plug. Reconnect carburetor and test run engine.

**IMPORTANT NOTE: Many test and leak detection procedures in general use for servicing gaseous fueled equipment specify a "soap bubble" test. Do not use to test closing of these highly sensitive demand type regulators. Soap bubble tension alone over regulator outlet is enough to block and shut off this regulator and such a test will prove inaccurate.**

REGULATOR ADJUSTMENT. Algas regulator (Fig. O159) is non-adjustable. It operates with an inlet pressure ranging from 6 ounces to 5 psi. It is no longer furnished on production engines. Because a number of units may still require service in the field, parts and overhaul kits are still obtainable. The Algas regulator features an optional solenoid primer as shown in Fig. O159 to assist in starting. It function is to hold regulator open during cranking. One electrical lead connects to starter solenoid switch on starter side; the other lead is grounded. Adjustment, for rapid start of a cold engine, should be performed when engine is hot and regulator, fuel line and carburetor are charged with gas. To adjust, loosen lock

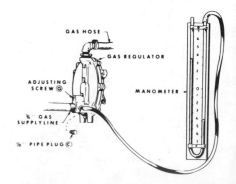

Fig. O161 – Typical connection of water column manometer to regulator. Note pressure is read as difference between tube levels expressed in inches of water. See text.

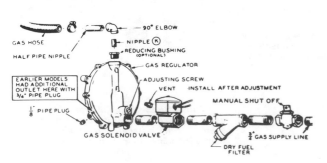

*Fig. O160 – Garretson regulator with installation fittings shown in normal arrangement. Gas solenoid valve shown is optional, but may be required by safety code in some areas. Refer to text.*

Illustrations courtesy of Onan Corp.

nut which secures primer stem to regulator cover and turn primer body inward (clockwise) for richer mixture. Primer is correctly set when slightly rough running and somewhat dark exhaust occur briefly when restarting the hot engine.

When a cold engine must be started and a solenoid primer is known to be out of adjustment, make coarse preliminary adjustment as follows:

Remove carburetor hose from regulator outlet and apply temporary battery voltage across solenoid primer leads.

Then, slowly turn primer body inward (clockwise) until valve opens and gas flow can be heard.

Reconnect primer leads in normal fashion and connect gas hose to regulator. If engine starts in 3 seconds or less adjustment is correct. If engine is slow to start, repeat procedure until cold engine with no gas in hose or carburetor can be rapidly started.

If engine cannot be started properly, test primer. To do so, remove primer from regulator cover and connect battery across solenoid leads while observing if primer plunger extends when power is switched on. If plunger is inoperative, check for mechanical interference or sticking in primer body and perform a continuity check to determine if there is an open circuit in windings or solenoid. Renew if defective.

Garretson regulator is shown in Fig. O160. This is the regulator furnished with current production engines. As a "demand type" regulator, it performs no metering function in regard to gas flow but is only open or closed. Maximum allowable inlet pressure is 8 ounces and minimum is 2 ounces. If supply line pressure exceeds 8 ounces (13.8 inches) a primary regulator should be installed in supply line to reduce working pressure. It is generally recognized that two-stage regulation is advisable in most

gas systems. An appliance type primary regulator installed ahead of secondary (final) regulator is excellent insurance against unsafe increase of pressure, regardless of fluctuations in pressure from gas supply or source. Such primary regulators should be set to deliver controlled pressure at the same level as operating pressure of secondary regulator. Secondary regulator should be adjusted so it will close off gas flow to carburetor at supply line pressure when there is no demand. This prevents gas leaks (seepage) when engine is not running and provides maximum regulator sensitivity. Factory setting is for operation between 2 and 4 ounces (3½ to 7 inches) of pressure. If pressure in gas supply line is from 4 to 8 ounces, readjust at lockout screw as in Fig. O162. Test of regulator shut-off by use of manometer is as previously covered under REGULATOR TESTING.

PRIMARY REGULATORS. When gas source pressures range from 6 ounces to 4 pounds, a pressure-reducing primary regulator is available. Engine model LK uses part number 148P33 which has a maximum gas flow capacity of 190 cubic feet per hour (CFH). CCK series and models JB and NH use No. 148P23 which delivers 330 CFH. For model JC use No. 148P34 which has a capacity of 680 CFH. These regulators all deliver an outlet pressure of 11 inches (water column). These engine models equipped with Garretson regulators in gas line to carburetor accept an inlet pressure of from 2 to 8 ounces (3.5-13.8 inch water column).

LPG VAPORIZER. A vaporizer, sometimes referred to as a converter (converts liquid to vapor) must be used in a system designed for LPG fuel. Its function is to offset cooling by expansion of liquid petroleum gas as it vaporizes from its liquid state. Vaporizer is normally installed in the path of the warm air flow off an air-cooled engine's cooling system to provide warming effect needed.

Fig. O163 shows a cross-section of a typical vaporizer with vaporization process illustrated. Note unit also contains an adjustable pressure regulator for control of outlet gas pressure.

Adjustment procedure proposed by manufacturer recommends vaporizer-regulator be removed from LPG line and fitted to a compressed air source at its inlet port. Seventy five pounds air pressure is adequate. Fit a pressure gage of sufficient range to regulator outlet port and back knurled pressure adjusting screw out nearly all the way. Turn on air pressure and slowly screw adjuster down until test gage reads 7 psi. If pressure rises after being set to specification, regulator valve or diaphragm is leaking and overhaul will be necessary.

**NOTES ON GASEOUS FUEL OPERATION.** Vibration and dirt contamination are major problems in vapor fuel systems. Use of flexible lines which will not transmit engine vibration or shock to regulators or controls is essential. Fuel filters and strainers as installed or recommended by manufacturer must be used and maintained.

Local safety codes which pertain to dispersal of exhaust fumes and engine heat, venting of regulators and fuel storage areas and location of fuel tanks and service lines in relation to structures must be observed.

Many local codes require an electric shut-off valve in fuel line (see Fig. O160) which is usually a solenoid-operated gate valve so connected as to shut off gas supply when engine is halted. Some safety codes will accept the final regulator as an adequate shut-off valve.

Ensign regulator, though no longer furnished by manufacturer, may still be encountered on older production series engines. If a problem develops, contact local factory branch for advice.

Fire safety and electrical installation codes, NEC and local, must be meticulously observed.

Technical Bulletin, T-015, entitled "Use of Gaseous Fuels with Onan Elec-

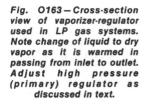

Fig. O162—Adjustment of regulator shut-off of Garretson regulator. Regulator should be set to close gas valve when there is no demand (vacuum) from carburetor. Refer to text.

*Fig. O163 — Cross-section view of vaporizer-regulator used in LP gas systems. Note change of liquid to dry vapor as it is warmed in passing from inlet to outlet. Adjust high pressure (primary) regulator as discussed in text.*

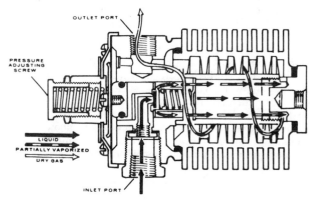

tric Generating Sets" was published by ONAN in August, 1972.

It should be kept in mind when fitting gaseous system fuel line pipe assemblies, strainers or regulators that high grade non-drying pipe joint compound be used on threads. Gaseous fuels have a severe drying effect and use of oil-base paint or other unsatisfactory substitute as a thread-sealant must be avoided or dangerous leaks may result.

## AIR COOLING SYSTEMS

**PRESSURE COOLING.** Most widely used conventional cooling system for these 1 to 4 cylinder models is referred to as pressure cooling. In this system, free air is drawn by flywheel rotation into engine sheet metal housing through flywheel grille opening and is forced through cylinder cooling fins and out through a rear or side aperture.

Larger, J-series engines may be equipped with a thermostat controlled shutter (Vernatherm) which allows engine compartment air to reach 120° F (49° C) before opening and becomes wide open at 140° F (60° C) for full ventilation of enclosure. Opening temperature of sensing element is not adjustable. To determine if this operating element is in working order, remove two screws which retain it to mounting bracket (note slotted holes for adjusting position) and test it by application of heat. Opening should begin at 120° F (49° C) and plunger should be fully extended at 140° F (60° C). Total movement should be at least 13/64-inch (5.16 mm). Reinstall so plunger when fully withdrawn into element body just touches roll pin as in Fig. O164 with shutter completely closed at ambient (free air) temperature.

If shutter operation is unsatisfactory, check for a weak shutter return (closing) spring and examine nylon shutter bearings for dirt or damage. Clean and renew as necessary.

**VACU-FLO COOLING.** This system is designed for cooling industrial power plants which are installed in a closed compartment. Note in Fig. O165 that flow of coolant air is drawn through engine shroud and cooling fins and forced out by flywheel blower through a vent or outside duct. Flow is in reverse direction from that of pressure-cooled engines. IMPORTANT: If flywheel or flywheel blower is renewed, be sure new part is correct for engine cooling system, whether pressure or Vacu-Flo.

Air volume requirement for proper cooling of these engines, expressed in cubic feet per minute is specified for each engine in factory-furnished operator's manual. Dependent upon

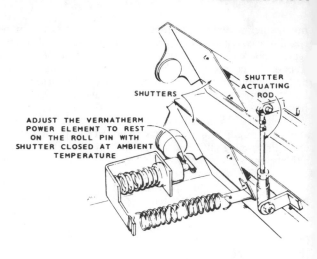

*Fig. O164 — View of thermostat controlled power shutter for pressure cooled engines. Note adjustment. Refer to text for details.*

engine size, this may range from 300 to 1600 cfm. Duct and vent sizes are detailed for each model and type of cooling system.

HIGH TEMPERATURE CUT-OFF. Some larger engine models were equipped with a high temperature safety switch for protection from overheating. Switch is normally closed, but opens to halt engine if compartment air temperature rises to 240° F (116° C) due to problem in cooling system caused by blockage or shutter failure to open. When engine compartment temperature drops to about 190° F (88° C) switch will automatically close and engine can be restarted.

**CLUTCH.** When optional Rockford clutches are furnished with these engines, an adaptor flange is fitted to engine output shaft for mounting clutch unit and a variety of housings are used dependent upon application and model of engine or clutch used. Refer to Fig. O172 for guidance in adjustment and proceed as follows:

Remove plate from top of housing and rotate engine manually until lock screw (1 – Fig. O172) is at top of ring (2) as shown. Loosen lock screw and turn adjusting ring clockwise (as facing through clutch toward engine) until toggles cannot be locked over center. Then, turn ring in reverse direction until toggles can just be locked over center by a very firm pull on operating lever. If a new clutch plate has been installed, slip under load to knock off "fuzz" and readjust. Lubricate according to instructions on unit plate.

**REDUCTION GEAR ASSEMBLIES.** Typical reduction gear unit is shown in Fig. O173. Ratio of 1:4 is common in industrial applications. Lubrication calls for use of SAE 50 motor oil or SAE 90 gear oil. Refer to instructions printed on gear case for guidance. In

most cases, a total of six plugs are fitted into case for lubricant fill or level check. Plug openings to be used are determined by positioning of gear box in relation to horizontal or vertical. It is recommended that square plug heads be cut off those plugs not to be used to fill, check

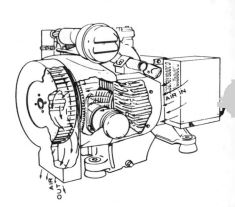

*Fig. O165 — Typical cooling air flow in Vacu-Flo system used for closed compartment installation.*

*Fig. O172 — Procedure for adjustment of Rockford clutch.*

1. Ring lock & screw
2. Adjuster ring
3. Clutch lever

or drain so as to eliminate chance of error by overfill or underfill. All parts shown are available for renewal if needed in overhaul.

**NOTE: In some installations, no shaft seal is fitted between engine crankcase and reduction gear housing. In these cases, with a common oil supply, engine oil lubricates gears and bearings of reduction gear unit and gear oil is not used. Be sure to check nameplate or operator's manual.**

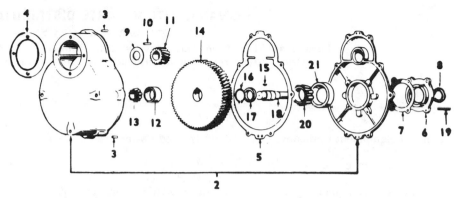

*Fig. O173 — Exploded view of typical reduction gear set. See text for service details.*

| | | |
|---|---|---|
| 2. Housing & cover | 7. Shims | 12. Bearing cup | 17. Bearing spacer |
| 3. Dowel pins (2) | 8. Oil seal | 13. Bearing cone | 18. Shaft |
| 4. Gasket (engine) | 9. Pinion washer | 14. Driven gear | 19. Key |
| 5. Cover gasket | 10. Pinion key | 15. Gear key | 20. Bearing cone |
| 6. Bearing retainer | 11. Pinion gear | 16. Snap ring | 21. Bearing cup |

## ONAN SERVICE TOOLS

Following special tool list indicates ONAN part number, tool name, use and application.

Refer to list of **ONAN CENTRAL WAREHOUSE DISTRIBUTORS** as a source of these tools.

**Valve tools,**
420-0311-VALVE SEAT
REMOVER . . . . . . . . . . . . . J Series,
NH, NHA, NHB,
NHC, NB, N52, T-260G

420-0274-CUTTER
BLADE . . . . . . . . . . . . . For 420-0311
Seat Remover

420-0349-SEAT
CUTTER . . . . . . . . . . . . . All Series

420-0351-CUTTER
BLADE . . . . . . . . . . . . . For 420-0349
Seat Cutter

420-0071-VALVE SEAT
DRIVER . . . . . . . ACK, CCK, CCKB,
CK, LK, LKB, BF,
BG, B43, B48

420-0270-VALVE SEAT
DRIVER . . . . . . . . . . . . . . . . J Series

420-0308-VALVE SEAT
DRIVER . . . . . . . . . NH, NHA, NHB,
NB, N52, T-260G

420-0310-VALVE SEAT
STAKER . . . . . . . . . . . . . (For Intake
Valve) NB,
NHA, NH, NHB,
NHC, N52, T-260G

420-0309-VALVE SEAT
STAKER . . . . . . . . . . . (For Exhaust
Valve) NB,
NHA, NH, NHB,
NHC, N52, T-260G

420-0305-VALVE GUIDE
HONE . . . . . . . . . . . . . . . . All Series

420-0363-REPLACEMENT STONES
(5/16 and 11/32-inch) . . . For 420-0305
Valve Guide Hone

420-0364-REPLACEMENT STONES
(3/8, 13/32 and 7/16-inch . . . . . . . For
420-0305 Valve
Guide Hone

420-0300-VALVE GUIDE
DRIVER . . . . . . . . . CCK, CCKB, LK,
LKB, NH, NHA,
NHB, NHC, N52, T-260G,
NB, BF, BG, B43,
B48, J Series

**Oil Seal Guides And Drivers,**
420-0387-BEARING
PLATE . . . . . . . . . . CK, ACK, CCK,
CCKB, LK, AJ,
AK, NH, NHA,
NHB, B48, T-260G

420-0250-BEARING
PLATE . . . . . . . . . . . . . . . . J Series
420-0389-GEAR COVER . . . . . . J Series
420-0313-GEAR
COVER . . . . . . . . . NH, NHA, NHB,
NHC, NB, N52, T-260G,
CCK, CCKB, BF,
BG, B43, B48

**Pullers,**
420-0100-FLYWHEEL
PULLER . . . . . . . . . . . . . . . . J Series
BH, CK, LK, LKB,
CCK, CCKB, NH,
NHA, NHB, NHC,
N52, NB, BF, BG,
B43, B48, T-260G

**Wrenchs,**
420-0169-CARBURETOR ADJUSTING
WRENCH . . . . . . . . . . BH, CK, CCK
420-0294-CARBURETOR ADJUSTING
WRENCH . . . . . . . . . . . . . . J Series,
CCKB, NH,
NHA, NHB, NHC,
NB, CCK, T-260G

## ONAN CENTRAL PARTS DISTRIBUTORS
(Arranged Alphabetically by States)

These franchised firms carry extensive stock of repair parts. Contact them for
name and address of nearest dealer who may have the parts you need.

Atchison Equipment Company, Inc.
Phone: (205) 591-2328
4724 First Avenue North
P.O. Box 2971
**Birmingham, Alabama 35212**

Delhomme Industries
Phone: (205) 473-6626
3422 Georgia Pacific Avenue
**Mobile, Alabama 36607**

Fremont Electric Company
Phone: (907) 277-9558
140 East Dowling Road
**Anchorage, Alaska 99502**

Harrison Industries of Arizona Inc.
Phone: (602) 243-7222
3502 East Broadway Road
**Phoenix, Arizona 85040**

Mecelec of Arkansas
Phone: (501) 490-1801
1701 Dixon Road
P.O. Box 9297
**Little Rock, Arkansas 72219**

Equipment Service Company
Phone: (213) 426-0311
3431 Cherry Avenue
P.O. Box 1307
**Long Beach, California 90801**

Cal-West Electric, Inc.
Phone: (916) 372-5522
3939 West Capitol Avenue
**West Sacramento, California 95691**

Cal-West Electric, Inc.
Phone: (415) 873-7710
1341 San Mateo Avenue
P.O. Box 2364
**San Francisco (South),
California 94080**

Equipment Service Company
Phone: (714) 562-2804
1954 Friendship Drive
**El Cajon, California 92020**

C.W. Silver Company, Inc.
Phone: (303) 399-7440
3945 East 50th Avenue
**Denver, Colorado 80216**

GLT Industries, Inc.
Phone: (203) 528-9944
29 Mascolo Road
P.O. Box 307
**South Windsor, Connecticut 06074**

Advanced Power Systems
Phone: (904) 355-4563
710 Haines Street
P.O. Box 38039
**Jacksonville, Florida 32206**

R. B. Grove, Inc.
Phone: (305) 854-5420
261 South West 6th Street
**Miami, Florida 33130**

Tampa Armature Works, Inc.
Phone: (305) 843-8250
3400 Bartlett Boulevard
**Orlando, Florida 32805**

Tampa Armature Works, Inc.
Phone: (813) 621-5661
440 South 78th Street
P.O. Box 3381
**Tampa, Florida 33601**

Blalock Machinery & Equipment
Company
Phone: (912) 436-1507
700 South Westover Road
**Albany, Georgia 31702**

Blalock Machinery & Equipment
Company
Phone: (404) 766-2632
5112 Blalock Industrial Boulevard
**College Park, Georgia 30349**

Atlas Electric Company, Inc.
Phone: (808) 524-5866
1151 Mapunapuna Street
**Honolulu, Hawaii 96819**

Power Systems, Division of E. C.
Distributing
Phone: (208) 342-6541
4499 Market Street
**Boise, Idaho 83705**

Power Systems, Division of E. C.
Distributing
Phone: (208) 234-2442
1060 South Main
**Pocatello, Idaho 83204**

Stannard Power Equipment Company
Phone: (312) 597-5500
4901 West 128th Place
**Alsip, Illinois 60658**

Service Automotive Warehouse, Inc.
Phone: (309) 794-0400
111 4th Avenue
**Rock Island, Illinois 61201**

Meco-Indiana, Inc.
Phone: (219) 262-4611
23900 County Road 6
**Elkhart, Indiana 46514**

Evansville Auto Parts, Inc.
Phone: (812) 425-8264
5 East Riverside Drive
**Evansville, Indiana 47713**

Meco-Indiana, Inc.
Phone: (317) 873-5005
5005 West 106th Street
**Zionsville, Indiana 46077**

Midwestern Power Products Company
Phone: (515) 278-5521
10100 Dennis Drive
**Des Moines, Iowa 50322**

Anderson Equipment Company, Inc.
Phone: (712) 255-8033
300 South Virginia Street
**Sioux City, Iowa 51102**

Mecelec of Kansas
Phone: (316) 522-4767
4631 Palisade Street
**Wichita, Kansas 67217**

Southern Power Systems
Phone: (502) 459-5060
2025 Old Shepherdsville Road
**Louisville, Kentucky 40218**

Delhomme Industries, Inc.
Phone: (318) 234-9837
337 Mecca Drive
**Lafayette, Louisiana 70505**

Delhomme Industries, Inc.
Phone: (318) 439-9700
2506 Elaine Street
**Lake Charles, Louisiana 70601**

Delhomme Industries, Inc.
Phone: (318) 365-5476
Northside Road
P.O. Box 266
**New Iberia, Louisiana 70560**

Menge Pump and Machinery
Company, Inc.
Phone: (504) 888-8830
2740 North Arnoult
P.O. Box 8210
**New-Orleans-Metairie,
Louisiana 70011**

Menge Pump & Machinery Company, Inc.
Phone: (318) 222-5781
1510 Grimmet Drive
P.O. Box 7323
**Shreveport, Louisiana 71107**

The Leen Company
Phone: (207) 989-7363
54 Wilson Street
**Brewer, Maine 04412**

The Leen Company
Phone: (207) 774-6266
366 West Commercial Street
**Portland, Maine 04101**

Curtis Engine & Equipment Company
Phone: (301) 633-5161
6120 Holabird Avenue
**Baltimore, Maryland 21224**

New England Engine Corporation
Phone: (617) 948-7331
RR 1
**Rowley, Massachusetts 01962**

Standby Power Inc.
Phone: (616) 949-7990
2745 South East 29th Street
**Grand Rapids, Michigan 49508**

Stanby Power Inc.
Phone: (313) 348-6400
4700 12 Mile Road
**Novi, Michigan 48050**

Flaherty Equipment Corporation
Phone: (612) 338-8796
2525 East Franklin Avenue
**Minneapolis, Minnesota 55406**

Interstate Detroit Diesel Allison Inc.
Phone: (612) 854-5511
2501 East 80th Street
**Minneapolis, Minnesota 55420**

Northstar Detroit Diesel Allison Inc.
Phone: (218) 749-4484
1921 West 16th Avenue
**Virginia, Minnesota 55792**

Menge Pump & Power Systems Division
Phone: (601) 969-9333
1327 South Gallatin
**Jackson, Mississippi 39202**

National Industrial Supply Company
Phone: (314) 621-0350
1100 Martin Luther King Drive
**St. Louis, Missouri 62201**

Comet Industries, Inc.
Phone: (816) 245-9400
4800 Deramus Avenue
**Kansas City, Missouri 64120**

Anderson Equipment Company, Inc.
Phone: (402) 558-1200
5532 Center Street
**Omaha, Nebraska 68106**

Anderson Industrial Engines
Phone: (402) 558-8700
2123 South 56th Street
**Omaha, Nebraska 68106**

Equipment Service Company of Nevada
Phone: (702) 382-3852
1916 Highland Avenue
**Las Vegas, Nevada 89102**

R. C. Equipment Company, Inc.
Phone: (609) 742-0220
522 South Broadway
**Gloucester City, New Jersey 08030**

GLT Industries, Inc.
Phone: (201) 767-9751
411 Clinton Avenue
**Northvale, New Jersey 07647**

C. W. Silver Company, Inc.
Phone: (505) 881-2454
4812 North East Jefferson
**Albuquerque, New Mexico 87109**

Power Plant Equipment Corporation
Phone: (518) 783-1991
6 Northway Lane
**Latham, New York 12110**

Ronco Communications & Electronics
Phone: (716) 424-3890
230 Metro Park
**Rochester, New York 14623**

Power Plant Equipment Corporation
Phone: (315) 475-7251
929 South Salina Street
**Syracuse, New York 13202**

Ronco Communications & Electronics, Inc.
Phone: (716) 873-0760
595 Sheridan Drive
**Tonawanda-Buffalo, New York 14150**

Owsley & Sons
Phone: (919) 668-2454
Interstate 40
P.O. Box 8627
**Greensboro, North Carolina 27410**

Interstate Detroit Diesel Allison, Inc.
Phone: (701) 258-2303
3801 Miriam Gateway
**Bismarck, North Dakota 58501**

Interstate Detroit Diesel Allison, Inc.
Phone: (701) 282-6556
3902 North 12th Avenue
**Fargo, North Dakota 58102**

Southern Ohio Power Systems, Inc.
Phone: (513) 821-6305
8148 Vine Street
**Cincinnati, Ohio 45216**

McDonald Equipment Company
Phone: (216) 951-8222
37200 Vine Street-Willoughby
**Cleveland, Ohio 44094**

Tuller Corporation
Phone: (614) 224-8246
947 West Goodale Boulevard
**Columbus, Ohio 43212**

G & R Equipment Company
Phone: (405) 685-5534
3826 Newcastle Road
**Oklahoma City, Oklahoma 73119**

Mechanical & Electrical Equipment Company
Phone: (918) 582-7777
712 Wheeling
P.O. Box 50323
Whittier Station
**Tulsa, Oklahoma 74150**

EC Distributing Division Electrical Construction
Phone: (503) 224-3623
2122 North West Thurman Street
**Portland, Oregon 97208**

A. F. Shane Company
Phone: (412) 781-8000
654 Alpha Drive RIDC Industrial Park
**Pittsburgh, Pennsylvania 15238**

Winter Engine Generator Service
Phone: (717) 848-3777
1600 Pennsylvania Avenue
**York, Pennsylvania 17404**

Owsley & Sons Inc.
Phone: (803) 548-3636
I-77 & SC Exit 72
Gold Hill Road
**Fort Mill, South Carolina 29715**

Hobbs Equipment Company
Phone: (615) 894-8400
6203 Provence Street
**Chattanooga, Tennessee 37421**

Hobbs Equipment Company
Phone: (615) 966-7550
10625 Lexington Drive
**Knoxville, Tennessee 37922**

Maritime & Industrial, Inc.
Phone: (901) 775-1204
P.O. Box 9397
292 East Mallory
**Memphis, Tennessee 38109**

Hobbs Equipment
Phone: (615) 244-4933
1327 Foster Avenue
**Nashville, Tennessee 37211**

Lightbourn Equipment Company
Phone: (214) 233-5151
P.O. Box 401870
13649 Beta Road
**Dallas, Texas 75240**

Harrison Equipment Company
Phone: (713) 879-2600
1100 West Airport Boulevard
**Houston, Texas 77001**

Power Support Systems
Phone: (915) 332-1429
P.O. Box 4417
3208 Kermit Highway
**Odessa, Texas 79760**

Lightbourn Equipment Company
Phone: (512) 333-7542
4260 Dividend Drive
**San Antonio, Texas 78219**

C. W. Silver Company, Inc.
Phone: (801) 355-5373
550 West 7th South
**Salt Lake City, Utah 84101**

T & L Electric Company
Phone: (802) 295-3114
Sykes Avenue
**White River Junction, Vermont 05001**

Curtis Engine & Equipment
Phone: (804) 627-9470
1114 Ballentine Boulevard
**Norfolk, Virginia 23504**

Owsley & Sons
Phone: (804) 275-2603
10300 Jefferson
Davis Highway
P.O. Box 34508
**Richmond, Virginia 23234**

Fremont Electric Company
Phone: (206) 633-2323
P.O. Box 31640
744 North 34th Street
**Seattle, Washington 98103**

Lay & Nord
Phone: (509) 453-5591
P.O. Box 472
511 South 3rd Street
**Yakima, Washington 98901**

Call Detroit Diesel Allison, Inc.
Phone: (304) 744-1511
P.O. Box 8245
Charleston Ordinance Center
**Charleston, West Virginia 25303**

Inland Diesel, Inc.
Phone: (414) 781-7100
13015 Custer Avenue
**Butler, Wisconsin 53007**

Morley-Murphy
Phone: (414) 499-3171
Box 3640
700 Morley Road
**Green Bay, Wisconsin 54303**

Clymar, Inc.
Phone: (414) 781-0700
N55 W13787 Oak Lane
Menomenee Falls
**Milwaukee, Wisconsin 53051**

Power Service Company
Phone: (307) 237-3773
P.O. Box 2880
5201 West Yellowstone Highway
**Casper, Wyoming 82606**

## CANADIAN DISTRIBUTORS

**Calgary, Alberta T2H 1J5
Simson-Maxwell**
6447-2nd Street S.E.

**South Edmonton, Alberta
T6H 1E7
Simson-Maxwell**
Box 4446
10375 59th Avenue

**Vancouver V6H 1A7, B.C.
Simson-Maxwell**
1380 W. 6th Avenue

Kenora, Ontario
**L.O.W.E. Power Systems, Ltd.**
P.O. Box 81

**Oakville, Ontario L6K 2H2
Hawker Siddeley Diesels & Electric**
355 Wyecroft Road

**Ottawa Ontario K1B 3V7
J.A. Faguy & Sons, Ltd.**
2544 Sheffield Road

**Thunder Bay, Ontario, P7B 5M5
Lake of the Woods Electric Ltd.**
1177 Roland Street
Postal Station F

**Montreal, Quebec H4T 1P3
J.A. Faguy & Sons, Ltd.**
750 Montee de Liesse

**Saskatoon, Saskatchewan S7K 1L2
Pryme Power & Diesels Ltd.**
5-2949 3rd Avenue North
Phone: (306) 665-8044

**Prince George, B.C.
Simson-Maxwell**
729 4th Avenue

**Winnipeg, Manitoba R3B 0A6
Kipp Kelly, Ltd.**
68 Higgins Avenue

**Fredericton, New Brunswick
Sansom Equipment, Ltd.**
Woodstock Rd., P.O. Box 1263

**Truro, Nova Scotia
Sansom Equipment Ltd.**
P.O. Box 152
80 Glenwood Drive

**Sault Ste. Marie, Ontario
Algonma Truck & Tractor Sales**
815 Great Northern Road

**Scarborouigh, Ontario
Total Power, Ltd.**
330 Nantucket Boulevard

# SUZUKI

| Model | No. Cyls. | Bore | Stroke | Displacement | Power Rating |
|---|---|---|---|---|---|
| SE300 | 1 | 44 mm (1.73 in.) | 40 mm (1.57 in.) | 61 cc (3.7 cu. in.) | 0.66 kW (0.9 hp) |

## ENGINE IDENTIFICATION

The 61 cc (3.7 cu. in.) engine used on Model SE300 generator is a four-stroke, single-cylinder, air-cooled, horizontal crankshaft engine. Rated output is 0.66 kW (0.9 hp) at 3600 rpm.

Engine serial number is located on the side of crankcase cover (Fig. SU60). Always furnish generator model number and engine serial number when ordering parts.

## MAINTENANCE

**SPARK PLUG.** Recommended spark plug is a NGK CM-6, or equivalent. Spark plug should be removed and cleaned and electrode gap set at 0.4-0.5 mm (0.016-0.020 inch) after every 100 hours of operation. Renew spark plug if electrode is burned or damaged.

**CAUTION: Caution should be exercised if abrasive type spark plug cleaner is used. Inadequate cleaning procedure may allow the abrasive cleaner to be deposited in engine cylinder accelerating wear and part failure.**

**CARBURETOR.** Model SE300 is equipped with a float type carburetor (Fig. SU61). Main fuel mixture is controlled by a fixed main jet (6—Fig. SU61). Main jet size is a number 52.5 and pilot jet (2) is a number 42.5.

Initial adjustment of idle fuel mixture screw (5) from a lightly seated position is 3/4 turn open.

Engine idle speed is 1000 rpm and idle speed adjustment should be made by adjusting the throttle stop screw (1). When making this adjustment, make certain

governor screw is loose so that the governor does not operate.

**AIR FILTER.** Model SE300 is equipped with a foam type air filter element. The air filter element should be removed and cleaned at 50 hour intervals. To clean element, wash in a nonflammable cleaning solvent and gently squeeze element dry. Oil element with clean engine oil and gently squeeze out the excess oil.

**GOVERNOR.** Model SE300 is equipped with a centrifugal governor which is located internally in engine with flyweight assembly attached to camshaft gear. Governed engine speed is 3600 rpm.

**IGNITION SYSTEM.** Model SE300 is equipped with a transistorized ignition system. There are no contact points used in this system, and no periodic maintenance or adjustments are required.

**VALVE ADJUSTMENT.** Valve clearance should be checked after every 300 hours of operation. To check clearance, remove tappet chamber cover and breather assembly from cylinder block. Turn crankshaft until piston is at top dead center on compression stroke, then use a feeler gage to measure clearance between end of valve stem and tappet (Fig. SU62). Correct valve stem clearance is 0.05-0.15 mm (0.002-0.006 in.) for intake and exhaust valves. Valve tappets

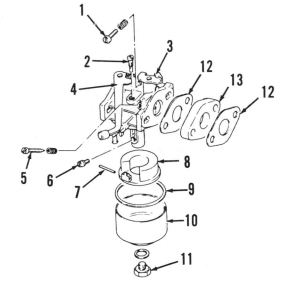

Fig. SU61-Exploded view of the BV13 float type carburetor used on Model SE300 generator.

1. Idle speed adjusting screw
2. Pilot jet
3. Throttle shaft
4. Choke shaft
5. Idle mixture screw
6. Main jet
7. Pin
8. Float
9. Gasket
10. Float bowl
11. Bolt
12. Gaskets
13. Spacer

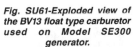

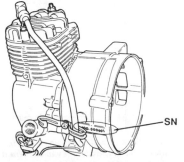

Fig. SU60-View showing location of engine serial number (SN) for Model SE300 generator.

Fig. SU62—Valve clearance should be checked with engine cold and with piston at top dead center on compression stroke. Use a feeler gage to measure clearance between valve stem end and tappet. Refer to text for adjustment procedure.

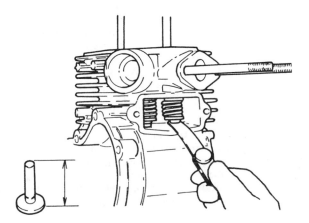

are available in a variety of lengths to adjust valve clearance. Refer to the REPAIRS section for service procedure.

**CYLINDER HEAD AND COMBUSTION CHAMBER.** Cylinder head, combustion chamber and piston should be cleaned and combustion deposits removed after every 300 hours of operation. Refer to REPAIRS section for service procedure. Cylinder compression reading should be 500 kPa (73 psi) or higher.

**LUBRICATION.** Model SE300 engine is lubricated by utilizing a dipper attached to connecting rod cap.

Oil level should be checked prior to each operating interval. Oil level should be maintained at top edge of reference marks on gage with oil fill plug just touching first threads. Do not screw oil fill plug and gage in when checking oil level.

Manufacturer recommends using SAE 5W-20 or 5W-30 oil if temperature is below -18° C (0° F); SAE 10W-30 or 10W-40 oil if temperature is -18° C (0° F) to 27° C (80° F); SAE 20W-40 or 20W-50 oil if temperature will be above 27° C (80° F).

Oil should be changed after the first 20 hours of operation and every 100 hours of operation thereafter. Crankcase capacity is 0.25 L (0.53 pt.).

**GENERAL MAINTENANCE.** Check and tighten all loose bolts, nuts or clamps prior to each day of operation. Check for fuel or oil leakage and repair if necessary.

Clean dust, dirt, grease or any foreign material from cylinder head and cylinder block cooling fins after every 100 hours of operation. Inspect fins for damage and repair if necessary.

## REPAIRS

**TIGHTENING TORQUES.** Recommended tightening torques are as follows:

Connecting rod . . . . . . . . . . . .4.8 N·m
(42 in.-lbs.)
Crankcase . . . . . . . . . . . . . . . .3.4 N·m
(30 in.-lbs.)
Cylinder head . . . . . . . . . .6.1-9.5 N·m
(54-84 in.-lbs.)
Rotor set bolt . . . . . . . . .15.7-22.5 N·m
(140-200 in.-lbs.)
Generator through-bolt . .8.8-11.6 N·m
(78-103 in.-lbs.)

**CYLINDER HEAD.** To remove cylinder head, first remove side cover, end cover and control panel. Disconnect generator electrical connector, ignition wires and ground wire. Disconnect fuel hose and remove the fuel tank. Remove

engine air cleaner. Remove the four nuts retaining engine and generator assembly to the bottom mounting plate. Remove engine cooling fan housing and cylinder air shroud. Loosen cylinder head bolts gradually in a crisscross pattern until all bolts are loose enough to remove by hand.

Remove spark plug and clean carbon and other combustion deposits from cylinder head. Check cylinder head and cylinder mating surfaces for distortion. If either is warped more than 0.030 mm (0.0012 inch), cylinder head or block must be renewed.

Reinstall cylinder head using a new head gasket. Tighten bolts evenly to specified torque in a crisscross pattern.

**CONNECTING ROD.** Connecting rod used in Model SE300 engine rides directly on the crankshaft crankpin journal.

To remove connecting rod, first remove cylinder head as outlined above. Remove air cleaner and muffler. Disconnect governor lever linkage from carburetor. Remove rotor retaining bolt and generator through-bolts. Remove generator stator assembly. Remove generator rotor from engine crankshaft using a suitable puller. Remove ignition coil from crankcase cover and remove Woodruff key from crankshaft. Unbolt and remove generator fan housing and engine crankcase cover assembly from cylinder block. Remove camshaft and governor flyweight assembly, mark tappet location and remove tappets. Remove connecting rod bolts, splasher plate and connecting rod cap. Push piston and connecting rod out the top cylinder block.

Remove piston pin retaining rings (4—Fig. SU63) and separate piston (2) from connecting rod (7) as required. Note that piston is installed on connecting rod so that arrow on top of piston is aligned with arrow on connecting rod.

Standard piston pin bore diameter in connecting rod is 13.006-13.014 mm (0.5120-0.5124 inch). Standard piston pin diameter is 12.995-13.000 mm (0.5115-0.5118 inch). If clearance between piston pin and connecting rod pin bore is 0.05 mm (0.002 inch) or more, renew pin and/or connecting rod.

Standard diameter of connecting rod crankshaft bearing bore is 18.013-18.027 mm (0.7092-0.7097 inch). Clearance between connecting rod bearing bore and crankshaft crankpin journal should be 0.013-0.035 mm (0.0005-0.0014 inch). If clearance exceeds 0.060 mm (0.0024 inch), renew connecting rod and/or crankshaft.

Side clearance between connecting rod and crankshaft should be 0.250-0.850 mm (0.0098-0.0335 inch). If clearance is 1.200 mm (0.047 inch) or more,

renew connecting rod and/or crankshaft.

When installing connecting rod and piston assembly, make certain that arrow marks (Fig. SU64) on piston crown and connecting rod point toward the valves. Be sure match marks on connecting rod and cap are aligned and are facing open side of cylinder block. Install oil slinger, new lock plate and then the connecting rod bolts. Tighten con-

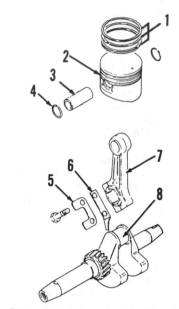

**Fig. SU63—Exploded view of piston and connecting rod assembly.**

| | |
|---|---|
| 1. Piston rings | 5. Lock plate |
| 2. Piston | 6. Oil slinger |
| 3. Piston pin | 7. Connecting rod |
| 4. Retaining rings | 8. Crankshaft |

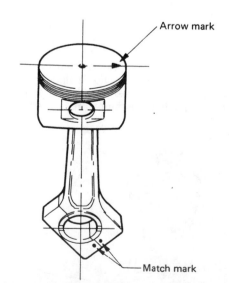

**Fig. SU64—Arrow on piston crown and connecting rod should point toward the valves. When installing connecting rod cap, make certain match marks are aligned.**

necting rod bolts to specified torque and bend lock plate against bolt heads.

**PISTON, PIN AND RINGS.** Refer to CONNECTING ROD paragraph for piston removal.

After separating piston and connecting rod, carefully remove rings and clean carbon and other combustion deposits from piston surface and ring lands.

**CAUTION: Extreme care should be exercised when cleaning ring lands. Do not damage squared edges or widen ring grooves. If ring lands are damaged, piston must be renewed.**

Inspect piston for excessive wear, scoring or other damage and renew as necessary. Measure piston skirt diameter at a point 90 degrees from piston pin bore and 15 mm (0.590 inch) up from bottom of skirt, and measure cylinder bore diameter at a point 20 mm (0.8 inch) below top surface of cylinder. Clearance between piston skirt and cylinder should be 0.030-0.050 mm (0.0012-0.0020 inch). If clearance exceeds 0.120 mm (0.0047 inch), renew piston and/or recondition cylinder bore.

Standard side clearance between piston rings and ring grooves in piston is 0.02-0.06 mm (0.0008-0.0024 inch). If ring to groove clearance exceeds 0.12 mm (0.0047 inch) for top ring or 0.10 mm (0.0039 inch) for second ring, renew piston.

Standard piston ring end gap is 0.15-0.35 mm (0.006-0.014 inch) for all rings. Maximum allowable ring end gap is 0.700 mm (0.0276 inch) for all rings.

Install rings on piston with the "R" stamped on ring toward top of piston and stagger ring end gaps around diameter of piston, however, do not position ring end gaps in line with piston pin center line.

Piston should be installed on connecting rod as outline in CONNECTING ROD paragraph.

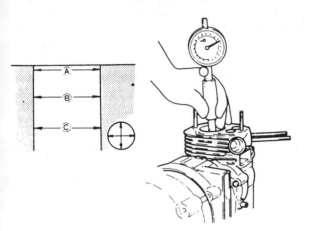

Fig. SU65—Cylinder bore diameter should be measured at the six locations illustrated. If measurements vary more than 0.10 mm (0.004 in.), cylinder bore must be reconditioned or cylinder renewed.

**CYLINDER/CRANKCASE ASSEMBLY.** The cylinder is an integral part of the crankcase casting.

Standard cylinder bore inside diameter is 44 mm (1.73 inch). Cylinder bore should be measured at six different locations as shown in Fig. SU65. If any of the measurements vary by 0.10 mm (0.004 inch) or more, cylinder should be reconditioned or renewed.

**CRANKSHAFT AND MAIN BEARINGS.** Crankshaft is supported by ball bearings (7—Fig. SU66) at each end. To remove crankshaft, first remove connecting rod and piston as outlined in CONNECTING ROD paragraph. Use a suitable puller to remove flywheel, then withdraw crankshaft from cylinder block.

Standard crankshaft crankpin journal diameter is 17.992-18.000 mm (0.7083-0.7087 inch). Specified clearance between crankpin and connecting rod bearing bore is 0.013-0.035 mm (0.0005-0.0014 inch) and wear limit is 0.060 mm (0.0024 inch). Ball bearing type main bearings should be a slight press fit in bores in crankcase and crankcase cover and on crankshaft journals.

Crankshaft oil seals (1—Fig. SU66) should be renewed using suitable seal installing tool. Pack space between the seal lips with grease before installing crankshaft.

When installing crankshaft in engine block, make certain that crankshaft and camshaft timing marks are aligned as shown in Fig. SU67.

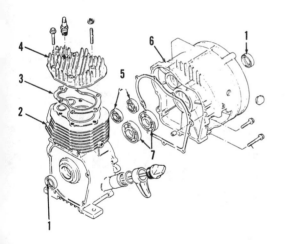

Fig. SU66—Camshaft and crankshaft are supported by ball bearings (5 and 7) located in cylinder block (2) and crankcase cover (6).

1. Crankshaft oil seals
2. Cylinder block
3. Head gasket
4. Cylinder head
5. Camshaft bearings
6. Crankcase cover
7. Crankshaft main bearings

Fig. SU67—When installing crankshaft and camshaft, make certain timing marks are aligned as shown.

**CAMSHAFT AND BEARINGS.** Camshaft is supported at each end in ball bearings (5—Fig. SU66). Refer to CONNECTING ROD paragraph for camshaft removal. Camshaft ball bearings should be a light press fit in camshaft bearing bores machined into crankcase and crankcase cover.

Standard camshaft lobe height is 9.974-10.014 mm (0.3927-0.3843 inch) for intake and exhaust lobes. If lobe height is 9.67 mm (0.380 inch) or less, renew camshaft. It is recommended that tappets be renewed if a new camshaft is installed.

Coat tappets with engine oil, then install in cylinder block before installing camshaft. Make certain governor sleeve (20—Fig. SU68) and governor flyweight assembly is correctly positioned on camshaft during reassembly. When installing camshaft, make certain camshaft gear and crankshaft gear timing marks are aligned as shown in Fig. SU67.

**GOVERNOR.** Refer to Fig. SU68 for an exploded view of the centrifugal flyweight governor assembly which is located on the camshaft gear. Make certain cotter pins retaining governor flyweights (19) to camshaft gear mounting plate (17) are bent to secure flyweights to mounting plate.

To adjust governor linkage, loosen clamp bolt retaining governor lever to governor shaft (22). Move governor lever manually to fully open carburetor throttle plate. Use a screwdriver to turn governor shaft fully clockwise, then tighten clamp bolt while holding governor shaft and lever in position.

**VALVE SYSTEM.** Clearance between valve stem and valve tappet (engine cold and piston at TDC on compression stroke) should be 0.05-0.15 mm (0.002-0.006 inch) for the intake and exhaust valve. Clearance is adjusted by installing tappets of different lengths. Tappets are identified by paint markings. Tappet, part number 12891-90100 has no paint marking and is 25.4 mm (1 inch) long. Tappet, part number 12891-90111 has a blue paint mark and is 25.33 mm (0.9973 inch) long. Tappet, part number 12891-90121 has a red paint mark and is 25.47 mm (1.0027 inch) long. It may be necessary to grind the valve stem end to obtain the correct valve stem clearance if tappets fail to do so.

Valve face angle is 45 degrees and desired valve seating contact width is 0.70-0.80 mm (0.028-0.031 inch). Valve seat should be ground at the 46 degree and 15 degree angles shown in Fig. SU69.

Inspect valves for wear, distortion or other damage. Renew valve if valve

head thickness (T—Fig. 70) after grinding is less than 0.30 mm (0.012 inch). Renew valve if valve head radial runout exceeds 0.03 mm (0.0012 inch). Measure valve stem outside diameter and valve guide inside diameter and refer to the following specifications:

Valve Stem OD
 Intake . . . . . . . . . . .5.460-5.475 mm
  (0.2150-0.2156 in.)
 Exhaust . . . . . . . . .5.450-5.465 mm
  (0.1970-0.2152 in.)
Valve Guide ID
 Intake . . . . . . . . . . .5.500-5.512 mm
  (0.2165-0.2170 in.)
 Exhaust . . . . . . . . .5.500-5.512 mm
  (0.2165-0.2170 in.)

Stem-to-Guide Clearance
 Intake-standard . . . .0.025-0.052 mm
  (0.0010-0.0020 in.)
 Wear limit . . . . . . . . . . . . .0.080 mm
  (0.003 in.)
 Exhaust-standard . .0.035-0.052 mm
  (0.0014-0.0020 in.)
 Wear limit . . . . . . . . . . . . .0.100 mm
  (0.004 in.)

Standard intake and exhaust valve spring free length is 18.50-19.50 mm (0.728-0.768 inch). If valve spring free length is 17.50 mm (0.689 inch) or less, renew valve spring.

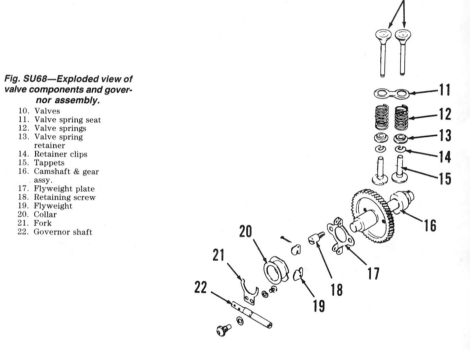

**Fig. SU68—Exploded view of valve components and governor assembly.**

10. Valves
11. Valve spring seat
12. Valve springs
13. Valve spring retainer
14. Retainer clips
15. Tappets
16. Camshaft & gear assy.
17. Flyweight plate
18. Retaining screw
19. Flyweight
20. Collar
21. Fork
22. Governor shaft

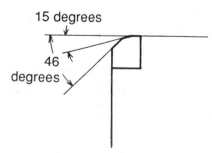

**Fig. SU69—Intake and exhaust valve seats are ground at 46 degrees with a clearance angle of 15 degrees ground at the top. Desired seat width is 0.70-0.80 mm (0.028-0.031 in.).**

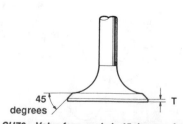

**Fig. SU70—Valve face angle is 45 degrees for intake and exhaust. Renew valve if head thickness (T) is less than 0.30 mm (0.012 in.).**

# SUZUKI

| Model | No. Cyls. | Bore | Stroke | Displacement | Power Rating |
|-------|-----------|------|--------|--------------|--------------|
| SE500 | 1 | 45 mm (1.77 in.) | 40 mm (1.57 in.) | 63.6 cc (3.88 cu. in.) | 0.75 kW (1.0 hp) |
| SE500A | 1 | 45 mm (1.77 in.) | 40 mm (1.57 in.) | 63.6 cc (3.88 cu. in.) | 0.75 kW (1.0 hp) |

## ENGINE IDENTIFICATION

The 63.6 cc (3.88 cu. in.) engines used on Models SE500 and SE500A generators are four-stroke, single-cylinder, air-cooled, horizontal crankshaft engines. Rated output is 0.75 kW (1.0 hp) at 3600 rpm.

Engine serial number is located on the side of crankcase cover (Fig. SU80). Always furnish generator model number and engine serial number when ordering parts.

## MAINTENANCE

**SPARK PLUG.** Recommended spark plug is a NGK BPM-6A, or equivalent. Spark plug should be removed and

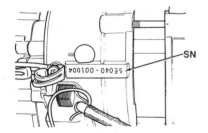

*Fig. SU80—View showing location of engine serial number for Model SE500 generator.*

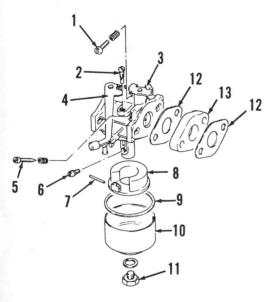

cleaned and electrode gap set at 0.6-0.7 mm (0.024-0.028 inch) after every 100 hours of operation. Renew spark plug if electrode is burned or damaged.

**CAUTION: Caution should be exercised if abrasive type spark plug cleaner is used. Inadequate cleaning procedure may allow the abrasive cleaner to be deposited in engine cylinder accelerating wear and part failure.**

**CARBURETOR.** All models are equipped with a float type carburetor (Fig. SU81). Main fuel mixture is controlled by a fixed main jet (6). Main jet size is a number 52.5 and pilot jet (2) is a number 42.5.

Initial adjustment of idle fuel mixture screw (5) from a lightly seated position is 3/4 turn open.

Engine idle speed is 1000 rpm and idle speed adjustment should be made by adjusting the throttle stop screw (1). When making this adjustment, make certain governor screw is loose so that the governor does not operate.

**AIR FILTER.** All models are equipped with a foam type air filter element. The air filter element should be removed and cleaned at 50 hour intervals. To clean element, wash in a nonflammable

*Fig. SU81—Exploded view of the BV13 float type carburetor.*

1. Idle speed adjusting screw
2. Pilot jet
3. Throttle shaft
4. Choke shaft
5. Idle mixture screw
6. Main jet
7. Pin
8. Float
9. Gasket
10. Float bowl
11. Bolt
12. Gaskets
13. Spacer

cleaning solvent and gently squeeze element dry. Oil element with clean engine oil and gently squeeze out the excess oil.

**GOVERNOR.** All models are equipped with a centrifugal governor which is located internally in engine with flyweight assembly attached to camshaft gear. Governed engine speed is 3600 rpm.

**IGNITION SYSTEM.** All models are equipped with a transistorized ignition system. There are no contact points used in this system, and no periodic maintenance or adjustments are required.

**VALVE ADJUSTMENT.** Valve clearance should be checked after every 300 hours of operation. To check clearance, remove tappet chamber cover and breather assembly from cylinder block. Turn crankshaft until piston is at top dead center on compression stroke, then use a feeler gage to measure clearance between end of valve stem and tappet (Fig. SU82). Correct valve stem clearance is 0.05-0.15 mm (0.002-0.006 inch) for intake and exhaust valves. Valve tappets are available in a variety of lengths to adjust valve clearance. Refer to the REPAIRS section for service procedure.

**CYLINDER HEAD AND COMBUSTION CHAMBER.** Cylinder head, combustion chamber and piston should be cleaned and combustion deposits removed after every 300 hours of operation. Refer to REPAIRS section for service procedure. Cylinder compression reading should be 500 kPa (73 psi) or higher.

**LUBRICATION.** Engine on all models is lubricated by utilizing a dipper attached to connecting rod cap.

Oil level should be checked prior to each operating interval. Oil level should be maintained at top edge of reference marks on gage with oil fill plug just touching first threads. Do not screw oil fill plug and gage in when checking oil level.

Manufacturer recommends using SAE 5W-20 or 5W-30 oil if temperature is below -18° C (0° F); SAE 10W-30 or 10W-40 oil if temperature is -18° C (0° F) to 27° C (80° F); SAE 20W-40 or 20W-50 oil if temperature will be above 27° C (80° F).

Oil should be changed after the first 20 hours of operation and every 100 hours of operation thereafter. Crankcase capacity is 0.35 L (0.75 pt.).

**GENERAL MAINTENANCE.** Check and tighten all loose bolts, nuts or clamps prior to each day of operation. Check for fuel or oil leakage and repair if necessary.

Clean dust, dirt, grease or any foreign material from cylinder head and cylinder block cooling fins after every 100 hours of operation. Inspect fins for damage and repair if necessary.

## REPAIRS

**TIGHTENING TORQUES.** Recommended tightening torques are as follows:

| | |
|---|---|
| Connecting rod | 4 N·m (35 in.-lbs.) |
| Crankcase | 6 N·m (53 in.-lbs.) |
| Cylinder head | 10 N·m (88 in.-lbs.) |
| Rotor set bolt | 10 N·m (88 in.-lbs.) |
| Generator through-bolt | 10 N·m (88 in.-lbs.) |

**CYLINDER HEAD.** To remove cylinder head, first remove side cover, end covers and control panel. Disconnect generator electrical connector, ignition wires and ground wire. Disconnect fuel hose and remove fuel tank. Remove the air cleaner. Remove recoil starter, then remove the four nuts retaining engine and generator assembly to the bottom mounting plate. Remove cooling fan cover and cylinder head cover shroud. Clean engine to prevent entrance of foreign material. Loosen cylinder head bolts gradually in a crisscross pattern until all bolts are loose enough to remove by hand.

Remove spark plug and clean carbon and other combustion deposits from cylinder head. Place cylinder head on a flat surface and check entire sealing surface for distortion. If warpage exceeds 0.030 mm (0.0012 inch), cylinder head must be renewed. Slight scratches may be repaired by lapping cylinder head with 400 grit emery paper on a flat surface. Use a straightedge to check gasket surface of cylinder for distortion also. Renew cylinder if warped more than 0.030 mm (0.0012 inch).

Reinstall cylinder head using a new head gasket. Tighten bolts evenly to specified torque in a crisscross pattern.

**CONNECTING ROD.** The connecting rod rides directly on the crankshaft connecting rod journal.

To remove connecting rod, first remove cylinder head as outlined above. Disconnect governor lever linkage from carburetor. Remove generator rotor retaining bolt and generator through-bolts. Withdraw stator assembly, then use a suitable puller to remove rotor. Remove ignition coil, Woodruff key from crankshaft, and crankcase cover retaining bolts. Tap crankcase cover with a soft hammer to remove cover from cylinder block. Remove camshaft and governor flyweight assembly, mark tappet location and remove tappets. Remove connecting rod bolts, oil slinger and connecting rod cap. Push piston and connecting rod out the top of cylinder block.

Remove piston pin retaining rings and separate piston from connecting rod as required. Note that piston is installed on connecting rod so that arrow on top of piston is aligned with arrow on connecting rod.

Standard piston pin bore diameter in connecting rod is 13.006-13.014 mm (0.5120-0.5124 inch). Standard piston pin diameter is 12.995-13.000 mm (0.5115-0.5118 inch). If clearance between piston pin and connecting rod pin bore exceeds 0.05 mm (0.002 inch), renew pin and/or connecting rod.

Standard diameter of connecting rod crankshaft bearing bore is 18.013-18.027 mm (0.7092-0.7097 inch). Clearance between connecting rod bearing bore and crankshaft crankpin journal should be 0.013-0.035 mm (0.0005-0.0014 inch). If clearance is 0.060 mm (0.0024 inch) or more, renew connecting rod and/or crankshaft.

Side clearance between connecting rod and crankshaft should be 0.250-0.850 mm (0.0098-0.0335 inch). If clearance exceeds 1.20 mm (0.047 inch), renew connecting rod and/or crankshaft.

When installing connecting rod and piston assembly, make certain that arrow marks (Fig. SU84) on piston crown and connecting rod point toward the valves. Be sure match marks on connecting rod and cap are aligned and are facing open side of cylinder block. Install oil slinger, new lock plate and then the connecting rod bolts. Tighten connecting rod bolts to specified torque and bend lock plate against bolt heads.

**PISTON, PIN AND RINGS.** Refer to CONNECTING ROD paragraph for piston removal.

After separating piston and connecting rod, carefully remove rings and clean carbon and other combustion deposits from piston surface and ring lands.

CAUTION: Extreme care should be exercised when cleaning ring lands. Do not damage squared edges or widen ring grooves. If ring lands are damaged, piston must be renewed.

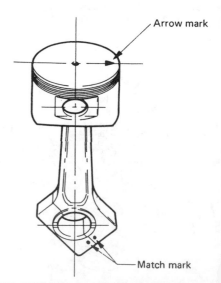

Fig. SU84—Arrow on piston crown and connecting rod should point toward the valves. When installing connecting rod cap, make certain match marks are aligned.

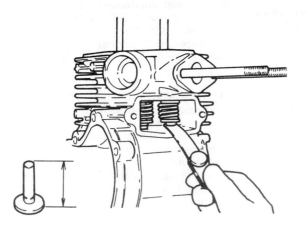

Fig. SU82—Valve clearance should be checked with engine cold and with piston at top dead center on compression stroke. Use a feeler gage to measure clearance between valve stem end and tappet. Refer to text for adjustment procedure.

Inspect piston for excessive wear, scoring or other damage and renew as necessary. Standard piston diameter, measured 90 degrees from piston pin bore and 15 mm (0.59 inch) up from bottom of skirt, is 43.965-43.970 mm (1.7309-1.7311 inch). Clearance between piston skirt thrust surface and cylinder should be 0.030-0.050 mm (0.0012-0.0020 inch). If clearance is greater than 0.12 mm (0.0047 inch), renew piston and/or recondition cylinder bore.

Clearance between piston pin and pin bore in piston should be 0.006-0.019 mm (0.00025-0.00075 inch). If clearance is greater than specified, renew piston pin and/or piston.

Standard side clearance between rings and ring grooves in piston is 0.02-0.06 mm (0.0008-0.0024 inch). If side clearance between a new ring and ring groove in piston is 0.12 mm (0.0047 inch) or more, renew piston.

Standard piston ring end gap is 0.15-0.35 mm (0.006-0.014 inch) for all rings. Maximum allowable ring end gap is 0.70 mm (0.0275 inch).

Install rings on piston with the "R" stamped on ring toward top of piston and stagger ring end gaps around diameter of piston, however, do not position ring end gaps in line with piston pin center line.

Piston should be installed on connecting rod as outline in CONNECTING ROD paragraph. Lubricate piston and cylinder with engine oil prior to installing piston in cylinder.

**CYLINDER/CRANKCASE ASSEMBLY.** The cylinder is an integral part of the crankcase casting.

Standard cylinder bore inside diameter is 45 mm (1.77 inch). Cylinder bore should be measured at six different locations as shown in Fig. SU85. If any of the measurements vary by 0.10 mm (0.004 inch) or more, recondition or renew cylinder.

**CRANKSHAFT AND MAIN BEARINGS.** Crankshaft (9—Fig. SU86) is supported by ball bearings (7) at each end. To remove crankshaft, first remove connecting rod and piston assembly as outlined in CONNECTING ROD paragraph. Use a suitable puller to remove flywheel, then withdraw crankshaft from cylinder block.

Standard crankshaft crankpin journal diameter is 17.992-18.000 mm (0.7083-0.7087 inch). Crankpin clearance in connecting rod bearing bore should be 0.013-0.035 mm (0.0005-0.0014 inch) with wear limit of 0.060 mm (0.0024 inch). Ball bearing type main bearings should be a slight press fit in bores in cylinder block and crankcase cover and on crankshaft journals.

Crankshaft oil seals (4—Fig. SU86) should be renewed using suitable seal installing tool. Pack space between the seal lips with grease before installing crankshaft.

When installing crankshaft in engine block, make certain crankshaft and camshaft timing marks are aligned as shown in Fig. SU87.

*Fig. SU85—Cylinder bore diameter should be measured at the six locations illustrated. If measurements vary more than 0.10 mm (0.004 in.), cylinder bore must be reconditioned or cylinder renewed.*

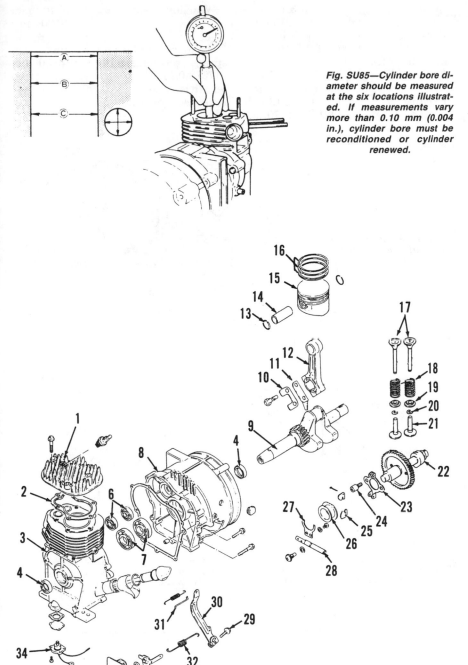

*Fig. SU86—Exploded view of engine assembly and governor components.*

| | | |
|---|---|---|
| 1. Cylinder head | 11. Oil slinger | 21. Tappets |
| 2. Head gasket | 12. Connecting rod | 22. Camshaft & gear assy. |
| 3. Cylinder block | 13. Retaining ring | 23. Flyweight bracket |
| 4. Crankshaft oil seal | 14. Piston pin | 24. Retaining screw |
| 5. Piston | 15. Piston | 25. Governor flyweights |
| 6. Camshaft bearings | 16. Piston rings | 26. Sleeve |
| 7. Crankshaft main bearings | 17. Valves | 27. Fork |
| 8. Crankcase cover | 18. Valve springs | 28. Governor shaft |
| 9. Crankshaft | 19. Spring retainers | 29. Clamp bolt |
| 10. Lock plate | 20. Retaining clips | |

30. Governor lever
31. Governor-to-carburetor link
32. Governor spring
33. Frequency adjustment knob & screw
34. Low oil level alarm sender (SE500A only)

**CAMSHAFT AND BEARINGS.** Camshaft (22—Fig. SU86) is supported at each end in ball bearings (6). Refer to CONNECTING ROD paragraph for camshaft removal. Camshaft ball bearings should be a light press fit on camshaft and in camshaft bearing bores machined into crankcase and crankcase cover.

Standard camshaft lobe height is 9.974-10.014 mm (0.3927-0.3943 inch) for intake and exhaust lobes. If lobe height is 9.670 mm (0.3810 inch) or less, renew camshaft. It is recommended that tappets be renewed if a new camshaft is installed.

Coat tappets with engine oil and position in cylinder block before installing camshaft. Make certain governor sleeve and governor assembly is correctly positioned on camshaft during reassembly. When installing camshaft, make certain camshaft gear and crankshaft gear timing marks are aligned as shown in Fig. SU87.

**GOVERNOR.** Refer to Fig. SU86 for an exploded view of the centrifugal flyweight governor assembly which is located on the camshaft gear. Make certain cotter pins retaining governor flyweights (25) to camshaft gear mounting plate (23) are bent to secure flyweights to mounting plate.

To adjust governor linkage, loosen clamp bolt (29) retaining governor lever (30) to governor shaft (28). Move governor lever manually to fully open carburetor throttle plate. Use a screwdriver to turn governor shaft fully clockwise, then tighten clamp bolt while holding governor shaft and lever in position.

**VALVE SYSTEM.** Clearance between valve stem and valve tappet (engine cold and piston at TDC on compression stroke) should be 0.05-0.15 mm (0.002-0.006 inch) for the intake and exhaust valve. Clearance is adjusted by installing tappets of different lengths. Tappets are identified by paint markings. Tappet, part number 12891-90100 has no paint marking and is 25.4 mm (1 inch) long. Tappet, part number 12891-90111 has a blue paint mark and is 25.33 mm (0.9973 inch) long. Tappet, part number 12891-90121 has a red paint mark and is 25.47 mm (1.0027 inch) long. It may be necessary to grind the valve stem end to obtain the correct valve stem clearance if tappets fail to do so.

Valve face angle is 45 degrees and desired valve seating contact width is 0.70-0.80 mm (0.028-0.031 inch). Valve seat should be ground at the 46 degree and 15 degree angles shown in Fig. SU89.

Inspect valves for wear, distortion or other damage. Renew valve if valve head thickness (T—Fig. 90) after grind-

ing is less than 0.30 mm (0.012 inch). Renew valve if valve head radial runout exceeds 0.03 mm (0.0012 inch). Measure valve stem outside diameter and valve guide inside diameter and refer to the following specifications:

Valve Stem OD
   Intake . . . . . . . . . . .5.460-5.475 mm
          (0.2150-0.2156 in.)
   Exhaust . . . . . . . . . .5.450-5.465 mm
          (0.1970-0.2152 in.)
Valve Guide ID
   Intake . . . . . . . . . . .5.500-5.512 mm
          (0.2165-0.2170 in.)
   Exhaust . . . . . . . . .5.500-5.512 mm
          (0.2165-0.2170 in.)

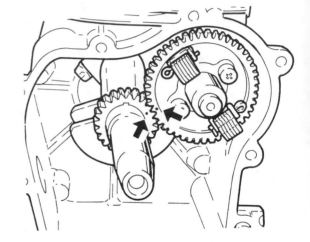

Fig. SU87—When installing crankshaft and camshaft, make certain that timing marks are aligned as shown.

Stem-to-Guide Clearance
   Intake-standard . . . .0.025-0.052 mm
          (0.0010-0.0020 in.)
   Wear limit . . . . . . . . . . . . .0.080 mm
          (0.003 in.)
   Exhaust-standard . .0.035-0.052 mm
          (0.0014-0.0020 in.)
   Wear limit . . . . . . . . . . . . .0.100 mm
          (0.004 in.)

Standard intake and exhaust valve spring free length is 18.50-19.50 mm (0.7283-0.7677 inch). If valve spring free length is 17.50 mm (0.6890 inch) or less, renew valve spring.

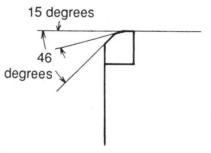

Fig. SU89—Intake and exhaust valve seats are ground at 46 degrees with a clearance angle of 15 degrees ground at the top. Desired seat width is 0.70-0.80 mm (0.028-0.031 in.).

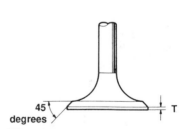

Fig. SU90—Valve face angle is 45 degrees for intake and exhaust. Renew valve if head thickness (T) is less than 0.30 mm (0.012 in.).

# SUZUKI

| Model | No. Cyls. | Bore | Stroke | Displacement | Power Rating |
|-------|-----------|------|--------|--------------|--------------|
| GY15D | 1 | 45 mm (1.77 in.) | 37 mm (1.45 in.) | 58 cc (3.54 cu. in.) | 0.75 kW (1.0 hp) |
| GY20D | 1 | 52 mm (1.69 in.) | 37 mm (1.45 in.) | 78 cc (4.76 cu. in.) | 1.02 kW (1.4 hp) |

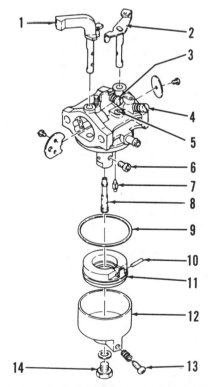

**Fig. SU100—Exploded view of fuel shut-off valve and fuel filter assembly. Filter element should be cleaned after every 100 hours of operation.**

1. Fuel shut-off valve
2. Strainer screen
3. Filter element
4. Sediment bowl
5. Retaining nut

## ENGINE IDENTIFICATION

The Suzuki GY15D and GY20D engines are four-stroke, single-cylinder, air-cooled horizontal crankshaft engines. Rated output for Model GY15D is 0.75 kW (1.0 hp) at 3600 rpm and rated output for Model GY20D is 1.02 kW (1.4 hp) at 3600 rpm.

## MAINTENANCE

**SPARK PLUG.** Recommended spark plug is a NGK BPM-4A, Champion CJ-8, or equivalent. Spark plug should be removed and cleaned and electrode gap set at 0.6-0.7 mm (0.024-0.028 inch) after every 100 hours of operation. Renew spark plug if electrode is excessively worn, burned or damaged.

**CAUTION: Caution should be exercised if abrasive type spark plug cleaner is used. Inadequate cleaning procedure may allow the abrasive cleaner to be deposited in engine cylinder accelerating wear and part failure.**

**AIR FILTER.** All models are equipped with a foam type air filter element. The air filter element should be removed and cleaned at 50 hour intervals. To clean element, wash in a nonflammable cleaning solvent and gently squeeze element dry. Oil element with clean engine oil and gently squeeze out the excess oil.

**FUEL FILTER.** A combination fuel shut-off valve (1—Fig. SU100), fuel filter (3) and sediment bowl (4) assembly is located in the fuel line between fuel tank and carburetor. The sediment bowl should be checked for accumulation of water or sediment prior to operating engine. If water or dirt is visible, close fuel shut-off valve, unscrew bowl retaining nut and remove and clean sediment bowl. Fuel filter element should be removed, inspected and cleaned after every 100 hours of operation.

**CARBURETOR.** All models are equipped with a float type carburetor (Fig. SU101). Main fuel mixture is controlled by a fixed main jet (6). Main jet size for Model GY15D is a number 52.5 and main jet size for Model GY20D is a number 57.5. Pilot jet (5) size for Model GY15D is a number 40 and pilot jet size for Model GY20D is a number 42.5.

Engine idle speed is 2650-2750 rpm. Adjust the idle stop screw (3) to obtain correct engine idle speed. Initial adjustment of idle fuel mixture screw (4) from a lightly seated position is 1-1/4 turn open.

**GOVERNOR.** All models are equipped with a centrifugal governor which is located internally in engine, with flyweight assembly attached to a gear which is driven by the camshaft gear. To adjust the governor, position the

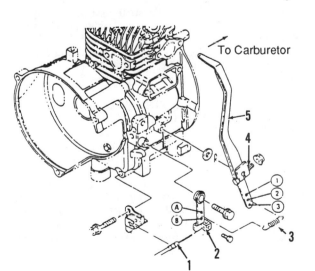

To Carburetor

**Fig. SU101—Exploded view of the float type carburetor used on all models.**

1. Choke lever
2. Throttle lever
3. Idle adjusting screw
4. Idle mixture screw
5. Pilot jet
6. Main jet
7. Needle valve
8. Main nozzle
9. Gasket
10. Pin
11. Float
12. Float bowl
13. Drain screw
14. Bolt

**Fig. SU102—Exploded view of governor linkage. Refer to text for adjustment procedure.**

1. Throttle cable
2. Speed control lever
3. Governor spring
4. Clamp bolt
5. Governor lever

speed control lever (2—Fig. SU102) at maximum speed position and tighten the wing nut to hold this position. Loosen the governor lever clamp bolt (4) and nut so governor lever (5) is released from the governor shaft. Turn the governor shaft clockwise as far as possible, hold shaft in this position and tighten the governor lever clamp bolt and nut. Note that governor spring (3) should be installed in hole (B) in speed control lever and in hole (2) in governor lever for 3600 rpm (60 Hz) operation. Maximum idling (no-load) speed is 3750-3850 rpm.

**IGNITION SYSTEM.** All models are equipped with a transistorized ignition system. There are no contact points in this system, and no periodic maintenance or adjustments are required.

**VALVE ADJUSTMENT.** Valve clearance should be checked after every 300 hours of operation. To check valve clearance, remove tappet chamber cover and breather assembly. Rotate crankshaft to position piston at top dead center on compression stroke. Measure clearance between valve stem end and tappet using a feeler gage. Correct valve stem clearance is 0.15 mm (.006 inch) for intake and exhaust valves. Adjust valve stem clearance by grinding valve stem to increase clearance or grinding valve seat deeper to decrease clearance. Refer to the REPAIRS section for service procedure.

**CYLINDER HEAD AND COMBUSTION CHAMBER.** Cylinder head, combustion chamber and piston should be cleaned and combustion deposits removed after every 300 hours of operation. Refer to REPAIRS section for service procedure.

**LUBRICATION.** Engine on all models is lubricated by utilizing a dipper attached to connecting rod cap.

Oil level should be checked prior to each operating interval. Oil level should be maintained at top edge of reference marks on gage with oil fill plug just touching first threads. Do not screw oil fill plug and gage in to check oil level.

Manufacturer recommends using an oil with service classification SE or SF. Use SAE 30 oil if temperature is 20° C (68° F) or above. SAE 20 or SAE 10W-30 oil is recommended if temperature is between 10°-19° C (44°-67° F). If temperature is below 10° C (44° F), SAE 10 or 10W-30 oil is recommended.

Oil should be changed after the first 20 hours of operation and every 50 hours of operation thereafter. Crankcase capacity for Model GY15D is 0.22 L (0.46 pt.) and crankcase capacity for Model GY20D is 0.29 L (0.61 pt.).

**GENERAL MAINTENANCE.** Check and tighten all loose bolts, nuts or clamps prior to each day of operation. Check for fuel or oil leakage and repair if necessary.

Clean dust, dirt, grease or any foreign material from cylinder head and cylinder block cooling fins after every 100 hours of operation. Inspect fins for damage and repair if necessary.

## REPAIRS

**TIGHTENING TORQUES.** Recommended tightening torques are as follows:

Connecting rod . . . . . . . . .4.0-4.5 N·m (35-40 in.-lbs.)
Crankcase cover . . . . . . .10.8-12.7 N·m (98-115 in.-lbs.)
Cylinder head . . . . . . . . .14.7-16.6 N·m (130-147 in.-lbs.)
Flywheel nut . . . . . . . . .24.5-29.4 N·m (18-21 ft.-lbs.)
Spark plug. . . . . . . . . . . .17.6-21.5 N·m (155-190 in.-lbs.)

**CYLINDER HEAD.** To remove cylinder head, first remove generator end covers, side cover and control panel. Disconnect generator wires and ignition wires. Disconnect fuel line and remove fuel tank. Remove intake manifold, carburetor and air cleaner assembly. Remove crankcase breather assembly. Remove recoil starter and cooling fan housing. Loosen cylinder head bolts in a crisscross pattern and remove cylinder head from cylinder.

Clean all carbon deposits from cylinder head, being careful not to scratch gasket mating surface. To check cylinder head for distortion, place cylinder head on a flat surface and check entire sealing surface using a feeler gage. If warpage exceeds 0.030 mm (0.0012 inch), cylinder head must be renewed. Slight warpage may be repaired by lapping cylinder head using 400 grit emery paper on a flat surface. Use a straightedge to check gasket surface of cylinder also.

Reinstall cylinder head using a new head gasket. Tighten bolts evenly to specified torque in a crisscross pattern.

**CONNECTING ROD.** Connecting rod used for all models rides directly on the crankshaft connecting rod journal. Piston is installed on connecting rod so that arrow on top of piston is aligned with arrow on connecting rod.

Standard diameter of connecting rod big end bearing bore is 16.000-16.011 mm (0.6300-0.6303 inch) for Model GY15D and 18.000-18.018 mm (0.7087-0.7094 inch) for Model GY20D. Clearance between connecting rod bearing bore and crankshaft crankpin journal should be

0.025-0.046 mm (0.0010-0.0018 inch) for Model GY15D and 0.020-0.051 mm (0.0008-0.0020 inch) for Model GY20D. If clearance is greater than 0.10 mm (0.004 inch) on either model, renew connecting rod and/or crankshaft.

Standard clearance between piston pin and bore in small end of connecting rod is 0.010-0.026 mm (0.0004-0.0010 inch). If clearance between piston pin and connecting rod pin bore is 0.10 mm (0.004 inch) or more, renew pin and/or connecting rod.

Side clearance between connecting rod and crankshaft should be 0.15-0.75 mm (0.006-0.030 inch) for all models. If clearance is 1.0 mm (0.040 inch) or more, renew connecting rod and/or crankshaft.

Connecting rod and connecting rod cap match marks should be aligned, and connecting rod is installed so that the side of the connecting rod with the match marks is toward open side of engine. Tighten connecting rod bolts to specified torque.

**PISTON, PIN AND RINGS.** All models are equipped with a three ring piston (Fig. SU103).

After separating piston and connecting rod, carefully remove rings and clean carbon and other combustion deposits from piston surface and ring lands.

**CAUTION: Extreme care should be exercised when cleaning ring lands. Do not damage squared edges or widen ring grooves. If ring lands are damaged, piston must be renewed.**

Standard piston diameter, measured 90 degrees from piston pin bore and 15 mm (0.59 inch) up from the bottom of the skirt, is 44.95-44.97 mm (1.7697-1.7705 inch) for Model GY15D and 51.94-51.96 mm (2.0449-2.0457 inch) for Model GY20D. If piston diameter is 44.75 mm (1.762 inch) or less for Model GY15D, renew piston. If piston diameter is 51.75 mm (2.037 inch) or less for Model GY20D, renew piston.

Clearance between piston skirt and cylinder should be 0.029-0.074 mm

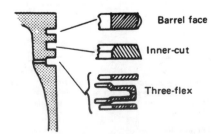

*Fig. SU103—View of three ring piston used on all models.*

(0.0012-0.0029 inch) for Model GY15D or 0.04-0.09 mm (0.0016-0.0035 inch) for Model GY20D. On all models, if clearance is 0.30 mm (0.012 inch) or more, renew piston and/or recondition cylinder bore.

Piston pin standard diameter is 9.994-10.000 mm (0.3935-0.3937 inch) for Model GY15D and 10.992-11.000 mm (0.4328-0.4330 inch) for Model GY20D. Clearance between piston pin and piston pin bore in piston should be 0.0-0.015 mm (0.0-0.0006 inch) for Model GY15D and 0.0-0.019 mm (0.0-0.0007 inch) for Model GY20D. On all models, if clearance is greater than 0.03 mm (0.0012 inch), renew piston pin and/or piston.

Standard side clearance between rings and ring grooves in piston is 0.030-0.065 mm (0.0012-0.0026 inch) for Model GY15D or 0.040-0.085 mm (0.0016-0.0033 inch) for Model GY20D. On all models, renew piston if ring side clearance exceeds 0.17 mm (0.0067 inch).

On all models, ring end gap should be 0.15-0.35 mm (0.006-0.014 inch) for top and second ring. Specified end gap for oil ring is 0.30-0.90 mm (0.012-0.035 inch). Maximum allowable ring end gap is 1.0 mm (0.039 inch) for top and second ring and 2.0 mm (0.079 inch) for oil ring.

When installing oil ring, position end gaps of top rail, bottom rail and spacer as shown in Fig. SU104. Be sure that split ends of spacer ring do not overlap. Install second and top rings on piston with the "R" stamped on ring toward top of piston. End gap of top ring should be positioned on side of piston facing the valves and end gap of second ring should be positioned 180 degrees from top ring.

Piston should be installed on connecting rod as outlined in CONNECTING ROD paragraph. Lubricate piston, rings and cylinder with engine oil prior to installing piston in cylinder block.

**CYLINDER/CRANKCASE ASSEMBLY.** The cylinder is an integral part of the crankcase casting.

Standard cylinder bore inside diameter for Model GY15D is 45.000-45.025 mm (1.7716-1.7726 inch), and for Model GY20D is 52.000-52.020 mm (2.047-2.048 inch). Cylinder bore should be measured in six different locations as shown in Fig. SU105. If cylinder bore inside diameter is greater than 45.15 mm (1.7776 inch) for Model GY15D or 52.75 mm (2.0374 inch) for Model GY20D, recondition or renew cylinder.

**CRANKSHAFT AND MAIN BEARINGS.** Crankshaft is supported by ball bearings at each end. To remove crankshaft, first remove piston and connecting rod assembly as outlined in CONNECTING ROD paragraph. Remove flywheel, then withdraw crankshaft from cylinder block.

Standard diameter of main journals is 16.983-16.994 mm (0.6686-0.6690 inch) for all models. Crankshaft main bearings should be a slight press fit on crankshaft journals.

Standard crankshaft crankpin journal diameter is 15.965-15.975 mm (0.6285-0.6289 inch) for Model GY15D or 17.970-17.980 mm (0.7075-0.7079 inch) for Model GY20D.

When installing crankshaft in engine block, make certain crankshaft and camshaft timing marks are aligned as shown in Fig. SU106.

**CAMSHAFT AND BEARINGS.** Camshaft is supported at each end in ball bearings. Camshaft ball bearings should be a light press fit on camshaft and in camshaft bearing bores machined into crankcase and crankcase cover.

Standard camshaft intake and exhaust lobe height is 16.5 mm (0.6496 inch) for Model GY15D or 18.3 mm (0.7205 inch) for Model GY20D. Renew camshaft if lobe height is less than 16.2 mm (0.6378 inch) for Model GY15D or 18 mm (0.7087 inch) for Model GY20D. It is recommended that tappets also be renewed if a new camshaft is installed.

When installing camshaft, make certain camshaft gear and crankshaft gear timing marks are aligned as shown in Fig. SU106.

**GOVERNOR.** All models are equipped with a centrifugal flyweight governor assembly mounted on a gear which is driven by the camshaft gear. Refer to the GOVERNOR paragraph in MAINTENANCE section for governor adjustment.

**VALVE SYSTEM.** Specified clearance between end of valve stem and tappet is 0.15 mm (0.006 inch) for intake and exhaust valve for all models. Clearance should be measured with engine cold

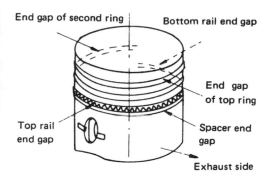

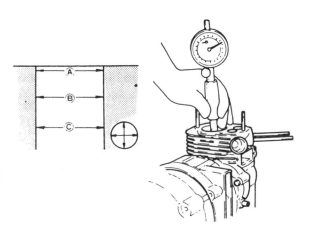

Fig. SU104—Stagger piston ring end gaps around piston as shown.

Fig. SU105—Cylinder bore diameter should be measured at the six locations illustrated. Refer to text for specifications.

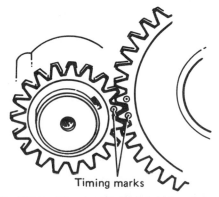

Timing marks

Fig. SU106—When installing crankshaft and camshaft, make certain timing marks are aligned as shown.

and piston at TDC on compression stroke. Adjust valve clearance by grinding valve stem to increase clearance or grinding valve seat deeper to decrease clearance.

Valve face angle is 45 degrees for intake and exhaust. Standard valve seat contact width is 0.70-0.80 mm (0.028-0.031 inch). Valve seats should be ground at 45 degree angle.

Inspect valves for excessive wear, distortion or other damage. Renew valve if head thickness (T—Fig. SU107) is less than 0.6 mm (0.024 inch). Renew valve if valve head radial runout exceeds 0.03 mm (0.0012 inch). Measure valve stem outside diameter and valve guide inside diameter and refer to the following specifications:

**Model GY15D**
Valve Stem OD—
　　Intake & Exhaust . . 5.445-5.460 mm
　　　　　　　　　　　　(0.2144-0.2150 in.)

Wear Limit . . . . . . . . . . . . . 5.35 mm
　　　　　　　　　　　　(0.2106 in.)
Valve-to-Guide Clearance—
　　Intake . . . . . . . . . . . 0.040-0.075 mm
　　　　　　　　　　　　(0.0016-0.0030 in.)
　　Wear Limit . . . . . . . . . . . . . 0.20 mm
　　　　　　　　　　　　(0.008 in.)
　　Exhaust . . . . . . . . . 0.040-0.085 mm
　　　　　　　　　　　　(0.0016-0.0033 in.)
　　Wear Limit . . . . . . . . . . . . . 0.20 mm
　　　　　　　　　　　　(0.008 in.)

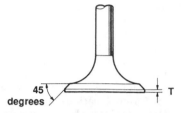

Fig. SU107—Valve face angle is 45 degrees on intake and exhaust valves. Minimum valve head thickness (T) is 0.60 mm (0.024 in.).

**Model GY20D**
Valve Stem OD—
　　Intake . . . . . . . . . . . 5.445-5.460 mm
　　　　　　　　　　　　(0.2144-0.2150 in.)
　　Wear Limit . . . . . . . . . . . . . 5.35 mm
　　　　　　　　　　　　(0.2106 in.)
　　Exhaust . . . . . . . . . 5.425-5.440 mm
　　　　　　　　　　　　(0.2136-0.2142 in.)
　　Wear Limit . . . . . . . . . . . . . 5.35 mm
　　　　　　　　　　　　(0.2106 in.)
Valve-to-Guide Clearance—
　　Intake . . . . . . . . . . . 0.040-0.075 mm
　　　　　　　　　　　　(0.0016-0.0030 in.)
　　Wear Limit . . . . . . . . . . . . . 0.20 mm
　　　　　　　　　　　　(0.008 in.)
　　Exhaust . . . . . . . . . 0.060-0.105 mm
　　　　　　　　　　　　(0.0024-0.0041 in.)
　　Wear Limit . . . . . . . . . . . . . 0.20 mm
　　　　　　　　　　　　(0.008 in.)

On Model GY15D, standard intake and exhaust valve spring free length is 18.00-19.00 mm (0.7087-0.7480 inch). On Model GY20D, standard intake and exhaust valve spring free length is 18.80-19.20 mm (0.740-0.756 inch). Renew valve spring if free length is not within specified range.

# SUZUKI

| Model | No. Cyls. | Bore | Stroke | Displacement | Power Rating |
|-------|-----------|------|--------|--------------|--------------|
| SE700 | 1 | 48 mm (1.89 in.) | 47 mm (1.85 in.) | 85 cc (5.19 cu. in.) | 1.2 kw (1.6 hp) |
| SE700A | 1 | 48 mm (1.89 in.) | 47 mm (1.85 in.) | 85 cc (5.19 cu. in.) | 1.2 kw (1.6 hp) |

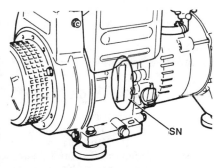

*Fig. SU110—View showing location of engine serial number (SN) for Model SE700 and SE700A generators.*

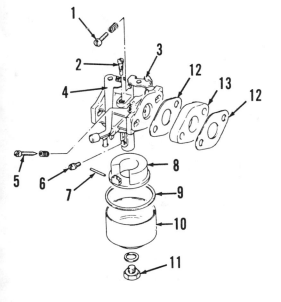

## ENGINE IDENTIFICATION

The 85 cc (5.19 cu. in.) engines used on Models SE700 and SE700A generators are four-stroke, single-cylinder, air-cooled, horizontal crankshaft engines. Rated output is 1.2 kw (1.6 hp) at 3600 rpm.

Engine serial number is located on the side of engine crankcase (Fig. SU110). Always furnish generator model number and engine serial number when ordering parts.

*Fig. SU111—Exploded view of the BV15-11 float type carburetor.*

1. Idle speed adjusting screw
2. Pilot jet
3. Throttle shaft
4. Choke shaft
5. Idle mixture screw
6. Main jet
7. Pin
8. Float
9. Gasket
10. Float bowl
11. Bolt
12. Gaskets
13. Spacer

*Fig. SU112—Valve clearance should be checked with engine cold and with piston at top dead center on compression stroke. Use a feeler gage to measure clearance between valve stem end and tappet. Refer to text for adjustment procedure.*

## MAINTENANCE

**SPARK PLUG.** Recommended spark plug is a NGK BPMR-6A, or equivalent. Spark plug should be removed and cleaned and electrode gap set at 0.6-0.7 mm (0.024-0.028 inch) after every 100 hours of operation. Renew spark plug if electrode is burned or damaged.

**CAUTION: Caution should be exercised if abrasive type spark plug cleaner is used. Inadequate cleaning procedure may allow the abrasive cleaner to be deposited in engine cylinder accelerating wear and part failure.**

**CARBURETOR.** All models are equipped with a BV15-11 float type carburetor (Fig. SU111). Main fuel mixture is controlled by a fixed main jet (6). Standard main jet size is number 57.5 and pilot jet (2) is number 45.

Initial adjustment of idle fuel mixture screw (5) from a lightly seated position is 7/8 turn open.

Engine idle speed is 1000 rpm and idle speed adjustment should be made by adjusting the throttle stop screw (1). When making this adjustment, make certain governor screw is loose so that the governor does not operate.

**AIR FILTER.** All models are equipped with a foam type air filter element. The air filter element should be removed and cleaned at 50 hour intervals. To clean element, wash in a nonflammable cleaning solvent and gently squeeze element dry. Oil element with clean engine oil and gently squeeze out the excess oil.

**GOVERNOR.** All models are equipped with a centrifugal governor which is located internally in engine with flyweight assembly attached to camshaft gear. Governed engine speed is 3600 rpm.

**IGNITION SYSTEM.** All models are equipped with a transistorized ignition system. There are no contact points used in this system, and no periodic maintenance or adjustments are required.

**VALVE ADJUSTMENT.** Valve clearance should be checked after every 300 hours of operation. To check clearance, remove tappet chamber cover and breather assembly from cylinder block. Turn crankshaft until piston is at top dead center on compression stroke, then use a feeler gage to measure clearance between end of valve stem and tappet (Fig. SU112). Correct valve stem clearance is 0.05-0.15 mm (0.002-0.006 inch) for intake and exhaust valves. Valve tappets are available in a variety of lengths to adjust valve clearance. Refer to the REPAIRS section for service procedure.

**CYLINDER HEAD AND COMBUSTION CHAMBER.** Cylinder head, combustion chamber and piston should be cleaned and combustion deposits removed after every 300 hours of operation. Refer to REPAIRS section for service procedure. Normal cylinder compression reading should be 570 kPa (83 psi) and minimum pressure is 480 kPa (70 psi).

**LUBRICATION.** Engine on all models is lubricated by utilizing a dipper attached to connecting rod cap.

Oil level should be checked prior to each operating interval. Oil level should be maintained at top edge of reference marks on gage with oil fill plug just touching first threads. Do not screw oil fill plug and gage in when checking oil level.

Manufacturer recommends using SAE 5W-20 or 5W-30 oil if temperature is below -18° C (0° F); SAE 10W-30 or 10W-40 oil if temperature is -18° C (0° F) to 27° C (80° F); SAE 20W-40 or 20W-50 oil if temperature will be above 27° C (80° F).

Oil should be changed after the first 20 hours of operation and every 100 hours of operation thereafter. Crankcase capacity is 0.5 L (0.52 qt.).

**GENERAL MAINTENANCE.** Check and tighten all loose bolts, nuts or clamps prior to each day of operation. Check for fuel or oil leakage and repair if necessary.

Clean dust, dirt, grease or any foreign material from cylinder head and cylinder block cooling fins after every 100 hours of operation. Inspect fins for damage and repair if necessary.

## REPAIRS

**TIGHTENING TORQUES.** Recommended tightening torques are as follows:

Connecting rod . . . . . . . . . . . . .5 N·m
(45 in.-lbs.)

Crankcase . . . . . . . . . . . . . . . .10 N·m
(88 in.-lbs.)
Cylinder head . . . . . . . . . . . . . .8 N·m
(70 in.-lbs.)
Flywheel bolt . . . . . . . . . . .30-40 N·m
(22-30 ft.-lbs.)
Rotor set bolt . . . . . . . . . . .16-23 N·m
(12-17 ft.-lbs.)
Generator through-bolt . . . . . .4-7 N·m
(35-62 in.-lbs.)

**CYLINDER HEAD.** To remove cylinder head, first remove transistor ignition cover and disconnect ignition wires. Disconnect generator electrical connector. Disconnect fuel hose, unbolt and remove fuel tank and control panel. Remove the cooling fan shroud and the air shroud from top of cylinder head. Loosen cylinder head bolts gradually in a crisscross pattern and remove cylinder head.

Remove spark plug and clean carbon and other combustion deposits from cylinder head. Place cylinder head on a flat surface and check entire sealing surface for distortion. If surface is warped more than 0.030 mm (0.0012 inch), cylinder head must be renewed. Slight scratches may be repaired by lapping cylinder

head with 400 grit emery paper on a flat surface. Use a straightedge to check gasket surface of cylinder for distortion also. Renew cylinder if warped more than 0.030 mm (0.0012 inch).

Reinstall cylinder head using a new head gasket. Tighten bolts evenly to specified torque in a crisscross pattern.

**CONNECTING ROD.** The connecting rod rides directly on the crankshaft crankpin journal.

To remove connecting rod, first remove cylinder head as outlined above. Remove muffler cover, muffler and heat shield. Disconnect governor lever linkage from carburetor and remove air cleaner and carburetor. Remove crankcase breather assembly. Remove generator rear cover. Remove brush holder retaining screws. Remove generator rotor retaining bolt and generator through-bolts. Withdraw stator assembly, then use a suitable puller to remove rotor. Remove ignition coil, Woodruff key from crankshaft, and generator fan housing. Unbolt and remove crankcase cover (1—Fig. SU113) from cylinder block. Remove camshaft (25) and governor flyweight assembly. Remove the tappets

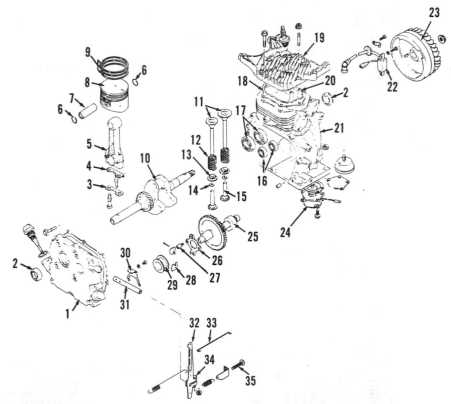

**Fig. SU113—Exploded view of engine assembly and governor components.**

| | | |
|---|---|---|
| 1. Crankcase cover | 11. Valves | 20. Valve guides |
| 2. Crankshaft oil seals | 12. Valve spring | 21. Cylinder block |
| 3. Lock plate | 13. Spring retainer | 22. Ignition coil |
| 4. Oil slinger | 14. Retainer clip | 23. Flywheel |
| 5. Connecting rod | 15. Tappet | 24. Low oil warning |
| 6. Retaining rings | 16. Camshaft bearings | sensor (SE700A) |
| 7. Piston pin | 17. Crankshaft bearings | 25. Camshaft & gear |
| 8. Piston | 18. Head gasket | assy. |
| 9. Piston rings | 19. Cylinder head | 26. Flyweight bracket |
| 10. Crankshaft | | 27. Screw |
| | | 28. Governor flyweight |
| | | 29. Sleeve |
| | | 30. Fork |
| | | 31. Governor shaft |
| | | 32. Governor lever |
| | | 33. Link to carburetor |
| | | 34. Clamp bolt |
| | | 35. High speed adjustment bolt |

(15), marking them so they can be reinstalled in original location. Remove connecting rod bolts, oil slinger and connecting rod cap. Push piston and connecting rod out the top of cylinder block.

Remove piston pin retaining rings (6) and piston pin (7) to separate piston (8) from connecting rod (5). Note that piston is installed on connecting rod so that arrow on top of piston is aligned with arrow on connecting rod (Fig. SU114).

Standard piston pin bore diameter in connecting rod is 13.006-13.014 mm (0.5120-0.5124 inch). Standard piston pin

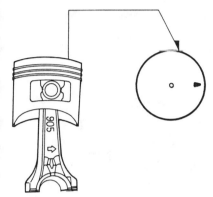

*Fig. SU114—Arrow on piston crown and connecting rod should point toward the valves.*

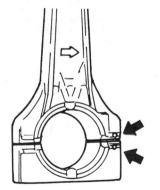

*Fig. SU115— When installing connecting rod cap, make certain match marks are aligned.*

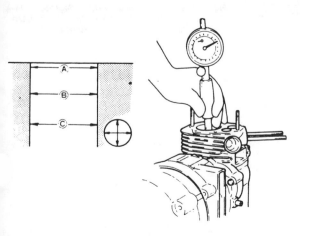

diameter is 12.995-13.000 mm (0.5115-0.5118 inch). If clearance between piston pin and connecting rod pin bore exceeds 0.05 mm (0.002 inch), renew pin and/or connecting rod.

Standard diameter of connecting rod crankshaft bearing bore is 20.005-20.015 mm (0.7876-0.7880 inch). Clearance between connecting rod bearing bore and crankshaft crankpin journal should be 0.005-0.025 mm (0.0002-0.0010 inch). If clearance is 0.060 mm (0.0024 inch) or more, renew connecting rod and/or crankshaft.

Side clearance between connecting rod and crankshaft should be 0.20-0.50 mm (0.008-0.020 inch). If clearance exceeds 1.20 mm (0.047 inch), renew connecting rod and/or crankshaft.

When installing connecting rod and piston assembly, make certain that arrow marks (Fig. SU114) on piston crown and connecting rod point toward the valves. Be sure match marks on connecting rod and cap (Fig. SU115) are aligned and are facing open side of cylinder block. Install oil slinger, new lock plate and then the connecting rod bolts. Tighten connecting rod bolts to specified torque and bend lock plate against bolt heads.

**PISTON, PIN AND RINGS.** Refer to CONNECTING ROD paragraph for piston removal.

After separating piston and connecting rod, carefully remove rings and clean carbon and other combustion deposits from piston surface and ring lands.

**CAUTION: Extreme care should be exercised when cleaning ring lands. Do not damage squared edges or widen ring grooves. If ring lands are damaged, piston must be renewed.**

Inspect piston for excessive wear, scoring or other damage and renew as necessary. Standard piston skirt diameter, measured 90 degrees from piston

*Fig. SU116—Cylinder bore diameter should be measured at the six locations illustrated. If measurements vary more than 0.10 mm (0.004 in.), cylinder bore must be reconditioned or cylinder renewed.*

pin bore and 15 mm (0.59 inch) up from bottom of skirt, is 47.960-47.975 mm (1.8882-1.8888 inch). Clearance between piston skirt and cylinder should be 0.025-0.055 mm (0.0010-0.0022 inch). If clearance exceeds 0.12 mm (0.0047 inch), renew piston and/or recondition cylinder bore.

Standard side clearance between piston rings and ring grooves in piston is 0.02-0.06 mm (0.0008-0.0024 inch). To check side clearance, position ring in piston groove and use a feeler gage to measure clearance between ring and piston ring land. If side clearance between a new ring and ring groove in piston is greater than 0.12 mm (0.005 inch) for top ring or 0.10 mm (0.004 inch) for second ring, renew piston.

Standard piston ring end gap is 0.15-0.35 mm (0.006-0.014 inch) for all rings. Maximum allowable ring end gap is 0.70 mm (0.0275 inch).

Install rings on piston with the "R" stamped on ring toward top of piston and stagger ring end gaps around diameter of piston, however, do not position ring end gaps in line with piston pin center line.

Piston should be installed on connecting rod as outline in CONNECTING ROD paragraph. Lubricate piston and cylinder with engine oil prior to installing piston in cylinder. Be sure that arrow marks on piston crown and connecting rod point toward the valves.

**CYLINDER/CRANKCASE ASSEMBLY.** The cylinder is an integral part of the crankcase casting.

Standard cylinder bore inside diameter is 48 mm (1.890 inch). Cylinder bore should be measured at six different locations as shown in Fig. SU116. If any of the measurements vary by 0.10 mm (0.004 inch) or more, recondition or renew cylinder.

**CRANKSHAFT AND MAIN BEARINGS.** Crankshaft (10—Fig. SU113) is supported by ball bearings (17) at each end. To remove crankshaft, first remove connecting rod and piston assembly as outlined in CONNECTING ROD paragraph. Use a suitable puller to remove flywheel, then withdraw crankshaft from cylinder block.

Standard crankshaft crankpin journal diameter is 19.990-20.000 mm (0.7870-0.7874 inch). Crankpin clearance in connecting rod bearing should be 0.005-0.025 mm (0.0002-0.0010 inch) with wear limit of 0.060 mm (0.0024 inch). Ball bearing type main bearings should be a slight press fit in bores in cylinder block and crankcase cover and on crankshaft journals.

Crankshaft oil seals (2—Fig. SU113) should be renewed using suitable seal installing tool. Pack space between the seal lips with grease before installing crankshaft.

When installing crankshaft in engine block, make certain crankshaft and camshaft timing marks are aligned as shown in Fig. SU117.

**CAMSHAFT AND BEARINGS.** Camshaft (25—Fig. SU113) is supported at each end in ball bearings (16). Refer to CONNECTING ROD paragraph for camshaft removal. Camshaft ball bearings should be a light press fit on camshaft and in camshaft bearing bores machined into crankcase and crankcase cover.

Standard camshaft lobe height is 19.967-20.027 mm (0.7861-0.7885 inch) for intake and exhaust lobes. If lobe height is 19.963 mm (0.7860 inch) or less, renew camshaft. It is recommended that tappets be renewed if a new camshaft is installed.

Coat tappets with engine oil and position in cylinder block before installing camshaft. Make certain governor sleeve and governor assembly is correctly positioned on camshaft during reassembly. When installing camshaft, make certain camshaft gear and crankshaft gear timing marks are aligned as shown in Fig. SU117.

**GOVERNOR.** Refer to Fig. SU113 for an exploded view of the centrifugal flyweight governor assembly which is located on the camshaft gear. Make certain cotter pins retaining governor flyweights (28) to camshaft gear mounting plate (26) are bent to secure flyweights to mounting plate.

To adjust governor linkage, loosen clamp bolt (34) retaining governor lever (32) to governor shaft (31). Move governor lever manually to fully open carburetor throttle plate. Use a screwdriver to turn governor shaft fully clockwise, then tighten clamp bolt while holding governor shaft and lever in position.

**VALVE SYSTEM.** Clearance between valve stem and valve tappet (engine cold and piston at TDC on compression stroke) should be 0.05-0.15 mm (0.002-0.006 inch) for the intake and exhaust

valve. Clearance is adjusted by installing tappets of different lengths. Tappets are identified by paint markings. Tappet, part number 12891-90500 has no paint marking and is 32.97-33.03 mm (1.298-1.300 inch) long. Tappet, part number 12891-90510 has a blue paint mark and is 32.90-32.96 mm (1.295-1.298 inch) long. Tappet, part number 12891-90520 has a red paint mark and is 33.04-33.10 mm (1.301-1.303 inch) long. It may be necessary to grind the valve stem end to obtain the correct valve stem clearance if tappets fail to do so.

Valve face angle is 45 degrees and desired valve seating contact width is 0.70-0.80 mm (0.028-0.031 inch). Valve seat should be ground at the 46 degree and 15 degree angles shown in Fig. SU118.

Inspect valves for wear, distortion or other damage. Renew valve if valve head thickness (T—Fig. 119) after grinding is less than 0.30 mm (0.012 inch). Renew valve if valve head radial runout exceeds 0.03 mm (0.0012 inch). Measure valve stem outside diameter and valve guide inside diameter and refer to the following specifications:

Valve Stem OD
Intake . . . . . . . . . . . .5.460-5.475 mm
(0.2150-0.2156 in.)
Exhaust . . . . . . . . . .5.445-5.460 mm
(0.2144-0.2150 in.)
Valve Guide ID
Intake . . . . . . . . . . . .5.500-5.512 mm
(0.2165-0.2170 in.)
Exhaust . . . . . . . . . .5.500-5.512 mm
(0.2165-0.2170 in.)
Stem-to-Guide Clearance
Intake-standard . . . .0.025-0.052 mm
(0.0010-0.0020 in.)
Wear limit . . . . . . . . . . . . .0.080 mm
(0.003 in.)
Exhaust-standard . .0.040-0.067 mm
(0.0016-0.0026 in.)
Wear limit . . . . . . . . . . . . .0.100 mm
(0.004 in.)

Standard intake and exhaust valve spring free length is 18.50-19.50 mm (0.7283-0.7677 inch). If valve spring free length is 17.50 mm (0.6890 inch) or less, renew valve spring.

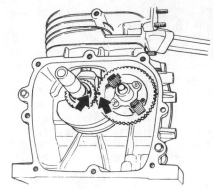

Fig. SU117—When installing crankshaft and camshaft, make certain that timing marks are aligned as shown.

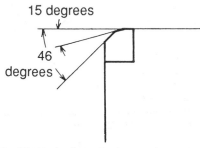

Fig. SU118—Intake and exhaust valve seats are ground at 46 degrees with a clearance angle of 15 degrees ground at the top. Desired seat width is 0.70-0.80 mm (0.028-0.031 in.).

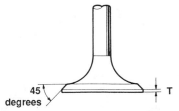

Fig. SU119—Valve face angle is 45 degrees for intake and exhaust. Renew valve if head thickness (T) is less than 0.30 mm (0.012 in.).

# TECUMSEH 4-STROKE

| Model | Bore | Stroke | Displacement |
|---|---|---|---|
| LAV25, LAV30, TVS75, H25, H30 prior to 1983 | 2.313 in. (58.74 mm) | 1.844 in. (46.84 mm) | 7.75 cu. in. (127 cc) |
| H30 after 1982, LV35, LAV35, TVS90, H35 prior to 1983, ECH90 | 2.500 in. (63.50 mm) | 1.844 in. (46.84 mm) | 9.05 cu. in. (148 cc) |
| H35 after 1982 | 2.500 in. (63.50 mm) | 1.938 in. (49.23 mm) | 9.51 cu. in. (156 cc) |
| LAV40, V40 external ignition, TVS105, HS40, ECV105 | 2.625 in. (66.68 mm) | 1.938 in. (49.23 mm) | 10.49 cu. in. (172 cc) |
| TNT100, TVS100, ECV100 | 2.625 in. (66.68 mm) | 1.844 in. (46.84 mm) | 9.98 cu. in. (164 cc) |
| V40, VH40, H40, HH40 | 2.500 in. (63.50 mm) | 2.250 in. (57.15 mm) | 11.04 cu. in. (181 cc) |
| ECV110 | 2.750 in. (69.85 mm) | 1.938 in. (49.23 mm) | 11.50 cu. in. (189 cc) |
| LAV50, ECV120, TNT120, TVS120, HS50 | 2.812 in. (71.43 mm) | 1.938 in. (49.23 mm) | 12.04 cu. in. (197 cc) |
| H50, HH50, V50, VH50, TVM125 | 2.625 in. (66.68 mm) | 2.250 in. (57.15 mm) | 12.18 cu. in. (229 cc) |
| H60, HH60, V60, VH60, TVM140 | 2.625 in. (66.68 mm) | 2.500 in. (63.50 mm) | 13.53 cu. in. (222 cc) |

## ENGINE IDENTIFICATION

Engines must be identified by the complete model number, including the specification number in order to obtain correct repair parts. These numbers are located on the name plate and/or tags that are positioned as shown in Fig. TE1 or Fig. TE1A. It is important to transfer identification tags from the original engine to replacement short block assemblies so unit can be identified when servicing.

If selecting a replacement engine and model or type number of the old engine is not known, refer to chart in Fig. TE1B and proceed as follows:

1. List the corresponding number which indicates the crankshaft position.

2. Determine the horsepower needed.

3. Determine the primary features needed. (Refer to the Tecumseh Engines Specification Book No. 692531 for specific engine variations.)

4. Refer to Fig. TE1C for Tecumseh engine model number and serial number interpretation.

The number following the letter code is the horsepower or cubic inch displacement. The number following the model number is the specification number. The last three digits of the specification number indicate a variation to the basic engine specification.

## MAINTENANCE

**SPARK PLUG.** Spark plug recommendations are shown in the following chart.

14 mm—3/8 inch
   reach . . . . . . . . . . . . . . Champion J8
18 mm—1/2 inch
   reach . . . . . . . . . . . . Champion D16
                 or MD16
All 7/8 inch reach . . . . Champion W18

**CARBURETOR.** Several different carburetors are used on these engines. Refer to the appropriate paragraph for model being serviced.

**Tecumseh Diaphragm Carburetor.** Refer to model number stamped on the carburetor mounting flange and to Fig. TE2 for exploded view of Tecumseh diaphragm type carburetor.

Initial adjustment of idle mixture and main fuel mixture screws from a lightly seated position is one turn open. Clockwise rotation leans mixture and counterclockwise rotation richens mixture.

Final adjustments are made with engine at operating temperature and running. Operate engine at rated speed and adjust main fuel mixture screw (14) for smoothest engine operation. Operate engine at idle speed and adjust idle mixture screw (10) for smoothest engine idle. If engine does not accelerate smoothly, slight adjustment of main fuel mixture screw may be required. Engine idle speed should be approximately 1800 rpm.

The fuel strainer in the fuel inlet fitting can be cleaned by reverse flushing with compressed air after the inlet needle and seat (19—Fig. TE2) are removed. The inlet needle seat fitting is metal with a neoprene seat, so the fitting (and enclosed seat) should be removed before

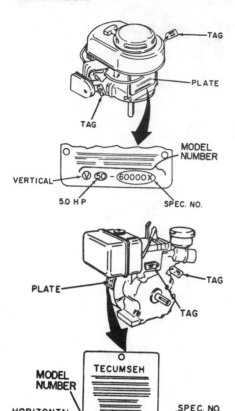

Fig. TE1—Tags and plates used to identify engine model will most often be located in one of the positions shown.

carburetor is cleaned with a commercial solvent. The stamped line on carburetor throttle plate should be toward top of carburetor, parallel with throttle shaft and facing outward as shown in Fig. TE3. Flat side of choke plate should

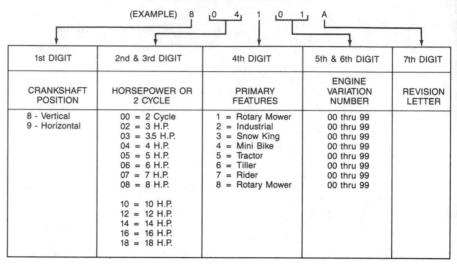

### TECUMSEH SERVICE NUMBER SYSTEM

(EXAMPLE)  8  0  4  1  0  1  A

| 1st DIGIT | 2nd & 3rd DIGIT | 4th DIGIT | 5th & 6th DIGIT | 7th DIGIT |
|---|---|---|---|---|
| CRANKSHAFT POSITION | HORSEPOWER OR 2 CYCLE | PRIMARY FEATURES | ENGINE VARIATION NUMBER | REVISION LETTER |
| 8 - Vertical<br>9 - Horizontal | 00 = 2 Cycle<br>02 = 3 H.P.<br>03 = 3.5 H.P.<br>04 = 4 H.P.<br>05 = 5 H.P.<br>06 = 6 H.P.<br>07 = 7 H.P.<br>08 = 8 H.P.<br><br>10 = 10 H.P.<br>12 = 12 H.P.<br>14 = 14 H.P.<br>16 = 16 H.P.<br>18 = 18 H.P. | 1 = Rotary Mower<br>2 = Industrial<br>3 = Snow King<br>4 = Mini Bike<br>5 = Tractor<br>6 = Tiller<br>7 = Rider<br>8 = Rotary Mower | 00 thru 99<br>00 thru 99<br>00 thru 99<br>00 thru 99<br>00 thru 99<br>00 thru 99<br>00 thru 99<br>00 thru 99 | |

Fig. TE1B—Reference chart used to select of identify replacement engines.

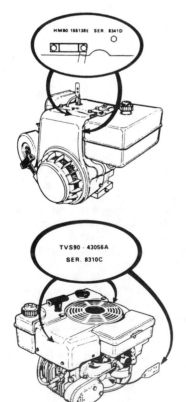

Fig. TE1C—Chart showing engine model number and serial number interpretation.

Fig. TE1A—Locations of tags and plates used to identify later model engines.

**V** - Vertical Shaft
**LAV** - Lightweight Aluminum Vertical
**VM** - Vertical Medium Frame
**TVM** - Tecumseh Vertical (Medium Frame)
**VH** - Vertical Heavy Duty (Cast Iron)
**TVS** - Tecumseh Vertical Styled
**TNT** - Toro N' Tecumseh
**ECV** - Exclusive Craftsman Vertical
**OVM** - Overhead Valve Vertical Medium Frame
**H** - Horizontal Shaft
**HS** - Horizontal Small Frame
**HM** - Horizontal Medium Frame
**HHM** - Horizontal Heavy Duty (Cast Iron) Medium Frame
**HH** - Horizontal Heavy Duty (Cast Iron)
**ECH** - Exclusive Craftsman Horizontal

### (EXAMPLE)

**TVS90-43056A** is the model and specification number

**TVS** - Tecumseh Vertical Styled
**90** - Indicates a 9 cubic inch displacement
**43056A** - is the specification number used for properly identifying the parts of the engine

**8310C** is the serial number

**8** - first digit is the year of manufacture (1978)
**310** - indicates calendar day of that year (310th day or November 6, 1978)
**C** - represents the line and shift on which the engine was built at the factory.

**SHORT BLOCKS.** New short blocks are identified by a tag marked SBH (Short Block Horizontal) or SBV (Short Block Vertical). Original model tags of an engine should always be transferred to a new short block for correct parts identification.

Illustrations courtesy of Tecumseh Products Company

be toward the fuel inlet fitting side of carburetor. Mark on choke plate should be parallel to shaft and should face IN-WARD when choke is closed. Diaphragm (21—Fig. TE2) should be installed with rounded head of center rivet up toward the inlet needle (19) regardless of size or placement of washers around the rivet.

On carburetor Models 0234-252, 265, 266, 269, 270, 271, 282, 293, 303, 322, 327, 333, 334, 344, 345, 348, 349, 350, 351, 352, 356, 368, 371, 374, 378, 379, 380, 404 and 405, or carburetors marked with an "F" as shown in Fig. TE2A, gasket (20—Fig. TE2) must be installed between diaphragm (21) and cover (22). All other models are assembled as shown, with gasket (20) between diaphragm (21) and carburetor body (6).

**Tecumseh Standard Float Carburetor.** Refer to Fig. TE4 for exploded view of Tecumseh standard float type carburetor.

Initial adjustment of idle mixture and main fuel mixture screws from a lightly seated position is one turn open. Clockwise rotation leans mixture and counterclockwise rotation richens mixture.

Fuel adjustments are made with engine at operating temperature and running. Operate engine at rated speed and adjust main fuel mixture screw (34) for

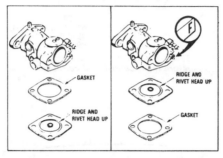

Fig. TE2A—Illustration showing correct position of diaphragm on carburetors unmarked and marked with a "F."

smoothest engine operation. Operate engine at idle speed and adjust idle mixture screws for smoothest engine idle. If engine does not accelerate smoothly, slight adjustment of main fuel mixture screw may be required.

Carburetor must be disassembled and all neoprene or Viton rubber parts removed before carburetor is immersed in cleaning solvent. Do not attempt to reuse any expansion plugs. Install new plugs if any are removed for cleaning.

A resilient tip on fuel inlet needle is used on some carburetors (Fig. TE5). The soft tip contacts the seating surface machined into the carburetor body.

On some carburetors, the fuel inlet valve needle seats against a Viton seat which must be removed before cleaning. The seat can be removed by blowing compressed air in from the fuel inlet fitting or by using a hooked wire. The grooved face of valve seat should be IN toward bottom of bore and the valve

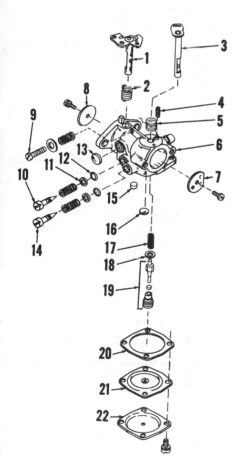

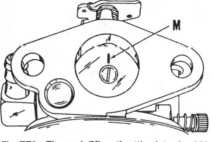

Fig. TE3—The mark (M) on throttle plate should be parallel to the throttle shaft and outward as shown. Some models may also have mark at 3 o'clock position.

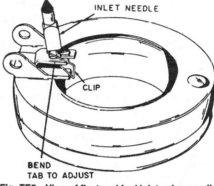

Fig. TE5—View of float and fuel inlet valve needle. The valve needle shown is equipped with a resilient tip and a clip. Bend tab shown to adjust float height.

Fig. TE2—Exploded view of typical Tecumseh diaphragm carburetor.

1. Throttle shaft
2. Return spring
3. Choke shaft
4. Choke stop spring
5. Return spring
6. Carburetor body
7. Choke plate
8. Throttle plate
9. Idle speed screw
10. Idle mixture screw
11. Washers
12. "O" ring
13. Welch plug
14. Main fuel mixture screw
15. Cup plug
16. Welch plug
17. Inlet needle spring
18. Gasket
19. Inlet needle & seat
20. Gasket
21. Diaphragm
22. Cover

Fig. TE4—Exploded view of standard Tecumseh float type carburetor.

1. Idle speed screw
2. Throttle plate
3. Return spring
4. Throttle shaft
5. Choke stop spring
6. Choke shaft
7. Return spring
8. Fuel inlet fitting
9. Carburetor body
10. Choke plate
11. Welch plug
12. Idle mixture needle
13. Spring
14. Washer
15. "O" ring
16. Ball plug
17. Welch plug
18. Pin
19. Cup plugs
20. Bowl gasket
21. Inlet needle seat
22. Inlet needle
23. Clip
24. Float shaft
25. Float
26. Drain stem
27. Gasket
28. Bowl
29. Gasket
30. Bowl retainer
31. "O" ring
32. Washer
33. Spring
34. Main fuel needle

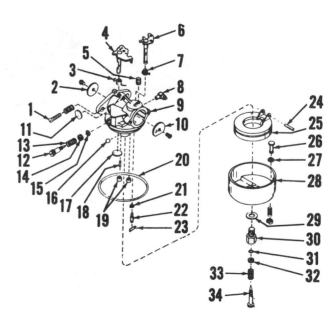

needle should seat on smooth side of the Viton seat (Fig. TE7).

A Viton seat contained in a brass seat fitting is used on some models. Use a 10-24 or 10-32 tap to pull the seat and fitting from carburetor bore as shown in Fig. TE6. Use a flat, close fitting punch to install new seat and fitting.

Install the throttle plate (2—Fig. TE4) with the two stamped marks out and at 12 and 3 o'clock positions. The 12 o'clock line should be parallel with the throttle shaft and toward top of carburetor. Install choke plate (10) with flat side down toward bottom of carburetor. Float height should be 0.200-0.220 inch (5.1-5.6 mm), measured from surface of carburetor body to float with carburetor body inverted as shown in Fig. TE8. Float height may also be set using Tecumseh float gage 670253A as shown in Fig. TE8A. Remove float and bend tab at float hinge to change float setting. The fuel inlet fitting (8—Fig. TE4) is pressed into body on some models. Start fitting into body, then apply a light coat of Loctite sealant to shank and press fitting into position. The flat on fuel bowl should be under the fuel inlet fitting. Refer to Fig. TE9.

Be sure to use correct parts when servicing the carburetor. Some fuel bowl gaskets are square section, while others are round. The bowl retainer screw (30—Fig. TE4) contains a drilled passage for fuel to the high speed metering needle (34). A diagonal port through one side of the bowl retainer is used on carburetors with external vent. The port is through both sides on models with internal vent (Fig. TE10).

**Tecumseh "Automagic" Float Carburetor.** Refer to Fig. TE11 and note carburetor can be identified by the absence of the idle mixture screw, choke and main mixture screw. Refer to Fig. TE12 for operating principles.

Float height setting is 0.210 inch (5.33 mm) measured from rim of carburetor body to surface of float as shown in Fig. TE8. Float may also be set using Tecumseh float gage 670253 as shown in Fig.

TE8A. Bend float tab to adjust float height.

Service procedures for "Automagic" carburetor are the same as for standard Tecumseh float carburetors.

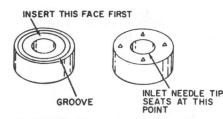

Fig. TE7—The Viton seat used on some Tecumseh carburetors must be installed correctly to operate properly. All-metal needle is used with the Viton seat.

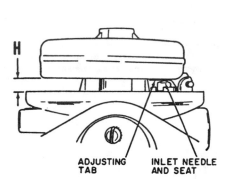

Fig. TE8—Distance (H) between carburetor body and float with body inverted should be 0.200-0.220 inch (5.1-5.6 mm).

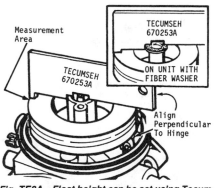

Fig. TE8A—Float height can be set using Tecumseh float tool 670253A as shown.

**Carter Carburetor.** Refer to Fig. TE13 for identification and exploded view of Carter carburetor.

Initial adjustment of idle mixture screw and main fuel mixture screws from a lightly seated position is 1-1/2 turns open for idle mixture screw and 1-3/4 turns open for main fuel mixture screw. Clockwise rotation leans mixture and counterclockwise rotation richens mixture.

Final adjustments are made with engine at operating temperature and running. Operate engine at rated speed and adjust main fuel mixture screw (4) for smoothest engine operation. Operate engine at idle speed and adjust idle mixture screw (9) for smoothest engine idle.

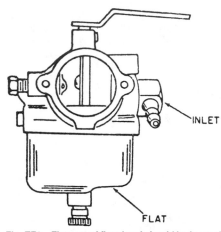

Fig. TE9—Flat part of float bowl should be located under the fuel inlet fitting.

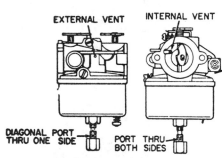

Fig. TE10—The bowl retainer contains a drilled fuel passage which is different for carburetors with external and internal fuel bowl vent.

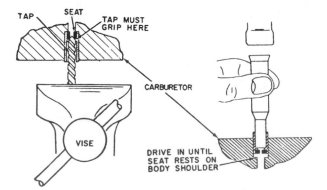

Fig. TE6—A 10-24 or 10-32 tap is used to pull the brass seat fitting and fuel inlet valve seat from some carburetors. Use a close fitting flat punch to install new seat and fitting.

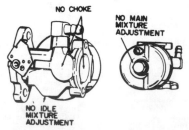

Fig. TE11—Tecumseh "Automagic" carburetor is similar to standard float models, but can be identified by absence of choke and adjusting needles.

If engine acceleration is not smooth, main fuel mixture screw may have to be adjusted slightly.

To check float level, invert carburetor body and float assembly. There should be 11/64 inch (4.37 mm) clearance between free side of float and machined surface of body casting. Bend float lever tang to provide correct measurement.

**Marvel-Schebler.** Refer to Fig. TE14 for identification and exploded view of Marvel-Schebler Series AH carburetor.

Initial adjustment of idle mixture screw and main fuel mixture screw from a lightly seated position is one turn open for idle mixture screw and 1-1/4 turn open for main fuel mixture screw. Clockwise rotation leans mixture and counterclockwise rotation richens mixture.

Final adjustments are made with engine at operating temperature and running. Operate engine at rated speed and adjust main fuel mixture screw (13) for smoothest engine operation. Operate engine at idle speed and adjust idle fuel mixture needle (18) for smoothest engine idle. Adjust idle speed screw (1) to obtain 1800 rpm.

To adjust float level, invert carburetor throttle body and float assembly. Float clearance should be 3/32 inch (2.38 mm) between free end of float and machined surface of carburetor body. Bend tang on float which contacts fuel inlet needle as necessary to obtain correct float level.

**Tillotson Type "MT" Carburetor.** Refer to Fig. TE15 for identification and exploded view of Tillotson type MT carburetor.

Initial adjustment of idle mixture screw and main fuel mixture screw from a lightly seated position is 3/4 turn open for idle mixture screw (21) and one turn open for main fuel mixture screw (20). Clockwise rotation leans fuel mixture

and counterclockwise rotation richens fuel mixture. Final adjustments are made with engine at operating temperature and running. Operate engine at rated speed and adjust main fuel mix-

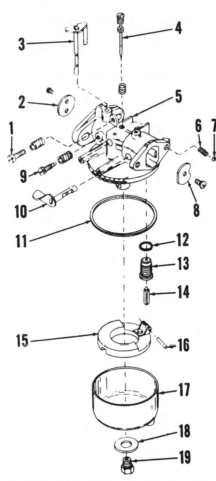

**Fig. TE13—Exploded view of typical Carter N carburetor.**

| | |
|---|---|
| 1. Idle speed screw | 10. Choke shaft |
| 2. Throttle plate | 11. Bowl gasket |
| 3. Throttle shaft | 12. Gasket |
| 4. Main adjusting screw | 13. Inlet valve seat |
| 5. Carburetor body | 14. Inlet valve |
| 6. Choke shaft spring | 15. Float |
| 7. Ball | 16. Float shaft |
| 8. Choke plate | 17. Fuel bowl |
| 9. Idle mixture screw | 18. Gasket |
| | 19. Bowl retainer |

ture screw for smoothest engine operation. Operate engine at idle speed and adjust idle mixture screw for smoothest engine idle. Adjust throttle stop screw (1) to obtain 1800 rpm. If engine acceleration is not smooth, main fuel mixture screw may have to be adjusted slightly.

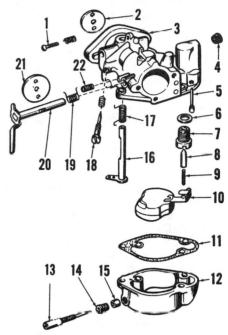

**Fig. TE14—Exploded view of Marvel-Schebler carburetor.**

| | |
|---|---|
| 1. Idle speed screw | 13. Main adjusting screw |
| 2. Throttle plate | 14. Retainer |
| 3. Carburetor body | 15. Packing |
| 4. Fuel screen | 16. Throttle shaft |
| 5. Float | 17. Throttle spring |
| 6. Gasket | 18. Idle mixture screw |
| 7. Inlet valve seat | 19. Choke spring |
| 8. Inlet valve | 20. Choke shaft |
| 9. Spring | 21. Choke plate |
| 10. Float | 22. Choke ratchet spring |
| 11. Gasket | |
| 12. Fuel bowl | |

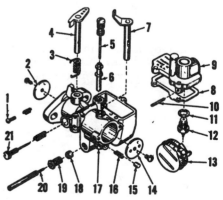

**Fig. TE15—Exploded view of typical Tillotson MT carburetor.**

| | |
|---|---|
| 1. Idle speed screw | 12. Inlet needle & seat |
| 2. Throttle plate | 13. Float |
| 3. Throttle spring | 14. Choke plate |
| 4. Throttle shaft | 15. Friction pin |
| 5. Idle tube | 16. Choke shaft spring |
| 6. Main nozzle | 17. Carburetor body |
| 7. Choke shaft | 18. Pacing |
| 8. Gasket | 19. Retainer |
| 9. Bowl cover | 20. Main adjusting screw |
| 10. Float shaft | 21. Idle mixture screw |
| 11. Gasket | |

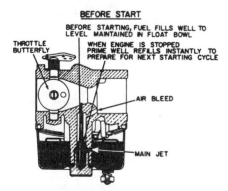

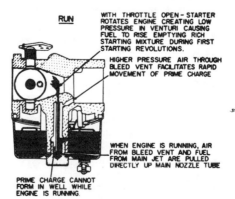

**Fig. TE12—The "Automagic" carburetor provides a rich starting mixture without using a choke plate. Mixture will be changed by operating with a dirty air filter or by incorrect float setting.**

To check float setting, invert carburetor throttle body and measure distance from top of float to carburetor body float bowl mating surface. Distance should be 1-13/32 inch (35.72 mm). Carefully bend float tang which contacts fuel inlet needle as necessary to obtain correct float level.

**Tillotson Type "E" Carburetor.** Refer to Fig. TE16 for identification and exploded view of Tillotson type E carburetor.

Initial adjustment of idle mixture screw and main fuel mixture screw from a lightly seated position, is ¾ turn open for idle mixture screw (18) and 1 turn open for main fuel mixture screw (14). Clockwise rotation leans mixture and counterclockwise rotation richens mixture.

Final adjustments are made with engine at operating temperature and running. Operate engine at rated speed and adjust main fuel mixture screw for smoothest engine operation. Operate engine at idle speed and adjust idle mixture screw for smoothest engine idle. Adjust throttle stop screw (8) to obtain desired idle speed. If engine does not accelerate smoothly, it may be necessary to adjust main fuel mixture screw slightly.

To check float setting, invert carburetor throttle body and measure distance from top of float at free end to float bowl mating surface on carburetor as shown in Fig. TE17. Distance should be 1-5/16 inch (27.38 mm). Gasket should not be installed when measuring float height. Carefully bend float tang which contacts fuel inlet needle as necessary to obtain correct float level.

**Walbro Carburetor.** Refer to Fig. TE18 for identification and exploded view of Walbro carburetor.

Initial adjustment of idle mixture and main fuel mixture screws from a lightly seated position is one turn open. Clockwise rotation leans fuel mixture and counterclockwise rotation richens fuel mixture.

Final adjustments are made with engine at operating temperature and running. Operate engine at rated speed and adjust main fuel mixture screw (33) for smoothest engine operation. Operate engine at idle speed and adjust idle mixture screw for smoothest engine idle. Adjust throttle stop screw (7) so engine idles at 1800 rpm.

To check float level, invert carburetor throttle body and measure clearance between free end of float and machined surface of carburetor. Clearance should be 1/8 inch (3.18 mm). Refer to Fig. TE19 Carefully bend float tang which contacts fuel inlet needle as necessary to obtain correct float setting.

**NOTE: If carburetor has been disassembled and main nozzle (19—Fig. TE18) removed, do not reinstall the original equipment nozzle. Install a new service nozzle. Refer to Fig. TE20 for differences between original and service nozzles.**

**PNEUMATIC GOVERNOR.** Some engines are equipped with a pneumatic (air vane) type governor. Engine speed under load should be 3600 rpm. On engines not equipped with slide control (Fig. TE23), obtain desired speed by varying the tension on governor speed

regulating spring by moving speed adjusting lever (Fig. TE21) in or out as required. If engine speed fluctuates due to governor hunting or surging, move carburetor end of governor spring into next outer hole on carburetor throttle arm shown in Fig. TE22.

A too-lean mixture or friction in governor linkage will also cause hunting or unsteady operation.

On engines with slide control carburetors, speed adjustment is made with engine running. Remove slide control cover (Fig. TE23) marked "choke, fast, slow, etc." Lock the carburetor in high speed "Run" position by matching the hole in slide control member (A) with hole (B) closest to choke end of brack-

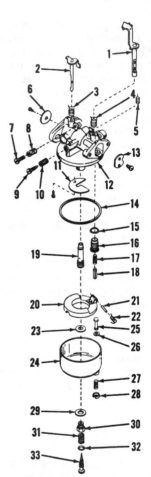

*Fig. TE18—Exploded view of Walbro LMG carburetor.*

1. Choke shaft
2. Throttle shaft
3. Throttle return spring
4. Choke return spring
5. Choke stop spring
6. Throttle plate
7. Idle speed stop screw
8. Spring
9. Idle mixture screw
10. Spring
11. Baffle
12. Carburetor body
13. Choke plate
14. Bowl gasket
15. Gasket
16. Inlet valve seat
17. Spring
18. Inlet valve
19. Main nozzle
20. Float
21. Float shaft
22. Spring
23. Gasket
24. Bowl
25. Drain stem
26. Gasket
27. Spring
28. Retainer
29. Gasket
30. Bowl retainer
31. Spring
32. "O" ring
33. Main fuel screw

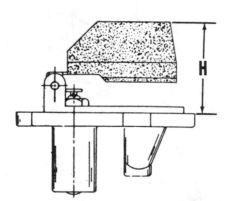

*Fig. TE17—Float height (H) should be measured as shown for Tillotson type "E" carburetors. Gasket should not be installed when measuring.*

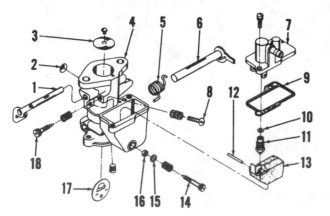

*Fig. TE16—Exploded view of typical Tillotson E carburetor.*

1. Throttle shaft
2. Welch plug
3. Throttle plate
4. Carburetor body
5. Return spring
6. Choke shaft
7. Bowl cover
8. Idle speed stop screw
9. Bowl gasket
10. Gasket
11. Inlet needle & seat
12. Float shaft
13. Float
14. Main fuel needle
15. Washer
16. Packing
17. Choke plate
18. Idle mixture screw

Illustrations courtesy of Tecumseh Products Company

et. Temporarily hold in this position by inserting a tapered punch or pin of suitable size into the hole.

At this time the choke should be wide open and the choke activating arm (E) should be clear of choke lever (F) by 1/64 inch (0.40 mm). Obtain clearance gap by bending arm (E). To increase engine speed, bend slide arm at point (D) outward from engine. To decrease speed, bend arm inward toward engine.

Make certain on models equipped with wiper grounding switch that wiper (G) touches slide (A) when control is moved to "Stop" position.

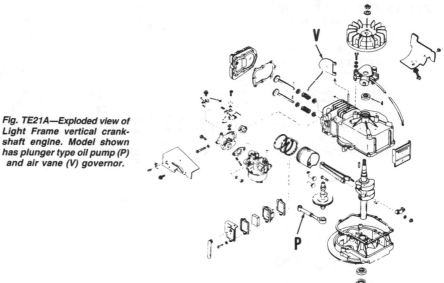

Fig. TE21A—Exploded view of Light Frame vertical crankshaft engine. Model shown has plunger type oil pump (P) and air vane (V) governor.

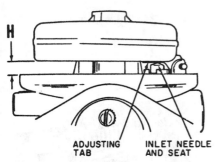

Fig. TE19—Float height (H) should be measured as shown on Walbro float carburetors. Bend the adjusting tab to adjust height.

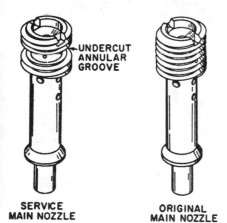

Fig. TE20—The main nozzle originally installed is drilled after installation through hole in body. Service main nozzles are grooved so alignment is not necessary.

Refer to Figs. TE24, TE25, TE26 and TE27 for assembled views of pneumatic governor systems.

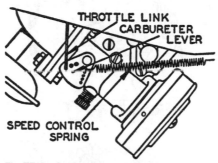

Fig. TE22—Outer spring anchorage holes in carburetor throttle lever may be used to reduce speed fluctuations or surging.

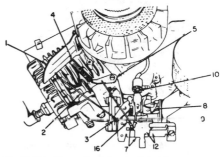

Fig. TE24—View of one type of speed control with pneumatic governor. Refer also to Figs. TE25, TE26 and TE27.

1. Air vane
2. Throttle control link
3. Carburetor throttle lever
4. Governor spring
5. Governor spring linkage
7. Speed control lever
8. Carburetor choke lever
9. Choke control
10. Stop switch
12. Alignment holes
16. Idle speed stop screw
18. High speed stop screw (Fig. TE27)

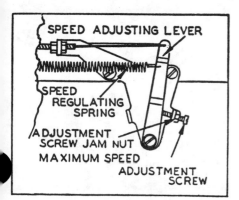

Fig. TE21—Speed adjusting screw and lever. The no-load speed should not exceed 3800 rpm.

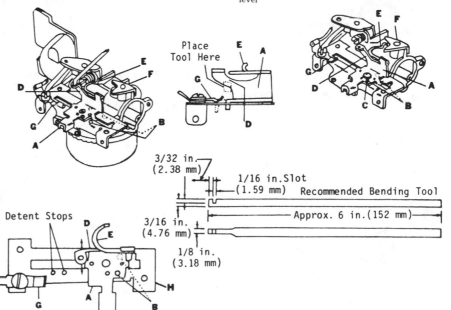

Fig. TE23—When carburetor is equipped with slide control, governed speed is adjusted by bending slide arm at point "D." Bend arm outward from engine to increase speed.

**MECHANICAL GOVERNOR.** Some engines are equipped with a mechanical (flyweight) type governor. To adjust the governor linkage, refer to Fig. TE28 and loosen governor lever screw. Twist protruding end of governor shaft counterclockwise as far as possible on vertical crankshaft engines; clockwise on horizontal crankshaft engines. On all models, move the governor lever until carburetor throttle shaft is in wide open position, then tighten governor lever clamp screw.

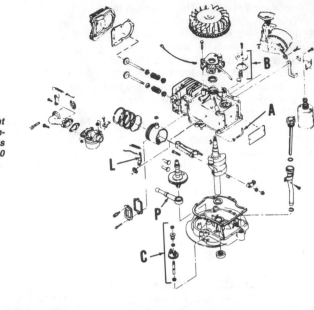

*Fig. TE27A—View of Light Frame vertical crankshaft engine typical of Models ECV100, ECV105, ECV110 and ECV120.*

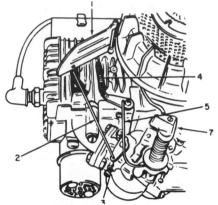

*Fig. TE25—Pneumatic governor linkage.*

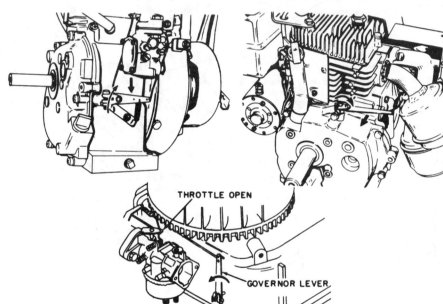

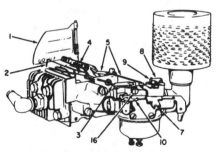

*Fig. TE26—Pneumatic governor linkage.*

*Fig. TE28—Views showing location of mechanical governor lever and direction to turn when adjusting position of governor shaft.*

THROTTLE OPEN

GOVERNOR LEVER

SCREW

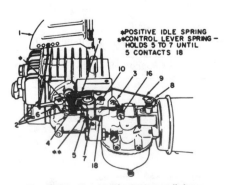

\*POSITIVE IDLE SPRING
\*CONTROL LEVER SPRING—
HOLDS 5 TO 7 UNTIL
5 CONTACTS 18

*Fig. TE27—Pneumatic governor linkage.*

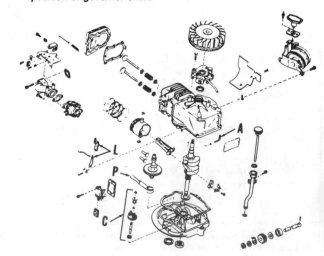

*Fig. TE28A—View of Light Frame vertical crankshaft engine.*

     Illustrations courtesy of Tecumseh Products Company

Binding or worn governor linkage will result in hunting or unsteady engine operation. An improperly adjusted carburetor will also cause a surging or hunting condition.

Refer to Figs. TE29 through TE47 for views of typical mechanical governor speed control linkage installations. On some models the governor gear shaft is renewable. New governor gear shaft must be pressed into bore in cover until the correct amount of the shaft protrudes. Refer to illustration in Fig. TE50.

**IGNITION SYSTEM.** A magneto ignition system with breaker points or capacitor-discharge ignition (CDI) may have been used according to model and application. Refer to appropriate paragraph for model being serviced.

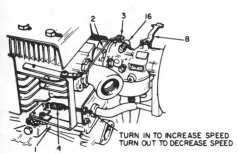

Fig. TE29—View of mechanical governor with one type of constant speed control. Refer also to Fig. TE30 through TE43 for other mechanical governor installations.

1. Governor lever
2. Throttle control
3. Carburetor throttle lever
4. Governor spring
8. Choke lever
16. Idle speed stop screw

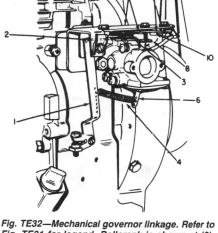

Fig. TE32—Mechanical governor linkage. Refer to Fig. TE31 for legend. Bellcrank is shown at (6).

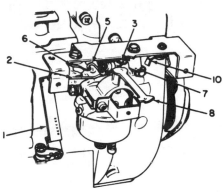

Fig. TE35—Mechanical governor linkage. Governed speed of engine is increased by closing loop in linkage (5); decrease speed by spreading loop. Refer to Fig. TE33 for legend.

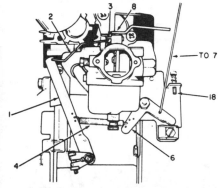

Fig. TE30—Mechanical governor linkage. Refer to Fig. TE29 for legend except for the following:

6. Bellcrank
7. Speed control lever
18. High speed stop screw

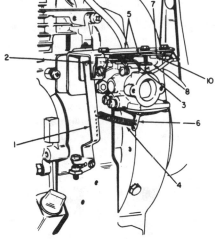

Fig. TE33—Mechanical governor linkage. Control cover is raised to view underside.

1. Governor lever
2. Throttle control link
3. Carburetor throttle lever
4. Governor spring
5. Governor spring linkage
7. Speed control lever
8. Choke lever
9. Choke control
10. Stop switch
12. Alignment holes
16. Idle speed stop screw
18. High speed stop screw

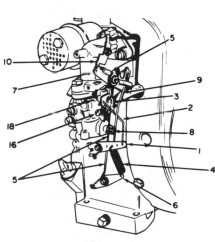

Fig. TE36—Mechanical governor linkage.

1. Governor lever
2. Throttle control link
3. Throttle lever
4. Governor spring
5. Governor spring linkage
6. Bellcrank
7. Speed control lever
8. Choke lever
9. Choke control link
10. Stop switch
16. Idle speed stop screw
18. High speed stop screw

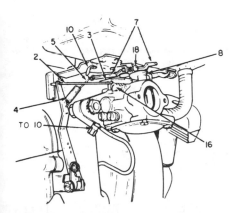

Fig. TE31—Mechanical governor linkage.

1. Governor lever
2. Throttle control link
3. Carburetor throttle
4. Governor spring
5. Governor spring linkage
7. Speed control lever
8. Choke lever
10. Stop switch
16. Idle speed stop screw
18. High speed stop screw

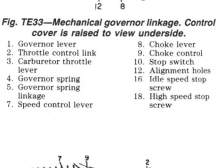

Fig. TE34—Mechanical governor linkage with control cover raised. Refer to Fig. TE33 for legend.

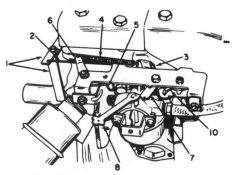

Fig. TE37—View of control linkage used on some engines. To increase governed engine speed, close loop (5); to decrease speed, spread loop. Refer to Fig. TE36 for legend.

Illustrations courtesy of Tecumseh Products Company

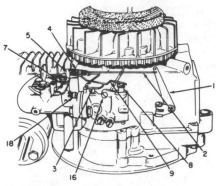

**Fig. TE38—View of mechanical governor control linkage. Governor spring (4) is hooked onto loop in link (2).**

1. Governor lever
2. Throttle control link
3. Throttle lever
4. Governor spring
5. Governor spring linkage
7. Speed control lever
8. Choke lever
9. Choke control link
16. Idle speed stop screw
18. High speed stop screw

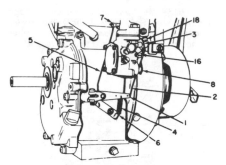

**Fig. TE39—Linkage for mechanical governor. Refer to fig. TE38 for legend. Bellcrank is shown at (6).**

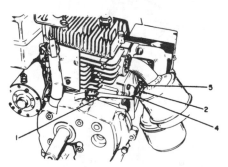

**Fig. TE40—Mechanical governor linkage. Refer to Fig. TE38 for legend.**

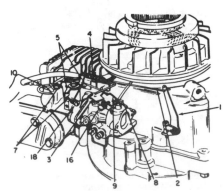

**Fig. TE41—Mechanical governor linkage. Refer to Fig. TE38 for legend.**

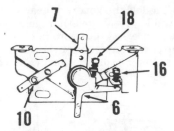

**Fig. TE42—Models with "Automagic" carburetor use control shown.**

6. Governor spring
7. Control lever
10. Idle speed stop screw
16. Idle speed stop screw
18. High speed stop screw

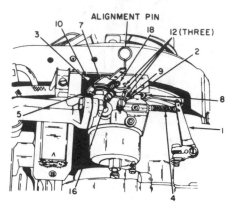

**Fig. TE43—On model shown, adjust location of cover so control lever is aligned with high speed slot and holes (12) are aligned.**

1. Governor lever
2. Throttle control link
3. Throttle lever
4. Governor spring
5. Governor spring linkage
7. Control lever
8. Choke lever
9. Choke linkage
10. Stop switch
12. Alignment holes
16. Idle speed stop screw
18. High speed stop screw

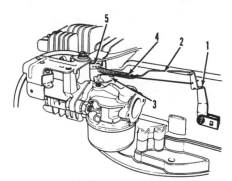

**Fig. TE44—View of governor linkage on TNT 100 engines. TNT120 engines are similar. Refer also to Fig. TE45 for models with primer carburetor.**

1. Governor lever
2. Throttle link
3. Throttle lever
4. Governor spring
5. Control lever

**Fig. TE46—View of governor linkage on TVS engines with adjustable carburetor.**

1. Governor lever
2. Throttle link
4. Governor spring
6. Idle mixture screw
7. Main mixture screw
8. Idle speed screw

**Breaker-Point Ignition System.** Breaker-point gap at maximum opening should be 0.020 inch (0.51 mm) for all models. Marks are usually located on stator and mounting post to facilitate timing (Fig. TE51).

Ignition timing can be checked and adjusted to occur when piston is at specific location (BTDC) if marks are missing. Refer to the following specifications for recommended timing.

| Models | Piston Position BTDC |
| --- | --- |
| HS40, LAV40, TVS105, ECV100, ECV105, ECV110, ECV120, TNT100, TNT120 | 0.035 in. (0.89 mm) |

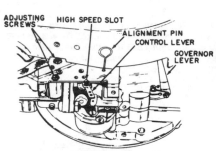

**Fig. TE45—View of governor linkage on TNT100 engine with primer carburetor. Refer to Fig. TE44 for parts identification.**

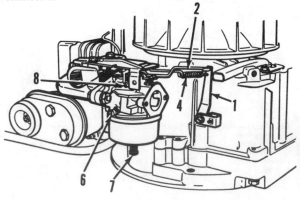

Illustrations courtesy of Tecumseh Products Company

V40, VH40, LAV50,
TVS120, H40, HH40, HS50 . . 0.050 in.
(1.27 mm)
LAV25, LAV30, LV35,
LAV35, TVS90, H25, H30,
H35, TVS75, ECH90 . . . . . . . 0.065 in.
(1.65 mm)
V50, VH50, V60, VH60,
H50, HH50, H60, HH60,
TVM125, TVM140 . . . . . . . . . 0.080 in.
(2.03 mm)

Some models may be equipped with the coil and laminations mounted outside the flywheel. Engines equipped with breaker-point ignition have the ignition points and condenser mounted under the flywheel, and the coil and laminations mounted outside the flywheel. This system is identified by the round shape of the coil and a stamping "Gray Key" in the coil to identify the correct flywheel key.

The correct air gap setting between the flywheel magnets and the coil laminations is 0.0125 inch (0.32 mm). Use Tecumseh gage 670297 or equivalent thickness plastic strip to set gap as shown in Fig. TE54.

**Solid-State Ignition System.** The Tecumseh solid-state ignition system does not use ignition breaker points. The only moving part of the system is the rotating flywheel with the charging magnets. As the flywheel magnet passes the input coil (2—Fig. TE52), a low voltage ac current is induced into input coil. Ac current passes through the rectifier (3), which converts this current to dc current. It then travels to the capacitor (4) where it is stored. The flywheel rotates approximately 180 degrees to position (1B). As it passes trigger coil (5), it induces a very small electric charge into the coil. This charge passes through resistor (6) and turns on the SCR (silicon controlled rectifier) switch (7). With the SCR switch closed, low voltage cur-

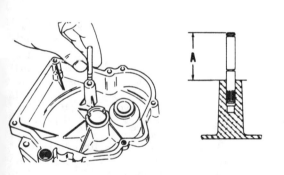

Fig. TE47—View of governor linkage on TVS engines with primer carburetor.
1. Governor lever
2. Throttle link
3. Throttle lever
4. Governor spring

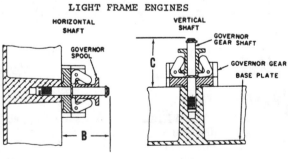

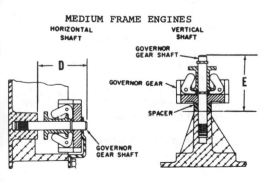

LIGHT FRAME ENGINES

MEDIUM FRAME ENGINES

Fig. TE50—The governor gear shaft must be pressed into bore until the correct amount of shaft protrudes (A, B, C, D or E). Refer also to illustrations for correct assembly of governor gear and associated parts.
B. 1-5/16 inches (33.34 mm)
C. 1-5/16 inches (33.34 mm)
D. 1-3/8 inches (34.92 mm)
E. 1-19/32 inches (40.48 mm)

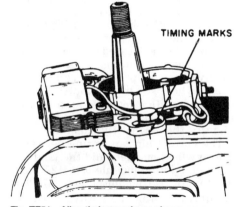

Fig. TE51—Align timing marks as shown on magneto ignition system.

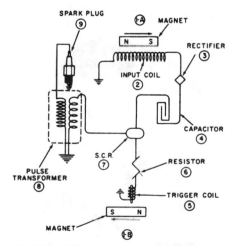

Fig. TE52—Wiring diagram of solid-state ignition system for models using ignition components shown in Fig. TE53.

rent stored in capacitor (4) travels to pulse transformer (8). Voltage is stepped up instantaneously and current is discharged across the electrodes of spark plug (9), producing a spark.

Some units are equipped with a second trigger coil and resistor set to turn the SCR switch on at a lower rpm. This second trigger pin is closer to the flywheel and produces a spark at TDC for easier starting. As engine rpm increases, the first (shorter) trigger pin picks up the small electric charge and turns the SCR switch on, firing the spark plug earlier (before TDC).

If system fails to produce a spark to the spark plug, first check high tension lead (Fig. TE53). If condition of high tension lead is questionable, renew pulse transformer and high tension lead assembly. Check low tension lead and renew if insulation is faulty. The magneto charging coil, electronic triggering system and mounting plate are available only as an assembly. If necessary to renew this assembly, place unit in position on engine. Start retaining screws, turn mounting plate counterclockwise as far as possible, then tighten retaining screws to 5-7 ft.-lbs. (7-10 N·m).

Engines with solid-state (CDI) ignition have all the ignition components sealed in a module and located outside the flywheel. There are no components under the flywheel except a spring clip to hold the flywheel key in position. This system is identified by the square shape module and a stamping "Gold Key" to identify the correct flywheel key.

The correct air gap setting between the flywheel magnets and the laminations on ignition module is 0.0125 inch (0.32 mm). Use Tecumseh gage 670297 or equivalent thickness plastic strip to set gap as shown in Fig. TE54.

**LUBRICATION.** Vertical crankshaft engines may be equipped with a barrel and plunger type oil pump or a gear-driven rotor type oil pump. Horizontal crankshaft engines may be equipped with a gear-driven rotor type pump or with a dipper type oil slinger attached to the connecting rod.

Oil level should be checked after every five hours of operation. Maintain oil level at lower edge of filler plug or at "FULL" mark on dipstick.

Manufacturer recommends oil with an API service classification SE or SF. Use SAE 30 or SAE 10W-30 motor oil for temperatures above 32° F (0° C). Use SAE 5W-30 or SAE 10W for temperatures below 32° F (0° C). Manufacturer explicitly states: DO NOT USE SAE 10W-40 motor oil.

Oil should be changed after the first two hours of engine operation (new or rebuilt engine) and after every 25 hours of operation thereafter.

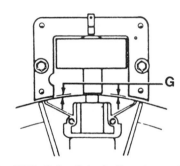

*Fig. TE54—Set coil lamination air gap (G) at 0.0125 inch (0.32 mm) at locations shown.*

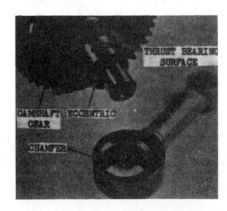

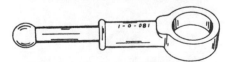

*Fig. TE55—Various types of barrel and plunger oil pumps have been used. Chamfered face of collar should be toward camshaft if drive collar has only one chamfered side. If drive collar has a flat boss, the boss should be next to the engine crankcase cover (mounting flange), away from camshaft gear.*

### REPAIRS

**TIGHTENING TORQUES.** Recommended tightening torque specifications are as follows:

Carburetor to intake
  pipe . . . . . . . . . . . . . . . . 48-72 in.-lbs.
                     (5-8 N·m)
Connecting rod nuts
(except Durlok nuts):
  1.7 hp, 2.5 hp, 3.0 hp,
  ECH90, ECV100 . . . . . . 65-75 in.-lbs.
                     (7-9 N·m)
  4 hp & 5 hp light frame
  models, ECV105, ECV110,
  ECV120 . . . . . . . . . . . . . 80-95 in.-lbs.
                    (9-11 N·m)
  5 hp medium frame,
  6 hp . . . . . . . . . . . . . . 86-110 in.-lbs.
                  (10-12 N·m)
Connecting rod screws
(Durlok):
  5 hp light frame . . . . 110-130 in.-lbs.
                  (12-15 N·m)
  5 hp medium
  frame, 6hp . . . . . . . . . 130-150 in.-lbs.
                  (15-17 N·m)
  All other models . . . . . 95-110 in.-lbs.
                  (11-12 N·m)
Cylinder head . . . . . . . 160-200 in.-lbs.
                  (18-23 N·m)
Flywheel:
  Light frame . . . . . . . . . . 30-33 ft.-lbs.
                  (41-45 N·m)
  Medium frame . . . . . . . 36-40 ft.-lbs.
                  (49-54 N·m)
  External ignition . . . . . 33-36 ft.-lbs.
                  (45-50 N·m)
Gear reduction cover . . 75-110 in.-lbs.
                  (9-12 N·m)
Gear reduction
  housing . . . . . . . . . . . 100-144 in.-lbs.
                  (11-16 N·m)
Intake pipe to cylinder . . 72-96 in.-lbs.
                  (8-11 N·m)
Magneto stator . . . . . . . . 40-90 in.-lbs.
                  (5-10 N·m)
Mounting flange . . . . . . . 75-110 in.-lbs.
                  (9-12 N·m)
Spark plug . . . . . . . . . . . . 21-30 ft.-lbs.
                  (28-41 N·m)

**FLYWHEEL.** On models so equipped, disengage flywheel brake as outlined in FLYWHEEL BRAKE section. If flywheel has tapped holes, use a suitable puller to remove flywheel. If no holes are present, screw a knock-off nut onto crankshaft so there is a small gap between nut and flywheel. Gently pry against bottom of flywheel while tapping sharply on nut.

After installing flywheel, tighten flywheel nut to torque listed in TIGHTENING TORQUE table.

**CONNECTING ROD.** Piston and connecting rod assembly is removed from cylinder head end of engine. Before

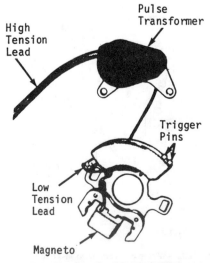

*Fig. TE53—Drawing of solid-state ignition system components used on some models.*

removing piston, remove any carbon or ring ridge from top of cylinder to prevent ring breakage. The aluminum alloy connecting rod rides directly on crankshaft crankpin.

Refer to the following table for standard crankpin journal diameter (if engine has external ignition, check list for a different specification than engines not so equipped):

| Model | Diameter |
|---|---|
| LAV25, LAV30, TVS75, H25, H30, LV35, LAV35, TVS90, H35 (prior 1983), ECH90, TNT100, ECV100, TVS100 | 0.8610-0.8615 in. (21.869-21.882 mm) |
| H35 (after 1982), LAV40, TVS105, V40 (external ignition), HS40, ECV105, ECV110, ECV120, TNT120, LAV50, HS50, TVS120 | 0.9995-1.0000 in. (25.390-25.400 mm) |
| V40, VH40, H40, HH40, H50, HH50, V50, VH50, TVM125, TVM140, V60, VH60, H60, HH60 | 1.0615-1.0620 in. (26.962-26.975 mm) |

Standard inside diameter for connecting rod crankpin bearing is listed in the following table (if engine has external ignition, check list for a different specification than engines not so equipped):

| Model | Diameter |
|---|---|
| LAV25, LAV30, TVS75, H25, H30, LV35, LAV35, TVS90, H35 (prior 1983), ECH90, TNT100, ECV100, TVS100 | 0.8620-0.8625 in. (21.895-21.908 mm) |
| H35 (after 1982), LAV40, TVS105, V40 (external ignition), HS40, ECV105, ECV110, ECV120, TNT120, LAV50, HS50, TVS120 | 1.0005-1.0010 in. (25.413-25.425 mm) |
| V40, VH40, H40, HH40, H50, HH50, V50, VH50, TVM125, TVM140, V60, VH60, H60, HH60 | 1.0630-1.0636 in. (27.000-27.013 mm) |

Connecting rod bearing-to-crankpin journal clearance should be 0.0005-0.0015 inch (0.013-0.038 mm) for all models.

When installing connecting rod and piston assembly, align the match marks on connecting rod and cap as shown in Fig. TE58. On some models, the piston pin hole is offset in piston and arrow on top of piston should be toward valves. On all engines, the match marks on connect-

ing rod and cap must be toward power takeoff (pto) end of crankshaft. Lock plates, if so equipped, for connecting rod cap retaining screws should be renewed each time cap is removed.

**PISTON, PIN AND RINGS.** Aluminum alloy pistons are equipped with two compression rings and one oil control ring.

Piston ring end gap is 0.007-0.017 inch (0.18-0.43 mm) on models without external ignition and displacement of 12.04 cu. in. (197 cc) or less. If engine is equipped with an external ignition, or displacement is 12.18 cu. in. (229 cc) or greater, piston ring end gap should be 0.010-0.020 inch (0.25-0.50 mm).

Piston skirt-to-cylinder clearances are listed in the following table (if engine has external ignition, check list for a different specification than engines not so equipped):

| Model | Clearance |
|---|---|
| H25, H30 (prior 1983), LAV25, LAV30, TVS75 | 0.0025-0.0043 in. (0.064-0.110 mm) |

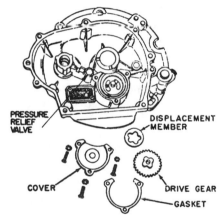

Fig. TE56—View of typical gear-driven rotor type oil pump disassembled.

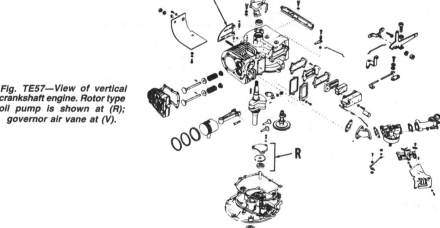

Fig. TE57—View of vertical crankshaft engine. Rotor type oil pump is shown at (R); governor air vane at (V).

| Model | Clearance |
|---|---|
| V40, VH40, H40, HH40 | 0.0055-0.0070 in. (0.140-0.178 mm) |
| ECV110 | 0.0045-0.0060 in. (0.114-0.152 mm) |
| H50, H60, HH50, HH60, TVM125, TVM140, V50, V60 | 0.0035-0.0050 in. (0.089-0.127 mm) |
| H50 (external ignition), H60 (external ignition), TVM125 (external ignition), TVM140 (external ignition) | 0.0030-0.0048 in. (0.076-0.123 mm) |
| HH50, VH50, HH60, VH60 | 0.0015-0.0055 in. (0.038-0.140 mm) |

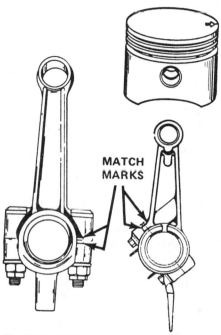

Fig. TE58—Match marks on connecting rod and cap should be aligned and should be toward pto end of crankshaft.

ECH90, ECV100, ECV105,
ECV120, H30 (after 1982)
H35 (prior 1983),
HS40, HS50, LAV35,
LAV40, LAV50, LV35,
TNT100, TNT120, TVS90
TVS100, TVS105, TVS120,
V40 (external
   ignition) . . . . . . . . .0.0040-0.0058 in.
        (0.102-0.147 mm)

Standard piston diameters measured at piston skirt 90 degrees from piston pin bore are listed in the following table (if engine has external ignition, check list for a different specification than engines not so equipped):

| Model | Diameter |
|---|---|
| LAV25, LAV30, TVS75, H25, H30 (prior 1983) | 2.3092-2.3100 in. (58.654-58.674 mm) |
| H40, HH40, V40, VH40 | 2.4945-2.4950 in. (63.360-63.373 mm) |
| LV35, LAV35, H30 (after 1982), H35, TVS90, ECH90 | 2.4952-2.4960 in. (63.378-63.398 mm) |
| ECV100, ECV105, HS40, LAV40, TNT100, TVS100, TVS105, V40 (external ignition) | 2.6202-2.6210 in. (66.553-66.573 mm) |
| HH50 (external ignition), HH60 (external ignition), VH50 (external ignition), VH60 (external ignition) | 2.6205-2.6235 in. (66.561-66.637 mm) |
| H50, HH50, V50, VH50, TVM125, TVM140, H60, HH60, V60, VH60 | 2.6210-2.6215 in. (66.573-66.586 mm) |
| TVM125 (external ignition), TVM140 (external ignition) | 2.6212-2.6220 in. (66.578-66.599 mm) |
| ECV110 | 2.7450-2.7455 in. (69.723-69.736 mm) |
| HS50, LAV50, TVS120, ECV120, TNT120 | 2.8072-2.8080 in. (71.303-71.323 mm) |

Standard ring side clearance in ring grooves is listed in the following table (if engine has external ignition, check list for a different specification than engines not so equipped):

| Model | Clearance |
|---|---|
| ECH90, H25, H30 H35, LAV25, LAV30, LAV35, LV35, TVS75, TVS90: | |

Compression rings . . .0.002-0.005 in.
        (0.05-0.13 mm)
Oil control ring . . .0.0005-0.0035 in.
        (0.013-0.089 mm)
ECV100, ECV105, ECV120,
H40, H50 (external ignition),
H60 (external ignition),
HH40, HH50 (external ignition),
HH60 (external ignition),
HS40, HS50,
HS50 (external ignition),
LAV40, LAV50,
LAV50 (external ignition),
TNT100, TNT120,
TVM125 (external ignition),
TVM140 (external ignition),
TVS100, TVS105, TVS120,
TVS120 (external ignition),
V40, V40 (external ignition),
VH40, VH50 (external ignition),
VH60 (external ignition):
  Compression rings . . .0.002-0.005 in.
        (0.05-0.13 mm)
  Oil control ring . . . . .0.001-0.004 in.
        (0.03-0.10 mm)
ECV110:
  Compression rings . . .0.002-0.004 in.
        (0.05-0.10 mm)
  Oil control ring . . . . .0.001-0.002 in.
        (0.03-0.05 mm)
H50, H60, HH50,
HH60, TVM125, TVM140,
V50, V60, VH50,
VH60:
  Compression rings . . .0.002-0.004 in.
        (0.05-0.10 mm)
  Oil control ring . . . . .0.002-0.004 in.
        (0.05-0.10 mm)

*Fig. TE59—View of Medium Frame engine with vertical crankshaft. Some models use plunger type oil pump (P); others are equipped with rotor type oil pump (R).*

Refer to CONNECTING ROD section for correct piston to connecting rod assembly and correct piston installation procedure. Piston pin should be a tight push fit in piston pin bore and connecting rod pin bore and is retained by snap rings at each end of piston pin bore. Install marked side of piston rings up. Stagger ring end gaps equally around circumference of piston during installation, however, on models with a "trenched" cylinder (see Fig. TE60), position ring end gaps so they are not in trenched area. This will lessen the possibility of the ring end snagging during installation of piston.

**CYLINDER AND CRANKCASE.** Cylinder and crankcase are an integral casting on all models. Cylinder should be honed and fitted to nearest oversize for

*Fig. TE60—Cylinder block on models H50, HH50, V50, VH50, TVM125 and TVM140 has been "trenched" to improve fuel flow and power.*

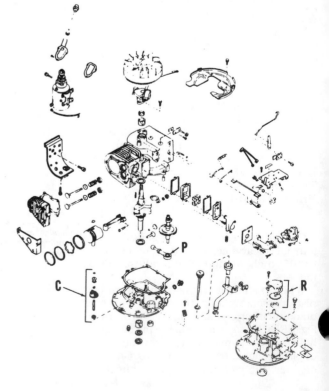

which piston and ring set are available if cylinder is scored, tapered or out-of-round more than 0.005 inch (0.13 mm).

Standard cylinder bore diameters are listed in the following table (if engine has external ignition, check list for a different specification than engines not so equipped):

| Model | Diameter |
|---|---|
| LAV25, LAV30, TVS75, H25, H30 (prior 1983) | 2.3125-2.3135 in. (58.738-58.763 mm) |
| ECH90, H30 (after 1982), H35, H40, HH40, LAV35, LV35, TVS90, V40, VH40 | 2.5000-2.5010 in. (63.500-63.525 mm) |
| ECV100, ECV105, H50, H60, HH50, HH60, HS40, LAV40, TNT100, TVM125, TVS100, TVS105, V40 (external ignition), V50, VH50 | 2.6250-2.6260 in. (66.675-66.700 mm) |
| ECV110 | 2.7500-2.7510 in. (69.850-69.875 mm) |
| LAV50, TVS120, HS50, ECV120, TNT120 | 2.8120-2.8130 in. (71.425-71.450 mm) |

Refer to PISTON, PIN AND RINGS section for correct piston-to-cylinder block clearance. Note also that cylinder block used on Models H50, HH50, V50, VH50, TVM125 and TVM140 has been "trenched" to improve fuel flow and power (Fig. TE60).

**CRANKSHAFT, MAIN BEARINGS AND SEALS.** Refer to CONNECTING ROD section for standard crankshaft crankpin journal diameters.

Crankshaft main bearing journals on some models ride directly in the aluminum alloy bores in the cylinder block and the crankcase cover (mounting flange). Other engines were originally equipped with renewable steel backed bronze bushings and some were originally equipped with a ball type main bearing at the pto end of crankshaft.

Standard main bearing bore diameters for main bearings are listed in the following table (if engine has external ignition, check list for a different specification than engines not so equipped):

| Model | Diameter |
|---|---|
| LAV25, LAV30 TVS75, LV35, LAV35, TVS90, ECV100, H25, H30 (prior 1983), ECH90 | 0.8755-0.8760 in. (22.238-22.250 mm) |
| ECV105, ECV110, ECV120, H35 (after 1982), H40, H50, H60, HH40, HH50, HH60, HS40, HS50, LAV40, LAV50, TNT120, TVM125, TVM140, TVS105, TVS120, V40, V50, V60, VH40, VH50, VH60 | 1.0005-1.0010 in. (25.413-25.425 mm) |
| ECV100 (external ignition), H30 (after 1982), TVS75 (external ignition), TVS90 (external ignition), TNT100, TVS100: | |
| Crankcase side | 1.0005-1.0010 in. (25.413-25.425 mm) |
| Cover (flange) side | 0.8755 0.8760 in. (22.238-22.250 mm) |

Standard diameters for crankshaft main bearing journals are shown in the following table (if engine has external ignition, check list for a different specification than engines not so equipped):

| Model | Diameter |
|---|---|
| LAV25, LAV30 TVS75, LV35, LAV35, TVS90, ECV100, H25, H30 (prior 1983), ECH90 | 0.8735-0.8740 in. (22.187-22.200 mm) |
| ECV105, ECV110, ECV120, H35 (after 1982), H40, H50, H60, HH40, HH50, HH60, HS40, HS50, LAV40, LAV50, TNT120, TVM125, TVM140, TVS105, TVS120, V40, V50, V60, VH40, VH50, VH60 | 0.9985-0.9990 in. (25.362-25.375 mm) |
| ECV100 (external ignition), H30 (after 1982), TVS75 (external ignition), TVS90 (external ignition), TNT100, TVS100: | |
| Crankcase side | 0.9985-0.9990 in. (25.362-25.375 mm) |
| Cover (flange) side | 0.8735-0.8740 in. (22.187-22.200 mm) |

Main bearing clearance should be 0.0015-0.0025 inch (0.038-0.064 mm). Crankshaft end play should be 0.005-0.027 inch (0.13-0.69 mm) for all models.

On Models H30 through HS50, it is necessary to remove a snap ring (Fig. TE61) which is located under oil seal. Note oil seal depth before removal of seal, then remove seal and snap ring.

Crankcase cover may now be removed. When installing cover, press new seal in to same depth as old seal before removal.

Ball bearing should be inspected and renewed if rough, loose or damaged. Bearing must be pressed on or off of crankshaft journal using a suitable press or puller.

Always note oil seal depth and direction before removing oil seal from crankcase or cover. New seals must be pressed into seal bores to the same depth as old seal before removal on all models.

When installing crankshaft, align crankshaft and camshaft gear timing marks as shown in Fig. TE65.

**CAMSHAFT.** The camshaft and camshaft gear are an integral part which rides on journals at each end of camshaft. Camshaft on some models also has a compression release mechanism mounted on camshaft gear which lifts exhaust valve at low cranking rpm to reduce compression and aid starting (Fig. TE63).

When removing camshaft, align timing marks on camshaft gear and crankshaft gear to relieve valve spring pressure on camshaft lobes. On models with compression release, it is necessary to rotate crankshaft three teeth past the aligned position to allow compression release mechanism to clear the exhaust valve tappet.

Renew camshaft if lobes or journals are worn or scored. Spring on compression release mechanism should snap weight against camshaft. Compression release mechanism and camshaft are serviced as an assembly only.

Standard camshaft journal diameter is 0.6230-0.6235 inch (15.824-15.837 mm) for Models V40, VH40, H40, HH40, H50, HH50, V50, VH50, H60, HH60, V60, VH60, TVM125 and TVM140, and

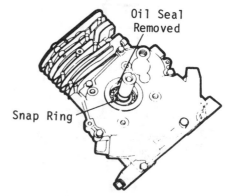

Fig. TE61—On Models H30 through H50 it is necessary to remove snap ring under oil seal before removing crankcase cover.

0.4975-0.4980 inch (12.637-12.649 mm) for all other models.

On models equipped with the barrel and plunger type oil pump, the pump is operated by an eccentric on the camshaft. Refer to OIL PUMP paragraph.

When installing camshaft, align crankshaft and camshaft timing marks as shown in Fig. TE65.

**OIL PUMP.** Vertical crankshaft engines may be equipped with a barrel and plunger type oil pump or a gear-driven rotor type oil pump. Horizontal crankshaft engines may be equipped with a gear-driven rotor type pump or with a dipper type oil slinger attached to the connecting rod.

The barrel and plunger type oil pump is driven by an eccentric on the camshaft. Chamfered side of drive collar (Fig. TE55) should be toward engine lower cover. Oil pumps may be equipped with two chamfered sides, one chamfered side or with flat boss as shown. Be sure installation is correct.

On engines equipped with gear-driven rotor oil pump (Fig. TE56), check drive gear and rotor for excessive wear and other damage. End clearance of rotor in pump body should be within limits of 0.006-0.007 inch (0.15-0.18 mm) and is controlled by cover gasket. Gaskets are available in a variety of thicknesses.

On all models with oil pump, be sure to prime pump during assembly to ensure immediate lubrication of engine.

**VALVE SYSTEM.** Clearance between valve tappet and valve stem (engine cold) is 0.010 inch (0.25 mm) for intake and exhaust valves for Models V50, VH50, H50, HH50, V60, VH60, H60, HH60, TVM125 and TVM140. Valve tappet clearance for all other models is 0.008 inch (0.20 mm) for intake and exhaust valves. Check clearance with piston at TDC on compression stroke. Grind valve stem end as necessary to obtain specified clearance.

Valve face angle is 45 degrees and valve seat angle is 46 degrees except on very early production 2.25 horsepower models which have 30 degree valve face and seat angles.

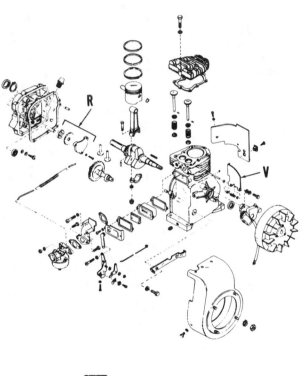

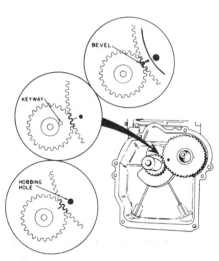

Fig. TE62—View of Light Frame engine with horizontal crankshaft. Air vane (V) type governor and rotor type oil pump (R) are used on model shown.

Fig. TE65—The camshaft and crankshaft must be correctly timed to ensure valves open at correct time. Different types of marks have been used, but marks should be aligned when assembling.

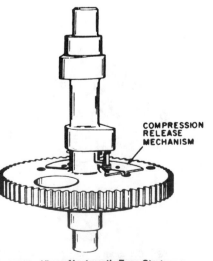

Fig. TE63—View of Instamatic Ezee-Start compression release camshaft.

Fig. TE64—View of Light Frame horizontal crankshaft engine with mechanical governor and splash lubrication. Governor centrifugal weights are shown at (C) and lubrication dipper at (D).

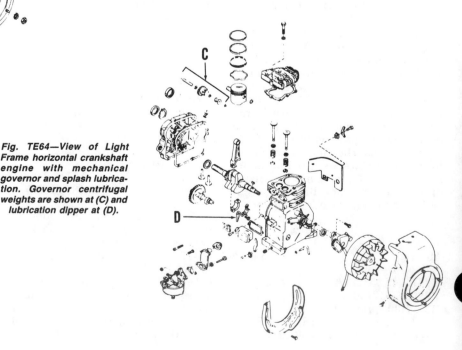

Valve seat width should be 0.042-0.052 inch (1.07-1.32 mm) for Models V40, VH40, H40, HH40, H50, HH50, V50, VH50, H60, HH60, V60, VH60, TVM125 and TVM140; and 0.035-0.045 inch (0.89-1.14 mm) for all other models.

Valve stem guides are cast into cylinder block and are not renewable. If excessive clearance exists between valve stem and guide, guide should be reamed and a new valve with an oversize stem installed.

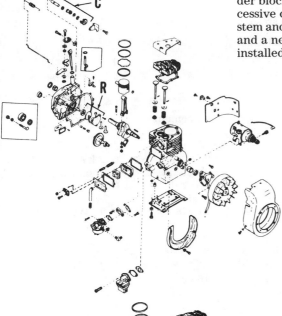

*Fig. TE66—View of Medium Frame horizontal crankshaft engine with mechanical governor (C). An oil dipper for splash lubrication is cast onto the connecting rod cap instead of using the rotor type oil pump (R).*

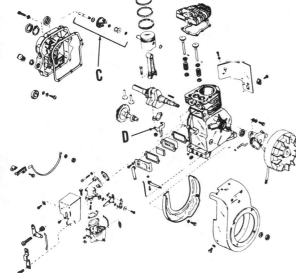

*Fig. TE67—View of Light Frame horizontal crankshaft engine with mechanical governor and splash lubrication.*

# TECUMSEH

**TECUMSEH PRODUCTS COMPANY**
**Grafton, Wisconsin 53024**

| Model | No. Cyls. | Bore | Stroke | Displacement | Power Rating |
|---|---|---|---|---|---|
| VM70* | 1 | 3.062 in. (77.8 mm) | 2.531 in. (64.3 mm) | 18.65 cu. in. (305.7 cc) | 7 hp. (5.2 kW) |
| VM70** | 1 | 2.94 in. (74.61 mm) | 2.531 in. (64.3 mm) | 17.16 cu. in. (281 cc) | 7 hp. (5.2 kW) |
| HH70 | 1 | 2.75 in. (69.9 mm) | 2.531 in. (64.3 mm) | 15 cu. in. (246.8 cc) | 7 hp. (5.2 kW) |
| VH70 | 1 | 2.75 in. (69.9 mm) | 2.531 in. (64.3 mm) | 15 cu. in. (246.8 cc) | 7 hp. (5.2 kW) |
| HM70* | 1 | 3.062 in. (77.8 mm) | 2.531 in. (64.3 mm) | 18.65 cu. in. (305.7 cc) | 7 hp. (5.2 kW) |
| HM70** | 1 | 2.94 in. (74.61 mm) | 2.531 in. (64.3 mm) | 17.16 cu. in. (281 cc) | 7 hp. (5.2 kW) |
| TVM170 | 1 | 2.94 in. (74.61 mm) | 2.531 in. (64.3 mm) | 17.16 cu. in. (281 cc) | 7 hp. (5.2 kW) |
| V80 | 1 | 3.062 in. (77.8 mm) | 2.531 in. (64.3 mm) | 18.65 cu. in. (305.7 cc) | 8 hp. (6 kW) |
| VM80*** | 1 | 3.062 in. (77.8 mm) | 2.531 in. (64.3 mm) | 18.65 cu. in. (305.7 cc) | 8 hp. (6 kW) |
| VM80**** | 1 | 3.125 in. (79.38 mm) | 2.531 in. (64.3 mm) | 19.4 cu. in. (318 cc) | 8 hp. (6 kW) |
| HM80*** | 1 | 3.062 in. (77.8 mm) | 2.531 in. (64.3 mm) | 18.65 cu. in. (305.7 cc) | 8 hp. (6 kW) |
| HM80**** | 1 | 3.125 in. (79.38 mm) | 2.531 in. (64.3 mm) | 19.4 cu. in. (318 cc) | 8 hp. (6 kW) |
| HH80† | 1 | 3.00 in. (76.2 mm) | 2.75 in. (69.9 mm) | 19.4 cu. in. (318.7 cc) | 8 hp. (6 kW) |
| HH80†† | 1 | 3.313 in. (84.15 mm) | 2.75 in. (69.9 mm) | 23.75 cu. in. (388.6 cc) | 8 hp. (6 kW) |
| VH80 | 1 | 3.313 in. (84.15 mm) | 2.75 in. (69.9 mm) | 23.75 cu. in. (388.6 cc) | 8 hp. (6 kW) |
| TVM195 | 1 | 3.125 in. (79.38 mm) | 2.531 in. (64.3 mm) | 19.4 cu. in. (318 cc) | 8 hp. (6 kW) |
| HM100††† | 1 | 3.187 in. (80.95 mm) | 2.531 in. (64.3 mm) | 20.2 cu. in. (330.9 cc) | 10 hp. (7.5 kW) |
| HM100†††† | 1 | 3.313 in. (84.15 mm) | 2.531 in. (64.3 mm) | 21.82 cu. in. (357.6 cc) | 10 hp. (7.5 kW) |
| VM100 | 1 | 3.187 in. (80.95 mm) | 2.531 in. (64.3 mm) | 20.2 cu. in. (330.9 cc) | 10 hp. (7.5 kW) |
| VH100 | 1 | 3.313 in. (84.15 mm) | 2.75 in. (69.9 mm) | 23.75 cu. in. (388.6 cc) | 10 hp. (7.5 kW) |
| HH100 | 1 | 3.313 in. (84.15 mm) | 2.75 in. (69.9 mm) | 23.75 cu. in. (388.6 cc) | 10 hp. (7.5 kW) |
| TVM220 | 1 | 3.313 in. (84.15 mm) | 2.531 in. (64.3 mm) | 21.82 cu. in. (357.6 cc) | 10 hp. (7.5 kW) |
| HH120 | 1 | 3.5 in. (88.9 mm) | 2.875 in. (72.9 mm) | 27.66 cu. in. (452.5 cc) | 12 hp. (9 kW) |

    \* VM70 or HM70 models prior to type letter A
   \*\* VM70 or HM70 models, type letter A and after
  \*\*\* VM80 or HM80 models prior to type letter E
 \*\*\*\* VM80 or HM80 models, type letter E and after
    † Prior to type letter B
   †† Type letter B and after
  ††† Early HM100 models
 †††† Late HM100 models

Medium and heavy frame, horizontal or vertical crankshaft, four cycle, one cylinder engines are covered in this section.

Care must be taken to correctly identify engine model as outlined in **TECUMSEH ENGINE IDENTIFICATION INFORMATION** section on page 441.

Connecting rod for all models rides directly on crankshaft crankpin journal. Lubrication is provided by splash lubrication for some models or a barrel and plunger type oil pump is used on some models.

A variety of ignition systems, magneto, battery or solid state, have been used. Refer to appropriate paragraph for model being serviced.

Tecumseh or Walbro float type carburetor is used according to model and application.

Refer to **TECUMSEH ENGINE IDENTIFICATION INFORMATION** section on page 441 for correct engine identification and always give complete engine model, specification and serial numbers when ordering parts or service material.

## MAINTENANCE

**SPARK PLUG.** Recommended spark plug is Champion J8 or equivalent. Electrode gap is 0.030 inch (0.762 mm) for all models.

**CARBURETOR.** Either a Tecumseh or Walbro float type carburetor is used. Refer to appropriate paragraph for model being serviced.

TECUMSEH CARBURETOR. Refer to Fig. T1 for exploded view of Tecumseh float type carburetor and location of fuel mixture adjustment screws.

For initial carburetor adjustment for VM70, VM80, VM100, HM70, HM80 and HM100 models, open idle mixture and main fuel mixture screws 1½ turns each.

For initial carburetor adjustment for VH70 model, open idle mixture screw 1 turn and main fuel mixture screw 1¼ turns.

For initial carburetor adjustment for HH80, HH100, HH120 and VH100 models, open idle mixture screw 1¼ turns and main fuel mixture screw 1¾ turns.

Make final adjustment on all models with engine at normal operating temperature and running. Place engine under load and adjust main fuel mixture screw for leanest mixture that will allow satisfactory acceleration and steady governor operation. Set engine at idle speed, no load and adjust idle mixture screw to obtain smoothest idle operation.

As each adjustment affects the other, adjustment procedure may have to be repeated.

To check float level, invert carburetor throttle body and float assembly. Float setting should be 7/32-inch (5.556 mm) measured between free end of float and rim on carburetor body.

Adjust float by carefully bending float lever tang that contacts inlet valve.

Refer to Fig. T1 for exploded view of carburetor during disassembly. When reinstalling Viton inlet valve seat, grooved side of seat must be installed in bore first so inlet valve will seat against smooth side (Fig. T2). Some later models have a Viton tipped inlet needle

(Fig. T3) and a brass seat.

Install throttle plate (2–Fig. T1) with the two stamped lines facing out and at 12 and 3 o'clock position. The 12 o'clock line should be parallel to throttle shaft and to top of carburetor.

Install choke plate (10) with flat side towards bottom of carburetor.

Fuel fitting (8) is pressed into body. When installing fuel inlet fitting, start fitting into bore; then, apply a light coat of "Loctite" 271 to shank and press fitting into place.

WALBRO CARBURETOR. Refer to Fig. T4 for exploded view of Walbro float type carburetor and location of fuel mixture adjustment screws.

For initial carburetor adjustment for VM80 and HM80 models, open idle mixture screw 1¾ turns and main fuel mixture screw 2 turns.

For initial carburetor adjustment for all remaining models, open idle mixture and main fuel mixture adjustment screws 1 turn each.

Make final adjustment on all models with engine at normal operating temperature and running. Place engine under load and adjust main fuel mixture screw for leanest mixture that will allow satisfactory acceleration and steady governor operation. Set engine at idle speed, no load and adjust idle mixture screw to obtain smoothest idle operation.

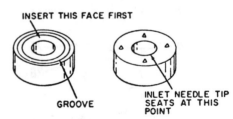

Fig. T2—Viton seat used on some Tecumseh carburetors must be installed correctly to operate properly. All metal needle is used with seat shown.

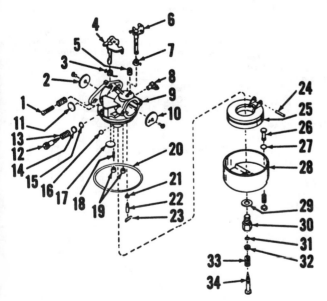

**Fig. T1—Exploded view of Tecumseh carburetor.**

1. Idle speed screw
2. Throttle plate
3. Return spring
4. Throttle shaft
5. Choke stop spring
6. Choke shaft
7. Return spring
8. Fuel inlet fitting
9. Carburetor body
10. Choke plate
11. Welch plug
12. Idle mixture needle
13. Spring
14. Washer
15. "O" ring
16. Ball plug
17. Welch plug
18. Pin
19. Cup plugs
20. Bowl gasket
21. Inlet needle seat
22. Inlet needle
23. Clip
24. Float shaft
25. Float
26. Drain stem
27. Gasket
28. Bowl
29. Gasket
30. Bowl retainer
31. "O" ring
32. Washer
33. Spring
34. Main fuel needle

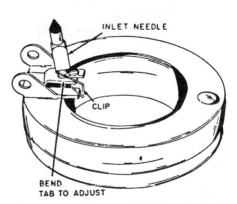

Fig. T3—View of float and fuel inlet valve needle. Valve needle shown is equipped with resilient tip and a clip. Bend tab shown to adjust float level.

As each adjustment affects the other, adjustment procedure may have to be repeated.

To check float level, invert carburetor throttle body and float assembly (Fig. T5). Float setting, dimension (H) should be 1/8-inch (3.175 mm) for horizontal crankshaft engines and 3/32-inch (2.381 mm) for vertical crankshaft engines.

Adjust float by carefully bending float lever tang that contacts inlet valve.

Float drop (travel) for all models should be 9/16-inch (14.288 mm) and is adjusted by carefully bending limiting tab on float.

**GOVERNOR.** A mechanical flyweight type governor is used on all models. Governor weight and gear assembly is driven by camshaft gear and rides on a renewable shaft which is pressed into engine crankcase or crankcase cover.

If renewal of governor shaft is necessary, press governor shaft in until shaft end is located as shown in Fig. T7, T8, T9 or T10.

To adjust governor lever position on vertical crankshaft models, refer to Fig. T11. Loosen clamp screw on governor lever. Rotate governor lever shaft counter-clockwise as far as possible. Move governor lever to left until throttle is fully open, then tighten clamp screw.

On horizontal crankshaft models, loosen clamp screw on lever, rotate governor lever shaft clockwise as far as possible. See Fig. T12. Move governor lever clockwise until throttle is wide open, then tighten clamp screw.

For external linkage adjustments, refer to Figs. T13 and T14. Loosen screw (A), turn plate (B) counter-clockwise as far as possible and move lever (C) to left until throttle is fully open. Tighten screw (A). Governor spring must be hooked in hole (D) as shown. Adjusting screws on bracket shown in Figs. T13 and T14 are used to adjust fixed or variable speed settings.

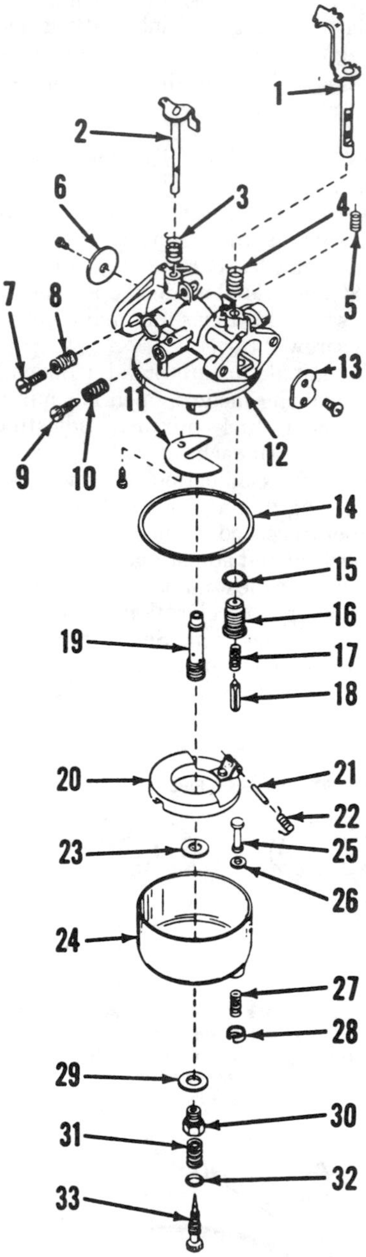

*Fig. T4 — Exploded view of Walbro carburetor.*

1. Choke shaft
2. Throttle shaft
3. Throttle return spring
4. Choke return spring
5. Choke stop spring
6. Throttle plate
7. Idle speed stop screw
8. Spring
9. Idle mixture needle
10. Spring
11. Baffle
12. Carburetor body
13. Choke plate
14. Bowl gasket
15. Gasket
16. Inlet valve seat
17. Spring
18. Inlet valve
19. Main nozzle
20. Float
21. Float shaft
22. Spring
23. Gasket
24. Bowl
25. Drain stem
26. Gasket
27. Spring
28. Retainer
29. Gasket
30. Bowl retainer
31. Spring
32. "O" ring
33. Main fuel adjusting needle

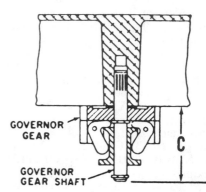

*Fig. T8 — Governor gear and shaft installation on VH80 and VH100 models. Dimension (C) is 1 inch (25.4 mm).*

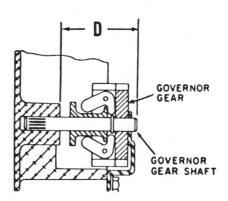

*Fig. T9 — Correct installation of governor shaft, gear and weight assembly on HH70, HM70, HM80 and HM100 models. Dimension (D) is 1-3/8 inch (34.925 mm) on HM70, HM80 and HM100 models or 1-17/64 inch (32.147 mm) on HH70 model.*

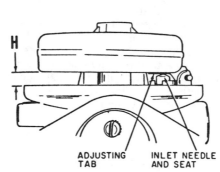

*Fig. T5 — Float height (H) should be measured as shown on Walbro float carburetors. Bend adjusting tab to adjust height.*

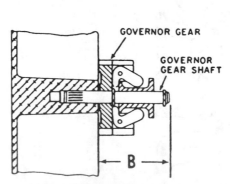

*Fig. T7 — View showing installation of governor shaft and governor gear and weight assembly on HH80, HH100 and HH120 models. Dimension (B) is 1 inch (25.4 mm).*

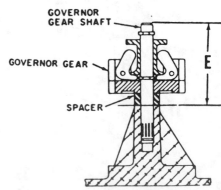

*Fig. T10 — Governor gear and shaft installation on VH70, VM70, VM80, VM100, TVM170, TVM195 and TVM220 models. Dimension (E) is 1-19/32 inch (40.481 mm).*

**IGNITION SYSTEM.** A variety of ignition systems, magneto, battery or solid state, have been used. Refer to appropriate paragraph for model being serviced.

**MAGNETO IGNITION.** Tecumseh flywheel type magnetos are used on some models. On VM70, HM70, VM80, HM80, VM100, HM100, HH70 and VH70 models, breaker points are enclosed by the flywheel. Breaker point gap must be adjusted to 0.020 inch (0.508 mm). Timing is correct when timing mark on stator plate is in line with mark on bearing plate as shown in Fig. T15. If timing marks are defaced, points should start to open when piston is 0.085-0.095 inch (2.159-2.413 mm) BTDC.

Breaker points on HH80, VH80, HH100, VH100 and HH120 models are located in crankcase cover as shown in Fig. T16. Timing should be correct when points are adjusted to 0.020 inch (0.508 mm). To check timing with a continuity light, refer to Fig. T17. Remove "pop" rivets securing identification plates to blower housing. Remove plate to expose a 1¼ inch hole. Connect continuity light to terminal screw (78–Fig. T16) and suitable engine ground. Rotate engine clockwise until piston is on compression stroke and timing mark is just below stator laminations as shown in Fig. T17. At this time, points should be ready to open and continuity light should be on.

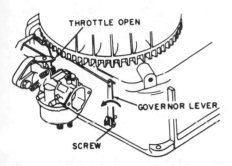

Fig. T11—*When adjusting governor linkage on VH70, VM70, VM80, VM100, TVM170, TVM195 or TVM220 models, loosen clamp screw and rotate governor lever shaft and lever counter-clockwise as far as possible.*

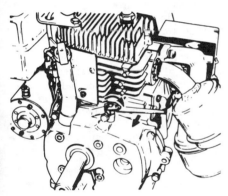

Fig. T12—*On HH70, HM70, HM80 and HM100 models, rotate governor lever shaft and lever clockwise when adjusting linkage.*

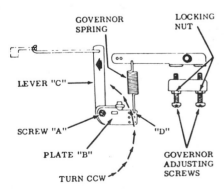

Fig. T13—*External governor linkage on VH80 and VH100 models. Refer to text for adjustment procedure.*

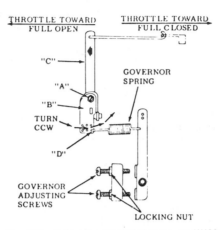

Fig. T14—*External governor linkage on HH80, HH100 and HH120 models. Refer to text for adjustment procedure.*

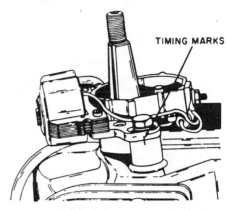

Fig. T15—*On VM70, VH70, HM70, HH70, VM80, HM80, VM100 and HM100 equipped with magneto ignition, adjust breaker point gap to 0.020 inch (0.508 mm) and align timing marks as shown.*

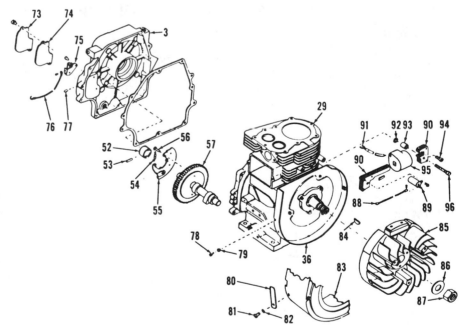

Fig. T16—*Exploded view of magneto ignition components used on HH80, HH100 and HH120 models. Timing advance and breaker points used on engines equipped with battery ignition are identical.*

|  |  |  |  |
|---|---|---|---|
| 3. Crankcase cover | 57. Camshaft assy. | 80. Ground switch | 88. Condenser wire |
| 29. Cylinder block | 73. Breaker box cover | 81. Screw | 89. Condenser |
| 36. Blower air baffle | 74. Gasket | 82. Washer | 90. Armature core |
| 52. Breaker cam | 75. Breaker points | 83. Blower housing | 91. High tension lead |
| 53. Push rod | 76. Ignition wire | 84. Flywheel key | 92. Washer |
| 54. Spring | 77. Pin | 85. Flywheel | 93. Spacer |
| 55. Timing advance weight | 78. Screw | 86. Washer | 94. Screw |
| 56. Rivet | 79. Clip | 87. Nut | 95. Coil |
|  |  |  | 96. Screw |

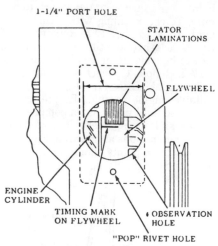

*Fig. T17—On HH80, HH100 and HH120 models, remove identification plate to observe timing mark on flywheel through the 1¼-inch hole in blower housing.*

*Fig. T19—View of solid state ignition system used on some models not equipped with flywheel alternator.*

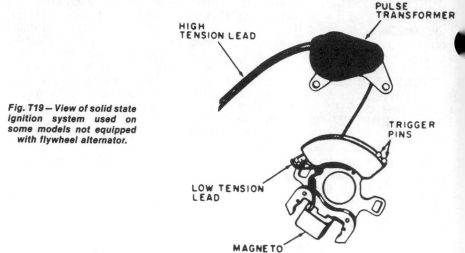

Rotate flywheel until mark just passes under edge of laminations. Points should open and light should be out. If not, adjust points slightly until light goes out. Points are actuated by push rod (53–Fig. T16) which rides against breaker cam (52). Breaker cam is driven by a tang on advance weight (55). When cranking, spring (54) holds advance weight in retarded position (TDC). At operating speeds, centrifugal force overcomes spring pressure and weight moves cam to advance ignition so spark occurs when piston is at 0.095 inch (2.413 mm) BTDC.

An air gap of 0.006-0.010 inch (0.1524-0.254 mm) should be between flywheel and stator laminations. To adjust gap, turn flywheel magnet into position under coil core. Loosen holding

screws and place shim stock or feeler gage between coil and magnet. Press coil against gage and tighten screws.

**BATTERY IGNITION.** Models HH80, HH100 and HH120 may be equipped with a battery ignition system. Coil and condenser are externally mounted while points are located in crankcase cover. See Fig. T18. Points should be adjusted to 0.020 inch (0.508 mm). To check timing, disconnect primary wire between coil and points and follow the same procedure as described in **MAGNETO IGNITION** section.

**SOLID STATE IGNITION (WITHOUT ALTERNATOR).** Tecumseh solid state ignition system shown in Fig. T19

may be used on some models not equipped with flywheel alternator. This system does not use ignition breaker points. The only moving part of system is rotating flywheel with charging magnets. As flywheel magnet passes position (1A–Fig. T20), a low voltage AC current is induced into input coil (2). Current passes through rectifier (3) converting this current to DC. It then travels to capacitor (4) where it is stored. Flywheel rotates approximately 180° to position (1B). As it passes trigger coil (5), it induces a very small electric charge into coil. This charge passes through resistor (6) and turns on SCR (silicon controlled rectifier) switch (7). With SCR switch closed, low voltage current stored in capacitor (4) travels to pulse transformer (8). Voltage is step-

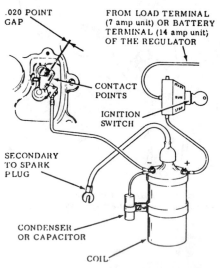

*Fig. T18—Typical battery ignition wiring diagram used on some HH80, HH100 and HH120 engines.*

*Fig. T20—Diagram of solid state ignition system used on some models.*

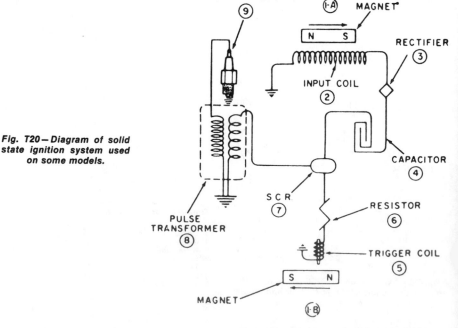

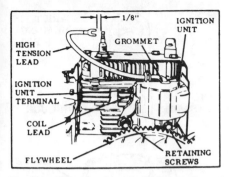

*Fig. T21 — View of solid state ignition unit used on some models equipped with flywheel alternator. System should produce a good blue spark 1/8-inch (3.175 mm) long at cranking speed.*

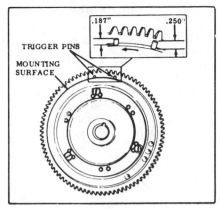

*Fig. T23 — Remove flywheel and drive trigger pins in or out as necessary until long pin is extended 0.250 inch (6.35 mm) and short pin is extended 0.187 inch (4.763 mm) above mounting surface.*

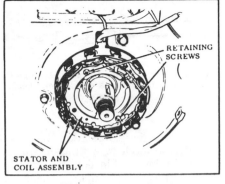

*Fig. T25 — Ignition generator coil and stator serviced only as an assembly.*

ped up instantaneously and current is discharged across electrodes of spark plug (9), producing a spark before TDC.

Some units are equipped with a second trigger coil and resistor set to turn SCR switch on at a lower rpm. This second trigger pin is closer to flywheel and produces a spark at TDC for easier starting. As engine rpm increases, first (shorter) trigger pin picks up small electric charge and turns SCR switch on, firing spark plug BTDC.

If system fails to produce a spark to spark plug, first check high tension lead (Fig. T19). If condition of high tension lead is questionable, renew pulse transformer and high tension lead assembly. Check low tension lead and renew if insulation is faulty. Magneto charging coil, electronic triggering system and mounting plate are available only as an assembly. If necessary to renew this assembly, place unit in position on engine. Start retaining screws, turn mounting plate counter-clockwise as far as possible, then tighten retaining screw to a torque of 5-7 ft.-lbs. (7-10 N·m).

SOLID STATE IGNITION (WITH ALTERNATOR). Tecumseh solid state ignition system used on some models equipped with flywheel alternator does not use ignition breaker points. The only

moving part of system is rotating flywheel with charging magnets and trigger pins. Other components of system are ignition generator coil and stator assembly, spark plug and ignition unit.

The long trigger pin induces a small charge of current to close SCR (silicon controlled rectifier) switch at engine cranking speed and produces a spark at TDC for starting. As engine rpm increases, first (shorter) trigger pin induces current which produces a spark when piston is 0.095 inch (2.413 mm) BTDC.

Test ignition system by holding high tension lead 1/8-inch (3.175 mm) from spark plug (Fig. T21), crank engine and check for a good blue spark. If no spark is present, check high tension lead and coil lead for loose connections or faulty insulation. Check air gap between trigger pin and ignition unit as shown in Fig. T22. Air gap should be 0.006-0.010 inch (0.1524-0.254 mm). To adjust air gap, loosen the two retaining screws and move ignition unit as necessary, then tighten retaining screws.

NOTE: Long trigger pin should extend 0.250 inch (6.35 mm) and short trigger pin should extend 0.187 inch (4.763 mm), measured as shown in Fig. T23. If not, remove flywheel and drive pins in or out as required.

Remove coil lead from ignition terminal and connect an ohmmeter as shown in Fig. T24. If series resistance test of ignition generator coil is below 400 ohms, renew stator and coil assembly (Fig. T25). If resistance is above 400 ohms, renew ignition unit.

LUBRICATION. On VH70, VM70, VM100, TVM170, TVM195 and TVM220 models, a barrel and plunger type oil pump (Fig. T26 or T27) driven by an eccentric on camshaft, pressure lubricates upper main bearing and connecting rod journal. When installing early type pump (Fig. T26), chamfered side of drive collar must be against thrust bearing surface on camshaft gear. When installing late type pump, place side of drive collar with large flat surface shown in Fig. T27 away from camshaft gear.

An oil slinger (59 – Fig. T28), installed on crankshaft between gear and lower bearing is used to direct oil upward for

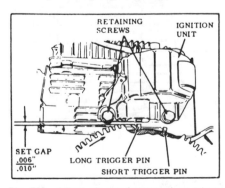

*Fig. T22 — Adjust air gap between long trigger pin and ignition unit to 0.006-0.010 inch (0.1524-0.254 mm).*

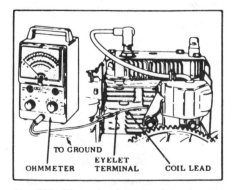

*Fig. T24 — View showing an ohmmeter connected for resistance test of ignition generator coil.*

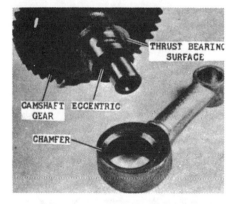

*Fig. T26 — View of early type oil pump used on VH70, VM70 and VM80 models. Chamfered face of drive collar should be toward camshaft gear.*

Fig. T27—Install late type oil pump so large flat surface on drive collar is away from camshaft gear.

complete engine lubrication on VH80 and VH100 models. A tang on slinger hub, when inserted in slot in crankshaft gear, correctly positions slinger on crankshaft as shown in Fig. T28.

Splash lubrication system on all other models is provided by use of an oil dipper on connecting rod. See Figs. T30 and T31.

Oils approved by manufacturer must meet requirements of API service classification SE or SF.

Use an SAE 30 oil for temperatures above 32° F (0° C) and SAE 5W20 or 5W30 oil for temperatures below 32° F (0° C).

Check oil level at 5 hour intervals or before initial start-up of engine. Maintain oil level at edge of filler hole or at "FULL" mark on dipstick, as equipped.

Recommended oil change interval for all models is every 25 hours of normal operation.

Fig. T28—Oil slinger (59) on VH80 and VH100 models must be installed on crankshaft as shown.

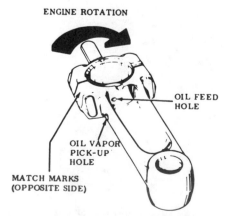

Fig. T29—Connecting rods used on VH80 and VH100 have two oil holes.

**REPAIRS**

**TIGHTENING TORQUE.** Recommended tightening torque specifications are as follows:

**Cylinder head—**
 VM70, HM70, VM80, HM80, VM100, HM100, HH70, VH70, TVM170, TVM195, TVM220 ...... 170 in.-lbs. (19 N·m)
 HH80, VH80, HH100, VH100, HH120 ......... 200 in.-lbs. (23 N·m)

**Connecting rod—**
 VM70, HM70, VM80, HM80, VM100, HM100, HH70, VH70, TVM170, TVM195, TVM220 ...... 120 in.-lbs. (13 N·m)
 HH80, VH80, HH100, VH100, HH120 ......... 110 in.-lbs. (12 N·m)

**Crankcase cover—**
 All models ........... 65-110 in.-lbs. (5-12 N·m)

**Bearing retainer—**
 HH80, VH80, HH100, VH100, HH120 ....... 65-110 in.-lbs. (5-12 N·m)

**Ball bearing retainer nut—**
 All models so equipped ........... 15-22 in.-lbs. (1-3 N·m)

**Flywheel nut—**
 Light frame ........ 400-440 in.-lbs. (45-50 N·m)
 Medium frame ...... 430-500 in.-lbs. (49-57 N·m)
 Heavy frame ....... 600-660 in.-lbs. (68-75 N·m)

**CYLINDER HEAD.** When removing cylinder head, note location of different length cap screws for aid in correct reassembly. Always install new head gasket and tighten cap screws evenly in sequence shown in Figs. T32, T33, T34

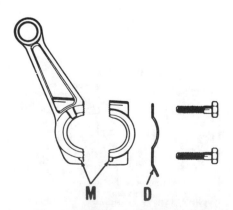

Fig. T30—Connecting rod assemble used on VH70, VM70, VM80, VM100, HH70, HM70, HM80, HM100, TVM170, TVM195 and TVM220 models. Note position oil dipper (D) and match marks (M).

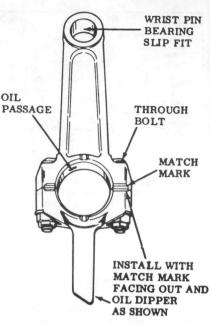

Fig. T31—Connecting rod assembly used on HH80, HH100 and HH120 models.

or T35. Refer to **TIGHTENING TORQUE** section for correct torque specifications for model being serviced.

**CONNECTING ROD.** Piston and connecting rod assembly is removed from cylinder head end of engine. Connecting rod rides directly on crankpin journal of crankshaft.

Standard crankpin journal diameter is 1.1865-1.1870 inches (30.14-30.15 mm) for VM70, HM70, VM80, HM80, VM100, HM100, HH70, VH70, TVM170, TVM195 and TVM220 models and 1.3750-1.3755 inches (34.93-34.94 mm) for all remaining models.

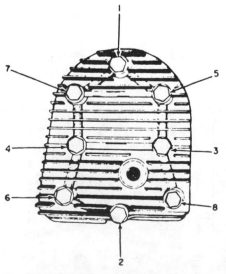

Fig. T32—On VM70, HM70, VH70 and HH70 models, tighten cylinder head cap screws evenly to a torque of 170 in.-lbs. (19 N·m) using tightening sequence shown.

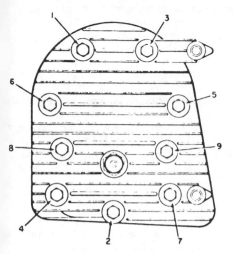

*Fig. T33 — Tighten cylinder head cap screws on HM80, VM80, HM100, VM100, TVM170, TVM195 and TVM220 models in sequence shown to 170 in.-lbs. (19N·m) torque.*

Connecting rod to crankpin journal running clearance should be 0.002 inch (0.0508 mm) for all models.

Connecting rods are equipped with match marks which must be aligned and face pto end of crankshaft after installation. See Figs. T29, T30 or T31.

Connecting rod is available in a variety of sizes for undersize crankshafts, as well as standard.

**PISTON, PIN AND RINGS.** Aluminum alloy piston is fitted with two compression rings and one oil control ring.

Piston skirt diameter and piston skirt to cylinder wall clearance is measured at bottom edge of skirt at a right angle to piston pin for all models.

Piston skirt diameter for HH70 and VH70 models is 2.7450-2.7455 inches (69.72-69.74 mm) and skirt to cylinder bore clearance is 0.0045-0.0060 inch (0.1143-0.1524 mm).

Piston skirt diameter for HH80 model prior to type letter B, VM70 and HM70 models with type letter A and after and TVM170 models is 2.9325-2.9335 inches (74.49-74.51 mm) and skirt to cylinder bore clearance is 0.004-0.006 inch (0.1016-0.1524 mm).

Piston skirt diameter for VM70 and HM70 models prior to type letter A, V80 model, VM80 and HM80 models prior to type letter E is 3.0575-3.0585 inches (77.66-77.69 mm) and skirt to cylinder bore clearance is 0.0035-0.0055 inch (0.0889-0.1397 mm).

Piston skirt diameter for VM80 and HM80 models with type letter E or after and TVM195 models is 3.1195-3.1205 inches (79.24-79.26 mm) and skirt to cylinder bore clearance is 0.004-0.006 inch (0.1016-0.1524 mm).

Piston skirt diameter for HM100 (early production) and VM100 models is 3.1817-3.1842 inches (80.815-80.823 mm) and skirt to cylinder bore clearance is 0.0028-0.0063 inch (0.0711-0.1600 mm).

Piston skirt diameter for HH80 models with type letter B and after, VH80, VH100, HM100 (late production), HH100 and TVM220 models is 3.308-3.310 inches (84.02-84.07 mm) and skirt to cylinder bore clearance is 0.002-0.005 inch (0.0508-0.1270 mm).

Piston skirt diameter for HH120 model is 3.4950-3.4970 inches (88.77-88.82 mm) and skirt to cylinder bore clearance is 0.003-0.006 inch (0.0762-0.1524 mm).

Side clearance of top ring in piston groove for VM70 (prior to type letter A), V80 model and VM80 model (prior to

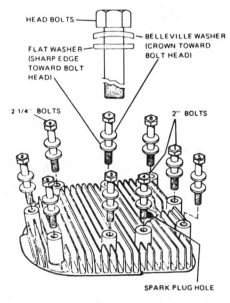

*Fig. T35 — Flat washers and Belleville Washers are used on cylinder head cap screws on late HH80, HH100, HH120 and all VH80 and VH100 engines. Tighten cap screws in sequence shown to a torque of 200 in.-lbs. (23 N·m).*

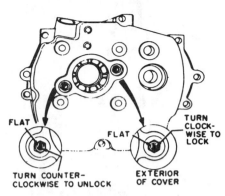

*Fig. T36 — View showing bearing locks on HM70, HH70, HM80 and HM100 models equipped with ball bearing main bearings. Locks must be released before removing crankcase cover. Refer to Fig. T37 for interior view of cover and locks.*

type letter E) is 0.003-0.004 inch (0.0762-0.1016 mm), for HM and TVM170 models is 0.0028-0.0051 inch (0.0711-0.1296 mm), for VM80 (type letter E and after), HM80, TVM195, HM100, VM100, TVM220, H80, VH100, HH100 and VH80 models is 0.002-0.005 inch (0.0508 0.1270 mm).

Standard piston pin diameter is 0.625-0.6254 inch (15.88-15.89 mm) for VM70, HM70, VM80, HM80, HH70, VH70, TVM170 and TVM195 models and 0.6873-0.6875 inch (17.457-17.462 mm) for all other models.

Piston pin clearance should be 0.0001-0.0008 inch (0.0025-0.0203 mm) in connecting rod and 0.0002-0.0005 inch (0.0051-0.0127 mm) in piston. If excessive clearance exists, both piston and pin must be renewed as pin is not available separately.

Pistons should be assembled on connecting rod so arrow on top of piston is pointing toward carburetor side of engine after installation.

Pistons and rings are available in a variety of oversizes as well as standard.

**CYLINDER.** If cylinder is scored or excessively worn, or if taper or out-of-round exceeds 0.004 inch (0.1016 mm), cylinder should be rebored to nearest

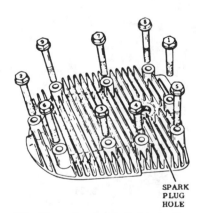

*Fig. T34 — View showing cylinder head cap screw tightening sequence used on early HH80, HH100 and HH120 engines. Tighten cap screw to 200 in.-lbs. (23 N·m) torque. Note type and length of cap screws.*

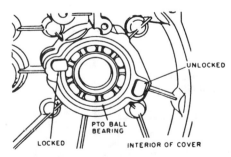

*Fig. T37 — Interior view of crankcase cover and ball bearing locks used on HM70, HH70, HM80 and HM100 models.*

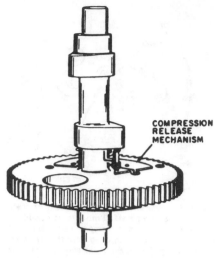

Fig. T38—View of Insta-matic Ezee-Start compression release camshaft assembly used on all models except HH80, HH100 and HH120.

oversize for which piston and rings are available.

Standard cylinder bore for HH70 and VH70 models is 2.750-2.751 inches (69.850-69.854 mm).

Standard cylinder bore for HH80 (prior to type letter B), VM70 and HH70 models with type letter A and after and TVM170 models is 2.9375-2.9385 inches (74.61-74.64 mm).

Standard cylinder bore for VM70 and HM70 models prior to type letter A, VM80 and HM80 models prior to type letter E and V80 models is 3.062-3.063 inches (77.78-77.80 mm).

Standard cylinder bore for VM80 and HM80 models with type letter E or after and TVM195 models is 3.125-3.126 inches (79.38-79.40 mm).

Standard cylinder bore for HM100 (early production) and VM100 models is 3.187-3.188 inches (80.95-80.98 mm).

Standard cylinder bore for HH80 (type letter B and after), VH80, HM100 (late production), VH100, HH100 and TVM220 models is 3.312-3.313 inches (84.13-84.15 mm).

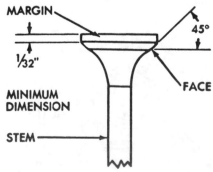

Fig. T39—Valve face angle should be 45°. Minimum valve head margin is 1/32-inch (0.794 mm).

Standard cylinder bore for HH120 model is 3.500-3.501 inches (88.90-88.93 mm).

**CRANKSHAFT AND MAIN BEARINGS.** Crankshaft main journals ride directly in aluminum alloy bearings in crankcase and mounting flange (engine base) on vertical crankshaft engines or in two renewable steel backed bronze bushings. On some horizontal crankshaft engines, crankshaft rides in a renewable sleeve bushing at flywheel end and a ball bearing or bushing at pto end. Models HH80, VH80, HH100, VH100 and HH120 are equipped with taper roller bearings at both ends of crankshafts.

Standard main bearing bore diameter for bushing type main bearings for VM70, VH70 and HH70 models should be 1.0005-1.0010 inches (25.41-25.43 mm) for either bearing.

Standard main bearing bore diameter for bushing type main bearings for all remaining models should be 1.0005-1.0010 inches (25.41-25.43 mm) for main bearing in cylinder block section and 1.1890-1.1895 inches (30.20-30.21 mm) for main bearing in cover section.

Normal running clearance of crankshaft journals in aluminum bearings or bronze bushings is 0.0015-0.0025 inch (0.038-0.064 mm). Renew crankshaft if main journals are more than 0.001 inch (0.025 mm) out-of-round or renew or regrind crankshaft if connecting rod journal is more than 0.0005 inch (0.0127 mm) out-of-round.

Check crankshaft gear for wear, broken teeth or loose fit on crankshaft. If gear is damaged, remove from crankshaft with an arbor press. Renew gear pin and press new gear on shaft making certain timing mark is facing pto end of shaft.

On models equipped with ball bearing at pto end of shaft, refer to Figs. T36 and T37 before attempting to remove crankcase cover. Loosen locknuts and rotate protruding ends of lock pins counter-clockwise to release bearing and remove cover. Ball bearing will remain on crankshaft. When reassembling, turn lock pins clockwise until flats on pins face each other, then tighten locknuts to 15-22 in.-lbs. (2-3 N·m) torque.

Crankshaft end play for all models except those with taper roller bearing main bearings is 0.005-0.027 inch (0.1270-0.6858 mm) and is controlled by varying thickness of thrust washers between crankshaft and cylinder block or crankcase cover.

Crankshaft main bearing preload for all models with taper roller bearing main bearings is 0.001-0.007 inch (0.0254-0.1778 mm) and is controlled by varying thickness of shim gasket (32—Fig. T42) on all models.

Taper roller bearings are a press fit on crankshaft and must be renewed if removed. Heat bearings in hot oil to aid installation.

Fig. T40—Exploded view of typical vertical crankshaft engine. Renewable bushings (13 and 36) are not used on VM70, VM80 and VM100 models.

1. Cylinder head
2. Head gasket
3. Exhaust valve
4. Intake valve
5. Pin
6. Spring cap
7. Valve spring
8. Spring cap
9. Cylinder block
10. Magneto
11. Flywheel
12. Oil seal
13. Crankshaft bushing
14. Breather assy.
15. Carburetor
16. Intake pipe
17. Top compression ring
18. Second compression ring
19. Oil ring expander
20. Oil control ring
21. Piston pin
22. Piston
23. Retaining ring
24. Connecting rod
25. Thrust washer
26. Crankshaft
27. Thrust washer
28. Rod cap
29. Rod bolt lock
30. Camshaft assy.
31. Valve lifters
32. Oil pump
33. Gasket
34. Mounting flange (engine base)
35. Oil screen
36. Crankshaft bushing
37. Oil seal
38. Spacer
39. Governor shaft
40. Governor gear assy.

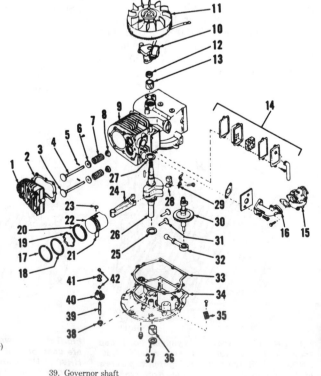

Illustrations courtesy of Tecumseh Products Company

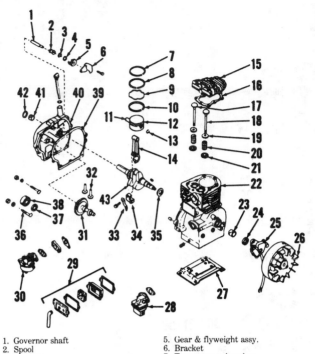

1. Governor shaft
2. Spool
3. Washer
4. Retaining ring

5. Gear & flyweight assy.
6. Bracket
7. Top compression ring
8. Second compression ring

*Fig. T41—Exploded view of typical horizontal crankshaft engine. Engines may be equipped with crankshaft bushing (41) or ball bearing (38) at pto end of shaft.*

9. Oil ring expander
10. Oil control ring
11. Piston pin
12. Piston
13. Retaining ring
14. Connecting rod
15. Cylinder head
16. Head gasket
17. Exhaust valve
18. Intake valve
19. Spring cap
20. Valve spring
21. Spring retainer
22. Cylinder block
23. Crankshaft bushing
24. Oil seal
25. Magneto
26. Flywheel
27. Mounting plate
28. Fuel pump
29. Breather assy.
30. Carburetor
31. Camshaft assy.
32. Valve lifter
33. Rod bolt lock
34. Rod cap
35. Thrust washer
36. Bearing lock pin
37. Thrust washer
38. Ball bearing
39. Gasket
40. Crankcase cover
41. Bushing
42. Oil seal
43. Crankshaft

HH80, VH80, HH100, VH100, HH120, (Both journals) . . 1.1865-1.1870 inches (30.14-30.15 mm)
VM80, HM80, HM100, VM100, TVM170, TVM195, TVM220
Flywheel end . . . . . 0.9985-0.9990 inch (25.36-25.38 mm)
Pto end . . . . . . . . 1.1870-1.1875 inches (30.15-30.16 mm)

**Crankpin journal diameter—**
HH80, VH80
HH100, HH120 . 1.3750-1.3755 inches (34.93-34.94 mm)

All other models . . . . . . . . 1.1865-1.1870 inches (30.14-30.15 mm)

Connecting rods are available in a variety of sizes for reground crankshaft crankpin journals.

When reinstalling crankshaft, make certain timing marks on camshaft and crankshaft are properly aligned.

**CAMSHAFT.** Camshaft and camshaft gear are an integral part which rides on journals at each end of camshaft. Camshaft journal diameter is 0.6235-0.6240 inch (15.84-15.85 mm) and journal to bearing running clearance should be 0.003 inch (0.0762 mm).

Renew camshaft if gear teeth are worn or if journal or lobe surfaces are worn or damaged.

Camshaft equipped with compression release mechanism (Fig. T38) is easier to remove and install if crankshaft is rotated three teeth past aligned position which allows compression release mechanism to clear exhaust valve lifter. Compression release mechanism parts should work freely with no binding or sticking. Parts are not serviced separate from camshaft assembly.

Some models are equipped with an automatic timing advance mechanism (52 through 56—Fig. T43) which must work freely with no binding or sticking. Renew damaged parts as necessary.

Make certain timing marks on camshaft gear and crankshaft gear are aligned after installation.

**VALVE SYSTEM.** Valve seats are machined directly into cylinder block assembly. Seats are ground at a 45° angle and should not exceed 3/64-inch (1.191 mm) in width.

Valve face is ground at a 45° angle and margin should not be less than 1/32-inch (0.794 mm). See Fig. T39. Valves are available with 1/32-inch oversize stem for use with oversize valve guide bore.

Valve guides are not renewable. If guides are excessively worn, ream to 0.3432-0.3442 inch (8.72-8.74 mm) and install valve with 1/32-inch (0.794 mm) oversize valve stem. Drill upper and

Bronze or aluminum bushing type main bearings are renewable and finish reamers are available from Tecumseh.

Standard crankshaft journal diameters are as follows:

**Main journal diameter—**
VM70, HM70, HH70, VH70, (Both journals) . . . . 0.9985-0.9990 inch (25.36-25.38 mm)

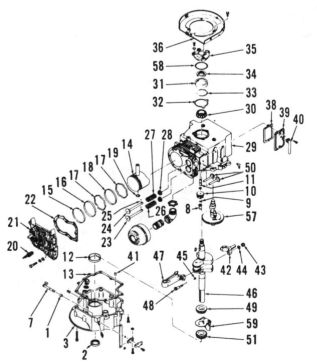

*Fig. T42—Exploded view of VH80 or VH100 model vertical crankshaft engine.*

12. Bearing cup
13. Gasket
14. Piston & pin assy.
15. Top compression ring
16. Second compression ring
17. Ring expander
18. Oil control ring
19. Retaining ring
20. Spark plug
21. Cylinder head
22. Head gasket
23. Exhaust valve
24. Intake valve
25. Pin
26. Exhaust valve spring
27. Intake valve spring
28. Spring cap
29. Cylinder block
30. Bearing cone
31. Bearing cup
32. Shim gasket
33. Steel washer (0.010 inch)
34. Oil seal
35. Bearing retainer cap
36. Blower air baffle
38. Gasket
39. Breather
40. Breather tube
42. Rod cap
43. Self locking nut
44. Washer
45. Crankshaft gear pin
46. Crankshaft
47. Connecting rod
48. Rod bolt
49. Crankshaft gear
50. Valve lifters
51. Bearing cone
57. Camshaft assy.
58. "O" ring
59. Oil slinger

1. Governor arm bushing
2. Oil seal
3. Mounting flange (engine base)
7. Governor arm

8. Thrust spool
9. Snap ring
10. Governor gear & weight assy.
11. Governor shaft

**467**

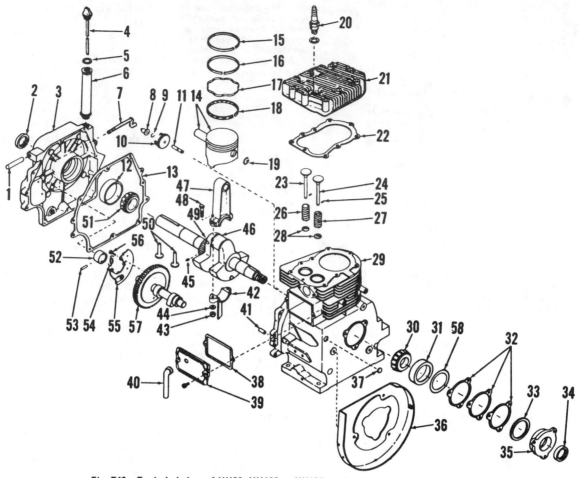

**Fig. T43 — Exploded view of HH80, HH100 or HH120 model horizontal crankshaft engine.**

1. Governor arm bushing
2. Oil Seal
3. Crankcase cover
4. Dipstick
5. Gasket
6. Oil filler tube
7. Governor arm
8. Thrust spool
9. Snap ring
10. Governor gear & weight assy.
11. Governor shaft
12. Bearing cup
13. Gasket
14. Piston & pin assy.
15. Top compression ring
16. Second compression ring
17. Oil ring expander
18. Oil control ring
19. Retaining ring
20. Spark plug
21. Cylinder head
22. Head gasket
23. Exhaust valve
24. Intake valve
25. Pin
26. Exhaust valve spring
27. Intake valve spring
28. Spring cap
29. Cylinder block
30. Bearing cone
31. Bearing cup
32. Shim gaskets
33. Steel washer (0.010 inch)
34. Oil seal
35. Bearing retainer cap
36. Blower air baffle
37. Plug
38. Gasket
39. Breather assy.
40. Breather tube
41. Dowel pin
42. Rod cap
43. Self locking nut
44. Washer
45. Crankshaft gear pin
46. Crankshaft
47. Connecting rod
48. Rod bolt
49. Crankshaft gear
50. Valve lifters
51. Bearing cone
52. Breaker cam
53. Push rod
54. Spring
55. Timing advance weight
56. Rivet
57. Camshaft assy.
58. "O" ring

lower valve spring caps as necessary for oversize valve stem.

Valve lifters should be identified before removal so they may be reinstalled in their original positions. Some models use lifters of different length. Short lifter is installed at intake position and longer lifter is installed at exhaust position.

Valve tappet gap (cold) is measured with piston at TDC on compression stroke. Correct clearance for all models is 0.010 inch (0.254 mm).

Adjust clearance by squarely grinding valve stem to increase clearance or grinding seat deeper to decrease clearance.

**DYNA-STATIC BALANCER.** Dyna-Static engine balancer used on some models consists of counterweighted

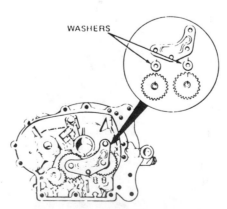

**Fig. T44 — View showing Dyna-Static balancer gears installed in models so equipped. Note location of washers between gears retaining bracket.**

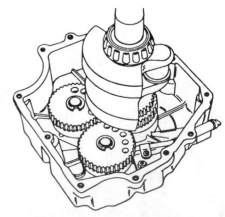

**Fig. T45 — View showing Dyna-Static balancer gears installed in HH80, HH100 or HH120 models. Note gear retaining snap rings. Refer to Fig. T44 also.**

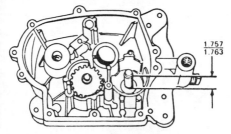

Fig. T46—On HM80, VM80, HM100 and VM100 models, balancer gear shafts must be pressed into cover or engine base so a distance of 1.757-1.763 inches (44.52-44.78 mm) exists between shaft bore boss and edge of step cut as shown.

MEASURE FROM COVER BOSS TO RING GROOVE OUTER EDGE

Fig. T47—On HH80, HH100 and HH120 models, press balancer gear shafts into cover until 1.7135-1.7185 inches (43.52-43.64 mm) exists between cover boss and outer edge of snap ring groove as shown.

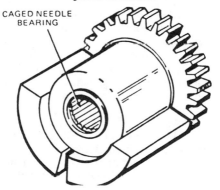

CAGED NEEDLE BEARING

Fig. T48—Using tool number 670210, press new needle bearing into HM80, VM80, HM100 or VM100 models balancer gears until bearing cage is flush to 0.015 inch (0.381 mm) below edge of bore.

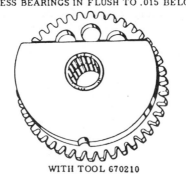

PRESS BEARINGS IN FLUSH TO .015 BELOW

WITH TOOL 670210

Fig. T49—Using tool number 670210, press new needle bearings into HH80, HH100 and HH120 models balancer gears until bearing cage is flush to 0.015 inch (0.381 mm) below edge of bore.

Fig. T50—To time engine balancer gears, remove pipe plugs and insert alignment tool number 670240 through crankcase cover (HM80 and HM100) or engine base (VM80 and VM100) and into slots in balancer gears. Refer also to Fig. T52.

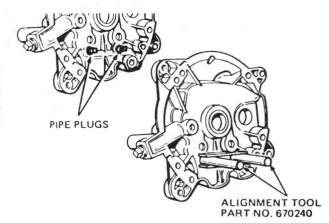

PIPE PLUGS

ALIGNMENT TOOL PART NO. 670240

gears driven by crankshaft gear to counteract unbalance caused by counterweights on crankshaft.

Counterweight gears on medium frame models are held in position on shafts by a bracket bolted to crankcase or engine base (Fig. T44). Snap rings are used on heavy frame models to retain counterweight gears on shafts which are pressed into crankcase cover (Fig. T47).

Renewable balancer gear shafts are pressed into crankcase cover or engine base. On medium frame models, press shafts into cover or engine base until a distance of 1.757-1.763 inches (44.52-44.78 mm) exists between shaft bore boss and edge of step cut on shafts as shown in Fig. T46. Heavy frame model shafts should be pressed into cover until a distance of 1.7135-1.7185 inches (43.52-43.64 mm) exists between cover boss and outer edge of snap ring groove as shown in Fig. T47.

All balancer gears are equipped with renewable cage needle bearings. Using tool number 670210, press new bearings into gears until cage is flush to 0.015 inch (0.381 mm) below edge of bore.

When reassembling engine, balancer gears must be timed with crankshaft for correct operation. Refer to Figs. T50

and T51 and remove pipe plugs. Insert alignment tool number 670240 through crankcase cover or engine base of medium frame models and into timing slots in balancer gears. On heavy frame models use timing tool number 670239. On all models, rotate engine until piston is at TDC on compression stroke and install cover and gear assembly while tools retain gears in correct position. See Figs. T52 and T53.

When correctly assembled, piston should be on TDC and weights on balancer gears should be in directly opposite position. See Figs. T52 and T53.

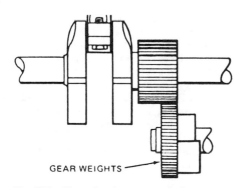

GEAR WEIGHTS

Fig. T52—View showing correct balancer gear timing to crankshaft gear on HM80, VM80, HM100 and VM100 models. With piston at TDC, weights should be directly opposite.

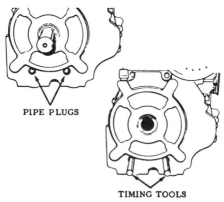

PIPE PLUGS

TIMING TOOLS

Fig. T51—To time balancer gears on HH80, HH100 and HH120 models, remove pipe plugs and insert timing tools number 67039 through crankcase cover and into timing slots in balancer gears. Refer also to Fig. T53.

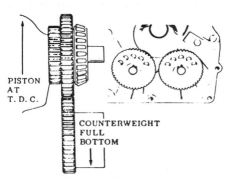

PISTON AT T.D.C.

COUNTERWEIGHT FULL BOTTOM

Fig. T53—On HH80, HH100 and HH120 models, balancer gears are correctly timed to crankshaft when piston is at TDC and weights are at full bottom position.

# TECUMSEH

## TECUMSEH PRODUCTS COMPANY
### Grafton, Wisconsin 53024

| Model | No. Cyls. | Bore | Stroke | Displacement | Power Rating |
|-------|-----------|------|--------|--------------|--------------|
| OH120 | 1 | 3.125 in. (79.375 mm) | 2.75 in. (69.85 mm) | 21.1 cu. in. (155.8 cc) | 12 hp. (8.9 kW) |
| OH140 | 1 | 3.312 in. (84.125 mm) | 2.75 in. (69.85 mm) | 23.7 cu. in. (175 cc) | 14 hp. (10.4 kW) |
| OH150 | 1 | 3.500 in. (88.90 mm) | 2.875 in. (72.625 mm) | 27.66 cu. in. (203.2 cc) | 15 hp. (11.2 kW) |
| OH160 | 1 | 3.500 in. (88.90 mm) | 2.875 in. (72.625 mm) | 27.66 cu. in. (203.2 cc) | 16 hp. (11.9 kW) |
| OH180 | 1 | 3.625 in. (92.075 mm) | 2.875 in. (72.625 mm) | 30 cu. in. (218 cc) | 18 hp. (13.4 kW) |
| HH140 | 1 | 3.312 in. (84.125 mm) | 2.75 in. (69.85 mm) | 23.7 cu. in. (175 cc) | 14 hp. (10.4 kW) |
| HH150 | 1 | 3.500 in. (88.90 mm) | 2.875 in. (72.625 mm) | 27.66 cu. in. (203.2 cc) | 15 hp. (11.2 kW) |
| HH160 | 1 | 3.500 in. (88.90 mm) | 2.875 in. (72.625 mm) | 27.66 cu. in. (203.2 cc) | 16 hp. (11.9 kW) |

Engines in this section are one cylinder, four cycle, horizontal crankshaft models. Crankshaft is supported at each end in tapered roller bearings and intake and exhaust valves are located in cylinder head assembly. Cylinder and crankcase are a single cast iron unit.

Connecting rod rides directly on crankshaft crankpin journal and all models are splash lubricated by an oil dipper connected to connecting rod cap.

All models are equipped with a solid state ignition system and an alternator battery charging system is available.

A Walbro LM model float type carburetor is used on all models.

Refer to **TECUMSEH ENGINE IDENTIFICATION INFORMATION** section on page 441 for engine identification. Always give model, serial and specification numbers when ordering parts or service material.

## MAINTENANCE

**SPARK PLUG.** Recommended spark plug is Champion L7 or equivalent. Electrode gap should be set at 0.030 inch (0.762 mm) for all models.

Walbro LM model float type carburetor is used on all models. For exploded view and location of adjustment mixture needles, refer to Fig. T55.

For initial carburetor adjustment, open idle mixture screw and main fuel mixture screw one turn. Make final adjustments with engine at normal operating temperature and running. Place engine under load at rated engine speed and adjust main fuel mixture screw for leanest setting that will allow satisfactory acceleration and steady governor operation. Set engine at idle speed, no load and adjust idle mixture screw for smoothest idle operation. Adjust idle speed stop screw (8 – Fig. T55) so engine idles at 1200 rpm.

As each adjustment affects the other, adjustment procedures may have to be repeated.

To check float level, refer to Fig. T56. Invert carburetor throttle body and float assembly. A distance of 0.275-0.315 inch (6.99-8.00 mm) should exist between float and center boss as shown.

Adjust float by carefully bending float lever tang that contacts inlet valve.

Refer to Fig. T55 for exploded view of carburetor during disassembly. When reinstalling Viton inlet valve seat, grooved side of seat must be installed in bore first so inlet valve will seat against smooth side (Fig. T57).

**GOVERNOR.** A mechanical flyweight type governor is used on all models and governor gear, flyweights and shaft are serviced only as an assembly. Refer to Fig. T59 for view showing governor assembly installed in crankcase. Governor gear is driven by camshaft gear.

To adjust external governor linkage, refer to Fig. T60. Loosen screw (A), turn plate (B) counter-clockwise as far as possible and move governor lever (C) to left until throttle is in wide open position. Tighten screw (A). Governor spring must be hooked in hole (D) as shown. Adjusting screws on bracket are used to adjust fixed or variable speed settings. Engine high idle speed should not exceed 3600 rpm.

**IGNITION SYSTEM.** Either a solid state ignition system without alternator or a solid state ignition system which incorporates an alternator charging system may be used. Refer to appropriate paragraph for model being serviced.

**SOLID STATE IGNITION (WITHOUT ALTERNATOR).** Tecumseh solid state ignition system shown in Fig. T61 is used on models not equipped with flywheel alternator. This system does not use ignition breaker points. The only moving part of system is rotating flywheel with charging magnets. As flywheel magnet passes position (1A – Fig. T62), a low voltage AC current is induced into input coil (2). Current passes through rectifier (3) converting this current to DC. It then travels to capacitor (4) where it is stored. Flywheel rotates approximately 180° to position (1B). As it passes trigger coil (5), it induces a very small electric charge into coil. This charge passes through resistor (6) and turns on SCR (silicon controlled rectifier) switch (7).

With SCR switch closed, low voltage current stored in capacitor (4) travels to pulse transformer (8). Voltage is stepped up instantaneously and current is discharged across electrodes of spark plug (9), producing a spark before "top dead center".

Units may be equipped with a second trigger coil and resistor set to turn SCR switch on at a lower rpm. This second trigger pin is closer to flywheel and produces a spark at TDC for easier starting. As engine rpm increases, first (shorter) trigger pin turns SCR switch on, firing spark plug before "top dead center".

If system fails to produce a spark at spark plug, first check high tension lead (Fig. T61). If condition of high tension lead is questionable, renew pulse transformer and high tension lead assembly. Check low tension lead and renew if insulation is faulty. Magneto charging coil, electronic triggering system and mounting plate are available only as an assembly. If necessary to renew this assembly, place unit in position on engine. Start retaining screws, turn mounting plate counter-clockwise as far as possible, then tighten retaining screws to 5-7 ft.-lbs. (7-10 N·m) torque.

SOLID STATE IGNITION (WITH ALTERNATOR). Tecumseh solid state ignition system used on models equipped with flywheel alternator does not use ignition points. The only moving part of system is rotating flywheel with charging magnets and trigger pins. Other components of system are ignition generator coil and stator assembly, spark plug and ignition unit.

Long trigger pin induces a small charge of current to close SCR (silicon controlled rectifier) switch at engine

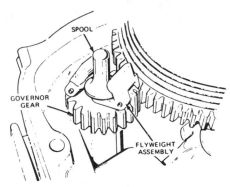

Fig. T59 — View showing governor assembly installed in crankcase. Governor gear is driven by camshaft gear.

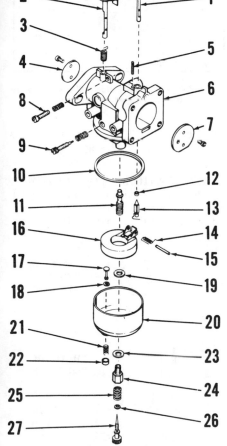

Fig. T55 — Exploded view of typical Walbro carburetor used on all models.

| | |
|---|---|
| 1. Choke shaft | 15. Float shaft |
| 2. Throttle shaft | 16. Float |
| 3. Throttle return spring | 17. Drain stem |
| 4. Throttle plate | 18. Gasket |
| 5. Choke stop spring | 19. Gasket |
| 6. Carburetor body | 20. Bowl |
| 7. Choke plate | 21. Spring |
| 8. Idle speed stop screw | 22. Retainer |
| 9. Idle mixture needle | 23. Gasket |
| 10. Bowl gasket | 24. Bowl retainer |
| 11. Main nozzle | 25. Spring |
| 12. Inlet valve seat | 26. "O" ring |
| 13. Inlet valve | 27. Main fuel adjusting |
| 14. Float spring | needle |

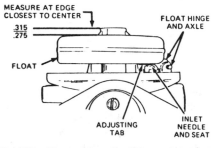

Fig. T56 — Float setting should be measured as shown. Bend adjusting tab to adjust float setting

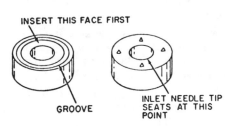

Fig. T57 — Viton inlet fuel valve seat must be installed grooved side first.

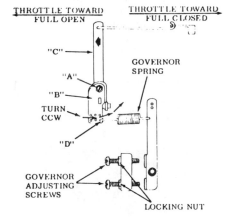

Fig. T60 — Typical external governor linkage. Refer to text for adjustment procedures.

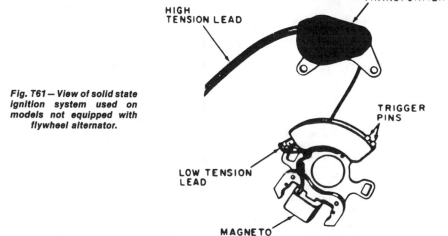

Fig. T61 — View of solid state ignition system used on models not equipped with flywheel alternator.

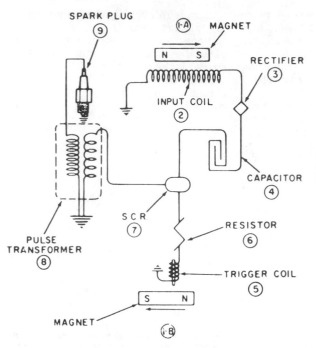

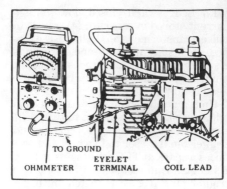

Fig. T62—Operational diagram of solid state ignition system used on some models.

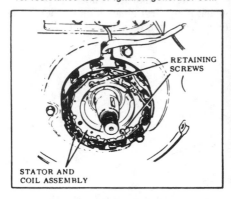

Fig. T66—View showing ohmmeter connected for resistance test of ignition generator coil.

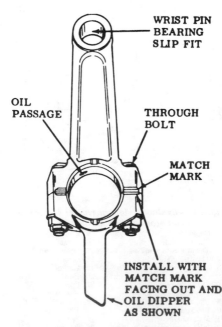

Fig. T67—Ignition generator coil and stator is serviced only as an assembly.

cranking speed and produces a spark at TDC for starting. As engine rpm increases, first (shorter) trigger pin induces current which produces a spark when piston is BTDC.

Test ignition system by holding high tension lead 1/8-inch (3.175 mm) from

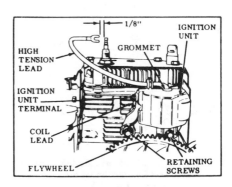

Fig. T63—View of solid state ignition unit used on models equipped with flywheel alternator. System should produce a good blue spark 1/8-inch long at cranking speed.

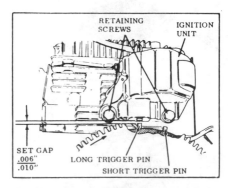

Fig. T64—Adjust air gap between long trigger pin and ignition unit to 0.006-0.010 inch (0.1524-0.254 mm).

spark plug (Fig. T63), crank engine and check for a good blue spark. If no spark is present, check high tension lead and coil lead for loose connections or faulty insulation. Check air gap between long trigger pin and ignition unit as shown in Fig. T64. Air gap should be 0.006-0.010 inch (0.1524-0.254 mm). To adjust air gap, loosen two retaining screws and move ignition unit as necessary, then tighten retaining screws.

**NOTE: Long trigger pin should extend 0.250 inch (6.35 mm) and short trigger pin should extend 0.187 inch (4.75 mm), measured as shown in Fig. T65.**

Remove coil lead from ignition terminal and connect an ohmmeter as shown in Fig. T66. If series resistance test of ignition generator coil is below 400 ohms, renew stator and coil

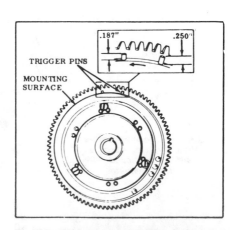

Fig T65—If trigger pin extension is incorrect, remove flywheel and drive pins in or out, as necessary, until long pin is extended 0.250 inch (6.35 mm) and short pin is extended 0.187 inch (4.75 mm) above mounting surface.

assembly (Fig. T67). If resistance is above 400 ohms, renew ignition unit.

**LUBRICATION.** Splash lubrication on all models is provided by an oil dipper on connecting rod cap (Fig. T68).

Oils approved by manufacturer must

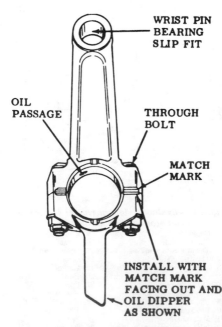

Fig. T68—Connecting rod assembly used on all models. Note oil dipper on rod cap.

meet requirements of API service classifications SE or SF.

Use SAE 30 oil for operating temperatures above 32° F (0° C) and SAE 10W oil for temperatures below 32° F (0° C).

## REPAIRS

**TIGHTENING TORQUES.** Recommended tightening torques are as follows:

Cylinder head bolts . . . . . . . 15-20 ft.-lbs.
(20-27 N·m)
Connecting rod . . . . . . . . . . . 7-9 ft.-lbs.
(10-12 N·m)
Crankcase cover . . . . . . . . . . 5-9 ft.-lbs.
(7-12 N·m)
Bearing retainer . . . . . . . . . . 5-9 ft.-lbs.
(7-12 N·m)
Flywheel nut . . . . . . . . . . . 50-55 ft.-lbs.
(68-75 N·m)
Spark plug . . . . . . . . . . . . . 18-23 ft.-lbs.
(24-31 N·m)
Stator mounting . . . . . . . . . . 5-7 ft.-lbs.
(7-10 N·m)
Carburetor to inlet pipe . . . . . 4-5 ft.-lbs.
(5-7 N·m)
Inlet pipe to head . . . . . . . . . . 5-8 ft.-lbs.
(7-11 N·m)
Rocker arm housing to head . . 7-8 ft.-lbs.
(10-11 N·m)
Rocker arm shaft screw . . . 15-18 ft.-lbs.
(20-24 N·m)
Rocker arm cover . . . . . . . . . . 1-2 ft.-lbs.
(1-3 N·m)

**CYLINDER HEAD.** To remove cylinder head, unbolt and remove blower housing, valve cover and breather assembly. Turn crankshaft until piston is at "top dead center" on compression stroke (Fig. T69), loosen locknuts on rocker arms and back off adjusting screws. Remove snap rings from rocker shaft and remove rocker arms. Using valve spring compressor tool (special tool number 670237) as shown in Fig. T70, remove valve retainers. Remove

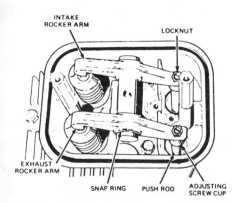

Fig. T69—View showing rocker arms used on all models. Slotted adjusting screws were used on early production engines. Later engines have adjusting nut on screw below rocker arm.

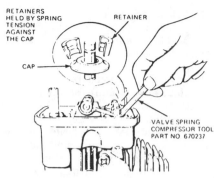

Fig. T70—Use tool No. 670237 to compress valve springs while removing retainers.

upper spring cap, valve spring, lower spring cap and "O" ring from each valve. Remove the three cap screws, washers and "O" rings from inside rocker arm housing and carefully lift off housing. Push rods and push rod tubes can now be withdrawn. Unbolt and remove carburetor and inlet pipe assembly from cylinder head. Remove cylinder head cap screws and lift off cylinder head, taking care not to drop intake and exhaust valves.

Always use new head gasket when reinstalling cylinder head and make certain Belleville washers and flat washers are properly installed (Fig. T71). Note location of the two short cap screws (Fig. T72) for correct installation. Tighten cylinder head cap screws evenly to 15-20 ft.-lbs. (20-27 N·m) torque using sequence shown in Fig. T72. Place new "O" rings on push rod tubes and install push rods and tubes. Install rocker arm housing and using new "O" rings on the three mounting cap screws, tighten cap screws to 7-8 ft.-lbs. (10-11 N·m) torque. Install new "O" ring, lower spring cap, valve spring and upper spring cap on each valve. Use valve spring compressor (special tool number 670237) to compress valve springs and install retainers. Install rocker arms and secure them with snap rings. Refer to **VALVE SYSTEM** paragraph for valve adjustment procedure and after correctly adjusting valves, install rocker arm cover, breather assembly, carburetor and inlet pipe assembly and blower housing.

**CONNECTING ROD.** Aluminum alloy connecting rod rides directly on

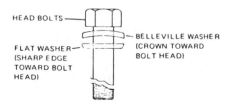

Fig. T71—Install Belleville washer and flat washer on cylinder head cap screws as shown.

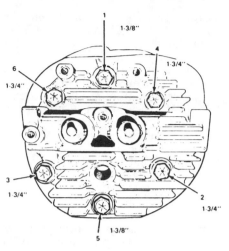

Fig. T72—Tighten cylinder head cap screws evenly to 15-20 ft.-lbs. (20-27 N·m) torque using tightening sequence shown. Note location of different length cap screws.

crankshaft crankpin journal for all models.

Piston and connecting rod assembly is removed from above after removing rocker arm housing, cylinder head, crankcase cover and connecting rod cap.

Standard crankpin journal diameter is 1.3750-1.3755 inches (34.925-34.938 mm) and connecting rods are available in a variety of sizes for undersize crankshafts as well as standard.

When installing piston and connecting rod assembly, make certain match marks on connecting rod and rod cap (Fig. T68) are aligned and marks are facing pto end of shaft. Always renew self-locking nuts on connecting rod bolts and tighten nuts to 7-9 ft.-lbs. (10-12 N·m) torque.

**PISTON, PIN AND RINGS.** Aluminum alloy piston for all models is fitted with two compression rings and one oil control ring. Piston should be renewed if visibly scored or damaged.

Piston skirt clearance in cylinder, measured at thrust side of piston just below oil control ring should be 0.006-0.008 inch (0.1524-0.2032 mm) for OH140 and HH140 models and

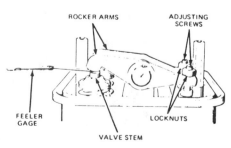

Fig. T73—Use a feeler gage when adjusting valve tappet gap. Refer to text for adjustment procedure.

0.010-0.012 inch (0.254-0.305 mm) for HH150, HH160, OH120, OH140, OH160 and OH180 models.

Compression ring groove width in piston is 0.095-0.096 inch (2.41-2.44 mm) for all models except OH180 model and oil control ring groove width in piston is 0.188-0.189 inch (4.78-4.80 mm) for all models except OH180 model.

Compression ring groove width in piston is 0.0955-0.0965 inch (2.43-2.45 mm) and oil control ring groove width in piston is 0.1880-0.1885 inch (4.7752-4.7879 mm) for OH180 model.

Ring side clearance in groove should be 0.0015-0.0035 inch (0.0381-0.0889 mm) for all models and piston should be renewed if side clearance exceeds 0.006 inch (0.1524 mm).

Standard piston pin diameter is 0.6876-0.6880 inch (17.465-17.475 mm) for all models except OH180.

Standard piston pin diameter is 0.7810-0.7812 inch (20.07-20.12 mm) for OH180 model.

Pin clearance to connecting rod pin bore should be 0.0001-0.0008 inch (0.0025-0.0203 mm). Pin and/or piston should be renewed if clearance is excessive.

Ring end gap should be 0.010-0.020 inch (0.254-0.508 mm) for all models. Rings should be installed as shown in Fig. T74 and ring end gaps should be staggered at 90° intervals around diameter of piston after installation.

Connecting rods on all models should be installed with match marks aligned and facing pto end of crankshaft.

Piston should be assembled to connecting rod so arrow on piston is pointing to carburetor side of engine after installation.

A variety of oversize pistons and rings are available for oversize bores for all models.

**CYLINDER.** Standard cylinder bore is 3.125-3.126 inches (79.375-79.400

mm) for OH120 model, 3.312-3.313 inches (84.125-84.150 mm) for HH140 and OH140 models, 3.500-3.501 inches 88.90-88.93 mm) for HH150, OH150 HH160 and OH160 models and 3.625-3.626 inches (92.075-92.100 mm) for OH180 model.

If cylinder is scored or if taper or out-of-round exceeds 0.004 inch (0.1016 mm), cylinder should be rebored to nearest oversize for which piston and rings are available.

**CRANKSHAFT.** Crankshaft for all models is supported at each end in tapered roller bearings and crankshaft bearings should have 0.001-0.007 inch (0.0254-0.1778 mm) preload. Preload is controlled by varying thickness of shim (17 – Fig. T76) on all models.

Bearings are a press fit on crankshaft and should be renewed if removed or if rough or damaged.

Standard crankpin journal diameter

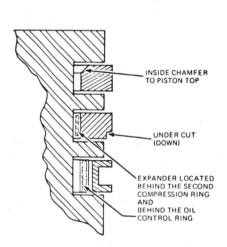

**Fig. T74 – Cross-sectional view showing correct installation of piston rings.**

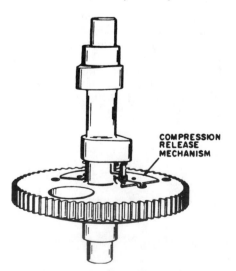

**Fig. T75 – View of Insta-matic Ezee-Start compression release camshaft assembly used on all models.**

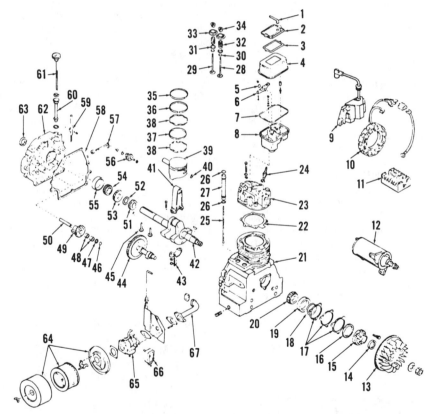

**Fig. T76 – Exploded view of basic engine used on all models.**

| | | |
|---|---|---|
| 1. Breather tube | 18. "O" ring | 35. Top compression ring |
| 2. Breather | 19. Bearing cup | 36. Second compression |
| 3. Gasket | 20. Bearing cone | ring |
| 4. Valve cover | 21. Cylinder block | 37. Oil control ring |
| 5. Snap ring | 22. Head gasket | 38. Ring expanders |
| 6. Rocker arm (2 used) | 23. Cylinder head | 39. Piston & pin assy. |
| 7. Seal ring | 24. Spark plug | 40. Retaining ring |
| 8. Rocker arm housing | 25. Push rod | 41. Connecting rod |
| 9. Ignition unit | 26. "O" ring | 42. Crankshaft |
| 10. Stator assy. | 27. Push rod tube | 43. Rod cap |
| 11. Regulator rectifier | 28. Intake valve | 44. Camshaft assy. |
| 12. Starter motor | 29. Exhaust valve | 45. Valve lifters |
| 13. Flywheel | 30. "O" ring | 46. Snap ring |
| 14. Oil seal | 31. Lower spring cap | 47. Thrust washer |
| 15. Bearing retainer cap | 32. Valve spring | 48. Needle bearing |
| 16. Steel washer | 33. Upper valve cap | 49. Dyna-Static balancer |
| 17. Shim gaskets | 34. Valve retainers | 50. Balancer shaft |

| |
|---|
| 51. Crankshaft gear |
| 52. Spacer |
| 53. Balancer drive gear |
| 54. Bearing cone |
| 55. Bearing cup |
| 56. Governor assy. |
| 57. Governor arm |
| 58. Gasket |
| 59. Oil filler tube extension |
| 60. Oil filler tube |
| 61. Dipstick |
| 62. Crankcase cover |
| 63. Oil seal |
| 64. Air cleaner assy. |
| 65. Carburetor |
| 66. Fuel pump |
| 67. Inlet pipe |

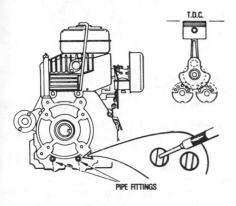

Fig. T77—View showing location of pipe plugs covering timing alignment holes on models with side-by-side Dyna-Static balancer gears. Note location of notches in balancer gears when properly installed.

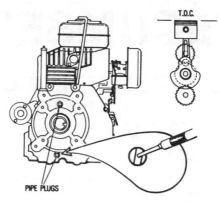

Fig. T78—View showing location of pipe plugs covering timing alignment holes on models with above and below crankshaft gear balance gears. Note location of notches in balancer gears when properly installed.

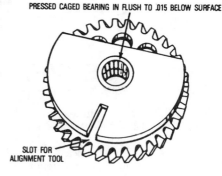

Fig. T79—Using tool number 670210, press new needle bearing into balancer gears until bearing cage is flush to 0.015 inch (0.381 mm) below edge of bore. Note alignment notch at lower side of balancer.

for all models is 1.3750-1.3755 inches (34.925-34.938 mm). Connecting rods are available in a variety of sizes for undersize crankshafts as well as standard.

Crankshaft should be renewed or reground if crankpin journal is tapered or worn over 0.002 inch (0.0508 mm) or is out-of-round more than 0.005 inch (0.0127 mm).

When installing crankshaft, align timing mark on crankshaft gear (chamfered tooth) with timing mark on camshaft gear.

Crankshaft oil seals should be installed flush to 0.025 inch (0.635 mm) below surface, with lips on seals facing inward.

**CAMSHAFT.** Camshaft and camshaft gear are an integral part which ride on journals at each end of camshaft. Camshaft is equipped with a compression release mechanism (Fig. T75). Check compression release parts for binding, excessive wear or other damage. If any parts are excessively worn or damaged, renew complete camshaft assembly. Parts are not serviced separately for compression release mechanism.

Renew camshaft if gear teeth are excessively worn or bearing surfaces or lobes are worn or scored. Camshaft lobe nose-to-heel diameter should be 1.3117-1.3167 inches (33.317-33.444 mm) for all models.

Diameter of camshaft journals is 0.6235-0.6240 inch (15.837-15.850 mm) for all models.

Maximum allowable clearance between camshaft journal and bearing bore is 0.003 inch (0.0762 mm).

When installing camshaft, align timing mark on camshaft gear with timing

mark (chamfered tooth) on crankshaft gear.

**DYNA-STATIC BALANCER.** Dyna-Static engine balancer consists of counterweighted gears, mounted side-by-side (Fig. T77) or above and below crankshaft gear (Fig. T78) and driven by crankshaft to counteract unbalance caused by counterweights on crankshaft. Balancer gears are held in position on balancer shafts by snap rings. Renewable balancer shafts are pressed into cover until a distance of 1.7135-1.7185 inches (43.52-43.65 mm) exists between boss on cover and outer edge of snap ring groove on shafts.

Balancer gears are equipped with renewable caged needle bearings (Fig. T79). Using tool number 670210, press new bearings into balancer gears until bearing cage is flush to 0.015 inch (0.381 mm) below edge of bore.

When reassembling engine, balancer gears must be timed with crankshaft for correct operation. To time balancer gears, refer to Fig. T77 or T78 and remove pipe plugs. Rotate crankshaft so piston is at TDC and install cover and weight assembly so slots in weights are visible through pipe plug openings. See Fig. T77 and T78. When correctly assembled, piston should be exactly at TDC and weights should be at full bottom position.

**VALVE SYSTEM.** Seats are machined directly into cylinder head surface. Seats should be ground at a 46° angle and seat width should be 0.042-0.052 inch (1.067-1.321 mm).

Valves should be ground at a 45° angle and valve should be renewed if margin is 0.060 inch (1.524 mm) or less.

Valves are available with 1/32-inch (0.794 mm) oversize stems for installation in guides which are badly worn. Guides must be reamed to proper dimension.

Standard valve guide inside diameter is 0.312-0.313 inch (7.925-7.950 mm) for all models. Guides may be reamed to 0.343-0.344 inch (8.71-8.74 mm) for use with 1/32-inch (0.794 mm) oversize valve stems.

To renew valve guides, remove and submerge head in large pan of oil. Heat on a hot plate until oil begins to smoke, about 15-20 minutes. Remove head from pan and place head on arbor press with valve seats facing up. Use a drift punch ½-inch in diameter to press guides out.

**CAUTION: Be sure to center punch. DO NOT allow punch to contact head when pressing guides out.**

To install new guides, place guides in freezer or on ice for 30 minutes prior to installation. Submerge head in pan of oil. Heat on hot plate until oil begins to smoke, about 15-20 minutes. Remove head and place, gasket surface down, on a 6 x 12 inch piece of wood. Using snap rings to locate both guides, insert silver color guide in intake side and brass colored guide in exhaust side. It may be necessary to use a rubber or rawhide mallet to fully seat snap rings. **DO NOT** use metal hammer or guide damage will result. Allow head to cool and reface both valve seats.

Recommended valve tappet gap (cold) for all models is 0.005 inch (0.127 mm) for intake valves and 0.010 inch (0.254 mm) for exhaust valves.

Valves must be adjusted with engine at TDC on compression stroke. Refer to **Fig. T73** for view of rocker arm adjusting screw.

# TECUMSEH

## SERVICING TECUMSEH ACCESSORIES

Some Tecumseh engines may be equipped with 12 volt electrical systems. Refer to the following paragraphs for servicing Tecumseh electrical units and 12 volt Delco-Remy starter-generator used on some models.

**12 VOLT STARTER MOTOR (BENDIX DRIVE TYPE).** Refer to Fig. T82 for exploded view of 12 volt starter motor and Bendix drive unit used on some engines. To identify starter, refer to service number stamped on end cap. When assembling starter motor use spacers (15) of varying thicknesses to obtain an armature end play of 0.005-0.015 inch (0.127-0.381 mm). Tighten armature nut (1) to 100 in.-lbs. (11 N·m) torque on motor numbers 29965, 32468, 32468A, 32468B and 33202, to 130-150 in.-lbs. (15-17 N·m) torque on motor number 32817. Tighten through-bolts to 30-35 in.-lbs. (3-4 N·m) torque on motor numbers 29965, 32468, 32468A, 32468B and 33202, to 35-44 in.-lbs. (4-5 N·m) torque on motor number 32817 and to 45-50 in.-lbs. (5-6 N·m) torque on motor number 32510.

To perform no-load test for starter motors 29965, 32468 and 32468A, use a fully charged 6 volt battery. Maximum current draw should not exceed 25 amps at 6 volts. Minimum rpm is 6500.

No-load test for starter motors 32468B, 33202, 32510 and 32817 must be performed with a 12 volt battery. Maximum current draw should not exceed 25 amps at 11.5 volts. Minimum rpm is 8000.

**ALTERNATOR CHARGING SYSTEMS.** Flywheel alternators are used on some engines for the charging system. Generated alternating current is converted to direct current by two rectifiers on rectifier panel (Fig. T83 and T84) or regulator-rectifier (Fig. T85).

System shown in Fig. T83 has a maximum charging output of about 3 amps at 3600 rpm. No current regulator is used on this low output system. Rectifier panel includes two diodes (rectifiers) and a 6 amp fuse for overload protection.

System shown in Fig. T84 has a maximum output of 7 amps. To prevent overcharging battery, a double pole switch is used in low output position to reduce output to 3 amps for charging battery. Move switch to high output position (7 amps) when using accessories.

System shown in Fig. T85 has a max-

imum output of 7 amps on engines of 7 hp.; 10 or 20 amps on engines of 8 hp. and larger. This system uses a solid state regulator-rectifier which converts generated alternating current to direct current for charging battery. Regulator-rectifier also allows only required amount of current flow for existing battery conditions. When battery is fully charged, current output is decreased to prevent overcharging battery.

**TESTING.** On models equipped with rectifier panel (Figs. T83 or T84), remove rectifiers and test them with

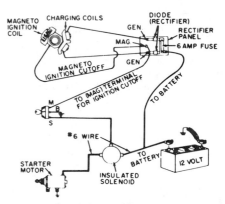

Fig. T83 — Wiring diagram of typical 3 amp alternator and rectifier panel charging system.

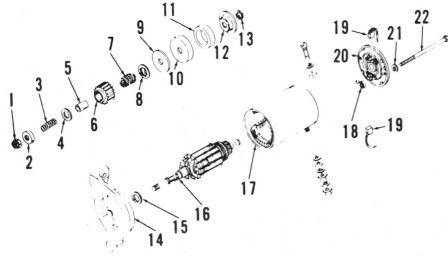

Fig. T82 — Exploded view of typical 12 volt starter motor assembly. Spacer (15) is available in different thicknesses to adjust armature end play.

| | | |
|---|---|---|
| 1. Nut | 8. Stop washer | 16. Armature |
| 2. Pinion stop | 9. Thrust washer | 17. Frame & field coil assy. |
| 3. Anti-drift spring | 10. Cushion cup | 18. Brush spring |
| 4. Washer | 11. Rubber cushion | 19. Brushes |
| 5. Anti-drift sleeve | 12. Thrust washer | 20. End cap |
| 6. Pinion gear | 13. Thrust bushing | 21. Washer |
| 7. Screw shaft | 14. Drive end cap | 22. Bolt |
| | 15. Spacer washer | |

Fig. T84 — Wiring diagram of typical 7 amp alternator and rectifier panel charging system. The double pole switch in one position reduces output to 3 amp for charging or increases output to 7 amp in other position to operate accessories.

either a continuity light or an ohmmeter. Rectifiers should show current flow in one direction only. Alternator output can be checked using an induction ammeter over positive lead wire to battery.

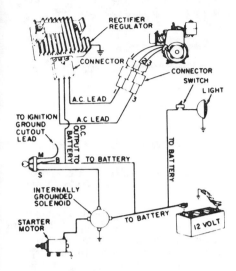

*Fig. T85 — Wiring diagram of typical 7, 10 or 20 amp alternator and regulator-rectifier charging systems.*

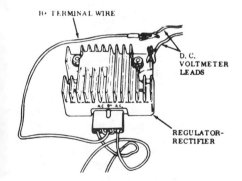

*Fig. T86 — Connect DC voltmeter as shown when checking regulator-rectifier.*

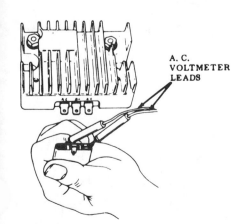

*Fig. T87 — Connect AC voltmeter to AC leads as shown when checking alternator coils.*

On models equipped with regulator-rectifier (Fig. T85), check system as follows: Disconnect B+ lead and connect a DC voltmeter as shown in Fig. T86. With engine running near full throttle, voltage should be 14.0-14.7. If voltage is above 14.7 or below 14.0 but above 0, regulator-rectifier is defective. If voltmeter reading is 0, regulator-rectifier or alternator coils may be defective. To test alternator coils, connect an AC voltmeter to AC leads as shown in Fig. T87. With engine running at near full throttle, check AC voltage. If voltage is less than 20.0 volts on 10 amp or 32.0 volts on 20 amp system, alternator is defective.

**MOTOR-GENERATOR.** Combination motor-generator (Fig. T88) functions as a cranking motor when starting switch is closed. When engine is operating and starting switch is open, unit operates as a generator. Generator output and circuit voltage for battery and various accessories are controlled by current-voltage regulator.

To determine cause of abnormal operation, motor-generator should be given a "no-load" test or a "generator output" test. Generator output test can be performed with a motor-generator on or off engine. No-load test must be made with motor-generator removed from engine.

Motor-generator test specifications are as follows:

**Motor-Generator Delco-Remy No. 1101980.**
Brush spring tension, oz. . . . . . . . .24-32

Field draw,
  Amperes . . . . . . . . . . . . . . .1.52-1.62
  Volts . . . . . . . . . . . . . . . . . . . . . . . . .12
Cold output,
  Amperes . . . . . . . . . . . . . . . . . . . . . .12
  Volts . . . . . . . . . . . . . . . . . . . . . . . . .14
  Rpm . . . . . . . . . . . . . . . . . . . . . . .4950
No-load test,
  Amperes (max.) . . . . . . . . . . . . . . . .18
  Volts . . . . . . . . . . . . . . . . . . . . . . . . .11
  Rpm (min.) . . . . . . . . . . . . . . . . .2500
  Rpm (max.) . . . . . . . . . . . . . . . . .2900

**CURRENT-VOLTAGE REGULA-TORS.** Two types of current-voltage regulators are used with motor-generator system. One is a low output unit which delivers a maximum of 7 amps. High output unit delivers a maximum of 14 amps.

Low output (7 amp) unit is identified by its four connecting terminals (three on one side of unit and one on underside of regulator). Battery ignition coil has a 3 amp draw. This leaves a maximum load of 4 amps which may be used on accessory lead.

High output (14 amp) unit has only three connecting terminals (all on side of unit). So with a 3 amp draw for battery ignition coil, a maximum of 11 amps can be used for accessories.

Regulator service test specifications are as follows:

**Regulator Delco-Remy No. 1118988 (7 amp)**
Ground polarity . . . . . . . . . . . .Negative
Cut-out relay,
  Air gap . . . . . . . . . . . . . . . . .0.020 inch
                    (0.508 mm)

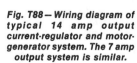

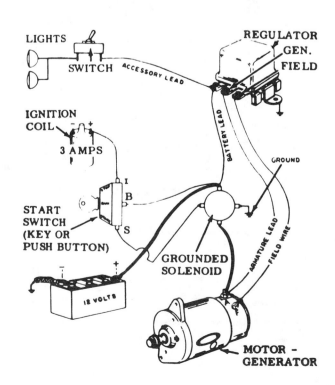

*Fig. T88 — Wiring diagram of typical 14 amp output current-regulator and motor-generator system. The 7 amp output system is similar.*

Point gap . . . . . . . . . . . . . . 0.020 inch
(0.508 mm)
Closing voltage, range . . . . . 11.8-14.0
Adjust to . . . . . . . . . . . . . . . . . . 12.8
Voltage regulator,
Air gap . . . . . . . . . . . . . . . . 0.075 inch
(1.9 mm)
Setting volts, range . . . . . . . 13.6-14.5
Adjust to . . . . . . . . . . . . . . . . . . 14.0

## Regulator Delco-Remy No. 1119207 (14 amp)

Ground polarity . . . . . . . . . . . Negative
Cut-out relay,
Air gap . . . . . . . . . . . . . . . . 0.020 inch
(0.508 mm)
Point gap . . . . . . . . . . . . . . 0.020 inch
(0.508 mm)
Closing voltage, range . . . . . 11.8-13.5
Adjust to . . . . . . . . . . . . . . . . . . 12.8
Voltage regulator,
Air gap . . . . . . . . . . . . . . . . 0.075 inch
(1.9 mm)
Voltage setting:
14.4-15.5 @ 65° (18° C)
14.2-15.2 @ 85° (29° C)
14.0-14.9 @ 105° (41° C)
13.8-14.8 @ 125° (52° C)
13.5-14.3 @ 145° (63° C)
13.1-13.9 @ 165° (74° C)
Current regulator,
Air gap . . . . . . . . . . . . . . . . 0.075 inch
(1.9 mm)
Current setting . . . . . . . . . . . . . 13-15

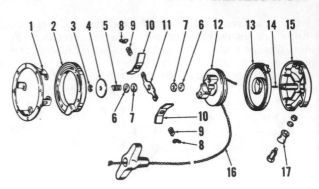

Fig. T90 — Exploded view of typical friction shoe rewind starter assembly.

1. Mounting flange
2. Flange
3. Retaining ring.
4. Washer
5. Spring
6. Slotted washer
7. Fibre washer
8. Spring retainer
9. Spring
10. Friction shoe
11. Actuating lever
12. Rotor
13. Rewind spring
14. Centering pin
15. Cover
16. Rope
17. Roller

## WIND-UP STARTER

**RATCHET STARTER.** On models equipped with ratchet starter, refer to Fig. T89 and proceed as follows: Move release lever to "RELEASE" position to remove tension from main spring. Remove starter assembly from engine. Remove left hand thread screw (26), retainer hub (25), brake (24), washer (23) and six starter dogs (22). Note position of starter dogs in hub (21). Remove hub (21), washer (20), spring and housing (12), spring cover (18), release gear (17) and retaining ring (19) as an assembly. Remove retaining ring, then carefully separate these parts.

**CAUTION: Do not remove main spring from housing (12). The spring and housing are serviced only as an assembly.**

Remove snap rings (16), spacer washers (29), release dog (14), lock dog (15) and spring (13). Winding gear (8), clutch (4), clutch spring (5), bearing (6) and crank handle (2) can be removed after first removing the retaining screw and washers (10, 30 and 9).

Reassembly procedure is reverse of disassembly. Centering pin (27) must align screw (26) with crankshaft center hole.

## REWIND STARTERS

**FRICTION SHOE TYPE.** To disassemble starter, refer to Fig. T90 and proceed as follows: Hold starter rotor (12) securely with thumb and remove the four screws securing flanges (1 and 2) to cover (15). Remove flanges and release thumb pressure enough to allow spring to rotate pulley until spring (13) is unwound. Remove retaining ring (3), washer (4), spring (5), slotted washer (6) and fibre washer (7). Lift out friction shoe assembly (8, 9, 10 and 11), then remove second fibre washer and slotted washer. Withdraw rotor (12) with rope from cover and spring. Remove rewind spring from cover and unwind rope from rotor.

When reassembling, lubricate rewind spring, cover shaft and center bore in rotor with a light coat of Lubriplate or equivalent. Install rewind spring so windings are in same direction as removed spring. Install rope on rotor, then place rotor on cover shaft. Make certain inner and outer ends of spring are correctly hooked on cover and rotor. Preload rewind spring by rotating rotor two full turns. Hold rotor in preload position and install flanges (1 and 2). Check sharp end of friction shoes (10) and sharpen or

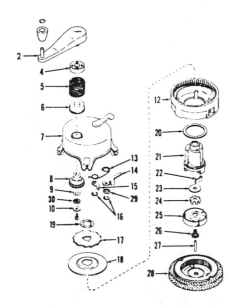

Fig. T89 — Exploded view of ratchet starter used on some engines.

| | |
|---|---|
| 2. Handle | 18. Spring cover |
| 4. Clutch | 19. Retaining ring |
| 5. Clutch spring | 20. Hub washer |
| 6. Bearing | 21. Starter hub |
| 7. Housing | 22. Starter dog |
| 8. Wind gear | 23. Brake washer |
| 9. Wave washer | 24. Brake |
| 10. Clutch washer | 25. Retainer |
| 12. Spring and housing | 26. Screw (left hand thread) |
| 13. Release dog spring | 27. Centering pin |
| 14. Release dog | 28. Hub and screen |
| 15. Lock dog | 29. Spacer washers |
| 16. Dog pivot retainers | 30. Lock washer |
| 17. Release gear | |

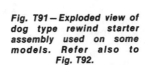

Fig. T91 — Exploded view of dog type rewind starter assembly used on some models. Refer also to Fig. T92.

1. Cover
2. Keeper
3. Recoil spring
4. Pulley
5. Spring
6. Dog
7. Brake spring
8. Retainer
9. Screw
10. Centering pin
11. Sleeve
12. Nut
13. Washer
14. Cup
15. Screen

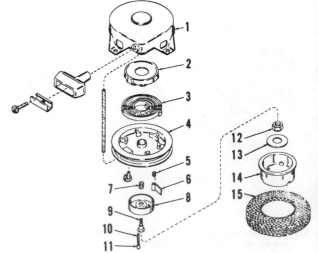

Illustrations courtesy of Tecumseh Products Company

renew as necessary. Install washers (6 and 7), friction shoe assembly, spring (5), washer (4) and retaining ring (3). Make certain friction shoe assembly is installed properly for correct starter rotation. If properly installed, sharp ends of friction shoes will extend when rope is pulled.

Remove brass centering pin (14) from cover shaft, straighten pin if necessary, then reinsert pin ⅓ of its length into cover shaft. When installing starter on engine, centering pin will align starter with center hole in end of crankshaft.

**DOG TYPE.** Two dog type starters may be used as shown in Fig. T91 and Fig. T92. Disassembly and assembly is similar. To disassemble starter shown in Fig. T91, remove starter from engine and while holding pulley remove rope handle. Allow recoil spring to unwind. Remove starter components in order shown in Fig. T91 noting position of dog (6) and direction spring (3) is wound. Be careful when removing recoil spring (3). Reassemble by reversing disassembly procedure. Turn pulley six turns before passing rope through cover so spring (3), is preloaded. Tighten retainer screw (9)

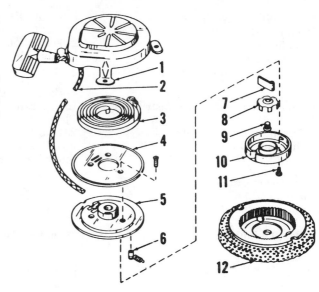

**Fig. T92 – Exploded view of dog type rewind starter assemby used on some models. Some units use three starter dogs (7).**

1. Cover
2. Rope
3. Rewind spring
4. Pulley half
5. Pulley half and hub
6. Retainer spring
7. Starter dog
8. Brake
9. Brake screw
10. Retainer
11. Retainer screw
12. Hub & screen assy.

to 45-55 in.-lbs. (5-6 N·m) torque.

To disassemble starter shown in Fig. T92, pull starter rope until notch in pulley half (5) is aligned with rope hole in cover (1). Hold pulley and prevent from rotating. Engage rope in notch and allow pulley to slowly rotate so recoil

spring will unwind. Remove components as shown in Fig. T92. Note direction recoil spring is wound being careful when removing spring from cover. Reassemble by reversing disassembly procedure. Preload recoil spring by turning pulley two turns with rope.

## TECUMSEH CENTRAL WAREHOUSE DISTRIBUTORS

(Arranged Alphabetically by States)

**These franchised firms carry extensive stocks of repair parts. Contact them for name of dealer in their area who will have replacement parts.**

Charlie C. Jones Battery &
Electric Co., Inc.
Phone: (602) 272-5621
2440 West McDowell Road
P.O. Box 6654
**Phoenix, Arizona 85005**

Pacific Power Equipment Company
Phone: (415) 692-1094
1565 Adrian Road
**Burlingame, California 94010**

Pacific Power Equipment Company
Phone: (303) 371-4081
500 Oakland Street
**Denver, Colorado 80239**

Spencer Engine Incorporated
Phone: (813) 253-6035
1114 West Cass Street
P.O. Box 2579
**Tampa, Florida 33601**

Sedco Incorporated
Phone: (404) 925-4706
1414 Red Plum Road NW
**Norcross, Georgia 30093**

Small Engine Clinic
Phone: (808) 488-0711
98019 Kam Highway
**Honolulu, Hawaii 96701**

Industrial Engine & Parts
Phone: (312) 927-4100
1133 West Pershing Road
**Chicago, Illinois 60609**

Medart Engines & Parts of Kansas
Phone: (913) 888-8828
15500 West 109th Street
**Lenexa, Kansas 66219**

Grayson Company of Louisiana
Phone: (318) 222-3211
100 Fannin
P.O. Box 206
**Shreveport, Louisiana 71102**

W. J. Connell Company
Phone: (617) 543-3600
65 Green Street
Route 106
**Foxboro, Massachusetts 02035**

Carl A. Anderson Inc. of Minnesota
Phone: (612) 452-2010
2737 South Lexington
**Eagan, Minnesota 55121**

Medart Engines & Parts
Phone: (314) 343-0505
100 Larkin William Industrial Ct.
**Fenton, Missouri 63026**

Original Equipment Incorporated
Phone: (406) 245-3081
905 Second Avenue North
Box 2135
**Billings, Montana 59103**

Carl A. Anderson Incorporated
Phone: (404) 339-4944
7410 "L" Street
P.O. Box 27139
**Omaha, Nebraska 68127**

E. J. Smith & Sons Company
Phone: (704) 394-3361
4250 Golf Acres Drive
P.O. Box 668887
**Charlotte, North Carolina 28266**

Uesco Warehouse Incorporated
Phone: (701) 237-0424
715 25th Street North
P.O. Box 2904
**Fargo, North Dakota 58108**

Gardner Engine & Parts Distribution
Phone: (614) 488-7951
1150 Chesapeake Avenue
**Columbus, Ohio 43212**

Mico Incorporated
Phone: (918) 627-1448
7450 East 46th Place
P.O. Box 470324
**Tulsa, Oklahoma 74147**

Brown & Wiser, Incorporated
Phone: (503) 692-0330
9991 South West Avery Street
**Tualatin, Oregon 97062**

Sullivan Brothers Incorporated
Phone: (215) 942-3686
Creek Road & Langoma Avenue
P.O. Box 140
**Elverson, Pennsylvania 19520**

Pitt Auto Electric Company
Phone: (412) 766-9112
2900 Stayton Street
**Pittsburgh, Pennsylvania 15212**

Locke Auto Electric Service
Incorporated
Phone: (605) 336-2780
231 North Dakota Avenue
P.O. Box 1165
**Sioux Falls, South Dakota 57101**

Medart Engines & Parts of Memphis
Phone: (901) 774-6371
674 Walnut Street
**Memphis, Tennessee 38126**

Engine Warehouse Incorporated
Phone: (713) 937-4000
7415 Empire Central Drive
**Houston, Texas 77040**

Frank Edwards Company
Phone: (801) 363-8851
100 South 300 West
P.O. Box 2158
**Salt Lake City, Utah 84110**

R B I Corporation
Phone: (804) 798-1541
101 Cedar Run Drive
Lake-Ridge Park
**Ashland, Virginia 23005**

BITCO Western Incorporated
Phone: (206) 682-4677
4030 1st Avenue South
P.O. Box 24707
**Seattle, Washington 98124**

Wisconsin Magneto Incorporated
Phone: (414) 445-2800
4727 North Teutonia Avenue
P.O. Box 09218
**Milwaukee, Wisconsin 53209**

## CANADIAN DISTRIBUTORS

Suntester Equipment (Central) Ltd.
Phone: (403) 453-5791
13315 146th Street
**Edmonton, Alberta, Canada T5L 4S8**

Suntester Equipment (Central) Ltd.
Phone: (416) 624-6200
5466 Timberlea Boulevard
**Mississauga, Ontario, Canada
L4W 2T7**

# WISCONSIN

TELEDYNE TOTAL POWER
3409 Democrat Road
Memphis, TN 38181

| Model | No. Cyls. | Bore | Stroke | Displacement | Power Rating |
|-------|-----------|------|--------|--------------|--------------|
| ADH | 1 | 2.75 in. (69.8 mm) | 3.25 in. (82.55 mm) | 19.3 cu. in. (308.7 cc) | 5.1 hp. (3.8 kW) |
| AE, AEH, AEHS, AEN, AENL, AENS | 1 | 3.0 in. (76.2 mm) | 3.25 in. (82.55 mm) | 23.0 cu. in. (376.5 cc) | 9.2 hp. (6.9 kW) |
| AFH | 1 | 3.25 in. (82.55 mm) | 4.0 in. (101.6 mm) | 33.2 cu. in. (543.8 cc) | 7.2 hp. (5.4 kW) |
| AGH, AGND | 1 | 3.5 in. (88.90 mm) | 4.0 in. (101.6 mm) | 38.5 cu. in. (630.7 cc.) | 12.5 hp. (9.3 kW) |
| AHH | 1 | 3.625 in. (92.08 mm) | 4.0 in. (101.6 mm) | 41.3 cu. in. (676.6 cc) | 9.2 hp. (6.9 kW) |
| AK | 1 | 2.875 in. (73.025 mm) | 2.75 in. (69.8 mm) | 17.8 cu. in. (292.3cc) | 4.1 hp. (3.1 kW) |
| AKS | 1 | 2.875 in. (73.025 mm) | 2.75 in. (69.8 mm) | 17.8 cu. in. (292.3 cc) | 4.7 hp. 3.5 kW) |
| AKN | 1 | 2.875 in. (73.025 mm) | 2.75 in. (69.8 mm) | 17.8 cu. in. (292.3 cc) | 6.2 hp. (4.6 kW) |

Engines in this section are four-cycle, one cylinder, horizontal crankshaft engines. Crankshaft is supported at each end by taper roller bearings.

Connecting rod rides directly on crankpin journal and shims between rod cap and rod provide running clearance adjustment. Pressure spray or splash lubrication is provided by a plunger type oil pump or a dipper located on end of connecting rod cap. Pump is driven off of camshaft lobe.

Ignition system consists of either a battery type system or one of a variety of magneto systems.

Float type carburetors from a variety of manufacturers are available. A mechanical fuel pump is available for most models.

Engine identification information is located on engine instruction plate and engine model, specification and serial number are required when ordering parts or service material.

## MAINTENANCE

**SPARK PLUG.** Recommended spark plug is Champion D16J or equivalent. Electrode gap should be 0.030 inch (0.762 mm).

**CARBURETOR.** A variety of float type carburetors are used as listed. Marvel Schebler VH53 and VH70 models are replacements for older VH12 and VH14 models.

Marvel Schebler VH53 model is used on AK, AKS and AKN engines, TSX147 model is used on AHH engines and TSX676 model is used on AGND engines.

Stromberg OH ⅝ model is used on AK, AKS and AKN engines, UC ¾ model is used on ADH, AE, AEH, AEHS and AFH engines, UC 7/8 model is used on AGH and AHH engines and UR ¾ model is used on ADH, AE, AEH, AEHS and AENL engines.

Zenith 87B5 and 87BY6 models were used on AKN, AK and AKS engines, 161-7 model is used on AE, AEH, AEHS, AFH, AGH, AHH, AEN and AENS engines and 68-7 model is used on some AENL and AGND engines.

For initial carburetor adjustment open idle mixture screw 1-¼ turns and main fuel mixture screw, if so equipped, 1-¼ turns. Make final adjustments with engine at normal operating temperature and running. Place engine under load and adjust main fuel mixture screw for leanest setting that will allow satisfac-

tory acceleration and steady governor operation. Set engine at idle speed, no load and adjust idle mixture screw for smoothest idle operation.

As each adjustment affects the other, adjustment procedure may have to be repeated.

Recommended float level for Marvel Schebler VH12 and VH14 models is ½-inch (12.7 mm) and ¼-inch (6.35 mm) for VH53, VH70 and TSX models.

Recommended float level for Zenith 87B5 and 87BY6 models is 31/32-inch (24.606 mm) and 1-5/32 inch (29.369 mm) for 161-7 and 68-7 models.

Recommended float level for Stromberg models is ½ to 9/16-inch (12.7 to 14.3 mm).

**GOVERNOR.** Flyweight type governors are used on all models.

On Models AK, AKN and AKS flyweights are attached to camshaft and operate a sleeve and shaft arrangement (Fig. W1) to control engine speed.

Governor flyweights on remaining models are attached to a gear assembly which is driven by camshaft gear and actuates an internal and external lever arrangement to control engine speed.

Speed changes on all models are obtained by using springs of varying tension and locating spring or governor rod in alternate holes of governor control lever.

Governor-to-carburetor rod should be adjusted so it just will enter hole in governor lever when governor weights are "in" and carburetor throttle is wide open.

**IGNITION SYSTEM.** Either a battery type ignition system with breaker points located in a distributor or a magneto type ignition system are used according to model and application.

BATTERY IGNITION. Breaker points on battery system are located inside a distributor which is driven by camshaft gear. Condenser is mounted externally on distributor body. Point gap is 0.020 inch (0.508 mm) for both type Auto-Lite distributors. See Fig. W5.

To correctly time distributor to engine during installation, remove screen over flywheel air intake, make certain piston is at "top dead center" on compression stroke (Fig. W3) and install distributor so points are just beginning to open. Connect timing light to engine, start and run engine at rated speed and rotate distributor as necessary to set timing to specifications as follows:

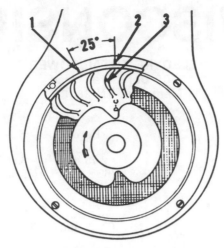

**Fig. W3 — Ignition timing marks for a typical "A" series engine. Piston is at top dead center position when leading edge of flywheel vane (3), marked "X" and "DC" is in register with static ignition timing mark (2).**

| | |
|---|---|
| ADH, AE, AEH, AEHS, AEN, AENS, AFH, AGH, AHH | 25° BTDC |
| AK, AKN, AKS | 28° BTDC |
| AENL, AGND | 20° BTDC |

MAGNETO IGNITION. Eisemann and Edison magnetos were obsoleted and replaced by Fairbanks-Morse or Wico magnetos.

Breaker point gap for all models is 0.015 inch (0.381 mm).

To correctly time magneto to engine during installation, remove screen over flywheel air intake opening and timing hole plug from crankcase. Make certain piston is at "top dead center" on compression stroke and leading edge of flywheel vane (3-Fig. W3) marked "X" and "DC" is in register with timing mark (2) on air shroud. Install magneto so marked tooth (Figs. W4 or W6) on magneto drive gear is visible through port in timing gear cover as shown. If timing is correct, impulse coupling will trip when crankshaft keyway is up as engine is being cranked.

Connect timing light and set timing to specifications listed for battery ignition system.

**LUBRICATION.** A plunger type oil pump driven off of a camshaft lobe may either pressure spray lubricate (Fig. W7) connecting rod and internal engine components or maintain proper oil level in an internal oil trough below connecting rod which enables dipper on rod cap to splash lubricate (Fig. W8) internal engine components.

Oil pump plunger to bore clearance should be 0.003-0.007 inch (0.0762-0.1778 mm) and plunger should be renewed if clearance exceeds 0.008 inch (0.203 mm).

Maintain oil level at full mark on dipstick or at top of filler plug but do not overfill. High quality detergent oil having API classification "SE" or "SF" is

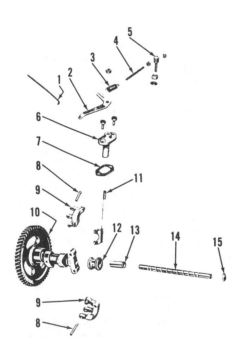

**Fig. W1 — Exploded view of governor assembly showing component parts and their relative positions.**

| | |
|---|---|
| 1. Governor control rod | 9. Flyweight |
| 2. Governor lever | 10. Camshaft & gear assy. |
| 3. Governor spring | 11. Governor yoke |
| 4. Adjusting screw | 12. Thrust sleeve |
| 5. Pin | 13. Spacer |
| 6. Support bracket | 14. Support pin |
| 7. Gasket | 15. Core plug |
| 8. Roll pin | |

**Fig. W4 — Timing gears and marks of AEN, AENL and AENS models shown in register. With marks on gears (3 and 5) aligned as shown at (4), "X" marked magneto gear tooth must show through port (2) when piston is at TDC.**

1. Magneto gear
2. Timing port
3. Camshaft gear
4. Timing marks
5. Crankshaft gear

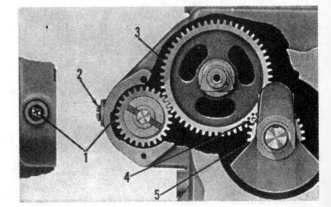

**Fig. W5 — Distributor used on Models AEN, AENL and AENS with battery ignition is driven from camshaft gear.**

3. Drive gear
4. Screw clamp
5. Distributor housing
6. Condenser
7. Breaker points

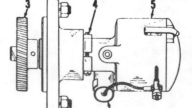

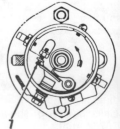

Illustrations courtesy of Teledyne Total Power

recommended. Recommended oil change interval is every 50 hours of normal operation.

Use SAE 30 oil when operating temperatures are above 40° F (4° C), SAE 20 oil for temperatures between 40° F (4° C) and 5° F (-15° C) and SAE 10 oil if temperature is below 5° F (-15° C).

## REPAIRS

**TIGHTENING TORQUES.** Recommended torque specifications are as follows:

Cylinder head –
AK, AKS & AKN . . . 14-18 ft.-lbs.
(19-24 N·m)
All other
models . . . . . . . . . 26-32 ft.-lbs.
(35-43 N·m)

Connecting rod –
AFH, AGH,
AHH, AGND . . . . . . . . 32 ft.-lbs.
(43 N·m)
All other
models . . . . . . . . . . . 18 ft.-lbs.
(24 N·m)

Cylinder to block (where applicable)
AGND . . . . . . . . . . . 40-50 ft.-lbs.
(54-68 N·m)
All other
models . . . . . . . . 62-78 ft.-lbs.
(84-106 N·m)

Main bearing plate –
AK, AKS & AKN . . . 14-18 ft.-lbs.
(19-24N·m)
All other
models . . . . . . . . . 26-32 ft.-lbs.
(35-43 N·m)

**CONNECTING ROD.** Connecting rod and piston assemblies can be removed from above after removing cylinder head and engine base.

Connecting rod rides directly on crankpin journal and rods are available for 0.010, 0.020 and 0.030 inch (0.254, 0.508 and 0.762 mm) undersize crankshafts.

Rod to journal running clearance should be 0.001-0.0018 inch (0.0254-0.0457 mm) for AEN, AENS and AENL models and 0.0007-0.002 inch (0.0178-0.0508 mm) for all other models. Clearance is controlled to some degree by varying thickness of shims between connecting rod and cap.

Connecting rod side clearance should be 0.004-0.013 inch (0.1016-0.3302 mm) for AEN, AENS and AENL models and 0.004-0.011 inch (0.1016-0.2794 mm) for all other models.

Connecting rod on all models must be installed with oil hole in rod cap facing oil pump side of engine for proper lubrication.

Crankpin standard journal diameter is 1.00 inch (25.4 mm) for AK, AKS and AKN models, 1.1255 inches (28.59 mm) for AE, AEH, AEN, AEHS, AENS, ADH and AENL models, 1.375 inches (34.925 mm) for AFH, AGH and AHH models and 1.750 inches (44.45 mm) for AGND models.

**PISTON, PIN AND RINGS.** Piston may be of three ring or four ring design according to model and application. Pistons and rings are available in a variety of oversizes as well as standard.

Piston ring end gap for all models should be 0.012-0.022 inch (0.3048-0.5588 mm) and ring side clearance should be 0.002-0.003 inch (0.0508-0.0762 mm).

Early models were equipped with a split skirt piston and split should be toward valve side of engine after installation. Later model cam ground piston has a wide and a narrow thrust face on piston skirt and wide side must be toward thrust side (side opposite valve) of engine after installation.

Split skirt piston should have 0.0045-0.005 inch (0.1143-0.1270 mm) clearance in cylinder measured 90° from piston pin at bottom edge of piston skirt and cam ground piston should have 0.003-0.0035 inch (0.0762-0.0889 mm) clearance in cylinder measured along thrust surface of skirt.

Floating type piston pin is retained by snap rings in bore of piston and should

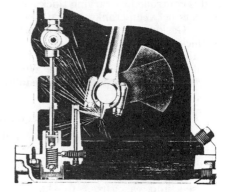

*Fig. W7 – View of pressure spray lubricating system used on some models.*

have 0.0002-0.0008 inch (0.0051-0.0203 mm) clearance in connecting rod.

**CYLINDER.** Cylinder bore section on ADH, AE, AEH, AEHS, AFH, AGH, AHH and AGND is separate from crankcase and may be renewed as an individual unit. Cylinder bore and crankcase for all other models are cast as a unit.

Cylinders for all models may be bored to oversize if worn or out-of-round more than 0.005 inch (0.127 mm).

Standard cylinder bore for ADH, AK, AKS and AKN is 2.750 inches (69.85 mm), for AEN, AENL and AENS bore is 3.0 inches (76.2 mm), for AE, AEHS and AFH bore is 3.250 inches (82.55 mm), for AGH and AGND bore is 3.5 inches (88.90 mm) and for AHH bore is 3.625 inches (92.08 mm).

Pistons are available in 0.005, 0.010, 0.020 and 0.030 inch (0.1270, 0.254, 0.508 and 0.762 mm) oversizes as well as standard.

**CRANKSHAFT AND MAIN BEARINGS.** Crankshaft for all models is supported at each end by taper roller bearings. Crankshaft end play should be 0.002-0.004 inch (0.0508-0.1016 mm)

*Fig. W6 – Placement of timing marks on Models AK, AKN, AKS and BKN when timing magneto to piston at top dead center.*

2. Crankshaft gear
3. Valve timing marks
4. Camshaft gear
5. Magneto gear
6. Magneto timing marks

*Fig. W8 – View of oil pump used to maintain correct oil level in oil trough of engines with splash lubricating system.*

and is controlled by varying thickness of gaskets between crankcase and bearing plate. Gaskets are available in a variety of thicknesses.

Main bearings are press fit on crankshaft. Refer to Fig. W4 or W6 for correct timing mark position for installation.

**CAMSHAFT.** The hollow camshaft turns on a stationary support pin (14-Fig. W1) and should have a running clearance of 0.001-0.0025 inch (0.0254-0.0635 mm). Support pin is pressed into case from flywheel end while camshaft and lifters are held in place. Camshaft plug (15) should be renewed.

Governor flyweights are attached to camshaft on some models and care should be taken to avoid damaging when camshaft is being installed.

**MAGNETO DRIVE GEAR.** On Models ADH, AE, AEH, AEHS, AFH, AGH and AHH magneto drive shaft should have a running clearance of 0.002-0.0035 inch (0.0508-0.0889 mm) in shaft bushing. Drive shaft end play should be 0.004-0.005 inch (0.1016-0.1270 mm) and is adjusted by varying thickness of gasket (2). Magneto drive gear is pressed on shaft so distance from coupling face of shaft to centerline of magneto mounting hole is 2.584 inches (65.6336 mm) as shown.

**VALVE SYSTEM.** Valve seats are renewable type inserts in cylinder block and seats should be ground to a 45° angle. Width should be 3/32-inch (2.381 mm).

Valve face is ground at 45° angle and stellite valves and seats and valve rotocaps are available for some models.

Valve stem clearance in block or guide should be 0.003-0.005 inch (0.0762-0.127 mm). AGND and AENL models have renewable guides. Valves with oversize stems are available for models which do not have renewable guides if stem clearance exceeds 0.006 inch (0.1524 mm).

Valve tappet gap (cold) for intake valve is 0.008 inch (0.2032 mm) and for exhaust valve is 0.016 inch (0.4064 mm). Gap is adjusted by turning adjusting screws on tappets if equipped or grinding end of valve stem if non-adjustable tappets are installed.

Illustrations courtesy of Teledyne Total Power

# WISCONSIN

**TELEDYNE TOTAL POWER**
3409 Democrat Road
Memphis, TN 38181

| Model | No. Cyls. | Bore | Stroke | Displacement | Power Rating |
|-------|-----------|------|--------|--------------|--------------|
| HBKN | 1 | 2.875 in. (73.03 mm) | 2.75 in. (69.85 mm) | 17.8 cu. in. (292.6 cc) | 7 hp. (5.2 kW) |
| HAENL | 1 | 3.0 in. (76.2 mm) | 3.25 in. (82.55 mm) | 23.0 cu. in. (376.5 cc) | 9.2 hp. (6.9 kW) |

Engines in this section are four-cycle, one cylinder, vertical crankshaft engines. Crankshaft is supported at each end by taper roller bearings.

Connecting rod rides directly on crankpin journal and shims between connecting rod cap and rod provide running clearance adjustment. Pressure spray lubrication is provided by a vane type oil pump mounted to and driven by crankshaft.

All models are equipped with a magneto ignition system.

Zenith float type carburetors are standard equipment for all models and a mechanical fuel pump is available.

Engine identification information is located on engine instruction plate and engine model, specification and serial number are required when ordering parts or service material.

## MAINTENANCE

**SPARK PLUG.** Recommended spark plug is Champion D16 or equivalent. Electrode gap should be 0.030 inch (0.762 mm).

**CARBURETOR.** Zenith 68-7 model (Fig. W9) carburetor is used on HAENL engines and Zenith 87B5 model (Fig. W10) carburetor is standard equipment for HBKN engines with a variety of other carburetors used for special applications. Refer to carburetor identification number for proper carburetor identification.

For initial adjustment on all models, open idle mixture screw and main fuel mixture screw (as equipped) 1-1/4 turns each. Make final adjustments with engine at normal operating temperature and running. Place engine under load and adjust main fuel mixture screw (as equipped) for leanest setting that will allow satisfactory acceleration and steady governor operation. Set engine at idle speed no load and adjust idle mixture screw for smoothest idle operation.

As each adjustment affects the other, adjustment procedure may have to be repeated.

To check float level, invert carburetor throttle body and float assembly (Fig. W13) and measure distance (A) from machined surface of throttle body to highest point on float. Model 87B5 carburetor float level should be 31/32-inch (24.606 mm) and 68-7 model carburetor float level should be 1-5/32 inch (29.369 mm). Adjust as necessary by bending float lever tang that contacts inlet valve.

**GOVERNOR.** Flyweight type governors are used on all models and flyweights are attached to camshaft and operate a sleeve and shaft arrangement (Fig. W14) to control engine speed.

Speed changes on all models is obtained by using springs (3) of varying tension, flyweights (9) of varying weights or locating spring or governor rod in alternate holes of governor control lever (2).

Governor-to-carburetor rod should be adjusted so it just will enter hole in governor lever when governor weights are "in" and carburetor throttle is wide open.

Camshaft must be removed from engine to service governor flyweights.

**IGNITION SYSTEM.** Fairbanks-Morse magnetos are used for all models and magneto is mounted in suspension below timing gear train. An oil drain

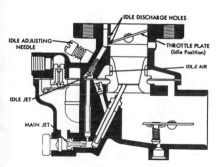

Fig. W9—Sectional view of Zenith 68-7 model carburetor showing location of idle adjusting screw and main fuel jet.

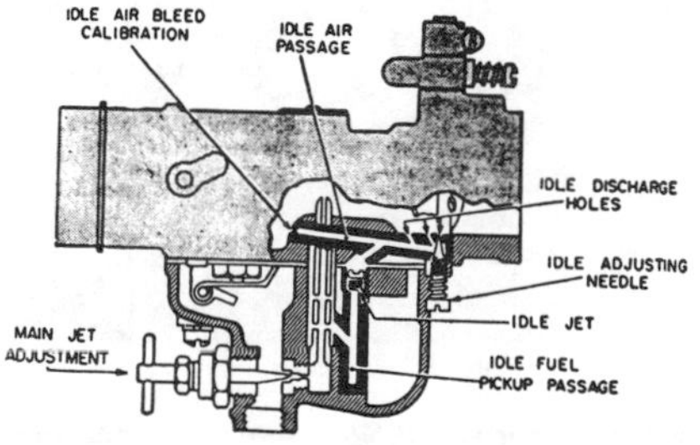

Fig. W10—Sectional view of Zenith 87B5 model carburetor showing location of idle adjusting screw and main fuel adjusting screw.

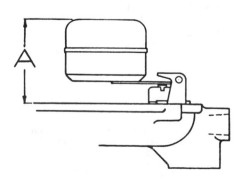

Fig. W13—Float level measurement for Zenith carburetors. On 87B5 model dimension "A" is 31/32-inch (24.6 mm). On 68-7 model, it is 1-5/32 inch (29.36 mm).

tube is connected to magneto body to allow excess oil to drain back into engine crankcase.

Magneto breaker points and condenser are accessible by removing magneto end cover. Breaker point gap should be adjusted to 0.015 inch (0.381 mm) clearance.

To correctly time magneto to HBKN engine, remove screen over flywheel air intake opening and timing hole plug from crankcase. Make certain piston is at "top dead center" on compression stroke and leading edge of flywheel vane (Fig. W17) marked "X" and "DC" is in register with timing mark (2) on air shroud. Install magneto so marked tooth (1) on magneto drive gear is visible through port in timing gear as shown. If timing is correct, impulse coupling will trip when crankshaft keyway is up as engine is being cranked.

To correctly time magneto to HAENL engine, remove screen over flywheel air intake opening and timing hole plug from crankcase. Make certain piston is at "top dead center" on compression stroke and leading edge of flywheel vane (Fig. W19) is in register with timing mark (3) on air shroud. Install magneto so marked tooth (1-Fig. W20) on magneto drive gear is visible through port in timing gear as shown.

To set running timing on either model,

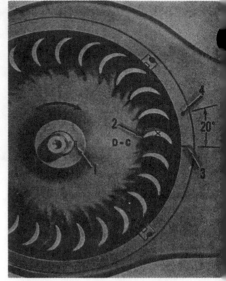

Fig. W17—Magneto timing for HBKN engines.
1. Magneto timing marks
2. Centerline mark
3. Running timing mark
4. Flywheel keyway

connect timing light, start and run engine at 1800 rpm and set timing on HBKN engines at 17° BTDC and set timing on HAENL engines 20° BTDC.

**LUBRICATION.** A vane type oil pump mounted to and driven by crankshaft is used to provide pressure spray lubrication on all models. Pump is located in engine adapter base (Fig. W21).

Pump vane retainer is held in position on crankshaft by a set screw which also holds pump body in place. A strap attached to engine bearing plate keeps oil pump body stationary.

Oil pressure relief valve reed should fit firmly against oil pressure relief hole. Renew reed if it is bent or out of shape.

Maintain oil level at filler plug but do not overfill. High quality detergent oil having API classification "SE" or "SF" is recommended. Recommended oil change interval is every 50 hours of normal operation.

Use SAE 30 oil when operating temperatures are above 40°F (4°C), SAE 20 oil for temperatures between 40°F (4°C) and 5°F (−15°C) and SAE

Fig. W19—View of HAENL model ignition timing marks lined up.
1. Keyway
2. Marked air vane
3. Centerline mark
4. Running timing mark

10 oil if temperature is below 5°F (−15°C).

## REPAIRS

**TIGHTENING TORQUES.** Recommended torque specifications are as follows:

Cylinder head—
HBKN . . . . . . . . . . . . . . 14-18 ft.-lbs.
(19-24 N·m)
HAENL . . . . . . . . . . . . . . 26-32 ft.-lbs.
(35-43 N·m)

Connecting rod—
HBKN . . . . . . . . . . . . . . 14-18 ft.-lbs.
(19-24 N·m)
HAENL . . . . . . . . . . . . . . . . . . 18 ft.-lbs.
(24 N·m)

Main bearing plate—
HBKN . . . . . . . . . . . . . . 14-18 ft.-lbs.
(19-24N·m)
HAENL . . . . . . . . . . . . 26-32 ft.-lbs.
(35-43 N·m)

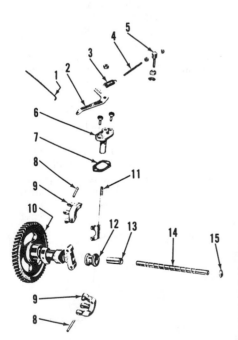

Fig. W14—Exploded view of governor assembly showing component parts and their relative positions.
1. Governor control rod
2. Governor lever
3. Governor spring
4. Adjusting screw
5. Pin
6. Support bracket
7. Gasket
8. Roll pin
9. Flyweight
10. Camshaft & gear assy.
11. Governor yoke
12. Thrust sleeve
13. Spacer
14. Support pin
15. Core plug

Fig. W18—Model HAENL timing marks shown in register.
1. Magneto timing mark
2. Timing port plug
3. Camshaft gear
4. Cam gear timing marks
5. Crankshaft gear timing mark

Fig. W20—Model HBKN timing marks shown in register.
1. Magneto timing marks
2. Crankshaft gear
3. Camshaft timing marks
4. Camshaft gear
5. Magneto gear

Crankcase cover plate—
HBKN ................6-8 ft.-lbs.
(8-11 N·m)
HAENL ................7-9 ft.-lbs.
(10-12 N·m)

Engine Adapter base—
All models.............24-26 ft.-lbs.
(33-35 N·m)

Spark plug—
All models.............25-30 ft.-lbs.
(34-40 N·m)

**CONNECTING ROD.** Connecting rod and piston assemblies can be removed from cylinder head surface end of block after removing cylinder head and side cover.

Connecting rod rides directly on crankpin journal and rods are available for 0.010, 0.020 and 0.030 inch (0.254, 0.508 and 0.762 mm) undersize crankshafts.

Rod to journal running clearance should be 0.0007-0.002 inch (0.0178-0.0508 mm) for all models and is controlled to some degree by varying thickness of shims between connecting rod and cap.

Connecting rod side clearance should be 0.004-0.010 inch (0.1016-0.254 mm) for HBKN engines and 0.006-0.013 inch (0.1524-0.3302 mm) for HAENL engines.

Connecting rod on all models must be installed with oil hole in rod cap facing oil pump side of engine for proper lubrication.

Crankpin standard diameter for HBKN engines is 1.000 inch (25.4 mm) and is 1.125 inches (31.75 mm) for HAENL engines.

**PISTON, PIN AND RINGS.** Pistons are equipped with two compression rings, one scraper ring and one oil control ring. Install rings as shown in Fig. W22. View "A" shows ring cross-section on HBKN engines and view "B" shows ring cross-section on HAENL engines.

HBKN engines operated at 3000 rpm or below require a different piston than HBKN engines operating at speed above 3000 rpm.

Ring end gap for all models should be 0.012-0.022 inch (0.3048-0.5588 mm) and side clearance of top ring in groove should be 0.002-0.0035 inch (0.0508-0.0889 mm). Side clearance of second or third ring should be 0.001-0.0025 inch (0.0254-0.0635 mm) and side clearance of oil control ring should be 0.0025-0.004 inch (0.0635-0.1016 mm). Ring gaps should be installed so they are at 90° intervals around piston.

Piston skirt clearance for HBKN engines operating at 3000 rpm or below should be 0.0055-0.006 inch (0.1397-0.1524 mm) and for HBKN engines operating above 3000 rpm clearance should be 0.006-0.0065 inch (0.1524-0.1651 mm).

Piston skirt clearance for all HAENL engines should be 0.003-0.0035 inch (0.0762-0.0889 mm).

Pistons and rings are available in 0.005, 0.010, 0.020 and 0.030 inch (0.1270, 0.254, 0.508 and 0.762 mm) oversizes as well as standard.

Floating type piston pin is retained by snap rings in each end of piston and pin clearance in connecting rod for HBKN engines should be 0.0002-0.0008 inch (0.0051-0.0203 mm) and for HAENL engines clearance should be 0.0005-0.001 inch (0.0127-0.0254 mm). Pins are available in a variety of oversizes as well as standard.

**CYLINDER.** Cylinder bore is an integral part of crankcase casting and if worn or out-of-round more than 0.005 inch (0.127 mm), cylinder should be bored to nearest size for which piston and rings are available.

If crankcase and cylinder renewal is required make certain number (Fig. W24) stamped as shown is added to basic part number shown in Wisconsin parts catalog when ordering new cylinder and crankcase.

**CRANKSHAFT AND MAIN BEARINGS.** Crankshaft on all models is supported at each end by taper roller bearings which are press fit on crankshaft. Races are driven into bearing plate or engine block. Crankshaft end play for HBKN engines should be 0.002-0.004 inch (0.0508-0.1016 mm), should be 0.001-0.003 inch (0.0254-0.0762 mm) for HAENL engines and is controlled on all

Fig. W22—Cross-section of piston rings as installed in vertical crankshaft engines. View "A" shows HBKN models and view "B" shows HAENL models.

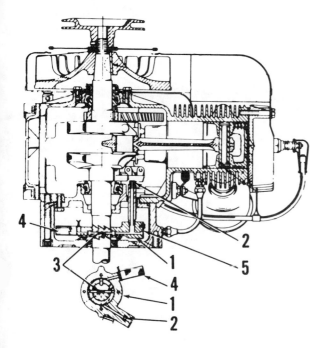

Fig. W21—Sectional view of engine oil pump as installed on vertical shaft models.

1. Pump body
2. Vertical spray jet
3. Vane assembly
4. Intake oil screen
5. Relief valve

Fig. W23—Crankshaft end play is controlled by number of gaskets (G) used under main bearing plate at pto end of crankcase.

models by varying number of gaskets between bearing plate and crankcase (Fig. W23).

HAENL engine is also equipped with a ball bearing at lower end of crankshaft which is mounted in engine adapter base below oil pump.

Install crankshaft so timing marks on crankshaft gear and camshaft gear are in register as shown in Figs. W18 or W20.

If necessary to renew crankshaft, refer to Fig. W25 for location of crankshaft part number.

**CAMSHAFT.** The hollow camshaft turns on a stationary support pin (14-Fig. 14) and should have a running clearance of 0.001-0.0025 inch (0.0254-0.0635 mm). Support pin is pressed into case from flywheel end while camshaft and lifters are held in place. Camshaft plug (15) should be renewed.

Governor flyweights are attached to camshaft on all models and care should be taken to avoid damaging when camshaft is being installed.

**VALVE SYSTEM.** Valve seats are renewable type inserts in cylinder block

Fig. W24 — Location of specification number (A) for special type crankcase and cylinder. If number is present, add to basic part number when renewing crankcase and cylinder.

and seats should be ground to 45° angle. Width should be 3/32-inch (2.381 mm).

Valve face is ground at 45° angle and stellite valves and seats and valve rotocaps are available for some models.

Valve stem clearance in guide should be 0.003-0.005 inch (0.0762-0.127 mm).

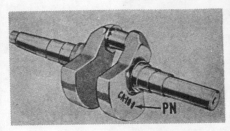

Fig. W25 — When renewing crankshaft, verify part number from crankshaft being removed from engine. Number appears on counterweight as shown at PN.

All models have renewable guides and valves with oversize stems are available for some models.

Valve tappet gap (cold) for HBKN engines should be 0.008 inch (0.2032 mm) for intake valve and 0.014 inch (0.3556 mm) for exhaust valve.

Valve tappet gap (cold) for HAENL engines should be 0.008 inch (0.2032 mm) for intake valve and 0.016 inch (0.4064 mm) for exhaust valve.

Valve gap is adjusted by turning adjusting screws on tappets if equipped or grinding end of valve stem if nonadjustable tappets are installed.

Illustrations courtesy of Teledyne Total Power

# WISCONSIN

**TELEDYNE TOTAL POWER**
3409 Democrat Road
Memphis, TN 38181

| Model | No. Cyls. | Bore | Stroke | Displacement | Power Rating |
|---|---|---|---|---|---|
| S-7D, HS-7D | 1 | 3 in. (76.2 mm) | 2.625 in. (66.68 mm) | 18.6 cu. in. (304.1 cc) | 7.25 hp. (5.4 kW) |
| S-8D, HS-8D | 1 | 3.125 in. (79.375 mm) | 2.625 in. (66.68 mm) | 20.2 cu. in. (330 cc) | 8.25 hp. (6.2 kW) |
| TR-10D | 1 | 3.125 in. (79.375 mm) | 2.625 in. (66.68 mm) | 20.2 cu. in. (330 cc) | 10 hp. (7.5 kW) |
| TRA-10D | 1 | 3.125 in. (79.375 mm) | 2.875 in. (73.03 mm) | 22.1 cu. in. (361.4 cc) | 10.1 hp. (7.8 kW) |
| S-10D | 1 | 3.25 in. (82.55 mm) | 3 in. (76.2 mm) | 24.9 cu. in. (407.8 cc) | 10.1 hp. (7.8 kW) |
| TRA-12D | 1 | 3.5 in. (88.9 mm) | 2.875 in. (73.03 mm) | 27.7 cu. in. (453 cc.) | 12 hp. (9.0 kW) |
| S-12D | 1 | 3.5 in. (88.9 mm) | 3 in. (76.2 mm) | 28.9 cu. in. (473 cc) | 12.5 hp. (9.3 kW) |
| S-14D | 1 | 3.75 in. (95.3 mm) | 3 in. (76.2 mm) | 33.1 cu. in. (543.5cc) | 14.1 hp. (10.5 kW) |

Engines in this section are four-cycle, one cylinder engines and all models except HS-7D and HS-8D are horizontal crankshaft engines. HS-7D and HS-8D engines have an adapter base for vertical mounting. Crankshaft for all models is supported at each end in taper roller bearings.

Connecting rod used in S-10D, S-12D and S-14D engines have renewable insert type bearings. Connecting rods on all other models ride directly on crankpin journal. All horizontal crankshaft models are splash lubricated by an oil dipper located on end of connecting rod and all vertical crankshaft models are pressure spray lubricated and have a plunger type oil pump located on lower end of crankshaft.

Various magneto, battery or solid-state electronic ignition systems are used according to model and application. Zenith or Walbro float type carburetor is used and a mechanical fuel pump is available.

Engine identification information is located on engine instruction plate and engine model, specification and serial number are required when ordering parts or service material.

## MAINTENANCE

**SPARK PLUG.** Recommended spark plug is Champion D16J or equivalent. Electrode gap should be 0.030 inch (0.762 mm).

**CARBURETOR.** A variety of float type carburetors are used as listed.

Zenith 72Y6 carburetor is standard on S-7D, S-8D, HS-7D and HS-8D engines. See Fig. W30.

Zenith 68-7 carburetor is standard on TR-10D and TRA-10D engines. See Fig. W31.

Walbro LME-35 carburetor is standard on TRA-12D engines. See Fig. W34.

1. Throttle body
2. Idle mixture needle
3. Idle speed adjusting screw
4. Throttle shaft
5. Needle valve seat
6. Needle valve
7. Throttle plate
8. Float pin
9. Float
10. Main fuel adjusting needle
11. Main jet
12. Fuel bowl
13. Choke plate
14. Choke shaft
15. Venturi
16. Idle tube
17. Gasket

Zenith 1408 carburetor is used on S-10D, S-12D and S-14D engines. See Fig. W35.

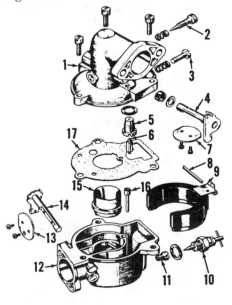

*Fig. W30—Exploded view of Zenith 72Y6 carburetor used on S-7D and S-8D engines and on HS-7D and HS-8D vertical shaft engines.*

For initial adjustment of Zenith 72Y6 carburetor (Fig. W30) open idle mixture screw (2) ½-turn and main fuel mixture screw (10) 2 turns.

For initial adjustment of Zenith 68-7 or Walbro LME-35 carburetor open idle mixture screw and main fuel mixture screw 1¼ turns. Refer to Fig. W31 or W34 for location of mixture screw of model being serviced.

For initial adjustment of Zenith 1408 carburetor (Fig. W35) open idle mixture screw (5) 1½ turns and main fuel mixture screw (13) 2¼ turns.

On all models make final adjustments with engine at normal operating temperature and running. Place engine under load and adjust main fuel mixture screw for leanest setting that will allow satisfactory acceleration and steady governor operation. Set engine at idle speed, no load and adjust idle mixture screw for smoothest idle operation.

As each adjustment affects the other, adjustment procedure may have to be repeated.

To check float level on all models, invert carburetor body and float assembly and refer to appropriate Figure for model being serviced. Adjust as necessary by bending float tang that contacts inlet valve.

**GOVERNOR.** Flyweight type governors driven by camshaft gear are used on all models.

Refer to Fig. W42 for S-7D, S-8D, HS-7D, HS-8D, TR-10D, TRA-10D and TRA-12D engine governor component parts and to Fig. W43 for S-10D, S-12D and S-14D engine governor component parts.

Major speed changes on all models is

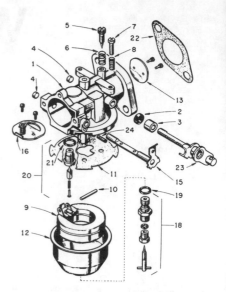

Fig. W35 — Exploded view of Zenith 1408 model carburetor which is used on S-10D, S12D and S-14D engines.

1. Carburetor body
2. Throttle shaft seal
3. Seal retainer
4. Cup plugs
5. Idle fuel needle
6. Spring
7. Idle speed stop screw
8. Spring
9. Float assy.
10. Float pin
11. Gasket
12. Fuel bowl
13. Throttle disc
15. Choke shaft
16. Choke disc
18. Main jet needle assy.
19. Washer
20. Inlet valve & seat assy.
21. Gasket
22. Gasket
23. Throttle shaft
24. Choke lever friction spring

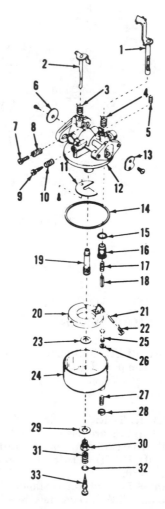

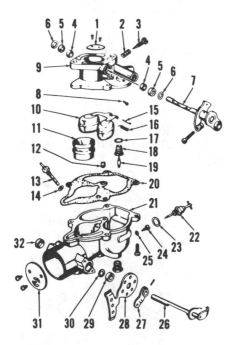

Fig. W31 — Zenith 68-7 model carburetor is used on TR-10D and TRA-10D engines.

1. Throttle plate
2. Spring
3. Idle mixture needle
4. Bushing
5. Seal
6. Retainer
7. Throttle shaft
8. Idle jet
9. Throttle body
10. Float
11. Venturi
12. Well vent
13. Discharge nozzle
14. Gasket
15. Float shaft
16. Float spring
17. Gasket
18. Inlet valve seat
19. Inlet valve
20. Gasket
21. Fuel bowl
22. Main fuel needle
23. Gasket
24. Main jet
25. Gasket
26. Choke shaft
27. Choke lever
28. Bracket
29. Retainer
30. Seal
31. Choke plate
32. Plug

Fig. W34 — Exploded view of Walbro LME-35 model carburetor used on TRA-12D engine.

1. Choke shaft
2. Throttle shaft
3. Throttle spring
4. Choke spring
5. Choke stop spring
6. Throttle plate
7. Idle speed screw
8. Spring
9. Idle mixture needle
10. Spring
11. Baffle
12. Carburetor body
13. Choke plate
14. Bowl gasket
15. Gasket
16. Inlet valve seat
17. Spring
18. Inlet valve
19. Main nozzle
20. Float
21. Float shaft
22. Spring
23. Gasket
24. Bowl
25. Drain stem
26. Gasket
27. Spring
28. Retainer
29. Gasket
30. Bowl retaier
31. Spring
32. "O" ring
33. Main fuel needle

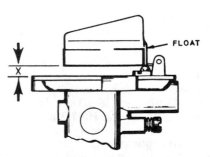

Fig. W36 — Zenith 72Y6 carburetor float must be parallel to casting. Dimension "X" should be the same, measured near hinge pin and at outer end of float.

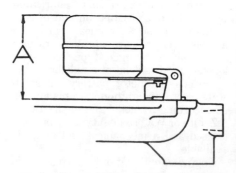

Fig. W37 — Zenith 68-7 carburetor float should be adjusted so dimension "A" is 1-5/32 inch (29.37 mm).

obtained by using springs of varying tension, flyweights of varying weights or locating spring or governor rod in alternate holes of governor control lever.

To correctly set governed speeds, use a tachometer to accurately record crankshaft rpm. Refer to appropriate table (I through IV) corresponding to engine model being serviced to determine proper governor lever hole for attaching governor spring to set required speed.

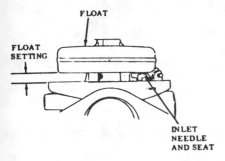

Fig. W38—Walbro LME-35 carburetor float should have 5/32-inch (3.97 mm) space between free end of float and gasket surface as shown.

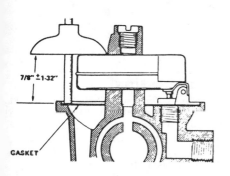

Fig. W39—Zenith 1408 carburetor float measurement should be 7/8-inch (22.23 mm) plus or minus 1/32-inch (0.794 mm) when measured as shown with gasket in place.

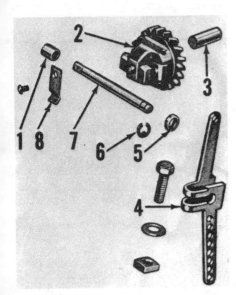

Fig. W42—Exploded view of governor assembly typical of that used on Models S-7D, HS-7D, S-8D, HS-8D, TR-10D, TRA-10D and TRA-12D.

1. Spacer
2. Gear/flyweight assy.
3. Governor shaft
4. Governor lever
5. Oil seal
6. Retaining ring
7. Fulcrum shaft
8. Vane

### Table I—Models S-7D, S-8D
**Engines to and including Serial Number 3909151**

| Desired Rpm Under Load | Hole Number | Adjust No-Load Rpm To: |
|---|---|---|
| 1600 | 3 | 1880 |
| 1700 | 3 | 1940 |
| 1800 | 3 | 1990 |
| 1900 | 3 | 2080 |
| 2000 | 4 | 2260 |
| 2100 | 4 | 2360 |
| 2200 | 4 | 2410 |
| 2300 | 4 | 2510 |
| 2400 | 4 | 2590 |
| 2500 | 4 | 2680 |
| 2600 | 5 | 2830 |
| 2700 | 5 | 2920 |
| 2800 | 5 | 2970 |
| *2900 | 5 | 3040 |
| *3000 | 6 | 3230 |
| *3100 | 6 | 3330 |
| *3200 | 6 | 3420 |
| *3300 | 6 | 3510 |
| *3400 | 6 | 3590 |
| *3500 | 7 | 3750 |
| *3600 | 7 | 3840 |

1600-2800 rpm, use 5-1/4 inch adjusting screw.
2900 rpm (*) and higher, use 5 inch adjusting screws.

**Engines beginning with Serial Number 3909152**

| Desired Rpm Under Load | Hole Number | Adjust No-Load Rpm To: |
|---|---|---|
| 1800 | 2 | 2030 |
| 1900 | 2 | 2125 |
| 2000 | 2 | 2220 |
| 2100 | 2 | 2320 |
| 2200 | 3 | 2430 |
| 2300 | 3 | 2520 |
| 2400 | 4 | 2690 |
| 2500 | 4 | 2720 |
| 2600 | 4 | 2845 |
| 2700 | 4 | 2930 |
| 2800 | 4 | 3010 |
| 2900 | 5 | 3150 |
| *3000 | 5 | 3230 |
| *3100 | 5 | 3300 |
| *3200 | 5 | 3350 |
| *3300 | 6 | 3575 |
| *3400 | 6 | 3650 |
| *3500 | 6 | 3750 |
| *3600 | 6 | 3800 |

1800-2900 rpm, use 5-5/8 inch adjusting screw.
3000 rpm (*) and higher, use 5-1/4 inch adjusting screw.

### Table II—Models HS-7D, HS-8D

| Desired Rpm Under Load | Hole Number | Adjust No-Load Rpm To: |
|---|---|---|
| 1800 | 4 | 2200 |
| 1900 | 4 | 2290 |
| 2000 | 4 | 2380 |
| 2100 | 4 | 2465 |
| 2200 | 4 | 2550 |
| 2300 | 5 | 2690 |
| 2400 | 5 | 2770 |
| 2500 | 5 | 2850 |
| 2600 | 6 | 3000 |
| 2700 | 6 | 3060 |
| 2800 | 6 | 3120 |
| 2900 | 6 | 3200 |
| *3000 | 6 | 3280 |
| *3100 | 6 | 3340 |
| *3200 | 6 | 3400 |
| *3300 | 7 | 3560 |
| *3400 | 7 | 3620 |
| *3500 | 7 | 3685 |
| *3600 | 7 | 3750 |

1800-2900 rpm, use 5-5/8 inch adjusting screw.
3000 rpm (*) and higher, use 5-1/4 inch adjusting screw.

### Table III—Models TRA-10D, TRA-12D, TR-10D
**Use Lever Spring Hole Number and Set No Load Rpm:**

| Desired Rpm Under Load | TRA-10D** | | TRA-12D | |
|---|---|---|---|---|
| 2000 | 1 | 2520 | 3 | 2230 |
| 2100 | 1 | 1580 | 4 | 2430 |
| 2200 | 1 | 2610 | 4 | 2515 |
| 2300 | 1 | 2690 | 4 | 2590 |
| 2400 | 1 | 2740 | 4 | 2660 |
| 2500 | 1 | 2800 | 4 | 2750 |
| 2600 | 1 | 2890 | 4 | 2810 |
| 2700 | 1 | 2935 | 5 | 3020 |
| 2800 | 2 | 3065 | 5 | 3100 |
| 2900 | 2 | 3160 | 5 | 3180 |
| *3000 | 3 | 3230 | 5 | 3260 |
| *3100 | 3 | 3300 | 5 | 3325 |
| *3200 | 3 | 3380 | 6 | 3535 |
| *3300 | 3 | 3460 | 6 | 3620 |
| *3400 | 4 | 3615 | 6 | 3700 |
| *3500 | 4 | 3690 | 6 | 3790 |
| *3600 | 5 | 3850 | 6 | 3860 |

**Applies to TR-10D engines Serial No. 3909152 and after.
2000-2900 rpm, TRA-10D uses 3-5/8 inch adjusting screw; TRA-12D uses 5-5/8 adjusting screw.
3000 rpm (*) and higher, TRA-10D uses 5 inch adjusting screw; TRA-12D uses 5-1/4 inch adjusting screw.

Table IV—Models S-10D, S-12D, S-14D

| Desired Rpm Under Load | Hole Number | Adjust No-Load Rpm To: |
|---|---|---|
| 1600 | 1 | 1760 |
| 1800 | 2 | 1975 |
| 1900 | 2 | 2040 |
| 2000 | 2 | 2120 |
| 2100 | 3 | 2260 |
| 2200 | 3 | 2340 |
| 2300 | 3 | 2400 |
| 2400 | 4 | 2580 |
| 2500 | 4 | 2650 |
| 2600 | 4 | 2720 |
| 2700 | 4 | 2810 |
| 2800 | 5 | 2910 |
| 2900 | 5 | 3010 |
| *3000 | 6 | 3150 |
| *3100 | 6 | 3230 |
| *3200 | 7 | 3360 |
| *3300 | 7 | 3455 |
| *3400 | 7 | 3520 |
| *3500 | 7 | 3590 |
| *3600 | 7 | 3680 |

1600-2900 rpm, use 3-15/16 inch adjusting screw.

3000 rpm (*) and higher, use 3-5/8 inch adjusting screws.

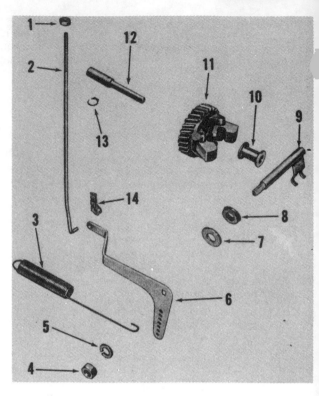

Fig. W43 — Governor mechanism as used on Models S-10D, S-12D and S-14D. Refer to text for service information.

1. Locknut
2. Throttle rod
3. Governor spring
4. Nut
5. Lockwasher
6. Governor lever
7. Flat washer
8. Oil seal
9. Cross (fulcrum) shaft
10. Thrust sleeve
11. Gear weight assy.
12. Shaft
13. Snap ring
14. Clip

**IGNITION SYSTEM.** Various magneto, battery or solid-state electronic ignition systems are used according to model and application. Refer to appropriate paragraph for model being serviced.

**MAGNETO.** Fairbanks-Morse or Wico magneto assembly located under flywheel is used (Fig. W47) and points and condenser are located in an external breaker box (Fig. W49 or W50) and are actuated by camshaft lobe via a short push rod.

Initial point gap is 0.018-0.020 inch (0.457-0.508 mm) and point gap is varied slightly to obtain correct timing as outlined in **BATTERY IGNITION** section.

**BATTERY IGNITION.** Battery ignition system uses a conventional ignition coil and points and condenser are the same as used for magneto system. Points and condenser location remains the same (Fig. W49 or W50) and initial point gap is 0.023 inch (0.5842 mm) and gap is varied slightly to obtain correct engine timing.

To set timing on either magneto system or battery system, position engine flywheel so timing marks appear in hole in flywheel shroud (Fig. W49) or align with timing pointer (Fig. W50), connect test light across points and adjust point gap so light just goes out.

Timing should be set at 15° BTDC on S-7D and HS-7D engines and 18° BTDC on all remaining models.

**SOLID-STATE ELECTRONIC IGNITION.** Breakerless capacitive-discharge (CD) ignition system is available for S-10D, S-12D and S-14D engines and is standard ignition system for TRA-12D engines.

No adjustments are possible and system has only three components which are magnet (part of flywheel), stator (containing trigger coil, rectifier diode and a silicone-controlled rectifier) located on bearing plate at flywheel side of engine and a special ignition coil.

If visual inspection of component parts fail to find possible cause of an ignition failure, a continuity test of ignition switch and coil should be made. If switch and coil are in working order but ignition system still fails, renew stator.

**LUBRICATION.** All horizontal crankshaft models are splash lubricated by an oil dipper attached to connecting rod cap. All vertical crankshaft models are pressure spray lubricated by a plunger type oil pump fitted to lower end of crankshaft.

Maintain oil level at full mark on dipstick or level with filler plug as equipped, but do not overfill. High quality detergent oil having API classification "SE" or "SF" is recommended. Recommended oil change interval is every 50 hours of normal operation.

Use SAE 30 oil when operating temperature is between 120°F (49°C)

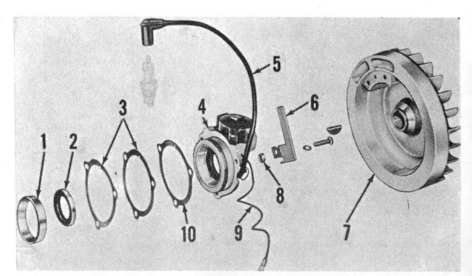

Fig. W47 — Typical flywheel magneto assembly. Note crankshaft end play is adjusted by shims (3) which are offered in a variety of thicknesses. Flywheel must be removed for access to magneto stator plate.

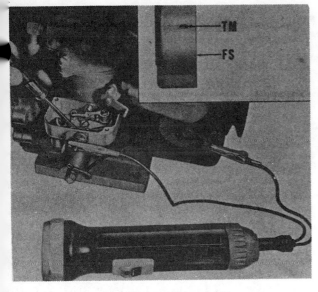

*Fig. W49 — Timing mark (TM) should appear in hole in flywheel shroud (FS) on S-7D and HS-7D engines as timing light goes out.*

and 40°F (4°C), SAE 20 oil between 40°F (4°C) and 15°F (−9°C), SAE 10 oil between 15°F (−9°C) and 0°F (−18°C) and SAE 5W-20 oil for below 0°F (−18°C) temperature.

## REPAIRS

**TIGHTENING TORQUES.** Recommended tightening torques are as follows:

**Models S-7D, S-8D, HS-7D and HS8D**

Gear cover . . . . . . . . . . . . . . . . .8 ft.-lbs.
(11N·m)
Stator plate. . . . . . . . . . . . . . . . .8 ft.-lbs.
(11 N·m)
Connecting rod . . . . . . . . . . . .18 ft.-lbs.
(24 N·m)
Spark plug . . . . . . . . . . . . . . . .29 ft.-lbs.
(39 N·m)
Cylinder head . . . . . . . . . . . . .18 ft.-lbs.
(24N·m)

**Models TR-10D, TRA-10D and TRA-12D**

Gear cover . . . . . . . . . . . . . . . . .8 ft.-lbs.
(11N·m)
Stator Plate . . . . . . . . . . . . . . . .8 ft.-lbs.
(11N·m)
Connecting rod . . . . . . . . . . . .22 ft.-lbs.
(30N·m)
Spark plug . . . . . . . . . . . . . . . .29 ft.-lbs.
(39 N·m)
Flywheel nut. . . . . . . . . . . . . . .55 ft.-lbs.
(75 N·m)
Cylinder head . . . . . . . . . . . . .18 ft.-lbs.
(24 N·m)

**Models S-10D, S-12D and S-14D**

Gear cover . . . . . . . . . . . . . . . .18 ft.-lbs.
(24 N−m)
Stator plate. . . . . . . . . . . . . . . .18 ft.-lbs.
(24 N·m)
Connecting rod . . . . . . . . . . . .32 ft.-lbs.
(43 N·m)
Spark plug . . . . . . . . . . . . . . . .29 ft.-lbs.
(39 N·m)

Flywheel nut. . . . . . . . . . . . . . .55 ft.-lbs.
(75 N·m)
Cylinder block nut . . . . . . . . . .50 ft.-lbs.
(68 N·m)
Cylinder head . . . . . . . . . . . . . .32 ft.-lbs.
(43 N·m)

**CYLINDER HEAD.** Always install a new head gasket when installing cylinder head. Note different lengths and styles of studs and cap screws for correct reinstallation and lightly lubricate threads.

Cylinder heads should be tightened in three equal steps to recommended torque using a criss-cross pattern working from the center out.

**CONNECTING ROD.** Connecting rod and piston are removed from cylinder head end of block after removing head and crankcase side cover. See Fig. W54 or W55.

Connecting rod for all models except S-10D, S-12D or S-14D ride directly on crankpin journal. S-10D, S-12D and S-14D models connecting rods have renewable insert type bearings.

Running clearance for insert type bearings should be 0.0005-0.0015 inch (0.0127-0.0381 mm). Running clearance for S-7D and HS-7D models rod should be 0.0012-0.002 inch (0.0305-0.0508 mm), for S-8D and HS-8D models clearance should be 0.0007-0.0015 inch (0.0178-0.0381 mm) and clearance for all TR and TRA models should be 0.0005-0.0015 inch (0.0127-0.0381 mm).

Side clearance for connecting rods

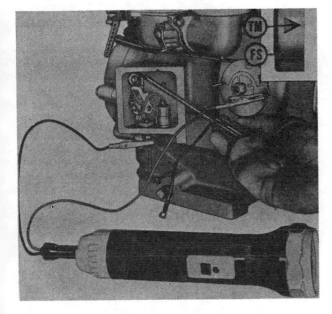

*Fig. W50-Timing mark (TM) should be aligned with timing pointer on flywheel shroud (FS) on all models except S-7D and HS-7D engines as timing light just goes out. Refer to Fig. W49 for S-7D and HS-7D engines.*

*Fig. W54 — Piston (P) is removed from top. Oil dipper (D) provides lubrication for horizontal crankshaft engines. Note placement of connecting rod index arrow at (A), location of governor shaft (GS) and that camshaft gear is fitted with a compression release (CR), typical of TR-TRA and larger models.*

Fig. W55—Open view of typical S-10D, S-12D or S-14D engine. Compare to Fig. W54, note difference in placement of governor shaft, style of oil dipper and that cylinder and crankcase are separate castings. Flywheel should be left in place to balance crankshaft when gear cover is removed.

with insert type bearing should be 0.004-0.013 inch (0.1016-0.3302 mm). Side clearance for S-7D, HS-7D, S-8D and HS-8D models should be 0.006-0.013 inch (0.1524-0.3302 mm) and side clearance for all TR and TRA models should be 0.009-0.016 inch (0.2286-0.4064 mm).

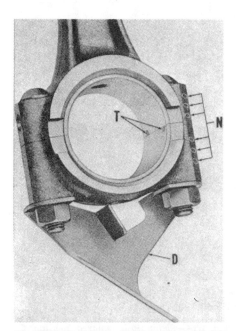

Fig. W56—Install connecting rod cap (S-10D, S-12D and S-14D) so tangs (T) of bearing inserts are on same side. Numbers (N) on rod end and cap should be aligned and installed toward gear cover side of crankcase. Oil dipper (D) open side faces outward.

Fig. W57—Sectional view showing proper arrangement of piston rings. In this typical view, top ring may not be chamfered inside as shown, however, all rings are marked with "TOP" or pit mark for correct installation.

When installing piston and connecting rod assembly on models which are not equipped with renewable bearing inserts align index arrows (A-Fig. W54) on rod end and cap. Arrows must face toward open end of crankcase. Horizontal shaft engines which are equipped with oil dipper (D) should have dipper installed so connecting rod cap screws are accessible from open end of crankcase. Refer to Fig. W56 when assembling cap to connecting rod on S-10D, S-12D and S-14D models and make certain fitting tangs (T) are on same side as shown. Stamped numbers (N) and oil dipper (D) should face open side of crankcase.

Standard connecting rod journal diameter for S-7D and HS-7D models should be 1.3750-1.3755 inches (34.93-34.94 mm), standard crankshaft diameter for S-8D, HS-8D, TR-10D, TRA-10D and TRA-12D should be 1.3755-1.3760 inches (34.94-34.95 mm) and standard diameter for S-10D, S-12D and S-14D models should be 1.4984-1.4990 inches (38.06-38.08 mm).

Connecting rods or bearing inserts are available in a variety of sizes to fit reground crankshafts.

Fig. W58—Location of specification number (A) on engines with cylinder cast as an integral part of crankcase.

Fig. W60—Locate timing mark (A) on camshaft gear between two marked teeth (B) of crankshaft gear. View is typical of S-7D, S-8D and all TR and TRA engines.

**PISTON, PIN AND RINGS.** Cam ground piston for all models is equipped with one chrome faced compression ring, one scraper ring and an oil control ring (Fig. W57). Top side of rings are marked for correct installation. Ring end gap for all models should be 0.010-0.020 inch (0.254-0.508 mm) and ring end gaps should be spaced at 120° intervals around piston. A variety of oversize pistons and rings are available.

Ring side clearance in groove for S-10D, S-12D and S-14D should be: Top and second ring, 0.002-0.004 inch (0.0508-0.1016 mm) and for oil control ring, 0.0015-0.0035 inch (0.0381-0.0889 mm).

Ring side clearance in groove for all remaining models should be: Top ring, 0.002-0.0035 inch (0.0508-0.0889 mm), second ring, 0.001-0.0025 inch (0.0254-0.0635 mm) and oil control ring, 0.002-0.0035 inch (0.0508-0.0889 mm).

Fig. W61—View of timing marks lined up in S-10D, S-12D and S-14D. In current prouction, camshaft gear will support a compression release as shown in Fig. W64. Camshaft thrust spring (S) and governor thrust sleeve (10) must be in place before installation of gear cover. Use heavy grease to hold camshaft thrust ball in cover hole during installation.

Piston to cylinder clearance should be measured at 90° angle to piston pin at lower edge of skirt. Clearance for S-7D, S-8D, HS-7D, HS-8D, TR-10D and TRA-10D models should be 0.004-0.0045 inch (0.1016-0.1143 mm), clearance for TRA-12D, S-10D and S-12D models should be 0.0025-0.003 inch (0.0635-0.0762 mm) and clearance for S-14D model should be 0.0025-0.004 inch (0.0635-0.1016 mm).

Floating type piston pin is retained by snap rings in bore of piston and should have 0.0002-0.0008 inch (0.0051-0.0203 mm) clearance in connecting rod of all models except S-10D, S-12D and S-14D. S-10D, S-12D and S-14D models use a bushing in connecting rod end and clearance in bushing should be 0.0005-0.0011 inch (0.0127-0.0279 mm).

Pins are available in a variety of over-sizes as well as standard. Connecting rod end or bushing and piston pin bore must be reamed to fit available oversize pin.

**CYLINDER.** Cylinder is an integral part of crankcase casting for S-7D, S-8D, HS-7D, HS-8D, TR-10D, TRA-10D and TRA-12D models (Fig. W58) but is a separate casting bolted to crankcase for S-10D, S-12D and S-14D models (Fig. W55).

Standard cylinder bore diameter for S-7D and HS-7D models should be 3 inches (76.2 mm), standard diameter for S-8D, HS-8D, TR-10D and TRA-10D models should be 3.125 inches (79.38 mm), standard diameter for S-10D model should be 3.25 inches (82.55 mm), standard diameter for TRA-12D and S-12D models should be 3.5 inches (88.9 mm) and standard diameter for S-14D model should be 3.75 inches (95.2 mm).

Cylinder should be bored to nearest oversize for which piston and rings are available if worn or out-of-round more than 0.005 inch (0.1270 mm).

Cylinder section of two piece assembly is bolted to crankcase section and one

cap screw is concealed within valve spring compartment (Fig. W67). Tighten concealed cap screw to 32 ft.-lbs. (43 N·m) torque and all remaining stud nuts to 42-50 ft.-lbs. (57-68 N·m) torque.

**CRANKSHAFT, MAIN BEARINGS AND SEALS.** Crankshaft on all models is supported at each end by taper roller bearings. Crankshaft end play should be 0.001-0.004 inch (0.0254-0.1016 mm) and is controlled by varying number of shims between crankcase and stator plate (main bearing support). See Fig. W47.

Main bearings are a press fit on crankshaft and bearing cups are pressed into stator plate and crankcase. Bearings and cups must be fully seated to correctly set crankshaft end play.

To properly time crankshaft to related engine components during installation, refer to Fig. W60 or W61 according to model being serviced.

Seals should be installed in stator plate and crankcase prior to crankshaft installation and seal protectors (Fig. W62 or W63) should be used to prevent seal damage during crankshaft installation.

**GOVERNOR GEAR AND WEIGHT ASSEMBLY.** Governor gear and weight assemblies rotate on a shaft which is a press fit in a bore of crankcase. Exploded views are shown in Figs. W42 and W43. On S-7D, S-8D, HS-7D, HS-8D and all TR and TRA models shaft (3-Fig. W42) has had its depth in block held by a snap ring beginning with production serial number 3090152. Models S-10D, S-12D and S-14D require end play of 0.003-0.005 inch (0.0762-0.1270 mm) be maintained on governor gear shaft between gear and its snap ring retainer. See Fig. W65 for measurement technique to be used on these models. Press-fit shaft is driven in or out of bore to make adjustment. On models with straight governor lever and governor gear mounted above cam gear,

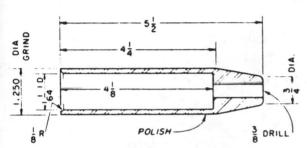

SLEEVE FOR ASSEMBLING GEAR COVER
WITH OIL SEAL ON TO CRANKSHAFT

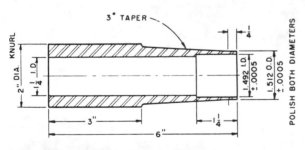

SLEEVE FOR ASSEMBLING GEAR COVER
WITH OIL SEAL, ON TO CRANKSHAFT.

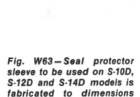

*Fig. W62—Dimensions of seal protector sleeve to be used on S-7D, S-8D, HS-7D, HS-8D, TR-10D, TRA-10D and TRA-12D models.*

*Fig. W63—Seal protector sleeve to be used on S-10D, S-12D and S-14D models is fabricated to dimensions shown.*

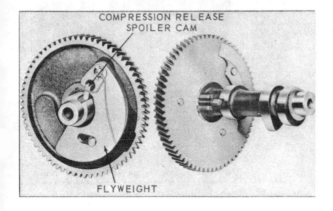

*Fig. W64—View of both sides of camshaft gear to show compression release assembly installed.*

*Fig. W65—Make certain snap ring is correctly seated and use feeler gage to measure end play of governor gear on shaft.*

note in Fig. W66 upper end of governor lever must tilt toward engine so governor vane will not be fouled or interfere with flyweights as gear cover is installed. On S-10D, S-12D and S-14D models, governor thrust sleeve must be in place as shown in Fig. W67 when gear cover is placed in position.

Camshafts on all models are supported at each end in unbushed bores in crankcase and gear cover. Camshaft end play is controlled by a thrust spring (Fig. W66 and W67) fitted into shaft hub which centers upon steel ball in a socket in gear cover. During assembly, ball is held in place by a coating of heavy grease. To remove or reinstall camshaft place block on its side as shown if Fig. W68 to prevent tappets from falling out.

On models equipped with a compression release type camshaft (Fig. W64), a spoiler cam holds exhaust valve slightly open during part of compression stroke while cranking. Reduced compression pressure allows for faster cranking speed with lower effort. When crankshaft reaches 650 rpm during cranking, centrifugal force swings flyweight on front of cam gear so as to turn spoiler cam to inoperative position allowing exhaust valve to seat and restore full compression. Whenever camshaft is removed, compression release mechanism should be checked for damage to spring or excessive wear on spoiler cam. Flyweight and spoiler cam must move easily with no binding.

See Figs. W60 and W61 to set timing marks in register during reassembly.

**VALVE SYSTEM.** Exhaust valve seats on all models are renewable insert type and intake seat may be machined directly into block or be of renewable insert type according to model and application. Seats should be ground to a 45° angle and width should not exceed 3/32-inch (2.381 mm).

Valve faces are ground at 45° angles and stellite exhaust valves and rotocaps are standard on all models.

Renewable guides are pressed in or out from top side of block. Tool, DF-72 guide driver is available from Wisconsin. Internal chamfered end of guide is installed toward tappet end of valve.

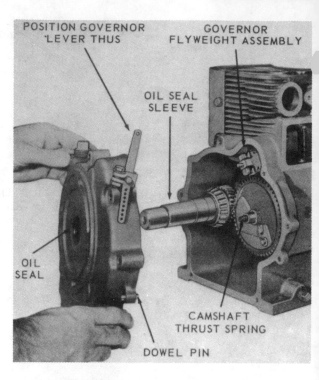

Fig. W66 — To install gear cover on all models where cylinder is an integral part of crankcase, governor assembly, camshaft thrust spring and oil seal protector must be in place. Make certain governor lever is tilted as shown and governor thrust ball is held in cover with heavy grease.

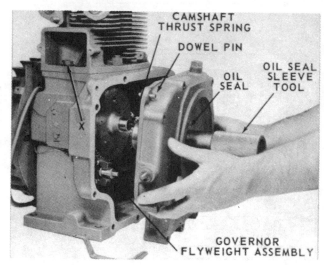

Fig. W67 — Installation of gear cover on engines where cylinder is separate from crankcase. Note cap screw (X) in valve compartment, referred to in text. Protect cover oil seal with sleeve tool as shown and make certain thrust sleeve is in place on governor shaft.

Inside diameter of valve guides should be 0.312-0.313 inch (7.93-7.95 mm) and valve stem diameter should be 0.310-0.311 inch (7.87-7.90 mm) for all intake valves and 0.309-0.310 inch (7.85-7.87 mm) for all exhaust valves except for S-10D, S-12D and S-14D exhaust valves which should be 0.308-0.309 inch (7.82-7.85 mm).

Maximum stem to guide clearance for all models except S-10D, S-12D and S-14D should be 0.006 inch (0.1524 mm). Clearance for S-10D and S-14D should be 0.007 inch (0.1778 mm).

Valve tappet gap (cold) for S-10D, S-12D and S-14D should be 0.007 inch

Fig. W68 — Place engine on its side as shown to prevent tappets dropping from block bores when camshaft is removed.

Illustrations courtesy of Teledyne Total Power

Fig. W69 — Reed type breather valve located in valve spring compartment of S-7D, S-8D, HS-7D, HS-8D, TR-10D, TRA-10D and TRA-12D models.

(0.1778 mm) for intake valves and 0.016 inch (0.4064 mm) for exhaust valves. Ad-

Fig. W70 — Breather valve used on S-10D, S-12D and S-14D models is located in valve spring compartment cover. Drain hole (H) must be kept open.

just by turning self-locking cap screw on tappet as requred.

Valve tappet gap (cold) for all remaining models should be 0.006 inch (0.1524 mm) for intake and 0.012 inch (0.3048 mm) for all exhaust valves except ones used on TRA-12D which should be 0.015 inch (0.3810 mm). Adjustment is made by carefully grinding valve stem end until required clearance is reached.

**BREATHER** A reed type breather valve (Fig. W69 or W70) is located in valve spring compartment. These reed valve assemblies should be kept clean and renewed whenever found to be inoperable.

If oil fouling occurs in ignition breaker box, condition of breather valve should be checked.

# WISCONSIN

## SERVICING WISCONSIN ACCESSORIES

### 12-VOLT MOTOR-GENERATOR

Combination motor-generator manufactured by Delco-Remy is used on some Wisconsin engines. Motor-generator functions as a cranking motor when starting switch is closed. When engine is operating and with starting switch open, unit operates as a generator. Generator output and circuit voltage for battery and various operating requirements are controlled by a current voltage regulator. See Figs. W130, W131 and W132.

Motor-generator belt tension should be adjusted until about 10 pounds (45 N) pressure applied midway between pulleys will deflect belt ½-inch (12.7 mm).

To determine cause of abnormal operation, motor-generator should be given a "no load" test or a "generator output" test. Generator output test can be performed with motor-generator on or off engine. No load test must be made with motor-generator removed from engine. Refer to Fig. W133 for exploded view of a typical motor-generator assembly. Parts are available from Wisconsin as well as authorized Delco-Remy service stations.

Motor-generator and regulator service test specifications are as follows:

#### Motor-Generator 1101696
Brush spring tension, oz. . . . . . . . 22-26
Field draw,
  Amperes . . . . . . . . . . . . . . . . .1.43-1.54
  Volts . . . . . . . . . . . . . . . . . . . . . . .12
Cold output,
  Amperes . . . . . . . . . . . . . . . . . . . .10
  Volts . . . . . . . . . . . . . . . . . . . . . . .14
  Rpm . . . . . . . . . . . . . . . . . . . . . .5750
No load test,
  Volts . . . . . . . . . . . . . . . . . . . . . . .11
  Amperes, max . . . . . . . . . . . . . . . .17
  Rpm, min. . . . . . . . . . . . . . . . . . .2350
  Rpm, max . . . . . . . . . . . . . . . . . .2850

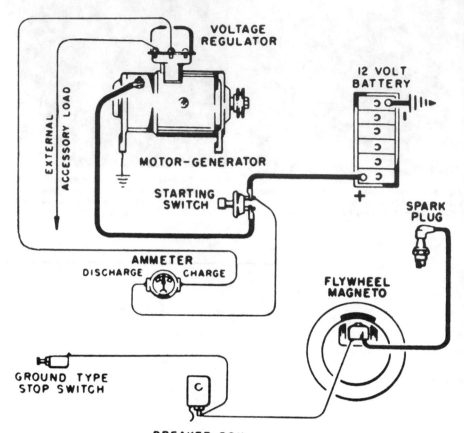

Fig. W130 — Typical wiring layout for combination motor-generator on engine equipped with flywheel magneto ignition.

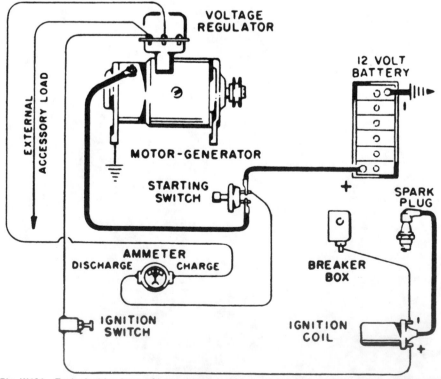

Fig. W131 — Typical wiring layout for combination motor-generator on engine equipped with battery ignition.

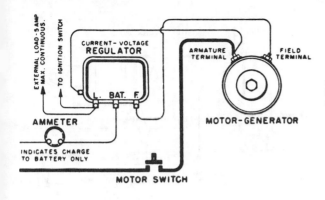

*Fig. W132—Wiring connections for current voltage regulator used with motor-generator.*

### Motor-Generator 1101870

Brush spring tension, oz . . . . . . . . 22-26
Field draw,
  Amperes . . . . . . . . . . . . . . . 1.52-1.62
  Volts . . . . . . . . . . . . . . . . . . . . . 12
Cold output,
  Amperes . . . . . . . . . . . . . . . . . . 12
  Volts . . . . . . . . . . . . . . . . . . . . . 14
  Rpm . . . . . . . . . . . . . . . . . . . . 4950
No load test,
  Volts . . . . . . . . . . . . . . . . . . . . . 11
  Amperes, max . . . . . . . . . . . . . . 18
  Rpm, min . . . . . . . . . . . . . . . . 2500
  Rpm, max . . . . . . . . . . . . . . . . 2900

### Motor-Generator 1101871 & 1101972

Brush spring tension, oz . . . . . . . . 24-32
Field draw,
  Amperes . . . . . . . . . . . . . . . 1.43-1.54
  Volts . . . . . . . . . . . . . . . . . . . . . 12
Cold output,
  Amperes . . . . . . . . . . . . . . . . . . 10
  Volts . . . . . . . . . . . . . . . . . . . . . 14
  Rpm . . . . . . . . . . . . . . . . . . . . 5450
No load test,
  Volts . . . . . . . . . . . . . . . . . . . . . 11
  Amperes, max . . . . . . . . . . . . . . 17
  Rpm, min . . . . . . . . . . . . . . . . 2500
  Rpm, max . . . . . . . . . . . . . . . . 3000

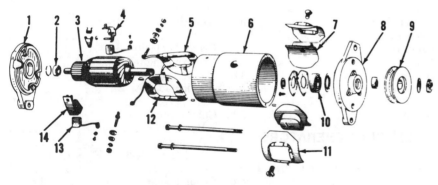

*Fig. W133—Exploded view of typical Delco-Remy motor-generator.*

1. Commutator end frame
2. Bearing
3. Armature
4. Ground brush holder
5. Field coil (L.H.)
6. Frame
7. Pole shoe
8. Drive end frame
9. Pulley
10. Bearing
11. Field coil insulator
12. Field coil (R.H.)
13. Brush
14. Insulated brush holder

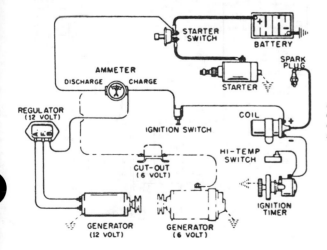

*Fig. W134—Typical wiring diagram for starting and charging system used on some Wisconsin one cylinder engines.*

### Regulators 1118791 & 1118985

Ground Polarity . . . . . . . . . . . . Positive
Cut-out relay,
  Air gap . . . . . . . . . . . . . . . . . . 0.020
    (0.508 mm)
  Point gap . . . . . . . . . . . . . . . . 0.020
    (0.508 mm)
  Closing voltage, range . . . . . 11.8-14.0
  Adjust to . . . . . . . . . . . . . . . . . 12.8
Voltage regulator,
  Air gap . . . . . . . . . . . . . . . . . . 0.075
    (1.905 mm)
  Setting voltage, range . . . . . 13.6-14.5
  Adjust to . . . . . . . . . . . . . . . . . 14.0

### Regulators 1118983 & 1118984

Ground polarity . . . . . . . . . . . Negative
Cut-out relay,
  Air gap . . . . . . . . . . . . . . . . . . 0.020
    (0.508 mm)
  Point gap . . . . . . . . . . . . . . . . 0.020
    (0.508 mm)
  Closing voltage, range . . . . . 11.8-14.0
  Adjust to . . . . . . . . . . . . . . . . . 12.8
Voltage regulator,
  Air gap . . . . . . . . . . . . . . . . . . 0.075
    (1.905 mm)
  Setting voltage, range . . . . . 13.6-14.5
  Adjust to . . . . . . . . . . . . . . . . . 14.0

## 6-VOLT STARTER MOTORS

Autolite and Prestolite 6-volt starter motors are used on some Wisconsin engines. Test specifications are as follows:

### Autolite MZ-4118, MZ-4175 & MZ-4184

Volts . . . . . . . . . . . . . . . . . . . . . . . 6
Brush spring tension, oz . . . . . . . . 42-53
No load test,
  Volts . . . . . . . . . . . . . . . . . . . . . 5
  Amperes . . . . . . . . . . . . . . . . . . 68
  Rpm . . . . . . . . . . . . . . . . . . . . 4000

### Autolite MAK-4008

Volts . . . . . . . . . . . . . . . . . . . . . . . 6
Brush spring tension, oz . . . . . . . . 38-61
No load test,
  Volts . . . . . . . . . . . . . . . . . . . . . 4
  Amperes . . . . . . . . . . . . . . . . . . 70
  Rpm . . . . . . . . . . . . . . . . . . . . 4700

### Autolite MDH-4001M

Volts . . . . . . . . . . . . . . . . . . . . . . . 6
Brush spring, tension, oz . . . . . . . . 42-66
No load test,
  Volts . . . . . . . . . . . . . . . . . . . . . 5
  Amperes . . . . . . . . . . . . . . . . . . 52
  Rpm . . . . . . . . . . . . . . . . . . . . 8100

### Prestolite MZ-4213

Volts . . . . . . . . . . . . . . . . . . . . . . . 6
Brush spring tension, oz . . . . . . . . 42-53
No load test,
  Volts . . . . . . . . . . . . . . . . . . . . 5.5
  Amperes . . . . . . . . . . . . . . . . . . 60
  Rpm . . . . . . . . . . . . . . . . . . . . 5000

## 12-VOLT STARTER MOTORS

Autolite, Prestolite and Delco-Remy 12-volt starters are used on some Wisconsin engines. Test specifications are as follows:

**Autolite MDL-6001**

Volts ........................12
Brush spring tension, oz. ........52-65
No load test,
   Volts ......................10
   Amperes ....................100
   Rpm ........................4700

**Autolite MDH-4002M**

Volts ........................12
Brush spring tension, oz ........42-66
No load test,
   Volts ......................10
   Amperes ....................38
   Rpm ........................10000

**Autolite MDL-6011**

Volts ........................12
Brush spring tension, oz. ........31-47
No load test,
   Volts ......................10
   Amperes ....................60
   Rpm ........................3200

**Prestolite MBG-4140**

Volts ........................12
Brush spring tension, oz. ....42-53
No load test,
   Volts ......................10
   Amperes ....................55
   Rpm ........................5200

**Prestolite MDY-7006**

Volts ........................12
Brush spring, tension, oz........42-53
No load test,
   Volts ......................10
   Amperes ....................60
   Rpm ........................4200

**Delco-Remy 1107246**

Volts ........................12
Brush spring tension, oz..........35
No load test,
   Volts ......................10.3
   Amperes ....................75
   Rpm ........................6900

### 6-VOLT GENERATORS

Autolite and Prestolite 6-volt generators are used on some Wisconsin engines. Test specifications are as follows:

**Autolite GAS-4301, GAS-4302, GAS-4303, GAS-4305 & GAS-4306**

Volts ........................6
Ground polarity .............Positive
Brush spring tension, oz. ........15-20
Field draw,
   Volts ......................5
   Amperes ...................3.4-3.8
Cold output,
   Volts ......................8
   Amperes ...................12.5

**Autolite GAS-4177**

Volts ........................6
Ground polarity .............Negative
Brush spring tension, oz. ........15-20
Field draw,
   Volts ......................5
   Amperes ...................3.4-3.8
Cold output,
   Volts ......................8
   Amperes ...................12.5

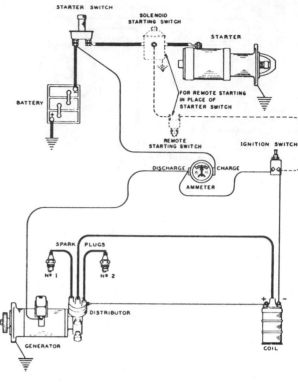

*Fig. W135 — Typical wiring diagram for starting and charging systems used on some Wisconsin two cylinder engines. Note distributor mounting on rear of gear driven generator.*

**Prestolite GAS-4301-1**

Volts ........................6
Ground polarity ...........Negative
Brush spring tension, oz .........15-20
Field draw,
   Volts ......................5
   Amperes .................3.4-3.8
Cold output,
   Volts ......................8
   Amperes.....................7.1

### 12-VOLT GENERATORS

Autolite, Prestolite and Delco-Remy 12-volt generators are used on some

Wisconsin engines. Test specifications are as follows:

**Autolite GJG-4001-M**

Volts ........................12
Ground polarity ...........Negative
Brush spring tension, oz. ........12-24
Field draw,
   Volts ......................10
   Amperes ..................1.7-1.9
Cold output,
   Volts ......................15
   Amperes ....................10

**Autolite GHH-6001-F**

Volts ........................12
Ground polarity .............Positive

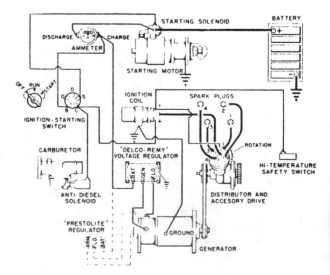

*Fig. W136 — Typical wiring diagram for starting and charging systems used on some Wisconsin V4 engines.*

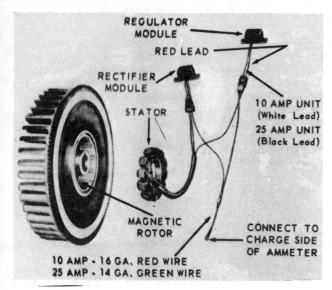

**Fig. W137—View of 10 or 25 amp flywheel alternator charging system used on some engines.**

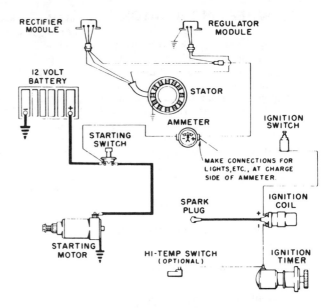

**Fig. W138—Typical wiring diagram for ignition, starting and alternator charging systems used on some single cylinder engines.**

Brush spring tension, oz . . . . . . . . .26-46
Field draw,
    Volts . . . . . . . . . . . . . . . . . . . . .10
    Amperes . . . . . . . . . . . . . . .1.6-1.7
Cold output,
    Volts . . . . . . . . . . . . . . . . . . . . .15
    Amperes . . . . . . . . . . . . . . . . ,,14.7

## Prestolite GJG-4001-MP
Volts . . . . . . . . . . . . . . . . . . . . . . .12
Ground polarity . . . . . . . . . . . .Positive
Brush spring tension, oz . . . . . . . .12-24
Field draw,
    Volts . . . . . . . . . . . . . . . . . . . . .10
    Amperes . . . . . . . . . . . . . . .1.7-1.9
Cold output,
    Volts . . . . . . . . . . . . . . . . . . . . .15
    Amperes . . . . . . . . . . . . . . . . . .10

## Prestolite GJG-4010-M
Volts . . . . . . . . . . . . . . . . . . . . . . .12
Ground polarity . . . . . . . . . . .Negative
Brush spring tension, oz. . . . . . .12-24
Field draw,
    Volts . . . . . . . . . . . . . . . . . . . . .10
    Amperes . . . . . . . . . . . . . . .1.7-1.9
Cold output,
    Volts . . . . . . . . . . . . . . . . . . . . .15
    Amperes . . . . . . . . . . . . . . . . . .10

## Prestolite GJY-7401-S
Volts . . . . . . . . . . . . . . . . . . . . . . .12
Ground polarity . . . . . . . . . . . .Positive
Brush spring tension, oz. . . . . . . .35-53
Field draw,
    Volts . . . . . . . . . . . . . . . . . . . . .10
    Amperes . . . . . . . . . . . . . .1.53-1.70
Cold output,
    Volts . . . . . . . . . . . . . . . . . . . . .15
    Amperes . . . . . . . . . . . . . . . . . .17

## Prestolite GJY-7401-SN
Volts . . . . . . . . . . . . . . . . . . . . . . .12
Ground polarity . . . . . . . . . . .Negative
Brush spring tension, oz. . . . . . . .35-53
Field draw,
    Volts . . . . . . . . . . . . . . . . . . . . .10
    Amperes . . . . . . . . . . . . . .1.53-1.70

Cold output,
    Volts . . . . . . . . . . . . . . . . . . . . .15
    Amperes . . . . . . . . . . . . . . . . . .17

## Delco-Remy 1102225
Volts . . . . . . . . . . . . . . . . . . . . . . .12
Ground polarity . . . . . . . . . . .Positive
Brush spring tension, oz . . . . . . . . .28
Field draw,
    Volts . . . . . . . . . . . . . . . . . . . . .12
    Amperes . . . . . . . . . . . . .1.67-1.79
Cold output,
    Volts . . . . . . . . . . . . . . . . . . . . .14

**Fig. W139—Typical wiring diagram for ignition, starting and alternator charging systems used on some four cylinder engines. Wiring diagram for two cylinder engines is similar.**

Amperes . . . . . . . . . . . . . . . . . . . .17

## Delco-Remy 1102343
Volts . . . . . . . . . . . . . . . . . . . . . . .12
Ground polarity . . . . . . . . . . .Negative
Brush spring tension, oz . . . . . . . . .28
Field draw,
    Volts . . . . . . . . . . . . . . . . . . . . .12
    Amperes . . . . . . . . . . . . .1.67-1.79
Cold output
    Volts . . . . . . . . . . . . . . . . . . . . .14
    Amperes . . . . . . . . . . . . . . . . . .17

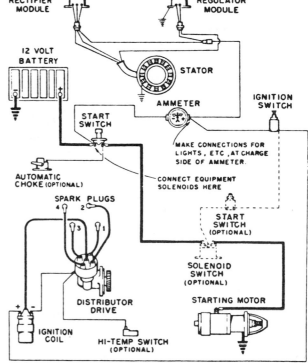

Illustrations courtesy of Teledyne Total Power

## FLYWHEEL ALTERNATOR

Beginning with serial number 5188288, some Wisconsin engines may be equipped with either a 10 amp or 25 amp flywheel alternator. See Fig. W137. To avoid possible damage to alternator system, the following precautions must be observed:

1. Negative post of battery must be connected to ground on engine.

2. Connect booster battery properly (positive to positive and negative to negative.)

3. Do not attempt to polarize alternator.

4. Do not ground any wires from stator or modules which terminate at connectors.

5. Do not operate engine with battery disconnected.

6. Disconnect battery cables when charging battery with a battery charger.

**OPERATION.** Alternating current (AC) produced by alternator is changed to direct current (DC) in rectifier module. See Figs. W138 and W139. Current regulation is provided by regulator module which "sense" counter-voltage created by battery to control or limit charging rate. No adjustments are possible on alternator charging system. Faulty components must be renewed. Refer to following trouble shooting paragraph to help pin point faulty component.

**TROUBLE SHOOTING.** Trouble conditions and their possible causes are as follows:

1. Full charge—no regulation. Could be caused by:
   a. Faulty regulator module
   b. Defective battery
2. Low or no charge. Could be caused by:
   a. Faulty windings in stator
   b. Faulty rectifier module
   c. Regulator module not properly grounded or regulator module defective.

If "full charge—no regulation" is the trouble, use a DC voltmeter and check battery voltage with engine operating at full rpm. If battery voltage is over 15.0 volts, regulator module is not functioning properly. If battery voltage is under 15.0 volts and over 14.0 volts, alternator, rectifier and regulator are satisfactory and battery is probably defective (unable to hold charge).

If "low" or "no charge" is the trouble, check battery voltage with engine operating at full rpm. If battery voltage is more than 14.0 volts, place a load on battery to reduce voltage to below 14.0 volts. If charge rate increases, alternator charging system is functioning properly and battery was fully charged. If charge increases, permanently install new rectifier module. If charge rate does not increase, stop engine and unplug all connectors between modules and stator. Start engine and operate at 2400 rpm. Using an AC voltmeter, check voltage between each black stator lead and ground. If either of the two voltage readings is zero or there is more than 10% difference between readings, stator is faulty and should be renewed.

# WISCONSIN ROBIN

**TELEDYNE TOTAL POWER**
3409 Democrat Road
Memphis, TN 38181

| Model | No. Cyls. | Bore | Stroke | Displacement | Power Rating |
|-------|-----------|------|--------|--------------|--------------|
| EY25W | 1 | 2.83 in. (71.88 mm) | 2.44 in. (61.98 mm) | 15.4 cu. in. (251.5 cc) | 6.5 hp. (4.9 kW) |
| EY27W | 1 | 2.91 in. (73.91 mm) | 2.44 in. (61.98 mm) | 16.26 cu. in. (265.5 cc) | 7.5 hp. (5.6 kW) |
| EY44W | 1 | 3.54 in. (89.92 mm) | 2.68 in. (68.07 mm) | 26.4 cu. in. (432.3 cc) | 10.5 hp. (7.8 kW) |
| EY21W | 2 | 2.95 in. (74.93 mm) | 2.76 in. (70.10 mm) | 37.74 cu. in. (618.2 cc) | 16.8 hp. (12.5 kW) |

EY25W, EY27W and EY44W models are four cycle, horizontal crankshaft, one cylinder engines and EY21W models are four cycle, horizontal crankshaft, two cylinder opposed engines. Crankshaft of one cylinder engines is supported at each end in ball bearings and crankshaft for two cylinder engine is supported by a ball bearing at front of shaft and by a roller bearing at rear of shaft.

Connecting rod of EY25W and EY27W model one cylinder engines and EY21W model two cylinder engine ride directly on crankshaft. Connecting rod for EY44W model is equipped with renewable insert type connecting rod bearings. All models are splash lubricated by a dipper attached to connecting rod cap.

One cylinder models have a magneto ignition system with points and condenser located beneath flywheel. Two cylinder models have battery type ignition with points and condenser located in an ignition timer located on top of engine.

A side draft, float type carburetor is used on one cylinder models and a downdraft, float type carburetor is used on two cylinder models. A mechanical type fuel pump is available for some models.

Engine model and specification number are stamped on name plate located on flywheel shroud and engine serial number for one cylinder models is stamped on crankcase base left mounting flange. Serial number for two

cylinder models is stamped on crankcase just below fuel pump.

Always give specification, model and

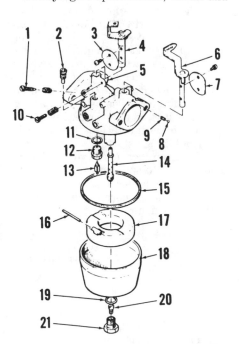

**Fig. WR1 — Exploded view of typical Mikuni float type carburetor used on EY25W, EY27W and EY44W models.**

1. Idle mixture needle
2. Idle jet
3. Throttle plate
4. Throttle shaft
5. Carburetor body
6. Choke shaft
7. Choke plate
8. Choke detent ball
9. Detent spring
10. Idle speed stop screw
11. Gasket
12. Fuel valve seat
13. Inlet fuel valve
14. Main nozzle
15. Bowl gasket
16. Float pin
17. Float
18. Fuel bowl
19. Gasket
20. Main fuel jet
21. Jet holder

serial number when ordering parts or service material.

## MAINTENANCE

**SPARK PLUG.** Recommended spark plug is Champion L86, or equivalent, for one cylinder models and Champion J8, or equivalent, for two cylinder models. Electrode gap is 0.020-0.025 inch (0.508-0.635 mm) for all models.

**CARBURETOR.** Mikuni float type carburetors are used on all models. Side draft style carburetor (Fig. WR1) is used on one cylinder models and downdraft style carburetor (Fig. WR2) is used on two cylinder models.

For initial carburetor adjustment, open idle mixture screw 1-5/8 turns for EY25W model, 1-1/2 turns for EY27W model, 1-3/4 turns for EY44W model and 1-1/4 turns for EY21W model.

Main fuel mixture is controlled by jet size and is non-adjustable.

Make final adjustment with engine at normal operating temperature and running. Set engine at idle speed (1000 rpm for EY21W model, 1200 rpm for EY44W model and 1250 rpm for EY25W and EY27W models) no load and adjust idle mixture screw to obtain smoothest idle operation and acceleration.

To check float level on one cylinder models, position carburetor as shown in Fig. WR3.

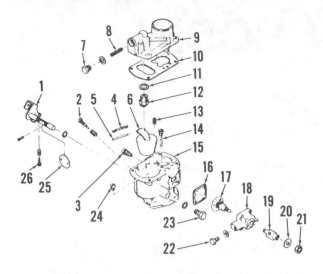

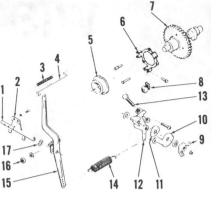

1. Throttle shaft
2. Idle mixture needle
3. Idle jet
4. Float pin retainer
5. Float pin
6. Float
7. Plug
8. Fuel screen
9. Upper body
10. Gasket
11. Gasket
12. Fuel inlet valve assy.
13. Idle air jet
14. Main nozzle
15. Lower body
16. Gasket
17. Starting valve
18. Valve housing
19. Starting lever
20. Washer
21. Nut.
22. Plug
23. Starting jet
24. Main fuel jet
25. Throttle plate
26. Idle speed stop screw

**NOTE: Needle valve is spring loaded and float tab should just contact needle valve pin, but should not compress spring.**

**Fig. WR5 — Exploded view of typical mechanical governor and linkage used on EY25W, EY27W and EY44W models.**

1. Governor lever shaft
2. Yoke
3. Link spring
4. Carburetor link
5. Thrust sleeve
6. Governor plate
7. Camshaft assy.
8. Flyweights (3 used)
9. Wing nut
10. Stop plate
11. Wave washer
12. Control lever
13. Speed stop screw
14. Governor spring
15. Governor lever
16. Clamp nut
17. Retaining ring

**VIEW LOOKING DOWN**
Carburetor Setting on Manifold Flange

FLOAT HINGE PIN

FLOAT

TAB JUST CONTACTING NEEDLE VALVE

0.842 - 0.921 For EY44W
0.910 - 0.980 For EY25W and EY27W

Fig. WR3 — With fuel bowl removed, stand carburetor on manifold flange and measure float setting as shown for EY25W, EY27W and EY44W models.

Use a depth gage to measure distance between body flange and free end of float (Fig. WR3). Distance should be 0.842-0.921 inch (21.39-23.39 mm) for EY44W model or 0.910-0.980 inch (23.11-24.89 mm) for EY25W and EY27W models.

Adjust float by bending float lever tab that contacts inlet valve.

To check float level on two cylinder models, refer to Fig. WR4 and note float setting will be within specified range if distance from top of float lever to bottom of float is 1-13/64 inches (30.56 mm) as shown. If necessary, bend float lever to obtain correct setting. Float lever must be parallel with float in area where fuel inlet valve contacts float lever.

Mikuni downdraft carburetor used on two cylinder model has a special starting fuel valve instead of a choke. Fuel from fuel bowl in lower body (15 – Fig. WR2) is metered through starting jet (23) and is mixed with air from inside vent. Mixture then flows into starting valve assembly where additional air from main air intake maintains a correct air/fuel ratio for easy starting.

**GOVERNOR.** Mechanical flyweight governor used on all models is mounted on and operated by camshaft gear (Fig. WR5 and WR6). Engine speed is controlled by tension on governor spring.

Before attempting to adjust governed speed on one cylinder models, synchronize governor linkage by loosening clamp nut (16-Fig. WR5) and turning governor lever (15) counter-clockwise until carburetor throttle plate is in wide open position. Insert screwdriver in slot in end of governor lever shaft (1) and rotate shaft counter-clockwise as far as possible. Tighten governor clamp nut.

No governor linkage adjustment is possible on two cylinder models as a slotted hole in governor lever link (Fig. WR6) fits on flat sides of governor shaft end.

To adjust all models for a particular loaded rpm, loosen wing nut (9-Fig. WR5) or lock knob (Fig. WR6). Start

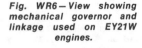

Fig. WR6 — View showing mechanical governor and linkage used on EY21W engines.

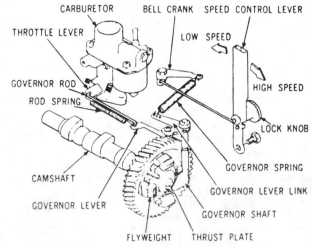

CARBURETOR    BELL CRANK    SPEED CONTROL LEVER
THROTTLE LEVER    LOW SPEED
GOVERNOR ROD    HIGH SPEED
ROD SPRING
LOCK KNOB
GOVERNOR SPRING
CAMSHAFT
GOVERNOR LEVER LINK
GOVERNOR SHAFT
GOVERNOR LEVER    THRUST PLATE
FLYWEIGHT

Fig. WR4 — Float setting on EY21W models will be correct if distance from top of float lever to bottom of float is 1-13/64 inches (30.56 mm). Refer to text.

engine and adjust stop screw at control lever to obtain correct no-load speed. If engine is to operate at a fixed speed, tighten wing nut or lock knob. For variable speed operation, do not tighten wing nut or lock knob.

For the following loaded engine speeds, adjust to the following no load speeds:

## MODEL EY25W

| Loaded Rpm | No-Load Rpm |
|---|---|
| 1800 | 2330 |
| 2000 | 2445 |
| 2200 | 2595 |
| 2400 | 2745 |
| 2600 | 2900 |
| 2800 | 3065 |
| 3000 | 3230 |
| 3200 | 3400 |
| 3400 | 3580 |
| 3600 | 3765 |

## MODEL EY27W

| Loaded Rpm | No-Load Rpm |
|---|---|
| 1800 | 2210 |
| 2000 | 2375 |
| 2200 | 2500 |
| 2400 | 2660 |
| 2600 | 2850 |
| 2800 | 3020 |
| 3000 | 3210 |
| 3200 | 3385 |
| 3400 | 3590 |
| 3600 | 3760 |

## MODEL EY44W

| Loaded Rpm | No-Load Rpm |
|---|---|
| 1800 | 2200 |
| 2000 | 2365 |
| 2200 | 2510 |
| 2400 | 2675 |
| 2600 | 2850 |
| 2800 | 3025 |
| 3000 | 3250 |
| 3200 | 3420 |
| 3400 | 3605 |
| 3600 | 3770 |

## MODEL EY21W

| Loaded Rpm | No-Load Rpm |
|---|---|
| 1800 | 2000 |
| 2000 | 2200 |
| 2200 | 2390 |
| 2400 | 2580 |
| 2600 | 2770 |
| 2800 | 2960 |
| 3000 | 3150 |
| 3200 | 3350 |
| 3400 | 3540 |
| 3600 | 3730 |
| 3800 | 3920 |

**IGNITION SYSTEM.** A flywheel type magneto ignition system is used on all one cylinder models and a battery type ignition sytem is used on all two cylinder models. Refer to appropriate type for model being serviced.

**MAGNETO IGNITION.** Flywheel type magneto used on all one cylinder models has points and condenser located beneath flywheel. Initial point gap for all models should be 0.014 inch (0.36 mm). See Fig. WR7.

To check and adjust engine timing, disconnect lead from shut-off switch. Connect one lead of a continuity light to disconnected lead and ground remaining lead from light to engine. Slowly turn flywheel in normal direction of rotation until light goes out. Immediately stop turning flywheel and check location of timing marks. Timing marks should be

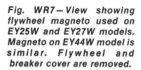

*Fig. WR7—View showing flywheel magneto used on EY25W and EY27W models. Magneto on EY44W model is similar. Flywheel and breaker cover are removed.*

*Fig. WR8—View showing 23°BTDC timing mark (M) on flywheel aligned with timing mark (D) on crankcase. Refer to text for timing procedure.*

aligned as shown in Fig. WR8. If timing mark (M) on flywheel is below timing mark (D) on crankcase, breaker point gap is too wide. If mark (M) is above mark (D), breaker point gap is too narrow. Carefully measure distance necessary to align the two marks, then remove flywheel and breaker point cover. Changing gap 0.001 inch (0.0254 mm) will change timing mark (M) position approximately 1/8-inch (3.175 mm). Reassemble and tighten flywheel nut to a torque of 44-47 ft.-lbs. (60-64 N·m) on EY25W and EY27W models or 68-72 ft.-lbs (92-98 N·m) on EY44W model.

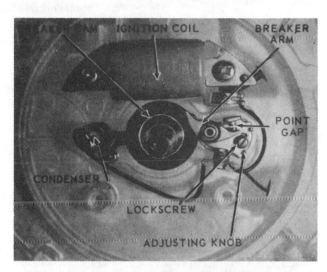

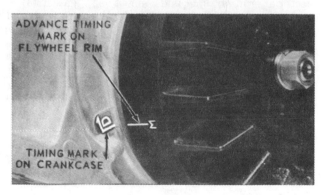

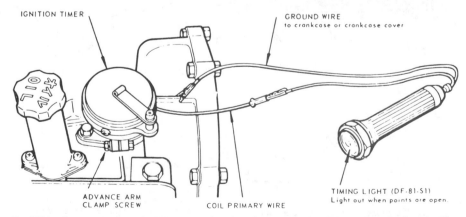

*Fig. WR9—Use continuity light (Wisconsin No. DF-81-S1 or equivalent) to check and adjust ignition timing on EY21W engines.*

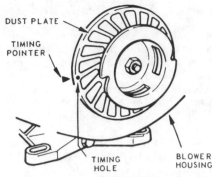

**Fig. WR10– View showing timing hole (8° BTDC static timing mark) in dust plate aligned with timing pointer on blower housing.**

**BATTERY IGNITION.** A 12 volt battery ignition system is standard on EY21W engines. Breaker points and condenser are located in ignition timer on top of engine (Fig. WR9). Timer is driven by a pinion on engine camshaft. Two lobe breaker cam rotates at half crankshaft speed and is equipped with a centrifugal spark advance mechanism. Initial point gap is 0.014 inch (0.36 mm).

Static timing is 8° BTDC and centrifugal advance mechanism starts at 600 rpm and continues to 18° BTDC at 2000 rpm and over.

To check and adjust ignition timing, make certain points are properly adjusted and disconnect primary wire connector at ignition timer and connect one lead of continuity light (Wisconsin No. DF-81-S1 or equivalent) to wire leading from timer. Connect remaining light lead to ground on engine (Fig. WR9). Slowly rotate crankshaft in normal direction of engine rotation until timing hole (8° BTDC static timing mark) in dust plate is aligned with timing pointer on blower housing as shown in Fig. WR10. Continuity light should just go out when timing marks are aligned. If

light goes out before or after timing marks line up, align timing marks and loosen timer advance arm screw (Fig. WR9). Rotate timer body slowly until breaker points are just beginning to open and light just goes out. Tighten clamp screw. Slowly rotate crankshaft in normal direction of rotation and recheck to make certain continuity light goes out just as timing marks are aligned.

**LUBRICATION.** Splash lubrication is provided by oil dippers attached to connecting rod caps (Figs. WR11, WR12 and WR13) for all models. Oil dippers on one cylinder models pick up oil directly from crankcase. Oil dipper on two cylinder models pick up oil from a trough located directly under crankpins. Trough oil level is maintained at a constant level by oil supplied by a small capacity oil pump located inside crankcase.

To remove oil pump, remove blower housing, blower and timing gear cover. Withdraw oil pump gear and shaft (2-Fig. WR14) with thrust washer (1). Unbolt and remove engine base, loosen pump adapter lockscrew (11) and remove oil pump assembly from inside crankcase. Remove the four screws and separate pump from adapter (3). Shaft (5), rotor (6), stator (7) and housing (8) are serviced as an assembly only.

When reassembling pump, use new gasket (4) and make certain shaft (5), rotor (6) and stator (7) turn free when screws are tightened securely.

Oils approved by manufacturer must meet requirements of API service classification SE or SF.

For one cylinder models use SAE 30 oil in temperatures above 40°F (4°C), SAE 20 oil in temperatures between 15° and 40°F (–9° and 4°C) and SAE 10W-30 oil in temperatures below 15° F (-9° C).

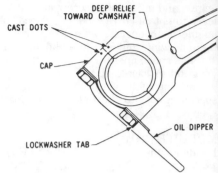

**Fig. WR13–Install connecting rods and rod caps with cast dots (match marks) together on EY21W engines.**

For two cylinder models use SAE 30 oil in temperatures above 60°F (16°C), SAE 10W-30 oil in temperatures between 0° and 60°F (–18° and 16°C) and SAE 5W-20 oil in temperatures below 0°F (–18°C).

Maintain oil level at "FULL" mark on dipstick. **DO NOT** overfill.

Recommended oil change interval for all models is every 50 hours of normal operation.

Crankcase oil capacity is 1.6 pints (0.76 L) for EY25W and EY27W models, 2.4 pints (1.14 L) for EY44W model and 5.25 pints (2.48 L) for EY21W model.

**CRANKCASE BREATHER.** A floating poppet type breather valve is located in breather plate at valve cover. A breather tube connects breather into air cleaner. Restricted or faulty breather is indicated when oil seeps from gasket surfaces and oil seals.

### REPAIRS

**TIGHTENING TORQUES.** Recom-

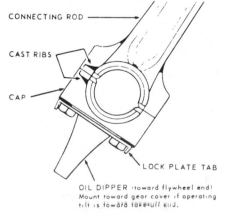

**Fig. WR11–Connecting rod and cap must be installed with cast ribs (match marks) together on EY25W and EY27W models.**

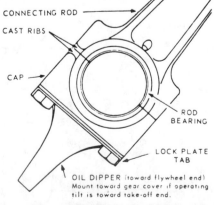

**Fig. WR12– Renewable insert type connecting rod bearings are used on EY44W models. Cast ribs (match marks) on connecting rod and cap must be installed together.**

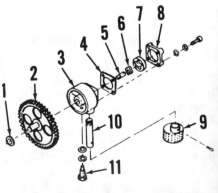

**Fig. WR14 – Exploded view of oil pump used on EY21W engines. Shaft (5), rotor (6), stator (7) and housing (8) are serviced as an assembly only.**

| | |
|---|---|
| 1. Thrust washer | |
| 2. Pump drive gear | 7. Stator |
| 3. Adapter | 8. Housing |
| 4. Gasket | 9. Oil strainer |
| 5. Shaft | 10. Oil intake pipe |
| 6. Rotor | 11. Lockscrew |

mended tightening torques are as follows:

Spark plug–
    EY25W, EY27W . . . . . . .24-27 ft.-lbs.
                       (33-37 N·m)
    EY44W. . . . . . . . . . . .18-22 ft.-lbs.
                       (24-30 N·m)
    EY21W. . . . . . . . . . . .22-29 ft.-lbs.
                       (30-39 N·m)

Cylinder head–
    EY25W, EY27W . . . . . . .25-27 ft.-lbs.
                       (34-37 N·m)
    EY44W. . . . . . . . . . . .36-37 ft.-lbs.
                       (49-50 N·m)
    EY21W . . . . . . . . . . . .33-36 ft.-lbs
                       (45-49 N·m)

Cylinder block–
    EY44W. . . . . . . . . . . . .25-29 ft.-lbs
                       (34-39 N·m)
    EY21W. . . . . . . . . . . . .23-26 ft.-lbs.
                       (31-35 N·m)

Connecting rod–
    EY25W, EY27W . . . . . . .15-18 ft.-lbs.
                       (20-24 N·m)
    EY44W, EY21W . . . . . . .18-22 ft.-lbs.
                       (24-30 N·m)

Flywheel nut–
    EY25W, EY27W . . . . . . .44-47 ft.-lbs.
                       (60-64 N·m)
    EY44W. . . . . . . . . . . .68-72 ft.-lbs.
                       (92-98 N·m)
    EY21W. . . . . . . . . . . .60-70 ft.-lbs.
                       (81-95 N·m)

Gear cover–
    EY25W,
    EY27W. EY44W . . . . . . .13-14 ft.-lbs.
                       (18-19 N·m)
    EY21W. . . . . . . . . . . .10-12 ft.-lbs.
                       (13-16 N·m)

Main bearing housing–
    EY21W. . . . . . . . . . . .10-12 ft.-lbs.
                       (13-16 N·m)

Engine base–
    EY21W. . . . . . . . . . . .22-27 ft.-lbs.
                       (30-37·m)

Camshaft nut–
    EY21W. . . . . . . . . . . .37-45 ft.-lbs.
                       (50-61 N·m)

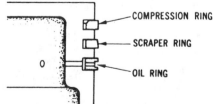

*Fig. WR15 – Install piston rings on piston as shown for EY25W, EY27W one cylinder engines and EY21W two cylinder engines.*

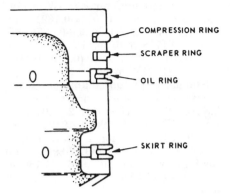

*Fig. WR16 – Install piston rings on piston as shown for EY44W one cylinder engines.*

**CYLINDER HEAD.** When removing cylinder head bolts from two cylinder models, note locations of bolts as they must be reinstalled in their original positions.

On all models, check cylinder head for distortion. If warpage is 0.006 inch (0.1524 mm) or more, renew cylinder head. Always use new head gasket when installing cylinder head. Tighten cylinder head retaining nuts or head bolts evenly, to correct specified torque.

**CONNECTING ROD.** On EY25W and EY27W models, connecting rod and piston assembly is removed from cylinder head end of block after cylinder head and gear cover are removed. Connecting rod to crankpin running clearance should be 0.0016-0.0026 inch (0.0406-0.0660 mm). Rod side clearance should be 0.004-0.012 inch (0.1016-0.3018 mm) with a maximum clearance of 0.039 inch (0.9906 mm). Connecting rod to piston pin clearance should be 0.0006-0.0014 inch (0.0152-0.0356 mm) with a maximum clearance of 0.0032 inch (0.0813 mm). When installing connecting rod in engine, make certain match marks (cast ribs) on rod and cap are together. Refer to Fig. WR11 for correct installation information. Tighten rod cap screws to 15-18 ft.-lbs. (20-24 N·m) torque.

On EY44W model, connecting rod and piston assembly is removed from cylinder head end of block after cylinder head and gear cover are removed. Connecting rod is equipped with renewable type bearings. Connecting rod bearing to crankpin running clearance should be 0.0016-0.0042 inch (0.0406-0.1067 mm). Rod side clearance should be

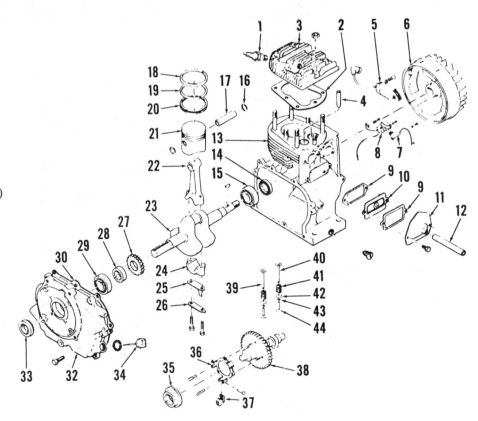

*Fig. WR17 – Exploded view of EY25W or EY27W basic engine assembly.*

| | | |
|---|---|---|
| 1. Spark plug | 12. Breather tube | 23. Crankshaft | 35. Governor sleeve |
| 2. Head gasket | 13. Cylinder block | 24. Rod cap | 36. Governor plate |
| 3. Cylinder head | 14. Oil seal | 25. Oil dipper | 37. Flyweight (3 used) |
| 4. Valve guide (2 used) | 15. Ball bearing | 26. Rod bolt lock plate | 38. Camshaft assy. |
| 5. Magneto coil | 16. Retaining ring | 27. Crankshaft gear | 39. Intake valve |
| 6. Flywheel | 17. Piston pin | 28. Adjusting collar | 40. Exhaust valve |
| 7. Condenser | 18. Compression ring | 29. Ball bearing | 41. Valve spring |
| 8. Breaker points | 19. Scraper ring | 30. Gasket | 42. Spring retainer |
| 9. Gaskets | 20. Oil control ring | 32. Gear cover | 43. Locks |
| 10. Breather plate | 21. Piston | 33. Oil seal | 44. Valve tappets |
| 11. Valve cover | 22. Connecting rod | 34. Dipstick | |

0.0079-0.0197 inch (0.2007-0.5004 mm) with a maximum clearance of 0.034 inch (0.8636 mm). Piston pin to connecting rod clearance should be 0.001-0.0019 inch (0.0254-0.0483 mm) with a maximum clearance of 0.0039 inch (0.0991 mm). Connecting rod bearing is available for a 0.010 inch (0.254 mm) undersize crankshaft crankpin journal. When installing connecting rod in engine, make certain match marks (cast ribs) on connecting rod and cap are together (Fig. WR12) and connecting rod is correctly installed. Use new lock plate and tighten connecting rod cap screws to 18-22 ft.-lbs. (24-30 N·m) torque.

On EY21W model, connecting rod and piston assemblies can be removed after removing engine base and cylinder heads. Identify each rod and piston unit so they can be reinstalled in their original position and do not intermix connecting rod caps. Connecting rods ride directly on crankpin journal and connecting rod to crankpin running clearance should be 0.0016-0.0032 inch (0.0406-0.0813 mm). Rod side clearance should be 0.004-0.020 inch (0.1016-0.508 mm) with a maximum clearance of 0.039 inch (0.9906 mm). Piston pin to connecting rod clearance should be 0.001-0.0018 inch (0.0254-0.0457 mm) with a maximum clearance of 0.0047 inch (0.1194 mm). Connecting rod is available for a 0.010 inch (0.254 mm) undersize crankpin journal. When reinstalling connecting rod in engine, make certain match marks (cast dots) on connecting rods and caps are together and that "V1" mark on top of pistons is toward blower end of engine. Refer to Fig. WR13 for correct connecting rod installation and tighten cap screws to 18-22 ft.-lbs (24-30 N·m) torque.

**PISTON, PIN AND RINGS.** On EY25W, EY27W and EY21W models, pistons are equipped with one compression ring, one scraper ring and one oil control ring (Fig. WR15). On EY44W models, a fourth ring is used on piston skirt (Fig. WR16).

Ring end gap should be 0.002-0.010 inch (0.05-0.25 mm) for EY25W model, 0.008-0.016 inch (0.20-0.40 mm) for EY27W model, 0.002-0.012 inch (0.05-0.30 mm) for EY21W model. Ring end gap for EY44W lower oil control ring should be 0.012-0.020 inch (0.30-0.50 mm) and 0.002-0.010 inch (0.05-0.25 mm) for all other rings.

On all models, if side clearance of new ring in piston ring groove is 0.006 inch (0.1524 mm) or more, renew piston. Stagger ring end gaps at 90° intervals around piston.

Recommended piston to cylinder bore clearance, measured at thrust face of piston, is 0.0024-0.0039 inch

(0.0610-0.0991 mm) for EY25W model, 0.0028-0.0052 inch (0.0711-0.1321 mm) for EY27W model, 0.006-0.007 inch (0.1524-0.1778 mm) for EY44W model and 0.0047-0.0071 inch (0.1194-0.1803 mm) for EY21W model.

Standard piston diameter when measured at skirt thrust face is 2.8315-2.8323 inches (71.92-71.94 mm) for EY25W model, 2.9090-2.9105 inches (73.89-74.13 mm) for EY27W model, 3.5382-3.5390 inches (89.87-89.89 mm) for EY44W model and 2.9468-2.9480 inches (74.85-74.88 mm) for EY21W model.

Pistons and rings are available in a variety of oversizes as well as standard.

Standard piston pin diameter is 0.6297-0.6300 inch (15.95-16.00 mm) for EY25W and EY27W models, 0.7870-0.7874 inch (19.99-20.00 mm) for EY44W model and 0.7081-0.7084 inch (17.986-17.993 mm) for EY21W model.

Piston pin to piston pin bore clearance should be 0.00035 inch (0.0089 mm) tight to 0.00039 inch (0.0099 mm) loose with a

maximum clearance of 0.0024 inch (0.061 mm) loose. Refer to **CONNECTING ROD** section for piston pin to connecting rod clearance.

To renew rings only on EY44W and EY21W models, engine need not be totally disassembled. Unbolt and remove cylinder block unit from crankcase to service piston and rings while connecting rod remains attached to crankshaft. When installing either piston on EY21W model, make certain identifying mark "V1" on top of piston is toward cooling blower end of engine.

**CYLINDER.** Aluminum crankcase and cylinder with cast-iron liner is a one piece unit on EY25W and EY27W models. Cast iron cylinder unit is removable from aluminum crankcase on EY44W and EY21W models.

On all models, if cylinder wall is scored or out-of-round more than 0.003 inch (0.0762 mm) or if cylinder bore taper ex-

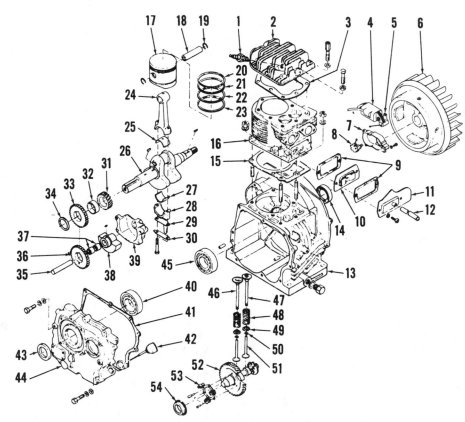

*Fig. WR18 — Exploded view of EY44W model basic engine assembly.*

1. Spark plug
2. Cylinder head
3. Head gasket
4. Magneto coil
5. Condenser
6. Flywheel
7. Cover
8. Breaker points
9. Gaskets
10. Breather plate
11. Valve cover
12. Breather tube
13. Crankcase
14. Oil seal
15. Cylinder gasket
16. Cylinder
17. Piston
18. Piston pin
19. Retaining ring
20. Compression ring
21. Scraper ring
22. Oil ring
23. Skirt ring
24. Connecting rod
25. Rod bearing (upper half)
26. Crankshaft
27. Rod. bearing (lower half)
28. Rod cap
29. Oil dipper
30. Rod bolt lock plate
31. Crankshaft gear
32. Spacer
33. Balancer drive gear
34. Adjusting collar
35. Balancer shaft
36. Balancer gear
37. Needle bearings
38. Balancer weight
39. Balancer housing
40. Ball bearing
41. Gasket
42. Dipstick
43. Oil seal
44. Gear cover
45. Ball bearing
46. Intake valve
47. Exhaust valve
48. Valve spring
49. Spring retainers
50. Locks
51. Valve tappets
52. Camshaft assy.
53. Governor plate & flyweights
54. Governor sleeve

ceeds 0.006 inch (0.1524 mm), cylinder should be rebored to nearest oversize for which piston and rings are available.

Standard cylinder bore is 2.8346-2.8354 inches (71.9988-72.0192 mm) for EY25W model, 2.9134-2.9141 inches (74.0003-74.0181 mm) for EY27W model, 3.5447-3.5456 inches (90.0354-90.0582 mm) for EY44W model and 2.9528-2.9539 inches (75.001-75.029 mm) for EY21W model.

When installing a cylinder unit on EY44W or EY21W models, use a new cylinder base gasket and tighten cylinder retaining nuts to specified torque.

**CRANKSHAFT AND MAIN BEARING.** EY25W, EY27W and EY44W models crankshaft is supported at each end in ball bearings. See 15 and 29, Fig. WR17 or 40 and 45, Fig. WR18.

EY21W model crankshaft is supported by a ball bearing (15-Fig. WR19) at front of crankshaft and a roller bearing (58) at rear of crankshaft.

On all models, ball bearings are a press fit on crankshaft and should be renewed if any indication of roughness, noise or excessive wear is found.

Crankshaft end play for EY25W and EY27W models should be 0.001-0.009 inch (0.0254-0.2286 mm) and is controlled by adjusting collar (28-Fig. WR17) located between crankshaft gear (27) and gear cover main bearing (29). Three lengths of adjusting collars, 0.740-0.748, 0.748-0.756 and 0.756-0.764 inch, are available. To determine correct length of adjusting collar with gear cover removed, refer to Fig. WR20 and measure distance (A) between machined surface

of crankcase face and end of crankshaft gear. Measure distance (B) between machined surface of gear cover and end of main bearing. Compressed thickness of gear cover gasket (C) is 0.007 inch (0.1778 mm). Select adjusting collar that is 0.001-0.009 inch (0.0254-0.2286 mm) less in length than total of A, B and C. Install adjusting collar on crankshaft with recessed side toward crankshaft gear. After reassembly, end play can be checked with a dial indicator.

Crankshaft end play for EY44W model should 0.001-0.008 inch (0.025-0.2032 mm) and is controlled by adjusting collar (34-Fig. WR18) located between balancer drive gear (33) and gear cover main bearing (40). Three lengths of adjusting collars, 0.0381-0.0443, 0.0443-0.0502 and 0.0502-0.0561 inch, are available. To

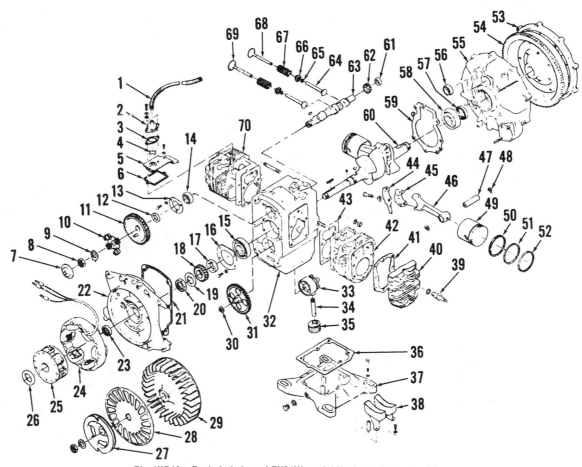

**Fig. WR19—Exploded view of EY21W model basic engine assembly.**

| | | | |
|---|---|---|---|
| 1. Breather tube | 15. Crankshaft bearing (front) | 29. Cooling blower | 44. Oil dipper | 57. Oil seal |
| 2. Adapter | 16. Bearing retainer | 30. Thrust washer | 45. Rod cap | 58. Crankshaft bearing (rear) |
| 3. Gasket | 17. Spacer | 31. Oil pump drive gear | 46. Connecting rod | 59. Gasket |
| 4. Breather valve | 18. Crankshaft gear | 32. Crankcase | 47. Piston pin | 60. Crankshaft, rod & piston assy. |
| 5. Valve cover | 19. Lock | 33. Oil pump assy. | 48. Retaining ring | |
| 6. Gasket | 20. Nut | 34. Oil suction tube | 49. Piston | 61. Spacer |
| 7. Governor thrust plate | 21. Gasket | 35. Oil strainer | 50. Oil control ring | 62. Timer drive gear |
| 8. Nut | 22. Gear cover | 36. Gasket | 51. Scraper ring | 63. Camshaft |
| 9. Lock | 23. Oil seal | 37. Engine base | 52. Compression ring | 64. Valve tappets |
| 10. Governor plate & flyweights | 24. Alternator stator | 38. Oil trough assy. | 53. Flywheel cover | 65. Locks |
| 11. Camshaft gear | 25. Rotor | 39. Spark plug | 54. Flywheel assy. | 66. Spring retainers |
| 12. Spacer | 26. Spacer | 40. Cylinder head | 55. Flywheel & bearing housing | 67. Valve springs |
| 13. Bearing retainer | 27. Pulley | 41. Head gasket | | 68. Exhaust valve (2 used) |
| 14. Camshaft bearing (front) | 28. Blower dust plate | 42. Cylinder | 56. Camshaft bearing (rear) | 69. Intake valve (2 used) |
| | | 43. Cylinder gasket | | 70. Cylinder & head assy. |

determine correct length of adjusting collar with gear cover removed, refer to Fig. WR21 and measure distance (A) between machined surface of crankcase face and face of balancer drive gear. Measure distance (B) between machined face of gear cover and end of main bearing. Compressed thickness of gear cover gasket (C) is 0.008 inch (0.2032 mm). Select adjusting collar that is 0.001-0.008 inch (0.0254-0.2032 mm) less in length than total of B plus C, minus distance A. Install adjusting collar on crankshaft with recessed side toward balancer drive gear. After reassembly, end play can be checked with a dial indicator.

Crankshaft end play for EY21W model should be 0.002-0.010 inch

(0.0508-0.254 mm) and is controlled by condition of front main (ball) bearing (15). If end play is excessive, renew front bearing.

Standard crankpin journal diameter is 1.1003-1.1008 inches (27.95-27.96 mm) for EY25W and EY27W models, 1.3754-1.3760 inches (34.94-34.95 mm) for EY44W model and 1.3750-1.3760 inches (34.93-34.95 mm) for EY21W model. If crankpin journal is worn or scored more than 0.0025 inch (0.0635 mm) on EY25W and EY27W models or 0.004 inch (0.1016 mm) for EY44W and EY21W models or if journal is tapered or out-of-round more than 0.0002 inch (0.0051 mm) on all models, crankshaft should be renewed or crankpin journal reground according to availability of oversize connecting rod or bearing in-

serts for model being serviced.

When reinstalling crankshaft in engine, make certain punch marked tooth on crankshaft gear is between two punch marked teeth on camshaft gear.

**CAMSHAFT.** EY25W, EY27W and EY44W models camshaft rides directly in bores machined into crankcase and gear cover. When removing camshaft assembly, lay engine on side to prevent tappets (44 – Fig. WR17 or 51 – Fig. WR18) from falling out. If valve tappets are removed they must be reinstalled in their original positions. Camshaft journal diameter is 0.5889-0.5893 inch (14.96-14.97 mm) for each end for EY25W and EY27W models and 0.7467-0.7472 inch (18.97-18.98 mm) for journal at gear cover end and 0.6682-0.6687 inch (16.98-16.99 mm) for journal at flywheel end for EY44W model.

EY44W model camshaft is equipped with an automatic compression release which at cranking speed, holds exhaust valve slightly open during first part of compression stroke. This reduces compression pressure and allows easier cranking of engine. As engine rpm increases, centrifugal action overcomes compression release mechanism and engine operates with normal compression.

When reinstalling camshaft install governor thrust sleeve (35-Fig. WR17) or (54-Fig. WR18) on governor flyweights. Slide camshaft into position making certain punch marked tooth on crankshaft gear is between two punch marked teeth on camshaft gear.

EY21W model camshaft (63-Fig. WR19) rides in a ball bearing at each end (14 and 56). To remove camshaft, remove blower housing, pulley (27), dust plate (28), blower (29), spacer (26), rotor (25), stator (24) and gear cover (22). Release valve springs and move tappets away from camshaft. Unbolt and remove fuel pump and ignition timer. Remove governor sleeve (7), nut (8) and lock (9). Use suitable puller to remove camshaft gear (11). Woodruff key and spacer (12). Unbolt retainer (13) and withdraw camshaft assembly. Check camshaft bearings (14 and 56) for excessive wear or other damage and renew as necessary. If valve tappets are removed they must be reinstalled in their original positions.

Reinstall camshaft by reversing removal procedure, making certain punch marked tooth on crankshaft gear is between two punch marked teeth on camshaft gear. Tighten camshaft gear retaining nut (8) to 40 ft.-lbs. (54 N·m) torque and secure with lock tab.

**VALVE SYSTEM.** Valve seats are ground at 45° angle and seat width

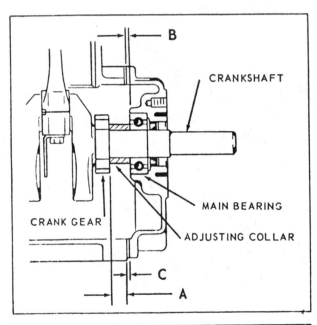

Fig. WR20—On EY25W and EY27W models, crankshaft end play is controlled by adjusting collar. Refer to text for procedure to determine length of collar to be used.

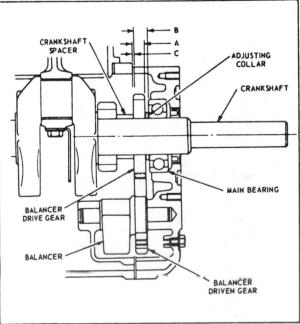

Fig. WR21—On EY44W model, crankshaft end play is controlled by adjusting collar. Adjusting collars are available in three lengths. Refer to text for procedure to determine correct length collar to be installed.

should be 3/64 to 1/16-inch (1.191 to 1.588 mm) for all models.

Valve faces are ground at 45° angle for all models and valve stem diameter is 0.273-0.274 inch (6.93-6.96 mm) for intake and exhaust valves for EY25W and EY27W models, 0.313-0.314 inch (7.95-7.98 mm) for intake valve and 0.310-0.311 inch (7.87-7.90 mm) for exhaust valves for EY44W model and 0.3118-0.3128 inch (7.93-7.95 mm) for intake valves and 0.3106-0.3114 inch (7.89-7.91 mm) for exhaust valve for EY21W model.

Valve stem to guide clearance should be 0.0015-0.004 inch (0.0381-0.1016 mm) for intake and exhaust valves for EY25W and EY27W models, 0.001-0.003 inch (0.0254-0.0762 mm) for intake valves and 0.004-0.006 inch (0.1016-0.1524 mm) for exhaust valves for EY44W model and 0.0022-0.0047 inch (0.0559-0.1194 mm) for intake valves and 0.0036-0.0059 inch (0.0914-0.1499 mm) for exhaust valves for EY21W model.

Valve guide bore diameters are 0.2755-0.2770 inch (7.00-7.04 mm) for EY25W and EY27W models, 0.315-0.316 inch (8.00-8.03 mm) for EY44W model and 0.315-0.3165 inch (8.00-8.04 mm) for EY21W model.

If valve stem to guide clearance is excessive for EY25W or EY27W models, renew valves and/or valve guides. If valve stem to guide clearance is excessive for EY44W or EY21W models, valve guides are not renewable and valves and/or cylinder unit must be renewed.

Valve spring free length should be 1.3582-1.4173 inches (34.50-36.00 mm) for EY25W and EY27W models and 1.750-1.811 inches (44.45-46.00 mm) for EY44W and EY21W models. Renew springs which are rusted, pitted or do not meet free length specifications.

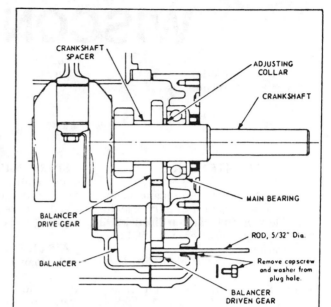

*Fig. WR22—Sectional view showing method of timing engine balancer to crankshaft. Refer to text.*

Valve tappets should have an operating clearance in crankcase bores of 0.001-0.0024 inch (0.0254-0.0610 mm) on all models.

Valve tappet gap (cold) for all models is 0.006-0.008 inch (0.1524-0.2032 mm). Adjustment is obtained by careful grinding of valve stem tips.

**ENGINE BALANCER.** Engine balancer on EY44W model offsets forces of reciprocating motion and reduces engine vibration to minimum. Balancer is driven by a gear on crankshaft and rotates in opposite direction of crankshaft. Balancer components, consisting of drive gear (33-Fig. WR18), shaft (35), driven gear (36), needle bearings (37), balance weight (38) and balancer housing (39), are not serviced separately. However, for cleaning and inspection balancer assembly can be disassembled. Refer to Fig. WR18.

When reinstalling gear cover and balancer assembly, time balancer driven gear (36) to drive gear (33) by rotating crankshaft until piston is at "top dead center". Refer to Fig. WR22, remove cap screw and washer from gear cover, insert a 5/32-inch rod through hole in gear cover and into hole in driven gear. Carefully install gear cover and balancer assembly allowing balancer gears to mesh. Tighten gear cover cap screws to 13-14 ft.-lbs. (18-19 N·m) torque and remove timing rod. Install cap screw and washer.

# WISCONSIN ROBIN

| Model | No. Cyls. | Bore | Stroke | Displacement |
|-------|-----------|------|--------|--------------|
| W1-145 | 1 | 63 mm | 46 mm | 143 cc |
| | | (2.48 in.) | (1.81 in.) | (8.73 cu. in.) |
| W1-185 | 1 | 67 mm | 52 mm | 183 cc |
| | | (2.64 in.) | (2.05 in.) | (11.2 cu. in.) |

## MAINTENANCE

**LUBRICATION.** Check engine oil level prior to each operating interval. Maintain oil level at full mark on dipstick. Oil should be changed after every 50 hours of operation.

Manufacturer recommends using oil with API service classification of SE or SF. Use SAE 30 oil when temperature is above 4° C (40° F), SAE 20 oil when temperature is between -9° C and 4° C (15° F and 40° F), and SAE 10W-30 oil when temperature is below -9° C (15° F).

**SPARK PLUG.** Recommended spark plug for all models is Champion L86, NGK B6HS or equivalent. Recommended spark plug electrode gap is 0.6-0.7 mm (0.024-0.027 inch).

**AIR FILTER.** The engine is equipped with a foam type air filter element. The air filter element should be removed and cleaned after every 50 hours of operation, or more often if operating in extremely dusty conditions. To clean element, wash in a nonflammable cleaning solvent and gently squeeze element dry. Oil element with clean engine oil and gently squeeze out the excess oil.

**CARBURETOR.** All models are equipped with a Mikuni float type carburetor. Fuel:air mixture at low speeds is controlled by the pilot jet (4—Fig. WR50), and at high speed by fixed main jet (15). There is no provision for adjustment of pilot jet or main jet on this carburetor. Engine idle speed is adjusted by turning throttle stop screw (12) as needed.

If carburetor related fuel system problems are encountered, disassemble carburetor and clean using suitable solvent and compressed air. Do not use drills or wire to clean fuel passages as calibration of carburetor fuel:air mixture will be affected if orifices of main jet or pilot jet are enlarged.

To check or adjust float level, remove float bowl (20). Place carburetor body on end (on manifold flange) so float pin is in vertical position. Move float to close inlet needle valve.

**NOTE: Needle valve is spring loaded. Float tab should just contact needle valve pin, but should not compress spring.**

Measure float setting as shown in Fig. WR51. Dimension "A" should be 12.5-14.5 mm (0.492-0.571 inch). Carefully bend tab on float lever to obtain correct setting.

**GOVERNOR.** All models are equipped with a centrifugal flyweight type governor. Governor assembly is located in the crankcase cover and is driven by the camshaft gear.

For 3600 rpm engine operation, be sure governor spring (6—Fig. WR52) is connected in hole "A" in speed control lever (8). To adjust external governor linkage, loosen clamp bolt (2) on governor lever (3) so governor lever can be moved independently of governor shaft (1). Move governor lever toward high speed position until throttle valve in carburetor is opened fully. While hold-

**Fig. WR50—Exploded view of carburetor used on all models.**

| | | | |
|---|---|---|---|
| 1. Screw | | 12. Throttle stop screw | |
| 2. Throttle plate | | 15. Main jet | |
| 3. Throttle shaft | | 16. Pin | |
| 4. Pilot jet | | 17. Fuel inlet valve | |
| 5. Choke shaft | | 18. Gasket | |
| 6. Screw | | 19. Float | |
| 7. Choke plate | | 20. Float bowl | |
| 10. Carburetor body | | 21. Gasket | |
| 11. Spring | | 22. Screw | |

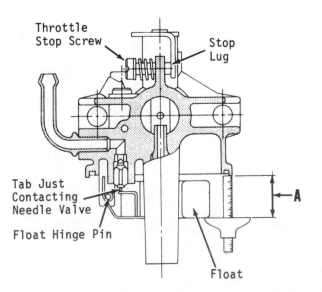

**Fig. WR51—Float height "A" should be 12.5-14.5 mm (0.492-0.571 in.). Bend float lever tab to adjust.**

ing lever in this position, insert screwdriver in slot at end of governor shaft and turn shaft clockwise as far as possible. Tighten governor lever clamp screw.

**IGNITION SYSTEM.** Early models are equipped with a breaker point type ignition system. Late models are equipped with a solid-state ignition system. Refer to the appropriate paragraphs for ignition type being serviced.

**Breaker-Point Ignition System.** Breaker points and condenser are located behind flywheel and the ignition coil is located outside flywheel. Specified gap between ignition coil and flywheel is 0.5 mm (0.020 inch). Initial breaker-point gap is 0.36 mm (0.014 inch). However, since specified engine timing of 23 degrees BTDC is regulated by the opening of breaker-points, a slight variation in point gap may be necessary to obtain specified timing.

To check and adjust engine timing, disconnect lead wire from engine shutoff switch. Connect test lead from a continuity light to lead wire and ground remaining test lead to engine. Slowly rotate flywheel in normal operating direction until light goes out (points open). If timing is correct, "P" mark (Fig. WR53) on flywheel should be aligned with "M" mark on crankcase. If timing marks are not aligned, adjust breaker-point gap until correct timing is obtained. Changing point gap 0.03 mm (0.001 inch) will change timing mark location on flywheel approximately 2 degrees or 3.18 mm (1/8 inch).

**Solid-State Ignition System.** Solid-state type ignition system does not have breaker points. There is no scheduled maintenance. Specified air gap between ignition coil and flywheel is 0.5 mm (0.020 inch).

**VALVE ADJUSTMENT.** Clearance between tappet and end of valve stem should be checked after every 500 hours of operation. To check clearance, remove tappet chamber cover (10—Fig. WR54) and breather plate (11). Rotate crankshaft to position piston at top dead center of compression stroke. Use a feeler gage to measure clearance between tappet and valve stem. Specified clearance with engine cold is 0.08-0.12 mm (0.003-0.005 inch) for intake and exhaust.

If clearance is smaller than specified, increase clearance by carefully grinding end of valve stem. To decrease clearance, renew valve and/or tappet. Refer to VALVE SYSTEM paragraph for valve removal and installation procedure.

**GENERAL MAINTENANCE.** Check and tighten all bolts, nuts or clamps prior to each operating interval. Clean dust, dirt or any other foreign material from cylinder head and cylinder block cooling fins after every 50 hours of normal operation.

Cylinder head should be removed and carbon and other combustion deposits removed after every 500 hours of operation.

*Fig. WR52—Exploded view of governor external linkage typical of all models. Refer to text for adjustment procedure.*

1. Governor shaft
2. Clamp bolt
3. Governor lever
4. Carburetor link
5. Link spring
6. Governor spring
7. High speed stop screw
8. Speed control plate
9. Spring plate
10. Stop plate

## REPAIRS

**TIGHTENING TORQUES.** Recommended tightening torque specifications are as follows:
Spark plug . . . . . . . . . . . . . 12-15 N·m
(9-11 ft.-lbs.)
Cylinder head . . . . . . . . . . 19-23 N·m
(14-16 ft.-lbs.)
Connecting rod bolts . . . . . . . 9-11 N·m
(7-8 ft.-lbs.)

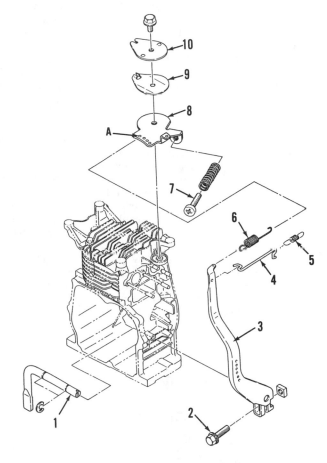

*Fig. WR53—On models equipped with breaker-point ignition, spark should occur when "P" mark on flywheel aligns with "M" mark on crankcase. Refer to text for adjustment procedure.*

Crankcase cover . . . . . . . . . . .8-9 N·m
(6-7 ft.-lbs.)
Flywheel nut . . . . . . . . . . . .60-64 N·m
(44-47 ft.-lbs.)

**CYLINDER HEAD.** To remove cylinder head, remove fuel tank, flywheel shroud, head cover and cylinder air baffle. Remove cylinder head retaining bolts and remove cylinder head and gasket.

Clean all carbon deposits from cylinder head. Use a straightedge and feeler gage to check cylinder head mounting surface for distortion. If cylinder head is warped more than 0.15 mm (0.006 inch), resurface head or renew as necessary.

Reinstall cylinder head using a new head gasket. Tighten head bolts evenly in a crisscross pattern to 19-23 N·m (14-16 ft.-lbs.).

**CONNECTING ROD.** To remove connecting rod and piston assembly, engine must be separated from generator unit as outlined in GENERATOR section. Remove cylinder head as previously outlined. Unbolt and remove crankcase cover (8—Fig. WR54) from cylinder block (3). Remove connecting rod cap (11—Fig. WR55) and withdraw piston and connecting rod assembly from top of cylinder block. Remove retaining ring (5), push out piston pin (6) and separate piston from connecting rod if necessary.

Connecting rod-to-crankpin clearance should be 0.038-0.064 mm (0.0015-0.0025 inch). If clearance exceeds 0.20 mm (0.008 inch), renew connecting rod and/or crankshaft.

Connecting rod-to-piston pin clearance should be 0.01-0.03 mm (0.0004-0.0012 inch). If clearance exceeds 0.12 mm (0.005 inch), renew piston pin and/or connecting rod.

Connecting rod side clearance on crankshaft should be 0.10-0.30 mm (0.004-0.012 inch). If side clearance exceeds 1.0 mm (0.040 inch), renew connecting rod and/or crankshaft.

When installing connecting rod and piston assembly, make certain match marks (cast ribs) on connecting rod and cap are adjacent to each other as shown in Fig. WR56. Install oil dipper (12—Fig. WR55) with offset facing flywheel end of engine. Install a new lock plate (13) and tighten connecting rod cap bolts to 9-11 N·m (7-8 ft.-lbs.).

**PISTON, PIN AND RINGS.** After removing piston rings and separating piston from connecting rod, carefully clean carbon and other deposits from piston surface and ring lands.

**CAUTION: Extreme care should be exercised when cleaning ring lands. Do not damage squared edges or widen ring grooves. If ring lands are damaged, piston must be renewed.**

Inspect piston for scoring or excessive wear and renew as necessary. Piston pin-to-piston bore fit should be 0.00 mm (0.00035 inch) tight to 0.010 (0.00039 inch) loose. Renew piston and pin if pin-to-piston clearance exceeds 0.06 mm (0.0024 inch) on all models. Refer to the following piston and ring specification data:

Piston-to-cylinder
  clearance—
Standard . . . . . . . . . .0.020-0.058 mm
(0.0008-0.0023 in.)
Wear limit . . . . . . . . . . . . . .0.25 mm
(0.010 in.)

Piston ring side clearance
  in groove (W1-145)—
Top ring . . . . . . . . . .0.090-0.135 mm
(0.0035-0.0053 in.)
Second ring . . . . . . .0.060-0.105 mm
(0.0024-0.0041 in.)
Oil ring . . . . . . . . . . .0.010-0.065 mm
(0.0004-0.0025 in.)
Wear limit, all rings . . . . . .0.15 mm
(0.006 in.)

Piston ring side clearance
  in groove (W1-185)—
Top ring . . . . . . . . . .0.050-0.095 mm
(0.0020-0.0037 in.)
Second ring . . . . . . .0.010-0.055 mm
(0.0004-0.0021 in.)
Oil ring . . . . . . . . . . .0.010-0.065 mm
(0.0004-0.0025 in.)
Wear limit, all rings . . . . . .0.15 mm
(0.006 in.)

**Fig. WR54—Exploded view of crankcase assembly typical of all models.**

1. Cylinder head
2. Head gasket
3. Cylinder block
4. Oil seal
5. Main bearings
6. Governor stub shaft
7. Oil seal
8. Crankcase cover
9. Oil drain plug
10. Tappet chamber cover
11. Breather plate

**Fig. WR55—Exploded view of crankshaft, connecting rod and piston assembly typical of all models.**

1. Compression ring
2. Scraper ring
3. Oil ring
4. Piston
5. Retaining rings
6. Piston pin
7. Connecting rod
8. Crankshaft
9. Crankshaft gear
10. Spacer
11. Rod cap
12. Oil dipper
13. Lock plate
14. Rod bolt

Illustrations courtesy of Teledyne Total Power

Piston ring end gap (W1-145)—
  All rings . . . . . . . . . . 0.20-0.40 mm
              (0.008-0.016 in.)
Piston ring end gap (W1-185)—
  All rings . . . . . . . . . . 0.05-0.25 mm
              (0.002-0.010 in.)

When assembling piston and rings, manufacturer's mark (if present) on top and second rings must face toward top of piston. Make certain that scraper ring is installed with notched side facing down (Fig. WR57), otherwise excessive oil consumption will result. It is recommended that piston pin retaining rings (5—Fig. WR55) be renewed whenever they are removed. Heating piston in warm water or oil will make installation of piston pin easier. Lubricate piston pin, piston and cylinder with oil prior to assembly. Install piston and connecting rod as outlined in CONNECTING ROD section.

**CYLINDER BLOCK.** Inspect cylinder bore for out-of-round, scoring, excessive wear or other damage. Cylinder can be rebored for installation of oversize piston and rings. Standard cylinder bore diameter for Model W1-145 is 63.000-63.020 mm (2.4803-2.4810 inches). If diameter exceeds 63.15 mm (2.486 inches), recondition or renew cylinder. Standard cylinder bore diameter for Model W1-185 is 67.000-67.0019 mm (2.6378-2.6385 inches). If diameter exceeds 67.15 mm (2.644 inches), recondition or renew cylinder.

**CRANKSHAFT.** The crankshaft is supported in two ball bearing type main bearings (5—Fig. WR54). Renew bearings if any indication of roughness, noise or excessive wear is noted.

Standard crankpin journal diameter for Model W1-145 is 23.950-23.963 mm (0.9429-0.9434 inch) and wear limit is 23.85 mm (0.939 inch).

Standard crankpin journal diameter for Model W1-185 is 25.950-25.963 mm (1.0216-1.0222 inch) and wear limit is 25.85 mm (1.018 inch).

On all models, specified clearance between crankpin and connecting rod is 0.037-0.063 mm (0.0015-0.0025 inch). If clearance exceeds 0.20 mm (0.008 inch), renew connecting rod and/or crankshaft.

Crankshaft end play should be 0.00-0.20 mm (0.000-0.008 inch) for all

models. End play is controlled by an adjusting collar or shim (10—Fig. WR55). Three thicknesses of adjusting collars are available for all models.

When reassembling engine, make certain that timing marks on camshaft gear and crankshaft gear are aligned. Be sure that crankshaft oil seals are installed with lip facing inward. Lubricate seals with engine oil prior to assembly.

**GOVERNOR.** On all models, the governor flyweight assembly (12—Fig. WR58) rotates on a stub shaft in the crankcase cover and is driven by the camshaft gear. Movement of governor weights is transmitted through external linkage to the carburetor throttle valve to maintain a relatively constant engine speed under various loads.

Governor assembly should be checked for free movement of governor weights. Governor should be renewed if weights (11), pivot pins (10) or sleeve (13) is worn or damaged. Adjust governor external linkage as outlined in MAINTENANCE section.

**CAMSHAFT.** On all models, the camshaft (9—Fig. WR58) rides directly in bores in crankcase and crankcase cover. When removing camshaft, position engine so tappets (8) fall away from cam lobes. Identify tappets as they are removed so they can be reinstalled in original positions if reused.

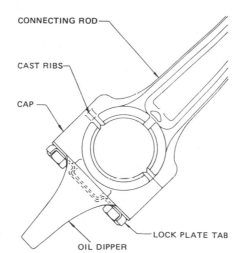

CONNECTING ROD

CAST RIBS

CAP

LOCK PLATE TAB

OIL DIPPER

**Fig. WR56—Cast ribs must be adjacent to each other for correct assembly of connecting rod and cap.**

Standard camshaft journal diameter for all models is 14.973-14.984 mm (0.5895-0.5899 inch) and wear limit is 14.95 mm (0.5886 inch). Inspect cam lobes and tappets for pitting, scoring or excessive wear and renew as necessary. On Model W1-145, specified camshaft lobe height is 24.85-25.05 mm (0.9783-0.9862 inch) and wear limit is 24.70 mm (0.9725 inch). On Model W1-185, specified camshaft lobe height is 28.70-28.90 mm (1.1299-1.1378 inch) and wear limit is 28.55 mm (1.124 inch). It is recommended that camshaft and tappets be renewed as a set.

When reinstalling camshaft, lubricate camshaft and tappets with engine oil. Make certain that timing marks on camshaft gear are aligned with timing mark on crankshaft gear.

**VALVE SYSTEM.** Valves can be removed after removing cylinder head and tappet chamber cover. Compress valve springs and remove spring retain-

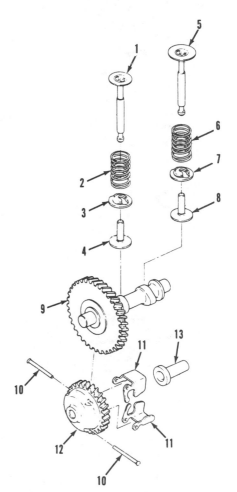

**Fig. WR58—Exploded view of governor flyweight assembly, camshaft and valve components.**

1. Exhaust valve
2. Spring
3. Retainer
4. Tappet
5. Intake valve
6. Spring
7. Retainer
8. Tappet
9. Camshaft & gear
10. Pins
11. Flyweights
12. Governor gear
13. Sleeve

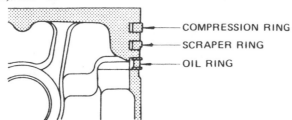

**Fig. WR57—Install piston rings as shown. Stagger ring end gaps at 90 degree intervals around piston.**

COMPRESSION RING

SCRAPER RING

OIL RING

ers (3 and 7—Fig. WR58), then lift valves out of cylinder block.

On all models, valve face and seat angles are 45 degrees for both intake and exhaust. Desired valve seat contact width is 1.2-1.5 mm (0.047-0.059 inch) and maximum allowable seat width is 2.5 mm (0.098 inch). Refer to the following valve specifications:

Valve stem diameter—
Intake . . . . . . . . . . . 6.460-6.475 mm
(0.2543-0.2549 in.)
Exhaust . . . . . . . . . . 6.422-6.444 mm
(0.2528-0.2537 in.)
Wear limit, all valves . . . . . 6.35 mm
(0.250 in.)
Valve stem-to-guide clearance—
Intake . . . . . . . . . . . 0.025-0.062 mm
(0.0010-0.0024 in.)
Exhaust . . . . . . . . . 0.056-0.100 mm
(0.0022-0.0039 in.)

Wear limit, intake & exhaust . . . . . . . . . . . . . . . . 0.30 mm
(0.012 in.)

Valve tappet gap (engine cold) should be 0.08-0.012 mm (0.003-0.005 inch) for both intake and exhaust. To increase clearance, carefully grind end of valve stem. To reduce clearance, grind valve seat deeper or renew valve and/or valve tappet.

# WISCONSIN ROBIN

| Model | No. Cyls. | Bore | Stroke | Displacement |
|-------|-----------|------|--------|--------------|
| EY20D | 1 | 67 mm | 52 mm | 183 cc |
|  |  | (2.64 in.) | (2.05 in.) | (11.17 cu. in.) |
| EY28D | 1 | 75 mm | 62 mm | 273 cc |
|  |  | (2.95 in.) | (2.44 in.) | (16.66 cu. in.) |
| EY40D | 1 | 84 mm | 70 mm | 388 cc |
|  |  | (3.31 in.) | (2.76 in.) | (23.68 cu. in.) |

## MAINTENANCE

**SPARK PLUG.** Spark plug should be removed and cleaned after every 50 hours of operation. Renew spark plug if excessively worn, burned or damaged.

Recommended spark plug is NGK BR6HS for Model EY20D, NGK BPR6HS for Model EY28D and NGK BPR4HS for Model EY40D. Spark plug electrode gap should be set to 0.6-0.7 mm (0.024-0.028 inch) on all models.

**AIR FILTER.** The engine is equipped with a foam type air filter element. The air filter element should be removed and cleaned after every 50 hours of operation, or more often if operating in extremely dusty conditions. To clean element, wash in a nonflammable cleaning solvent and gently squeeze element dry. Oil element with clean engine oil and gently squeeze out the excess oil.

**FUEL FILTER.** Engines are equipped with a filter screen and sediment bowl mounted on bottom of the fuel shut-off valve. Sediment bowl and screen should be removed and cleaned after every 200 hours of operation, or more often if water or dirt is visible in the sediment bowl.

**CARBURETOR.** All models use a float type side draft carburetor. Carburetor is accessible after removing the air filter.

Refer to Fig. R5-1 for exploded view of carburetor used on Model EY20D. The pilot jet (3) controls fuel supply at low engine speed. Main fuel mixture is controlled by a fixed main jet (18) for middle and high engine speeds. There is no provision for adjustment of pilot jet or main jet on this carburetor. Engine idle speed is adjusted by turning throttle stop screw (13) as necessary.

Refer to appropriate Fig. R5-2 or R5-3 for exploded view of carburetors used on Models EY28D and EY40D. The pilot jet (3) controls fuel supply at low speed and a fixed main jet (18) controls fuel mixture at middle and high engine speeds. Initial adjustment of pilot jet

screw (14) is 1-1/2 turns out from a lightly seated position. Final adjustment should be made with engine at operating temperature and running. Adjust pilot jet screw to obtain smooth idle operation and acceleration. On all models, adjust throttle stop screw (13) to obtain desired idle speed.

If carburetor related fuel system problems are encountered, disassemble carburetor and clean using suitable solvent and compressed air. Do not use drills or wire to clean fuel passages as calibration of carburetor fuel:air mixture will be affected if orifices of main jet or pilot jet are enlarged.

**GOVERNOR.** A mechanical flyweight governor is used. The governor unit (26—Fig. R5-9) on Models EY20D and EY28D rotates on a stub shaft (28) in crankcase cover and is driven by the camshaft gear (24). On Model EY40D, governor flyweight assembly is mounted on the camshaft gear (24—Fig. R5-10).

Adjustment of governor is made with carburetor and governor linkage installed. For 3600 rpm engine operation (60 Hz) on Models EY20D and EY28D, be sure that governor spring is connected to hole No. 1 (Fig. R5-5) in speed control lever (13). For 3600 rpm engine operation on Model EY40D, governor spring should be connected to hole No. 1 (Fig. R5-6) in governor lever (8) and hole "B" in speed control lever (13). Loosen governor lever clamp bolt (7). With engine stopped, move governor lever to fully open carburetor throttle valve and hold in this position. Use a screwdriver to rotate governor shaft (6) COUNTERCLOCKWISE (Models EY20D and EY28D) or CLOCKWISE (Model EY40D) as far as it will go. While holding parts in this position, tighten governor lever clamp bolt.

Maximum engine speed is adjusted by turning speed control lever adjusting screw (12—Fig. R5-5 or R5-6). Specified engine speed at no load is 3800 rpm for 60 Hz operation.

**IGNITION.** An electronic, breakerless ignition system is used on all models. No

periodic maintenance or adjustments are required.

**LUBRICATION.** Manufacturer recommends changing engine oil after first 20 hours of operation and every 100 hours thereafter. An oil drain plug is located in the bottom of engine crankcase. To ensure quick and complete draining, it is recommended that oil be drained while engine is warm. Crankcase capacity is approximately 0.6 liter (1.3 pints)

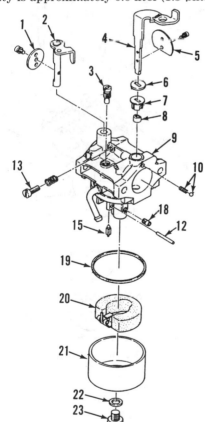

**Fig. R5-1—Exploded view of float type carburetor used on Model EY20D.**

| | |
|---|---|
| 1. Throttle valve | |
| 2. Throttle shaft | |
| 3. Pilot jet | 12. Pin |
| 4. Choke shaft | 13. Idle speed |
| 5. Choke valve | adjustment screw |
| 6. Seal | 15. Fuel inlet valve |
| 7. Retainer | 19. Gasket |
| 8. Ring | 20. Float |
| 9. Carburetor body | 21. Float bowl |
| 10. Choke detent spring | 22. Gasket |
| & ball | 23. Bolt |

for Model EY20D, 0.85 liter (1.8 pints) for Model EY28D or 1.2 liter (2.5 pints) for Model EY40D.

Use engine oil with API service classification SE or SF. SAE 10W-30 or SAE 10W-40 oil is recommended for use in all temperatures. If single viscosity oil is used, select appropriate viscosity for ambient temperature as shown in lubrication chart (Fig. R5-7).

**CLEANING CARBON.** Cylinder head should be removed and carbon cleaned from cylinder head and piston after every 500 hours of operation. Refer to CYLINDER HEAD paragraph for removal and installation procedure.

**VALVE ADJUSTMENT.** Clearance between tappet and end of valve stem should be checked after every 500 hours of operation. To check clearance, remove tappet chamber cover (14—Fig.

R5-8) and breather plate (12). Rotate crankshaft to position piston at top dead center of compression stroke. Use a feeler gage to measure clearance between tappet and valve stem. On Models EY20D and EY28D, specified clearance with engine cold is 0.08-0.12 mm (0.003-0.005 inch) for intake and exhaust. On Model EY40D, specified clearance with engine cold is 0.13-0.17 mm (0.005-0.007 inch) for intake and exhaust.

If clearance is smaller than specified, increase clearance by carefully grinding end of valve stem. To decrease clearance, renew valve and/or tappet. Refer to VALVE SYSTEM paragraph for valve removal and installation procedure.

## REPAIRS

**TIGHTENING TORQUES.** Recommended tightening torques are as follows:

Cylinder head—
EY20D . . . . . . . . . . . . . . .19-22 N·m
(14-16 ft.-lbs.)
EY28D . . . . . . . . . . . . . . .22-26 N·m
(16-19 ft.-lbs.)
EY40D . . . . . . . . . . . . . . .34-41 N·m
(25-30 ft.-lbs.)

Connecting rod bolts—
EY20D, EY28D . . . . . . . . .17-20 N·m
(13-15 ft.-lbs.)
EY40D . . . . . . . . . . . . . . .25-29 N·m
(19-21 ft.-lbs.)
Flywheel nut—
EY20D, EY28D . . . . . . . . .60-65 N·m
(44-48 ft.-lbs.)
EY40D . . . . . . . . . . . . . . .80-100 N·m
(59-73 ft.-lbs.)
Crankcase cover screws—
EY20D . . . . . . . . . . . . . . . .8-10 N·m
(6-7 ft.-lbs.)
EY28D, EY40D . . . . . . . . .17-19 N·m
(13-14 ft.-lbs.)

**CYLINDER HEAD.** To remove cylinder head, remove fuel tank, flywheel shroud, head cover and cylinder air baffle. Remove cylinder head retaining bolts and remove cylinder head and gasket.

Clean all carbon deposits from cylinder head. Use a straightedge and feeler gage to check cylinder head mounting surface for distortion. If cylinder head is warped more than 0.15 mm (0.006 inch), resurface head or renew as necessary.

Reinstall cylinder head using a new head gasket. Tighten head bolts evenly in a crisscross pattern to specified torque.

**CONNECTING ROD.** To remove connecting rod and piston assembly, engine must be separated from generator unit as outlined in GENERATOR section. Remove cylinder head as previously outlined. Unbolt and remove crankcase cover (3—Fig. R5-8) from cylinder block (7). Remove connecting rod cap (17—Fig. R5-9 or R5-10) and withdraw piston and connecting rod assembly from top of cylinder block. Remove retaining ring (4), push out piston pin (5) and separate piston from connecting rod if necessary.

Check connecting rod against the following specifications:

### Model EY20D

Connecting rod big end
inside diameter . .26.000-26.013 mm
(1.0236-1.0241 in.)
Connecting rod-to-crankpin
clearance . . . . . . . . .0.037-0.063 mm
(0.0015-0.0025 in.)
Wear limit . . . . . . . . . . . . . .0.20 mm
(0.008 in.)
Connecting rod small end
inside diameter .0.14.010-14.021 mm
(0.5516-0.5520 in.)
Wear limit . . . . . . . . . . . .14.08 mm
(0.5543 in.)
Piston pin-to-connecting
rod clearance . . . . . .0.010-0.029 mm
(0.0004-0.0012 in.)
Wear limit . . . . . . . . . . . .0.12 mm
(0.005 in.)

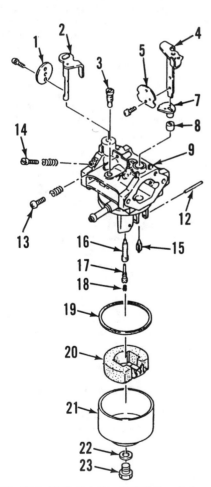

*Fig. R5-2—Exploded view of float type carburetor used on Model EY28D.*

| | |
|---|---|
| 1. Throttle valve | 14. Idle mixture |
| 2. Throttle shaft |     adjustment screw |
| 3. Pilot jet | 15. Fuel inlet valve |
| 4. Choke valve | 16. Nozzle |
| 5. Choke shaft | 17. Main jet pipe |
| 7. Retainer | 18. Main jet |
| 8. Ring | 19. Gasket |
| 9. Carburetor body | 20. Float |
| 12. Pin | 21. Float bowl |
| 13. Idle speed | 22. Gasket |
|     adjustment screw | 23. Bolt |

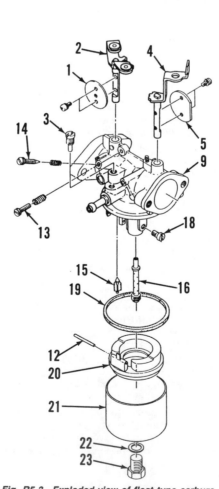

*Fig. R5-3—Exploded view of float type carburetor used on Model EY40D. Refer to Fig. R5-2 for legend.*

Connecting rod side
  clearance . . . . . . . . . .0.10-0.30 mm
                     (0.004-0.012 in.)
  Wear limit . . . . . . . . . . . . . .1.0 mm
                     (0.039 in.)

## Model EY28D

Connecting rod big end
  inside diameter . .28.000-28.013 mm
                     (1.1024-1.1029 in.)
  Wear limit . . . . . . . . . . . . .28.10 mm
                     (1.1063 in.)
Connecting rod-to-crankpin
  clearance . . . . . . . . .0.020-0.046 mm
                     (0.0008-0.0018 in.)
  Wear limit . . . . . . . . . . . . . .0.20 mm
                     (0.008 in.)
Connecting rod small end
  inside diameter . .16.010-16.021 mm
                     (0.6303-0.6307 in.)
  Wear limit . . . . . . . . . . . . .16.08 mm
                     (0.633 in.)
Piston pin-to-connecting
  rod clearance . . . . .0.010-0.029 mm
                     (0.0004-0.0011 in.)
  Wear limit . . . . . . . . . . . . . .0.12 mm
                     (0.0047 in.)
Connecting rod side
  clearance . . . . . . . . . .0.10-0.30 mm
                     (0.004-0.012 in.)
  Wear limit . . . . . . . . . . . . . .1.0 mm
                     (0.039 in.)

## Model EY40D

Connecting rod big end
  inside diameter . .34.000-34.016 mm
                     (1.3386-1.3392 in.)
  Wear limit . . . . . . . . . . . . .34.10 mm
                     (1.3425 in.)
Connecting rod-to-crankpin
  clearance . . . . . . . . .0.070-0.102 mm
                     (0.0028-0.0040 in.)
  Wear limit . . . . . . . . . . . . . .0.20 mm
                     (0.008 in.)
Connecting rod small end
  inside diameter . . . .20.02-20.03 mm
                     (0.7882-0.7886 in.)
  Wear limit . . . . . . . . . . . . .20.08 mm
                     (0.7905 in.)
Piston pin-to-connecting
  rod clearance . . . . .0.020-0.040 mm
                     (0.0008-0.0017 in.)
  Wear limit . . . . . . . . . . . . . .0.10 mm
                     (0.004 in.)
Connecting rod side
  clearance . . . . . . . . . .0.10-0.40 mm
                     (0.004-0.016 in.)
  Wear limit . . . . . . . . . . . . . .1.0 mm
                     (0.039 in.)

When installing connecting rod and piston assembly, make certain that match marks (case ribs) on connecting rod and cap are together as shown in Fig. R5-11 or R5-12. On Models EY20D and EY28D, install oil dipper (16—Fig. R5-9) so dipper is toward flywheel end

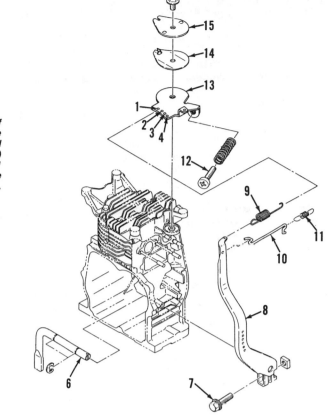

*Fig. R5-5—Exploded view of governor external linkage used on Models EY20D and EY28D. Governor spring (9) should be connected to No. 1 hole in speed control plate (13) for 3600 rpm operation.*

6. Governor shaft
7. Clamp bolt
8. Governor lever
9. Governor spring
10. Carburetor link
11. Link spring
12. High speed stop screw
13. Speed control plate
14. Spring washer
15. Stop plate

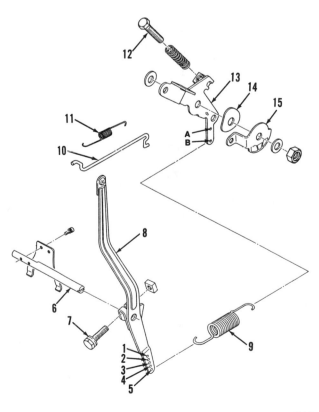

*Fig. R5-6—Exploded view of governor external linkage used on Model EY40D. For 3600 rpm operation, connect governor spring (9) in No. 1 hole in governor lever (8) and "B" hole in speed control plate (13). Refer to Fig. R5-5 for legend.*

of connecting rod cap. On Model EY40D, a dual tang oil dipper (16—Fig. R5-10) is used and can be installed either way. On all models, install a new lock plate (15) and tighten rod cap bolts evenly to specified torque. Turn crankshaft and check for free movement of connecting rod. Bend tabs on lock plate against hex head of rod bolts.

**PISTON, PIN AND RINGS.** After removing piston rings and separating piston from connecting rod, carefully clean carbon and other deposits from piston surface and ring lands.

CAUTION: Extreme care should be exercised when cleaning ring lands. Do not damage squared edges or widen ring grooves. If ring lands are damaged, piston must be renewed.

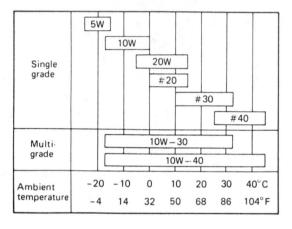

Fig. R5-7—Select engine oil viscosity according to ambient temperature.

Inspect piston for scoring or excessive wear and renew as necessary. Piston pin-to-piston bore fit should be 0.009 mm (0.00035 inch) tight to 0.010 mm (0.00039 inch) loose for Models EY20D and EY28D, or 0.011 mm (0.0004 inch) tight to 0.011 mm (0.0004 inch) loose for Model EY40D. Renew piston and pin if pin-to-piston clearance exceeds 0.06 mm (0.0024 inch) on all models. Refer to the following piston and ring specification data:

### Model EY20D

Piston skirt diameter—
  Standard . . . . . . . . . . 66.96-66.98 mm
    (2.6362-2.6370 in.)
  Wear limit . . . . . . . . . . . . . 66.88 mm
    (2.6331 in.)

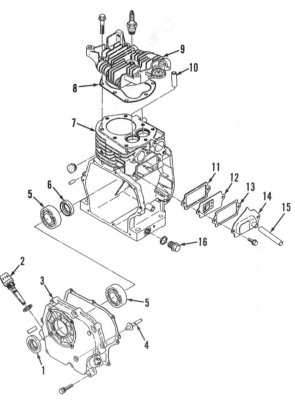

Fig. R5-8—Exploded view of cylinder block and cylinder head typical of Models EY20D and EY28D. Model EY40D is similar except governor stub shaft (4) is not used.

1. Oil seal
2. Dipstick
3. Crankcase cover
4. Governor stub shaft
5. Main bearings
6. Oil seal
7. Cylinder block
8. Head gasket
9. Cylinder head
10. Valve guide
11. Gasket
12. Breather plate
13. Gasket
14. Tappet chamber cover
15. Breather tube
16. Oil drain plug

Piston-to-cylinder clearance—
  Standard . . . . . . . . . 0.020-0.059 mm
    (0.0008-0.0023 in.)
  Wear limit . . . . . . . . . . . . . 0.25 mm
    (0.010 in.)
Piston ring side clearance in groove—
  Top ring . . . . . . . . . 0.050-0.095 mm
    (0.0020-0.0037 in.)
  Second ring . . . . . . . 0.010-0.055 mm
    (0.0004-0.0021 in.)
  Oil ring . . . . . . . . . . 0.010-0.065 mm
    (0.0004-0.0025 in.)
  Wear limit, all rings . . . . . . 0.15 mm
    (0.006 in.)
Piston ring end gap—
  All rings . . . . . . . . . . . 0.05-0.25 mm
    (0.004-0.012 in.)

### Model EY28D

Piston skirt diameter—
  Standard . . . . . . . . . 74.95-74.97 mm
    (2.9508-2.9515 in.)
  Wear limit . . . . . . . . . . . . 74.88 mm
    (2.9480 in.)
Piston-to-cylinder clearance—
  Standard . . . . . . . . . 0.030-0.069 mm
    (0.0012-0.0027 in.)
  Wear limit . . . . . . . . . . . . . 0.25 mm
    (0.010 in.)
Piston ring side clearance in groove—
  Top ring . . . . . . . . . 0.050-0.090 mm
    (0.0020-0.0035 in.)
  Second ring . . . . . . . 0.030-0.070 mm
    (0.0012-0.0028 in.)
  Oil ring . . . . . . . . . . 0.010-0.065 mm
    (0.0004-0.0025 in.)
  Wear limit, all rings . . . . . . 0.15 mm
    (0.006 in.)
Piston ring end gap—
  All rings . . . . . . . . . . . 0.10-0.30 mm
    (0.004-0.012 in.)

### Model EY40D

Piston skirt diameter—
  Standard . . . . . . . . . 83.91-83.94 mm
    (3.3035-3.3047 in.)
  Wear limit . . . . . . . . . . . . 83.84 mm
    (3.301 in.)
Piston-to-cylinder clearance—
  Standard . . . . . . . . . 0.060-0.112 mm
    (0.0024-0.0044 in.)
  Wear limit . . . . . . . . . . . . . 0.30 mm
    (0.012 in.)
Piston ring side clearance in groove—
  Top & second rings . 0.050-0.090 mm
    (0.0020-0.0035 in.)
  Oil ring . . . . . . . . . . . . 0.02-0.05 mm
    (0.0008-0.0021 in.)
  Wear limit, all rings . . . . . . 0.15 mm
    (0.006 in.)
Piston ring end gap—
  All rings . . . . . . . . . . . 0.20-0.40 mm
    (0.008-0.016 in.)

Illustrations courtesy of Teledyne Total Power

When assembling piston and rings, manufacturer's mark (if present) on top and second rings must face toward top of piston. Make certain that scraper ring is installed with notched side facing down (Fig. R5-13), otherwise excessive oil consumption will result. It is recommended that piston pin retaining rings (4—Fig. R5-9 or R5-10) be renewed whenever they are removed. Heating piston in warm water or oil will make installation of piston pin easier. Lubricate piston pin, piston and cylinder with oil prior to assembly. Install piston and connecting rod as outlined in CONNECTING ROD section.

**CYLINDER BLOCK.** Cylinder and crankcase are a one-piece casting on all models. Inspect cylinder for excessive wear, scoring or other damage. If cylinder is out-of-round more than 0.08 mm (0.003 inch) or tapered more than 0.15 mm (0.006 inch), cylinder should be rebored to nearest oversize for which piston and rings are available, or renew cylinder block if necessary.

Standard cylinder bore diameter is 67.000-67.019 mm (2.6378-2.6385 inches) for Model EY20D, 75.000-75.019 mm (2.9528-2.9535 inches) for Model EY28D and 84.000-84.022 mm (3.3071-3.3079 inches) for Model EY40D.

**CRANKSHAFT AND MAIN BEARINGS.** The crankshaft is supported at each end by ball type main bearings (5—Fig. R5-8). To remove crankshaft, separate engine from generator. Remove flywheel shroud. Remove flywheel nut and starter pulley. Use a suitable puller to remove flywheel from crankshaft taper. Remove Woodruff key from crankshaft. Remove crankcase cover and cylinder head. Remove connecting rod cap and push piston and connecting rod through top of cylinder. Slide camshaft out of cylinder block and remove tappets. Mark the tappets so they can be reinstalled in their original positions. Withdraw crankshaft from block.

Inspect crankshaft for excessive wear or damage and renew as necessary. Make certain that ball type main bearings turn smoothly. Bearings should be a light press fit on crankshaft journals. Renew oil seals (1 and 6) in crankcase and crankcase cover if necessary. Refer to the following crankshaft specifications:

**Model EY20D**

Crankpin diameter—
    Standard . . . . . . . 25.950-25.963 mm
                  (1.0216-1.0222 in.)
    Wear limit . . . . . . . . . . . . .25.85 mm
                  (1.018 in.)

Connecting rod-to-crankpin
  clearance—

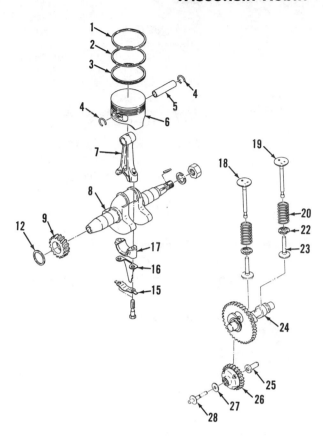

**Fig. R5-9—Exploded view of engine internal components typical of Models EY20D and EY28D.**

1. Compression ring
2. Scraper ring
3. Oil ring
4. Retaining rings
5. Piston pin
6. Piston
7. Connecting rod
8. Crankshaft
9. Crankshaft gear
12. Spacer
15. Lock plate
16. Oil dipper
17. Rod cap
18. Intake valve
19. Exhaust valve
20. Valve spring
22. Spring retainer
23. Tappet
24. Camshaft
25. Sleeve
26. Governor gear
27. Thrust washer
28. Stub shaft

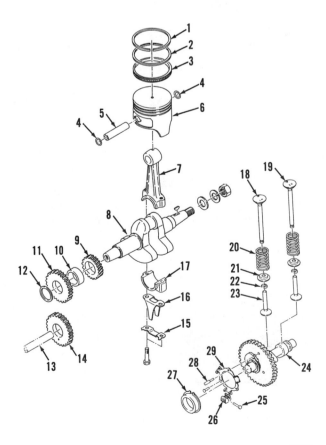

**Fig. R5-10—Exploded view of internal engine components used on Model EY40D.**

1. Compression ring
2. Scraper ring
3. Oil ring
4. Retaining rings
5. Piston pin
6. Piston
7. Connecting rod
8. Crankshaft
9. Crankshaft gear
10. Spacer
11. Balancer drive gear
12. Spacer
13. Balancer shaft
14. Balancer driven gear
15. Lock plate
16. Oil dipper
17. Rod cap
18. Intake valve
19. Exhaust valve
20. Valve spring
21. Spring retainer
22. Retainer lock
23. Tappet
24. Camshaft
25. Pin
26. Governor flyweight
27. Sleeve
28. Pin
29. Flyweight bracket

Standard . . . . . . . . 0.037-0.063 mm
(0.0015-0.0035 in.)
Wear limit . . . . . . . . . . . . 0.20 mm
(0.008 in.)
Side clearance between connecting
rod and crankshaft—
Standard . . . . . . . . . . 0.10-0.30 mm
(0.004-0.012 in.)
Wear limit . . . . . . . . . . . . . 1.0 mm
(0.039 in.)
Crankshaft end play . . . . . 0.0-0.2 mm
(0.000-0.008 in.)

## Model EY28D
Crankpin diameter—
Standard . . . . . . . 27.967-27.980 mm
(1.1011-1.1016 in.)

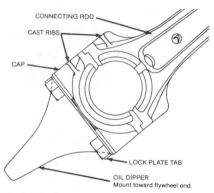

Fig. R5-11—Connecting rod and cap must be installed with cast ribs together on Models EY20D and EY28D.

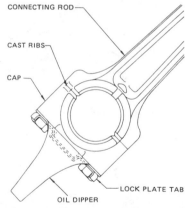

Fig. R5-12—Connecting rod and cap must be installed with cast ribs together on Model EY40D.

COMPRESSION RING
SCRAPER RING
OIL RING

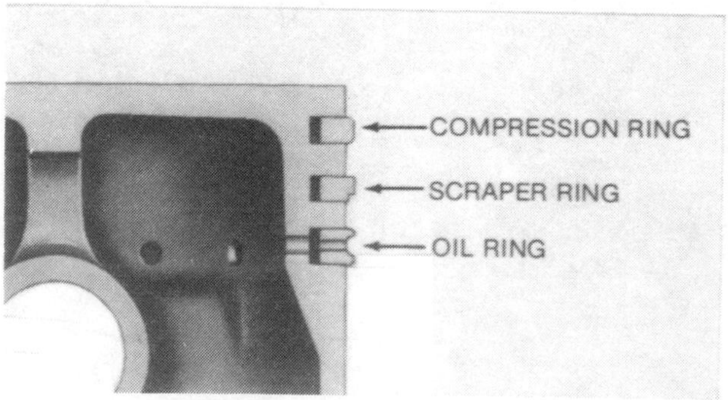

Fig. R5-13—Install piston rings on piston as shown.

Wear limit . . . . . . . . . . . 27.86 mm
(1.0969 in.)
Connecting rod-to-crankpin
clearance—
Standard . . . . . . . . . 0.020-0.046 mm
(0.0008-0.0018 in.)
Wear limit . . . . . . . . . . . . 0.20 mm
(0.008 in.)
Side clearance between connecting
rod and crankshaft—
Standard . . . . . . . . . . 0.10-0.30 mm
(0.004-0.012 in.)
Wear limit . . . . . . . . . . . . . 1.0 mm
(0.039 in.)
Crankshaft end play . . . . . 0.0-0.2 mm
(0.000-0.008 in.)

## Model EY40D

Crankpin diameter—
Standard . . . . . . . 33.914-33.930 mm
(1.3352-1.3358 in.)
Wear limit . . . . . . . . . . . 33.85 mm
(1.3327 in.)
Connecting rod-to-crankpin
clearance:
Standard . . . . . . . . . 0.070-0.102 mm
(0.0028-0.0040 in.)
Wear limit . . . . . . . . . . . . 0.20 mm
(0.008 in.)
Side clearance between connecting
rod and crankshaft—
Standard . . . . . . . . . . 0.10-0.40 mm
(0.004-0.016 in.)
Wear limit . . . . . . . . . . . . . 1.0 mm
(0.039 in.)
Crankshaft end play . 0.025-0.203 mm
(0.001-0.008 in.)

To reassemble, reverse the disassembly procedure while noting the following special instructions: Align timing marks on camshaft gear with timing mark on crankshaft gear (Fig. R5-14). Be careful not to damage lip of oil seal when installing crankcase cover. Check crankshaft end play using a dial indicator. If necessary, remove crankcase cover and install different thickness spacer (12—Fig. R5-9 or R5-10) to obtain recommended end play.

**BALANCER.** Engine balancer (Fig. R5-15) on Model EY40D offsets forces of reciprocating motion and reduces engine vibration. The balancer is driven by a gear on the crankshaft.

When reinstalling crankcase cover and balancer, time balancer driven gear to drive gear as follows: Rotate crankshaft until piston is at top dead center. Remove plug from bottom of crankcase cover and insert a 4 mm (5/32 inch) diameter rod through hole in cover and into hole in balancer gear (Fig. R5-15). Install cover and balancer assembly allowing balancer gears to mesh. Tighten crankcase cover cap screws to 17-19 N·m (12-13 ft.-lbs.) and remove timing rod.

**CAMSHAFT AND TAPPETS.** The camshaft (24—Fig. R5-9 or R5-10) and tappets (23) can be removed from cylinder block after removing crankcase cover as outlined in CRANKSHAFT section.

Inspect camshaft and tappets for excessive wear, scoring, pitting or other damage and renew as necessary. On Model EY20D, specified camshaft lobe height is 28.70-28.90 mm (1.1299-1.1378 inch) and wear limit is 28.55 mm (1.124 inch). On Model EY28D, specified camshaft lobe height is 30.7-30.9 mm (1.2087-1.2165 inch) for intake and exhaust. Wear limit is 30.55 mm (1.2027 inch). On Model EY40D, specified camshaft lobe height is 35.9-36.1 mm (1.4134-1.4213 inch) for intake and 35.4-35.6 mm (1.3937-1.4016 inch) for exhaust. Wear limit is 35.75 mm (1.4075 inch) for intake and 35.25 mm (1.3878 inch) for exhaust. It is recommended that camshaft and tappets be renewed as a set.

When installing camshaft, be sure to align timing marks on camshaft gear with timing mark on crankshaft gear (Fig. R5-14).

**GOVERNOR.** On Models EY20D and EY28D, the internal centrifugal flyweight governor assembly (25 through 28—Fig. R5-9) is mounted on a stub shaft in the crankcase cover. On Model EY40D, governor assembly (25 through 29—Fig. R5-10) is mounted on the camshaft gear. On all models, movement of governor weights is transmitted through external linkage to the carburetor throttle valve to maintain a relatively constant engine speed under various loads.

Governor assembly should be checked for free movement of governor weights. Governor should be renewed if weights, pivot pins or sleeve is worn or damaged. Adjust governor external linkage as outlined in MAINTENANCE section.

**VALVE SYSTEM.** Valves can be removed after removing cylinder head and tappet chamber cover (14—Fig. R5-8). Use a screwdriver or other suitable tool to compress valve spring, then slip spring retainer (22—Fig. R5-9 or R5-10) off end of valve stem and remove valve and spring.

Valve face and seat angle for both intake and exhaust valves is 45 degrees. The following specifications apply to valve guides and valves:

## Models EY20D, EY28D

Valve stem OD—
    Intake . . . . . . . . . . . 6.460-6.475 mm
        (0.2543-0.2549 in.)
    Exhaust . . . . . . . . . 6.422-6.444 mm
        (0.2528-0.2537 in.)
Valve guide ID . . . . . . . . 6.50-6.52 mm
        (0.2550-0.2568 in.)
Valve stem-to-guide clearance—
    Intake . . . . . . . . . . . 0.025-0.062 mm
        (0.0010-0.0024 in.)
    Exhaust . . . . . . . . . 0.056-0.100 mm
        (0.0022-0.0039 in.)
Wear limit—intake &
    exhaust . . . . . . . . . . . . . . . . 0.30 mm
        (0.012 in.)
Valve spring free length—
    Intake & exhaust . . . . . . . . . 37 mm
        (1.457 in.)
    Minimum allowable . . . . . . 35.5 mm
        (1.398 in.)

## Model EY40D

Valve stem OD—
    Intake . . . . . . . . . . . 7.945-7.970 mm
        (0.3128-0.3138 in.)
    Exhaust . . . . . . . . . . 7.910-7.930 mm
        (0.3114-0.3122 in.)
Valve guide ID . . . . . . . 8.00-8.036 mm
        (0.3150-0.3164 in.)
Valve stem-to-guide clearance—
    Intake . . . . . . . . . . . 0.030-0.091 mm
        (0.0012-0.0036 in.)
    Wear limit . . . . . . . . . . . . . . 0.15 mm
        (0.006 in.)
    Exhaust . . . . . . . . . 0.070-0.126 mm
        (0.0028-0.0050 in.)
    Wear limit . . . . . . . . . . . . . . 0.20 mm
        (0.008 in.)
Valve spring free length—
    Intake & exhaust . . . . . . . . . 46 mm
        (1.811 in.)
    Minimum allowable . . . . . . 44.45 mm
        (1.750 in.)

Specified valve clearance for both intake and exhaust valves with engine cold is 0.08-0.12 mm (0.003-0.005 inch) for Models EY20D and EY28D, and 0.13-

0.17 mm (0.005-0.007 inch) for Model EY40D. To check clearance, remove tappet chamber cover and breather plate. Turn crankshaft to position piston at top dead center of compression stroke. Use a feeler gage to measure clearance between end of valve stem and tappet. Valve clearance may be increased by

carefully grinding end of valve stem. To decrease clearance, renew valve and/or tappet. Note that valve must be renewed if distance from end of valve stem to spring retainer groove is less than 3.45 mm (0.136 inch) on Models EY20D and EY28D, or less than 4.8 mm (0.209 inch) on Model EY40D.

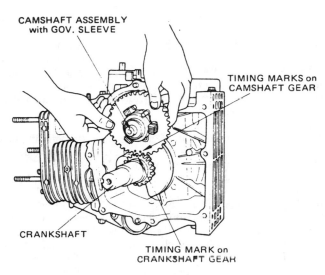

Fig. R5-14—Single marked tooth on crankshaft gear must be installed between two marked teeth on camshaft gear.

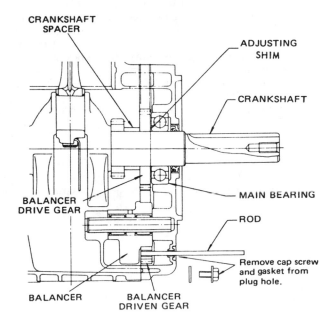

Fig. R5-15—Sectional view showing method of timing engine balancer to crankshaft on Model EY40D. On all models, crankshaft end play is adjusted by installing different thickness adjusting shim.

# WISCONSIN ROBIN

| Model | No. Cyls. | Bore | Stroke | Displacement |
|-------|-----------|------|--------|--------------|
| DY23D | 1 | 70 mm | 60 mm | 230 cc |
| | | (2.76 in.) | (2.36 in.) | (14.0 cu. in.) |
| DY27D | 1 | 75 mm | 60 mm | 265 cc |
| | | (2.95 in.) | (2.36 in.) | (16.2 cu. in.) |
| DY41D | 1 | 82 mm | 76 mm | 412 cc |
| | | (3.20 in.) | (3.10 in.) | (25.14 cu. in.) |

## MAINTENANCE

**LUBRICATION.** Manufacturer recommends changing the engine oil after the first 25 hours of operation and every 50 hours thereafter. All engines are equipped with a reusable oil filter element located in the bottom of crankcase cover. Oil filter should be removed and cleaned using a suitable solvent at each oil change interval.

Use good quality diesel engine oil having API service classification CC or CD. SAE 10W-30 oil is recommended for general all-temperature use. If single viscosity oil is used, refer to chart shown in Fig. R6-1 to select appropriate viscosity for the ambient temperature.

Crankcase capacity is approximately 0.9 liter (0.95 qt.) for Models DY23D and DY27D, and 1.1 liter (1.16 qt.) for Model DY41D. Do not screw oil dipstick into the crankcase when checking oil level on Models DY23D and DY27D.

**AIR FILTER.** Engines may be equipped with either a dual element dry type air filter assembly (Fig. R6-2) or a dual element oil bath type filter assembly (Fig. R6-3). On all models, air filter assembly should be cleaned after every 100 hours of operation, or more often if operating in extremely dusty conditions.

Dry type filter elements can be cleaned using soap and water. Do not wash elements with kerosene, gasoline or oil. Allow elements to dry thoroughly before reinstalling.

Oil bath type filter elements should be cleaned in kerosene or other suitable solvent. After cleaning, soak secondary element with engine oil. Refill oil bowl with engine oil of same viscosity as used in crankcase.

**FUEL FILTER AND BLEEDING FUEL SYSTEM.** A renewable fuel filter element (Fig. R6-4 and R6-5) is located in fuel line between the fuel tank and injection pump. Water and/or sediment should be drained from fuel filter after every 200 hours of operation. Fuel filter element should be renewed after every 500 hours of operation, or sooner if loss of engine power is evident.

Air must be bled from fuel system whenever engine has run out of fuel, fuel lines are disconnected or fuel filter element removed. To bleed fuel system, turn fuel shut-off valve to OPEN position. Loosen air vent screw (Fig. R6-6) on fuel injection pump and let fuel flow until free of air bubbles, then retighten vent screw. If engine fails to start after bleeding air at vent screw, loosen cap nut on high pressure fuel line at the injector. With speed control lever in high speed position, crank engine until fuel is discharged at the loosened cap nut. Retighten cap nut and start the engine.

**VALVE ADJUSTMENT.** Manufacturer recommends checking valve clearance after every 300 hours of operation.

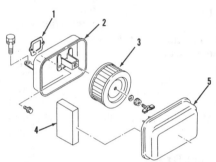

**Fig. R6-2—A dual-element dry type air cleaner is used on Models DY23D and DY27D.**

1. Gasket
2. Filter housing
3. Secondary element
4. Primary element
5. Cover

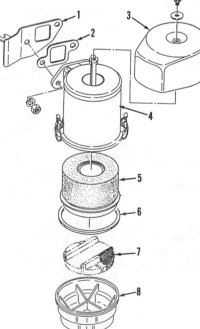

**Fig. R6-3—An oil bath type air cleaner is used on Model DY41D.**

1. Mounting bracket
2. Gasket
3. Cover
4. Filter housing
5. Secondary element
6. Gasket
7. Primary element
8. Oil cup

| | | | | | | | | |
|---|---|---|---|---|---|---|---|---|
| Single grade | 5W | | | | | | | |
| | | 10W | | | | | | |
| | | | 20W | | | | | |
| | | | #20 | | | | | |
| | | | | #30 | | | | |
| | | | | | #40 | | | |
| Multi-grade | | 10W–30 | | | | | | |
| | | 10W–40 | | | | | | |
| Ambient temperature | –20 | –10 | 0 | 10 | 20 | 30 | 40°C | |
| | –4 | 14 | 32 | 50 | 68 | 86 | 104°F | |

**Fig. R6-1—Select engine oil viscosity according to ambient temperature.**

Valve clearance may be checked and adjusted after removing rocker arm cover and turning crankshaft to position piston at top dead center of compression stroke. Use a feeler gage to measure clearance between rocker arm and valve stem as shown in Fig. R6-7. On all models, specified clearance with engine cold is 0.07-0.10 mm (0.003-0.004 inch) for intake and exhaust. To adjust clearance, loosen locknut and turn rocker arm adjusting screw as necessary.

## REPAIRS

**TIGHTENING TORQUES.** Refer to the following table for special tightening torques. All fasteners are metric.

### Models DY23D, DY27D
Cylinder head—
DY23D . . . . . . . . . . . . . . . 30-33 N·m
(22-24 ft.-lbs.)
DY27D . . . . . . . . . . . . . . . 33-35 N·m
(24-25 ft.-lbs.)
Connecting rod cap . . . . . . . 18-20 N·m
(13-14 ft.-lbs.)
Crankcase cover . . . . . . . . . . 17-19 N·m
(12.5-13.5 ft.-lbs.)
Flywheel nut . . . . . . . . . . . . 60-65 N·m
(44-47 ft.-lbs.)
Injector retainer nuts . . . . . . . 5-6 N·m
(3.5-4.5 ft.-lbs.)

### Model DY41D
Cylinder head . . . . . . . . . . . 33-35 N·m
(24-25 ft.-lbs.)
Connecting rod cap . . . . . . . 25-27 N·m
(18-19 ft.-lbs.)
Crankcase cover . . . . . . . . . 20-25 N·m
(15-18 ft.-lbs.)
Flywheel nut . . . . . . . . . . 200-215 N·m
(145-159 ft.-lbs.)
Injector retainer nuts . . . . . . 9-10 N·m
(6.5-7 ft.-lbs.)
Main bearing housing . . . . . 20-23 N·m
(15-17 ft.-lbs.)
Oil pump cover bolts . . . . . . . 8-10 N·m
(6-8 ft.-lbs.)

**CYLINDER HEAD AND VALVE SYSTEM.** To remove cylinder head, remove muffler and air cleaner. Disconnect injector high pressure line and fuel return line, and immediately cap all openings to prevent entry of dirt into fuel system. Remove injector from cylinder head. On Model DY41D, remove push rod tube retaining nuts. On all models, remove rocker arm cover, unscrew cylinder head retaining nuts and remove cylinder head.

On Models DY23D and DY27D, rocker shaft (9—Fig. R6-8) is retained in cylinder head by snap rings (10). On Model DY41D, rocker shaft (14—Fig. R6-9) is a light press fit in cylinder head. Use a brass drift and hammer to tap rocker shaft out of cylinder head. On all models, be sure to identify all parts as they are removed so they can be reinstalled in original positions if reused.

Inspect all parts for wear or damage. Use a straightedge and feeler gage to check cylinder head mounting surface for distortion. If cylinder head is warped more than 0.05 mm (0.002 inch), resurface or renew cylinder head as necessary. Valve face and seat angle for both valves is 45 degrees. Desired valve seat contact width is 1.5 mm (0.060 inch) and maximum allowable seat contact width is 2.2 mm (0.086 inch). Refer to the following valve system specifications:

### Models DY23D, DY27D
Valve stem OD—
Intake . . . . . . . . . . . 5.422-5.437 mm
(0.2135-0.2141 in.)
Exhaust . . . . . . . . . 5.402-5.417 mm
(0.2127-0.2133 in.)
Valve stem-to-guide
clearance—
Intake . . . . . . . . . . . 0.063-0.096 mm
(0.0025-0.0038 in.)
Exhaust . . . . . . . . . 0.083-0.116 mm
(0.0033-0.0046 in.)
Wear limit—intake &
exhaust . . . . . . . . . . . . . . . . . 0.30 mm
(0.012 in.)

Rocker shaft OD . . . 11.937-11.984 mm
(0.4714-0.4718 in.)
Wear limit . . . . . . . . . . . . . 11.92 mm
(0.4693 in.)
per arm bore . . . . . . 12.000-12.018 mm
(0.4724-0.4731 in.)
Wear limit . . . . . . . . . . . . . 12.07 mm
(0.4752 in.)
Rocker shaft-to-rocker
arm clearance . . . . . 0.016-0.045 mm
(0.0006-0.0018 in.)
Wear limit . . . . . . . . . . . . . 0.15 mm
(0.006 in.)
Valve spring free length—
Standard . . . . . . . . . . . . . . . 32.5 mm
(1.280 in.)
Minimum allowable . . . . . . . 31.0 mm
(1.220 in.)

### Model DY41D
Valve stem OD—
Intake . . . . . . . . . . . 6.922-6.937 mm
(0.2725-0.2731 in.)

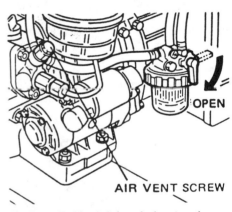

Fig. R6-6—To bleed air from fuel system, loosen air vent screw for fuel injection pump and allow fuel to flow until free of air bubbles.

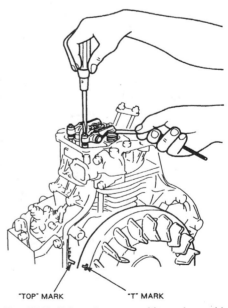

Fig. R6-7—Valve clearance with engine cold should be 0.07-0.10 mm (0.003-0.004 inch).

Fig. R6-4—A renewable fuel filter assembly is used on Models DY23D and DY27D.

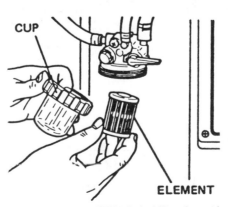

Fig. R6-5—On Model DY41D, fuel filter element is located inside a sediment bowl which is attached to fuel shut-off valve.

Exhaust . . . . . . . . . .6.902-6.917 mm
(0.2717-0.2723 in.)
Valve guide ID . . . . . . .7.000-7.015 mm
(0.2756-0.2762 in.)
Valve stem-to-guide
clearance—
Intake . . . . . . . . . . .0.063-0.093 mm
(0.0025-0.0036 in.)
Exhaust . . . . . . . . . .0.083-0.113 mm
(0.0033-0.0044 in.)
Wear limit—intake &
exhaust. . . . . . . . . . . . . . . . .0.30 mm
(0.012 in.)
Rocker shaft OD . . .11.966-11.984 mm
(0.4711-0.4718 in.)
Wear limit . . . . . . . . . . . .11.92 mm
(0.469 in.)
Rocker arm bore . . .12.000-12.018 mm
(0.4725-0.4731 in.)
Wear limit . . . . . . . . . . . . .12.07 mm
(0.475 in.)
Rocker shaft-to-rocker
arm clearance . . . . .0.016-0.052 mm
(0.0.0006-0.0020 in.)
Wear limit . . . . . . . . . . . . . .0.15 mm
(0.006 in.)
Valve spring free length . . . .36.5 mm
(1.437 in.)

Reassemble by reversing disassembly
procedure. All components should be installed in their original locations if reused. Install cylinder head and tighten
cylinder head nuts in three steps to final torque of 30-33 N·m (22-24 ft.-lbs.)
on Model DY23D or 33-35 N·m (24-25 ft.-
lbs.) on Models DY27D and DY41D. Adjust valve clearance as outlined in
MAINTENANCE section.

**CYLINDER.** On Models DY23D and
DY27D, the cylinder and crankcase are
an integral casting. Standard cylinder
bore diameter is 70.000-70.019 mm
(2.7559-2.7567 inches) for Model DY23D
and wear limit is 70.25 mm (2.7657
inches). On Model DY27D, standard bore
diameter is 75.000-75.019 mm (2.9528-
2.9535 inches) and wear limit is 75.25
mm (2.9626 inches). Cylinder may be rebored for installation of oversize piston
and rings if necessary.

On Model DY41D, the cylinder is a separate casting from crankcase. Cylinder
may be removed after removing cylinder head. Standard cylinder bore diameter is 82.00-82.019 mm (3.228-3.229
inches) and wear limit is 82.25 mm
(3.238 inches). Cylinder may be rebored
for installation of oversize piston and
rings if necessary.

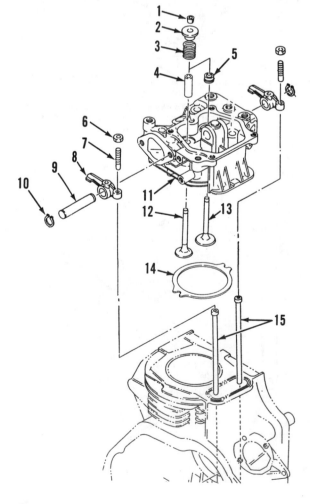

**Fig. R6-8—Exploded view of
cylinder head and valve components used on Models
DY23D and DY27D.**

1. Retainer lock
2. Spring retainer
3. Valve spring
4. Valve guide
5. Valve stem seal
6. Locknut
7. Adjusting screw
8. Rocker arm
9. Rocker shaft
10. Snap ring
11. Cylinder head
12. Exhaust valve
13. Intake valve
14. Spacer
15. Push rods

On all models, if cylinder, piston, connecting rod or crankshaft have been renewed, check and adjust clearance between piston and cylinder head as
follows: Position piston at top dead center and measure distance from top of
piston to top surface of cylinder as
shown in Fig. R6-10. On Models DY23D
and DY27D, select proper thickness cylinder head spacer (14—Fig. R6-8) so
clearance is 0.6-0.7 mm (0.024-0.027
inch). On Model DY41D, install correct
thickness spacer (19—Fig. R6-9) between cylinder and crankcase to obtain
clearance of 0.6-0.7 mm (0.024-0.027
inch).

**PISTON, PIN AND RINGS.** On
Models DY23D and DY27D, piston and
connecting rod are removed as an assembly after removing cylinder head,
crankcase cover and connecting rod cap.
On Model DY41D, piston can be separated from connecting rod after removing
the cylinder head and cylinder.

After separating piston from connecting rod and removing piston rings, carefully clean carbon and other deposits
from piston surface and ring lands.

**CAUTION: Extreme care should be exer-
cised when cleaning ring lands. Do not damage squared edges or widen ring grooves.
If ring lands are damaged, piston must be
renewed.**

Inspect piston for scoring or excessive
wear and renew as necessary. Standard
piston skirt diameter, measured at bottom of skirt and 90 degrees from piston
pin, is 69.961-69.981 mm (2.7544-2.7552
inches) for Model DY23D, 74.961-74.981
mm (2.9512-2.9520 inches) for Model
DY27D and 81.940-81.960 mm (3.226-
3.227 inches) for Model DY41D. Refer to
the following specifications:

**Models DY23D, DY27D**
Piston-to-cylinder
clearance . . . . . . . . .0.019-0.058 mm
(0.0.0008-0.0022 in.)
Wear limit . . . . . . . . . . . . . .0.15 mm
(0.006 in.)
Piston pin bore . . . .18.001-18.008 mm
(0.7087-0.7089 in.)
Wear limit . . . . . . . . . . . . .18.03 mm
(0.7098 in.)
Piston pin OD . . . . .18.000-18.006 mm
(0.7087-0.7089 in.)
Wear limit. . . . . . . . . . . . .17.980 mm
(0.7079 in.)
Piston ring side clearance
in ring groove—
Top & second rings...0.05-0.09 mm
(0.0020-0.0035 in.)
Wear limit . . . . . . . . . . . . . .0.15 mm
(0.006 in.)
Oil ring . . . . . . . . . .0.015-0.055 mm
(0.0006-0.0022 in.)

Wear limit ..............0.10 mm
(0.004 in.)
Piston ring end gap (DY23D)—
Top & second rings...0.30-0.50 mm
(0.012-0.020 in.)
Oil ring..............0.25-0.45 mm
(0.010-0.018 in.)
Piston ring end gap (DY27D)—
Top & second rings...0.10-0.30 mm
(0.004-0.012 in.)
Oil ring............0.25-0.45 mm
(0.010-0.018 in.)

## Model DY41D
Piston-to-cylinder
clearance .........0.060-0.102 mm
(0.0023-0.0040 in.)
Wear limit .............0.20 mm
(0.008 in.)
Piston pin bore...21.000-21.008 mm
(0.8268-0.8270 in.)
Wear limit ............21.03 mm
(0.828 in.)
Piston pin OD .....21.000-21.006 mm
(0.8268-0.0.8270 in.)
Wear limit ............20.98 mm
(0.826 in.)

Piston ring side clearance
in ring groove—
Top & second rings..0.05-0.09 mm
(0.0020-0.0035 in.)
Wear limit .............0.15 mm
(0.006 in.)
Oil ring .........0.015-0.055 mm
(0.0005-0.0021 in.)

Piston ring end gap—
Top & second rings..0.30-0.50 mm
(0.012-0.020 in.)
Oil ring............0.25-0.45 mm
(0.010-0.018 in.)

When installing top and second rings on piston, make sure that side of ring with "N" mark or punch mark faces up. See Fig. R6-11. Stagger ring end gaps around piston. Assemble piston on connecting rod so letter or arrow stamped on top of piston is toward side of connecting rod marked "FAN" (Fig. R6-12). Install new piston pin retaining rings. Lubricate piston and cylinder with engine oil prior to reassembly. Install piston and connecting rod so "FAN" mark

side of connecting rod is facing flywheel side of engine. Align match marks on connecting rod and cap and tighten rod cap bolts to specified torque.

**CONNECTING ROD.** On Models DY23D and DY27D, connecting rod and piston are removed as an assembly af-

ter first removing cylinder head, crankcase cover and connecting rod cap.

On Model DY41D, connecting rod can be removed after first removing cylinder head, cylinder, oil pan and connecting rod cap.

All models are equipped with renewable, precision insert type bearings in

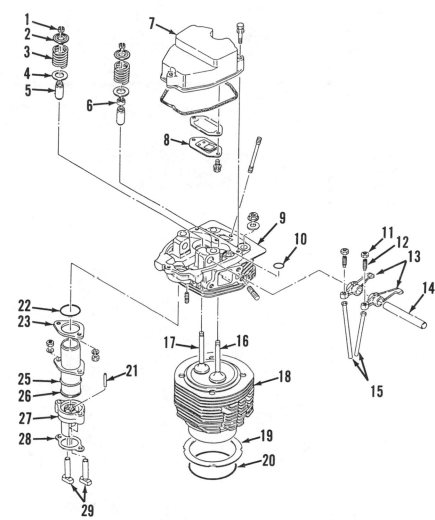

Fig. R6-9—Exploded view of cylinder head and valve components used on Model DY41D.

| | | |
|---|---|---|
| 1. Retainer lock | 8. Breather valve | 15. Push rods | 22. "O" ring |
| 2. Spring retainer | 9. Cylinder head | 16. Intake valve | 23. Retainer plate |
| 3. Valve spring | 10. "O" ring | 17. Exhaust valve | 25. "O" ring |
| 4. Washer | 11. Locknut | 18. Cylinder | 26. Push rod sleeve |
| 5. Valve guide | 12. Adjusting screw | 19. Spacer | 27. Tappet guide |
| 6. Valve stem seal | 13. Rocker arms | 20. "O" ring | 28. Gasket |
| 7. Rocker cover | 14. Rocker shaft | 21. Spring pin | 29. Tappets |

Fig. R6-10—Clearance between piston and top of cylinder should be checked whenever cylinder, piston, connecting rod or crankshaft is renewed. Refer to text.

Fig. R6-11—View of piston rings showing correct installation.

Top ring

Second ring

Oil ring

big end of connecting rod. Check connecting rod against the following specifications:

**Models DY23D, DY27D**
Connecting rod small end
bore . . . . . . . . . . . 18.013-18.034 mm
(0.7092-0.7100 in.)

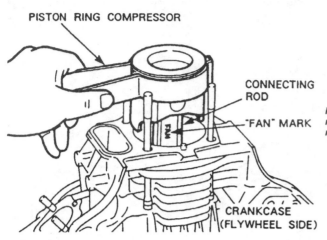

PISTON RING COMPRESSOR

CONNECTING ROD

"FAN" MARK

FAN

CRANKCASE (FLYWHEEL SIDE)

*Fig. R6-12—Connecting rod must be installed with side marked "FAN" toward flywheel end of engine.*

Wear limit . . . . . . . . . . . . 18.05 mm
(0.7106 in.)
Connecting rod-to-piston
pin clearance . . . . . . 0.007-0.034 mm
(0.0003-0.0013 in.)
Wear limit . . . . . . . . . . . . . 0.08 mm
(0.003 in.)
Crankshaft crankpin diameter—
Standard . . . . . . . 33.011-33.027 mm
(1.2996-1.3003 in.)
Wear limit . . . . . . . . . . . . 32.85 mm
(1.2933 in.)
Connecting rod-to-crankpin
clearance . . . . . . . . 0.023-0.081 mm
(0.0009-0.0032 in.)
Wear limit . . . . . . . . . . . . . 0.10 mm
(0.004 in.)
Connecting rod side clearance
on crankshaft . . . . . . . 0.10-0.30 mm
(0.004-0.012 in.)
Wear limit . . . . . . . . . . . . . 0.50 mm
(0.020 in.)

## Model DY41D
Connecting rod small end
bore . . . . . . . . . . . 21.013-21.034 mm
(0.8273-0.0.8281 in.)
Wear limit . . . . . . . . . . . . 21.05 mm
(0.829 in.)
Connecting rod-to-piston
pin clearance . . . . . . 0.007-0.034 mm
(0.0003-0.0013 in.)
Wear limit . . . . . . . . . . . . . 0.08 mm
(0.003 in.)
Crankshaft crankpin diameter—
Standard . . . . . . . 40.024-40.040 mm
(1.5758-1.5764 in.)
Wear limit . . . . . . . . . . . . 39.874 mm
(1.5698 in.)
Connecting rod-to-crankpin
clearance . . . . . . . . 0.010-0.068 mm
(0.0004-0.0026 in.)
Wear limit . . . . . . . . . . . . . 0.10 mm
(0.004 in.)
Connecting rod side clearance
on crankshaft . . . . . . . 0.07-0.33 mm
(0.003-0.013 in.)
Wear limit . . . . . . . . . . . . . 0.50 mm
(0.020 in.)

Reverse removal procedure to install connecting rod. Be sure that side of connecting rod marked "FAN" (Fig. R6-12) is facing flywheel side of engine. Align match marks on connecting rod and cap and tighten rod bolts to specified torque.

### Models DY23D, DY27D

**CRANKSHAFT AND CRANKCASE.** Crankshaft is supported by a ball type main bearing (12—Fig. R6-14) at flywheel end and a sleeve type bushing (15) at pto end. To remove crankshaft, drain engine oil. Remove blower housing, flywheel nut and starter pulley. Use a suitable puller to remove flywheel. Remove key from crankshaft. Remove

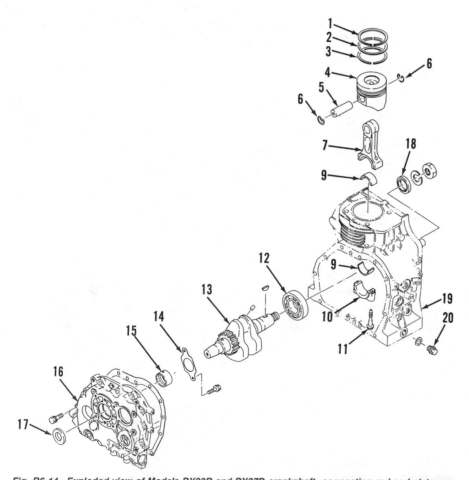

*Fig. R6-14—Exploded view of Models DY23D and DY27D crankshaft, connecting rod and piston assemblies.*

1. Compression ring
2. Scraper ring
3. Oil ring
4. Piston
5. Piston pin
6. Snap ring
7. Connecting rod
9. Bearing insert
10. Rod cap
11. Rod bolt
12. Main bearing
13. Crankshaft
14. Thrust washer
15. Main bearing
16. Crankcase cover
17. Oil seal
18. Oil seal
19. Cylinder block
20. Oil drain plug

cylinder head, crankcase cover and fuel injection pump. Remove camshaft and tappets. Mark the tappets so they can be reinstalled in original positions. Remove connecting rod and piston assembly. Use a plastic hammer to tap crankshaft out of crankcase. Remove main bearings and oil seals from crankcase and crankcase cover if necessary.

Inspect crankshaft for wear, scratches, scoring or other damage. Crankshaft main bearing (15—Fig. R6-14) in crankcase cover and connecting rod bearing insert (9) are available in undersizes as well as standard size. Refer to the following specifications:

Crankpin diameter—
  Standard . . . . . . . .33.011-33.027 mm
           (1.1.2996-1.3003 in.)
  Wear limit . . . . . . . . . . . . .32.85 mm
           (1.2933 in.)
Main journal diameter (pto end)—
  Standard . . . . . . .32.984-33.000 mm
           (1.2986-1.2992 in.)
  Wear limit . . . . . . . . . . . . .32.85 mm
           (1.2933 in.)
Clearance between main journal
  and main bearing (pto end)—
  Standard . . . . . . . . .0.014-0.086 mm
           (0.0006 0.0034 in.)

Wear limit . . . . . . . . . . . . .0.12 mm
           (0.0047 in.)

When reinstalling crankshaft, be careful not to damage lip of oil seals. Lubricate seals and crankshaft journals with engine oil. Align single timing mark on crankshaft gear between double timing marks on camshaft gear (Fig. R6-15). Crankshaft end play is adjusted by varying thickness of thrust washer (14—Fig. R6-14). Specified end play is 0.10-0.30 mm (0.004-0.012 inch). Tighten crankcase cover retaining screws evenly to 17-19 N·m (12.5-13.5 ft.-lbs.). Tighten flywheel retaining nut to 60-65 N·m (44-47 ft.-lbs.). Install fuel injection pump as outlined in INJECTION PUMP section.

## Model DY41D

**CRANKSHAFT AND CRANKCASE.** Crankshaft is supported by a ball type main bearing (12—Fig. R6-16) at flywheel end and by a sleeve type bushing (22) at timing gear end. To remove crankshaft, drain engine oil. Remove blower housing (28) and output shaft (31). Use suitable puller to remove flywheel (30) from crankshaft taper. Re-

move key from crankshaft. Remove fuel injection pump. Unbolt and remove timing gear/crankcase cover (21). Remove oil pan (19) and remove connecting rod cap (10). Remove cylinder head retaining nuts and push rod tube retaining nuts. Withdraw cylinder, piston and connecting rod assembly. Remove cap screws retaining main bearing housing (26), then remove crankshaft and bearing housing as an assembly from crankcase. Separate bearing housing and ball bearing from crankshaft.

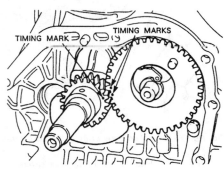

Fig. R6-15—When reassembling engine, make certain that timing marks on crankshaft gear and camshaft gear are aligned to ensure correct valve timing.

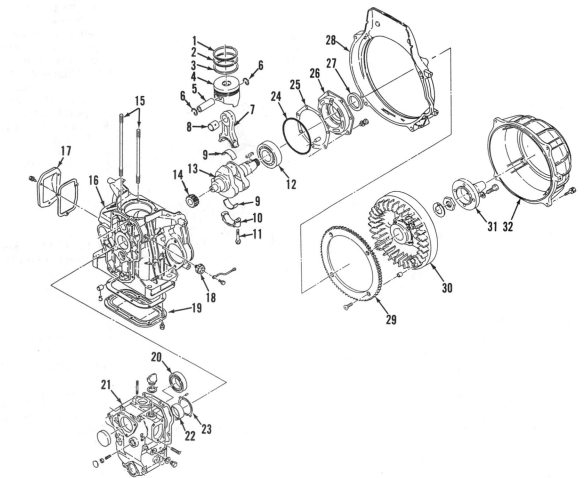

1. Compression ring
2. Scraper ring
3. Oil ring
4. Piston
5. Piston pin
6. Snap ring
7. Connecting rod
8. Bushing
9. Bearing insert
10. Rod cap
11. Rod bolt
12. Main bearing
13. Crankshaft
14. Crankshaft gear
15. Cylinder studs
16. Crankcase
17. Cover
18. Oil pressure sender
19. Oil pan
20. Camshaft bearing
21. Crankcase cover
22. Main bearing
23. Thrust washer
24. "O" ring
25. Shim
26. Main bearing
    housing
27. Oil seal
28. Blower housing
29. Ring gear (electric start)
30. Flywheel
31. Drive shaft
32. Adapter housing

Fig. R6-16—Exploded view of Model DY41D engine.

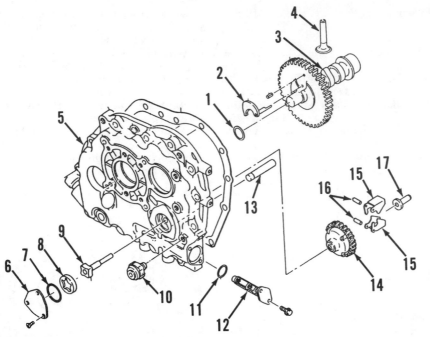

**Fig. R6-17—Exploded view of camshaft, oil pump and governor assemblies used on Models DY23D and DY27D.**

| | | |
|---|---|---|
| 1. Spacer | 5. Crankcase cover | 9. Pump inner rotor & governor shaft | 13. Oil suction pipe |
| 2. Compression release mechanism | 6. Oil pump cover | 10. Oil pressure sender | 14. Governor gear |
| 3. Camshaft & gear assy. | 7. "O" ring | 11. "O" ring | 15. Governor weights |
| 4. Tappet | 8. Oil pump outer rotor | 12. Oil filter | 16. Pins |
| | | | 17. Sleeve |

Inspect crankshaft for wear, scratches, scoring or other damage. Crankshaft main bearing (22—Fig. R6-16) and connecting rod bearing insert (9) are available in oversizes as well as standard size. Refer to the following specifications:

Crankpin diameter—
  Standard . . . . . . . 40.024-40.040 mm
            (1.5758-1.5764 in.)
  Wear limit . . . . . . . . . . . . . 39.874 mm
            (1.5698 in.)
Main journal diameter (timing gear end)—
  Standard . . . . . . 41.984-42.000 mm
            (1.6529-1.6535 in.)
  Wear limit . . . . . . . . . . . . . 41.85 mm
            (1.6476 in.)
Crankshaft-to-main bearing clearance (timing gear end)—
  Standard . . . . . . . . . 0.020-0.092 mm
            (0.0008-0.0036 in.)
  Wear limit . . . . . . . . . . . . . . 0.15 mm
            (0.006 in.)
Crankshaft end play . . . 0.10-0.20 mm
            (0.004-0.008 in.)

Lubricate crankshaft oil seal (27—Fig. R6-16) with oil or grease before installing main bearing housing (26) onto crankshaft. Assemble crankshaft with thrust washer (23) and main bearing housing into crankcase, then check crankshaft end play. Install suitable thickness shim (25) on main bearing housing to obtain desired end play. Tighten main bearing housing cap screws evenly to 20-23 N·m (15-17 ft.-lbs.). Align match marks on crankshaft gear and camshaft gear. Install crankcase cover and tighten cap screws to 20-25 N·m (15-18 ft.-lbs.). Install fuel injection pump as outlined in INJECTION PUMP section. Install cylinder, piston and connecting rod as previously outlined. Tighten flywheel retaining nut to 200-215 N·m (145-159 ft.-lbs.). Tighten output shaft cap screws to 55-70 N·m (40-50 ft.-lbs.).

**CAMSHAFT.** Refer to appropriate Fig. R6-17 or R6-18 for exploded view of camshaft and related components. To remove camshaft, first drain engine oil. Remove fuel injection pump. Remove rocker cover and loosen rocker arm adjusting screws. Unbolt and remove crankcase cover, then withdraw camshaft and tappets. Mark tappets so they can be reinstalled in their original positions.

Inspect camshaft and tappets for excessive wear, scoring, pitting or other damage. It is recommended that camshaft and tappets be renewed as a set. Refer to the following specifications:

## Models DY23D, DY27D
Camshaft lobe height—

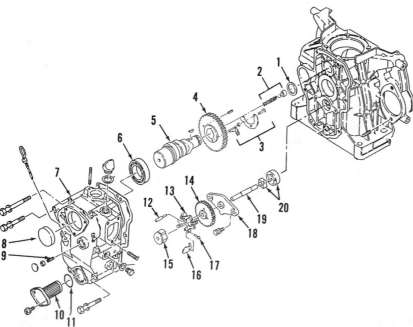

**Fig. R6-18—Exploded view of camshaft, oil pump and governor assemblies used on Model DY41D.**

| | | |
|---|---|---|
| 1. Spacer | 6. Ball bearing | 11. "O" ring | 16. Governor flyweight |
| 2. Check valve | 7. Crankcase cover | 12. Pin | 17. Pin |
| 3. Compression release mechanism | 8. Plug | 13. Flyweight bracket | 18. Oil pump cover |
| 4. Camshaft gear | 9. Smoke set screw | 14. Governor gear | 19. Oil pump/governor shaft |
| 5. Camshaft | 10. Oil filter | 15. Governor sleeve | 20. Oil pump rotors |

Intake & exhaust...32.13-32.23 mm
(1.2650-1.2689 in.)
Wear limit . . . . . . . . . . . .31.98 mm
(1.2591 in.)
Injection pump . . . .29.95-30.05 mm
(1.1791-1.1831 in.)
Wear limit . . . . . . . . . . . . .29.85 mm
(1.1752 in.)
Camshaft journal diameter—
Flywheel end. . . .24.934-24.947 mm
(0.9817-0.9822 in.)
Wear limit . . . . . . . . . . . .24.90 mm
(0.9803 in.)
Pto end . . . . . . . . .14.973-14.984 mm
(0.5895-0.5899 in.)
Wear limit . . . . . . . . . . . .14.90 mm
(0.5866 in.)
Camshaft end play . . . . .0.05-0.25 mm
(0.002-0.010 in.)
Tappet stem OD . . . . .7.960-7.975 mm
(0.3134-0.3140 in.)
Wear limit . . . . . . . . . . . .7.93 mm
(0.3122 in.)
Tappet guide ID . . . . .8.000-8.015 mm
(0.3150-0.3156 in.)
Wear limit . . . . . . . . . . . . .8.08 mm
(0.318 in.)

**Model DY41D**
Camshaft lobe height—
Intake & exhaust.36.445-36.545 mm
(1.4349-1.4388 in.)
Wear limit . . . . . . . . . . . .36.30 mm
(1.429 in.)
Injection pump . . . .39.95-40.05 mm
(1.5728-1.5768 in.)
Wear limit . . . . . . . . . . . .39.85 mm
(1.569 in.)
Camshaft journal diameter—
Flywheel end. . . .21.959-21.980 mm
(0.8645-0.8653 in.)
Crankcase cover
end . . . . . . . . .34.977-34.988 mm
(1.3770-1.3775 in.)
Camshaft end play . . . . .0.10-0.30 mm
(0.004-0.012 in.)
Tappet stem OD—
Standard . . . . . . . . .7.967-7.987 mm
(0.3137-0.3144 in.)
Wear limit . . . . . . . . . . . . .7.93 mm
(0.312 in.)
Tappet guide ID—
Standard . . . . . . . . . .8.00-8.015 mm
(0.3150-0.3155 in.)

Wear limit . . . . . . . . . . . . .8.08 mm
(0.318 in.)

Camshaft end play is adjusted by installing correct thickness shim spacer (1—Fig. R6-17 or R6-18). To adjust end play on Models DY23D and DY27D, temporarily install camshaft in crankcase and measure from machined surface of crankcase to camshaft shim surface. Measure from machined surface of crankcase cover to shim surface in the cover. The difference between the two measurements is camshaft end play. To adjust end play on Model DY41D, temporarily install camshaft in crankcase cover and measure from machined surface of cover to shim surface of camshaft. Measure from machined surface of crankcase to shim surface in crankcase. The difference between the two measurements is camshaft end play. Select and install correct thickness shim as needed.

When installing camshaft, be sure to align timing marks on camshaft gear and crankshaft gear. Install crankcase cover and tighten retaining cap screws evenly to 17-19 N·m (12.5-13.5 ft.-lbs.) on Models DY23D and DY27D or to 20-25 N·m (15-18 ft.-lbs.) on Model DY41D. Install injection pump as outlined in INJECTION PUMP section. Adjust valve clearance as outlined in MAINTENANCE section.

**GOVERNOR.** The centrifugal flyweight governor and governor gear assembly (Fig. R6-17 or R6-18) is mounted on engine oil pump shaft and is driven by the camshaft gear. Remove crankcase cover for access to governor assembly. Use a suitable puller to separate governor gear from oil pump shaft on Models DY23D and DY27D.

Inspect all parts for excessive wear or damage. Governor components must move freely for proper governor operation.

On Models DY23D and DY27D, governor gear and oil pump shaft must be renewed as a set. When reassembling, install oil pump shaft in crankcase cover and press governor gear onto shaft un-

til there is a clearance of 1 mm (0.040 inch) between gear hub and surface of cover as shown in Fig. R6-19.

On Model DY41D, governor and oil pump are available only as an assembly. When installing oil pump and governor, tighten oil pump cover cap screws to 8-10 N·m (6-7 ft.-lbs.).

**INJECTOR.** To remove injector, first clean dirt from injector, fuel lines and cylinder head. Disconnect fuel return line and high pressure line and cap or plug all openings. Remove retainer plate nuts and withdraw injector and copper washer from cylinder head.

A suitable test stand is required to check injector operation.

**WARNING: Fuel leaves the injector nozzle with sufficient force to penetrate the skin. When testing, keep yourself clear of nozzle spray.**

Attach injector to tester and operate pump lever a few quick strokes to purge air from injector and to make sure nozzle valve is not stuck. When operating properly during test, injector nozzle will emit a buzzing sound and cut off quickly with no leakage at seat. Spray pattern should be uniform and well atomized from the four spray holes.

Opening pressure should be 19125 kPa (2775 psi) for all models. Opening pressure is adjusted by varying the thickness of shims (3—Fig. R6-20).

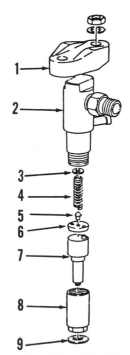

*Fig. R6-20—Exploded view of fuel injector typical of all models.*

1. Retainer bracket
2. Nozzle holder
3. Shim
4. Spring
5. Spring seat
6. Spacer
7. Nozzle valve assy.
8. Nozzle nut
9. Washer

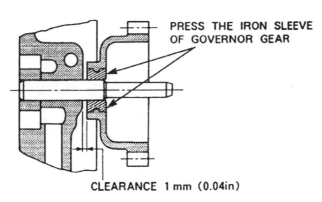

*Fig. R6-19—When installing governor gear on Models DY23D and DY27D, press gear onto shaft until there is clearance of 1 mm (0.040 in.) between gear and crankcase cover.*

PRESS THE IRON SLEEVE
OF GOVERNOR GEAR

CLEARANCE 1 mm (0.04in)

Operate tester to maintain a pressure 3450 kPa (500 psi) below opening pressure for 10 seconds. If a drop forms at nozzle tip, nozzle valve is not seating and injector must be overhauled or renewed.

To disassemble injector, remove nozzle holder nut (8) and separate components from nozzle holder. Thoroughly clean all parts in suitable solvent. Clean nozzle spray holes using cleaning wire slightly smaller in diameter than spray holes. Spray hole diameter is 0.22 mm (0.009 inch) on all models. Inspect all polished surfaces for corrosion or other damage and renew as necessary. Nozzle body (7) and needle must be renewed as a matched set.

When reassembling injector, make certain that all components are clean and wet with clean diesel fuel. Tighten nozzle holder nut to 30-40 N·m (22-29 ft.-lbs.). Retest injector as outlined

above. If nozzle spring (4) was renewed, adjust opening pressure to 19615-20595 kPa (2845-2987 psi). Opening pressure will decrease to 19125 kPa (2775 psi) after the new spring takes a set.

Install injector using a new copper seal washer (9). Tighten retaining nuts to 5-6 N·m (3.5-4.5 ft.-lbs.) on Models DY23D and DY27D or 9-10 N·m (6.5-7.5 ft.-lbs.) on Model DY41D.

**INJECTION PUMP.** The injection pump is actuated by a lobe on the engine camshaft. Pump timing is adjusted by installing different thickness pump mounting gasket (8—Fig. R6-21). Injection timing is 23 degrees BTDC on all models. Pump can be removed and installed without regard for position of crankshaft.

To remove injection pump, first clean dirt from pump, fuel lines and surrounding area. Disconnect fuel lines and im-

mediately cap or plug all openings. Remove pump retaining nuts and withdraw pump from engine.

Refer to Fig. R6-21 for an exploded view of fuel injection pump. The injection pump should be tested and overhauled by a shop qualified in diesel fuel injection pump repair.

To determine correct thickness of pump mounting gasket, measure the distance between face of cam base and pump mounting surface as shown in Fig. R6-22. Select gasket of proper thickness so distance is 65.95-66.05 mm (2.597-2.600 inches) for Models DY23D and DY27D or 75.95-76.05 mm (2.990-2.994 inches) for Model DY41D.

When installing injection pump, make sure that pump control rack engages governor lever correctly. Check governor linkage for smooth operation. Bleed air from fuel system as outlined in MAINTENANCE section.

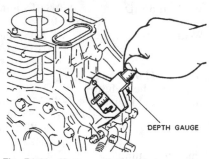

*Fig. R6-22—Measure distance between face of cam base to pump mounting surface on crankcase or crankcase cover to determine correct thickness of pump mounting gasket. Refer to text.*

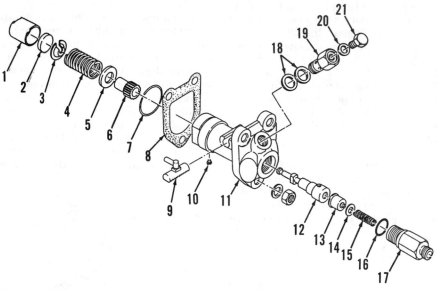

*Fig. R6-21—Exploded view of fuel injection pump used on Model DY41D. Injection pump used on Models DY23D and DY27D is similar.*

| | | |
|---|---|---|
| 1. Cam follower | 7. Retaining ring | 12. Plunger |
| 2. Spacer | 8. Gasket | 13. Delivery valve |
| 3. Spring retainer | 9. Control rack | 14. Gasket |
| 4. Spring | 10. Set screw | 15. Spring |
| 5. Spring seat | 11. Pump body | 16. "O" ring |
| 6. Control sleeve | | |

17. Delivery valve holder
18. Gaskets
19. Fuel inlet fitting
20. Gasket
21. Air vent screw

# WISCONSIN ROBIN

## SERVICING WISCONSIN ROBIN ACCESSORIES

### REWIND STARTER

**OVERHAUL.** To disassemble rewind starter, refer to Fig. WR23. Release spring tension by pulling rope handle until about 18 inches of rope extends from unit. Use thumb pressure against ratchet retainer to prevent reel from rewinding and place rope in notch in outer rim of reel. Release thumb pressure slightly and allow spring mechanism to slowly unwind. Twist loop of return spring and slip loop through slot in ratchet retainer. Refer to Fig. WR24 and remove nut, lockwasher, plain washer and ratchet retainer. Reel will completely unwind as these parts are removed. Remove compression spring, three ratchets and spring retainer washer. Slip fingers into two of the cavity openings in reel hub (Fig. WR25) and carefully lift reel from support shaft in housing.

**CAUTION: Take extreme care that power spring remains in recess of housing. Do not remove spring unless new spring is to be installed.**

If power spring escapes from housing, form a 4½ inch (114.3 mm) I.D. wire ring and twist ends together securely. Starting with the outside loop, wind spring inside ring in a counter-clockwise direction.

**NOTE: New power springs are secured in a similar wire ring for ease in assembly.**

Place spring assembly over recess in housing so hook in outer loop of spring is over tension tab in housing. Carefully press spring from wire ring and into recess of housing.

Using a new rope of same length and diameter as original, place rope in handle and tie a figure eight knot about 1-½ inches (38 mm) from end. Pull knot into top of handle. Install other end of rope through guide bushing of housing and through hole in reel groove. Pull rope out through cavity opening and tie a slip knot about 2-½ inches (64 mm) from end. Place slip knot around center bushing as shown in Fig. WR26 and pull knot tight. Stuff end of rope into reel cavity. Spread a film of light grease on power spring and support shaft. Wind rope ¼-turn clockwise in reel and place rope in notch on reel. Install reel on support shaft and rotate reel counterclockwise until tang on reel engages

hook on inner loop of power spring. Place outer flange of housing in a vise and use finger pressure to keep reel in housing. Then, by means of rope hooked in reel notch, preload power spring by turning reel 7 full turns counter clockwise. Remove rope from notch and allow reel to slowly turn clockwise as rope winds on pulley and handle returns to guide bushing on housing.

Install spring retaining washer (Fig. WR24), cup side up and place compression spring into cupped washer. Install return spring with bent end hooked into hole of reel hub. Place the three ratchets in position so they fit contour of

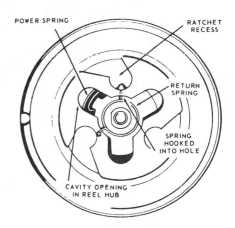

**Fig. WR25 — Use fingers in reel hub cavities to lift reel from support shaft.**

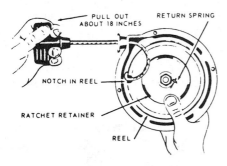

**Fig. WR23 — View showing method of releasing spring tension on rewind starter assembly.**

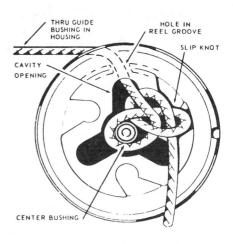

**Fig. WR26 — Install rope through guide bushing and hole in reel groove, then tie slip knot around center bushing.**

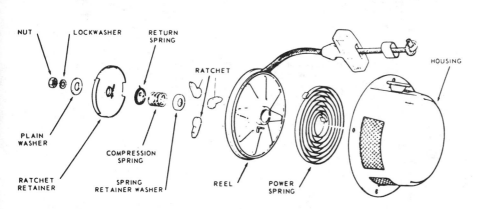

**Fig. WR24 — Exploded view of rewind starter assembly used on some Wisconsin Robin engines.**

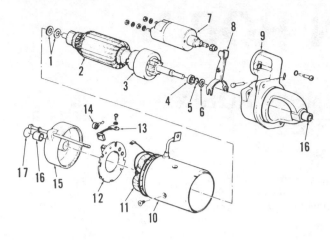

**Fig. WR27—Exploded view of 12 volt starter motor used on EY44W model.**

1. Thrust washers
2. Armature
3. Drive assy.
4. Pinion stop collar
5. Snap ring
6. Thrust washer
7. Solenoid
8. Shift lever
9. Drive end housing
10. Frame
11. Field coils
12. Brush holder plate
13. Brushes (2 used)
14. Brush spring
15. End cover
16. Bushings
17. Expansion plug

recesses. Mount ratchet retainer so loop end of return spring extends through slot. Rotate retainer slightly clockwise until ends of slots just begin to engage the three ratchets. Press down on retainer, install flat washer, lockwasher and nut, then tighten nut securely.

## 12-VOLT STARTER MOTORS

The 12 volt starting motors used on Models EY44W and EY21W are magnetic shift type in which the solenoid provides positive drive pinion engagement with ring gear. Two brush type starter shown in Fig. WR27 is used on Model EY44W. Refer to Fig. WR28 for the four brush starter used on Model EY21W. Disassembly and reassembly is conventional and procedure is obvious after examination of units and reference to Fig. WR27 or WR28. Parts are available from Wisconsin distributors or service centers.

## ALTERNATORS

On models equipped with an alternator, the following precautions must be observed:

A. Do not reverse battery connections. System is for negative ground only.

B. Connect booster batteries properly (positive to positive and negative to negative).

C. Do not attempt to polarize alternator.

D. Do not ground any wires from stator or rectifier which terminate at connectors.

E. Do not operate engine with battery disconnected.

F. Disconnect battery cables when charging battery with a battery charger.

**Fig. WR28—Exploded view of 12 volt starter motor used on EY21W model.**

1. End cover & brush holder assy.
2. Negative brush (2 used)
3. Positive brush (2 used)
4. Brush spring
5. Frame & field coil assy.
6. Bushing
7. Thrust washer
8. Armature
9. Center bearing
10. Drive assy.
11. Pinion stop collar
12. Snap ring
13. Thrust washer
14. Drive end housing
15. Bushing
16. Shift lever
17. Solenoid
18. Gasket
19. Dust plate

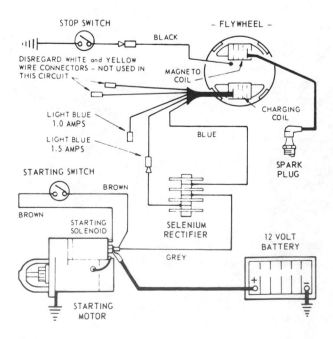

**Fig. WR29—Typical wiring diagram of EY44W model ignition system and flywheel alternator charging system.**

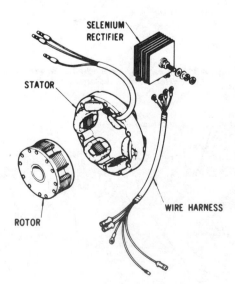

**Fig. WR30—Alternator stator and rotor and selenium rectifier used on EY21W model.**

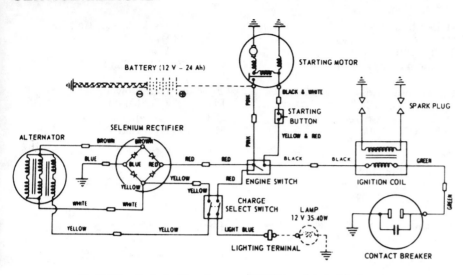

**Fig. WR31 — Typical wiring diagram of EY21W model electrical system.**

On Model EY44W, alternator charging coil (Fig. WR29) is located behind flywheel and rotating magnets are bolted to inner side of flywheel. A selenium rectifier is used to convert AC current to DC current to charge battery. If battery overcharging is indicated, disconnect light blue wire marked 1.5A from rectifier and connect light blue wire marked 1.0A. Alternator system is rated at 12 volts, 1.5 amps.

On model EY21W, alternator stator and rotor (Fig. WR30) are located behind cooling blower. AC current from alternator is converted to DC current by a selenium rectifier. A manual Hi-Lo charge select switch (Fig. WR31) is used to prevent overcharging battery. Alternator system is rated at 12 volts, 3 amps.

## TELEDYNE WISCONSIN CENTRAL PARTS DISTRIBUTORS

(Arranged Alphabetically by States)

**These franchised firms carry extensive stocks of repair parts. Contact them for name of dealer in their area who will have replacement parts.**

Parts Service Company
Phone: (205) 262-4485
12 Randolph Street
**Montgomery, AL 36101**

Sahlberg Equipment, Inc.
Phone: (907) 276-5494
1702 Ship Avenue
**Anchorage, AK 99501**

Southwest Products Corporation
Phone: (602) 269-3581
2949 North 30th Avenue
**Phoenix, AZ 85061**

Keeling Supply Company
Phone: (501) 945-4511
4227 East 43rd Street
**North Little Rock, AR 72115**

Lanco Engine Services, Inc.
Phone: (213) 772-2471
12915 Weber Way
**Los Angeles, CA 90250**

E.E. Ricter & Son, Inc.
Phone: (415) 658-1100
6598 Hollis Street
**San Francisco (Emeryville), CA 94608**

Central Equipment Company
Phone: (303) 388-3696
4477 Garfield Street
**Denver, CO 80216**

Highway Equipment & Supply Co.
Phone: (904) 783-1630
5366 Highway Avenue
**Jacksonville, FL 32206**

P.H. Neff & Sons, Inc.
Phone: (305) 592-5240
5295 N.W. 79th Avenue
**Miami FL 33144**

Highway Equipment & Supply Co.
Phone: (305) 843-6310
1016 West Church Street
**Orlando, FL 32805**

American Outdoor Corporation
Phone: (813) 522-5502
4475 28th Street North
**St. Petersburg, FL 33714**

Georgia Engine Sales & Service
Phone: (404) 446-1100
5715 Oak Brook Parkway
**Norcross, GA 30093**

Lanco Engine Service, Inc.
Phone: (808) 841-5896
3140 Koapaka Street
**Honolulu, HI 96819**

Teledyne Total Power
Phone: (208) 522-3872
1230 North Skyline Drive
**Idaho Falls, ID 83402**

Port Huron Machinery Company
Phone: (515) 266-1136
555 North East 16th
**Des Moines, IA 50308**

Harley Industries
Phone: (316) 262-5156
1607 Wabash
**Wichita, KS 67203**

Wilder Motor & Equipment Company
Phone: (502) 966-5141
4022 Produce Road
**Louisville, KY 40218**

Wm. F. Surgi Equipment Corporation
Phone: (504) 293-0556
5707 Siegen Lane
**Baton Rouge, LA 70814**

Wm. F. Surgi Equipment Corporation
Phone: (504) 733-0101
221 Laitram Lane
**Harahan, LA 70183-0715**

Wm. F. Surgi Corporation
Phone: (318) 233-6322
220 Industrial Parkway
**Lafayette, LA 70502**

Diesel Engine Sales Division
Phone: (617) 341-1760
199 Turnpike Street
**Staughton, MA 02072**

Engine Supply
Phone: (313) 349-9330
44455 Grand River Avenue
**Detroit (Novi), MI 48050**

Teledyne Total Power
Phone: (612) 425-7200
8575 County Road, 18
**Minneapolis (Osseo), MN 55429**

Allied Construction Equipment Co.
Phone: (314) 371-1818
4015 Forest Park Avenue
**St. Louis, MO 63108**

Central Motive Power, Inc.
Phone: (505) 884-2525
3740 Princeton Drive, North East
**Albuquerque, NM 87107**

John Reiner & Company
Phone: (201) 460-9444
145 Commerce Road
**Carlstadt, NJ 07072**

John Reiner & Company
Phone: (315) 474-5741
946 Spencer Street
**Syracuse, NY 13208**

King-McIver Sales Inc.
Phone: (919) 294-4600
6375 New Burnt Poplar Road
**Greensboro, NC 27420**

Northern Engine & Supply Company
Phone: (701) 232-3284
2710 3rd Avenue North
**Fargo, ND 58102**

Cincinnati Engine & Parts Co., Inc.
Phone: (513) 221-3525
2863 Stanton Avenue
**Cincinnati, OH 45206**

Allied Farm Equipment, Inc.
Phone: (614) 486-5283
1066 Kinnear Road
**Columbus, OH 43212**

Harley Industries
Phone: (405) 670-1341
1720 South Prospect
**Oklahoma City, OK 72129**

Harley Industries
Phone: (918) 672-9220
6845 East 41st Street
**Tulsa, OK 74145**

I.D. Inc.
Phone: (503) 646-8285
5500 South West Artic Drive
**Beaverton, OR 97075**

Jos. L. Pinto Inc.
Phone: (215) 747-3877
719 East Baltimore Pike
**Philadelphia, PA 19050**

Wilder Motor & Equipment Company
Phone: (803) 799-1220
1219 Rosewood Drive
**Columbia, SC 29201**

RCH Distributors, Inc.
Phone: (901) 345-2200
3150 Carrier Street
**Memphis, TN 38101**

Wilder Motor & Equipment Co., Inc.
Phone: (615) 329-2365
301 15th Avenue
**Nashville, TN 37203**

Harley Industries
Phone: (512) 851-1991
1918 Holley Road
**Corpus Christi, TX 78408**

Harley Industries
Phone: (214) 638-4504
8005 Sovereign Row
**Dallas, TX 75247**

Harley Industries
Phone: (713) 492-6445
17150 Park Row
**Houston, TX 77084**

Harley Industries
Phone: (915) 337-8676
3220 Kermit Highway
**Odessa, TX 79760**

Harley Industries
Phone: (512) 342-4255
8406 Speedway Drive
**San Antonio, TX 78320**

Teledyne Total Power
Phone: (703) 752-9395
1127 Ind. Pkwy.
**Fredricksburg, VA 22150**

Engine Sales & Service Co., Inc.
Phone: (304) 342-2131
601 Ohio Avenue
**Charleston, WV 25301**

Teledyne Total Power
Phone (414) 786-1600
2244 West Bluemound Road
**Milwaukee (Waukesha), WI 53187**

## CANADA

Mandem Div. of Asamera, Inc.
Phone: (306) 523-2631
3611 60th Street S. E.
**Calgary, Alberta T2A-2E5**

Mandem Div. of Asamera, Inc.
Phone: (403) 465-0244
5925 83rd Street
**Edmonton, Alberta T6E-4Y3**

Pacific Engines
Phone: (604) 254-0804
1391 William Street
**Vancouver, British Columbia V5L-2P6**

Mandem Div. of Asamera, Inc.
Phone: (204) 885-4440
21 Murray Park road
**Winnipeg, Manitoba R3J-3S2**

Mandem Div. of Asamera, Inc.
Phone: (506) 854-0982
146 Albert Street
**Moncton, New Brunswick E1C-1B2**

Mandem Div. of Asamera, Inc.
Phone: (416) 255-8158
3 Bestobell Road
**Toronto, Ontario L4W-1A4**

Mandem Div. of Asamera, Inc.
Phone: (514) 342-9233
8550 Delmeade Road
**Montreal, Quebec H4T-1L7**

Mandem Div. of Asamera, Inc.
Phone: (306)523-2631
1250 St. John Street
**Regina, Saskatchewan S4R-1R9**

Mandem Div. of Asamera, Inc.
Phone: (306) 244-1505
1729 Ontario Avenue
**Saskatoon, Saskatchewan S4R-1R9**

Mandem Div. of Asamera, Inc.
Phone: (403) 667-6939
114 Calcite Road
**Whitehorse, Yukon Y1A-4S2**